Student Solutions Manual to Accompany:

ORGANIC CHEMISTRY

Fifth Edition

Stanley H. Pine
Professor of Chemistry
California State University, Los Angeles

McGraw-Hill, Inc.
New York St. Louis San Francisco Auckland Bogotá
Caracas Lisbon London Madrid Mexico Milan
Montreal New Delhi Paris San Juan Singapore
Sydney Tokyo Toronto

Student Solutions Manual to Accompany:
ORGANIC CHEMISTRY

5 6 7 8 9 0 WHTWHT 9 2

ISBN 0-07-050119-X

The editor was Karen S. Misler;
the production supervisor was Leroy A. Young.
The Whitlock Press, Inc. was printer and binder.

Contents

TO THE STUDENT

In the preface to your textbook, I make the statement "organic chemistry is not a spectator sport". The point is that most students do not learn organic chemistry simply by listening to lectures. Though this is a part of the process, mastery of the subject results from the exercise of applying the concepts to problems of organic chemistry. By this approach students are able to find out what they do and what they do not know about the principles under study. Furthermore, the experiences associated with deciding what information is needed, then how it can be integrated to reach a solution, is the important process of problem solving applicable to any discipline.

Your textbook contains two types of problems: those that reinforce the learning of specific concepts and those that integrate many concepts. Problems of the first type are generally found throughout the body of the text, within individual sections of the chapters. You will derive the greatest benefit by working these problems as each is encountered in the text.

The problems at the ends of each chapter generally integrate many concepts from the current and other chapters. These are the problems which are particularly useful to reinforce your knowledge of organic chemistry while developing the very important skills of problem solving.

Much of the content of the organic chemistry course involves concepts of structure and of the bond making and bond breaking processes. It is therefore important that you use pen, pencil, (or computer display) to draw out the appropriate chemical structures so as to clearly visualize the relationships of atoms and the bonding changes that take place.

Often the process of solving a problem is more important than the final solution. However, when you complete the solution to a problem try to use some "common sense" to see if your answer is reasonable. The course in organic chemistry is not meant to be tricky or mysterious.

This "Student Solutions Supplement" is a tool to be used in the learning process. The Supplement should be consulted only after you have made a conscientious effort at solving a problem. Using it instead of solving the problems yourself almost always leads to a disappointing performance as the course progresses.

After working through a problem, check your answer against an acceptable solution found in the Supplement. You will find that many of the problems may have more than one acceptable solution. Only one is commonly given, but you should not hesitate to discuss alternate ideas with your instructor. If you find that your answer is in error, use this information in a constructive manner. Recall your thought processes to see how or why you missed the correct approach. This can often be a very important part of the learning sequence.

I hope that this Student Supplement will be a useful part of your learning resources and that it helps you to appreciate the excitement of organic chemistry.

Stanley H. Pine

2 Bonding in Organic Molecules

2-1

Methanol:

Bonded	2 x 5 =	10
Nonbonded	2 x 2 =	4
Electrons utilized		14

Carbon	1 x 4 =	4
Hydrogen	4 x 1 =	4
Oxygen	1 x 6 =	6
Available electrons		14

Ethene:

Bonded	6 x 2 =	12
Electrons utilized		12

Carbon	2 x 4 =	8
Hydrogen	4 x 1 =	4
Available electrons		12

Ethyne:

Bonded	5 x 2 =	10
Electrons utilized		10

Carbon	2 x 4 =	8
Hydrogen	2 x 1 =	2
Available electrons		10

2-2

a) 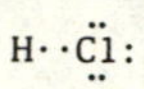

b) 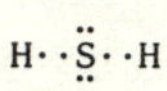

c) 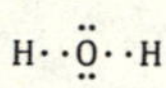

d) 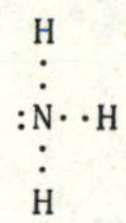

e) 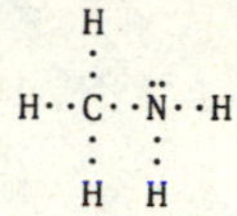

f) 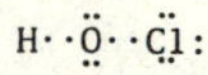

2-3

a) 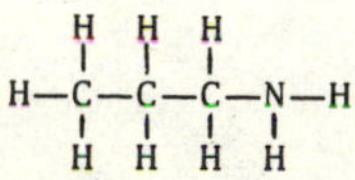 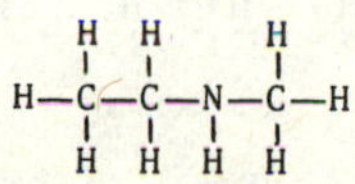 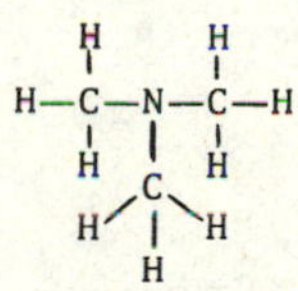 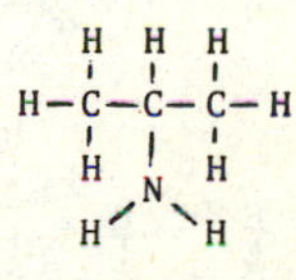

b) 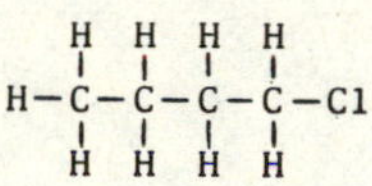 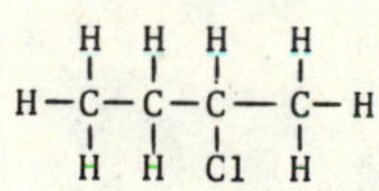

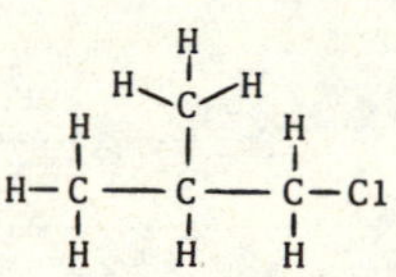 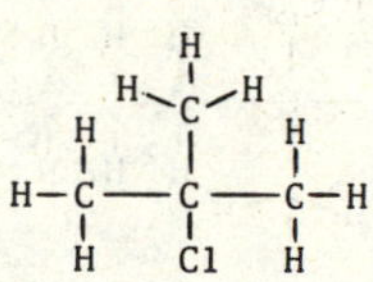

2-3 contd..

c)

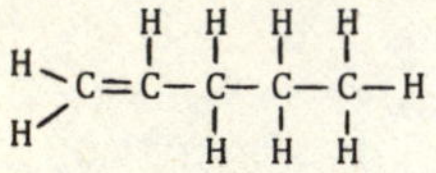
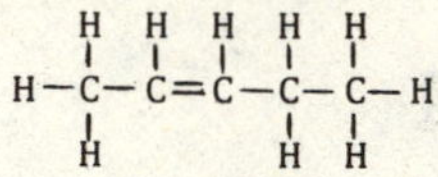
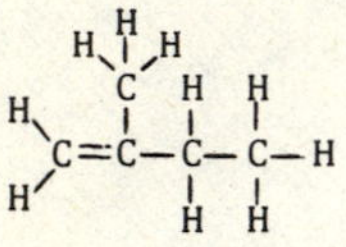
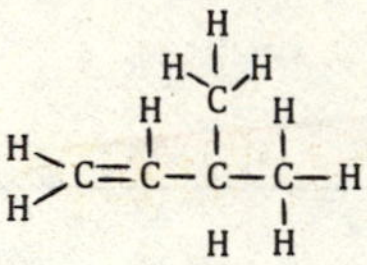
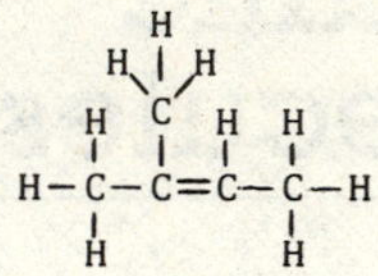
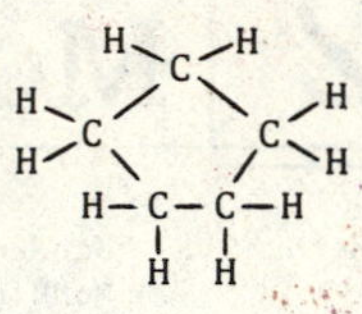
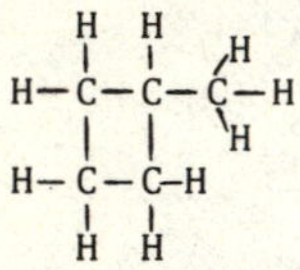
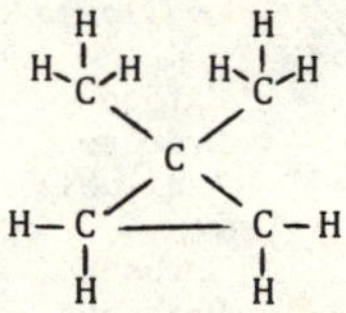
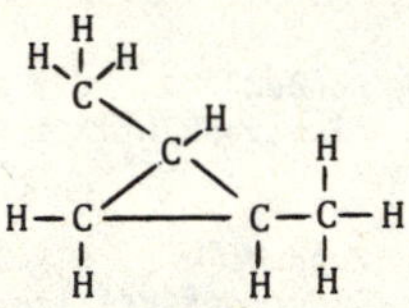

d)

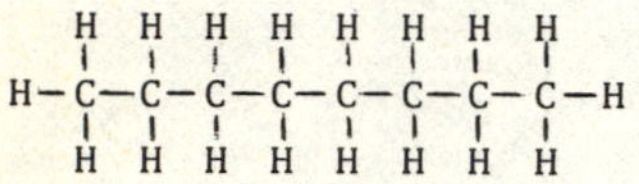
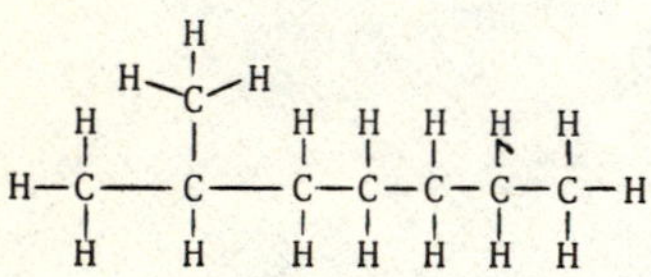
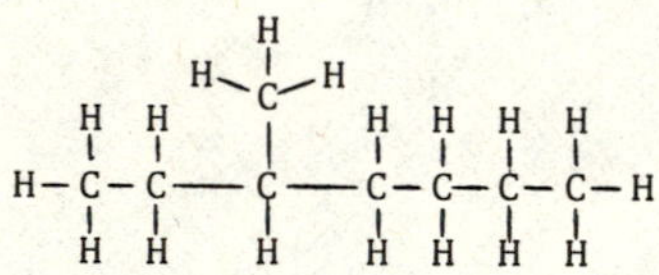
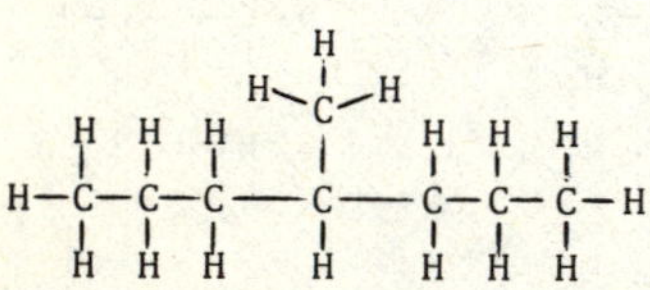
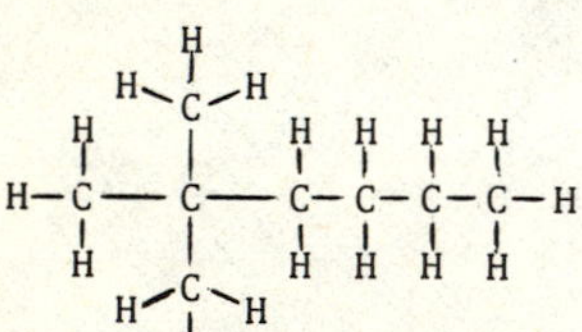
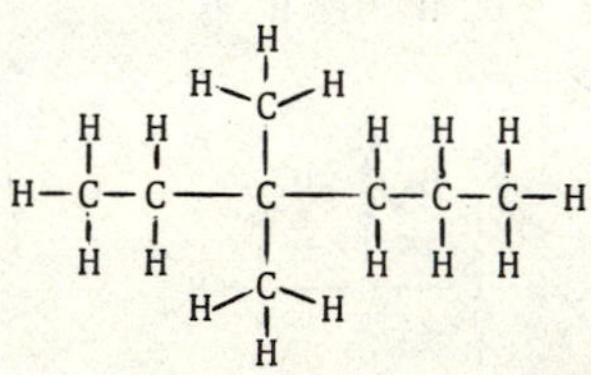

2-3 contd...

d) contd...

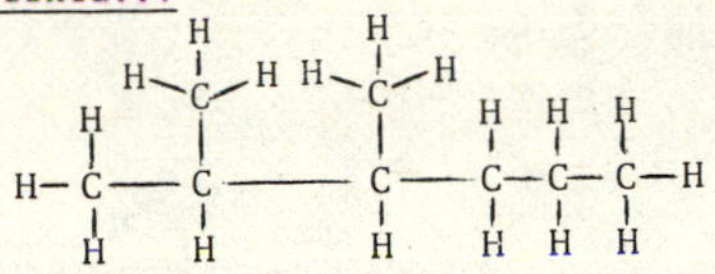

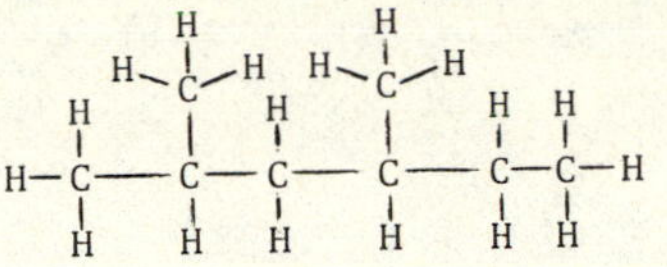

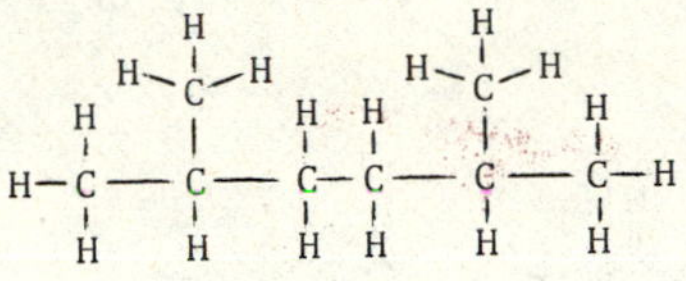

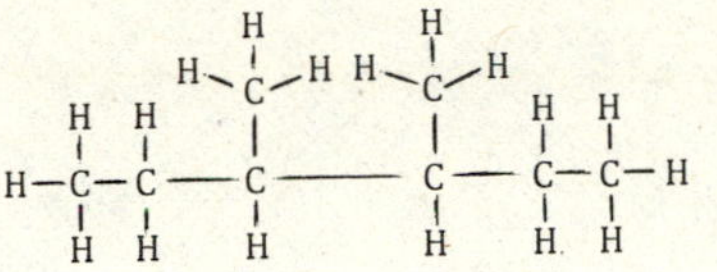

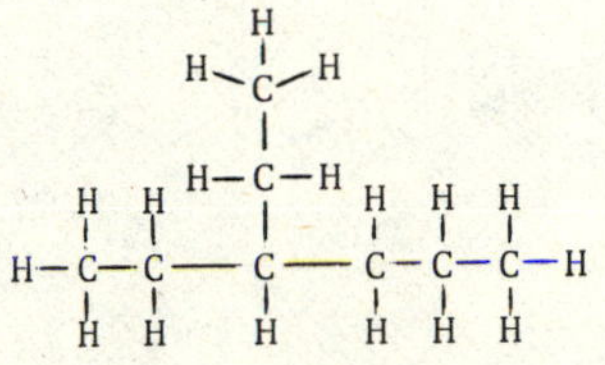

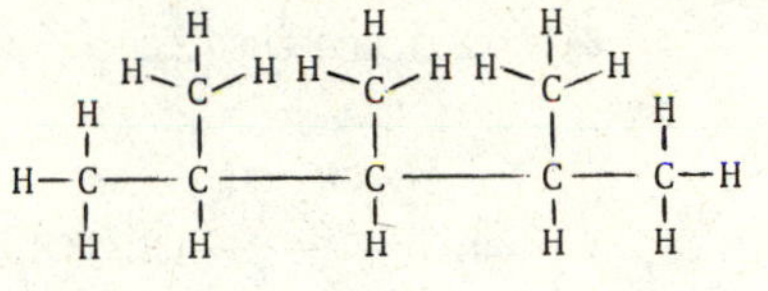

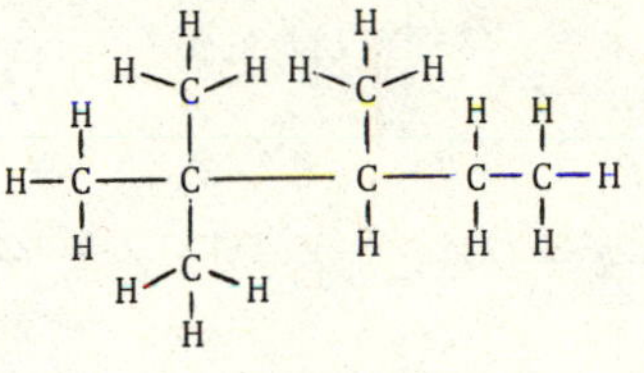

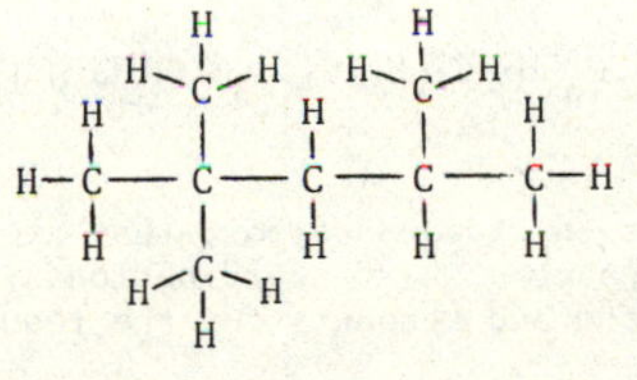

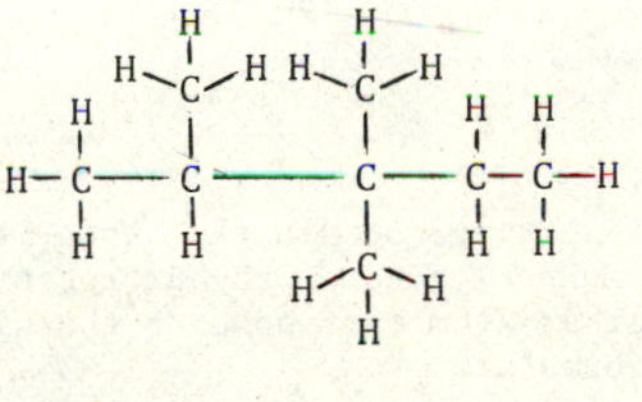

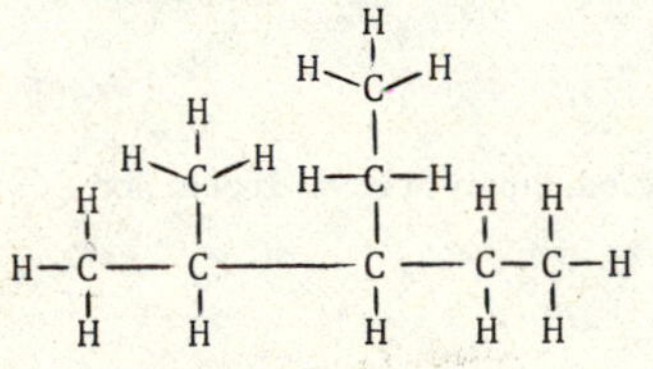

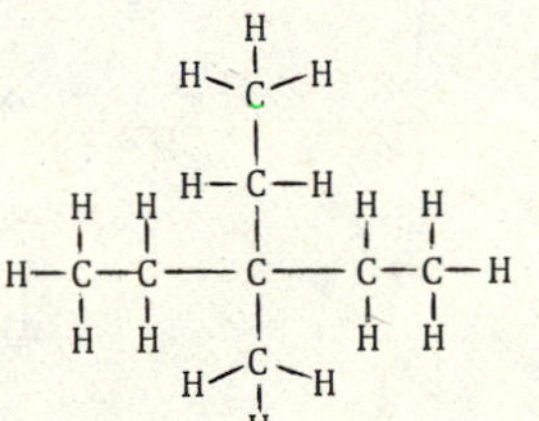

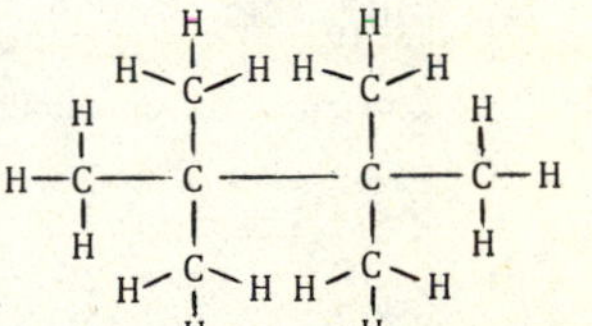

2-4

a) 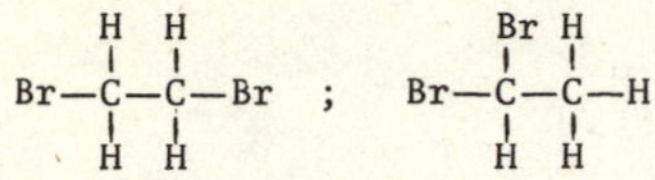

A B

b) 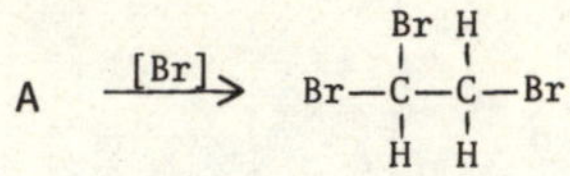

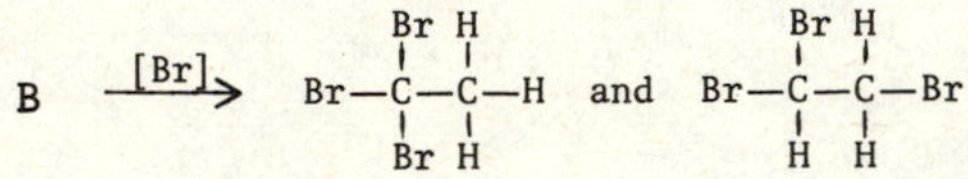

2-5

a) 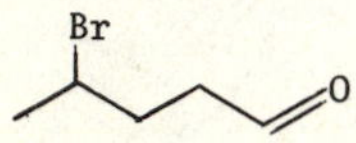

b) 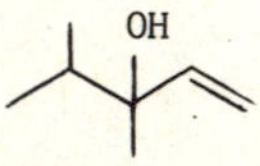

c) 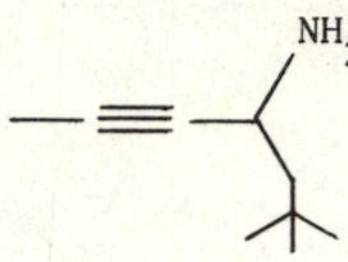

d) 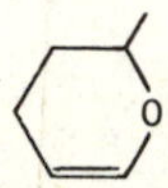

e) 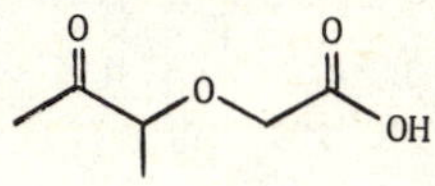

f) 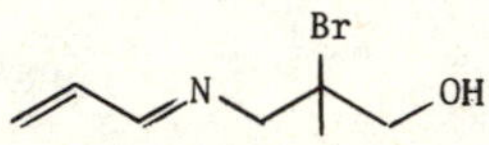

2-6

a) $(CH_3)_2CHCH_2CH_2OH$

b) 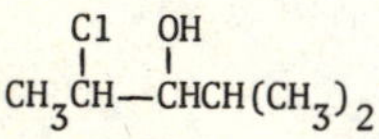

c) 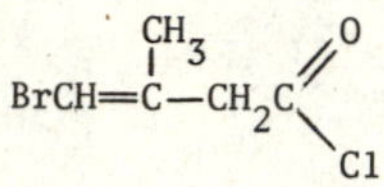

d) $(CH_3)_2C{=}CHCH_2NH_2$

e) 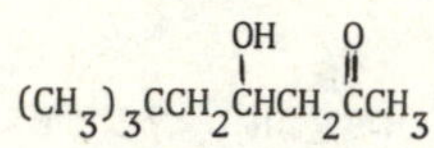

f) 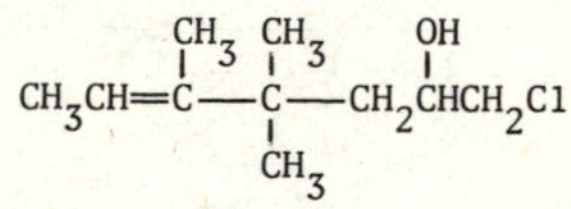

or

$CH_3CH{=}C(CH_3)C(CH_3)_2CH_2CH(OH)CH_2Cl$

2-7 A nitrogen atom is normally tricoordinate. When it replaces an atom on carbon, that atom, or another, can bond as a second ligand on the nitrogen. That leaves a third coordination site at which a hydrogen or another atom must bond to the nitrogen atom and accounts for the required extra H in the parent formula.

2-8

a) C_5H_{10} *IHD* = 1

One ring or one double bond.

b) $C_{10}H_{18}$ *IHD* = 2

Two double bonds; one triple bond; two rings; one ring and one double bond.

c) C_7H_{16} *IHD* = 0

Saturated noncyclic hydrocarbon.

d) C_8H_{12} *IHD* = 3

Three double bonds; three rings; one double bond and two rings; two double bonds and one ring; one triple bond and one ring; one triple bond and one double bond.

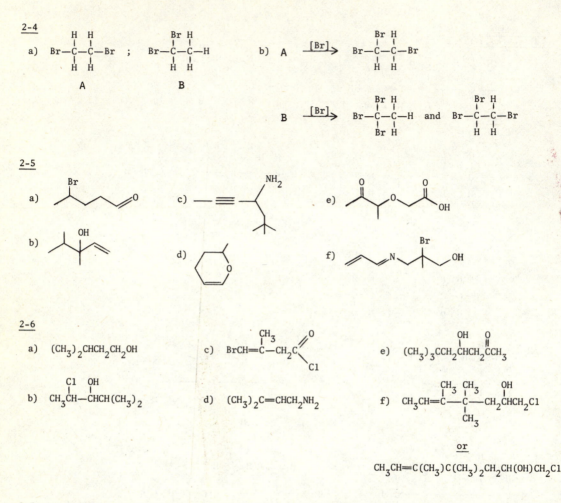

2-9

a) The nuclear charge increases as one moves across the periodic table, thus affinity for electrons increases.

b) As one moves down in the periodic table the outer (valence) electrons move further from the nuclear charge. Electron affinity decreases.

2-10

a) 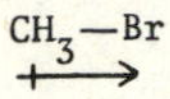 b) 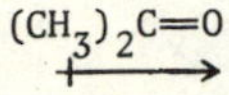 c) 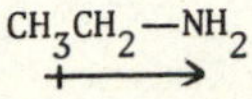 d) 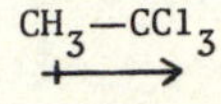

2-11

Bond dipole moments

Molecular dipole moment

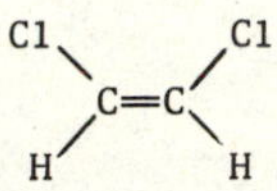

μ = 1 D

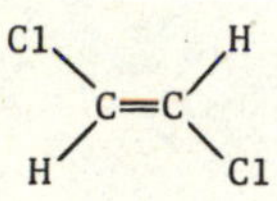

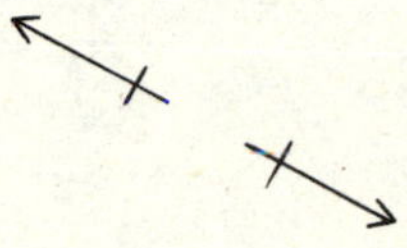

μ = 0

2-12

a) 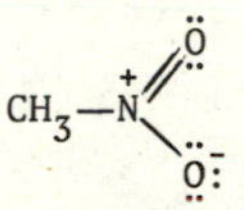

Nitrogen shares eight electrons, thus is assigned four of them. That is one less than the five associated with the free atom so that nitrogen is assigned +1. The singly bonded oxygen possesses six nonbonding electrons plus one from the shared pair. The total of seven is one more than is expected for oxygen giving it a -1 charge.

b) 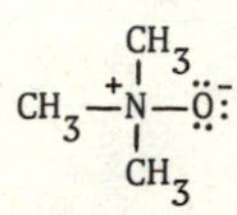

Nitrogen possesses four of the shared electrons resulting in a +1 charge. Oxygen possesses six nonbonding plus one shared electron and has one negative charge.

2-13

a) 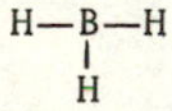 and 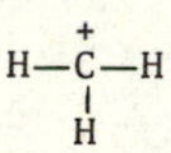

b) 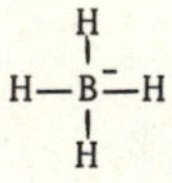 and 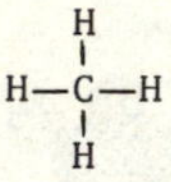 and 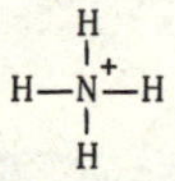

c) 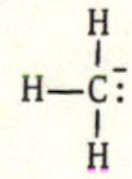 and 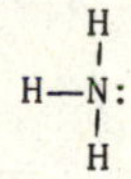 and 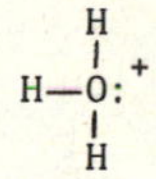

d) 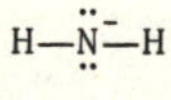 and 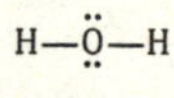

2-14

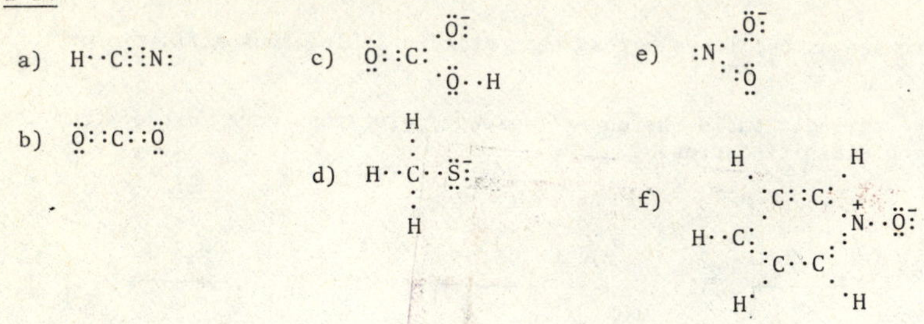

2-15

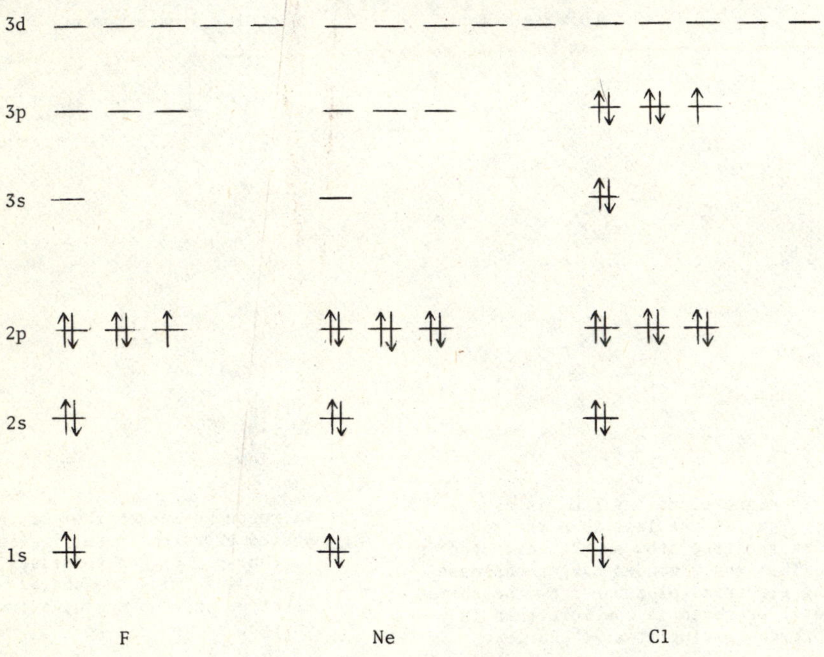

2-16

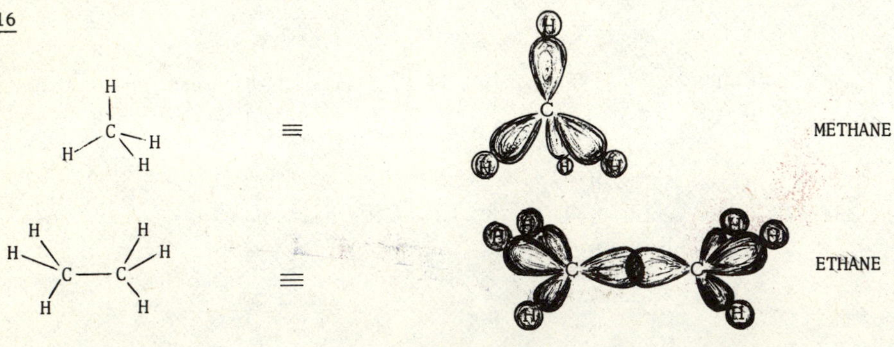

2-17

a) $>C{=}O$ $\equiv$ 

b) $-C{\equiv}N$ $\equiv$ 

2-18

sp^3 = 25%s + 75%p

sp^2 = 33%s + 67%p

sp = 50%s + 50%p

2-19

The 112° C-C-C angle is larger than tetrahedral and reflects less p-character than sp^3, therefore is assigned sp^{3-} (calculation predicts $sp^{2.7}$).
The 106° H-C-H angle is smaller than tetrahedral and indicates more p-character than sp^3, i.e., sp^{3+} (calculation predicts $sp^{3.3}$).

2-20

Fluorine is more electronegative than H so that electron-pair density is attracted away from the carbon atom along the C-F bond. VSEPR between the C-F bonds is therefore less than that between the C-H bonds. The F-C-F bond angle is consequently less than the H-C-H bond angle.

2-21

a) The ammonium cation has tetrahedral bond angles typical of a symmetrical tetracoordinate central atom. In NH_3 the nonbonding electron pair repels the N-H bonds more than N-H bonds repel each other.

b) H_2S possesses two nonbonding electron pairs which repel the electrons of the S-H bonds.

c) The central atoms, S and P are third row and thus larger than an N atom. The bonding electron pairs are further from each other so that their repulsions are less. Consequently, the nonbonding pairs which are closer to the nucleus have a greater repulsive effect. Only one nonbonding pair of electrons repels the P-H bonds in PH_3.

2-22

As the atoms attached to the trigonal carbon atom become more electronegative, electron density moves further away from the central carbon atom. The angles between the bonds become smaller because of decreasing repulsion close to the central carbon atom.

2-23

a) Electron repulsions decrease as the more electronegative atoms attract electron density away from the central atom. (See also prob. 2-22)

b) The nonbonding electron pairs on the carbonyl oxygen atom increase repulsions relative to the carbon-carbon double bond.

c) The molecule containing chlorine has intermediate angles even though chlorine is the largest atom. The size of the atoms (nonbonded repulsions) is not the major factor in this case.

2-24

a) i) Because repulsions involve two identical sets of bonds connected to the central carbon atom, the atoms attached to those bonds move as far from each other as possible. VSEPR therefore predicts a linear (180°) geometry.

ii) An orbital picture connects an *sp* hybrid at the center carbon atom to sp^2 orbitals of the two terminal carbons. The *sp* bonds are 180° apart. Overlap of the two sets of *p*-orbitals produces pi-orbitals in perpendicular planes.

b) Both hybrid orbital (sp^3) and VSEPR arguments predict a near tetrahedral geometry. VSEPR theory predicts greater repulsion by the electron pair resulting in H-C-H angles smaller than 109.5°. Hybrid orbital theory predicts more *s* character associated with the nonbonding electrons since they are closer to the nucleus as is an *s* orbital. Greater *p* character is thus expected in the C-H bonds and bond angles are less than tetrahedral.

2-25

a)

Bonds broken	kcal/mol	kJ/mol
H—H	104	436
C=C	146	610
	250 kcal/mol	1046 kJ/mol

Bonds formed		
C—C	83	347
2C—H	198	828
	281 kcal/mol	1175 kJ/mol

ΔH° = 250-281 = -31 kcal/mol
(= 1046-1175 = -129 kJ/mol)

∴ Reaction is exothermic.

b)

Bonds broken	kcal/mol	kJ/mol
C—O	86	359
H—Br	87	365
	173 kcal/mol	724 kJ/mol

Bonds formed		
C—Br	68	284
O—H	111	464
	179 kcal/mol	748 kJ/mol

ΔH° = 173-179 = -6 kcal/mol
(= 724-748 = -24 kJ/mol)

∴ Reaction is exothermic.

Contd....

2-25 Contd...

c)

Bonds broken	kcal/mol	kJ/mol
N—H	93	389
C—Br	68	284
	161 kcal/mol	673 kJ/mol

Bonds formed	kcal/mol	kJ/mol
C—N	73	305
H—Br	87	365
	160 kcal/mol	670 kJ/mol

$\Delta H° = 161-160 = 1$ kcal/mol
(= 673-670 = 3 kJ/mol)

∴ Reaction is endothermic.

2-26

The electronegative fluorine atoms attract electron density away from each carbon atom. The carbon-carbon distance consequently decreases due to the decreased electron-electron repulsion.

2-27

a) C_2H_6S: IHD = 0; saturated, acyclic.

CH_3SCH_3 ; CH_3CH_2SH

b) C_5H_{12}: IHD = 0; saturated, acyclic.

$CH_3CH_2CH_2CH_2CH_3$; $(CH_3)_2CHCH_2CH_3$; $(CH_3)_4C$

c) C_4H_8O: IHD = 1; one ring or one double bond.

$CH_3C(=O)CH_2CH_3$; $CH_3CH_2CH_2CHO$; $(CH_3)_2CHCHO$;

$CH_2{=}CHCH_2CH_2OH$; $CH_3CH{=}CHCH_2OH$; $CH_3CH_2CH{=}CHOH^*$; $CH_2{=}CHCH(OH)CH_3$; $CH_3CH{=}C(OH)CH_3{}^*$;

$CH_3CH_2C(OH){=}CH_2{}^*$; $(CH_3)_2C{=}CHOH^*$; $CH_2{=}C(CH_3)CH_2OH$; $CH_2{=}CHOCH_2CH_3$; $CH_2{=}CHCH_2OCH_3$;

$CH_3CH{=}CHOCH_3$; $CH_2{=}C(CH_3)OCH_3$

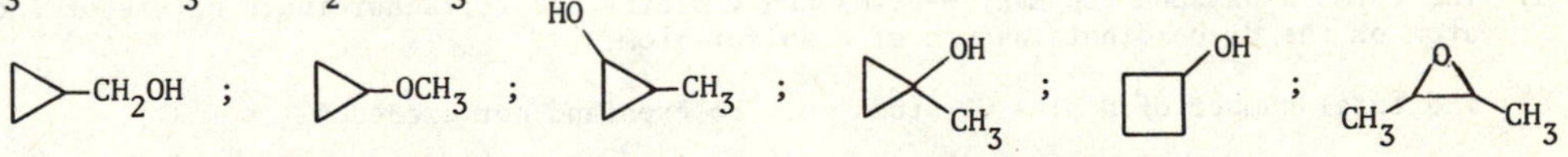

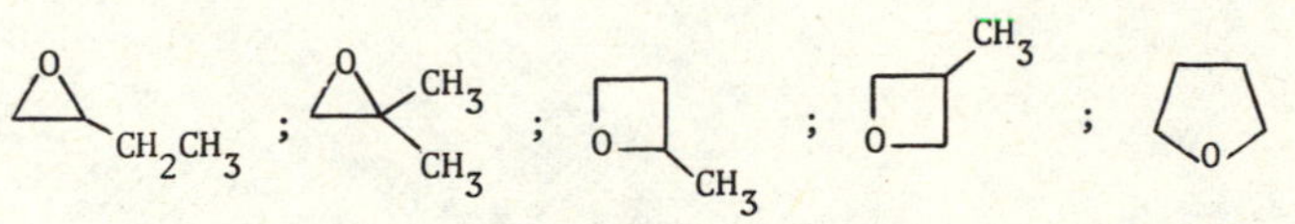

Contd....

2-27 Contd...

d) $C_3H_8O_2$: IHD = 0; saturated, acyclic.

$CH_3CH_2CH(OH)_2$; $CH_3\overset{OH}{C}HCH_2OH$; $HOCH_2CH_2CH_2OH$; $CH_3CH_2OCH_2OH$; $CH_3\underset{OH}{\overset{OH}{C}}CH_3$; $CH_3OCH_2CH_2OH$

$CH_3O\overset{OH}{C}HCH_3$; $CH_3OCH_2OCH_3$; $CH_3OOCH_2CH_3$; $CH_3CH_2CH_2OOH$; $(CH_3)_2CHOOH$

e) C_3H_9NO: IHD = 0; saturated, acyclic.

$CH_3CH_2\overset{OH}{C}HNH_2$; $CH_3\overset{OH}{C}HCH_2NH_2$; $HOCH_2CH_2CH_2NH_2$; $HOCH_2\overset{CH_3}{C}HNH_2$; $CH_3CH_2CH_2NHOH$;

$(CH_3)_2CHNHOH$; $CH_3CH_2OCH_2NH_2$; $CH_3OCH_2CH_2NH_2$; $CH_3O\overset{CH_3}{C}HNH_2$; $CH_3\overset{OH}{C}HNHCH_3$; $CH_3CH_2\overset{OH}{N}CH_3$;

$HOCH_2CH_2NHCH_3$; $CH_3CH_2NHCH_2OH$; $CH_3OCH_2NHCH_3$; $CH_3CH_2NHOCH_3$; $(CH_3)_2NCH_2OH$;

$(CH_3)_2NOCH_3$; $CH_3\underset{NH_2}{\overset{OH}{C}}CH_3$; $CH_3CH_2CH_2ONH_2$; $CH_3CH_2ONHCH_3$, $(CH_3)_2CHONH_2$

f) C_3H_6O: IHD = 1; one ring or double bond.

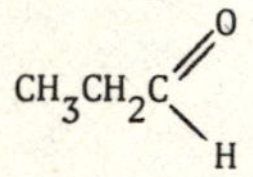 ; $CH_3\overset{O}{\overset{\|}{C}}CH_3$; $CH_2{=}CHOCH_3$; $CH_2{=}\overset{OH}{C}CH_3$* ; $CH_2{=}CHCH_2OH$; $CH_3CH{=}CHOH$* ;

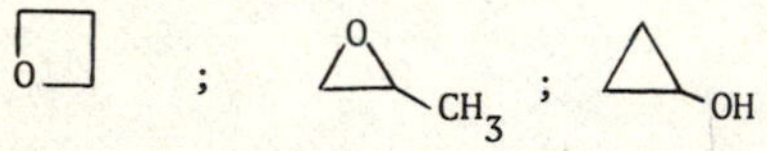

*These compounds actually exist as more stable tautomers (sec. 7-2C).

2-28

a) The formula requires one more or one less H-atom (or other single bonded atom) to accomodate the tetracoordinate carbon. The formula requires an even number of H's.

b) Amines require one more H-atom than the parent hydrocarbon; i.e., an odd number of H'swith an odd number of N-atoms.

c) The formula requires one more or one less H-atom; i.e., an even number of H's.

d) The formula has one too many H-atoms and violates the tetracoordinate nature of a carbon atom or the dicoordinate nature of a sulfur atom.

e) The total number of H plus Cl atoms must be even[and not exceed 6 $(2n + 2)$].

f) The total number of H plus Cl atoms must be odd[and not exceed 9 $(2n + 2)$].

2-29

The similar dipole moments suggest that only the alcohol portion of each molecule (—C—O—H—) is making a significant contribution to the net molecular dipole moment. Both the O—H and the C—O bond dipoles contribute to the molecular dipoles, but the O—H bond makes the greater contribution.

2-30

a) 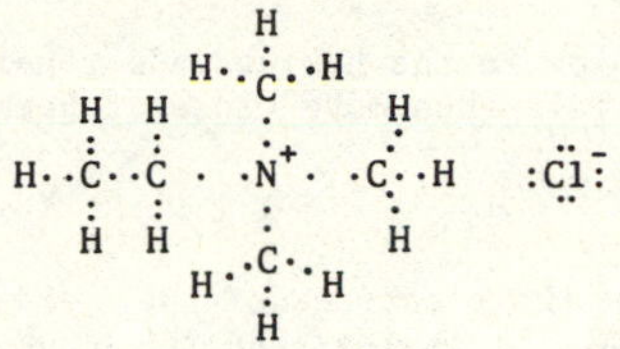

c) 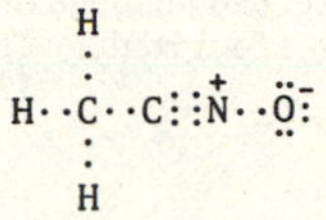

d) 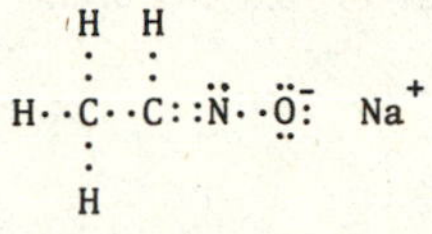

b) 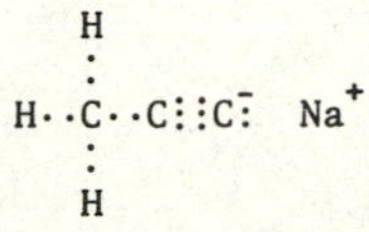

2-31

The VSEPR theory would suggest that the electronegative chlorine atoms attract electron density away from the carbon atoms and the Cl—C—Cl bond angles decrease from a trigonal geometry as repulsions near the carbons decrease.
In the hybrid orbital picture the decrease in electron density at carbon increases the *p*-character at that atom. A carbon hybrid orbital with greater *p*-character than that of an sp^2 orbital will have bond angles smaller than 120°.

2-32

a) sp^3 c) sp e) sp^2

b) sp^3 d) sp^3 f) sp^2

2-33

a) The angle is much less than the 109.5° predicted for sp^3 hybrid orbitals, thus is expected to be high in *p* character. The orbital is predicted to be between sp^4 and sp^5 in hybrid character.

b) Since the C-C-C component of the molecular orbitals is high in *p* character, the H-C-H part must be high in *s* character and thus possess an angle greater than 109.5°. This angle is actually found to be about 115°.

2-34

The electronegativities decrease in the order F > Cl > Br. Thus the shifting of electron density in the H—X bond decreases in that same order with an associated decrease in dipole moment.

2-35

i) If four sp^3 hybrid orbitals from the oxygen atom were used to accommodate the two hydrogen atoms and the two electron pairs, a tetrahedral angle of 109.5° would be the basis for prediction. Since the nonbonding electrons are closer to the nucleus, the orbital which they occupy would have more *s* character. The orbitals of the O-H bonds would thus have a higher *p* character and an angle smaller than 109.5°.

ii) The VSEPR argument assumes that the nonbonding pairs are closer to the nucleus and repel each other more than do the bonding electrons. The bonding orbitals thus move closer together than 109.5°.

2-36

Because the central sulfur atom is larger than an oxygen atom the electrons of the S-H bonds are farther apart than those of the O-H bonds in water. Repulsion is less and the bond angle is thus smaller.

2-37

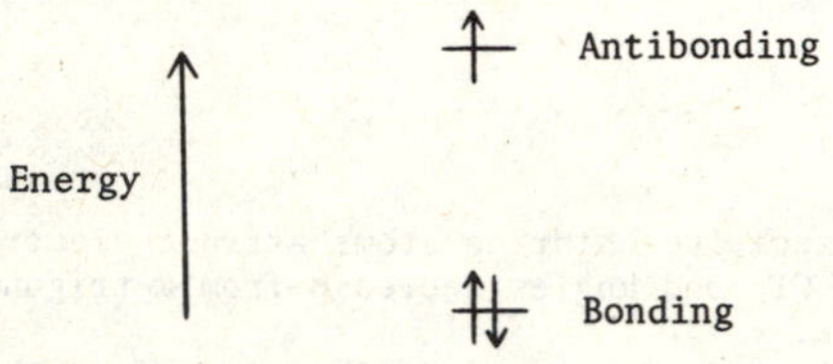

The Pauli exclusion principle requires that the third electron occupy the antibonding molecular orbital. That is a high energy configuration.

2-38 The C—Cl bond length decreases as the atoms substituted on carbon increase in electronegativity. Electron attraction toward the substituent gives the bond to that atom more *p*-character and leaves more *s*-character in the carbon-chlorine bond, thus shortening that bond. Another rationale is that the electronegativity difference generates a dipolar structure in which the opposotly charged atoms attract each other ($\overset{\delta+}{C}$——$\overset{\delta-}{Cl}$).

2-39

a)

Bonds broken	kcal/mol	kJ/mol
C—C	83	347
C—O	86	359
O—H	111	464
	280 kcal/mol	1170 kJ/mol

Bonds formed	kcal/mol	kJ/mol
C—H	99	414
C═O	192	803
	291 kcal/mol	1217 kJ/mol

$\therefore \Delta H°$ = 280 - 291 = -11 kcal/mol
(= 1170 - 1217 = -47 kJ/mol)

$\therefore$ Reaction is exothermic

Contd...

2-39 Contd...

b)

Bonds broken	kcal/mol	kJ/mol
2 C—N	146	610
N=N	100	418
	246 kcal/mol	1028 kJ/mol

Bonds formed		
C—C	83	347
N≡N	226	945
	309 kcal/mol	1292 kJ/mol

∴ $\Delta H° = 246 - 309 = -63$ kcal/mol
(= 1028 - 1292 = -264 kJ/mol)

∴ Reaction is exothermic

c)

Bonds broken	kcal/mol	kJ/mol
C—C	83	347
C—H	99	414
C—O	86	359
	268 kcal/mol	1120 kJ/mol

Bonds formed		
C=C	146	610
O—H	111	464
	257 kcal/mol	1074 kJ/mol

∴ $\Delta H° = 268 - 257 = 11$ kcal/mol
(= 1120 - 1074 = 46 kJ/mol)

∴ Reaction is endothermic

d)

Bonds broken	kcal/mol	kJ/mol
C—H	99	414
C—N	73	305
N=O	145	606
	317 kcal/mol	1325 kJ/mol

Bonds formed		
O—H	111	464
C=N	147	615
N—O	53	221
	311 kcal/mol	1300 kJ/mol

∴ $\Delta H° = 317 - 311 = 6$ kcal/mol
(= 1325 - 1300 = 25 kJ/mol)

∴ Reaction is endothermic

3 Classes and Nomenclature of Organic Compounds—Functional Groups

3-1

Addition of the remaining two carbons as an ethyl (C_2H_5) fragment would convert the potential butane parent to a pentane or hexane already represented.

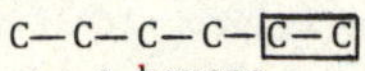

a hexane

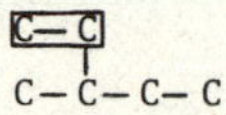

a pentane

3-2

The following responses make use of several commonly used representations for condensed structural formulas.

a) $CH_3CH(CH_3)CH(CH_3)CH(CH_3)CH_2CH_3$

b) 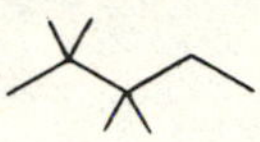

c) 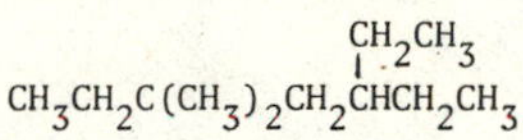

d) 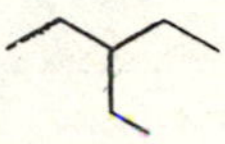

e) 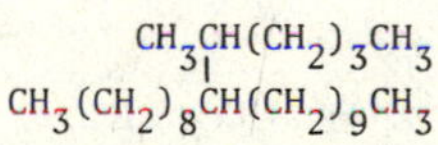

f) 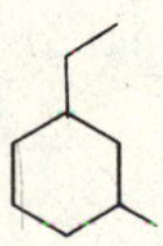

g) 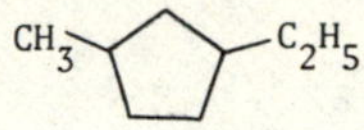

h) 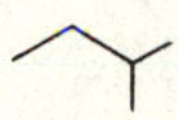

i) 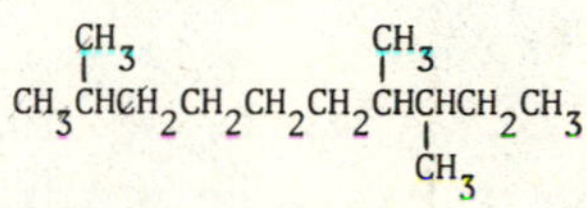

j) $CH_3CH_2CH(CH_3)C(CH_3)_2CH(CH_3)CH_2CH_3$

k) 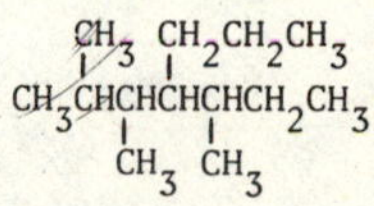

l) 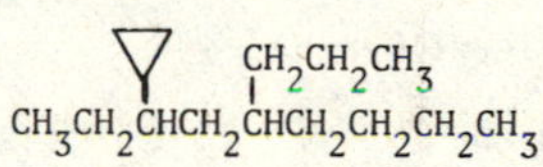

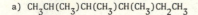

3-3

a) 3,4-Dimethylheptane

b) 5-(1-Methylpropyl)nonane or 5-*sec*-Butylnonane

c) 3-Methyl-1-(1-methylethyl)cyclohexane or 1-isopropyl-3-methylcyclohexane

d) 4-Ethyl-3,3-dimethyl-4-propyldecane

(Two decane chains are present. The parent chosen gives substituents the lowest numbers.)

e) 2,3-Dimethylpentane

f) 2,3,5-Trimethylhexane

g) (1,1-Dimethylethyl)cyclopentane <u>or</u> *tert*-Butylcyclopentane

h) 2,4-Dimethyl-3,3-di(1-methylethyl)pentane <u>or</u> 2,4-Dimethyl-3,3-diisopropylpentane

3-4

A, C, D, E, and H all are C_6H_{14} isomers.

3-5

a) C_nH_{2n} b) C_nH_{2n-2}; *IHD* = 2

3-6

a) Ethenyl b) 2-Propenyl c) 1-Methylethenyl

3-7

a) $CH_3\overset{CH_3}{\overset{|}{C}}{=}CHC(CH_3)_3$

b) $CH_2{=}CHCH_2CH_2CH{=}CH_2$

c) $CH_2{=}CH\overset{CH_2CH_3}{\overset{|}{C}H}CH_2CH_2CH_3$

d) $CH_2{=}C(CH_3)CH_2CH_3$

e) 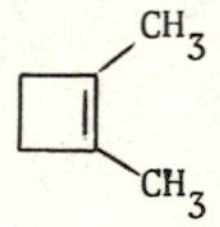

f) $CH_2{=}\overset{CH_3}{\overset{|}{C}}CH_2\underset{C_2H_5}{\underset{|}{\overset{CH_3}{\overset{|}{C}}}}CH_2CH_2CH_3$

3-8

a) 1-Ethyl-3,4-dimethylcyclohexene

b) 1,3-Dicyclobutylpropene

c) 4-Cyclopentyl-3-methyl-1,3-pentadiene

d) 2,2,5,5-Tetramethyl-3-hexene

e) 2-Isopropyl-3-(3,3-dimethylbutyl)-1,4-pentadiene <u>or</u>

3-(3,3-Dimethylbutyl)-2-(1-methylethyl)-1,4-pentadiene

f) 4-Vinyl-2,4-heptadiene <u>or</u> 4-Ethenyl-2,4-heptadiene

g) Butylbenzene

h) 1-Ethyl-3-methylbenzene <u>or</u> *m*-Ethyltoluene

3-9

a) 1,3-Hexadiene-5-yne

b) 4-Methyl-2-pentyne

c) 2,4,6-Octatriyne

d) Ethynylbenzene

3-10

a) 3-Methyl-1-butanol

b) 2-Ethyl-5,5-dimethyl-1-heptanol

c) 3-Methyl-5-hexyn-2-ol

d) 2-Cyclohexen-1-ol

e) 1,3-Propanediol

f) 1-Phenylethanol

3-11

The three hydroxy groups account for a significant degree of hydrogen bonding which increases intermolecular attractions and decreases volatility. Water solubility is due to the three hydroxy groups which make up a significant portion of the molecule and thus dominate over the hydrocarbon part of the molecule.

3-12

In higher molecular weight alcohols the functional group accounts for only a small part of the molecule and its associated physical properties. In these cases the hydrocarbon character dominates.

3-13

a) 2-Isopropyl-5-methylcyclohexanol <u>or</u> 2-(1-Methylethyl)-5-methylcyclohexanol

b) 3,7-Dimethyl-2,6-octadien-1-ol

3-14

Boiling points similar to those of similar molecular weight hydrocarbons suggest that hydrogen bonding association is not very important. Because the electrons on the larger sulfur are more diffuse, the atom is a weaker Lewis base than an oxygen atom.

3-15

a) 3-Methyl-1-butanethiol b) 2-Propene-1-thiol c) 1-Propanethiol

3-16

a) Dimethyl ether; Methoxymethane; 2-Oxapropane

b) *tert*-Butyl isopropyl ether; 2-*tert*-Butoxypropane; 2-Isopropoxy-2-methylpropane;

2,2,4-Trimethyl-3-oxapentane

c) Cyclohexyl methyl ether; Methoxycyclohexane

d) Methyl vinyl ether; Methyl ethenyl ether; Methoxyethene; 2-Oxa-3-butene (Ether takes priority over alkene)

3-17

Hydrogen bonding between ether molecules is not possible because no moderately acidic hydrogen atoms are present. Ether molecules are thus not strongly associated with each other and their boiling points are low. However, water is moderately soluble in ethers (and alcohols) because the ether oxygen atom can form hydrogen bonds to the water hydrogen atoms.

3-18

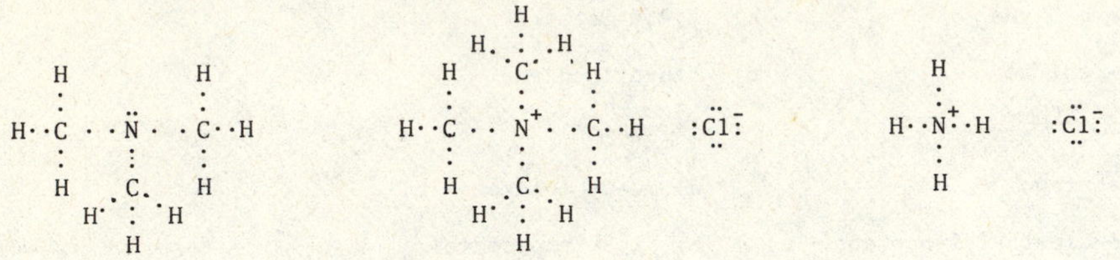

3-19

a) *tert*-Butylamine; 2-Methyl-2-propylamine; 2-Methyl-2-propanamine

b) *N*-Ethyl-*N*-methyl-1-propylamine; *N*-Ethyl-*N*-methyl-1-propanamine

c) 3-Butenylamine; 3-Butenamine

d) *N*,*N*-Diethylaniline; *N*,*N*-Diethylbenzenamine

3-20

The tertiary amine does not have a hydrogen atom on the nitrogen so that intermolecular hydrogen bonding is not possible.

3-21

a) 2-Bromo-2-methylpropane

b) 1,1,2-Trichloroethene

c) 3-Fluorocyclopentene

d) 1,3-Dichloro-5-methylbenzene; 3,5-Dichlorotoluene

e) 1,4-Diiodobutane

f) Tetrachloromethane

3-22

a) 1-Phenylethanone; Methyl phenyl ketone

b) 2-Chloro-3-pentanone; 1-Chloroethyl ethyl ketone

c) 1-Cyclohexyl-1-propanone; Cyclohexyl ethyl ketone

d) 5-Methyl-4-hexen-3-one; Ethyl 2-methyl-1-propenyl ketone

e) 3-Chloro-4-hydroxycyclohexanone.

3-23

a) 2-Methyl-2-butenal

b) 3-Chloro-2-methylpentanal

c) Ethanedial

d) Cyclobutanecarbaldehyde

e) 6-Bromo-2-oxocyclohexanecarbaldehyde (Aldehyde has priority over ketone which, in turn, has priority over bromine.)

3-24

a) 2-Methyl-3-phenylpentanoic acid

b) 2-Chlorobutanedioic acid

c) Cyclopentanecarboxylic acid

d) 2-Cyclopropylethanoic acid

e) 3-Cyclohexene-1,2-dicarboxylic acid

3-25

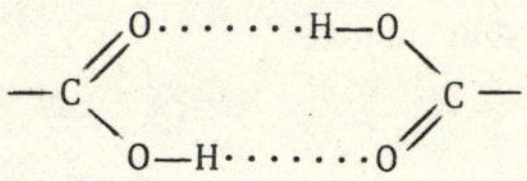

Double hydrogen bonding between two molecules is particularly favorable.

3-26

a) 2,2-Dimethylpropanoyl chloride

b) Methyl 3-chloro-3-methylbutanoate

c) 3-Cyclohexyl-2-methylpropanenitrile

d) *N*-Phenylmethanamide; *N*-Phenylformamide

e) Propanoic anhydride

f) Ethyl cyclobutanecarboxylate

g) *N*-Methylbenzamide; *N*-Methylphenylcarboxamide

h) 3-Phenylpentanenitrile

3-27

a) 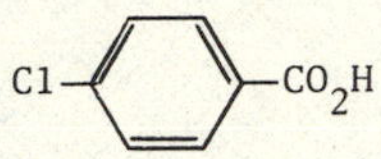

b) 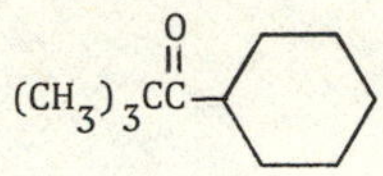

c) 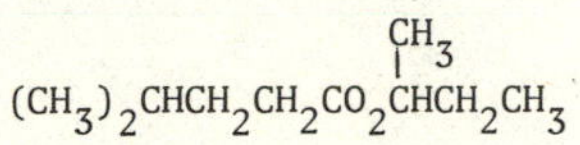

d) 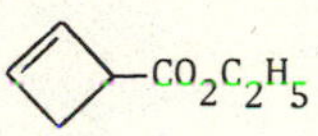

e) $CH_3OCH_2CH{=}CHCO_2H$

f) 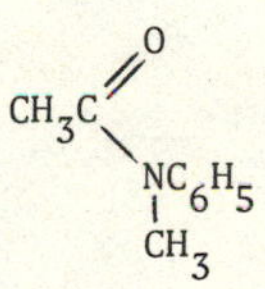

g) 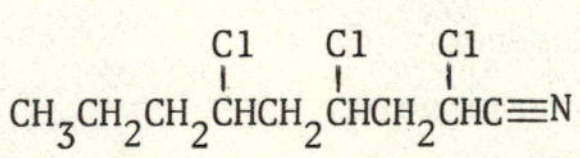

h) $(CH_3)_3CCO_2H$

i) 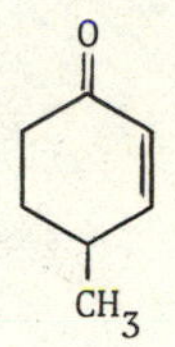

j) $(CH_3CH_2CH_2CH_2)_4N^+Br^-$

k) 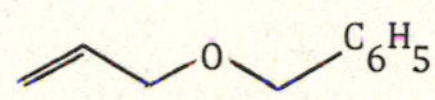

l) 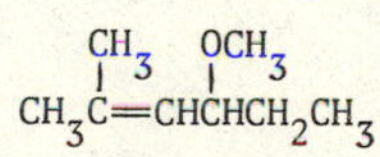

m) 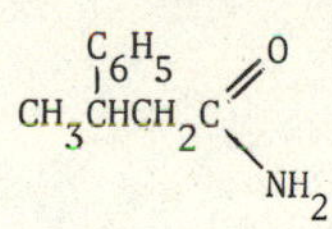

3-28

a) 3-Chloro-2,5-dimethyl-4-heptanone

b) 3-Amino-5-ethyl-1-hydroxybenzene

c) 2-Amino-4,4-dimethylpentanoic acid

d) 3-(1-Hydroxyethyl)-5-methylheptanal

e) 3-Bromo-5-ethylcyclopentanone

f) 5-Amino-2-methyl-2-cyclohexenone

g) 2-Ethyl-4-(*N*-ethylamino)-2-methylbutanal

h) 4-Hydroxy-5-methyl-2-hepten-6-ynoyl chloride

Contd...

3-28 Contd...

i) 3-Ethyl-3,5-dihydroxypentanamide

j) 3-Chloro-*N*-ethyl-3-methylpentanamide

k) 3-Hydroxy-4-methyl-4-pentenal

l) 3,6,9-Trimethyl-2,4,6,8-decatetraenoic acid

m) 3-Methyl-2-butanone

3-29

a) $C_6H_5CH_2CH_2OH$

b) 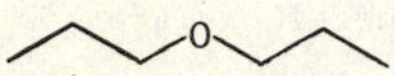

c) $CH_3\overset{\overset{O}{\|}}{C}CH_2CO_2H$

d) $CH_3SCH_2C_6H_5$

e) 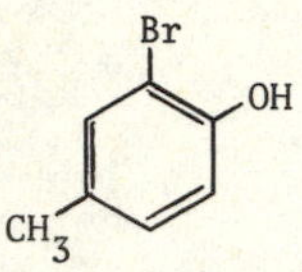

f) $[(CH_3)_2CH]_3N$

g) $CH_3\overset{\overset{OH}{|}}{C}HCH{=}CHCO_2H$

h) $(CH_3)_3CCH_2C(=O)NHCH_3$

i) $CH_2{=}CHCH{=}CHCH{=}C(CH_3)_2$

j) $CH_3CH_2CH_2\underset{\underset{C_2H_5}{|}}{\overset{\overset{CN}{|}}{C}}CH_2CH_2CO_2H$

k) 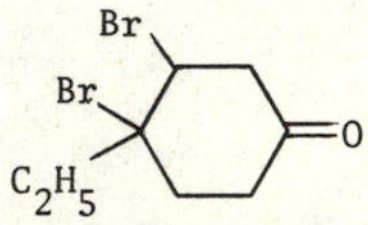

l) 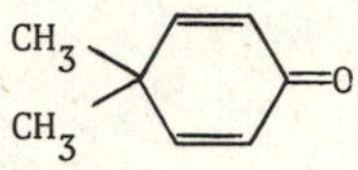

m) 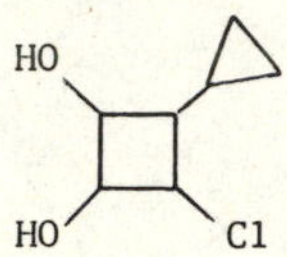

3-30

a) 4-Methyl-2-hexyne

b) 5-Methyl-3-phenyl-2-hexanol

c) Benzyl ethyl ether or 2-Oxa-1-phenylbutane

d) 5-Oxopentanoic acid

e) *N*,3,3-Trimethylbutanamine or 5,5-Dimethyl-2-azahexane

f) Chloromethylcyclopropane or Chlorocyclopropylmethane

g) 1-Chloro-4-methylbenzene

h) 4-Chloro-1-vinylbenzene or 4-Chloro-1-ethenylbenzene

i) 2-Isobutylcyclopentanone or 2-(2-Methylpropyl)cyclopentanone

j) 1,4-Cyclooctadiene

k) Cyclobutylethanoic acid

l) 3-Chloro-5-methyl-4-hexenoic acid

m) Methyl 3-chlorobutanoate

3-31

a) $CH_3C(=O)CH_2CH_3$
2-Butanone

b) $CH_3CH_2CH_2C(=O)H$
Butanal

$(CH_3)_2CHC(=O)H$
2-Methylpropanal

c) $CH_3CH_2C(=O)OCH_3$
Methyl propanoate

$CH_3C(=O)OCH_2CH_3$
Ethyl acetate
Ethyl ethanoate

$HC(=O)OCH_2CH_2CH_3$
Propyl formate
Propyl methanoate

$HC(=O)OCH(CH_3)_2$
Isopropyl formate
1-Methylethyl methanoate

d) $CH_3CH_2CH_2CH_2OH$
1-Butanol

$CH_3CH_2CH(OH)CH_3$
2-Butanol

$(CH_3)_2CHCH_2OH$
2-Methyl-1-propanol

$(CH_3)_3COH$
2-Methyl-2-propanol
tert-Butyl alcohol

e) $CH_3OCH_2CH_2CH_3$
Methyl propyl ether

$CH_3CH_2OCH_2CH_3$
Diethyl ether

$(CH_3)_2CHOCH_3$
Isopropyl methyl ether

3-32

a) CH_3OCH_3 and CH_3CH_2OH
Dimethyl ether   Ethanol

b) Ethanol

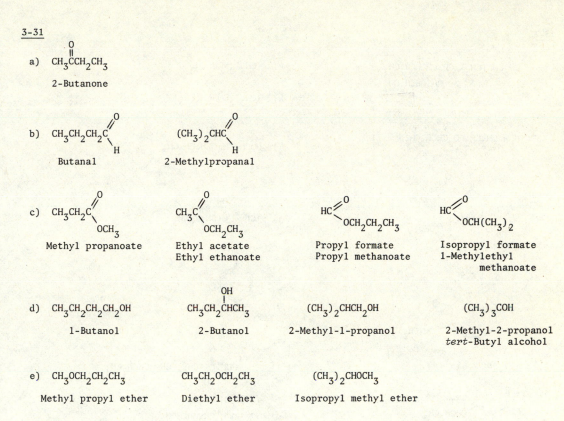

c) Ethanol

3-33

a) OH, C_2H_5
$CH_3CHC(CH_3)_2CH_2CHCHCH_2CH_3$
$CH_3CHCH_2CH_3$

b) CH_3
$C_6H_5CH_2CH_2CO_2CHCH_2CH_3$

c)
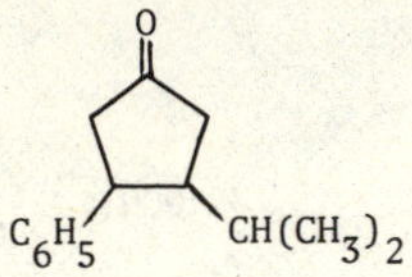

d)
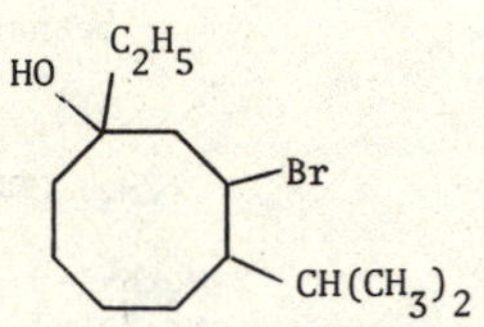

e)
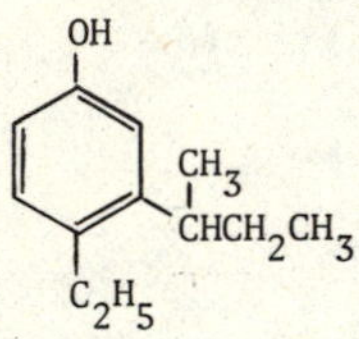

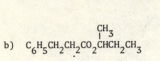

f) $(C_2H_5)_3\overset{+}{N}H \quad OH^-$

g) OH
$(CH_3)_2NCH_2CH_2CHCH_3$

h) $(CH_3)_2CHSCH(CH_3)_2$

i) OH
$HOCH_2CH_2CCH_2CH_2CH_3$
C_2H_5

j) OH
$(CH_3)_2C(CH_2)_5CH_3$

k)
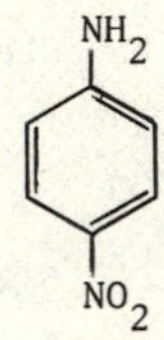

l) Cl, O
$CH_3CHCH_2CC(CH_3)_2CH{=}CHCH_2CH_3$

m) $ClCH_2CH_2CH_2CH_2$
$HOCH_2CH_2CHCHCH_3$
OH

n)
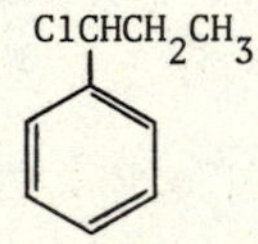

o)
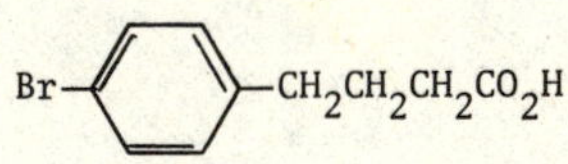

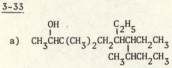

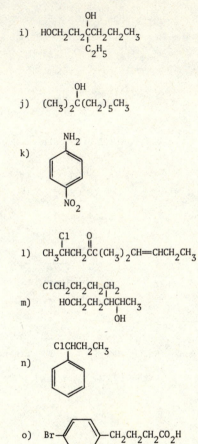

3-34

a) 5-Bromo-7-methyl-2,4,6-octatrienal

b) 3-Phenyl-2-butanone

c) 2,5-Dimethyl-4-hexen-3-one

d) 1,5-Diamino-2-pentanol

e) 5-*tert*-Butyl-3,5-nonanediol

f) 1-Ethoxy-2-propanol

g) 3-Bromo-2-ethyl-6-methylcyclohexanone

h) 3-(3-Bromopropyl)-1-ethyl-4-methylcyclohexene

i) 4-Chloro-3-(2-methylpropyl)-2-heptanone

j) Methyl 2-methoxy-6-methyl-3-cyclohexenecarboxylate

k) 3-(4-Chlorophenyl)-1,2-diphenyl-2-propen-1-one

l) 4-Bromo-5-hydroxy-2-pentanone

m) 3-Phenylpentanenitrile

n) 3-*tert*-Butyl-2,2,4,4-tetramethylpentane

o) 3-Methyl-2-oxa-4-phenylpentane

3-35

a) 4-*tert*-Butyl-1-methoxybenzene

b) 4-Chloro-2-methylpentanenitrile

c) 2-Methyl-4-nitro-2-pentanol

d) 2,4-Dimethyl-5-heptene-2,4-diol

e) 5-Amino-6-(1-methylpropyl)-2-cyclohexen-1-ol

f) Methyl 3-bromo-2-hydroxy-2-methylbutanoate

g) 4-Bromo-2-hydroxy-*N*,3-dimethyl-4-hepten-6-ynamide

h) 2-Bromo-2-methyl-3-cyclopentenone

i) 3-Chloro-4-oxo-5-propylcyclohexanecarbonitrile

j) 2-Butyl-4-chloro-5-methyl-2,4-hexadienoyl chloride

k) 1-Cyclopropyl-4-hydroxy-1-butanone

4 Characteristic Reactions of Organic Compounds

4-1

$CH_3\ddot{O}H + B^{-} \rightleftharpoons CH_3\ddot{O}^{-} + BH$ Methanol as an acid

$CH_3\ddot{O}H + HX \rightleftharpoons CH_3\overset{+}{\ddot{O}}H_2 + X^{-}$ Methanol as a base

4-2

A proton is a Lewis acid for it is an electron pair acceptor. It is not a proton donor.

4-3

a) $CH_3CO_2H + H_2O \overset{K_a}{\rightleftharpoons} CH_3CO_2^- + H_3O^+$

$$K_a = \frac{[CH_3CO_2^-][H_3O^+]}{[CH_3CO_2H]} = 1.76 \times 10^{-5}$$

$$pK_a = -\log 1.76 \times 10^{-5} = -(0.25 - 5) = 4.75$$

The concentration of H_3O^+ is determined by assuming that $[CH_3CO_2H] \approx 1$ for a 1N solution since the percent dissociation is small. Since $[H_3O^+] = [CH_3CO_2^-]$ in this case, $[H_3O^+]^2 = 1.76 \times 10^{-5}$ and $[H_3O^+] = 4.2 \times 10^{-3}$.

$\therefore pH = -\log 4.2 \times 10^{-3} = 2.4$

b) $\frac{4.2 \times 10^{-3}}{1} \times 100 = 0.4\ \%$

4-4

a) For the monoprotic acid HA, in aqueous media:

$$HA + H_2O \overset{K'_a}{\rightleftharpoons} H_3O^+ + A^-$$

$$K_a = \frac{[H_3O^+][A^-]}{[HA]}$$

$$-\log K_a = -\log[H_3O^+] - \log\frac{[A^-]}{[HA]}$$

$$= -\log[H_3O^+] + \log\frac{[HA]}{[A^-]}$$

$$\therefore pK_a = pH + \log\frac{[HA]}{[A^-]}$$

b) For a monoprotic acid at one-half neutralization $[HA] = [A^-]$ $\therefore \frac{[A^-]}{[HA]} = 1$

$K_a = [H_3O^+]$ and $\log K_a = \log[H_3O^+]$.

Since $pK_a = -\log K_a$ and $pH = -\log[H_3O^+]$, then $pK_a = pH$

4-5

a) Decreasing acidity: HI > HBr > HCl > HF

b) Decreasing acidity: $CH_3\overset{+}{O}H_2 > CH_3CO_2H > CH_3OH$

Protonated methanol is a very strong acid for it is the conjugate acid of a very weak base. The carboxylic acid is a weak acid. Methanol is a very weak acid since it is the conjugate acid of a strong base.

c) Ammonium salts are weak acids because they are the conjugate acids of strong bases. Amines possessing an N-H are very weak acids for they are the conjugate acids of very strong bases, $-\ddot{N}-$.

4-6

a)

$[HCl + H_2O]$ Reactants

ΔG° (arrow) $\qquad \Delta G^\circ < 0$

$[H_3O^+ + Cl^-]$ Products

b) $[CF_3CO_2^- + H_3O^+]$ Products $\qquad [CF_3CO_2H + H_2O]$ Reactants $\qquad \Delta G^\circ \approx 0$

Contd...

4-6 Contd...

c) $[C_6H_5O^- + H_3O^+]$ Products

ΔG° ↑ $\Delta G^\circ > 0$

$[C_6H_5OH + H_2O]$ Reactants

4-7

Base	Conjugate acid	pK_a
NaH	H-H	-
t-BuLi	$(CH_3)_3CH$	>50
n-BuLi	$CH_3(CH_2)_2CH_3$	≈50
$(iPr)_2NLi$	$(iPr)_2NH$	38
$NaNH_2$	NH_3	36
$(C_6H_5)_3CNa$	$(C_6H_5)_3CH$	32
$(CH_3)_3COK$	$(CH_3)_3COH$	18
C_2H_5ONa	C_2H_5OH	16
NaOH	HOH	15.7
Et_3N	$Et_3\overset{+}{N}H$	≈10
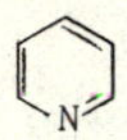	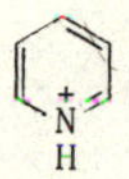	5.2
CH_3CO_2Na	CH_3CO_2H	4.8

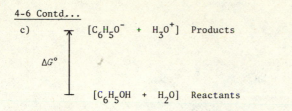

4-8

a) For the two reactants, consider dissociation of the acid and the conjugate acid of the base.

$$CH_3CO_2H \underset{}{\overset{K_a}{\rightleftharpoons}} CH_3CO_2^- + H^+ \qquad K_a = \frac{[CH_3CO_2^-][H^+]}{[CH_3CO_2H]} = 10^{-4.8}$$

$$C_2H_5\overset{+}{N}H_3 \underset{}{\overset{K'_a}{\rightleftharpoons}} C_2H_5NH_2 + H^+ \qquad K'_a = \frac{[C_2H_5NH_2][H^+]}{[C_2H_5\overset{+}{N}H_3]} = 10^{-10}$$

Contd...

4-8 Contd...

a) Contd...

For the total reaction

$$CH_3CO_2H + C_2H_5NH_2 \overset{K}{\rightleftharpoons} CH_3CO_2^- + C_2H_5\overset{+}{N}H_3$$

$$K = \frac{[CH_3CO_2^-][C_2H_5\overset{+}{N}H_3]}{[CH_3CO_2H][C_2H_5NH_2]} = \frac{K_a}{K'_a} = \frac{10^{-4.8}}{10^{-10}} = 10^{5.2}$$

The reaction is considered complete since $K > 10^2$. Note that the value of the equilibrium constant for these acid-base reactions is calculated from the ratio of the K_a value of the acid to the K_a value of the conjugate acid of the base.

b) $CH_3O^- + CH_3COCH_2CO_2CH_3 \rightleftharpoons CH_3OH + CH_3CO\bar{C}HCO_2CH_3$

$$K = \frac{10^{-11}}{10^{-15}} = 10^4 \quad \therefore \text{complete}$$

c) $CO_3^{2-} + HCN \rightleftharpoons HCO_3^- + CN^-$

$$K = \frac{10^{-9.1}}{10^{-10.2}} = 10^{1.1} \quad \therefore \text{only about 70\% complete}$$

d) 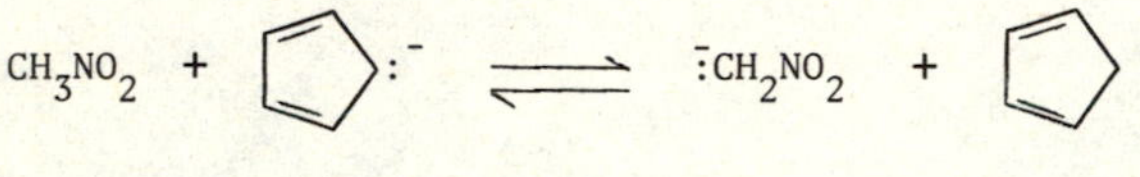

$$K = \frac{10^{-10.2}}{10^{-16}} = 10^{5.8} \quad \therefore \text{complete}$$

e) $HC{\equiv}CH + NH_2^- \rightleftharpoons HC{\equiv}C:^- + NH_3$

$$K = \frac{10^{-25}}{10^{-36}} = 10^{11} \quad \therefore \text{complete}$$

f) $HCO_3^- + HCO_2H \rightleftharpoons H_2CO_3 + HCO_2^-$

$$K = \frac{10^{-3.7}}{10^{-6.4}} = 10^{2.7} \quad \therefore \text{complete}$$

4-8 Contd...

g) $Cl_2CHCO_2H + C_6H_5NH_2 \rightleftharpoons Cl_2CHCO_2^- + C_6H_5\overset{+}{N}H_3$

$$K = \frac{10^{-1.3}}{10^{-4.6}} = 10^{3.3} \quad \therefore \text{complete}$$

h) 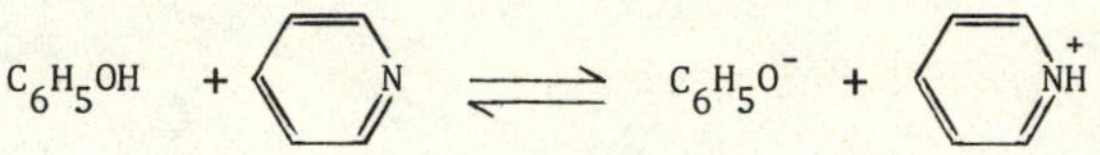

$$K = \frac{10^{-10}}{10^{-5.2}} = 10^{-4.8} \quad \therefore \text{equilibrium lies far to the left}$$

4-9

$CH_3(CH_2)_5CH{=}CH_2$
1-Octene

$CH_3(CH)_5CHBrCH_2Br$
1,2-Dibromooctane

$CH_3C{\equiv}CH$
Propyne

$CH_3CBr{=}CHBr$
1,2-Dibromopropene

$CH_3CBr_2{-}CHBr_2$
1,1,2,2-Tetrabromopropane

$CH_3CH_2CH{=}CH_2$
1-Butene

$CH_3CH_2CH_2CH_3$
Butane

1,3-Cyclohexadiene

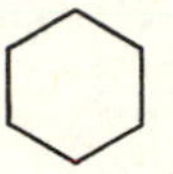
Cyclohexane

$CH_2{=}CH_2$
Ethene

CH_3CH_2OH
Ethanol

$CH_3CH{=}CH_2$
Propene

$CH_3CH(OH)CH_3$
2-Propanol

CH_3CH_2Cl
Chloroethane

4-10

a) Assuming that the carbon skeleton remains unchanged during the reactions, A must be a cyclobutyl ring with a one carbon side chain.
Since a C_5 parent would have H_{12}, the IHD=2 would be consistent with the cyclic structure plus one double bond. The addition of hydrogen and of bromine is also consistent with a double bond.
Three possible structures fit the data.

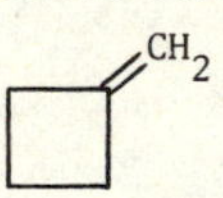

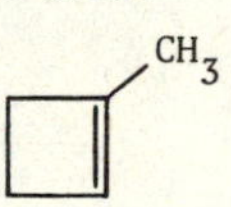

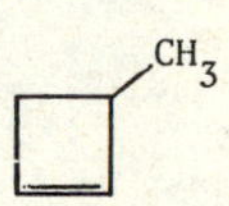

4-10 Contd...

b) Using one of the potential structures.

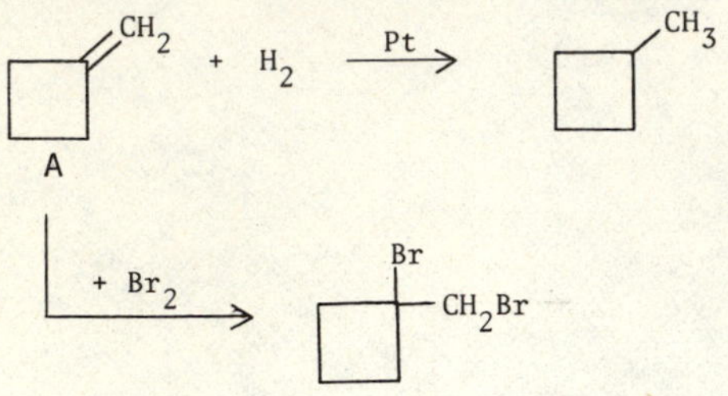

c) The data do not define a unique structure.

4-11

$CH_3CH_2CH_2CH_2OH$
1-Butanol

$CH_3CH_2CH{=}CH_2$
1-Butene

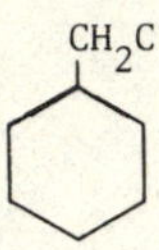

Chloromethylcyclohexane

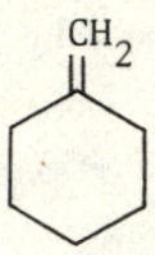

Methylenecyclohexane

OH
|
$CH_3CH_2CHCH_3$
2-Butanol

$CH_3CH{=}CHCH_3$
2-Butene

$CH_3CH_2CH{=}CH_2$
1-Butene

Br
|
$(CH_3)_2CCH_2CH_3$
2-Bromo-2-methylbutane

$(CH_3)_2C{=}CHCH_3$
2-Methyl-2-butene

CH_3
|
$CH_2{=}CCH_2CH_3$
2-Methyl-1-butene

4-12

The compound C_3H_8O has IHD=0 and is therefore saturated and acyclic. Two alcohols are possible.

$CH_3CH_2CH_2OH$ (A) or $CH_3CH(OH)CH_3$ (A′)

OH
|
CH_3CHCH_3
A′

Heating with acid dehydrates either alcohol to the same alkene, propene (B).

$$\text{A or A}' \xrightarrow[\Delta]{H^+} \underset{\text{B}}{CH_3CH{=}CH_2}$$

Propene adds Br_2 to give 1,2-dibromopropane (D).

$$CH_3CH{=}CH_2 + Br_2 \longrightarrow \underset{\text{D}}{CH_3CHBrCH_2Br}$$

Br
|
CH_3CHCH_2Br
D

Contd...

4-12 Contd...

The choice of which alcohol is correct depends on the fact (sec. 4-2A) that addition of water to propene gives C , the product in which a hydrogen atom from water has added to the alkene carbon atom already possessing the greater number of hydrogen atoms.

$$CH_3CH{=}CH_2 \quad + \quad H_2O \xrightarrow{H^+} \underset{A' \equiv C}{CH_3\overset{OH}{\overset{|}{C}}HCH_3}$$

Thus A is the original alcohol.

4-13

CH_3CH_2Br Bromoethane	CH_3CH_2OH Ethanol	$(CH_3)_3CCl$ 2-Chloro-2-methylpropane	$(CH_3)_3COH$ 2-Methyl-2-propanol
$CH_3CH_2CH_3$ Propane	$CH_3CH_2CH_2Cl$ 1-Chloropropane	$CH_3\overset{Cl}{\overset{\vert}{C}}HCH_3$ 2-Chloropropane	 Benzene
			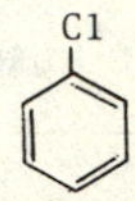 Chlorobenzene

4-14

a)

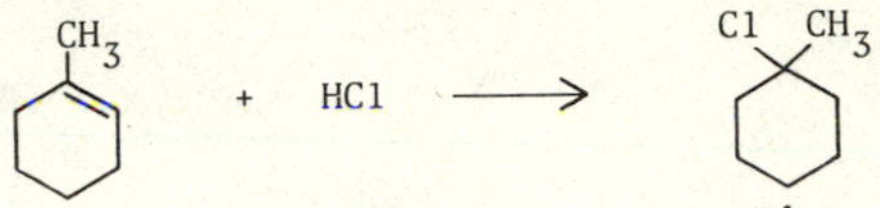

A

Addition

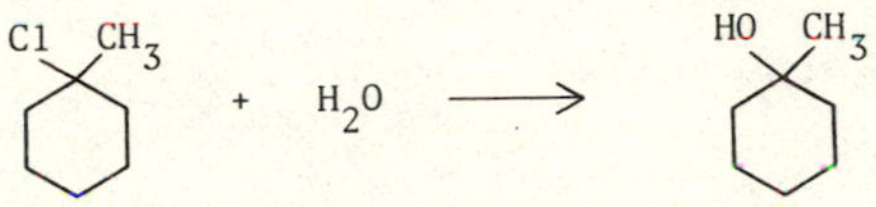

B

Substitution

4-15

a) Substitution

b) Elimination

c) Substitution

d) Addition

e) Elimination

f) Elimination

4-16

a) $\underset{-2}{C}H_3OH$

b) $\underset{-3}{CH_3}-\overset{O}{\overset{\|}{\underset{+2}{C}}}-\underset{-3}{CH_3}$

c) $\underset{-3}{CH_3}-\underset{-2}{CH_2}-\underset{+3}{C}(=O)OH$

d) $\underset{-2}{CH_3}-O-\underset{0}{CH}{=}\underset{-1}{CH}-\underset{+3}{C}{\equiv}N$

e) 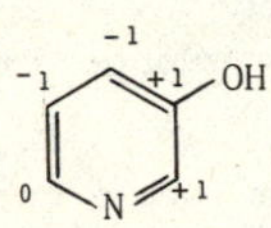

f) $H\underset{-1}{C}{\equiv}\underset{+1}{C}-NH-\overset{O}{\overset{\|}{\underset{+3}{C}}}-\underset{-1}{CH_2}Cl$

4-17

Change in oxidation state:

$C_6H_5-CH{=}CH_2$ (−1, −2) ⟶ $C_6H_5-C(=O)-C(=O)OH$ (+2, +3) (change = +8)

$3e^- + Mn^{+7} \longrightarrow Mn^{+4}$ (change = -3)

Adjusting stoichiometry and electron balance:

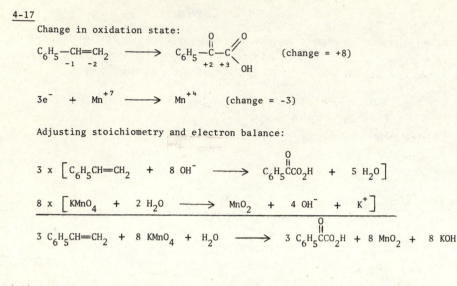

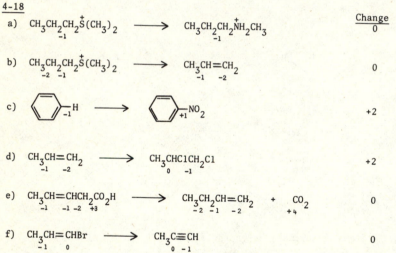

$3 \times [C_6H_5CH{=}CH_2 + 8\ OH^- \longrightarrow C_6H_5C(=O)CO_2H + 5\ H_2O]$

$8 \times [KMnO_4 + 2\ H_2O \longrightarrow MnO_2 + 4\ OH^- + K^+]$

$3\ C_6H_5CH{=}CH_2 + 8\ KMnO_4 + H_2O \longrightarrow 3\ C_6H_5C(=O)CO_2H + 8\ MnO_2 + 8\ KOH$

4-18

	Reaction	Change
a)	$CH_3CH_2CH_2\overset{+}{S}(CH_3)_2$ (−1) ⟶ $CH_3CH_2CH_2\overset{+}{N}H_2CH_3$ (−1)	0
b)	$CH_3CH_2CH_2\overset{+}{S}(CH_3)_2$ (−2, −1) ⟶ $CH_3CH{=}CH_2$ (−1, −2)	0
c)	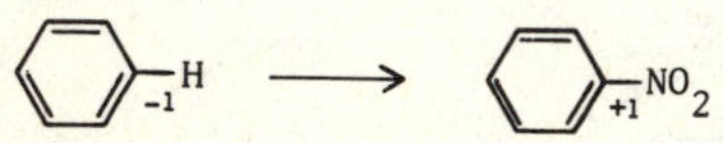	+2
d)	$CH_3CH{=}CH_2$ (−1, −2) ⟶ $CH_3CHClCH_2Cl$ (0, −1)	+2
e)	$CH_3CH{=}CHCH_2CO_2H$ (−1, −1, −2, +3) ⟶ $CH_3CH_2CH{=}CH_2$ (−2, −1, −2) + CO_2 (+4)	0
f)	$CH_3CH{=}CHBr$ (−1, 0) ⟶ $CH_3C{\equiv}CH$ (0, −1)	0

4-19

Weight of C in sample from CO_2 recovered $= \frac{12.011}{44.011} \times 8.80 = 2.402$ mg.

Weight of H in sample from H_2O recovered $= \frac{2.016}{18.016} \times 3.60 = 0.403$ mg.

$\%\ C = \frac{2.402}{6.00} \times 100 = 40.03$ mg/atms $C = \frac{40.03}{12.011} = 3.33$

$\%\ H = \frac{0.403}{6.00} \times 100 = 6.72$ mg/atms $H = \frac{6.72}{1.008} = 6.7$

% O (by difference) = 53.25 mg/atms $O = \frac{53.25}{16.00} = 3.33$

Contd...

4-19 Contd...

The atom ratio is 3.33 : 6.7 : 3.33 = 1 : 2 : 1. Empirical formula = CH_2O

The molecular formula of acetic acid is $C_2H_4O_2$.

4-20

a) Substitution
b) Addition
c) Substitution
d) Addition
e) Addition
f) Elimination
g) Addition
h) Substitution
i) Addition
j) Addition
k) Elimination

4-21

Change

a) $\underset{-4}{CH_4} + Cl_2 \longrightarrow \underset{-2}{CH_3Cl} + HCl$ +2

b) $\underset{-3}{CH_3}\underset{-2}{CH_2}\underset{-1}{CH}{=}\underset{-2}{CH_2} + HCl \longrightarrow \underset{-3}{CH_3}\underset{-2}{CH_2}\underset{0}{CHCl}\underset{-3}{CH_3}$ 0

(Total = -8) (Total = -8)

c) 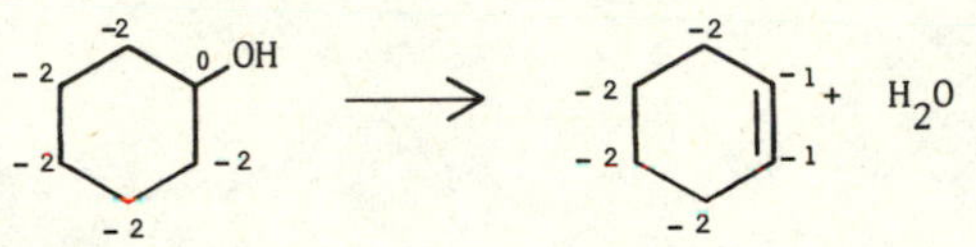 0

(Total = -10) (Total = -10)

d) $\underset{-3}{CH_3}\underset{-2}{CH_2}\underset{+3}{C}(=O)OH + \underset{-2}{CH_3OH} \longrightarrow \underset{-3}{CH_3}\underset{-2}{CH_2}\underset{+3}{C}(=O)\underset{-2}{OCH_3} + H_2O$ 0

(Total = -4) (Total = -4)

e) 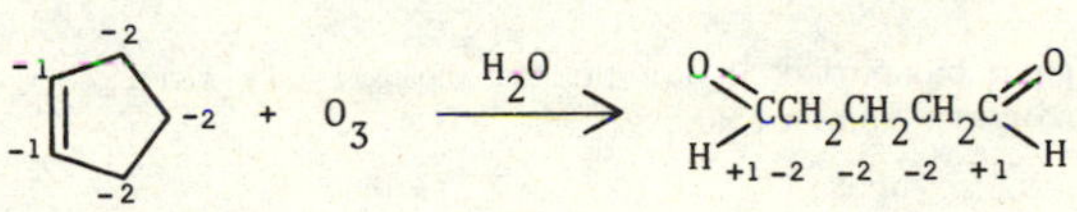 +4

(Total = -8) (Total = -4)

4-22

$pH = -\log[H_3O^+] = -\log 10^{-7} = 7$

$K_a = \dfrac{[H_3O^+][OH^-]}{[H_2O]} = \dfrac{[10^{-7}][10^{-7}]}{[55.5]^*} = 1.8 \times 10^{-16}$ $\therefore pK_a = -\log K_a = 15.7$

* 55.5 is the molarity of water

4-23

To remove >50% of one proton the conjugate acid of the base must have a pK_a value equal to or greater (be less acidic) than the substrate. The conjugate acids of the bases are:

CH_3CO_2H	pK_a = 4.8	C_2H_5OH	pK_a = 16	$(C_6H_5)_3CH$	pK_a = 32
$(C_2H_5)_3\overset{+}{N}H$	pK_a = 10	$(CH_3)_3COH$	pK_a = 18		

a) $C_6H_5N\underline{H}_2$; pK_a = 27 ∴ $(C_6H_5)_3C^-Na^+$

b) $CH_3CH{=}CHC\underline{H}_3$; pK_a = 35 ∴ None

c) $C_2H_5COC\underline{H}_3$; pK_a = 20 ∴ $(C_6H_5)_3C^-Na^+$

d) $(NC)_2C\underline{H}_2$; pK_a = 11.2 ∴ $(C_6H_5)_3C^-Na^+$, $(CH_3)_3CO^-Na^+$, $C_2H_5O^-Na^+$

e) 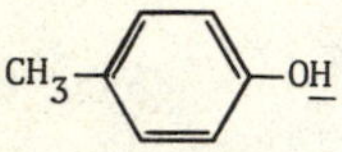

; pK_a ≳ 10 ..same as (d)

f) 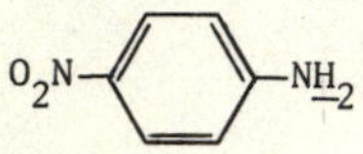

; pK_a = 18 ∴ $(C_6H_5)_3C^-Na^+$; $(CH_3)_3CO^-Na^+$

4-24

a) 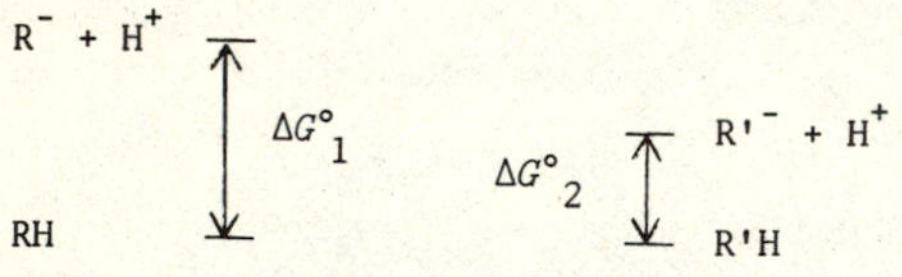

$\Delta G°_1$ more endergonic than $\Delta G°_2$
∴ R'H more acidic

b) 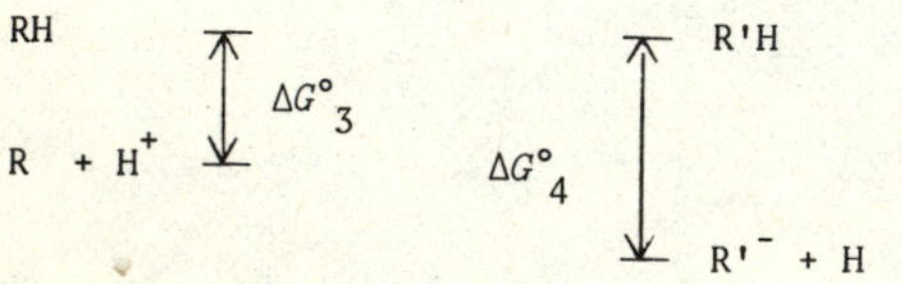

$\Delta G°_4$ more exergonic than $\Delta G°_3$
∴ R'H more acidic

Note that in both cases above the conjugate base of R'H has become relatively more favorable thermodynamically than the conjugate base of RH.

4-25

Sulfuric acid is a relatively stronger acid than acetic. Acid-base relationships are *relative* between the compounds being compared.

4-26

All of the acid dissociations are endergonic.

a) BH because ΔG_B has the smallest value

b) CH

c) AH

4-27

a) 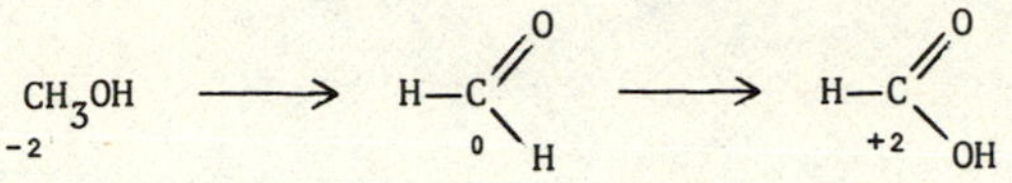

Each step is an oxidation. Total change at carbon = +4..

b) 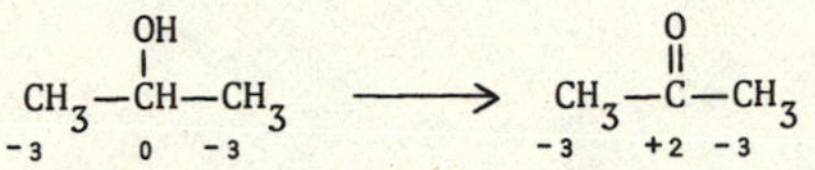

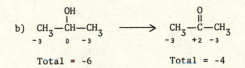

Total = -6 Total = -4

Total oxidation change = +2.

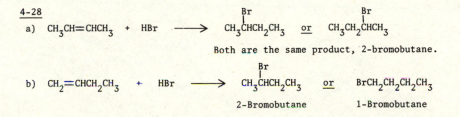

4-28

a) $CH_3CH=CHCH_3$ + HBr ⟶ $CH_3\overset{Br}{C}HCH_2CH_3$ or $CH_3CH_2\overset{Br}{C}HCH_3$

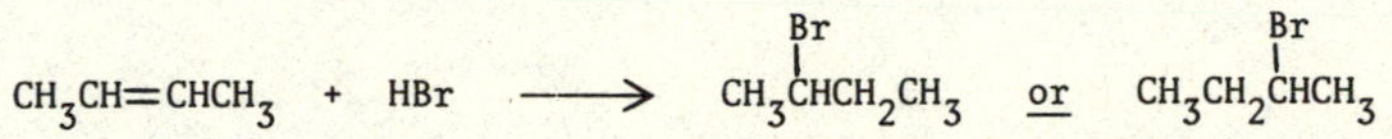

Both are the same product, 2-bromobutane.

b) $CH_2=CHCH_2CH_3$ + HBr ⟶ $CH_3\overset{Br}{C}HCH_2CH_3$ or $BrCH_2CH_2CH_2CH_3$

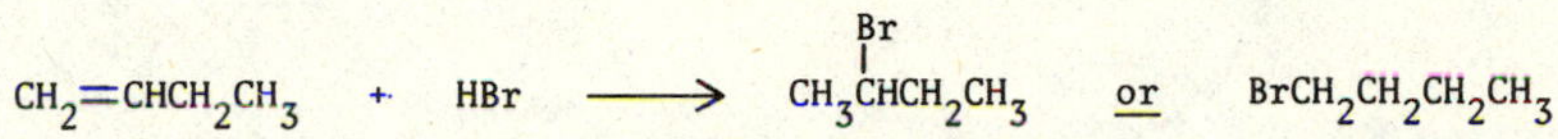

2-Bromobutane 1-Bromobutane

5 Reaction Mechanism

5-1

Nucleophiles: $:\ddot{I}:^-$; $H_2\ddot{O}$; $(CH_3)_3N:$; $:CN^-$

Electrophiles: $:\ddot{Cl}^+$; Li^+

5-2 and 5-3

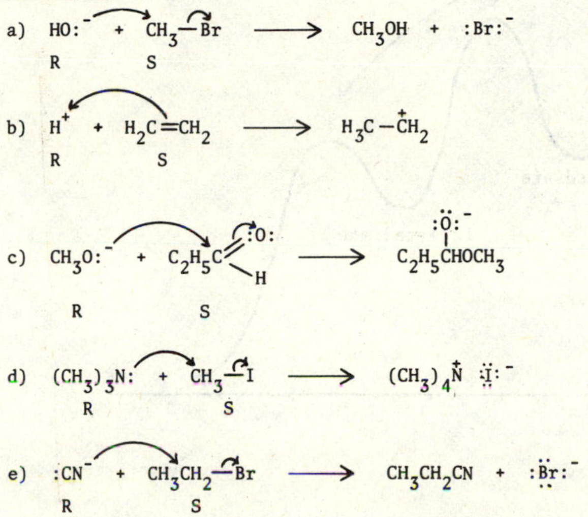

a) Nucleophilic substitution

b) Electrophilic addition

c) Nucleophilic addition

d) Nucleophilic substitution

e) Nucleophilic substitution

5-4

a) $RT = 1.99 \times 10^{-3} \times 298 = 0.6$ kcal/mol (2.5 kJ/mol)

b) Usually this difference is small relative to the magnitude of Ea (or $\Delta H^{\ddagger}$), thus $\Delta H^{\ddagger}$ is commonly equated to the measured Ea.

5-5

a)

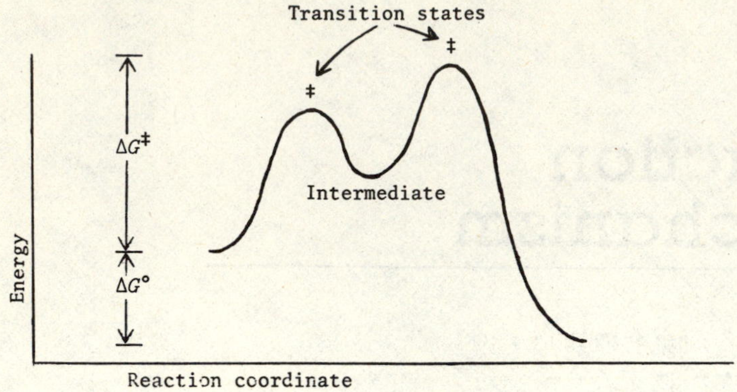

b)

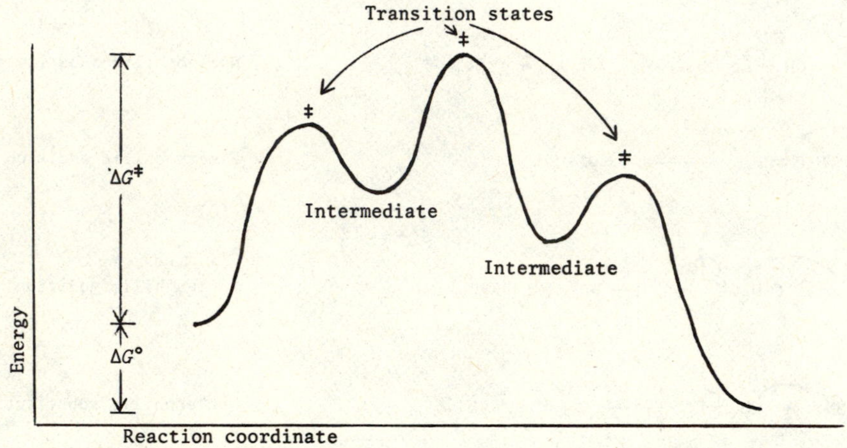

5-6

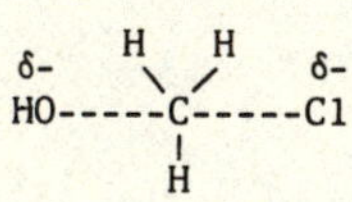

5-7

At the half life, $C_t = \frac{1}{2}C_o$

$\ln C_o/C_t = \ln 2 = 0.69$

$\therefore t_{\frac{1}{2}} = 0.69/k$

5-8

a) k_2 has the greater value.

b) Since the rate constant for the first step is smaller its energy of activation is greater and the step is slower.

5-9

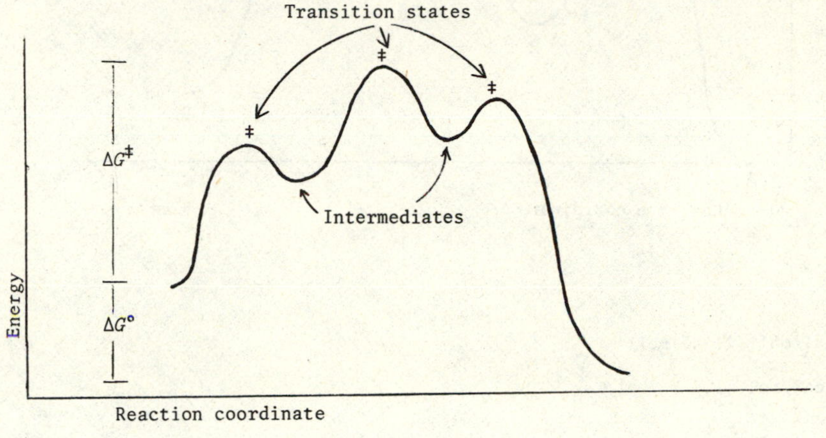

5-10

a) Nucleophilic substitution

b) Elimination (no electronic classification is commonly used)

c) Electrophilic substitution

d) Nucleophilic substitution

e) Nucleophilic addition

f) Electrophilic addition

g) Nucleophilic substitution

h) Nucleophilic substituion

5-11

a) + b)

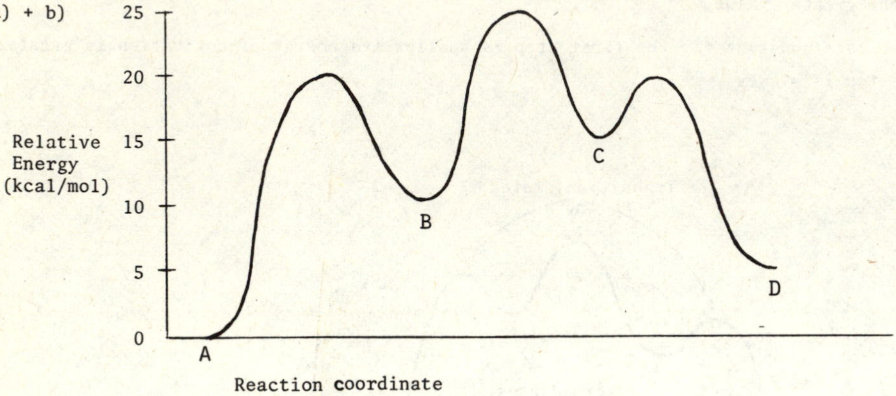

c) B $\longrightarrow$ C

d) ΔG° = 5 kcal/mol (21 kJ/mol)

$\Delta G^\ddagger$ = 25 kcal/mol (104 kJ/mol)

5-12

a) 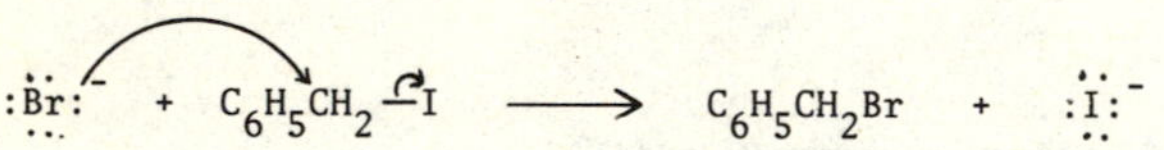

b) 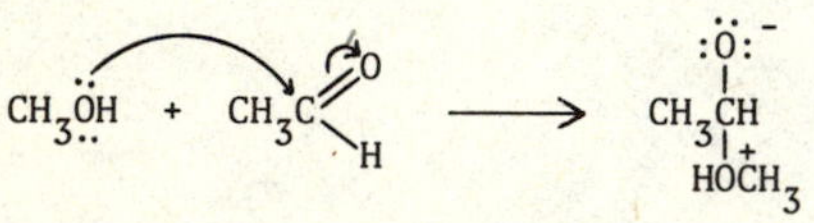

c) 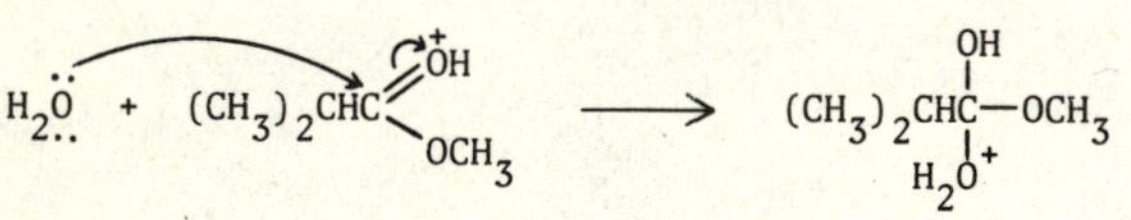

d) 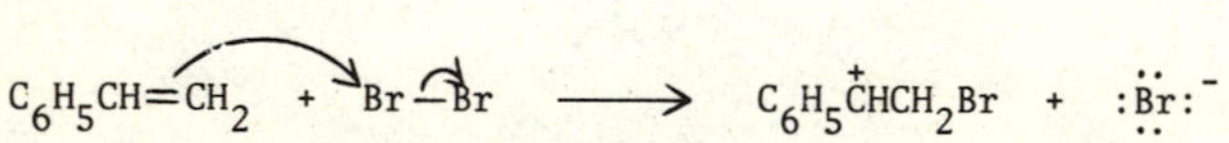

e) 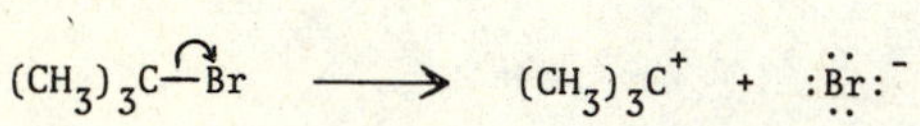

f) 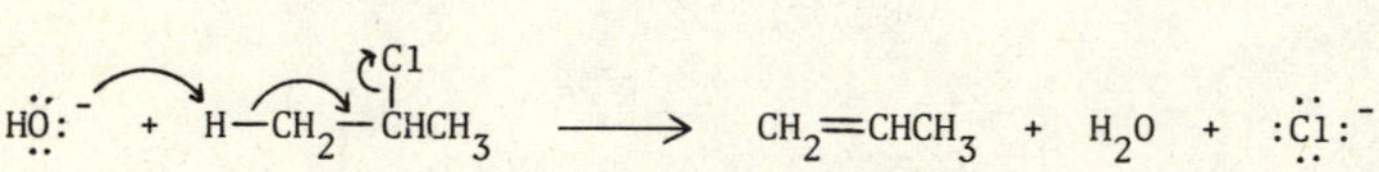

5-13

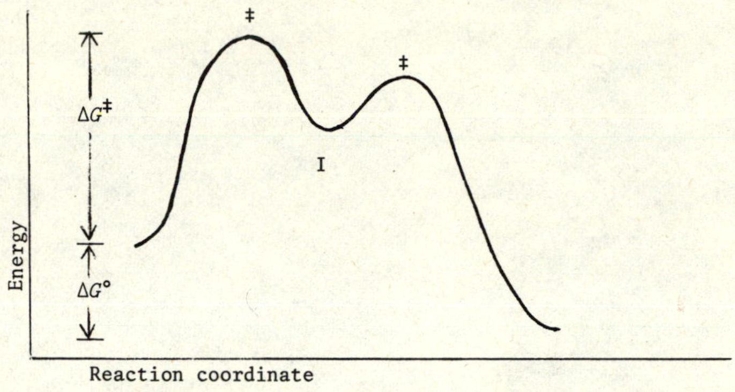

I ≡ 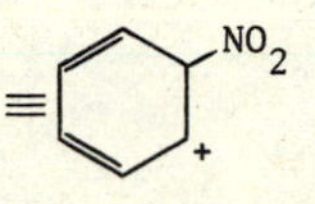

6 The Shapes of Molecules—Stereochemistry

6-1

The following three formulas represent one of the many possible spatial arrangements of 2,3-dichlorobutane.

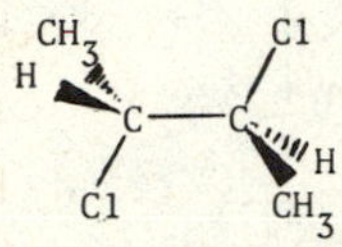

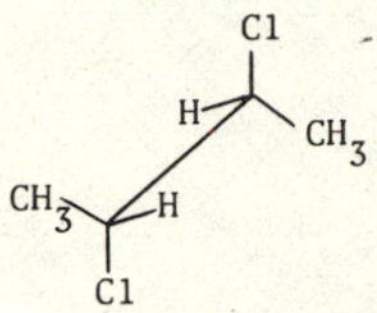

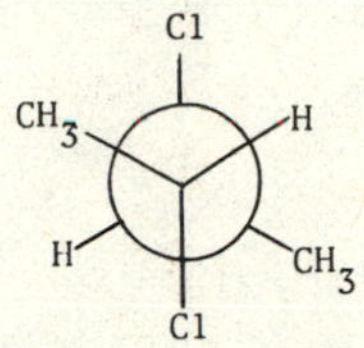

6-2

a)

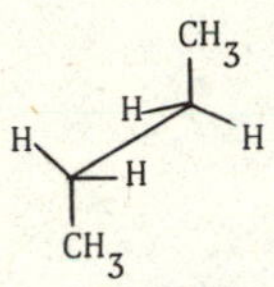

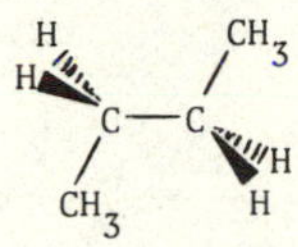

b)

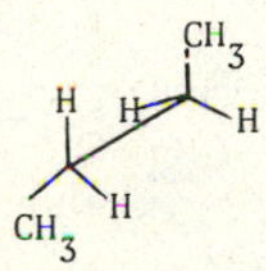

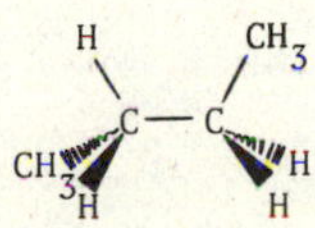

c)

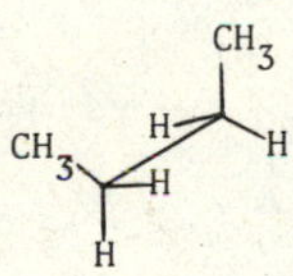

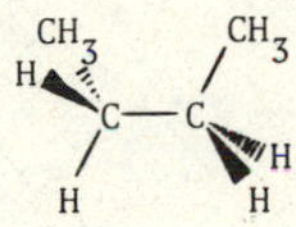

d)

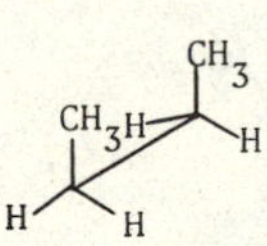

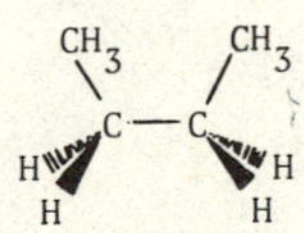

6-3

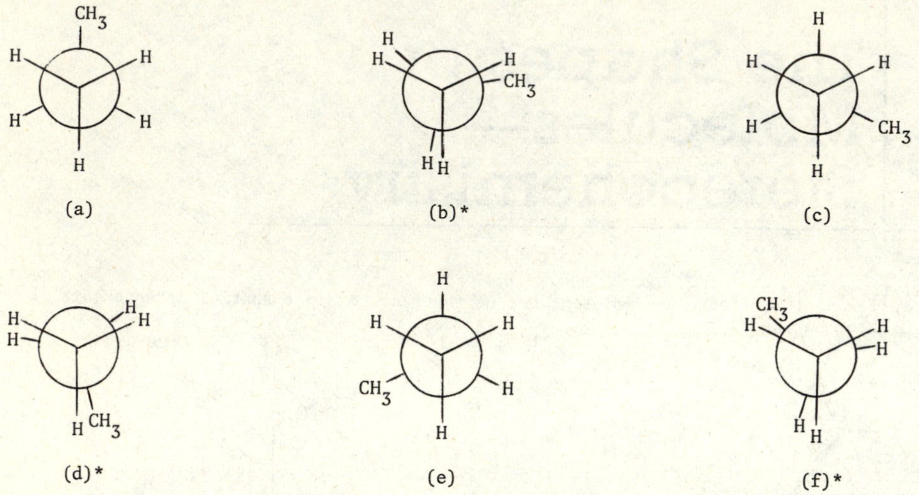

*Eclipsed rotamers slightly staggered for convenience in drawing.

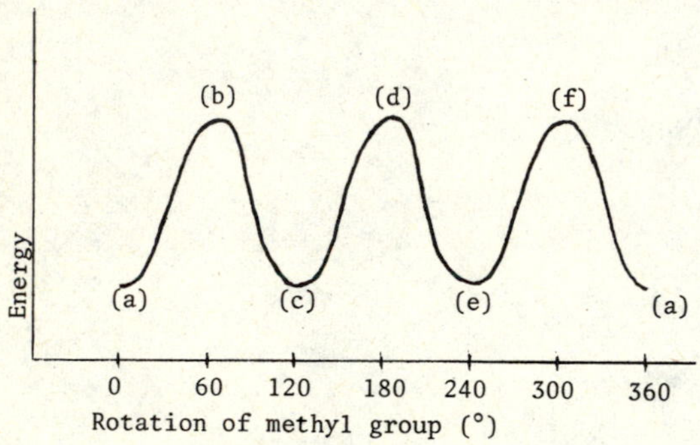

6-4

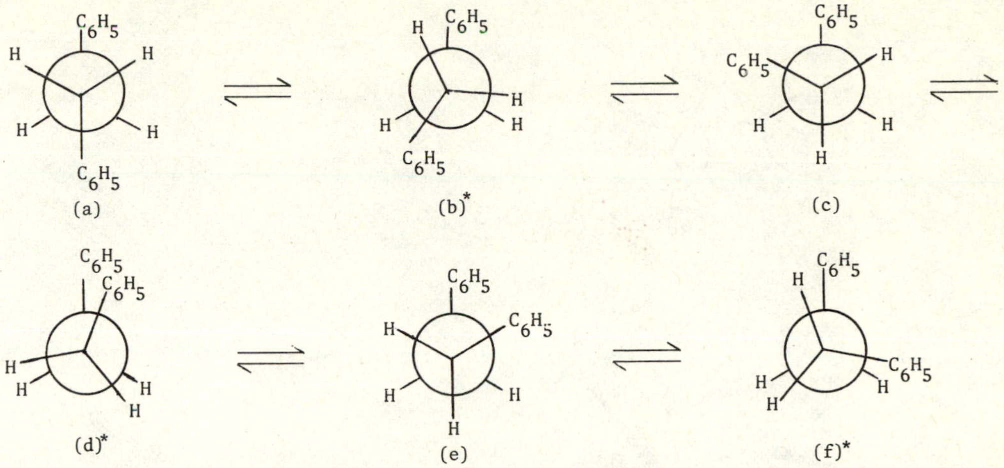

*Eclipsed rotamers slightly staggered for convenience in drawing

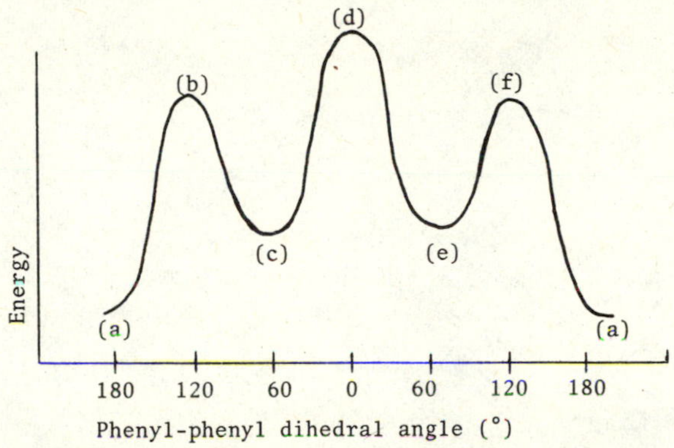

In contrast to propane where there are three identical staggered and three identical eclipsed conformations, 1,2-diphenylethane has several different conformational energies with a particularly unfavorable phenyl-phenyl eclipsing.

6-5

Chair cyclohexane is expected to have greater nonbonded repulsions than ethane because there are gauche interactions between the methylene (CH_2) groups (see fig. 6-9b). The gauche butane is probably a good model for one-half of the cyclohexane ring.

6-6

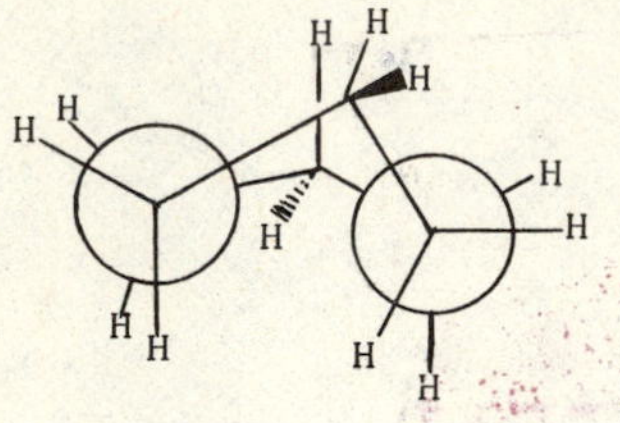

6-7

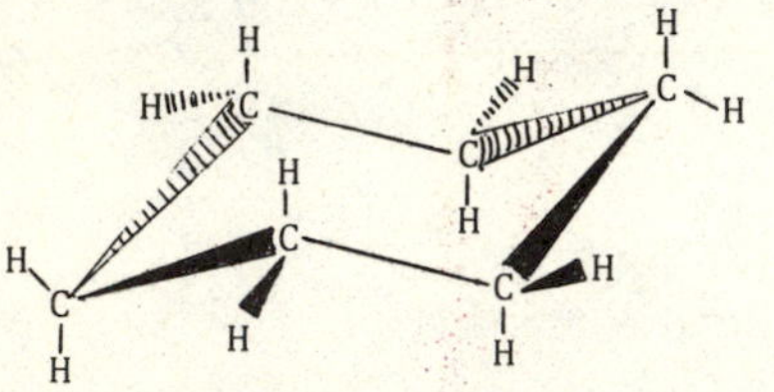

6-8

a)

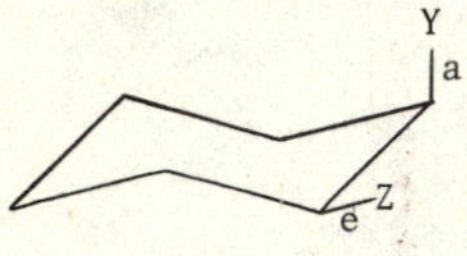

b)

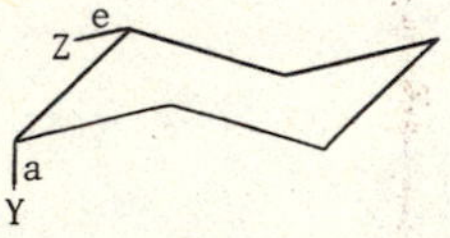

c) By convention the two carbon atoms depicted in the lower part of the drawing are in front of the plane of the two dimensional drawing surface and the upper two carbon atoms of the drawing are behind the plane (see solution to problem 6-7). Rotation of the (a) drawing moved the Z substituent from in front to behind the paper. Drawing of (a) and (b) are mirror images - enantiomers (sec. 6-5A).

6-9

Manipulation of a molecular model of cyclohexane allows you to examine the various nonbonded interactions.

6-10

A planar structure introduces angle strain at all carbon atoms whereas the proposed ring flip via the twist boat involves only minor angle strain as one side of the ring flips into

6-10 Contd...

the twist boat. Eclipsing strain would also be considerable because all twelve hydrogen atoms are eclipsed.

6-11

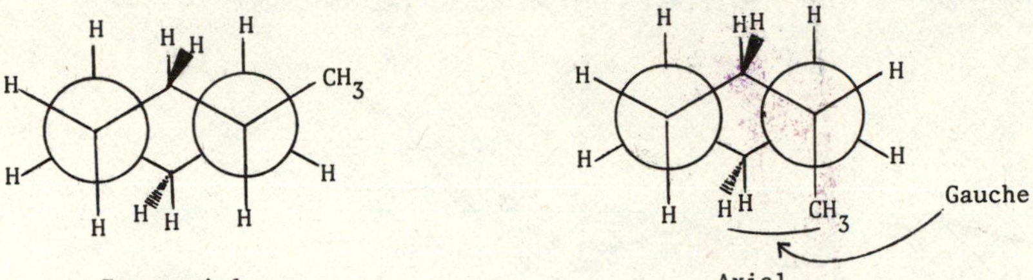

Equatorial   Axial

The axial conformation has unfavorable gauche methyl-methylene as well as 1,3-diaxial interactions.

6-12

The carbon-halogen bond lengths increase as the halogen atom becomes larger. The centers of the larger atoms are further away from other atoms in the molecule so that the effect of size is effectively decreased.

6-13

A planar molecule would have all hydrogen atoms eclipsed. The envelope shape enables the hydrogen atoms to be somewhat staggered.

6-14

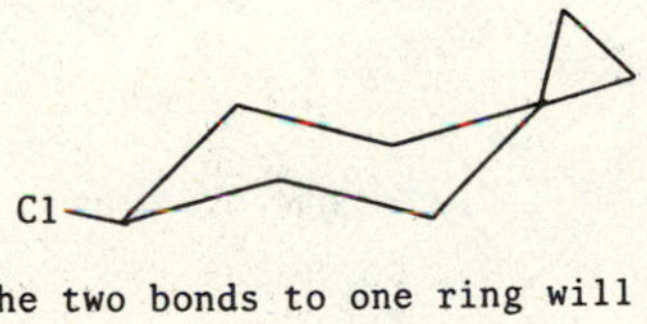

The two bonds to one ring will lie in a plane perpendicular to the plane containing the two bonds to the second ring.

6-15

a) 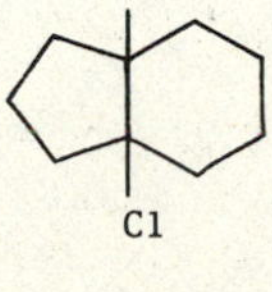

(b) 

(c) 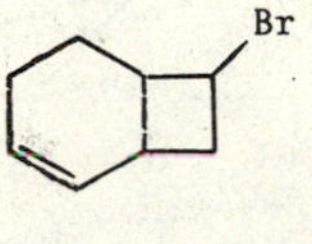

(d) 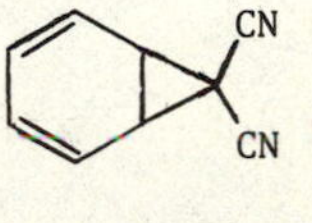

6-16

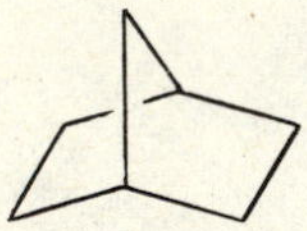

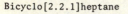

Bicyclo[2.2.1]heptane

Norbornane

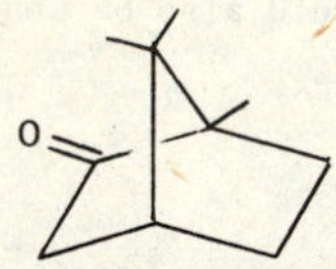

1,7,7-Trimethylbicyclo[2.2.1]-2-heptanone

Camphor

6-17

a)

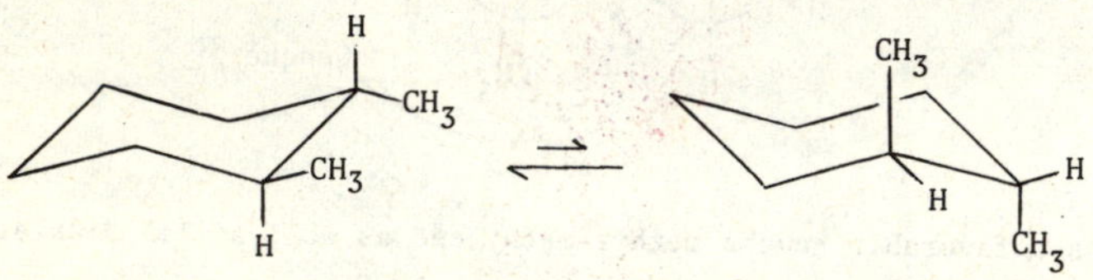

eq.-eq. ax.-ax.

b)

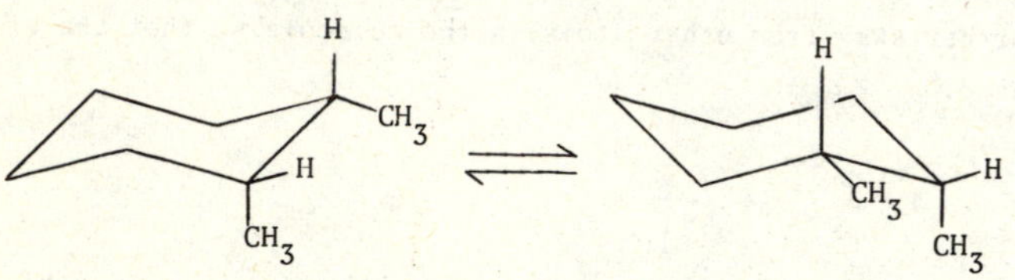

ax.-eq. eq.-ax.

c)

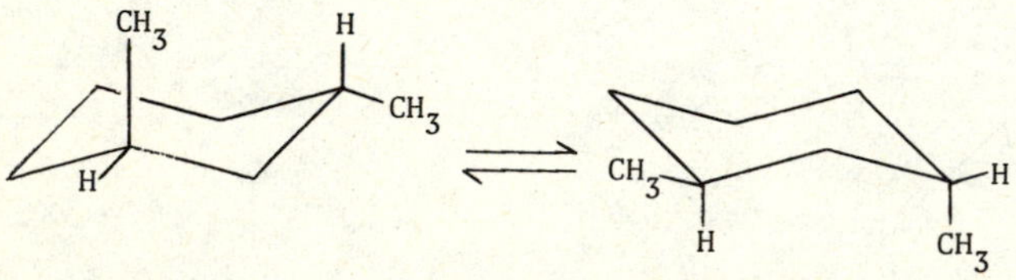

ax.-eq. eq.-ax.

d)

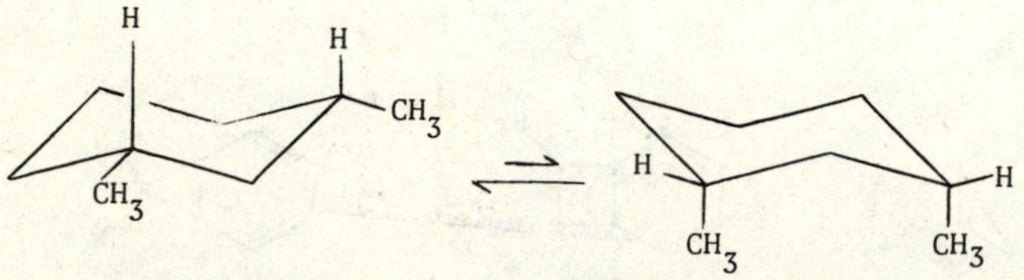

eq.-eq. ax.-ax.

6-18

a)

Isomer	Conformation	$\Delta G°$ (relative)	
trans	eq.-eq.	0	(no axial groups)
	ax.-ax.	3.4 kcal/mol (14.6 kJ/mol)	(two axial CH_3 groups)
cis	eq.-ax. or ax.-eq.	1.7 kcal/mol (7.3 kJ/mol)	(one axial CH_3 group)

b)

Isomer	Conformation	$\Delta G°$ (relative)	
trans	eq.-eq	0	(no axial groups)
	ax.-ax	2.2 kcal/mol (9.2 kJ/mol)	(one axial CH_3 and one axial Cl)
cis	ax.-eq.	1.7 kcal/mol (7.3 kJ/mol)	(one axial CH_3 group)
	eq.-ax.	0.5 kcal/mol (2.2 kJ/mol)	(one axial Cl group)

c)

Isomer	Conformation	$\Delta G°$ (relative)	
trans	eq.-ax.	0.5 kcal/mol (2 kJ/mol)	(one axial Br groups)
	ax.-eq.	2.2 kcal/mol (9.2 kJ/mol)	(one axial $(CH_3)_2CH$ group)
cis	eq.-eq.	0	(no axial groups)
	ax.-ax.	2.7 kcal/mol (11.2 kJ/mol)	(one axial Br and one axial $(CH_3)_2CH$)

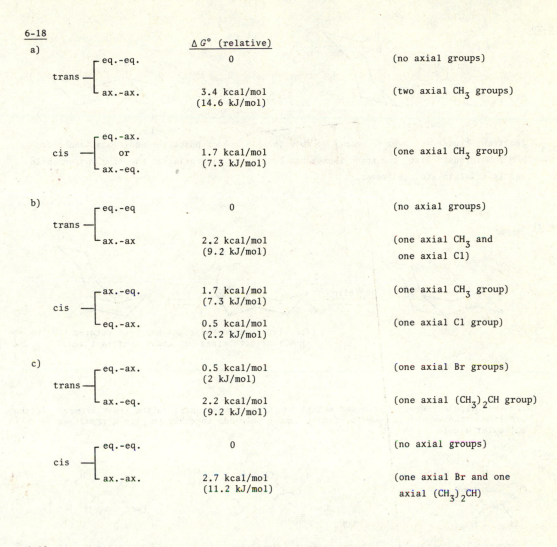

6-19

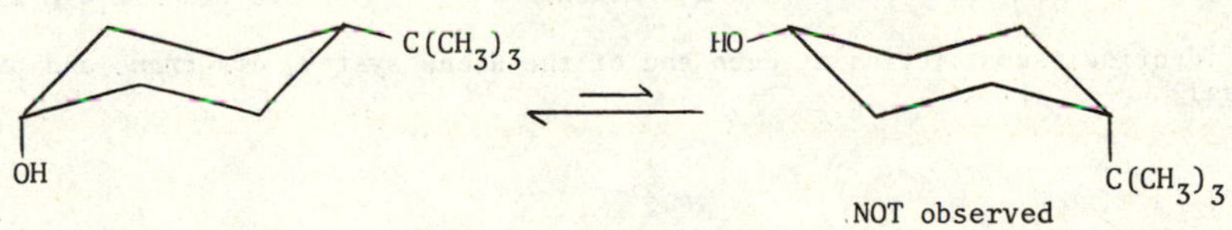

tert-Butyl is sterically unfavorable as an axial substituent, thus essentially always is equatorial. The cis isomer therefore must exist predominately in the conformation with an axial hydroxy and an equatorial *tert*-butyl group.

6-20

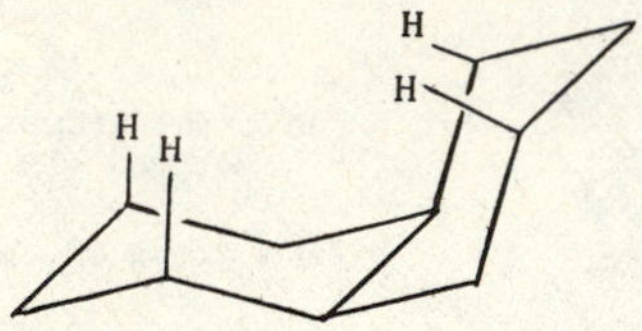

cis-Decalin

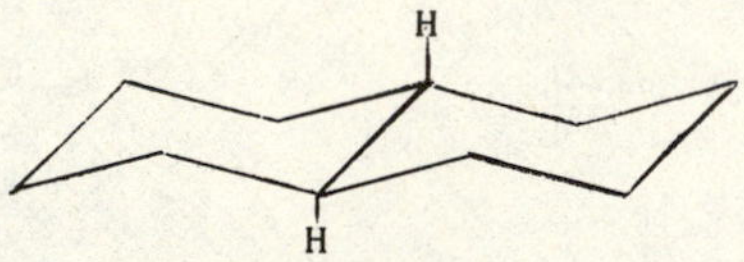

trans- Decalin

The trans isomer is lower in energy because the cis isomer possesses additional nonbonded H-H repulsions. Also, the trans isomer has both bonds equatorial at the ring fusion while one is axial in the cis isomer.

6-21

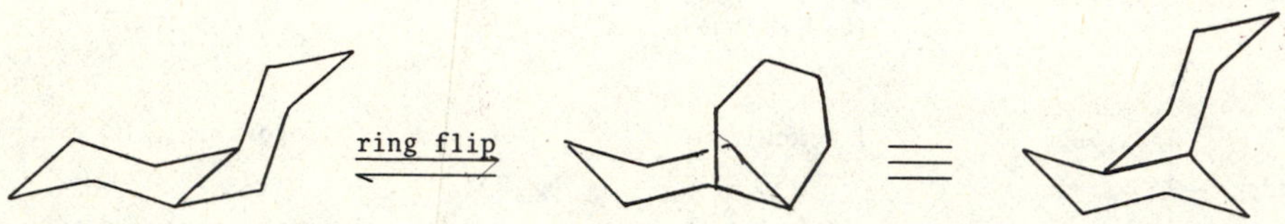

(The structure on the right has been rotated 120° to show that both rings are chair conformations.)

6-22

The bond dipole moments of the two methyl to carbon bonds cancel in the trans isomer. In the cis isomer the vector components of each bond dipole add together to give a resultant molecular dipole.

6-23

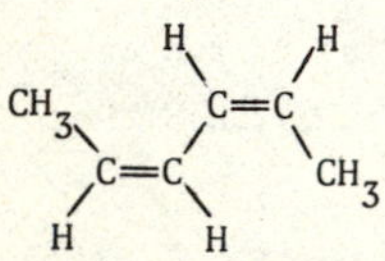

cis-cis-2,4-Hexadiene

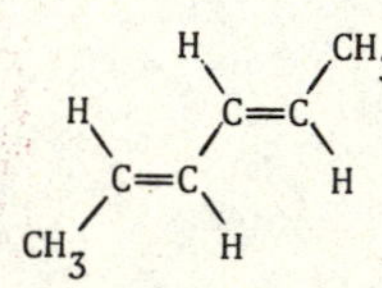

trans-trans-2,4-Hexadiene

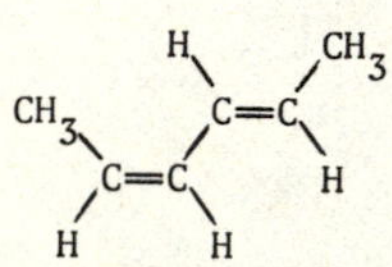

cis-trans-2,4-Hexadiene

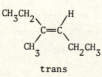

Because of identical substitution at each end of the diene system, *cis-trans* and *trans-cis* are identical.

6-24

a)

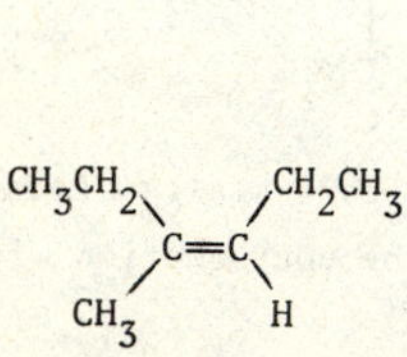

cis

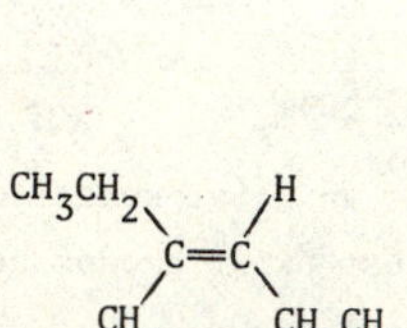

trans

Contd...

6-24 Contd...

b) 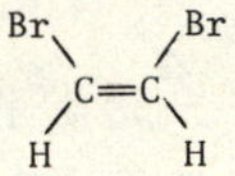 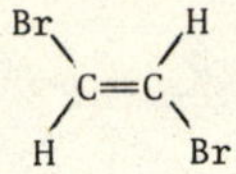

cis trans

c) 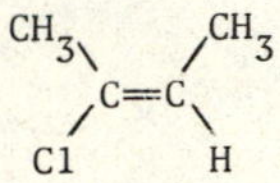 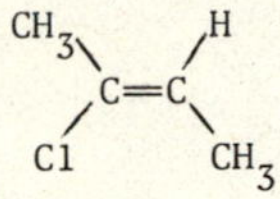

cis trans

d) 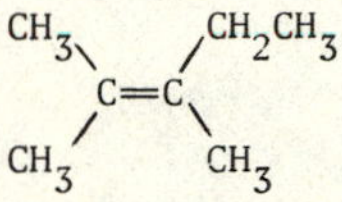

No geometrical isomers.

e) 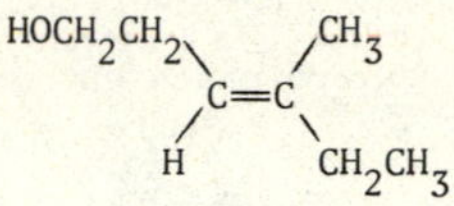 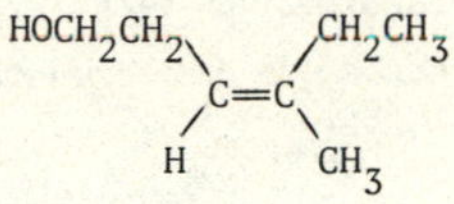

trans cis

f) 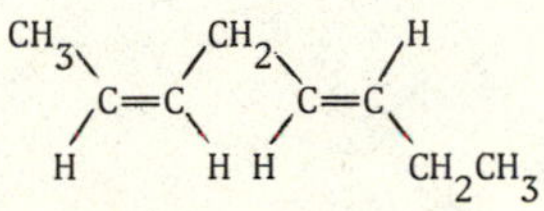 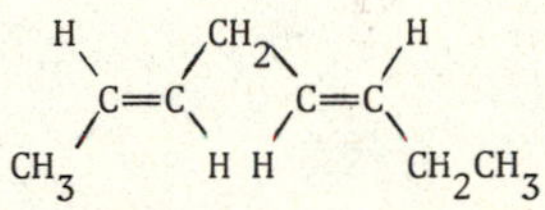

cis trans   trans trans

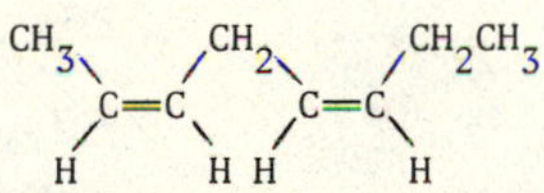 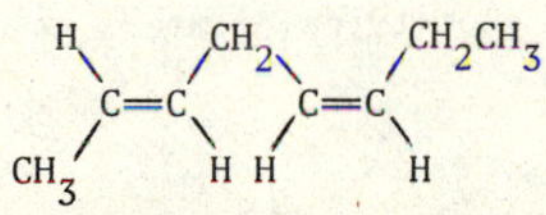

cis cis   trans cis

6-25

a) 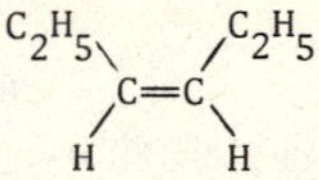 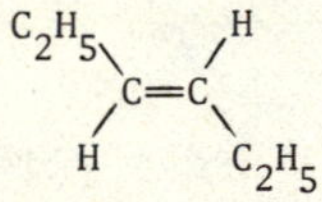

Z *E*

b) 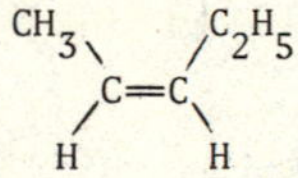 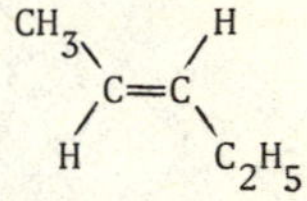

Z *E*

6-25 Contd...

c) 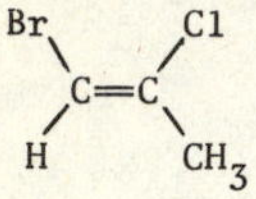

Z

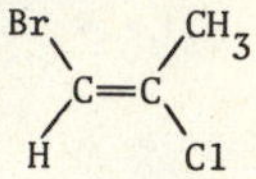

E

d) 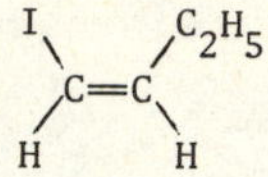

Z

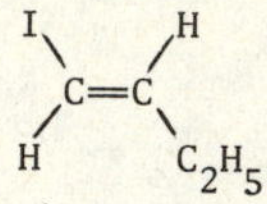

E

e) 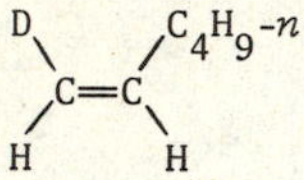

Z

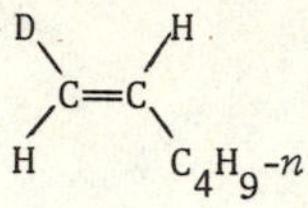

E

6-26

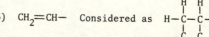

a) *Z*-3-Chloro-4-methyl-3-hexene
(Chloro takes precedence over methyl in numbering.)

b) *E*-1-Bromo-2-chloromethyl-4,4,4-trideuterio-3-methyl-2-butene
(The choice of parent name in this case is based on the longest carbon chain with the higher priority groups.)

c) *Z*-1-Ethyl-2-methylcyclooctene
(Numbering priority goes to the more complex group or by alphabetical order.)

d) *Z*-2-Ethyl-4-methyl-2-penten-1-ol

6-27

Because the terminal double bond has two identical substituents (H's) attached to one end and does not give rise to geometrical isomers.

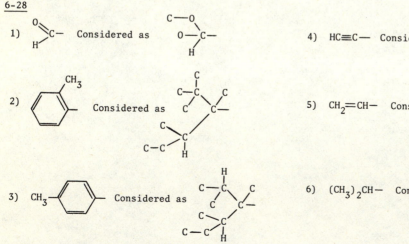

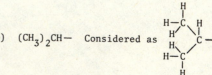

6-28

1) Considered as 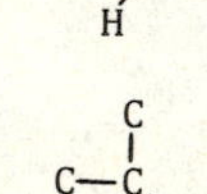

2) 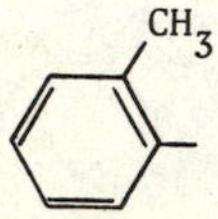 Considered as 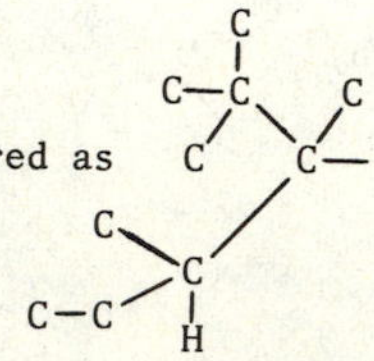

3) 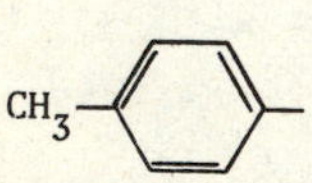 Considered as 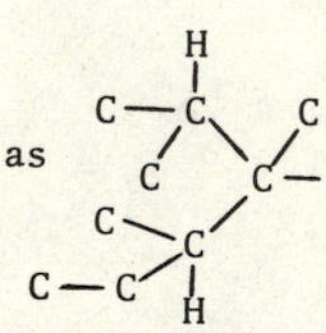

4) HC≡C— Considered as 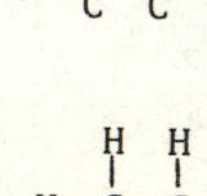

5) CH_2=CH— Considered as 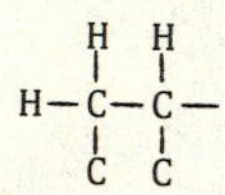

6) $(CH_3)_2CH$— Considered as 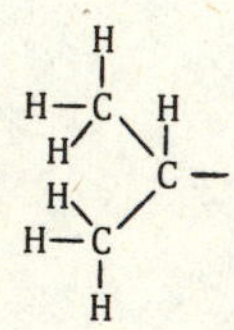

6-29

a) *Z*-3-Phenyl-3-hexene

b) *Z*-3-Hydroxymethyl-3-penten-2-one

c) *Z*-3-Cyclobutyl-4-(1-methylethyl)-3-heptene
or *Z*-3-Cyclobutyl-4-isopropyl-3-heptene

d) *E*-2-(4-Methylphenyl)-2-butenoic acid
or *E*-2-*p*-Tolyl-2-butenoic acid

6-30

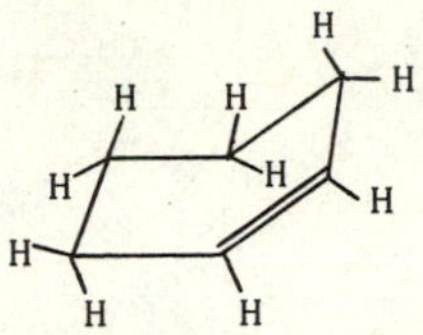

6-31

In all cases the single bonds on each end of the double bond must be highly strained for they lie in planes which are essentially perpendicular to each other.

6-32

a) See fig. 6-27 and compare molecular models.

b) 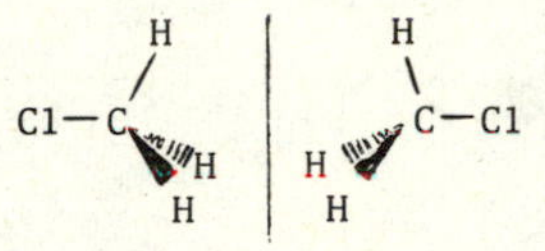 when turned around becomes 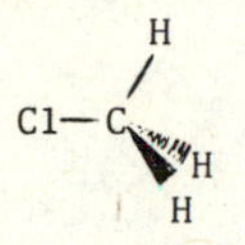

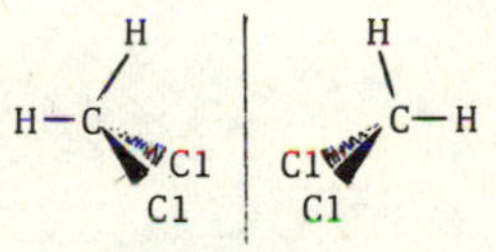 when turned around becomes 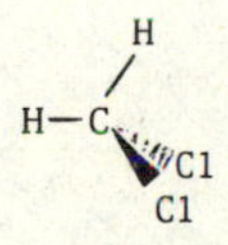

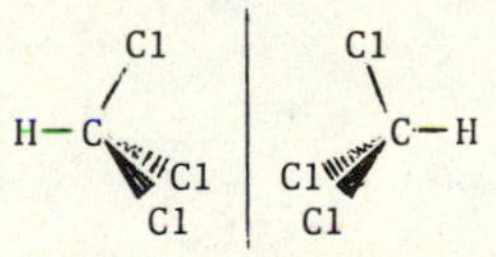 when turned around becomes 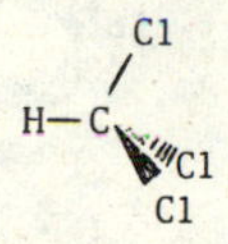

6-33

a) Achiral

b) Chiral at carbon atom 3

c) Achiral

d) Chiral at both carbon atoms 2 and 3

e) Achiral - possesses a plane of symmetry perpendicular to the molecular plane and passing through the Cl and Br atoms.

f) Chiral at carbon atom 4

g) Chiral at carbon atom 1 of the ethanol

h) Achiral - the molecule has plane of symmetry

6-34

a) 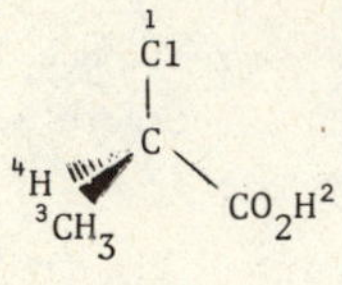

(*R*)

c) 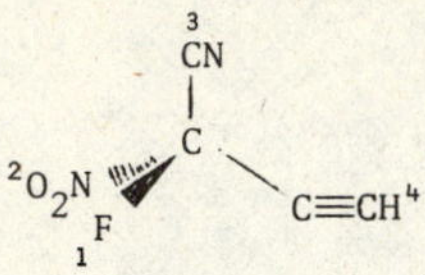

(*R*)

e) 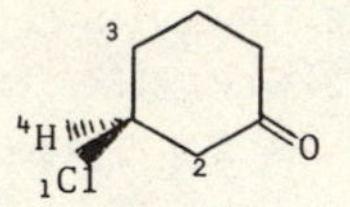

(*S*)

b) 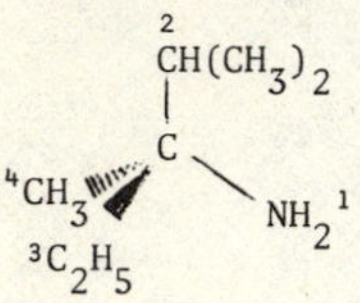

(*S*)

d) 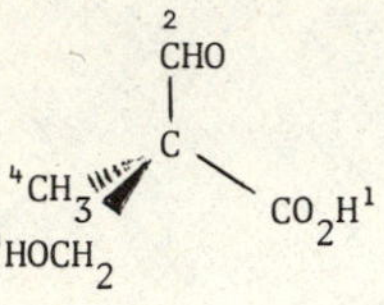

(*S*)

f) 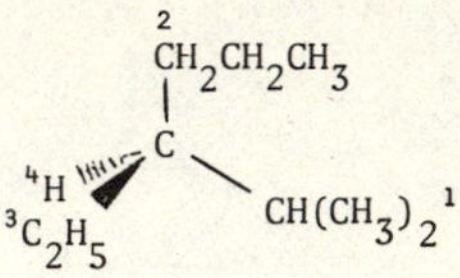

(*S*)

6-35

a) 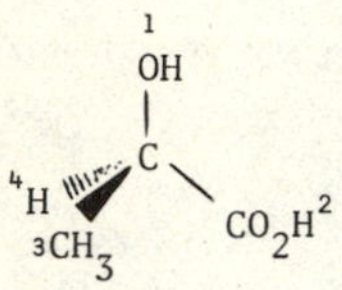

c) 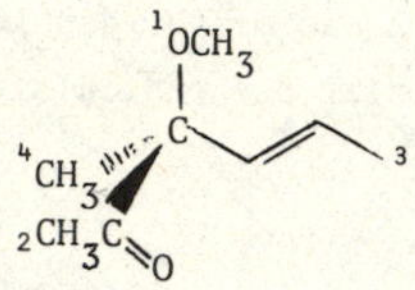

e) 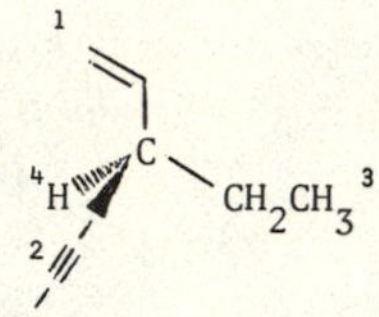

b) 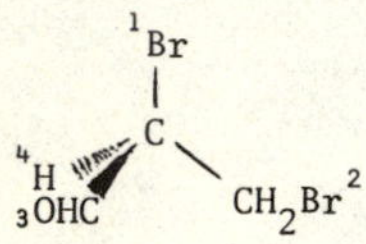

d) 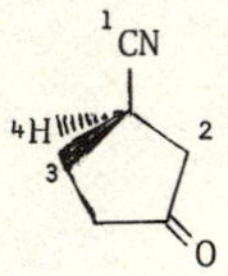

f) 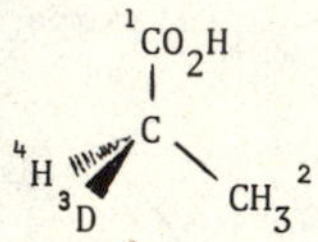

6-36

a) 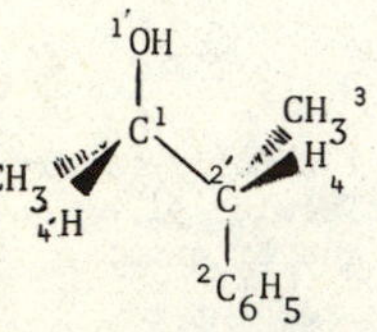

(*R*) (*R*)

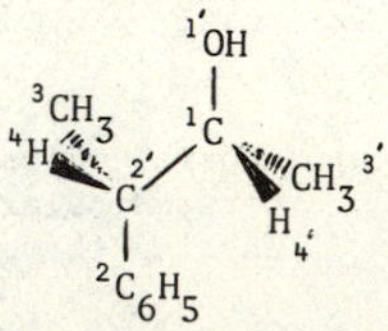

(*S*) (*S*)

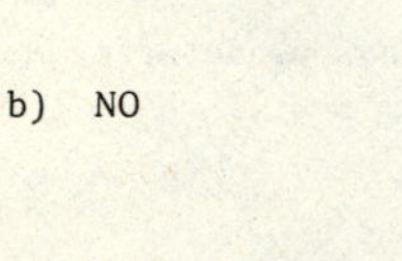

(*S*) (*R*)

(*S*) (*R*)

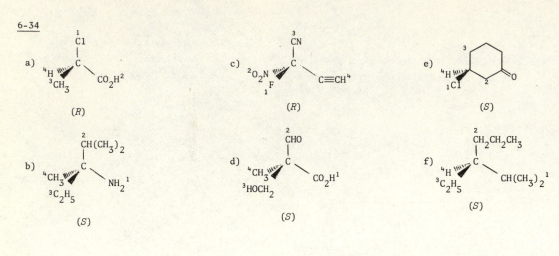

b) NO

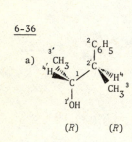

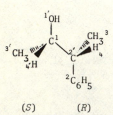

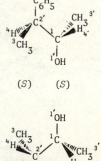

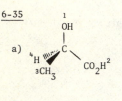

6-37

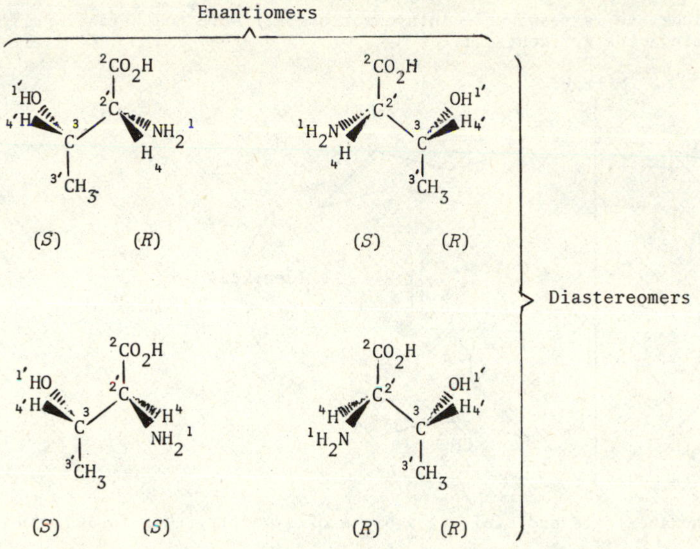

6-38

The configurations are (*R*,*R*) for (+)-tartaric acid and (*S*,*S*) for (-)-tartaric acid; that is configuration is identical at each carbon atom. For the *meso*-tartaric acid the configurations are (*R*,*S*) or (*S*,*R*). In the meso isomer the groups around both chiral atoms are identical. The chiral centers have a mirror image relationsbip so that the molecule has an internal plane of symmetry.

6-39

Compare the molecular models.

6-40

a) *trans*-1,3-Dimethylcyclohexane is resolvable into enantiomers. Note that either pair of ring-flip conformers are actually identical.

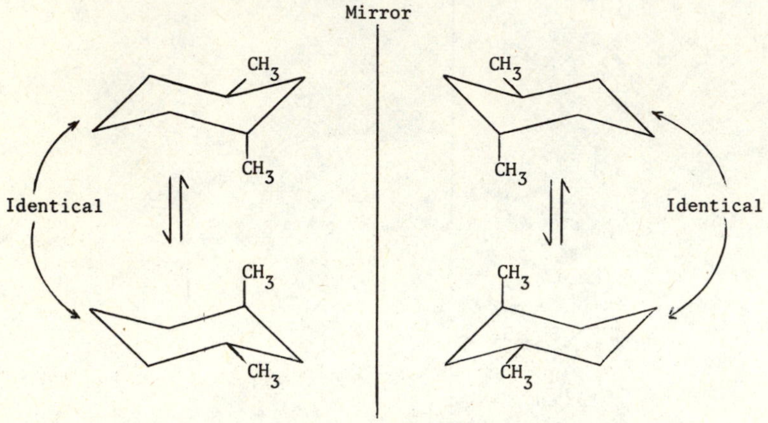

b) *cis*-1,3-Dimethylcyclohexanes (a,a and e,e) has a plane of symmetry, thus is achiral.

c) *cis*-3-Methylcyclohexanol is resolvable into enantiomers.

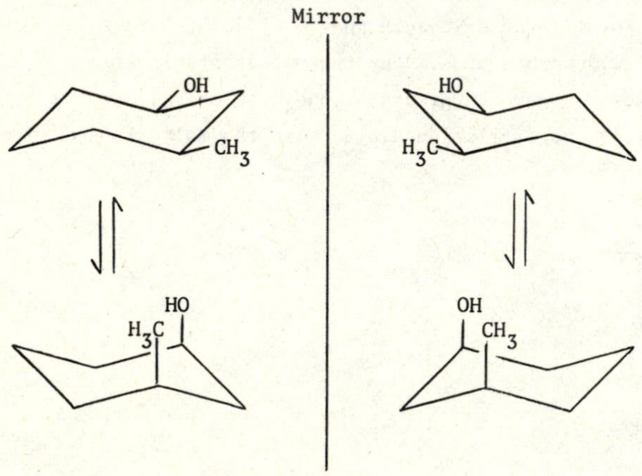

d) and e) 1,4-Disubstituted cyclohexanes have a plane of symmetry, thus are not resolvable.

6-41

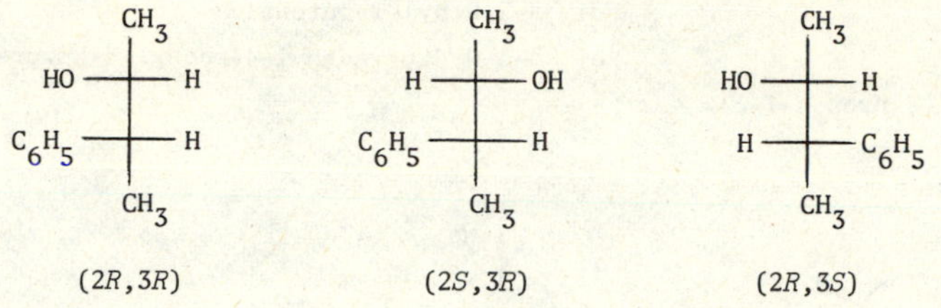

6-42

When the fourth priority group is on a vertical line it is, by Fischer projection convention, behind the plane of the two dimensional surface. Thus the relation of the other three groups is exactly the same as is derived from a three dimensional formula. When the fourth priority group is on a horizontal line it is above the plane of the surface and all rotational based assignments are reversed.

6-43

Follow the conventions for what a Fischer projection formula means in relation to a 3-dimensional molecule. The 3-D drawings can be moved out of the plane of the paper for comparison.

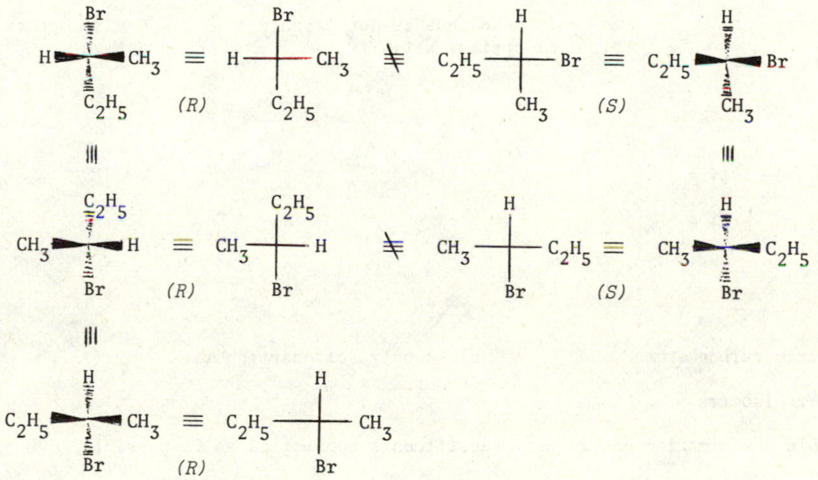

6-44

Spearmint: (5*R*)-2-Methyl-5-(1-methylethenyl)-2-cyclohexen-1-one

Caraway: (5*S*)-2-Methyl-5-(1-methylethenyl)-2-cyclohexen-1-one

6-45

a) 3-Methyl-1-hexene

b) *Z*-1-Methylcyclohexene

c) *E*-3-Fluoro-3-iodo-2-methyl-2-propen-1-ol

d) *Z*-2-Ethyl-2-butenal

e) *E*-3-Hydroxymethyl-4-phenyl-3-hexen-1-ol

6-46

a)

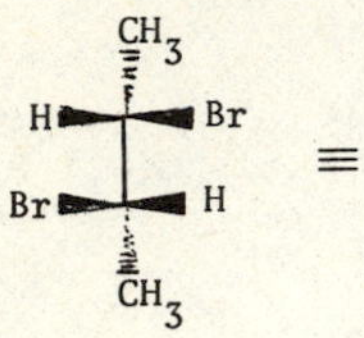

3-Dimensional drawing of Fischer projection

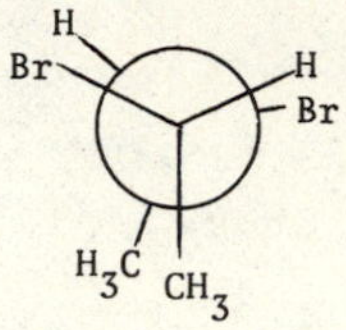

Comparable eclipsed Newman projection

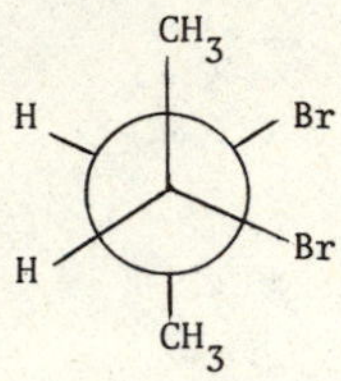

Most stable staggered rotamer with methyl groups trans
Can be optically active

b)

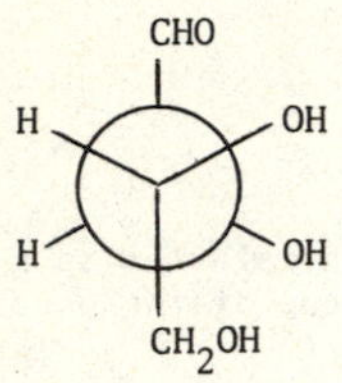

Can be optically active

c)

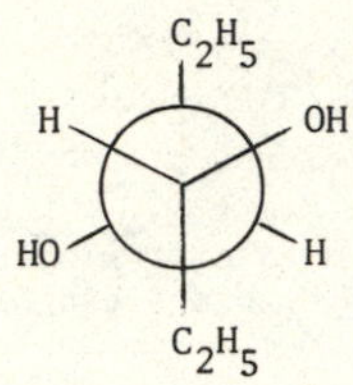

Is meso thus is not capable of optical activity.

d)

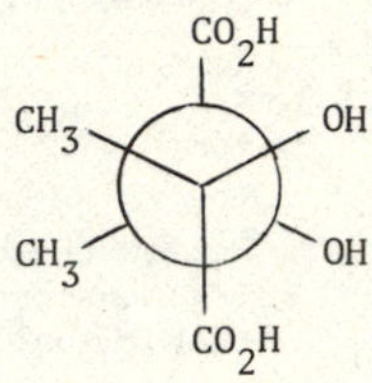

Can be optically active

6-47

a) *R* c) *R* e) *S*

b) *S* d) *S* f) *R*

6-48

a) Three asymmetric carbon atoms

$\therefore 2^3 = 8$ stereoisomers

b) 4 pairs of enantiomers

c) The most stable conformation has as many substituents equatorial as is possible.

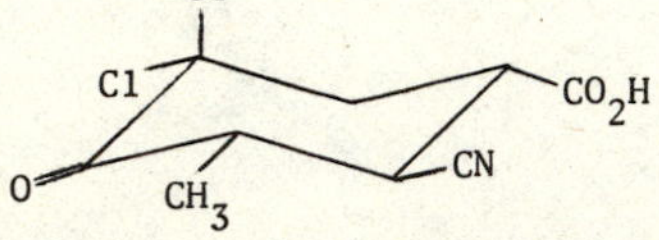

6-49

a)

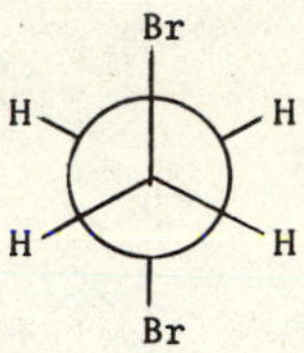

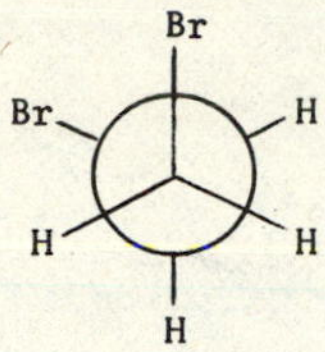

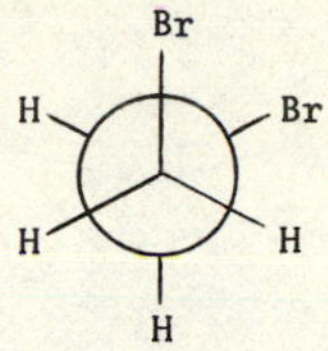

b) Anti Gauche Gauche

c)

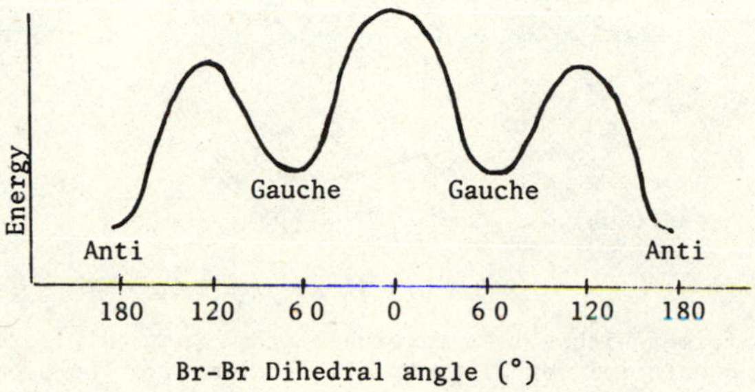

6-50

a) 2

b) Although four stereoisomers ($2^2 = 4$) might have been predicted only three actually exist; a pair of *dl* isomers and a meso form.

c) and d)

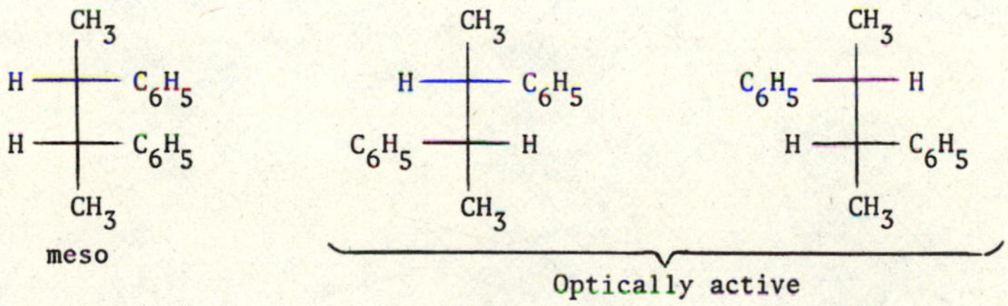

e) (*S*)–(*R*) (*S*)–(*S*) (*R*)–(*R*)

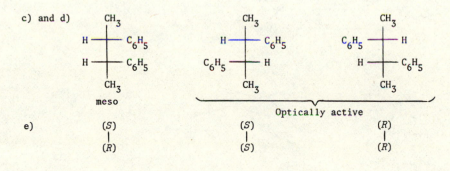

6-51

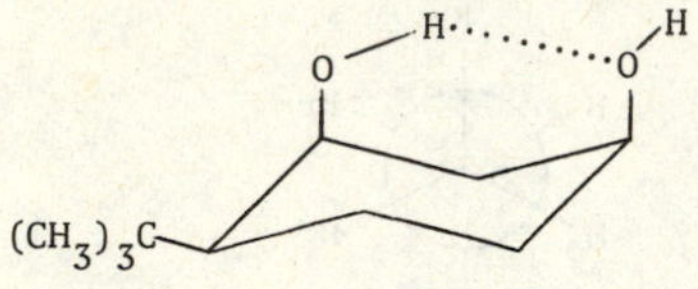

A

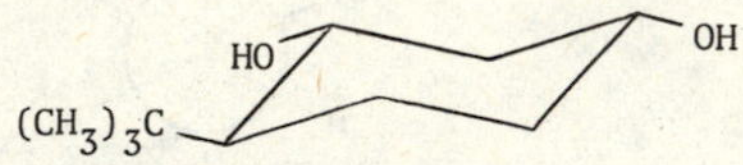

B

The *tert*-butyl group determines ring conformation by remaining equatorial. In A both hydroxy groups are axial and located in a 1,3-relationship favorable for interaction.

6-52

a)

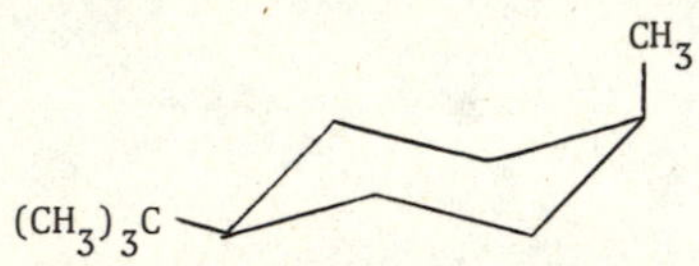

A

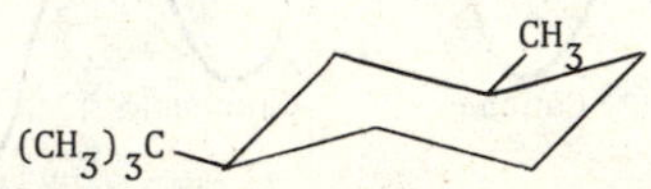

B

b) Ring flip of A would give the conformer with a less favorable axial *tert*-butyl group (>4 kcal/mol) but more favorable equatorial methyl group (1.7 kcal/mol). The process is endergonic by >2.3 kcal/mol.
Ring flip of B puts both groups in an axial position and is endergonic by >5.7 kcal/mol.

6-53

The more stable conformer is indicated. In some cases isomers are predicted to have identical conformational energies.

a)

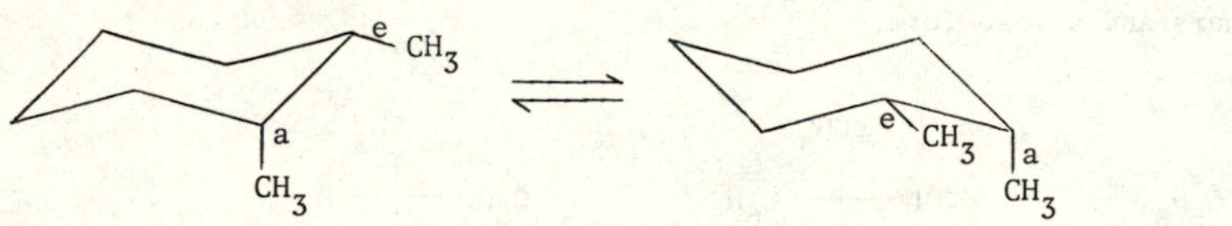

b)

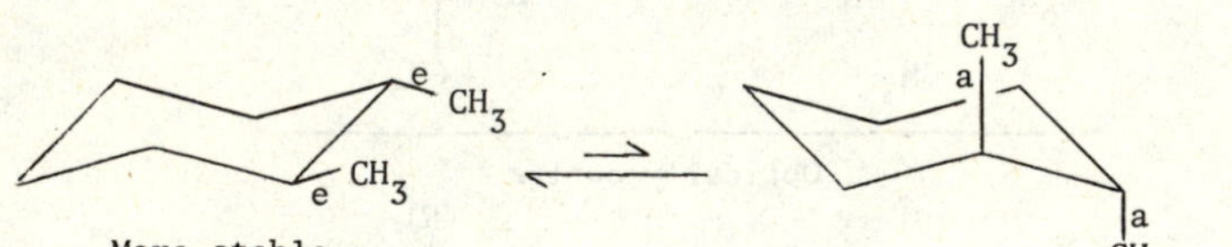

More stable

c)

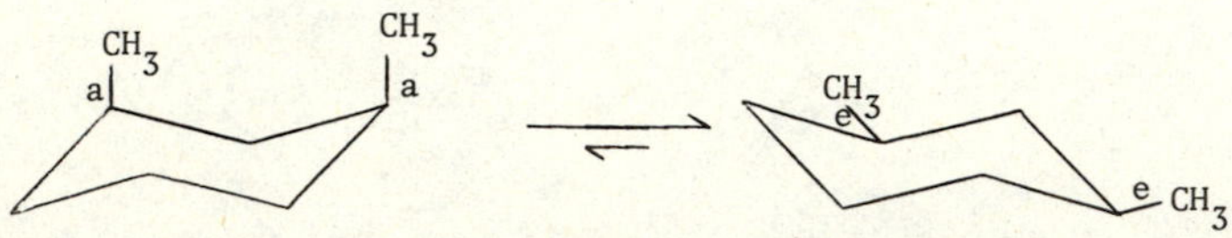

More stable

Contd...

6-53 Contd...

d)

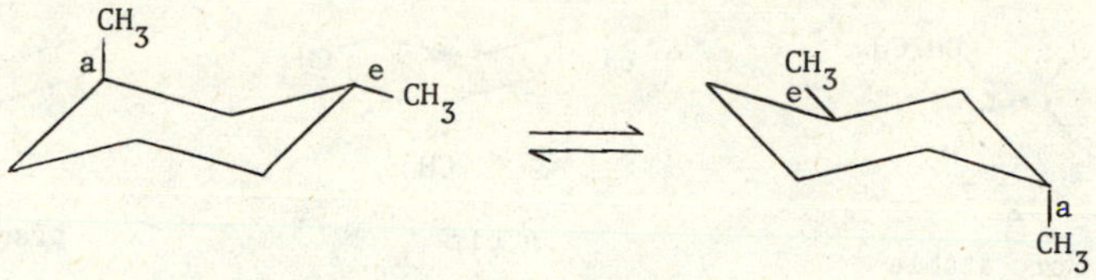

e)

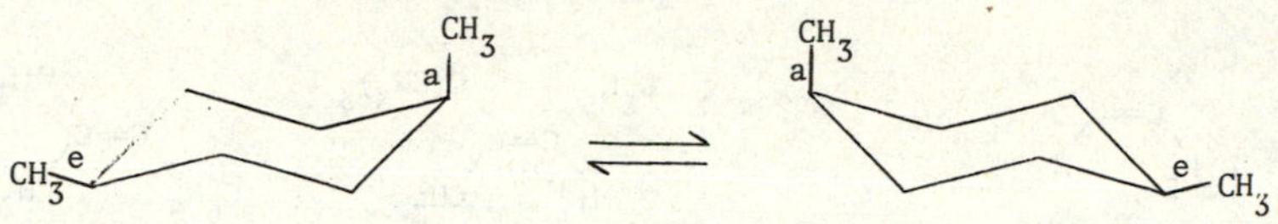

f)

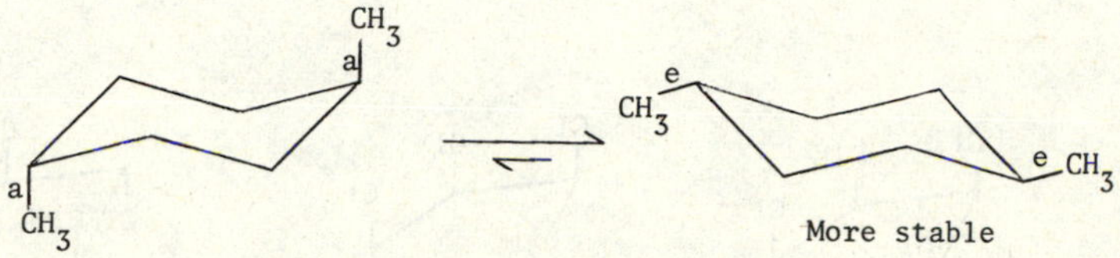

The predicted stability is f = c > b > e = d > a

6-54

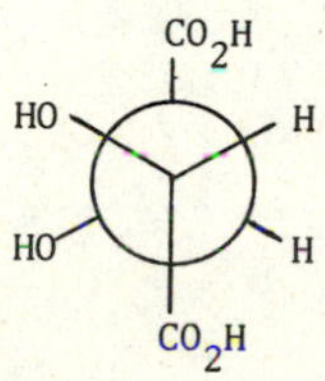

Most stable rotamer of (+)-tartaric acid

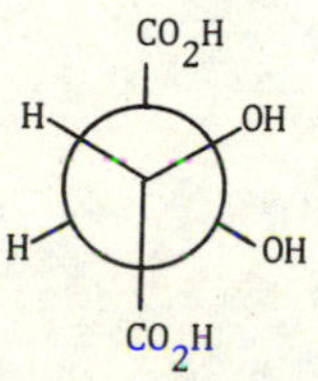

Most stable rotamer of (-)-tartaric acid

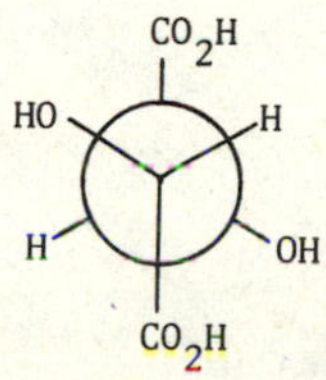

Most stable rotamer of meso tartaric acid

The meso isomer, with identical groups anti, is expected to be the most stable. The relative stabilities indicated for the three isomers are based on group sizes. Intramolecular hydrogen bonding between hydroxy, carboxy, and hydroxy-carboxy could alter those predictions.

6-55

a)

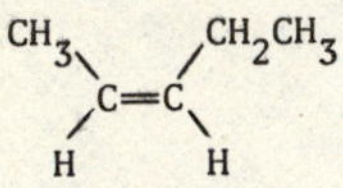

Z

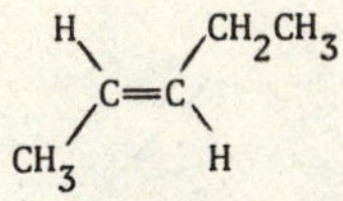

E

More stable

c)

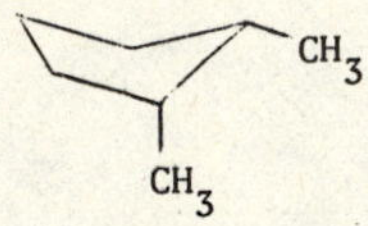

cis

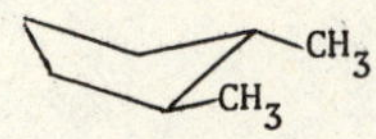

trans

More stable

b)

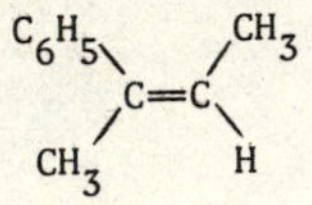

Z

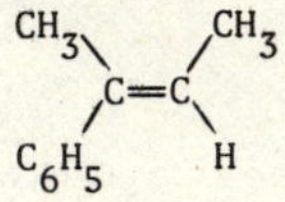

E

More stable

d)

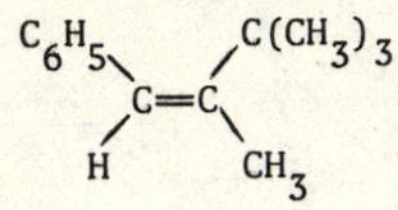

Z

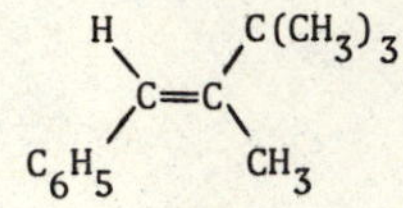

E

More stable

6-56

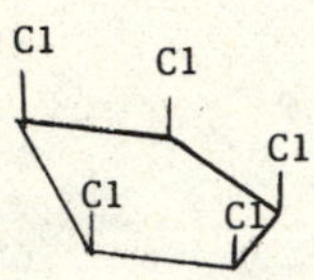

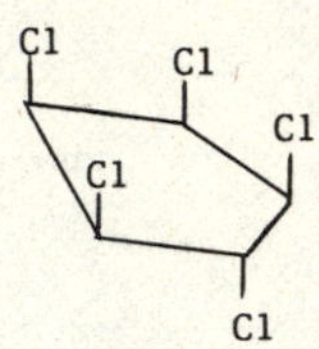

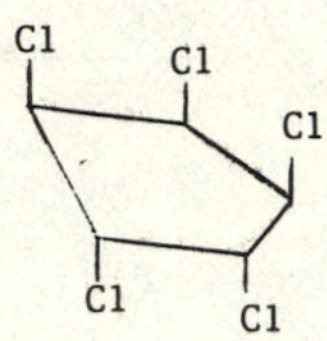

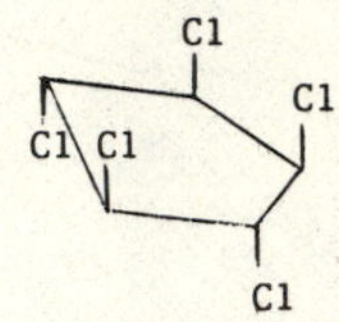

None are optically active because they all have a plane of symmetry,

6-57

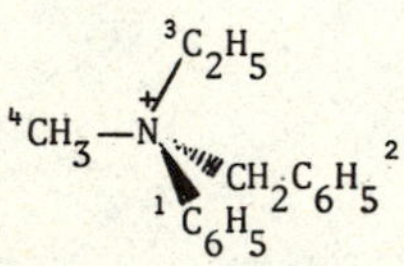

S

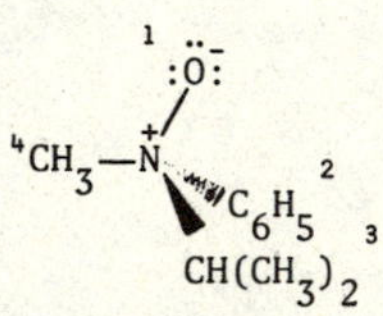

R

6-58

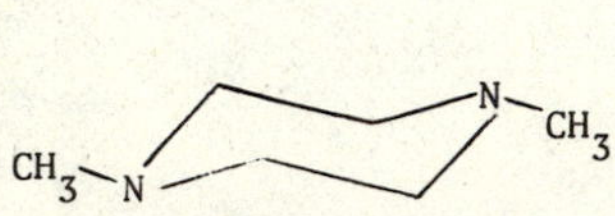

μ = 0

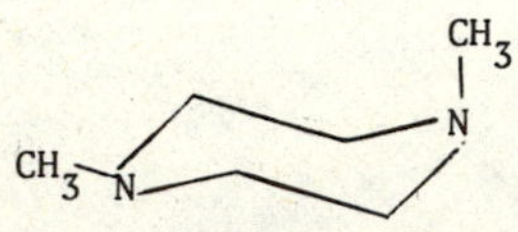

μ = 1.55 D
(calculated)

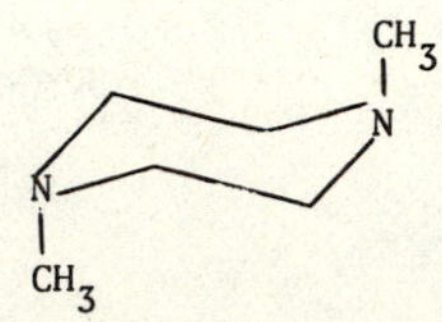

μ = 0

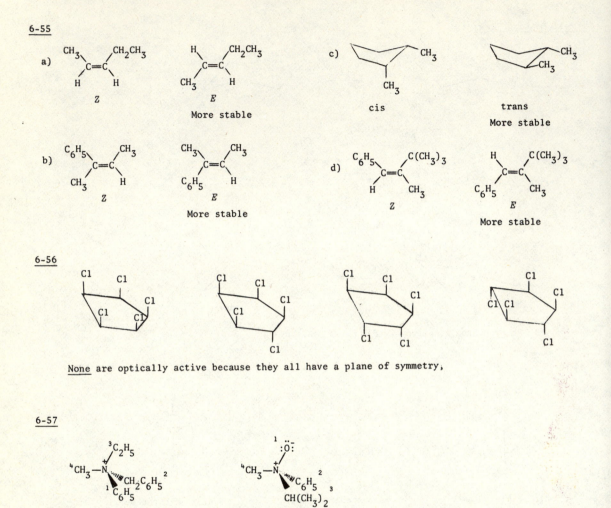

6-59

Ricinoleic acid has a *Z* and *E* geometrical isomer in addition to the asymmetric carbon atom. There are both *Z* and *E* pairs with (*R*) and (*S*) configurations at the asymmetric carbon.

6-60

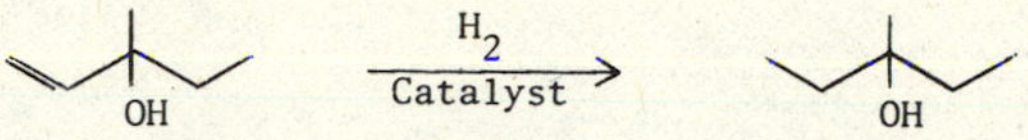

6-61

a)

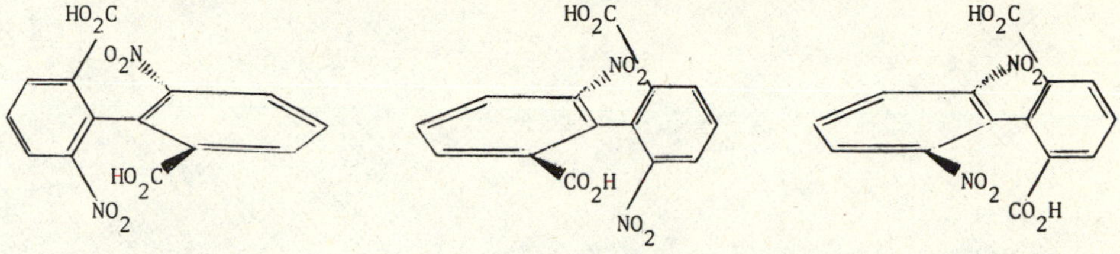

A B C

b)

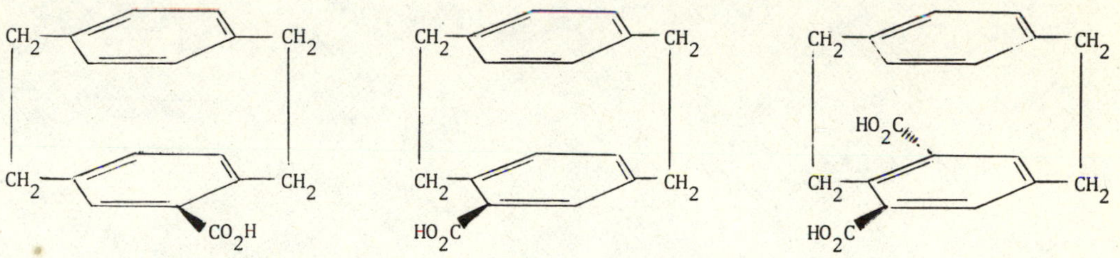

D E F

c)

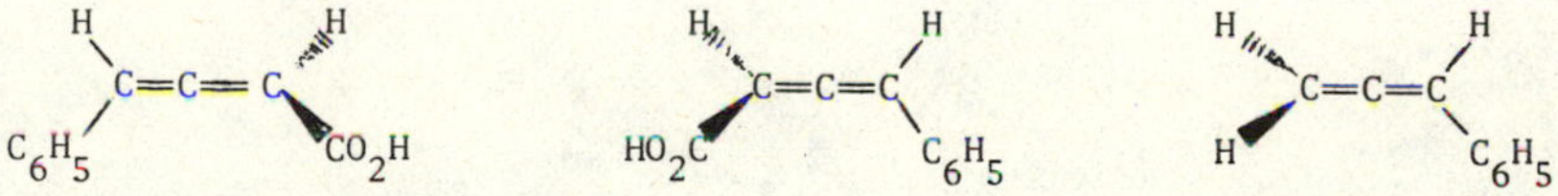

G H I

Free rotation about the aryl-aryl bond is hindered by the *ortho* substituents of A, B, and C.

Compounds A and B are enantiomers. Compound C is achiral because identical ortho substituents lead to a plane of symmetry.
Compounds D, E, and F (known as paracyclophanes) are not free to rotate because the carbon bridges connecting the aromatic rings are too short. Compounds D and E are enantiomers while F is achiral because it possesses a plane of symmetry.
Allenes G and H are chiral because of the perpendicular orientation of the two cumulative double bonds. Allene I is also rigid but has a plane of symmetry.

7 Structural Effects on Reactivity—Resonance

7-1 The effects of bond dipole interactions and of solvation are expected to be greater in the carboxlyate anions than in the acids. Nevertheless, even small differences in relative energies of the acids influence the overall acidities.

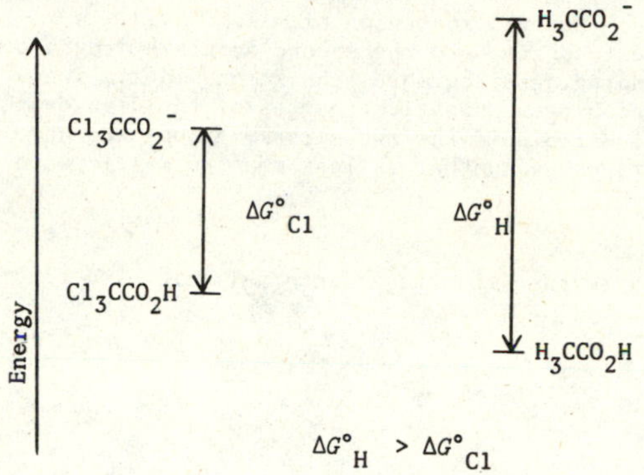

$$\Delta G°_H > \Delta G°_{Cl}$$

7-2 The inductive effect of the halogen atoms is related to their electronegativities. The most electronegative atom, fluorine, has the greatest acidifying influence.

7-3 The trifluoromethyl group stabilizes the anion of the conjugate base relative to the alcohol by inductive electron withdrawal.

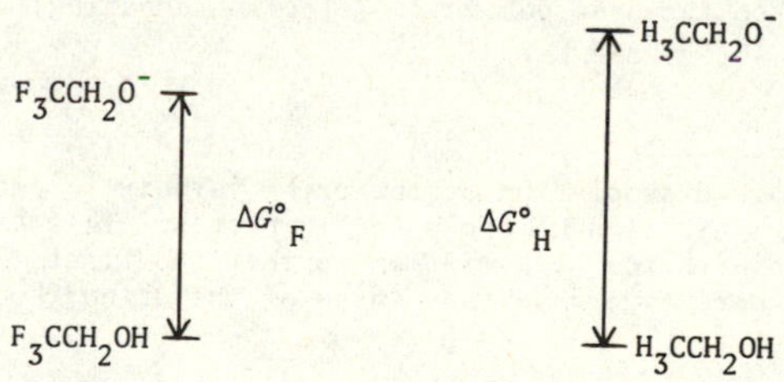

$$\Delta G°_H > \Delta G°_F$$

(Alcohols arbitrarily set as equal in energy)

7-4 The greater "s" character of an unsaturated carbon atom suggests that the electrons are attracted more closely to the nucleus. To the degree that this effect is transferred to adjacent groups, the unsaturated substituent will be electron withdrawing.

7-5

a) Oxalic ≡ Ethanedioic acid

b) Malonic ≡ Propanedioic acid

c) Succinic ≡ Butanedioic acid

d) Glutaric ≡ Pentanedioic acid

e) Maleic ≡ *Z*-Butenedioic acid

f) Fumaric ≡ *E*-Butenedioic acid

7-6 The two carboxy groups of maleic acid are closer together than those of fumaric. The electron withdrawing field effect of one carboxy on the other enhances the first ionization of maleic acid to a greater degree. But the monoanion of maleic acid. is closer to the undissociated carboxy group and tends to hold the second acidic hydrogen atom close to the molecule by intramolecular H-bonding. Furthermore, the second ionization of maleic acid is particularly unfavorable because the negative charges of the dianion that would be formed would repel each other. In fumaric acid the two carboxy groups are on opposite sides of the molecule. Their effects on each other is less than in maleic acid.

7-7 The carboxylic acid group provides a proton to the amine in an internal acid-base reaction.

$$H_2NCH_2CO_2H \rightleftharpoons H_3\overset{+}{N}CH_2CO_2^-$$

7-8 Solvation stabilization of the conjugate base from *tert*-butyl alcohol is hindered due to crowding by the three methyl groups. Dissociation is therefore relatively less favorable for *tert*-butyl alcohol than for methanol. In the gas phase solvation is not a factor. The gas phase data indicate that *tert*-butyl alcohol is intrinsically more acidic than methanol, that is, $-CH_3$ is more electron donating than $-C(CH_3)_3$.

7-9 The order reflects steric interactions between the alkyl groups of the acid and base. As more alkyl groups are brought close to each other the equilibrium constant for the acid-base reaction decreases. The result is opposite to the base enhancing (electron donating) inductive effects of methyl groups observed in the gas phase.

7-10 Intramolecular hydrogen bonding which favors dissociation of the ortho carboxylic acid by stabilizing the carboxylate anion, inhibits ionization of the phenolic proton. That kind of intramolecular association cannot take place with the para isomer so that the first ionization is less favorable and the second more favorable than those of the ortho isomer.

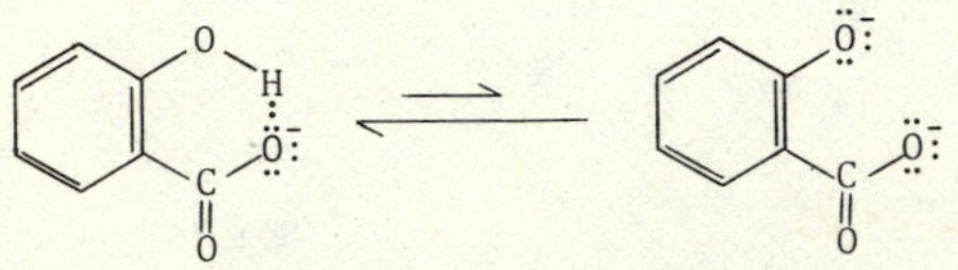

7-11

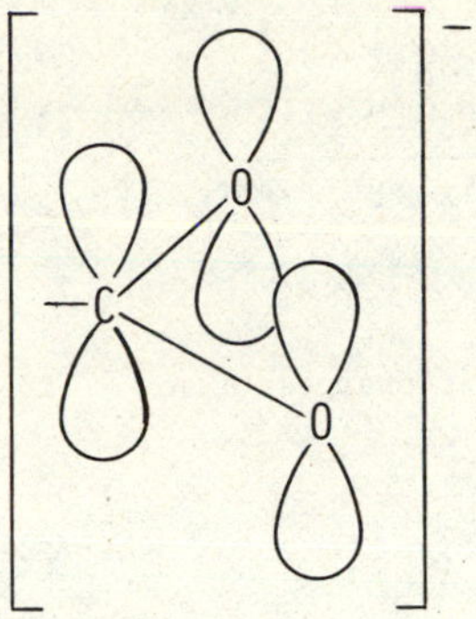

7-12

a) 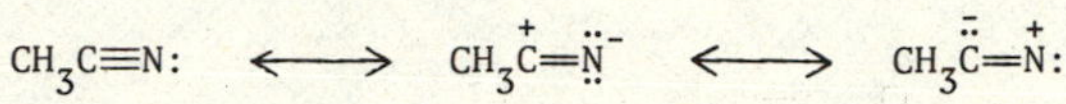

b) 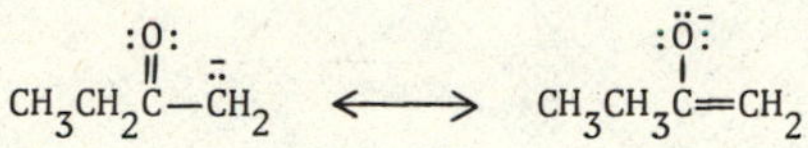

c) 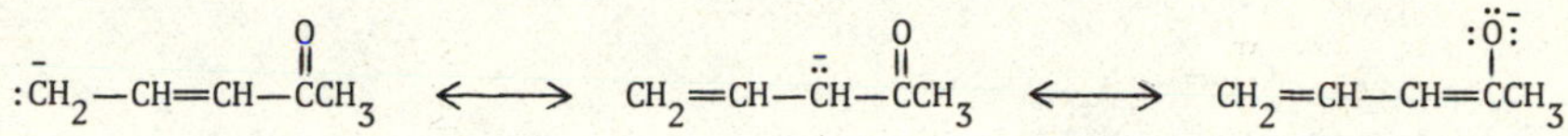

7-13

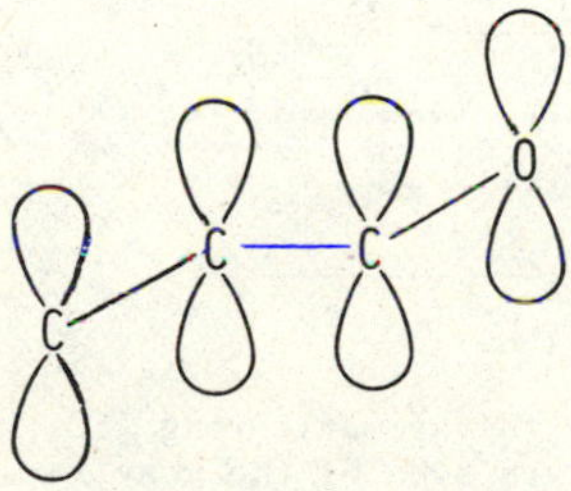

7-14

a) 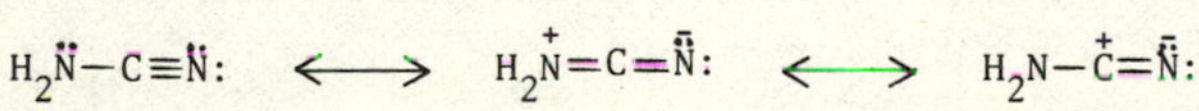

Most important

The noncharged structure is the most important.

b) 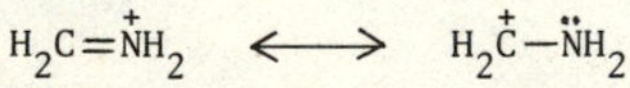

Most important

The structure in which the carbon atom has an electron octet is favored even though the more electronegative atom (nitrogen) has a positive charge.

c) 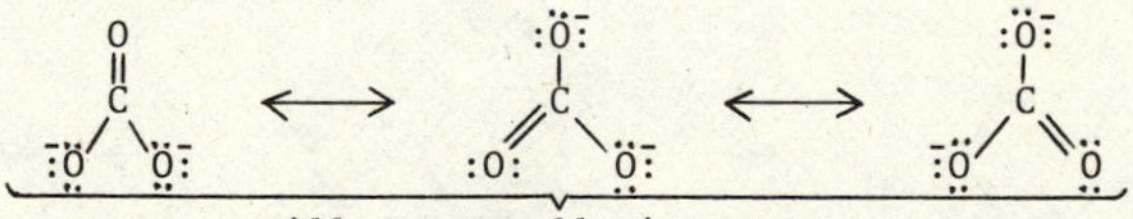

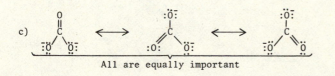

All are equally important

Contd...

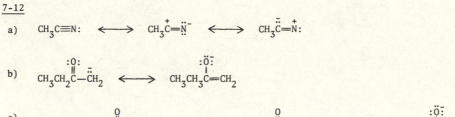

7-14 Contd...

d) 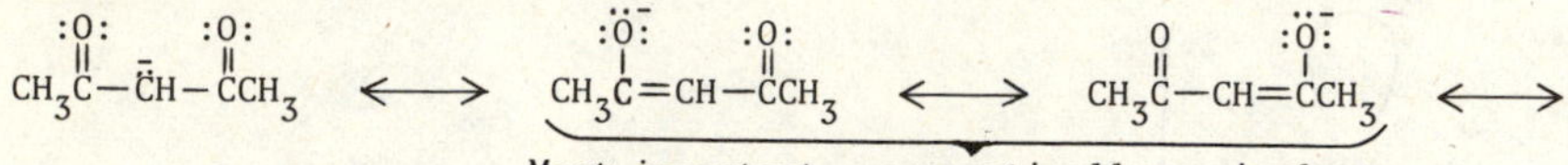

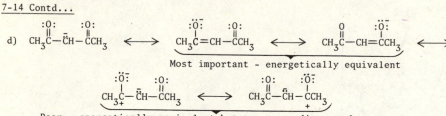

Most important - energetically equivalent

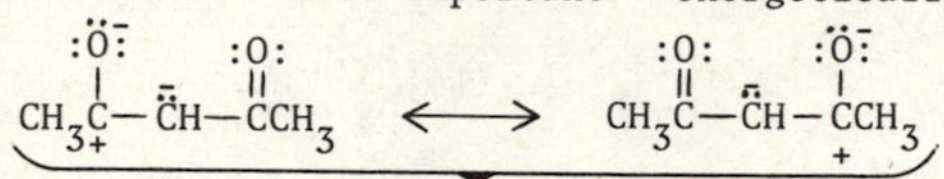

Poor - energetically equivalent but too many adjacent charges; also have lost one pi bond.

7-15

In structure B all atoms have an octet of electrons and an energetically more favorable pi bond is present.

7-16 Tautomers: a; d; e.

Resonance hybrids: b; c.

7-17

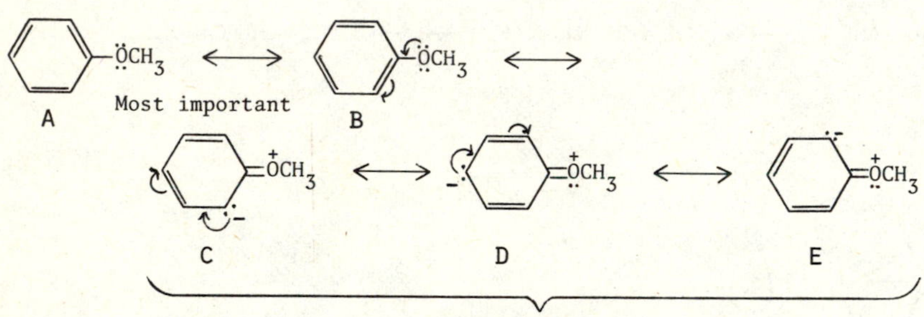

Structures A and B differ in the arrangement of bonds in the aromatic ring (i.e., they represent the two Kekulé forms). Structures C, D, and E are some of the many forms in which electrons are delocalized into the aromatic ring. Though not as important as A and B, the contributing structures C, D, and E are important in understanding the chemistry of this type of substituted aromatic.

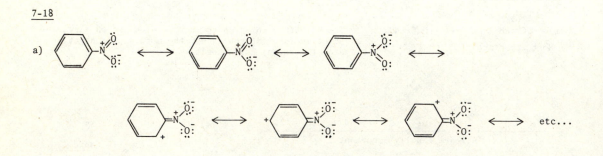

7-18

a) 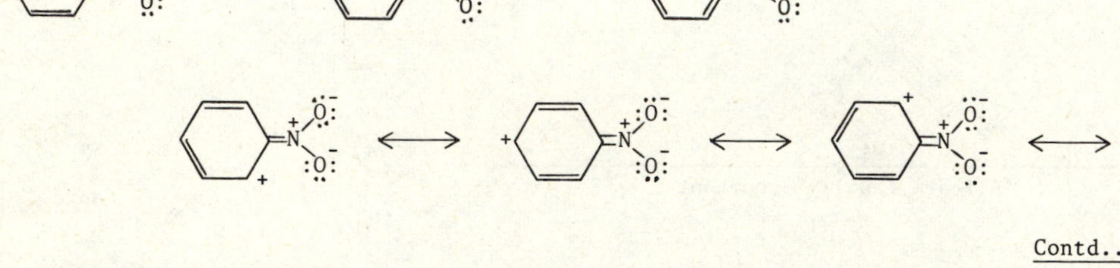 etc...

Contd...

7-18 Contd...

b) 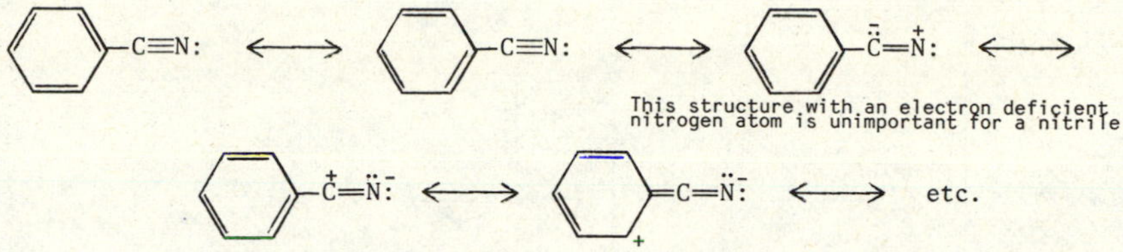

c) 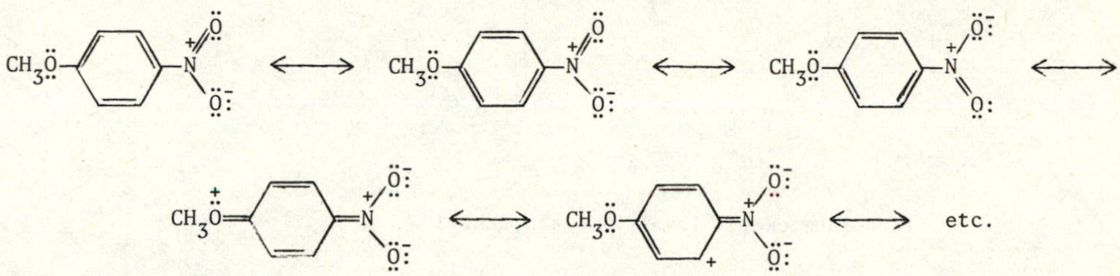

d) 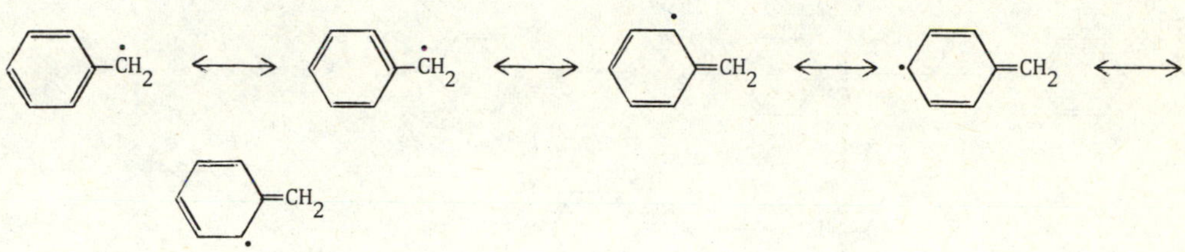

7-19

a) 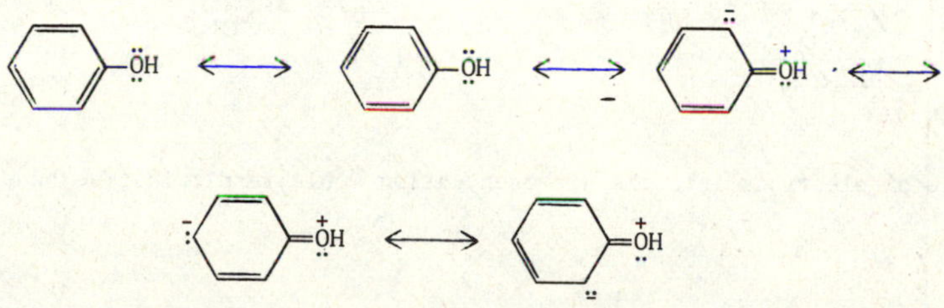

b) Acidity reflects the *difference* in energy between acid and conjugate base. The phenoxide must be resonance-stabilized relatively more than the phenol. With the exception of the two benzene Kekulé structures, phenol resonance structures involve generation and separation of opposite charges, thus resonance is not as important for the phenols as for the phenoxides.

Contd...

7-19 Contd...

c)

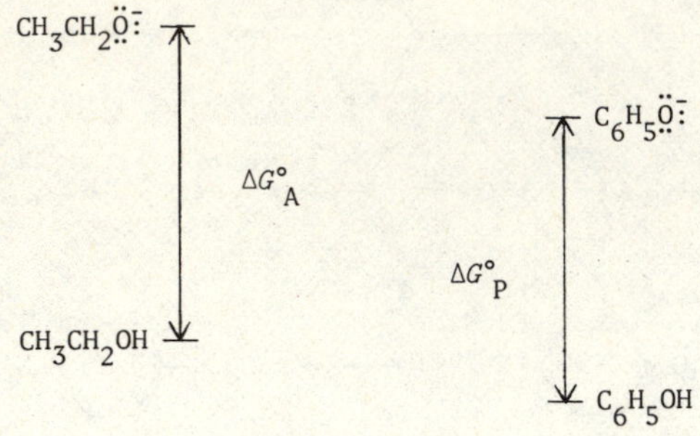

$$\Delta G°_A > \Delta G°_P$$

(Absolute energy levels are arbitrary)

7-20

a)

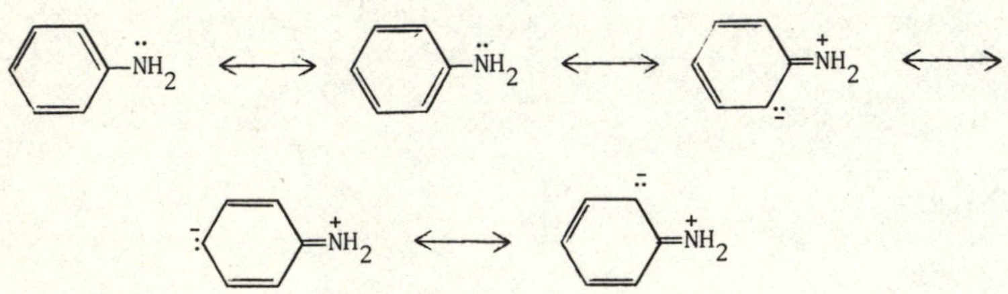

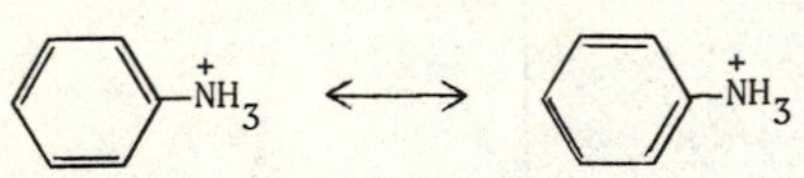

(Delocalization of the pi electrons into the nitrogen cation would result in five bonds to nitrogen.)

b)

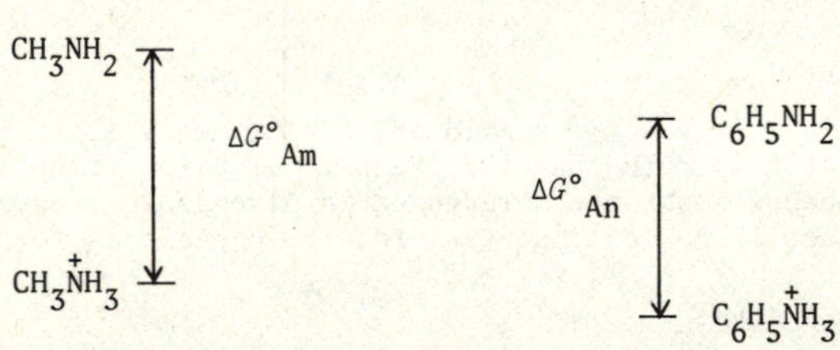

$$\Delta G°_{Am} > \Delta G°_{An}$$

(Absolute energy levels are arbitrary)

7-21 Resonance between the meta nitro group and the electrons on oxygen is not possible. Only an inductive influence can operate and thus the acid strengthening effect of nitro is relatively small in this case.

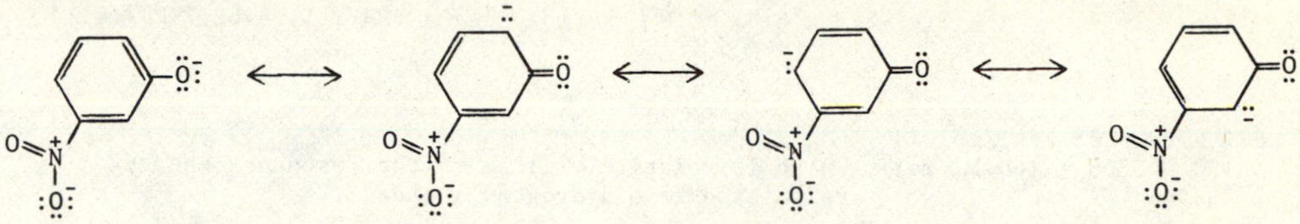

7-22 The carbonyl group in this hydroxy ketone is three atoms away from the acidic hydroxy. Inductive interactions are expected to be very small. However conjugation between the carbonyl and oxygen anion can stabilize the conjugate acid in a manner similar to that of a carboxylic acid. The pK_a value is thus similar to that of a carboxylic acid.

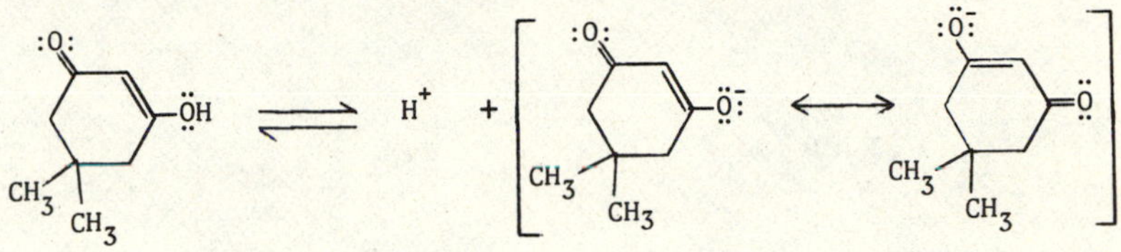

7-23 The smaller difference in dipole moments between the aromatic system ($\Delta\mu$ = 2.0) compared to the aliphatic system ($\Delta\mu$ = 4.8) can be attributed to contributions from delocalized structures which oppose the $\overset{+}{N}$-$\overset{-}{O}$ dipole.

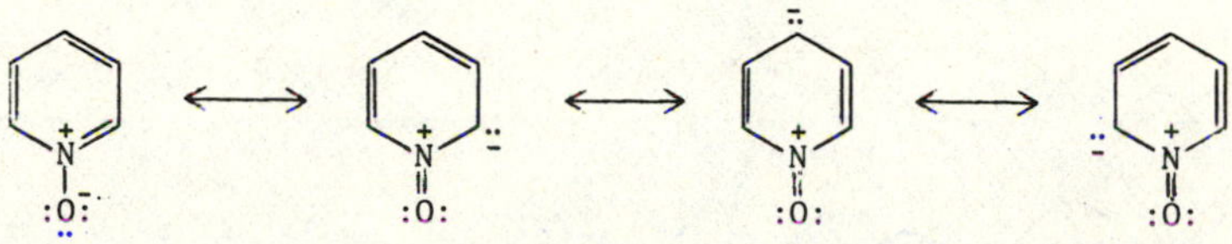

7-24

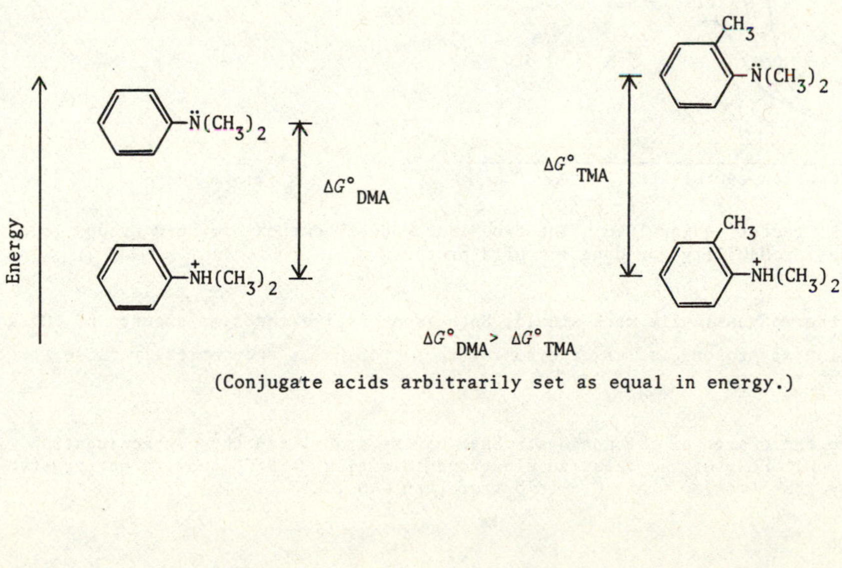

$\Delta G°_{DMA} > \Delta G°_{TMA}$

(Conjugate acids arbitrarily set as equal in energy.)

7-25 $\Delta K = 10^{11}$

$$\Delta\Delta G^\circ = -2.3\ RT\ \Delta \log K = -2.3 \times 1.99 \times 10^{-3}\ \text{kcal/mol - }K \times 298 \times 11 = 15.0\ \text{kcal/mol}$$

$$(= -2.3 \times 8.31 \times 10^{-3}\ \text{kJ/mol - }K \times 298 \times 11 = 62.7\ \text{kJ/mol})$$

7-26
a) 827 -789 = 38 kcal/mol
(3460 - 3300 = 160 kJ/mol), which is relatively close to the resonance energy calculated from hydrogenation data.

b)

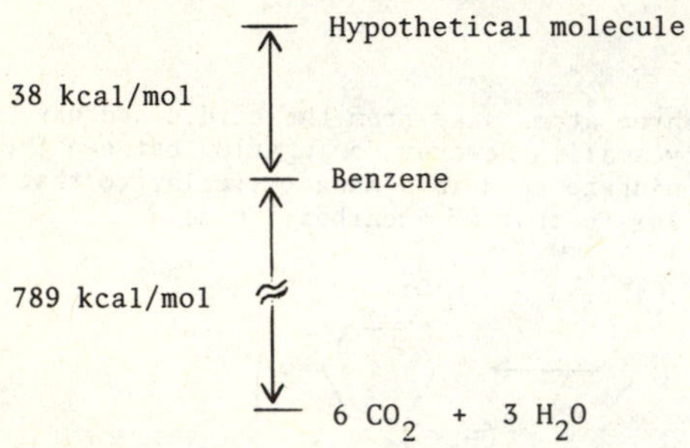

7-27 Using -30.3 kcal/mol (-126.8 kJ/mol) as the standard for the heat of hydrogenation of a terminal double bond in a C_4 hydrocarbon, the two double bonds of 1,3-butadiene without stabilization would be -60.6 kcal/mol(-253.6 kJ/mol).

Stabilization energy = 57.1 - (60.6) = 3.5 kcal/mol
= -238.9 - (-253.6) = 14.7 kJ/mol

7-28

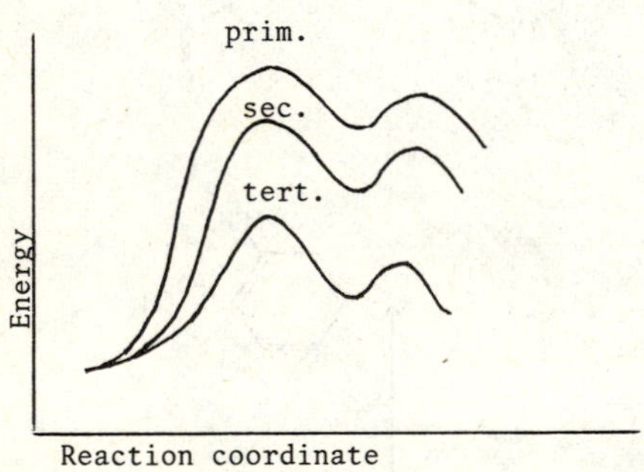

7-29 A statistical factor is involved. The two "independent" carboxylic acid groups lead to two times the probability for donation of a proton.

7-30 Propene is thermodynamically more stable. Both isomers give the same amounts of CO_2 and H_2O but combustion of propene is less exergonic, thus propene is energetically closer to the stable products.

7-31 In each case the charge of the conjugate base can be stabilized through conjugation with the carbonyl group. However the relative electronegativities O>N>C leads to better stabilization of charge by that atom so that $-\ddot{\underset{..}{O}}:^-$ is better than $-\ddot{\underset{..}{N}}H^-$, etc.

7-32

a) Alkyl groups are electron donating, thus destabilize a carbanion. The three alkyl groups of a tertiary carbanion will have a greater destabilizing effect than the two of a secondary carbanion, etc...

b) The benzene ring can delocalize the electrons of the carbanion and stabilize the species.

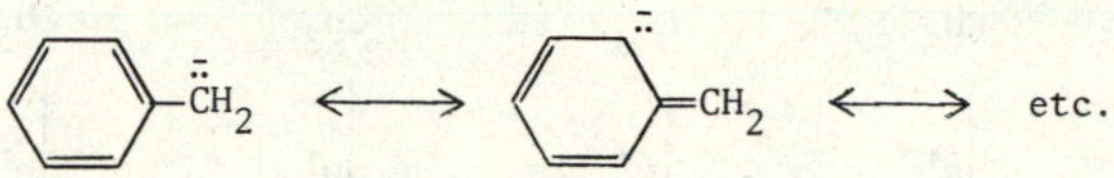

7-33 In each comparison, resonance structures can account for the differences in dipole moments.

a) $CH_2{=}CH{-}\ddot{Cl}:$ $\longleftrightarrow$ $\bar{C}H{-}CH{=}\overset{+}{Cl}:$

The dipole of this structure opposes the C-Cl dipole.

b)

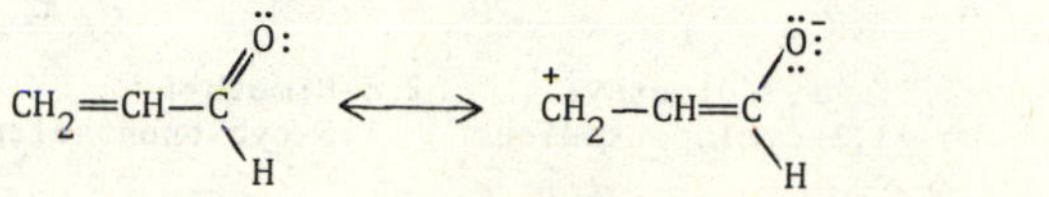

The greater charge separation in this structure enhances the total dipole moment.

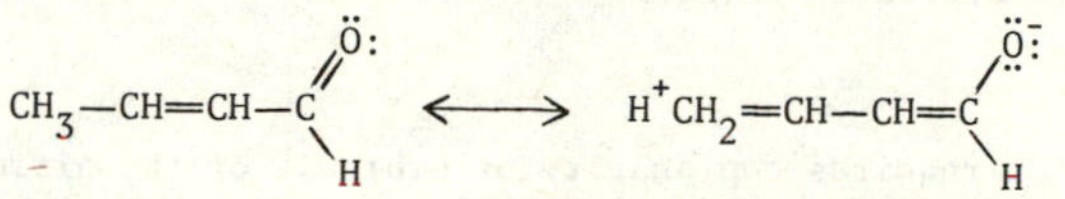

Methyl hyperconjugation enhances the total dipole moment even more than that of propenal.

c)

Interaction between two groups on the aromatic ring increases the total charge separation more than occurs in each monosubstituted molecule.

7-34 The *p*-orbitals of the two double bonds are perpendicular to each other, thus do not interact as is required for conjugation.

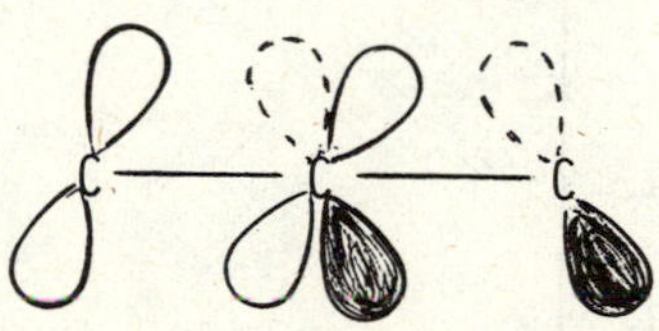

7-35 The base abstracts a proton to form a cyclopentadienyl anion. Protonation can take place at any carbon atom of the resonance stabilized system.

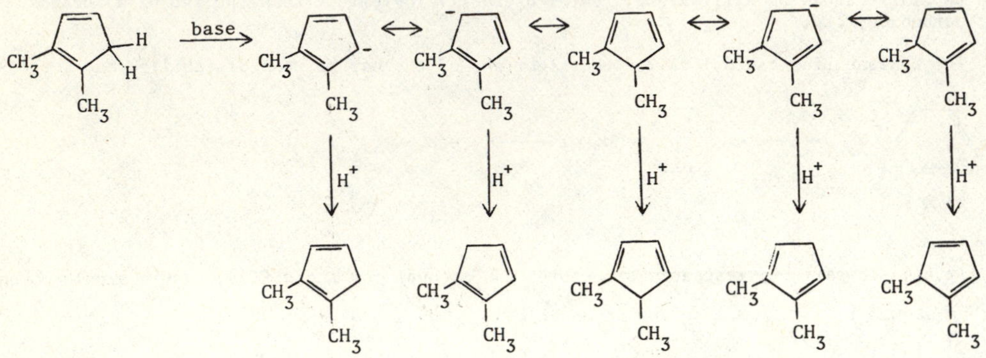

A — same as A — B — C — same as B

1,2-Dimethyl-1,3-cyclopentadiene (A)

1,5-Dimethyl-1,3-cyclopentadiene (B)

2,3-Dimethyl-1,3-cyclopentadiene (C)

7-36 Each successive negative charge decreases the stability of the conjugate base by electrostatic repulsions and therefore decreases acidity.

7-37 Resonance stabilization of the phenoxide requires coplanarity of orbitals of the nitro group, the benzene ring, and the phenoxide oxygen atom. A methyl group ortho to the nitro group sterically inhibits planarity. The nitro group must twist about the C-N bond and thereby the resonance stabilization of the phenoxide is reduced. The steric effect is increased by two ortho methyl groups.

7-38 The conjugate acid of guanidine has three energetically equivalent resonance structures and thus is stabilized more than most amines.

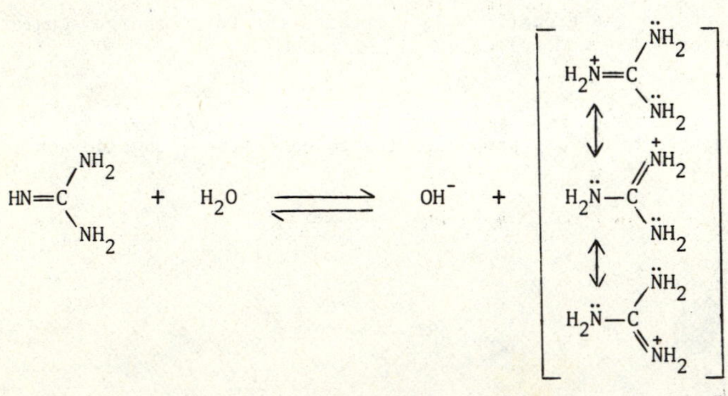

7-39

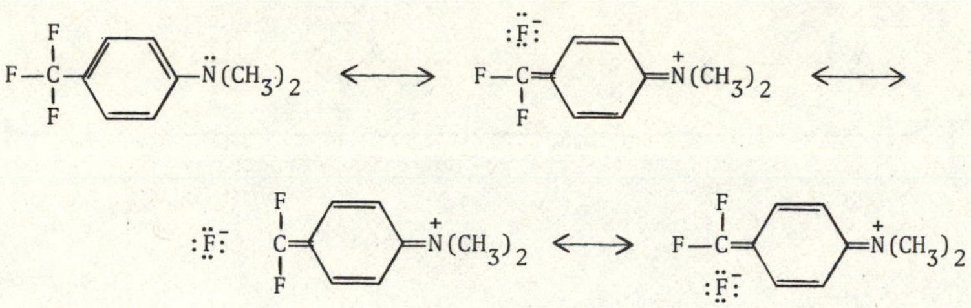

7-40

a) *p*-Cyanophenol; because of resonance stabilization of the conjugate base.

b) $CH_3CH{=}CHOH$; because of resonance stabilization of the conjugate base.

c) $NCCH_2CN$; because two groups provide resonance stabilization of the conjugate base.

d) 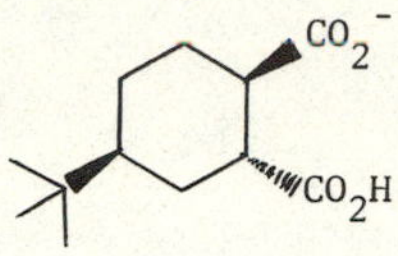 ; because the *t*-butyl group remains equatorial, the $-CO_2^-$ and $-CO_2H$ groups are trans-diaxial in this chair conformation. In the other isomer intramolecular H-bonding between the diequatorial groups decreases acidity.

e) CH_3CF_2OH; because the inductive electron-withdrawing effect by F falls off with distance.

f) 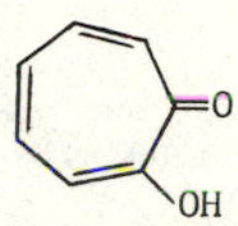 ; because the charge of the conjugate base is delocalized over a larger system.

g) 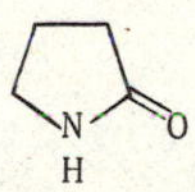 ; because of charge delocalization of the conjugate base by the carbonyl group.

7-41 They are structural isomers. Carbon atoms move and the sequence of covalently bonded atoms is different. Sigma bonds are made and broken.

7-42

a) 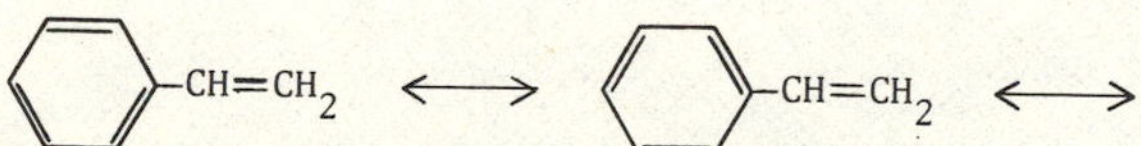

Most important - energetically equivalent

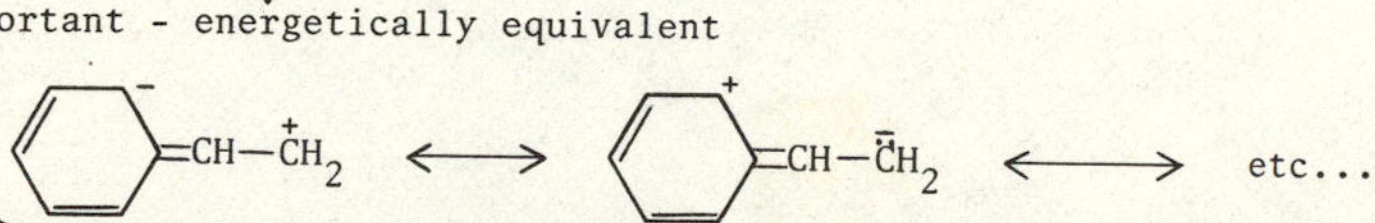

These, plus related ring-delocalized charged structures are not very important

Contd...

7-42 Contd...

b) 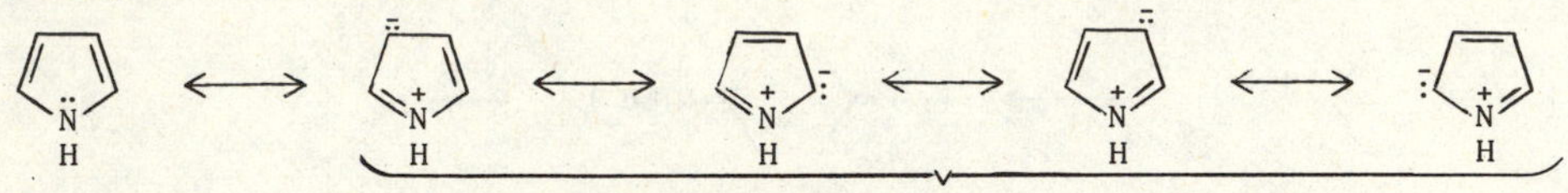

Most important

Less important - charge generated structures

c) 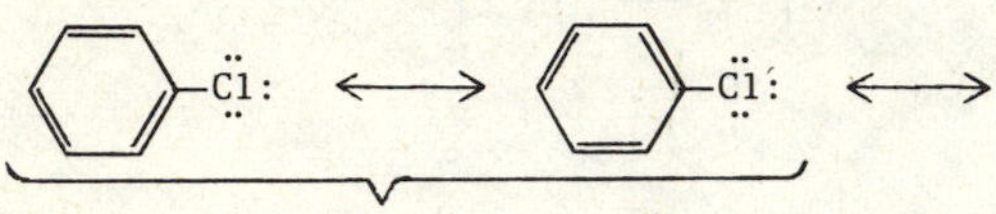

Most important - energetically equivalent

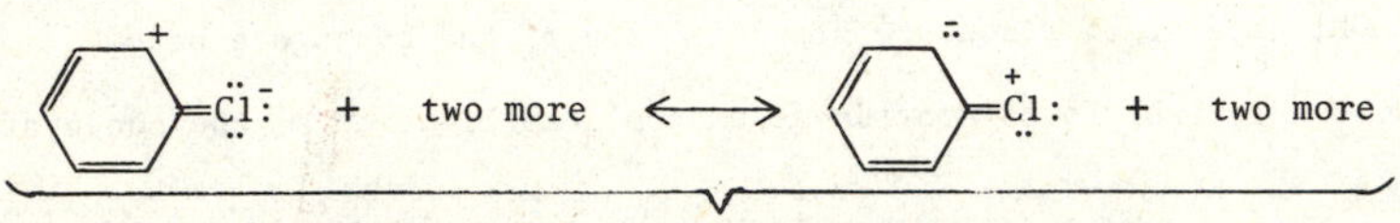

Less important - charge generated structures

7-43 The nitrogen atom of simple amides such as A is not basic because the electrons are delocalized by the carbonyl groups. In B, such conjugation would require that a double bond form at the bicyclic bridgehead. This would be very unfavorable energetically (Bredt's rule). The position of the carbonyl IR absorption of B is typical of an aldehyde or ketone rather than an amide.

7-44 Acidity properties suggest that A is a carboxylic acid and B is a phenol. The nmr spectra suggest that both are *p*-disubstituted aromatic compounds with deshielded, but unsplit methyl groups. The compounds are

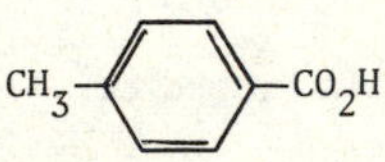

A

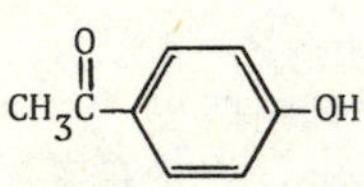

B

Note that the methyl groups are deshielded similarly by the carbonyl group or the aromatic ring.

8 Nucleophilic Additions to the Carbonyl Group—Aldehydes and Ketones

8-1

a) 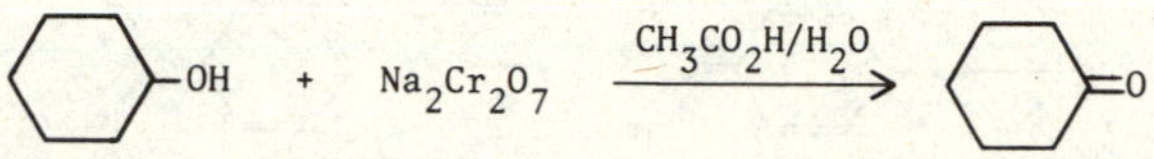

b) 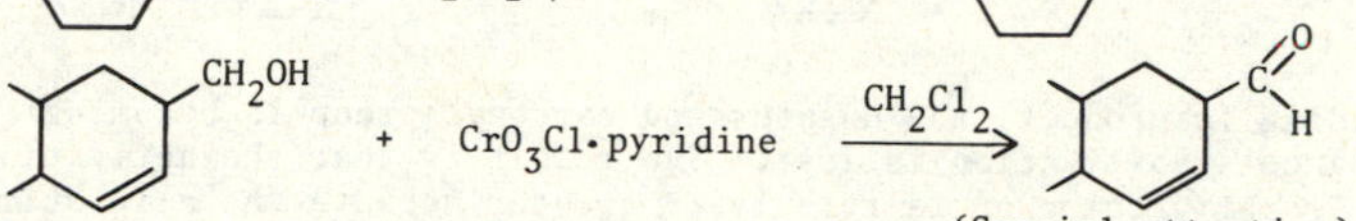

(Special attention)

c) $C_6H_5CH_2OH$ + $CrO_3Cl \cdot$pyridine $\xrightarrow{CH_2Cl_2}$ C_6H_5CHO

(Special attention)

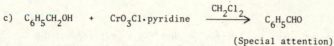

d) $C_6H_5CH_2CH(OH)CH_3$ + NaOCl $\xrightarrow{CH_3CO_2H/H_2O}$ $C_6H_5CH_2C(=O)CH_3$

e) 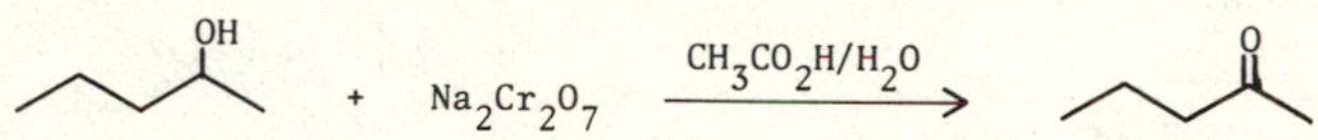

f) 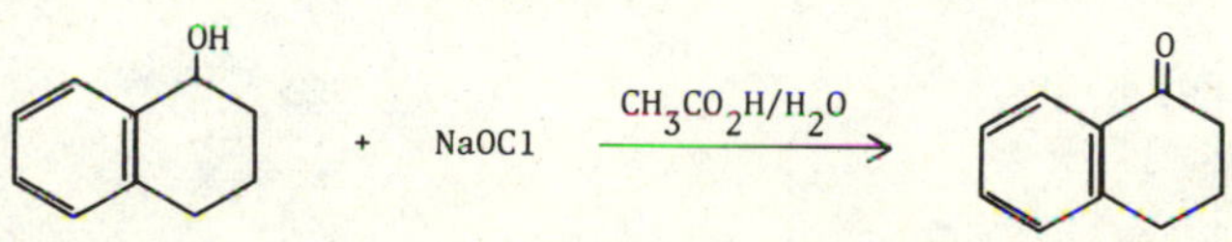

8-2

a) 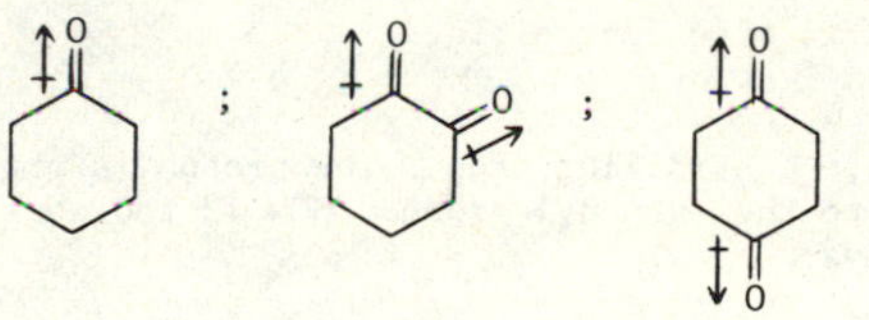

A B C

Contd...

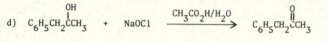

8-2 Contd...

b) Using vector addition of the carbonyl group dipoles gives:

A $\mu \approx 3.0$ D B $\mu \approx 5$ D C $\mu \approx 0$ D

8-3 No, for the carbonyl carbon atom would then possess ten electrons.

8-4 The aromatic aldehyde is stabilized more than the aliphatic aldehyde because of resonance with the benzene ring.

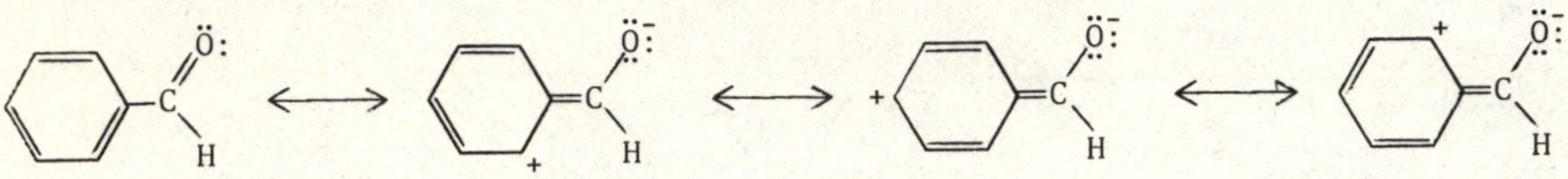

At the transition state leading to product, the unsaturated carbonyl group is becoming saturated. The extra resonance stabilization is lost. The result is that the activation energy for reaction of the aromatic compound is relatively greater because the aliphatic compound never had a significant resonance energy to lose.

8-5 The major difference between the two resonance hybrids is that the charge separated resonance structure of the ketone involves a secondary carbocation and that of the aldehyde a primary carbocation. Secondary carbocations are the more stable.

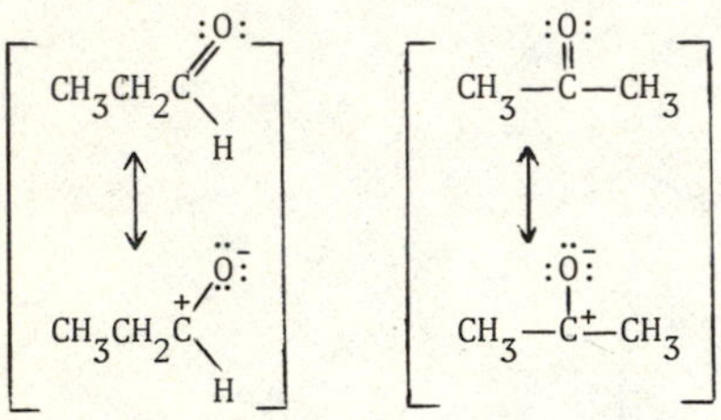

8-6

a) The addition of cyanide to the carbonyl group is rate controlling.

b) The addition of H^+ does not enter into the rate controlling step. The proton is provided by the weak acid HCN after cyanide has added to the carbonyl group. HCN is too weak of an acid to appreciably protonate the carbonyl oxygen atom.

8-7

a) $CH_3COCH_2CH_2CHO + HCN \longrightarrow CH_3COCH_2CH_2CH(OH)CN$ (The aldehyde group is more reactive.)

b) $C_6H_5COCH_2CH_2COCH_3 + HCN \longrightarrow C_6H_5COCH_2CH_2C(OH)(CN)CH_3$ (The resonance stabilized carbonyl group is less reactive.)

Contd...

8-7 Contd...

c)

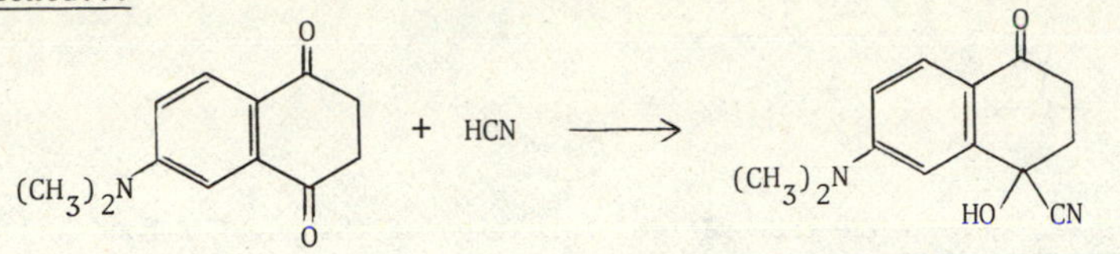

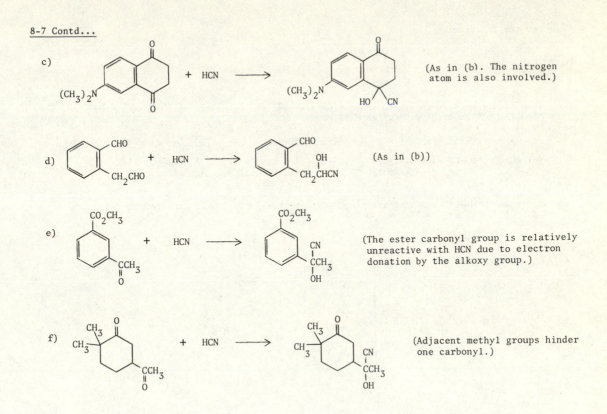

(As in (b). The nitrogen atom is also involved.)

d)

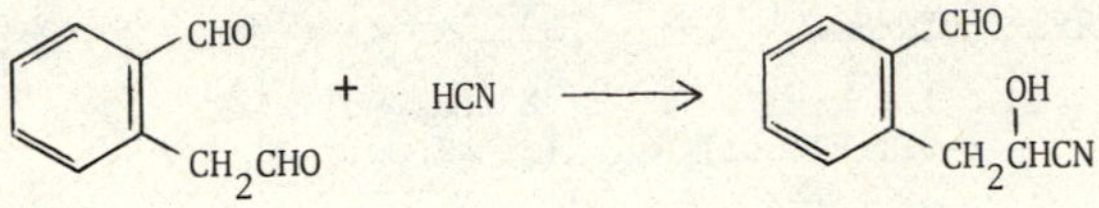

(As in (b))

e)

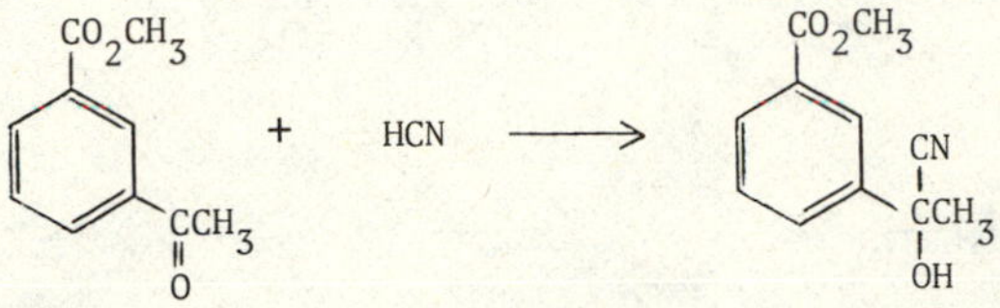

(The ester carbonyl group is relatively unreactive with HCN due to electron donation by the alkoxy group.)

f)

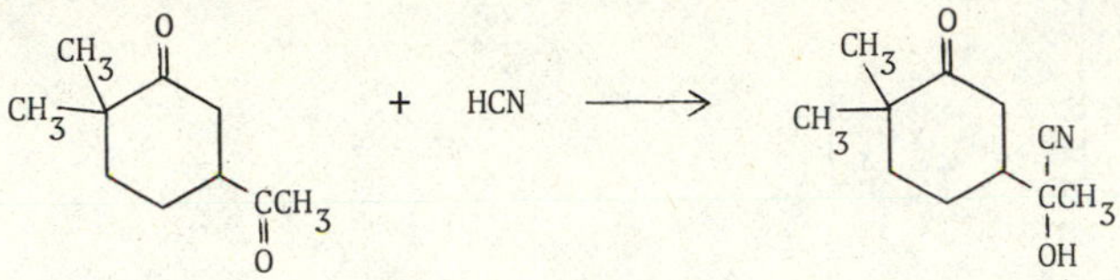

(Adjacent methyl groups hinder one carbonyl.)

8-8 Conjugation between the benzene ring and the carbonyl group stabilizes acetophenone relative to acetone.

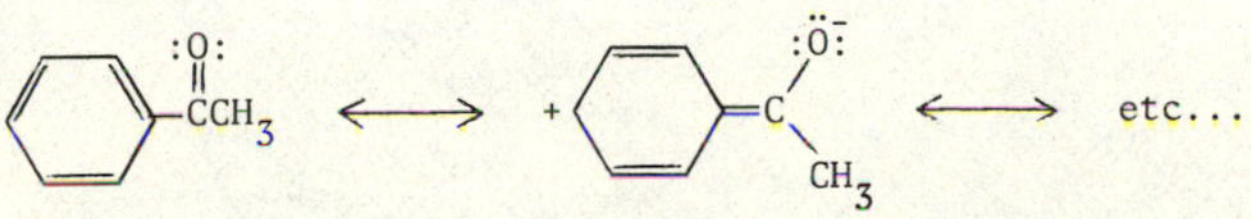

That interaction is not important in either of the cyanohydrin products. Acetophenone is stabilized sufficiently relative to its cyanohydrin that the addition reaction is slightly endergonic ($K < 1$). Resonance is not significant in acetone or its cyanohydrin so that this reaction is exergonic.

8-9 There is little difference between the stabilities of cyclopentanone and cyclohexanone. However the cyanohydrins differ in their nonbonded interactions. More eclipsing strain is present in the cyclopentanone cyanohydrin than in the completely staggered cyclohexanone cyanohydrin. Conversion of cyclohexanone to its cyanohydrin is therefore more exergonic.

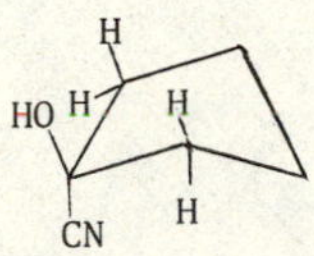

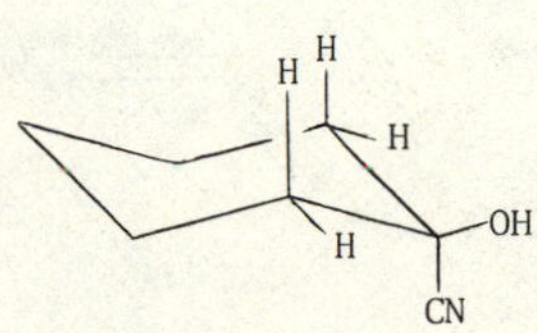

8-10 For the reaction

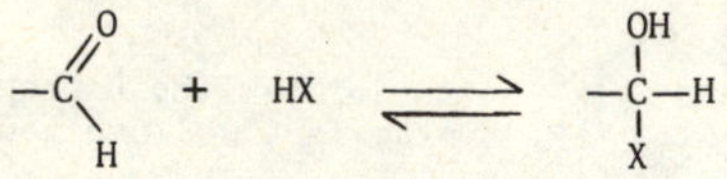

the difference between X = Cl and X = CN will reflect formation of the new C-X bond and the breaking of the original H-CN vs H-Cl bonds.

	Bonds broken		Bonds formed		$\Delta H°$
For HCN	H—CN	130 kcal/mol (544 kJ/mol)	C—CN	-122 kcal/mol (-511 kJ/mol)	8 kcal/mol (33 kJ/mol)
For HCl	H—Cl	103 kcal/mol (431 kJ/mol)	C—Cl	-84 kcal/mol (-352 kJ/mol)	19 kcal/mol (79 kJ/mol)

8-11

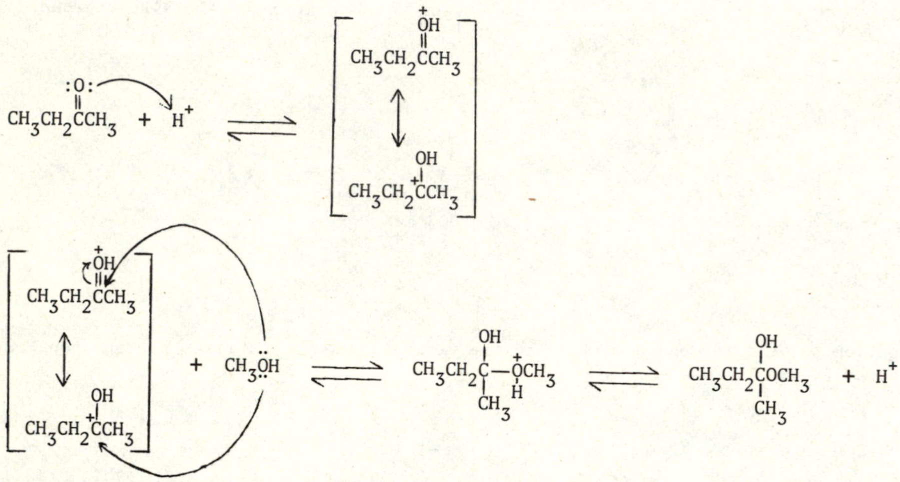

In base:

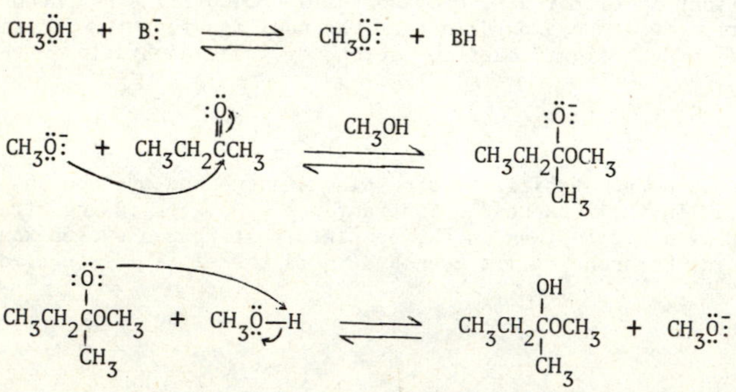

8-12

a) 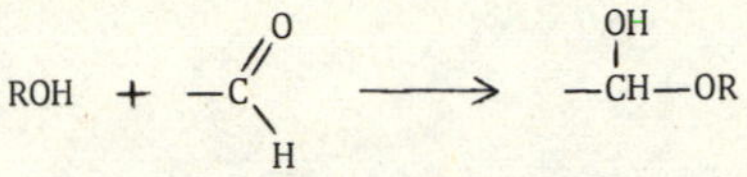

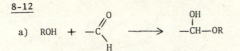

The change in bonds involves a loss of C=O and gain of two C—O bonds. Energies for cleavage of the alcohol O-H bond and formation of hemiacetal O-H bond cancel each other.

Bonds broken	Bonds formed	$\Delta H°$
C=O 176 kcal/mol (736 kJ/mol)	2 x C—O -172 kcal/mol (-718 kJ/mol)	4 kcal/mol (18 kJ/mol)

b) Since $\Delta G° = \Delta H° - T\Delta S°$, a more favorable entropy (more positive $\Delta S°$) makes the value of $\Delta G°$ more negative and energetically favorable. A less favorable (more negative) entropy leads to a relatively more positive value of $\Delta G°$. Since the $\Delta H°$ term is relatively small, the $T\Delta S°$ term can become the significant factor controlling $\Delta G°$.

8-13

a) $(CH_3)_2CHCH_2CH(OCH_3)_2$

b) 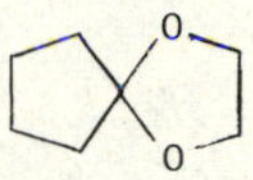

c) 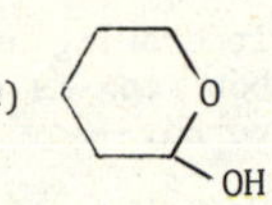

d) 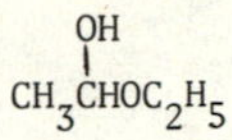

e) 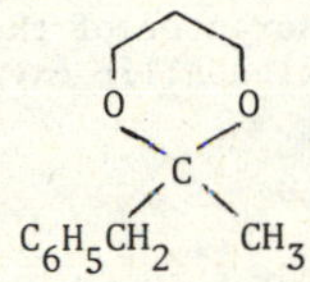

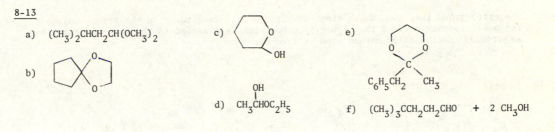

f) $(CH_3)_3CCH_2CH_2CHO + 2\ CH_3OH$

8-14

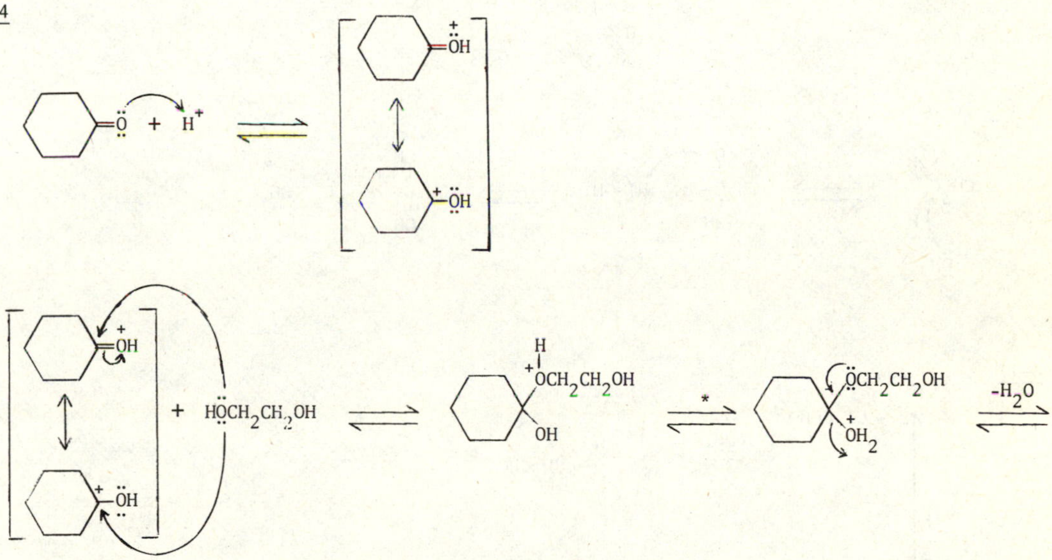

Contd...

8-14 Contd...

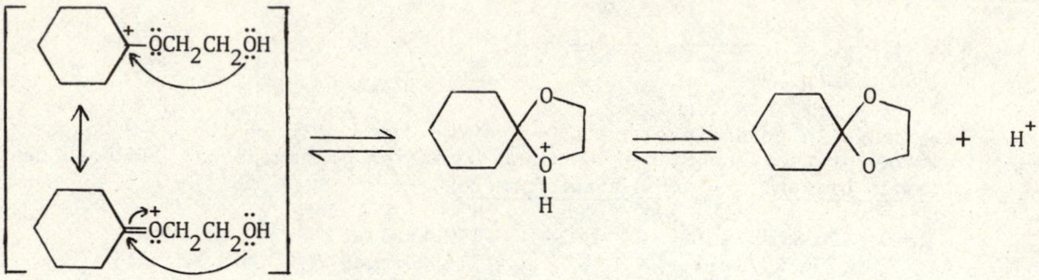

*We will often use an abbreviated notation that shows a proton moving from one atom to another in a single molecule. In solution the departing proton normally goes to solvent, then some other proton is donated back to the substrate by the solvent molecules.

8-15 The ester group functions as an electron-withdrawing substituent on the ketone carbonyl. The positive character of the carbonyl carbon atom is enhanced so the initial attack by an alcohols' nucleophilic oxygen atom is favored.

8-16 In acid:

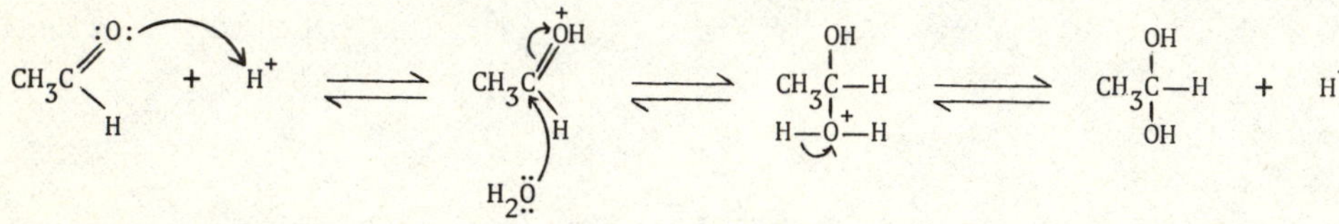

In base:

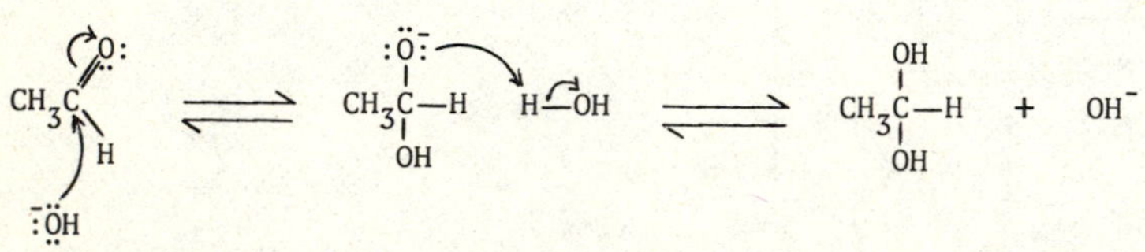

8-17

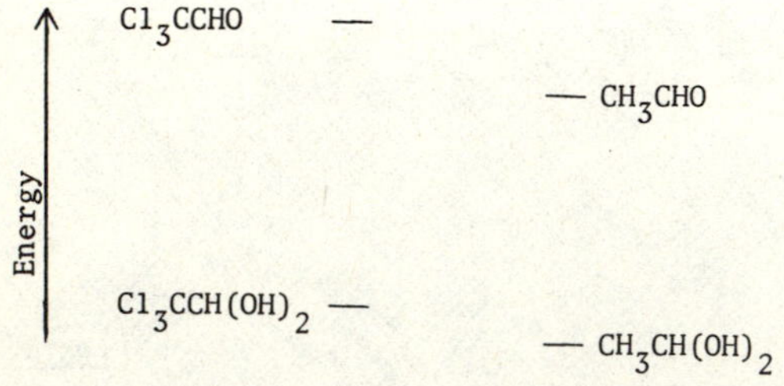

8-18 The *t*-butyl group is large and results in significant crowding when the addition of H_2O converts the trigonal carbonyl group to a tetrahedral configuration. The value of the equilibrium constant for that substrate is less than 1, while the less crowded acetaldehyde has an equilibrium constant slightly greater than 1.

8-19

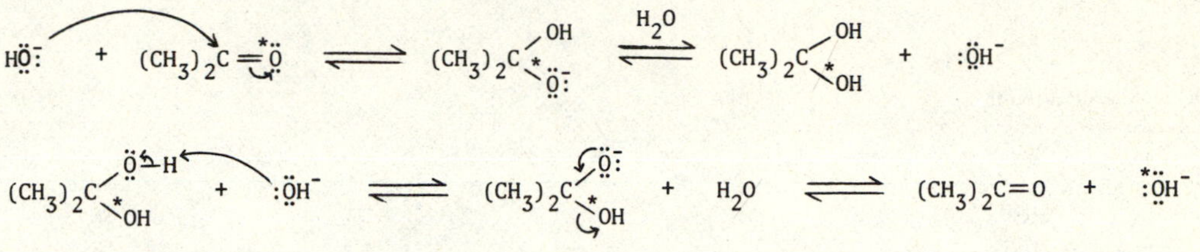

8-20

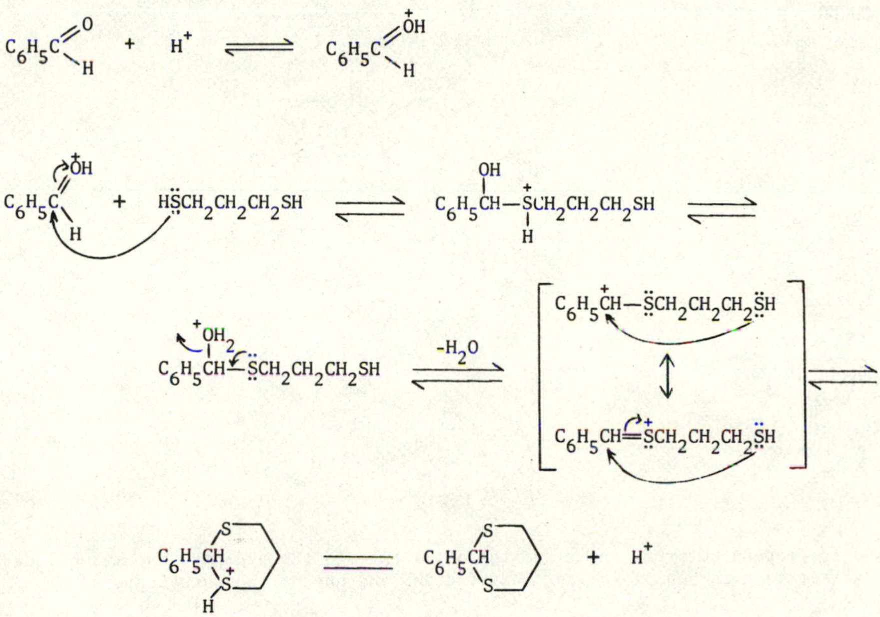

8-21

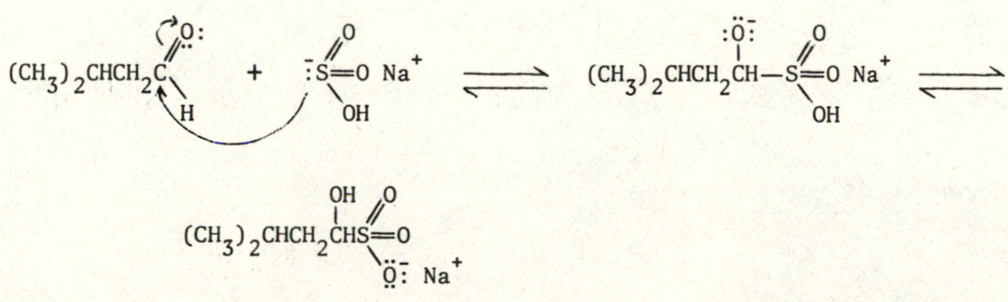

8-22 Alcohols are almost as acidic as water. The hydroxy proton, like that from water, readily reacts with $LiAlH_4$.

8-23 Base minimizes the availability of protons which would slowly react with the hydride from $NaBH_4$.

8-24

a) $CH_3CH_2CH_2CHO$

b) 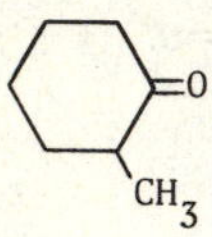

c) $CH_3CH{=}CHCHO$

d) $CH_3CH_2CH(CH_3)\overset{\overset{O}{\|}}{C}CH_3$

e) C_6H_5CHO

f) $(C_6H_5)_2CHCHO$

g) 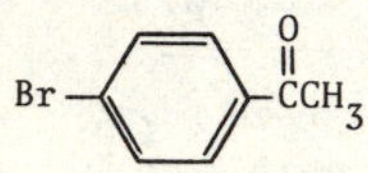

h) 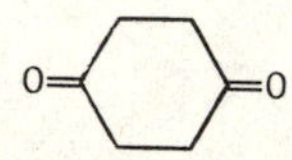

i) 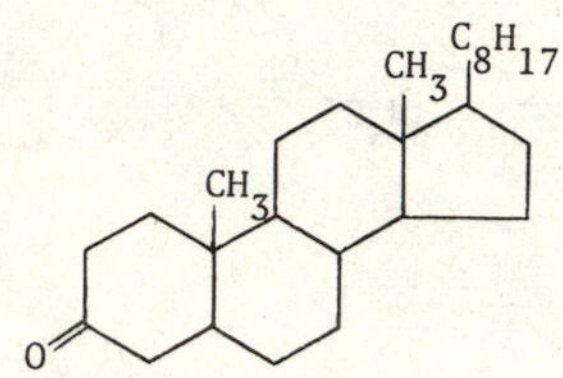

8-25

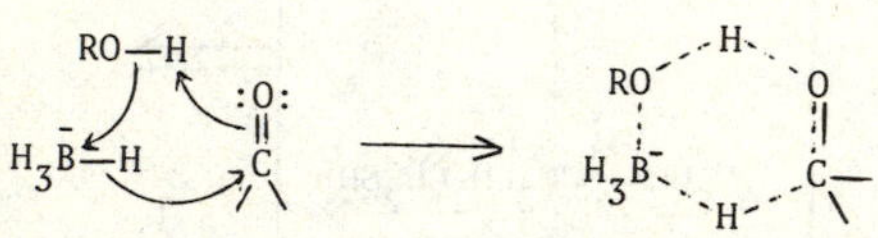

8-26 The plane is perpendicular to the ring and passes through the hydroxy, the 1 and 4 carbon atoms, the tertiary carbon of the *tert*-butyl group and one of the methyl groups.

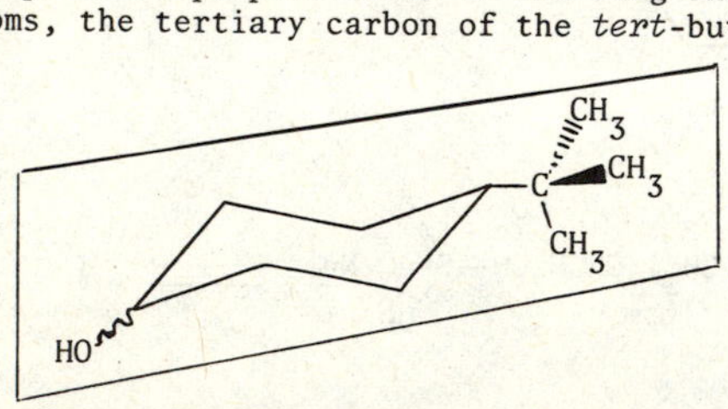

cis- or *trans*-4-*t*-Butylcyclohexanol

8-27

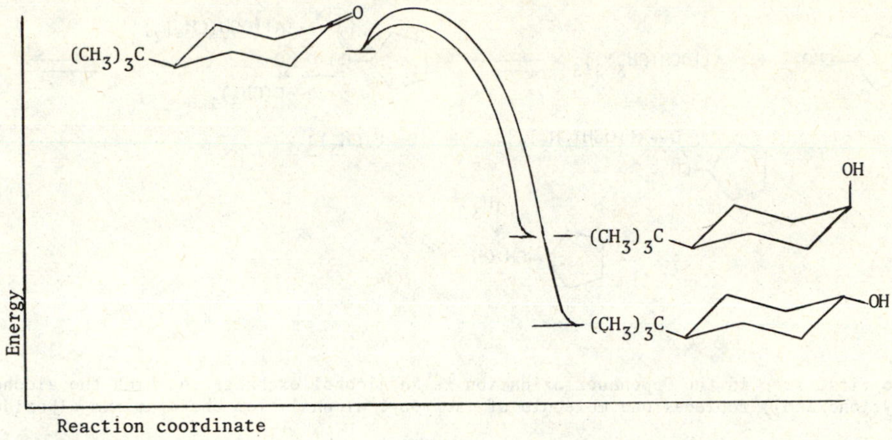

8-28

a)

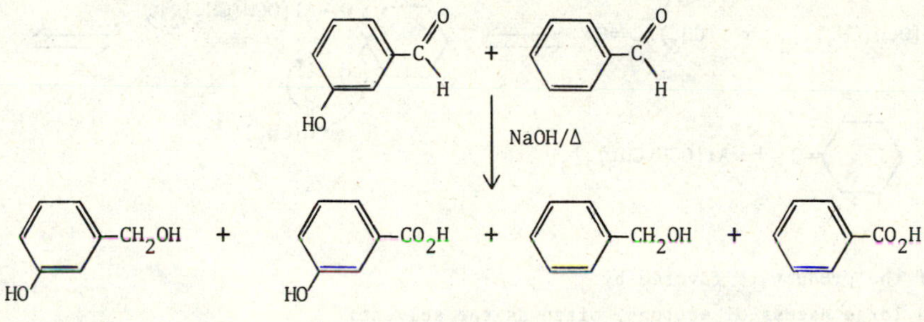

b)

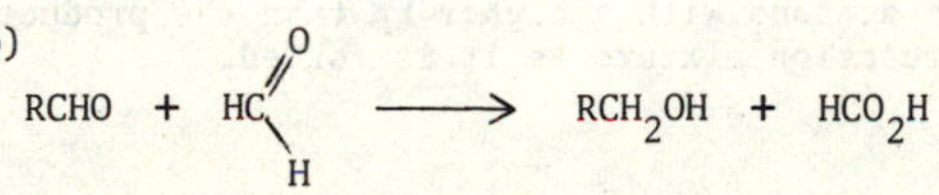

Formaldehyde is inexpensive and the formic acid produced is usually easily removed from the reaction mixture as the water-soluble salt, sodium formate.

8-29 The hydride transfer is rate controlling and requires a significant energy of activation ($\Delta G^{\ddagger}$).

8-30

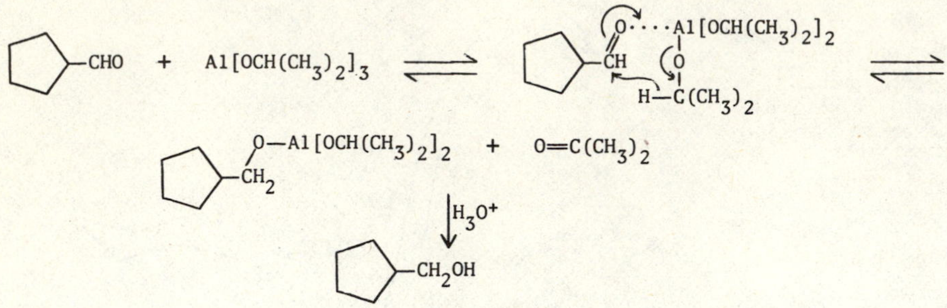

8-31

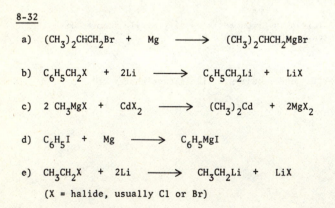

a) The first step in the Oppenauer oxidation is an alcohol exchange in which the alcohol (cyclohexanol) replaces one molecule of isopropyl alcohol from the aluminum alkoxide.

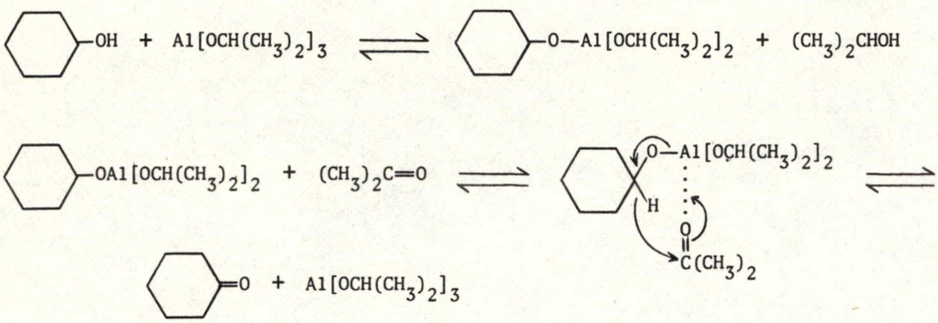

b) Formation of the product is favored by:

i) Use of a large excess of acetone, often as the solvent;

ii) Use of a ketone solvent-reactant other than acetone with a higher bp than the product so that product can be distilled from the reaction mixture as it is formed.

8-32

a) $(CH_3)_2CHCH_2Br + Mg \longrightarrow (CH_3)_2CHCH_2MgBr$

b) $C_6H_5CH_2X + 2Li \longrightarrow C_6H_5CH_2Li + LiX$

c) $2\ CH_3MgX + CdX_2 \longrightarrow (CH_3)_2Cd + 2MgX_2$

d) $C_6H_5I + Mg \longrightarrow C_6H_5MgI$

e) $CH_3CH_2X + 2Li \longrightarrow CH_3CH_2Li + LiX$

(X = halide, usually Cl or Br)

8-33

a) $(CH_3)_2CHCH_2CH_2OH$
3-Methyl-1-butanol
(Isoamyl alcohol)

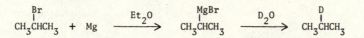

c) $(CH_3)_2C(OH)C\equiv CCH_3$
2-Methyl-3-pentyn-2-ol

b) $C_6H_5CH(OH)CH_2CH_3$
1-Phenyl-1-propanol

d) 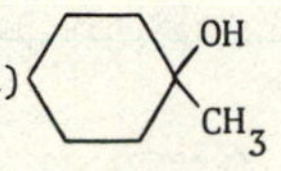
1-Methylcyclohexanol

8-34

$$CH_3CH(Br)CH_3 + Mg \xrightarrow{Et_2O} CH_3CH(MgBr)CH_3 \xrightarrow{D_2O} CH_3CH(D)CH_3$$

8-35 Both involve a 6-atom cyclic complex. The hydride comes from the atom beta to the metal.

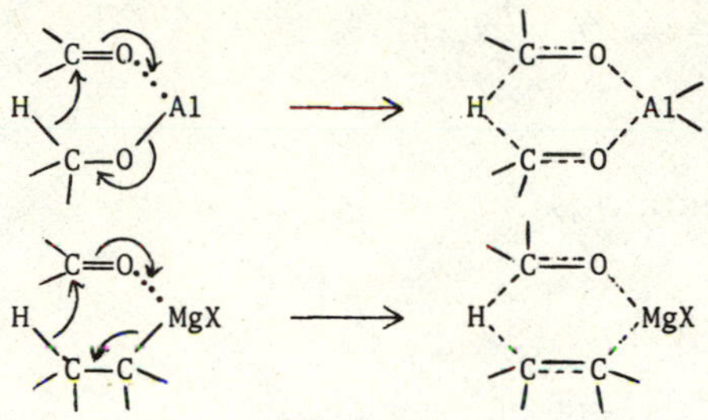

8-36

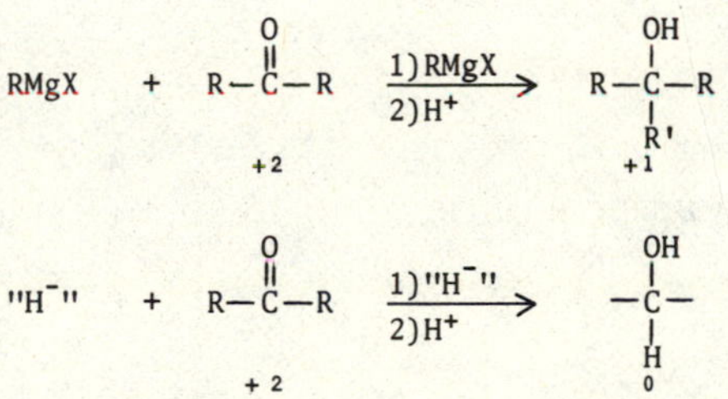

Oxidation change at carbon = -1

Oxidation change at carbon = -2

8-37 The first step in devising a synthetic sequence is the recognition of possible components (*synthons*) from which the product molecule (*the target*) can be constructed. In the Grignard reactions below, the carbinol carbon atom (the carbon atom bonded to hydroxy) must have originally been part of a carbonyl group so that some carbonyl compound is readily identified as one potential reactant. The second reactant, the Grignard reagent, must form a bond to this carbonyl carbon atom by addition of the nucleophilic carbon atom. Thus the original organohalogen compound is defined. As syntheses become more complex and multiple steps are involved, it is often useful to devise a sequence by working backwards, from product to reactants.

In each of the following examples, the Grignard reagent is prepared from the appropriate organohalogen compound and magnesium metal in diethyl ether or tetrahydrofuran (THF). THF is a higher boiling ether solvent commonly used when reaction is not favorable in diethyl ether.

a) $C_6H_5MgBr + (CH_3)_2C{=}O$ or $CH_3MgI + C_6H_5\overset{O}{\overset{\|}{C}}CH_3$

$$\xrightarrow[2)H_2O/NH_4Cl]{1)Et_2O} C_6H_5\overset{OH}{\overset{|}{C}}(CH_3)_2$$

b) $$(CH_3)_2CHCH_2MgBr + HCHO \xrightarrow[2)H_3O^+]{1)Et_2O} (CH_3)_2CHCH_2CH_2OH$$

c) $CH_3CH_2MgI + (CH_3)_2C{=}O$ or $CH_3MgI + CH_3\overset{O}{\overset{\|}{C}}CH_2CH_3$

$$\xrightarrow[2)H_2O/NH_4Cl]{1)Et_2O} (CH_3)_2\overset{OH}{\overset{|}{C}}CH_2CH_3$$

d) $$C_6H_5CH_2MgCl + HCHO \xrightarrow[2)H_3O^+]{1)Et_2O} C_6H_5CH_2CH_2OH$$

e) $$C_6H_5MgBr + C_6H_5\overset{O}{\overset{\|}{C}}C_6H_5 \xrightarrow[2)H_3O^+]{1)Et_2O} (C_6H_5)_3COH$$

f) $CH_3MgI + CH_3CH_2\overset{O}{\overset{\|}{C}}CH_2CH_2CH_3$

or

$CH_3CH_2MgI + CH_3\overset{O}{\overset{\|}{C}}CH_2CH_2CH_3$

or

$CH_3CH_2CH_2MgI + CH_3\overset{O}{\overset{\|}{C}}CH_2CH_3$

$$\xrightarrow[2)H_2O/NH_4Cl]{1)Et_2O} CH_3CH_2\underset{OH}{\underset{|}{\overset{CH_3}{\overset{|}{C}}}}CH_2CH_2CH_3$$

8-38

a) The imine double bond is part of a highly conjugated system.

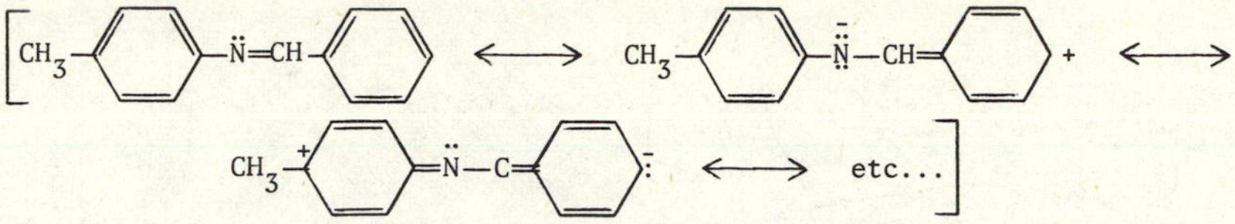

b) $p\text{-}CH_3C_6H_4\ddot{N}H_2 + C_6H_5CHO \rightleftharpoons p\text{-}CH_3C_6H_4\overset{+}{N}H_2CH(O^-)C_6H_5 \rightleftharpoons p\text{-}CH_3C_6H_4\ddot{N}HCH(OH)C_6H_5$

$p\text{-}CH_3C_6H_4\ddot{N}H\text{—}CH(OH)C_6H_5 \xrightleftharpoons{H_3O^+} p\text{-}CH_3C_6H_4\overset{+}{N}H{=}CHC_6H_5 + H_2O \rightleftharpoons$

$p\text{-}CH_3C_6H_4\ddot{N}{=}CHC_6H_5 + H_3O^+$

8-39

a) i) $p\text{-}CH_3OC_6H_4CHO + NH_2OH \longrightarrow p\text{-}CH_3OC_6H_4CH{=}NOH$

ii) 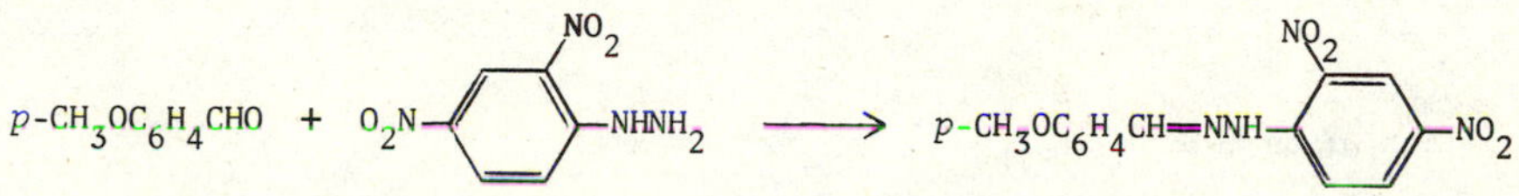

iii) $p\text{-}CH_3OC_6H_4CHO + H_2NNHCONH_2 \longrightarrow p\text{-}CH_3OC_6H_4CH{=}NNHCONH_2$

(The amide nitrogen atom is *not* the nucleophile because of delocalization of its electron pair by the carbonyl group.)

iv) $p\text{-}CH_3C_6H_4CHO + C_6H_5NHNH_2 \longrightarrow p\text{-}CH_3OC_6H_4CH{=}NNHC_6H_5$

b) $p\text{-}CH_3OC_6H_4CHO + \ddot{N}H_2OH \rightleftharpoons p\text{-}CH_3OC_6H_4CH(O^-)\text{—}\overset{+}{N}H_2OH \rightleftharpoons$

$p\text{-}CH_3OC_6H_4CH(OH)\text{—}NHOH \rightleftharpoons p\text{-}CH_3OC_6H_4CH{=}NOH + H_3O^+$

8-40

a) Enamine formation is reversible. Removal of water helps to drive the reaction toward completion.

b)

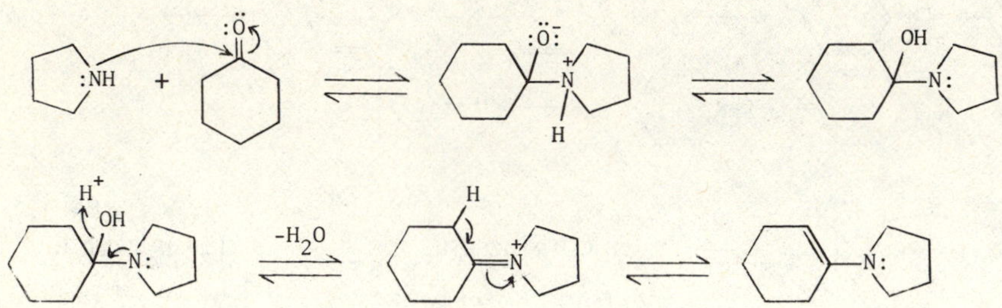

8-41

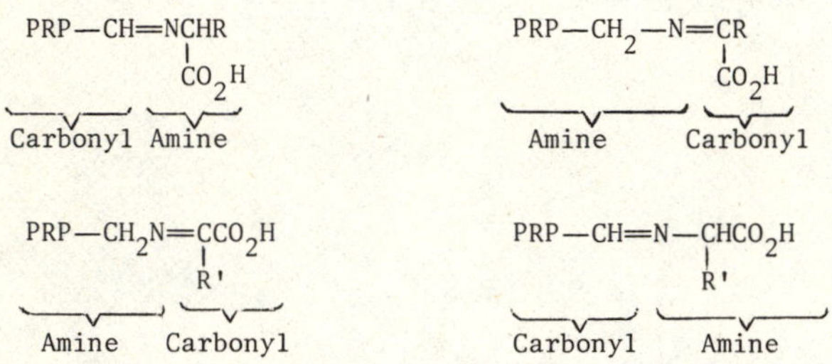

8-42

i) Formation of an iminium ion:

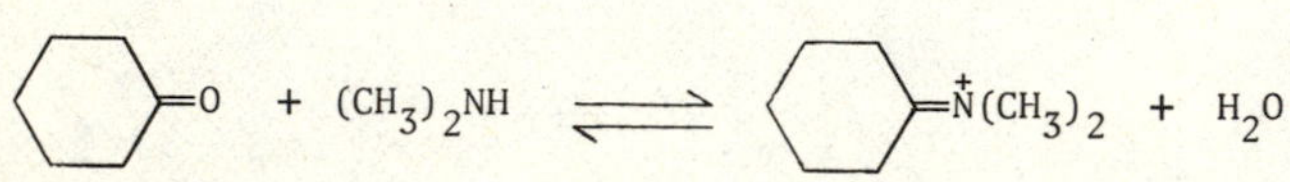

ii) Reduction:

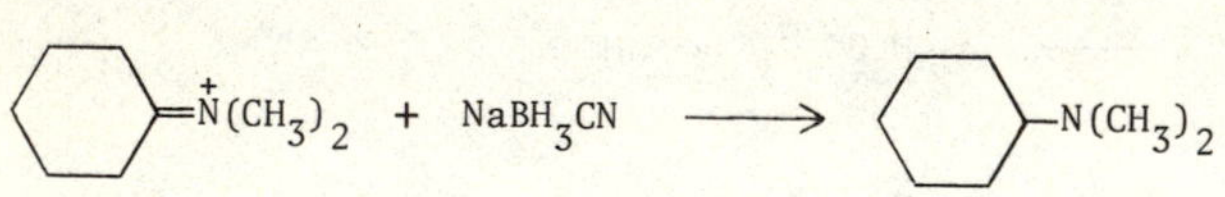

8-43 The nitrile-Grignard adduct forms a ketone only after the addition of water. Any unreacted Grignard reagent is also destroyed in the hydrolysis step.

8-44

a) $(CH_3)_2CHCH_2COC_6H_5$

b) p-$CH_3C_6H_4CH_2N(CH_3)_2$

c) $(CH_3CH_2)_2C(OH)CH_3$

d) $CH_3CH_2CH(OH)CH_2CH_3$

e) 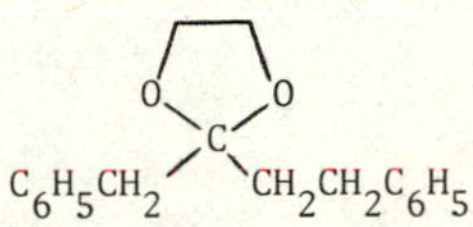

f) $C_6H_5CH{=}NOH$

g) $C_6H_5CH_2OH$ + CH_3CH_3

(Grignard reagents readily react with the acidic hydrogen atom of alcohols.)

h) 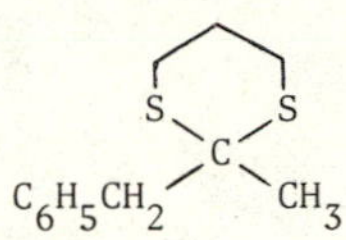

i) 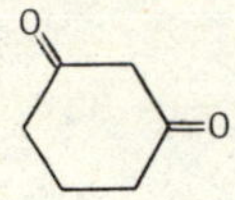 + $(CH_3)_2CH_2$

The 1,3-diketone has a pK_a value near 9. The Grignard reagent functions as a base to abstract an acidic proton rather than as a nucleophile.

j) $CH_3CH(OH)SO_3^-Na^+$

k) 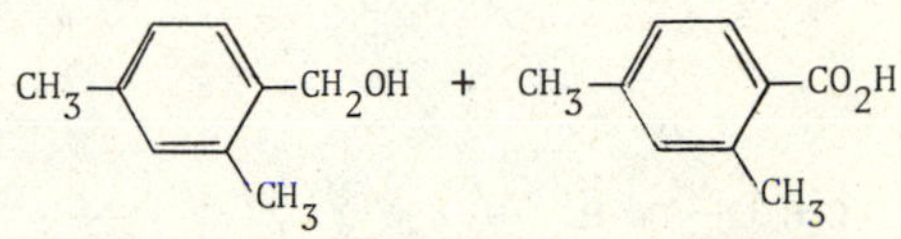

l) $C_6H_5CH(OH)C_6H_5$

m) $C_6H_5CO_2H$ + CH_4

n) $C_6H_5CH_2C(=O)CH_2CH_3$

8-45

a) Yes, because transfer of D shows that the source of D (or H) in the reduction is the carbinol D (or H) of the alkoxide group. (The carbon atom bonded to hydroxy is the carbinol carbon.)

b) The more favorable isomer is trans because the hydroxy and methyl groups are equatorial. The reaction is reversible so that the equilibrium favors the more stable stereoisomer.

8-46 The formation and hydrolysis of the acetal does not involve cleavage of the C-O bond in 2-octanol so that configuration at the chiral center does not change.

8-47

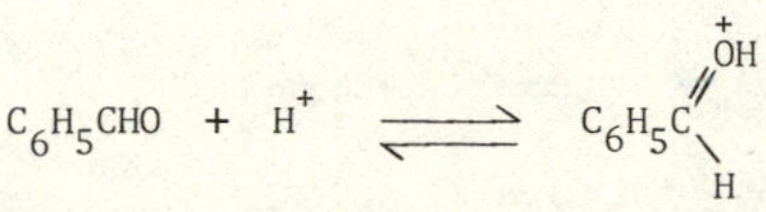

Contd...

8-47 Contd...

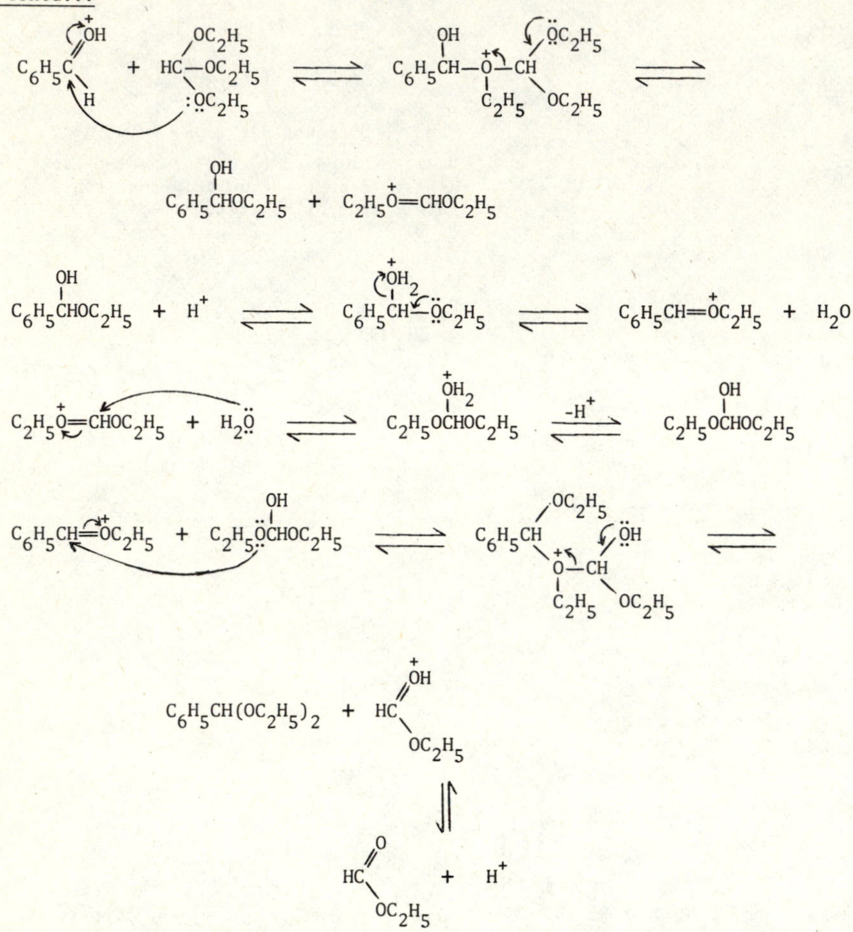

8-48

$C_6H_5CH{=}C(CH_3)CHO$

2-Methyl-3-phenylpropenal

(α-Methylcinnamaldehyde)

a) $C_6H_5CH{=}C(CH_3)CH_2OH$

b) $C_6H_5CH{=}C(CH_3)CH_2OH$ + $C_6H_5CH{=}C(CH_3)CO_2H$

c) $C_6H_5CH{=}C(CH_3)CH(OC_2H_5)_2$

d) $C_6H_5CH{=}C(CH_3)CH_2OH$

e) $C_6H_5CH{=}C(CH_3)CH(OH)CH_3$

f) 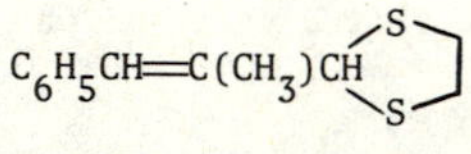

g) $C_6H_5CH{=}C(CH_3)CH{=}NNHC_6H_5$

h) $C_6H_5CH{=}C(CH_3)CH_2OH$

i) $C_6H_5CH{=}C(CH_3)CH(OC_2H_5)_2$

8-49

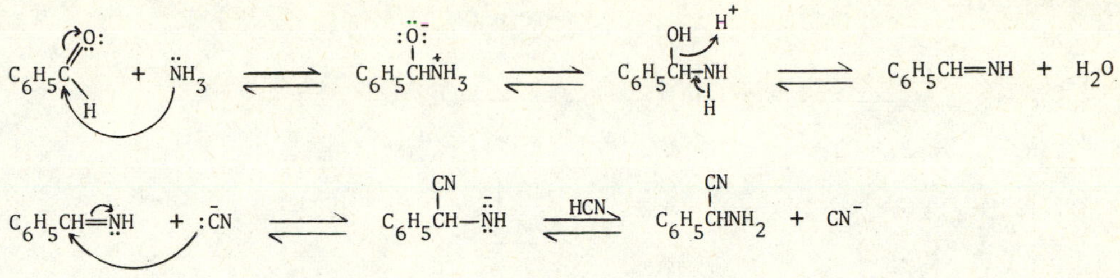

8-50

a) Cyclopropanone is a relatively strained compound. Addition to the carbonyl group converts the sp^2 hybridized carbon atom to sp^3 with an associated decrease in angle strain.

b) + c) The electron withdrawing groups (CF_3 or C=O) on either side of the central carbonyl destabilize the compound relative to the hydrate because of dipole repulsions. Conversion to the hydrate reduces those interactions. Another explanation is that the reverse reaction, loss of H_2O with the C—O bonding electrons is inhibited by the adjacent electron withdrawing groups.

8-51

a) 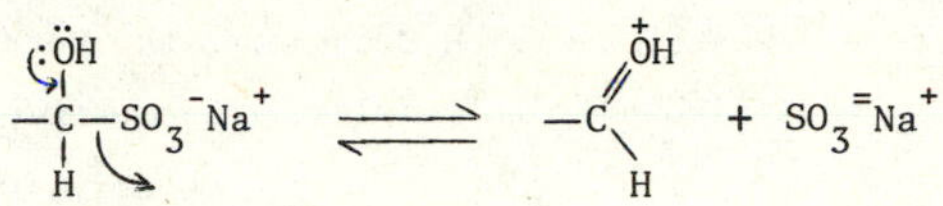

Bisulfite addition product

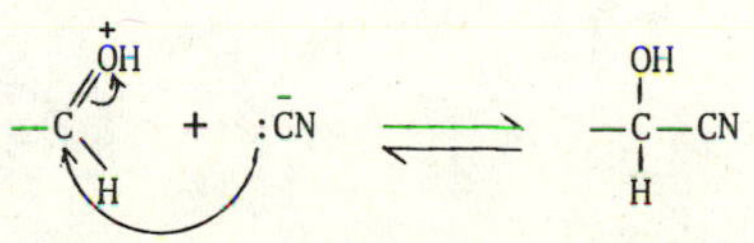

b) The result suggests that cyanohydrins are more stable than bisulfite addition products.

8-52

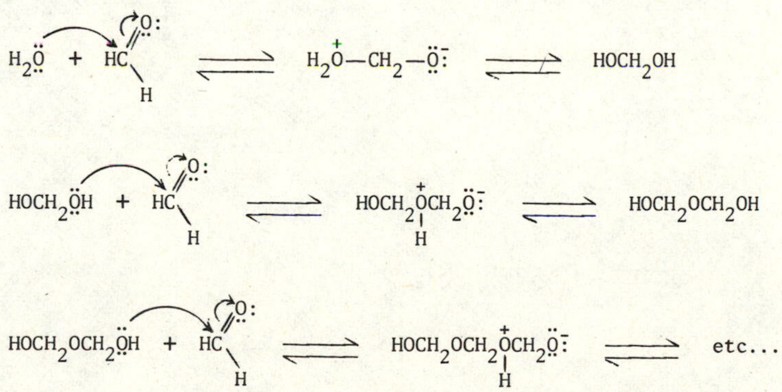

8-53

a) 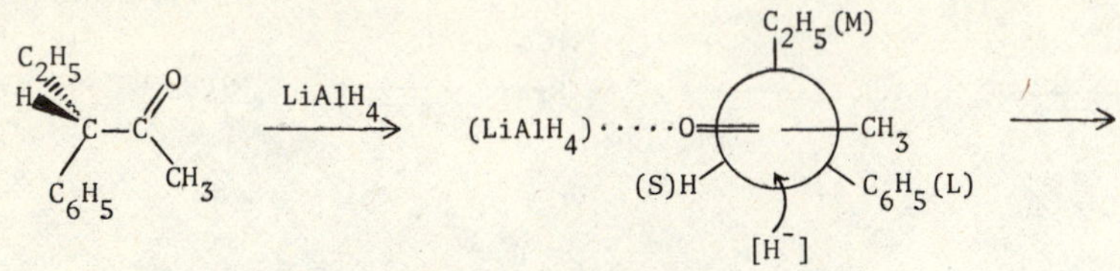

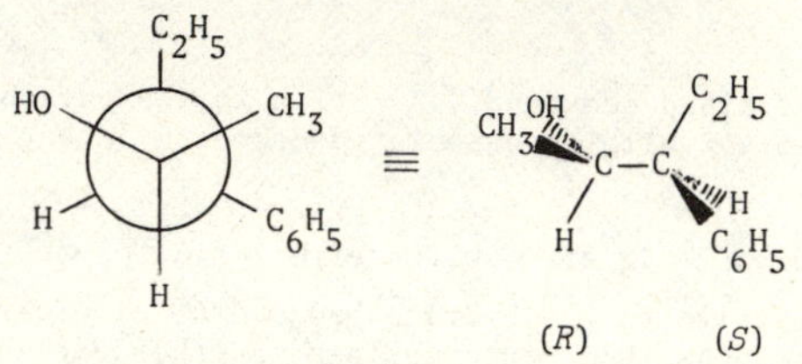

The carbonyl group is complexed with the $LiAlH_4$, thus is considered to be large. Sterically, it is best located between the small (S) and medium (M) groups. Hydride then attacks from the less hindered side of the C=O.

8-54 2,2-Dimethoxypropane is the ketal of acetone. Reaction with H_2O converts the ketal to acetone and methanol, both of which can be distilled away to complete this reaction.

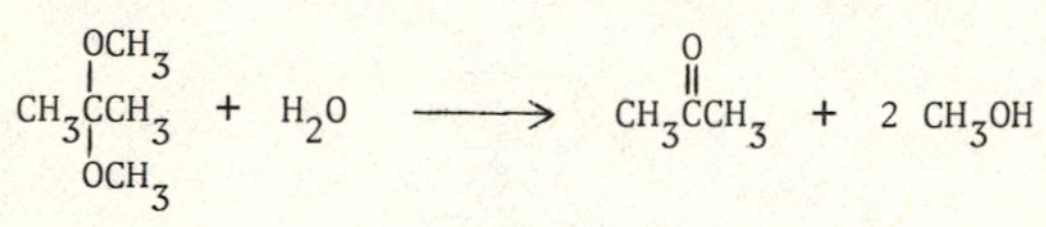

8-55

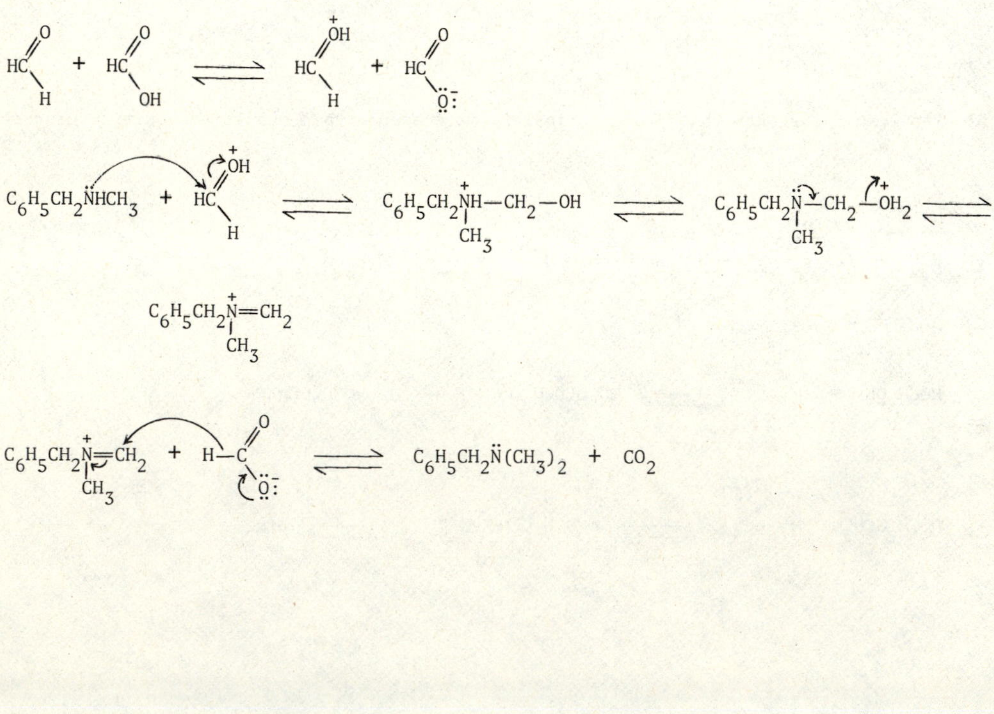

8-56

a) Reaction of base with the initially formed anion can produce the dianion.

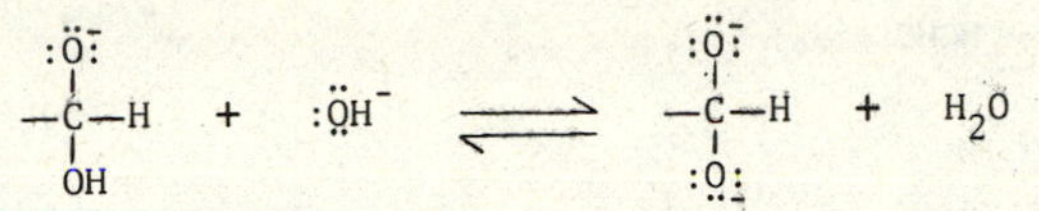

b) Loss of the hydride anion reduces the high negative charge present in the dianion molecule.

8-57

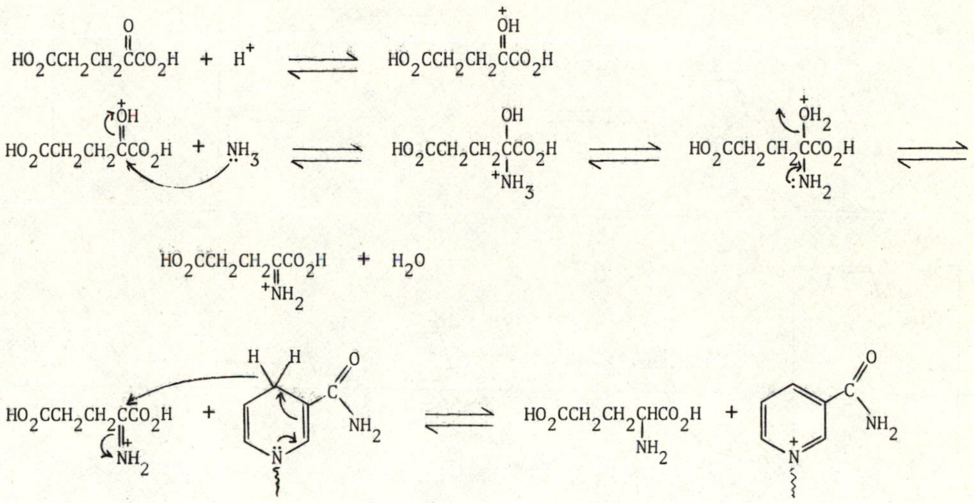

NADH

8-58

a) 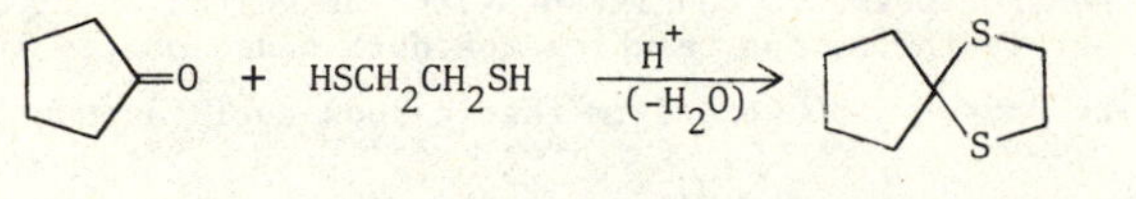

b) 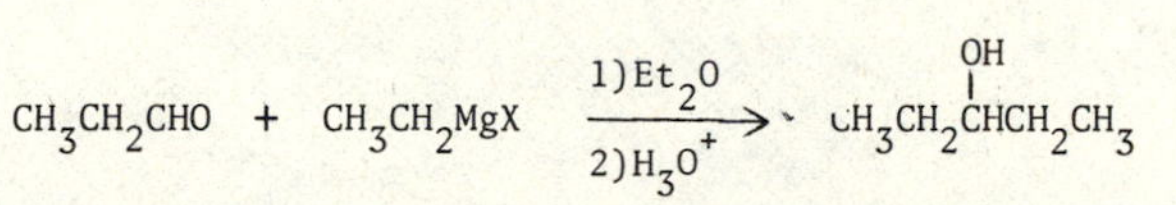

c) 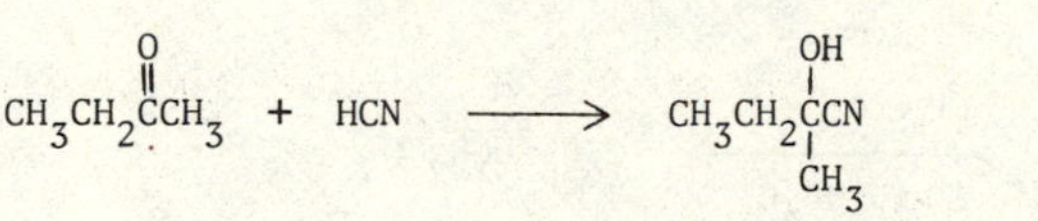

d)

e)

8-59

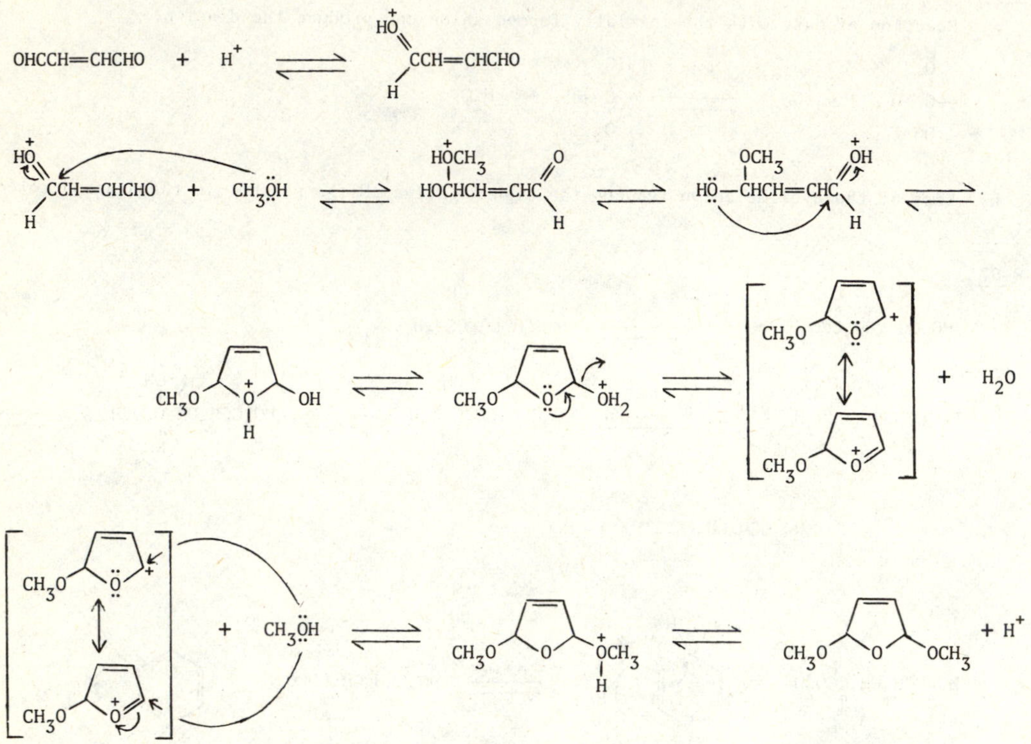

8-60 Elemental analysis provides the empirical formula of $C_9H_{12}O$ *IHD* = 4.

The IR spectrum shows a hydroxy group at 3300-3400cm^{-} and the ^{1}H-nmr spectrum shows C_6H_5—. The ^{1}H-nmr triplet (3) and quartet (2) at 1.8ppm are typical of CH_3CH_2—. The singlet at 2.0 is presumably the -OH proton. The triplet (1) at 4.5ppm corresponds to a deshielded —CH—CH_2—. This, with the molecular formula, suggests —CH—CH_2CH_3. The ^{13}C-nmr shows four different aromatic carbon atoms which is consistent with a monosubstituted benzene. Three additional carbon resonance peaks is consistent with the molecular formula and the deshielded peak at 70ppm is surely the carbon to which the deshielded proton in the ^{1}H-nmr spectrum is connected. The oxygen atom bonded to that carbon could account for those data.

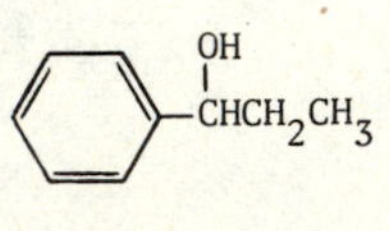

1-Phenyl-1-propanol

$$C_6H_5MgBr + CH_3CH_2CHO \xrightarrow[2)\,H_2O/NH_4Cl]{1)\,Et_2O} C_6H_5CH(OH)CH_2CH_3$$

8-61

a) $CH_3CH_2\overset{O}{\overset{\|}{C}}CH_2CH_2CH_2CH_3$

3-Heptanone

b) $CH_3CH_2CH_2\overset{O}{\overset{\|}{C}}CH_2CH_2CH_3$

4-Heptanone

These isomers can be differentiated by the number of different kinds of carbon atoms in their ^{13}C-nmr spectra. Both have a strongly deshielded carbonyl carbon atom. 3-Heptanone has six different additional carbon atoms (fig. 8-9a) while the symmetrical 4-heptanone has only three different additional carbon atoms (fig. 8-9b).

9 Nucleophilic Substitutions on the Carbonyl Group—The Carboxylic Acid Family

9-1

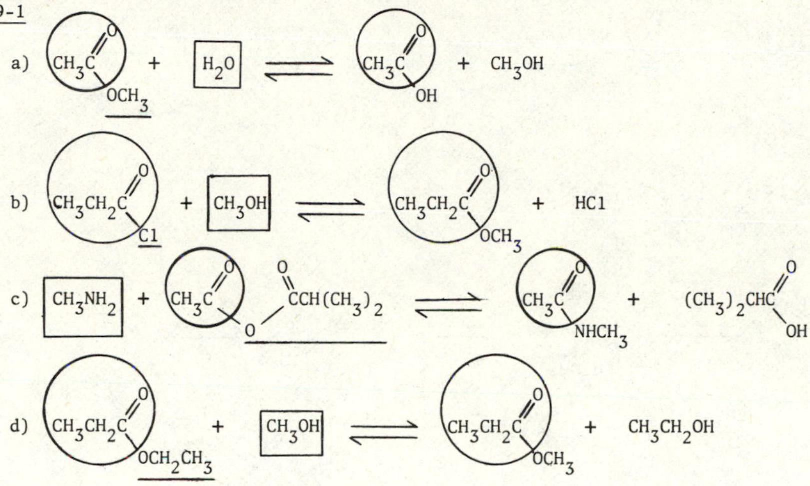

9-2

Good leaving groups	pK_a of conjugate acid	Groups converted to good leaving groups by protonation	pK_a of conjugate acid	Not normally leaving groups	pK_a of conjugate acid
$:\ddot{Cl}:^-$	-7.0	H_2O	-1.7	$H:^-$	-
$H\ddot{S}:^-$	7.0	ROH	~ -2	$\rangle C:^-$	~50
$RC(=\ddot{O})\ddot{O}:^-$	~5	NH_3	9.2		
		R_2NH	~ 10		

9-3

a) For the reaction

$$CH_3CO_2H + C_2H_5OH \overset{K}{\rightleftharpoons} CH_3CO_2C_2H_5 + H_2O$$

Let X = $[CH_3CO_2C_2H_5]$ and $[H_2O]$ at equilibrium

Then 1-X = $[CH_3CO_2H]$ and $[C_2H_5OH]$ at equilibrium

$$K = 4 = \frac{[X][X]}{[1-X][1-X]} \qquad X = 0.67 \text{ moles/l}$$

b) $$K = 4 = \frac{[X][X]}{[10-X][1-X]} \qquad X = 0.97 \text{ moles/l}$$

9-4

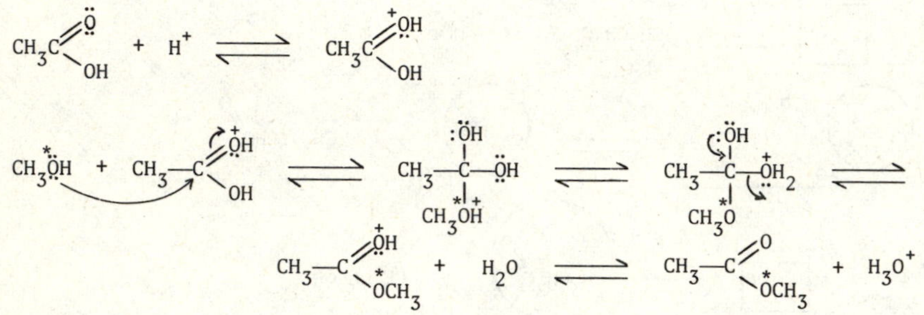

9-5

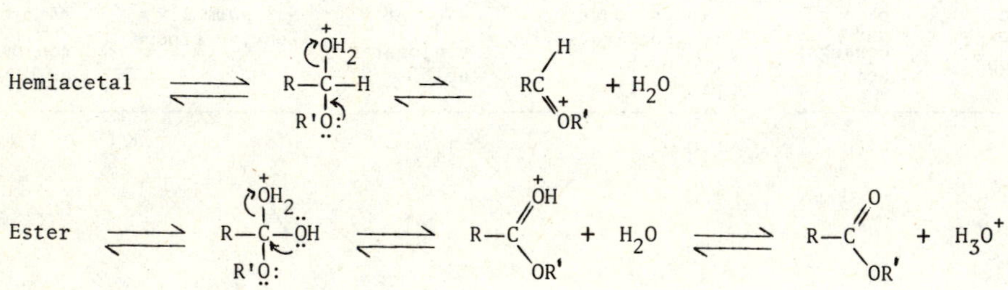

Loss of H_2O from the hemiacetal adduct gives an unstable oxonium ion. The ester adduct loses water to give a protonated ester which readily loses H^+ to give the stable product.

9-6

a) The nonbonding electrons of the carbonyl oxygen will be sp^2-like while those of the alcohol oxygen will be sp^3-like. The greater the "*s*" character, the closer and tighter the electrons are held to the nucleus and therefore the less tightly they hold a proton.

b) To the degree that structure B contributes to the resonance hybrid of the ester the alcohol oxygen atom is nonbasic. Hybrid orbital considerations also suggest that there is more "*s*" character at the alcohol oxygen of B. The electrons are held close to the nucleus and therefore the atom is less basic.

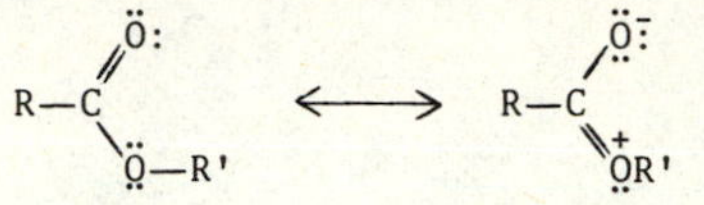

A B

Examination of the stability of each protonated species (the conjugate acids) leads to the same conclusions. The ester protonated at the carbonyl oxygen is stabilized by resonance. Similar stabilization in the alcohol oxygen protonated species would be very unlikely.

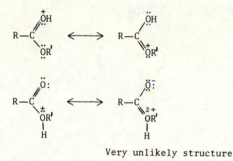

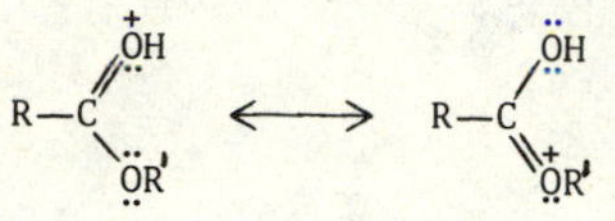

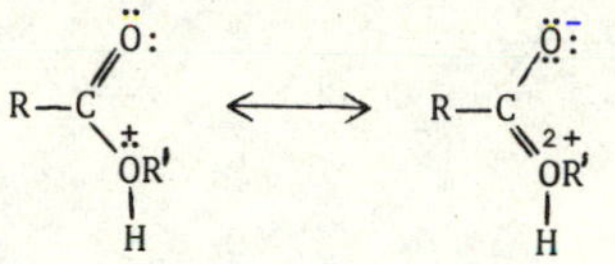

Very unlikely structure

9-7

a) $CH_3CH_2CH_2CO_2C_2H_5$

Ethyl butanoate

(Ethyl butyrate)

b) C_6H_5COCl

Benzoyl chloride

c) 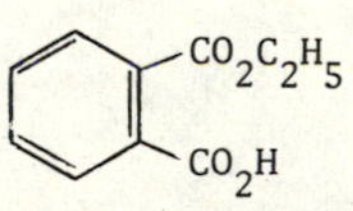

Ethyl hydrogen phthalate

d) $CH_3CO_2CH(CH_3)_2$

i-Propyl ethanoate

i-Propyl acetate

1-Methylethyl ethanoate

e) $(C_6H_5CO)_2O$

Benzoic anhydride

f) $C_6H_5CO_2C_6H_5$

Phenyl benzoate

9-8

i)

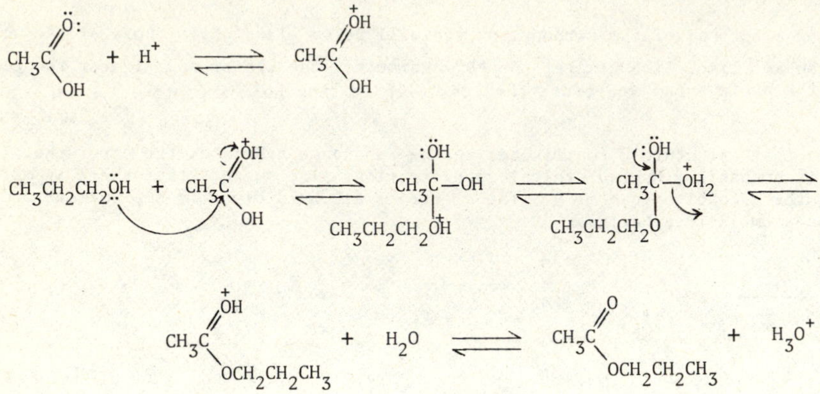

ii)

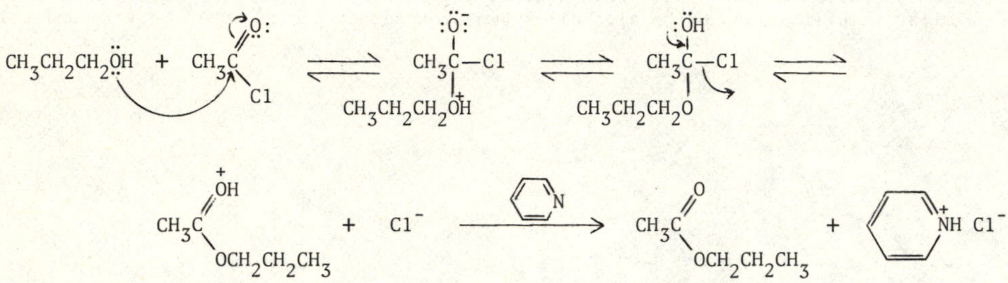

iii)

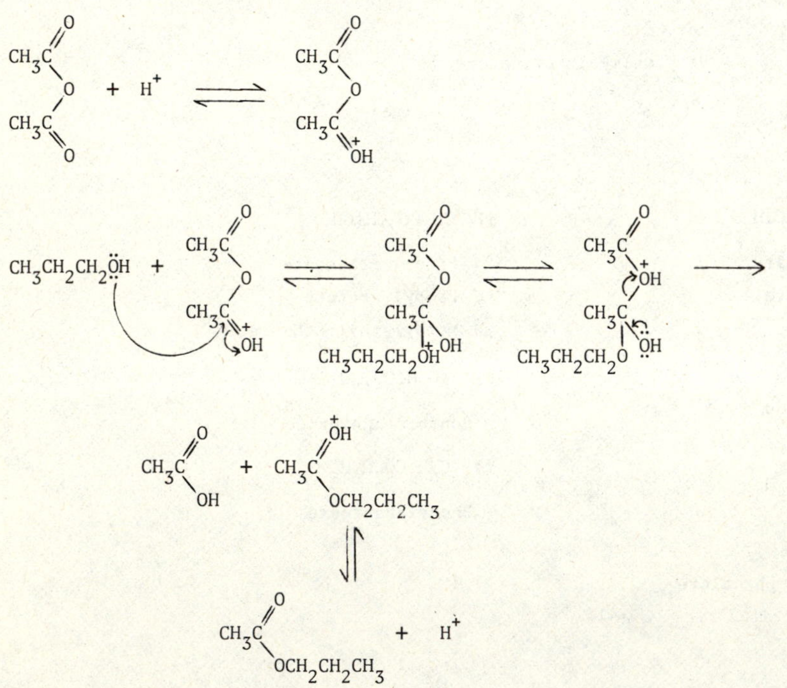

9-9

i) Use a large excess of the 1-butanol.

ii) Remove the low boiling methanol product by distillation.

9-10

i) Exchange

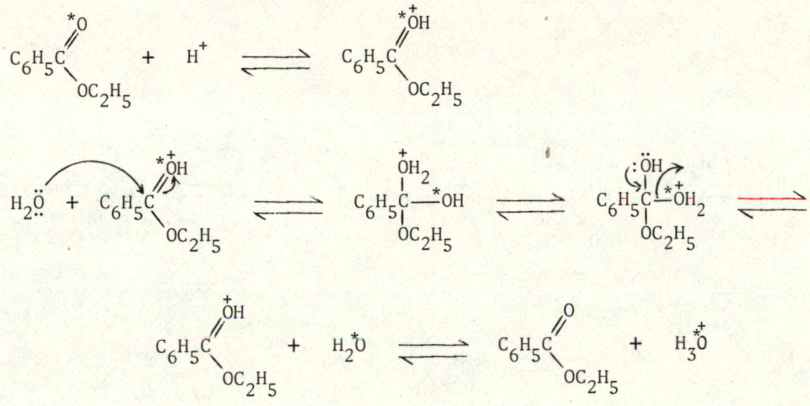

ii) Hydrolysis (or esterification)

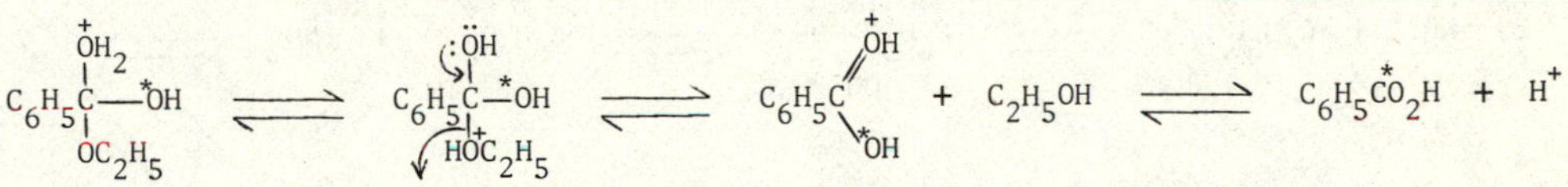

The exchange sequence must take place at a more rapid rate than does hydrolysis (or esterification).

9-11

a)

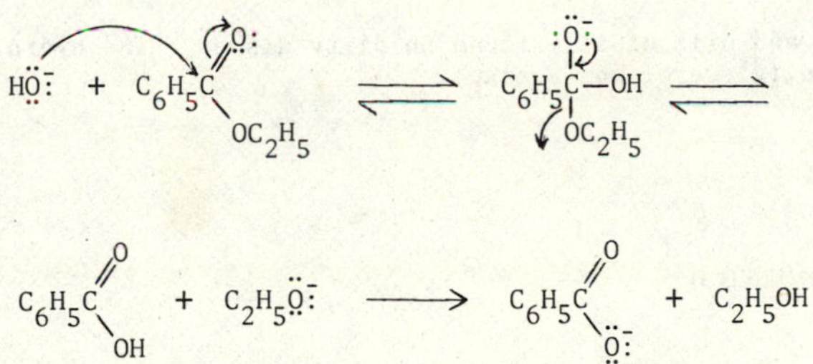

Contd...

9-11 Contd...

b) Base promoted esterification would involve addition of an alkoxide anion to the carboxylate anion. That would form an electrostatically unfavorable dianion.

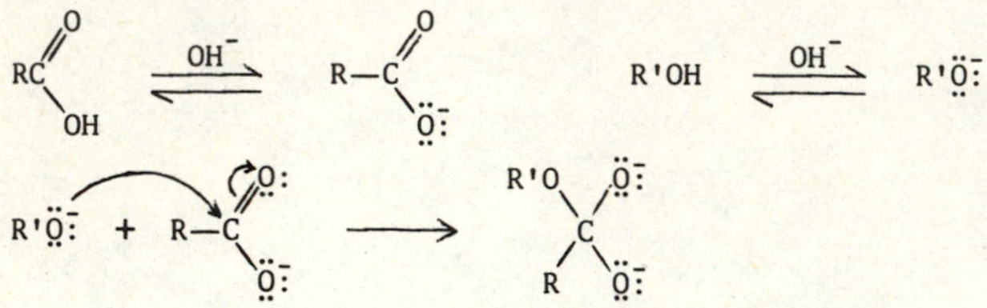

9-12

a)

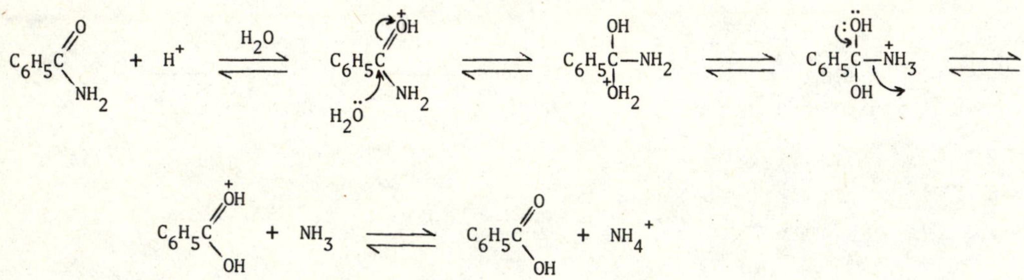

b)

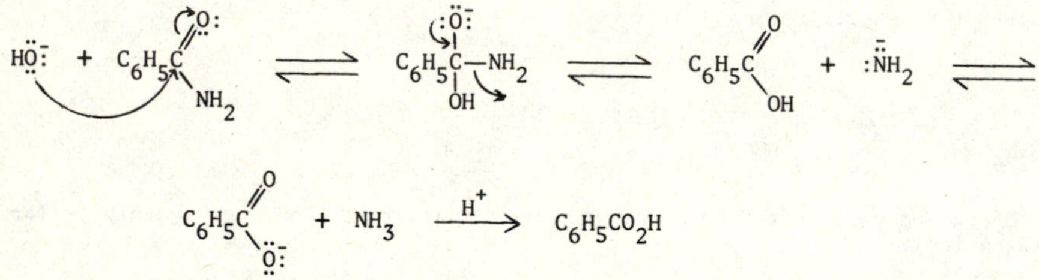

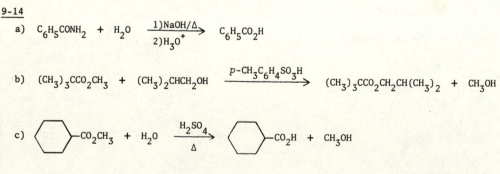

9-13 The strong bases hydrolyze the fats and oils usually found on dirty dishes. The hydrolysis products are usually water soluble (glycerol and a soap).

9-14

a) $C_6H_5CONH_2 + H_2O \xrightarrow[2)H_3O^+]{1)NaOH/\Delta} C_6H_5CO_2H$

b) $(CH_3)_3CCO_2CH_3 + (CH_3)_2CHCH_2OH \xrightarrow{p\text{-}CH_3C_6H_4SO_3H} (CH_3)_3CCO_2CH_2CH(CH_3)_2 + CH_3OH$

c)

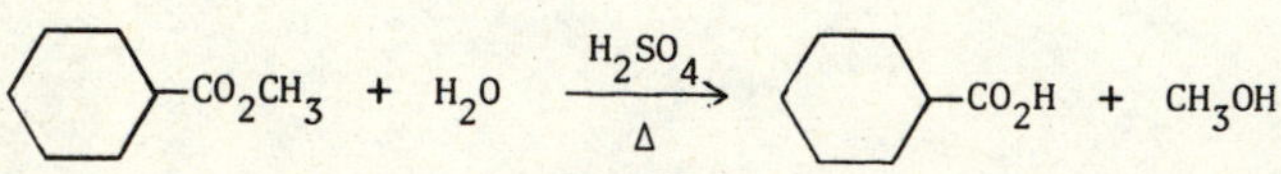

Contd...

9-14 Contd...

d) 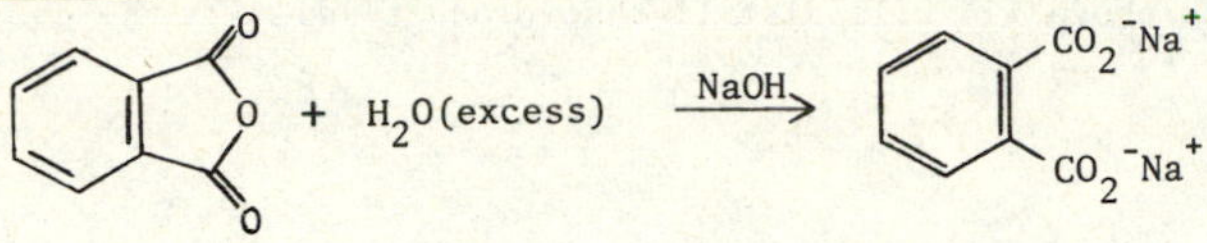

e) 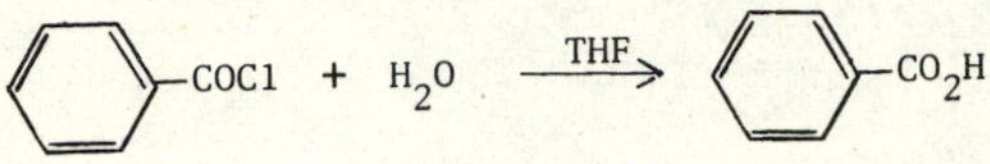

f) 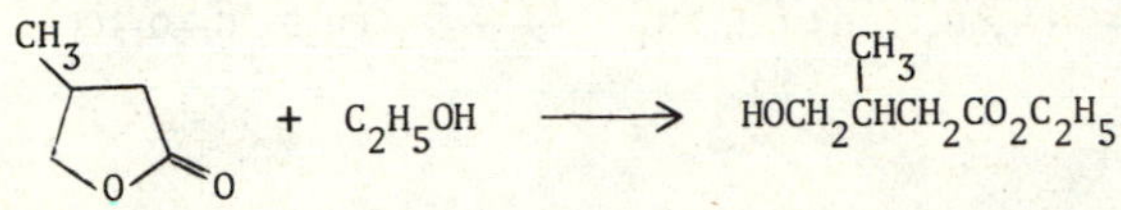

9-15 HCl liberated as reaction occurs converts unreacted amine to a non-nucleophilic ammonium salt.

9-16

a) Because an ammonium carboxylate is the salt of a weak acid and a weak base, a low equilibrium concentration of ammonia or amine and carboxylic acid can form. The ammonia or amine then functions as the nucleophile.

b) $CH_3CH_2CO_2^-\,\overset{+}{N}H_4 \rightleftharpoons CH_3CH_2CO_2H + NH_3$

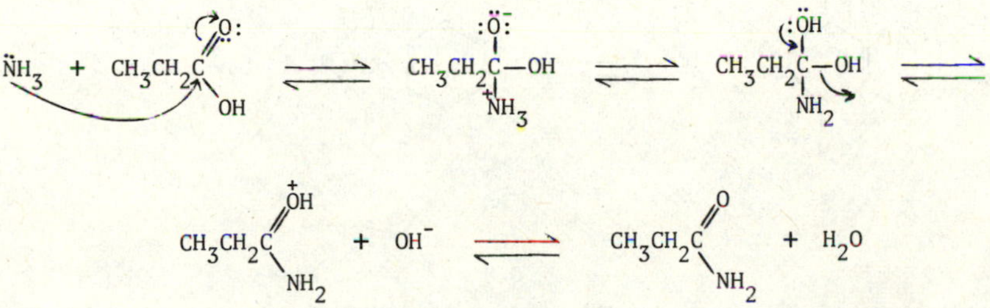

The unfavorable reaction is driven to completion by high temperature and loss of water.

9-17

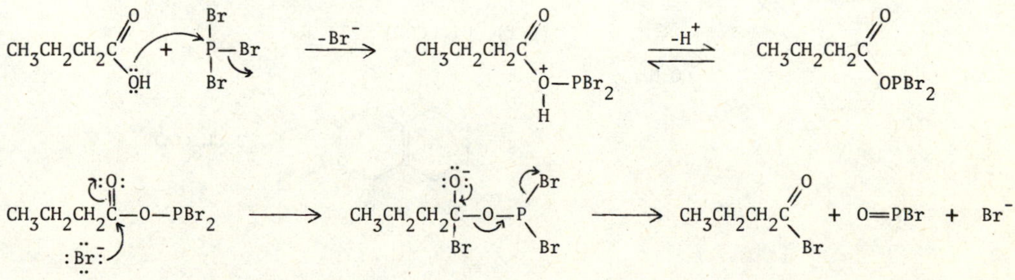

9-18 The desired acetyl chloride is the lowest boiling component in the reaction mixture. Warming the reaction mixture to above 51° will distill that product and shift the equilibrium concentrations.

9-19

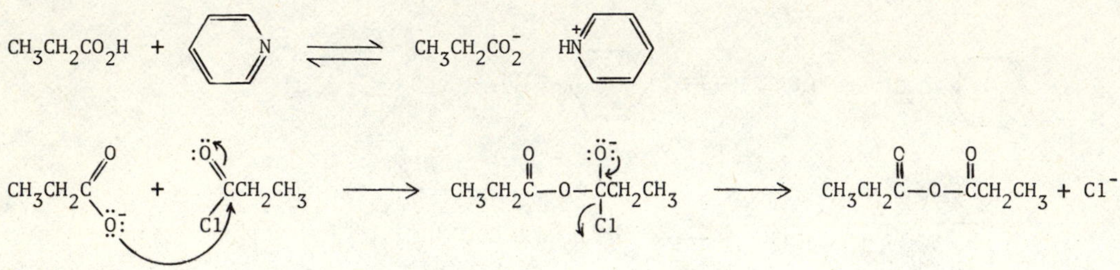

9-20

a) The entropy for cyclization of a single molecule to a nonstrained ring is more favorable than intermolecular reaction of two molecules.

b)

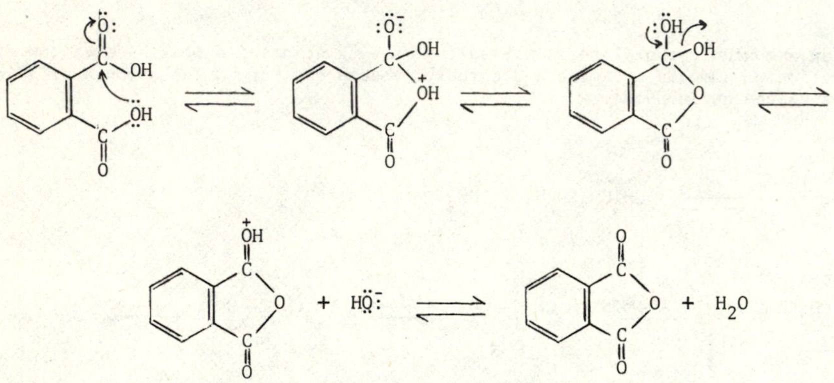

9-21

a) $(C_6H_5CH_2CO)_2O$

b) $C_6H_5CH_2COCl$

c) 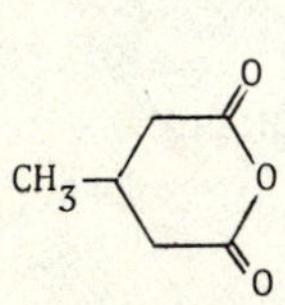 + 2 CH_3CO_2H

d) $CH_3\overset{O}{\overset{\|}{C}}O\overset{O}{\overset{\|}{C}}H$

e) $(CH_3)_2CHCH_2COCl$

f) Cl_3CCOCl

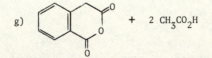

g) 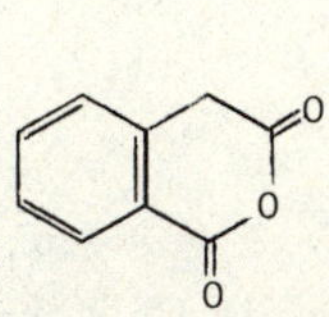 + 2 CH_3CO_2H

9-22

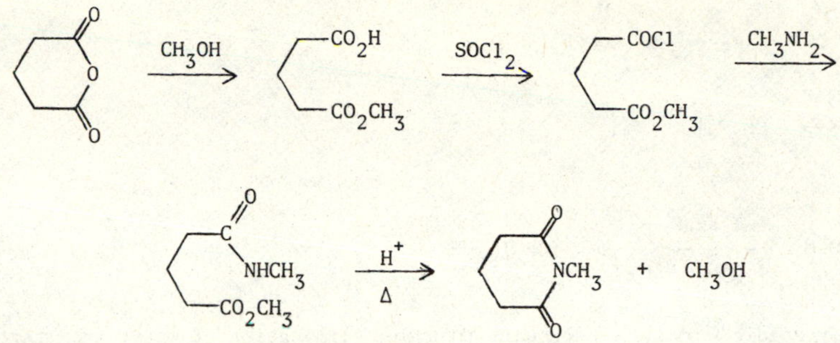

9-23 Addition of water could evolve hydrogen in a dangerously exothermic reaction if much $LiAlH_4$ remains. Ethyl acetate, an ester, is reduced to ethyl alcohol in a mild reaction to consume any excess $LiAlH_4$. Water can then be safely added.

9-24

a) $C_6H_5CH_2CO_2H$ + $LiAlH_4$ $\xrightarrow[2)\,H_3O^+]{1)\,THF}$ $\underline{C_6H_5CH_2CH_2OH}$

b) $\underline{C_6H_5CHO}$ + $NaBH_4$ $\xrightarrow[2)\,H_3O^+]{1)\,i\text{-}PrOH/H_2O/OH^-}$ $C_6H_5CH_2OH$

c) $\underline{(CH_3)_2CHCONHCH_3}$ + $(BH_3)_2$ $\xrightarrow[2)\,HCl/H_2O]{1)\,Diglyme}$ $(CH_3)_2CHCH_2NHCH_3$

d)

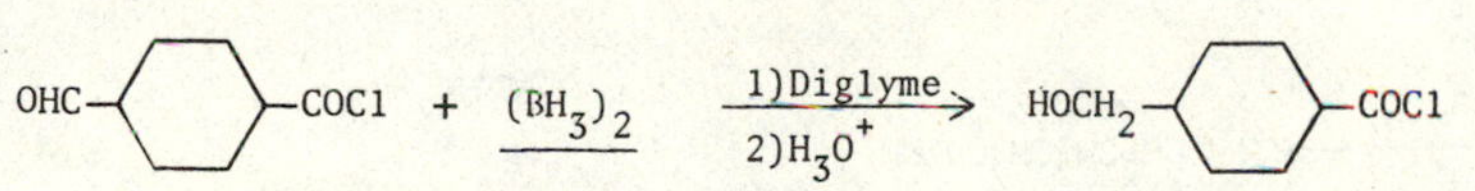

e) $C_6H_5CO_2CH(CH_3)_2$ + $\underline{LiAlH_4 \text{ or } (BH_3)_2}$ $\xrightarrow[2)\,H_3O^+]{1)\,THF}$ $C_6H_5CH_2OH$ + $(CH_3)_2CHOH$

f)

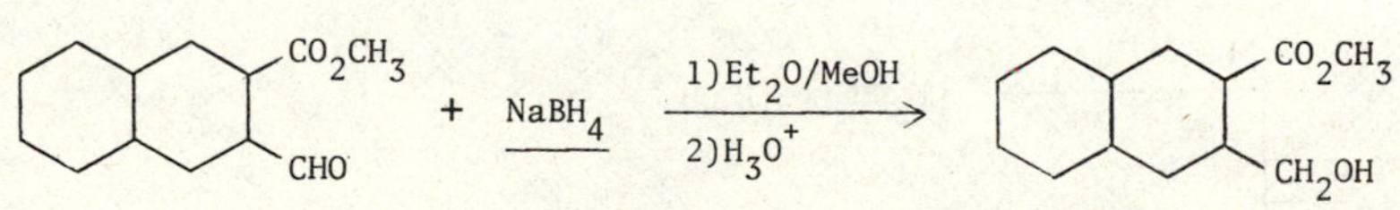

9-25

a) $$Cl-\overset{O}{\overset{\|}{C}}-Cl + 2\ CH_3OH \longrightarrow CH_3O\overset{O}{\overset{\|}{C}}OCH_3$$

b) $$CH_3I + Mg \xrightarrow{Et_2O} CH_3MgI$$

$$3\ CH_3MgI + CH_3O\overset{O}{\overset{\|}{C}}OCH_3 \xrightarrow[2)H_2O/NH_4Cl]{1)Et_2O} (CH_3)_3COH$$

Note that the weakly acidic workup is necessary to avoid elimination of water from tertiary alcohols.

9-26

a) $$C_6H_5CH_2CO_2C_2H_5 \xrightarrow[2)H_3O^+]{1)LiAlH_4} C_6H_5CH_2CH_2OH + C_2H_5OH$$

$$C_6H_5CH_2CO_2C_2H_5 \xrightarrow[2)H_2O/NH_4Cl]{1)2C_2H_5MgI} C_6H_5CH_2\overset{OH}{\overset{|}{C}}(C_2H_5)_2 + C_2H_5OH$$

b) $$CH_3CH_2\overset{O}{\overset{\|}{C}}CH_2CH_2CO_2CH_3 \xrightarrow[2)H_3O^+]{1)LiAlH_4} CH_3CH_2\overset{OH}{\overset{|}{C}}HCH_2CH_2CH_2OH + CH_3OH$$

$$CH_3CH_2\overset{O}{\overset{\|}{C}}CH_2CH_2CO_2CH_3 \xrightarrow[2)H_2O/NH_4Cl]{1)3C_2H_5MgI} (CH_3CH_2)_2\overset{OH}{\overset{|}{C}}CH_2CH_2\overset{OH}{\overset{|}{C}}(C_2H_5)_2$$

In both of these reactions the ketone carbonyl would react more rapidly, but it is unlikely that reaction could be stopped at only that one addition

c) 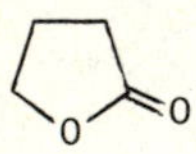

$$\xrightarrow[2)H_3O^+]{1)LiAlH_4} HOCH_2CH_2CH_2CH_2OH$$

$$\xrightarrow[2)H_2O/NH_4Cl]{1)2C_2H_5MgI} HOCH_2CH_2CH_2\overset{OH}{\overset{|}{C}}(C_2H_5)_2$$

d) $$CH_3CH_2CH_2CO_2H \xrightarrow[2)H_2O]{1)LiAlH_4} CH_3CH_2CH_2CH_2OH$$

$$CH_3CH_2CH_2CO_2H \xrightarrow[2)H_2O/NH_4Cl]{1)C_2H_5MgI} CH_3CH_2CH_2CO_2H + C_2H_6$$

9-27 The ketone and organometallic reagents are both so crowded that addition of another *t*-butyl reagent is sterically inhibited. The more reactive *tert*-butyllithium will convert the ketone to an alcohol.

9-28 When CO_2 is bubbled through the reaction mixture a large excess of Grignard reagent is present. Excess reagent adds to the initial adduct to form some tertiary alcohol. When the Grignard reagent is poured over dry ice, CO_2 is in excess and only one addition usually occurs.

9-29 Other approaches to the following syntheses are also possible.

a) $C_2H_5Br + Mg \xrightarrow{Et_2O} C_2H_5MgBr$

$2\ C_2H_5MgBr + HCO_2C_2H_5 \xrightarrow[2)H_3O^+]{1)Et_2O} CH_3CH_2CH(OH)CH_2CH_3$

b) $C_6H_5Br + Mg \xrightarrow{Et_2O} C_6H_5MgBr$

$2\ C_6H_5MgBr + CdCl_2 \longrightarrow (C_6H_5)_2Cd$

$(C_6H_5)_2Cd + C_6H_5COCl \xrightarrow[2)H_3O^+]{1)Et_2O} (C_6H_5)_2C{=}O$

c) $2\ C_2H_5MgBr$ + 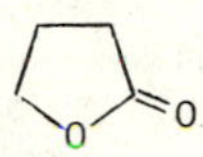 $\xrightarrow[2)H_2O/NH_4Cl]{1)THF} HOCH_2CH_2CH_2C(OH)(CH_2CH_3)_2$

d) $n\text{-}C_4H_9Br + Mg \xrightarrow{Et_2O} n\text{-}C_4H_9MgBr$

$2\ n\text{-}C_4H_9MgBr + (CH_3)_3CCH_2CO_2CH_3 \xrightarrow[2)H_3O^+]{1)Et_2O} (CH_3)_3CCH_2C(OH)(C_4H_9\text{-}n)_2$

9-30

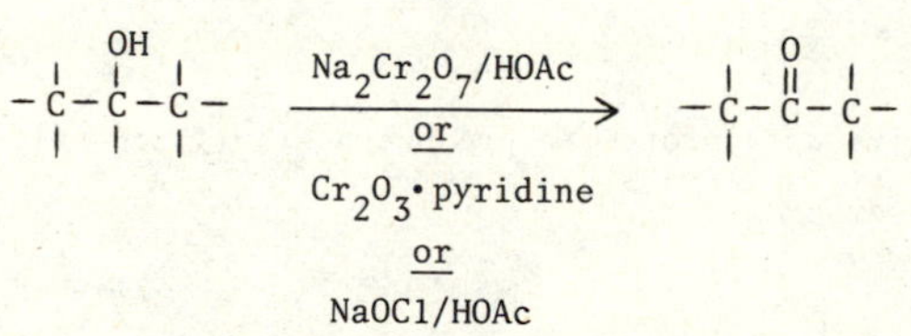

Oxidation

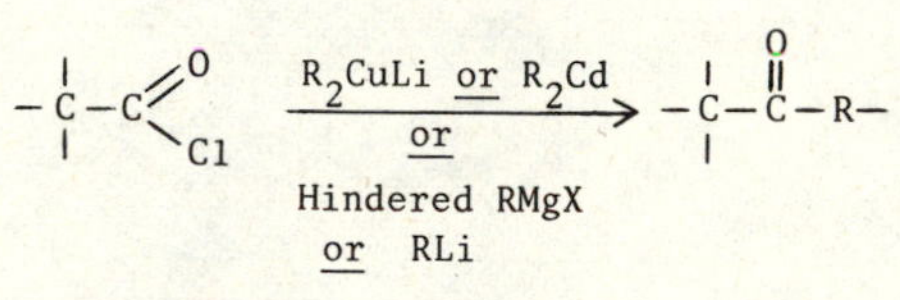

Substitution

Contd...

9-30 Contd...

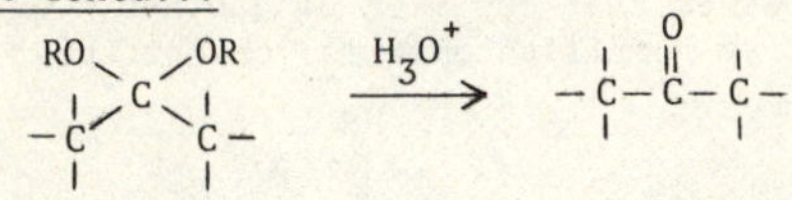

Hydrolysis

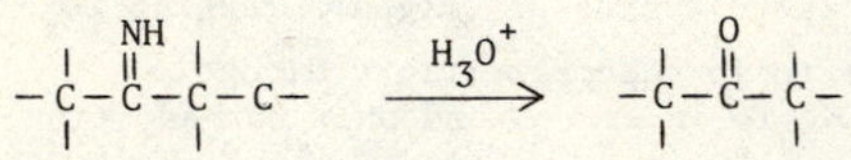

Hydrolysis

9-31

a) The departure of the alcohol must involve cleavage of the acyl-oxygen bond. If the bond between the asymmetric carbon atom and oxygen were broken (alkyl-oxygen cleavage) some loss of optical purity would be expected to occur.

b)

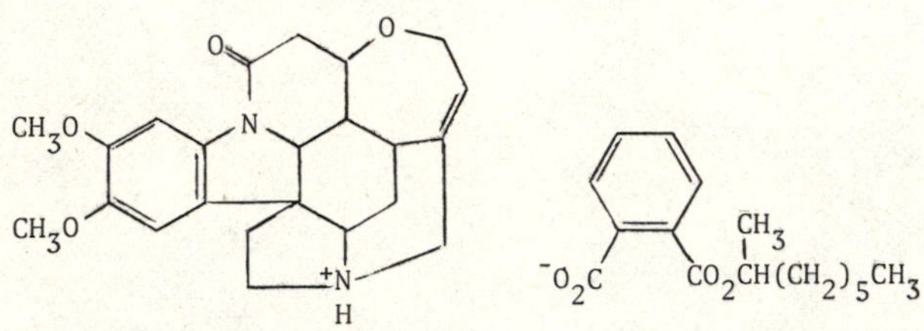

Brucine 2-octyl phthalate is the ammonium salt of the free carboxy group. Note that the amino nitrogen atom is the basic center. Amide nitrogen atoms are relatively nonbasic.

9-32 $C_6H_5SO_2Cl + RNH_2 \longrightarrow C_6H_5SO_2NHR \xrightarrow[H_2O]{NaOH} C_6H_5SO_2\ddot{N}R\ Na^+$

$C_6H_5SO_2Cl + R_2NH \longrightarrow C_6H_5SO_2NR_2$

$C_6H_5SO_2Cl + R_3N \longrightarrow$ No reaction

The sulfonamide of a primary amine has an acidic hydrogen atom which is abstracted by base to give a water soluble salt. The sulfonamide from the secondary amine is not capable of forming a salt and remains insoluble in water. The tertiary amine does not form a sulfonamide since one alkyl group would have to depart from the nitrogen atom.

9-33 One -OH group is not esterified, thus can give up a proton to produce a stabilized anion (conjugate base).

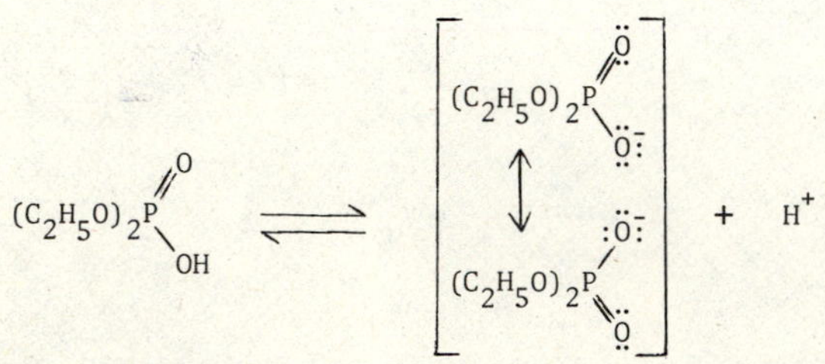

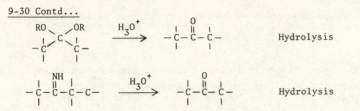

9-34 Absence of ^{18}O exchange suggests that departure of Cl^- after addition of water is faster than return to starting material, or that a direct displacement mechanism rather than an addition-elimination is taking place. The latter is generally accepted.

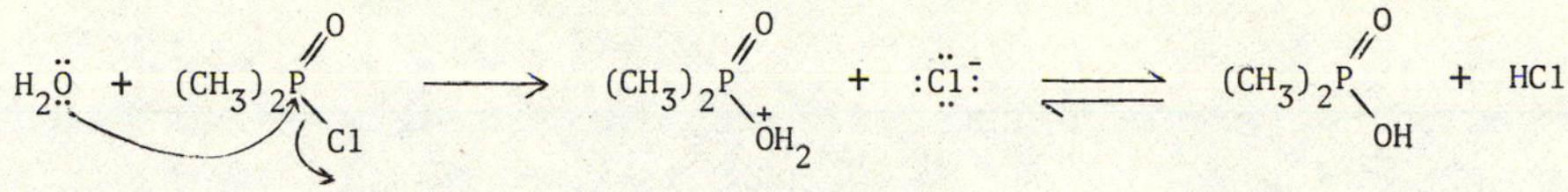

9-35

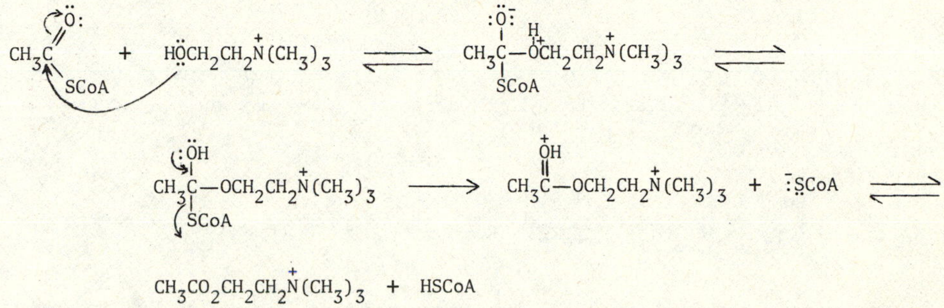

9-36

a) CH_3CO_2H + $(C_6H_5)_2CHOH$

b) 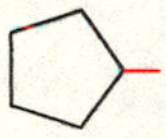 CO_2CH_3 + C_2H_5OH

c) $(CH_3)_2CHCH_2CO_2CH_2CH(CH_3)_2$

d) $CH_3CH_2C(=O)SCH_3$

e) p-$CH_3C_6H_4SO_3CH(CH_3)CH_2CH_3$

f) $CH_3CHOHCH_2CO_2C_2H_5$

g) $CH_2(CO_2H)_2$ + 2 C_2H_5OH

h) $CH_3CH_2C(=O)NHC_2H_5$

i) 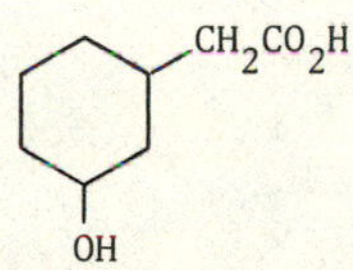

j) $C_6H_5C(=O)NH_2$ + $C_6H_5CO_2^-NH_4^+$

9-37 The *tert*-butyl group remains equatorial in both isomers and therefore determines configuration of the carboethoxy group. In the trans isomer the ester group is equatorial and it is axial in the cis isomer. The equatorial position is less crowded and can better accommodate the tetrahedral intermediate of ester hydrolysis.

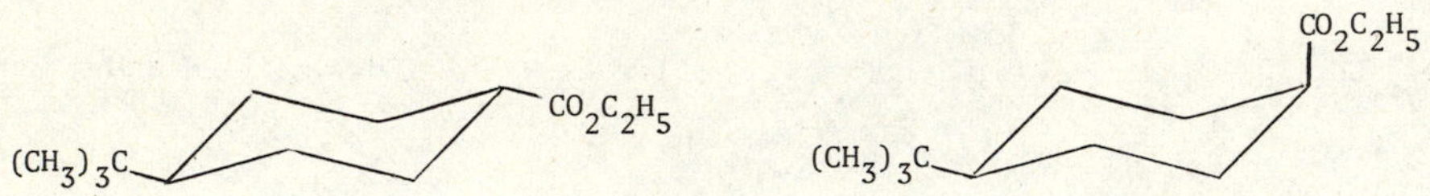

trans cis

9-38

a)

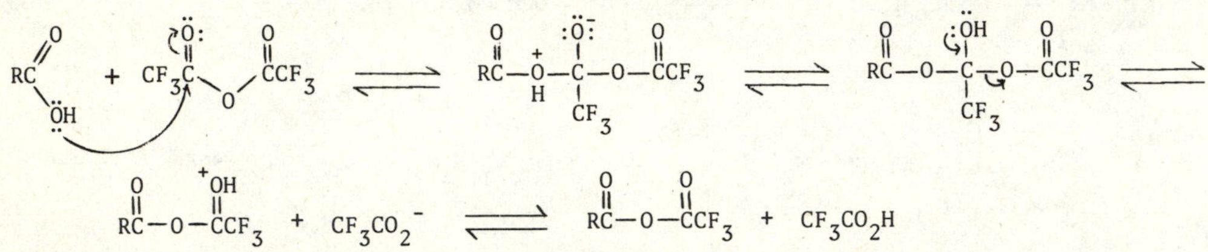

b) Trifluoroacetate is the conjugate base of a strong acid thus is a better leaving group than HO^- which must depart from the carboxylic acid.

9-39

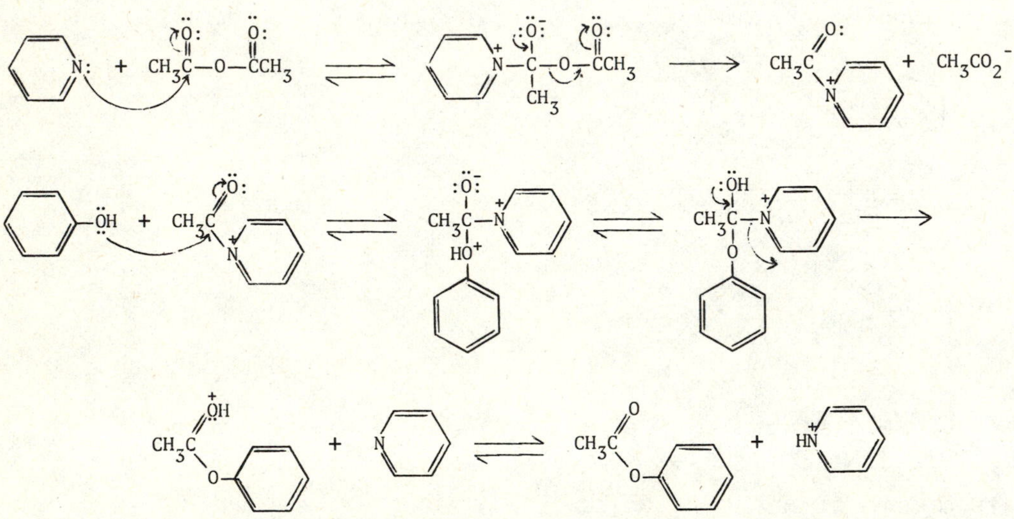

9-40

a) $CH_3CH_2CH_2CH_2CO_2H + NaOH \longrightarrow CH_3CH_2CH_2CH_2CO_2^- Na^+$

b) $CH_3CH_2CH_2CH_2CO_2H + SOCl_2 \longrightarrow CH_3CH_2CH_2CH_2COCl$

Contd...

9-40 contd...

c) $CH_3CH_2CH_2CH_2CO_2H$ + $SOCl_2$ $\xrightarrow{\text{excess } CH_3NH_2}$ $CH_3CH_2CH_2CH_2CONHCH_3$

$CH_3CH_2CH_2CH_2CO_2H$ + CH_3NH_2 $\xrightarrow{\Delta}$ $CH_3CH_2CH_2CH_2CONHCH_3$

d) $CH_3CH_2CH_2CH_2CO_2H + NH_3 \longrightarrow CH_3CH_2CH_2CH_2CO_2^{-}\overset{+}{N}H_4$

e) $CH_3CH_2CH_2CH_2CO_2H + CH_3OH\text{(excess)} \xrightarrow{H^+} CH_3CH_2CH_2CH_2CO_2CH_3$

9-41

a) 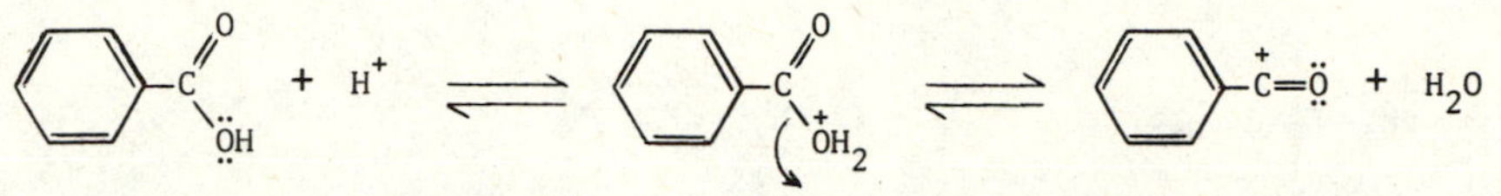

An acylium ion

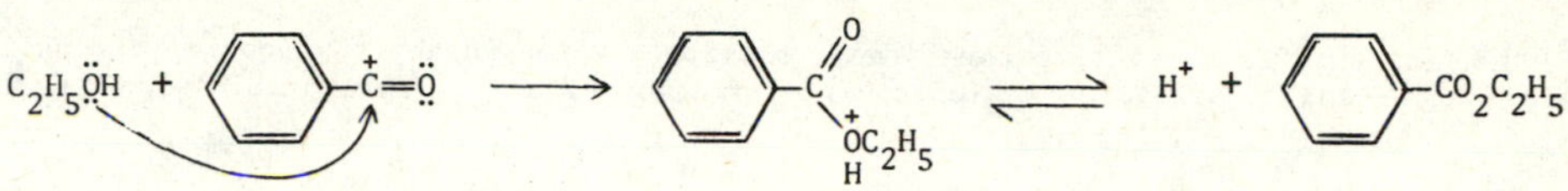

The intermediate acylium ion is a high energy species. It would only be expected in special cases of esterification.

b) The usual addition-elimination would involve a very crowded tetrahedral intermediate in this sterically hindered compound. The acylium ion pathway does not involve any significant increase in crowding and actually decreases steric interaction during the elimination step.

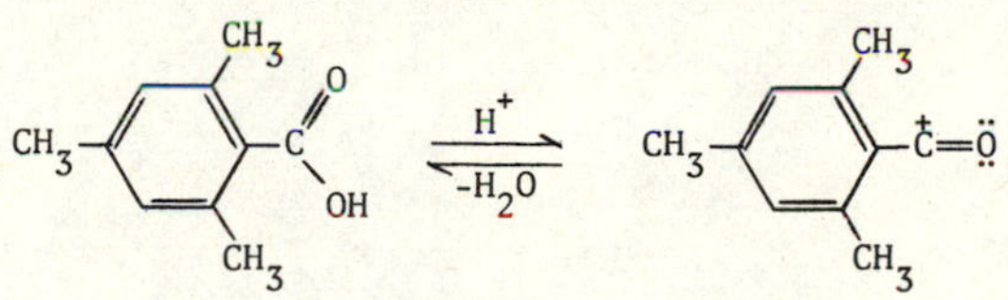

9-42

a) 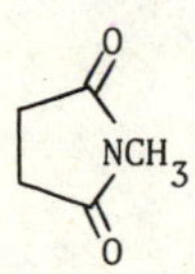

b) 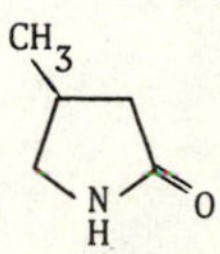

c) $CH_3CH_2CO_2H$

d) $CH_3CH_2CH_2CH_2\overset{O}{\overset{\|}{C}}CH_3$

e) $HO_2CCH_2CH_2CO_2CH_3$

f) 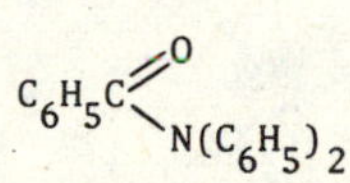

Contd...

9-42 Contd...

g) 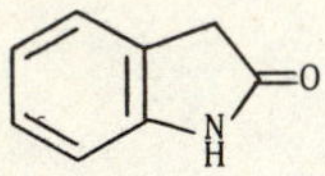

h) 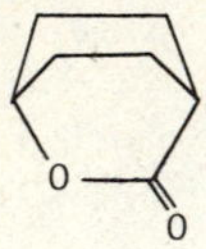

i) $CH_3(CH_2)_5CO_2CH_2C_6H_5$ + C_2H_5OH

j) $C_6H_5CO_2CH(CH_3)_2$ + $C_6H_5CO_2H$

k) $C_6H_5CH_2CH(CO_2CH_3)CH(OH)CH_3$

l) $(CH_3)_2CHCOC_4H_9$-*n*

9-43

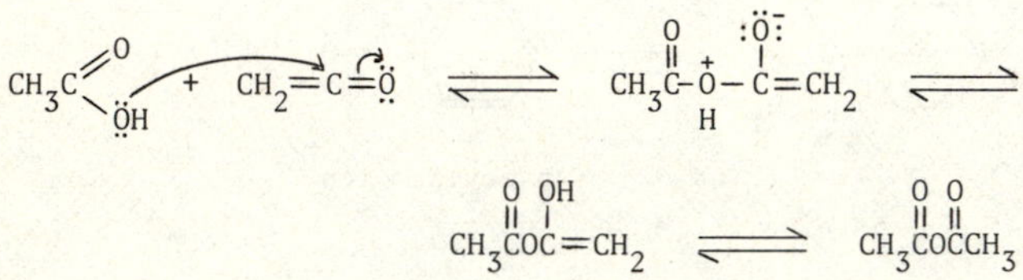

9-44 An initial acid-base reaction forms the carboxylate anion which is a poor electrophile. Addition of additional amide would require formation of a very unfavorable dianion.

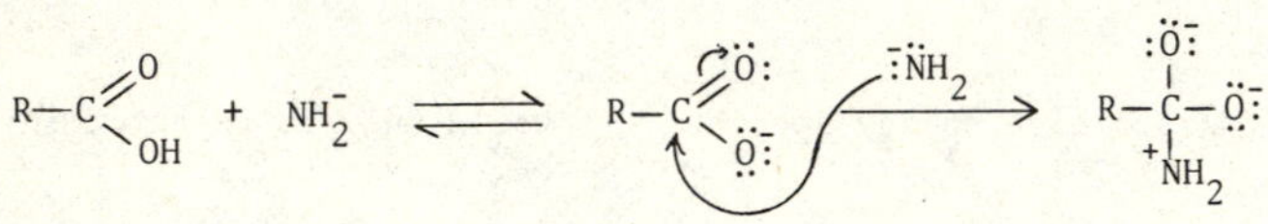

9-45

a) 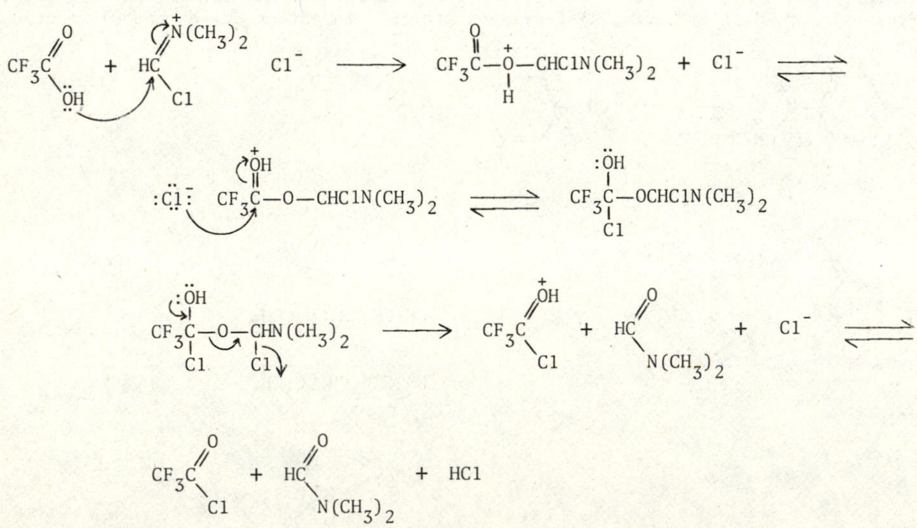

b) DMF is utilized, then regenerated. It is actually a catalyst.

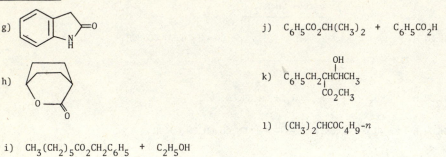

9-46 Esters hydrolyze (saponification) in aqueous base. If the ester is not soluble in the basic solution, the saponification reaction is markedly decreased.

9-47

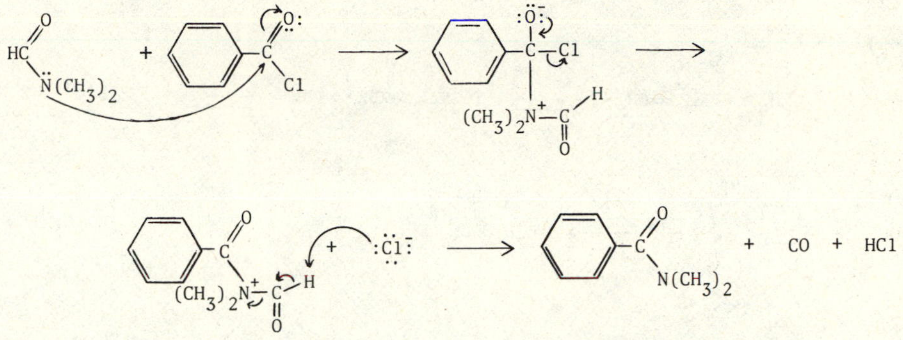

The sequence of bond breaking is driven to completion by the high temperature and loss of CO.

9-48

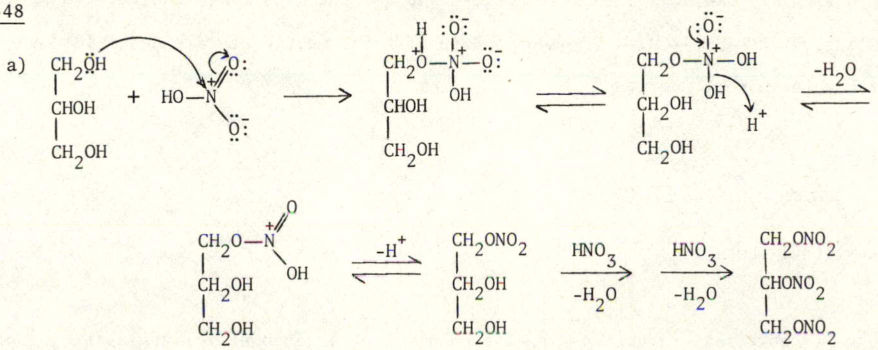

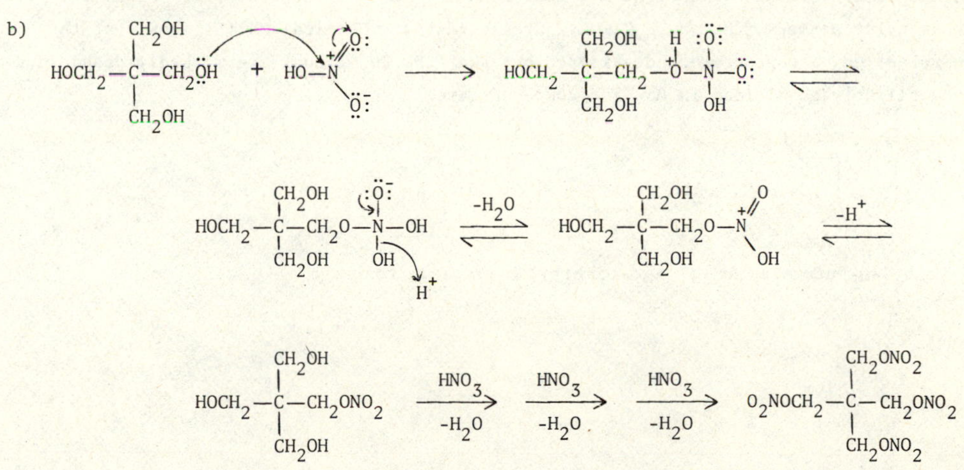

9-49

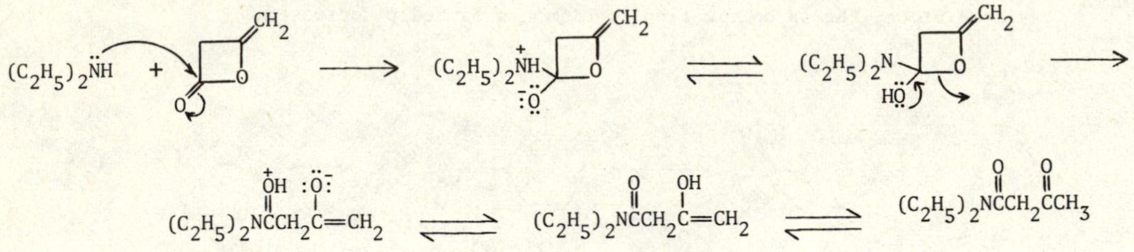

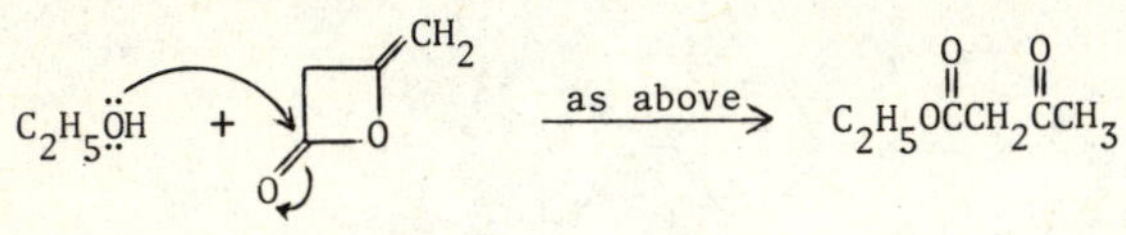

9-50 *IHD* = 5

The peak at 7.3 ppm with an area of 5 indicates a monosubstituted aromatic and accounts for *IHD* = 4. The singlet peak at 2.0 suggests $CH_3\overset{O}{\overset{\|}{C}}$— and accounts for *IHD* = 1. The two triplets of area = 2 suggest a —CH_2—CH_2— fragment. Both of those multiplets are deshielded so that the structure must be

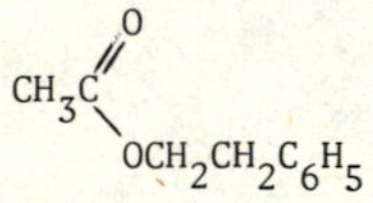

2-Phenylethyl acetate

9-51 The IR carbonyl absorptions suggest 4-ring lactones which can account for at least C_3O_2 of the molecular formula. The eight H's must actually be in a ratio of 6:2. Since they are both singlet peaks, $>CH_2$ and $>C(CH_3)_2$ are suggested. Chemical shifts show that the equivalent CH_3 groups are not deshielded but that the CH_2 groups are markedly deshielded in B and slightly deshielded in A. The compounds are

A B

β-Methyl-β-butyrolactone α,α-Dimethyl-β-propiolactone

9-52

a) *IHD* = 5. A disubstituted aromatic ring (nmr = 6.7 - 7.7) accounts for *IHD* = 4 and some type of conjugated carbonyl group (IR = 1680 cm^{-1}) accounts for *IHD* = 1. The nmr peak at 9.7 ppm which is lost on exchange with D_2O is consistent with N-H. Since the compound can be prepared from *p*-bromotoluene, one nmr peak at 2.8 or 2.9 ppm must be a methyl group on the aromatic ring. The other peak would be consistent with an $-NHCH_3$ since only one H is located on the nitrogen atom. The compound is

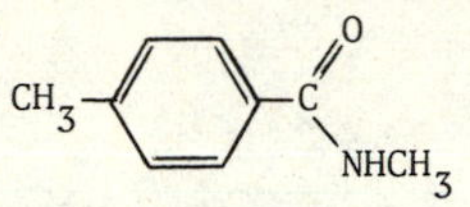

N,*p*-Dimethylbenzamide

b) 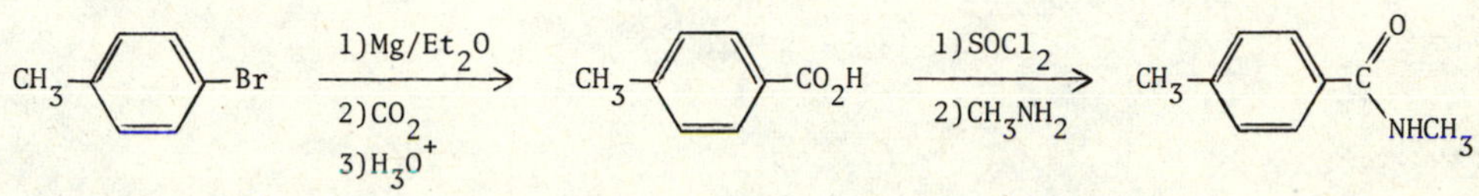

10 Nucleophilic Substitutions at Saturated Carbon

10-1

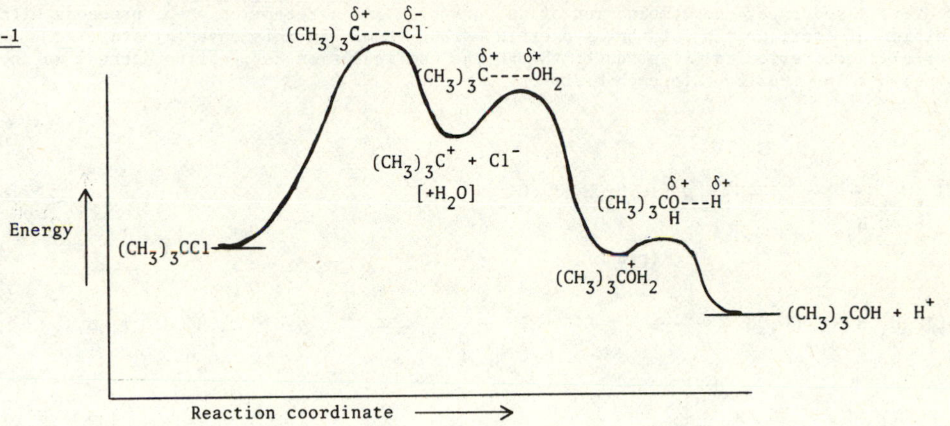

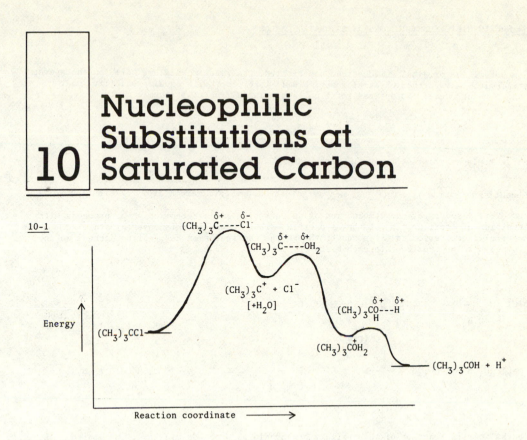

10-2 The rate-controlling step for the S_N1 reaction is a dissociation. The position of equilibrium can be shifted back to undissociated starting material by increasing the concentration of one of the products, in this case X^- (mass law effect). The leaving group X^-, and the added nucleophile both compete for the intermediate carbocation.

10-3 The reaction proceeds by an S_N1 mechanism. The reaction rate depends on the formation of the carbocation, not on the nature of the nucleophile.

10-4

a) Rate = $k_2[C_6H_5CH_2Br][N_3^-]$

b) The rate would double.

c) The rate would increase by four (2 x 2).

10-5 The ratio k_α/k_e would equal 1.
Consider two molecules of substrate. Their reaction with iodide is measured by k_e. When we assume an achiral intermediate, then one molecule is expected to react by retention and the other by inversion to give a racemic mixture. Thus the rate of reaction of those two molecules (k_e) is equal to the rate of racemization (k_α) of them also.

10-6 When 50% of the original 2-iodooctane has reacted, the reaction mixture will, on the average, be racemic. Any further chemical reaction will take place with equal probability on the *R* and *S* enantiomers. No optical rotation is generated from this racemic mixture.

10-7

a) Azide replaces chloride by an S_N2 mechanism, ∴inversion.

b) Reduction of azide to amine does not affect the asymmetric carbon atom, ∴retention.

The overall sequence, the combination of an inversion and a retention step, proceeds with inversion of configuration at the asymmetric carbon atom. In this case (*R*) starting material is converted to (*S*) product, though the configuration designation depends on the groups, not the reaction stereochemistry.

10-8

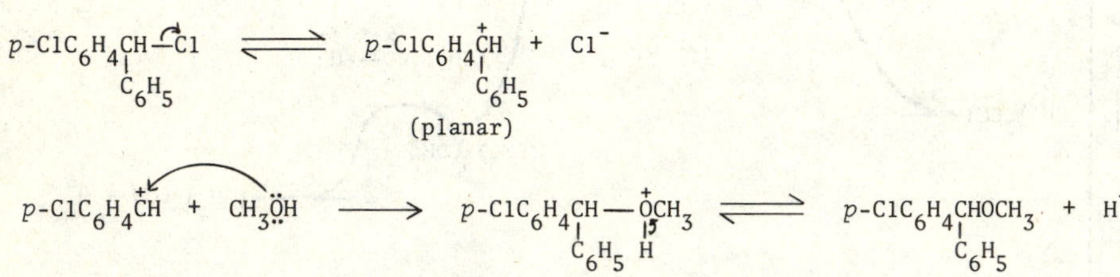

10-9 The solvolysis reaction of the tertiary organohalogen proceeds via an S_N1 mechanism. The polar solvent system enhances loss of Br^- to give a symmetrically solvated carbocation. The intermediate carbocation is easily attacked by the nucleophile H_2O from either side to give racemic product.

10-10 Substitution on this secondary benzylic substrate can proceed by an S_N1 or an S_N2 pathway. Solvolysis by methanol proceeds by the S_N1 mechanism and gives a mixture of inversion and racemization. With the good nucleophile methoxide, an S_N2 reaction takes place with inversion.

10-11

a) 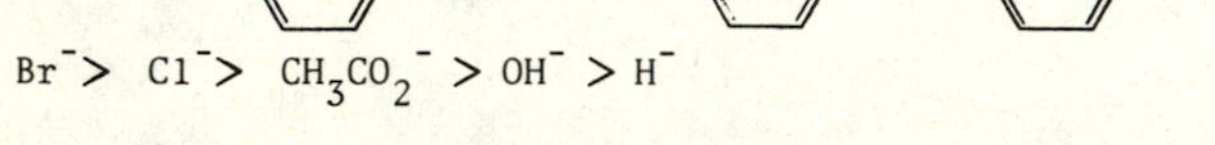

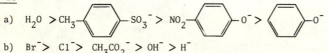

b) $Br^- > Cl^- > CH_3CO_2^- > OH^- > H^-$

10-12 Hydoxide is a poor leaving group. However, in the presence of acid the hydroxy group is protonated and the good leaving group, water, departs.

10-13

i) In the oxygen and nitrogen series anions are better nucleophiles than neutral species.

ii) Delocalization of electrons decreases nucleophilicity.

iii) Nitrogen is a better nucleophile than oxygen because oxygen is more electronegative and holds its electrons more tightly.

10-14 A nucleophile must have a nonbonding electron pair to share in bond making, that is, it must be a Lewis base. BF_4^- is negatively charged because of its bonded electron configuration. It does not have an available electron pair.

10-15 Protonation, an acid-base reaction takes place on the basic oxygen atom. Nucleophilic substitution occurs preferentially at the nucleophilic sulfur atom.

10-16

a) $(CH_3)_3P > (CH_3)_3B$ — The boron atom has no unshared electron pair. It is actually a Lewis acid.

b) $C_6H_5O^- > C_6H_5OH$ — Anions are usually better nucleophiles than analogous noncharged species.

c) $(CH_3)_2NH > CH_3NH_2$ — The electron donating methyl groups enhance nucleophilicity. Steric factors are less important in this specific reaction.

d) $p\text{-}CH_3C_6H_4O^- > p\text{-}NO_2C_6H_4O^-$ — The electron withdrawing nitro group delocalizes the electron pair and decreases nucleophilicity, whereas methyl donates electron density and increases nucleophilicity.

e) $CH_3SH > CH_3OH$ — The sulfur atom is the more nucleophilic.

f) $n\text{-}C_4H_9O^- > t\text{-}C_4H_9O^-$ — The tertiary anion leads to steric hindrance.

g) $H_3N > H_4\overset{+}{N}$ — Cations are not nucleophilic.

h) 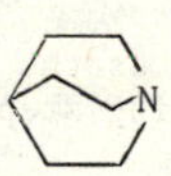 $> (C_2H_5)_3N$ — The alkyl groups are "tied back" in the bicyclic compound so that less steric interactions occur.

10-17 Steric hindrance to the S_N2 reaction accounts for the slower rates in the 2-substituted pyridines. The greater rate of the 4-methyl pyridine can be attributed to electron donation by the methyl group which enhances the nitrogen nucleophilicity.

10-18

a) 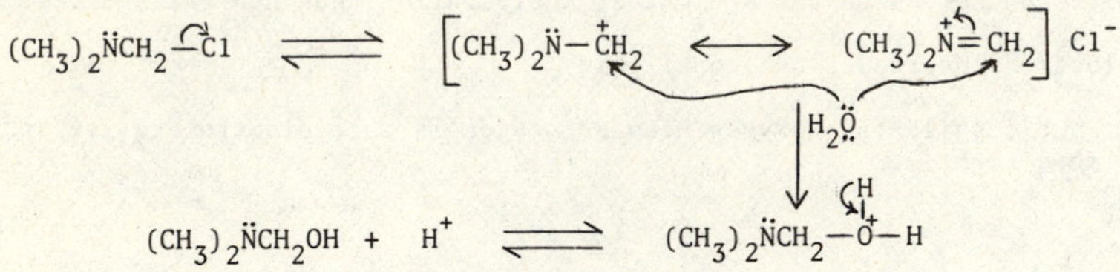

b) Resonance stabilization of the intermediate cation involves an important contributing structure in which the positive charge is located on the heteroatom. A nitrogen atom is less electronegative than oxygen and can better accommodate the positive charge.

10-19

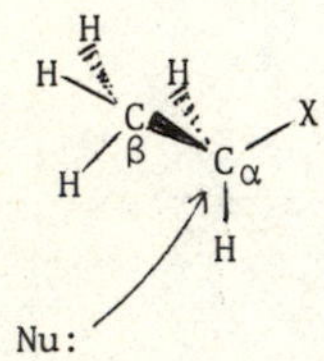

Haloethane

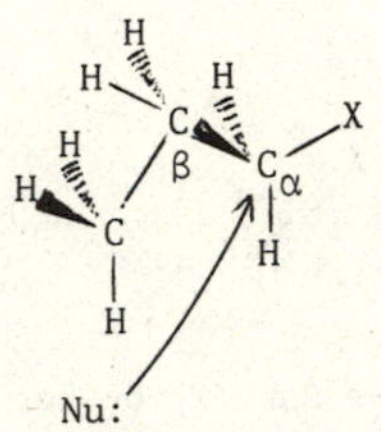

1-Halopropane

In the haloethanes, back-side attack of the nucleophile is sterically hindered by H-atoms on the β-carbon atom. With the propyl systems at least 1/3 of the rotamer populations involve blocking of the back-side of the reaction center by a methyl group on the β-carbon.

10-20

a) (p-MeO$C_6H_4)_3$CCl > Chlorodiphenylmethane > 2-Chloro-2,3-dimethylbutane > 3-Chloro-2-methylpropene > 2-Chloropropane.

b) Chloromethyl methyl ether > Chlorophenylmethane > Chloromethane > 2-Chloropropane > 1-Chloro-2,2-dimethylpropane.

10-21 A is a very crowded molecule. Dissociation into a carbocation relieves strain as the molecule changes from a tetrahedral to a trigonal geometry. The rate controlling step is favored by "steric acceleration".

10-22 The reactants and products will have essentially the same energies in the two solvent systems. The major difference will be in the transition states and intermediates where the charged species are more stable in the solvent containing a higher percentage of water.

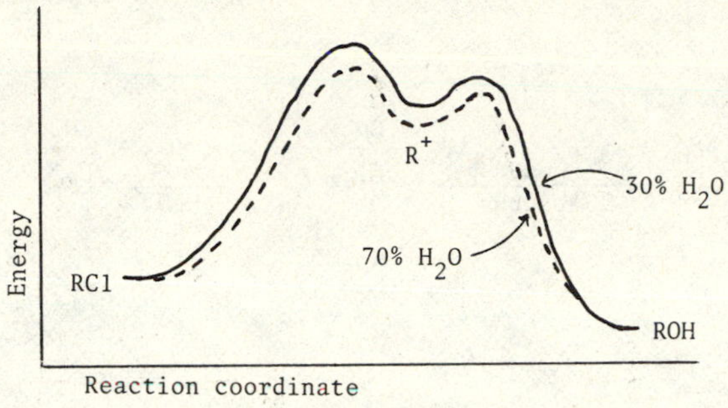

10-23 Water is a more polar solvent than methanol and dissociation to a more symmetrically solvated carbocation is favored. A higher degree of racemization results.

10-24

i) R—L ⟶ $\overset{\delta+}{R}$----$\overset{\delta-}{L}$ Formation of charge

ii) $Nu:^-$ + R—L ⟶ $\overset{\delta-}{Nu}$----R----$\overset{\delta-}{L}$ Dispersal of charge

iii) Nu: + R—L ⟶ $\overset{\delta+}{Nu}$----R----$\overset{\delta-}{L}$ Formation of charge

iv) $Nu:^-$ + $R—L^+$ ⟶ $\overset{\delta-}{Nu}$----R----$\overset{\delta+}{L}$ Destruction of charges

10-25 The hydration of F^- is the most exothermic. Hydration is very favorable so that it is difficult to desolvate the ion to free the nucleophile. Iodide is the least strongly solvated and therefore is most easily freed to function as a nucleophile.

10-26 Bond energies C—Cl = 81 kcal/mol (339 kJ/mol)

C—Br = 68 kcal/mol (284 kJ/mol)

Activation energies in the aprotic solvent DMSO reflect bond strengths since solvation is relatively unimportant. Bond strengths and the rate of bond formation (kinetics) are in the same order.

10-27 The S_N1 pathway becomes relatively more important in both cases as solvent becomes more polar and capable of enhancing an ionization pathway. The benzylic substrate can form the more stable carbocation, thus gives relatively more S_N1 reaction while S_N2 predominates with the secondary propyl substrate.

10-28 Substitution on chlorodiphenylmethane, a benzylic substrate, proceeds by an S_N1 mechanism. Thus the rate of reaction is essentially unaffected by the nature of the nucleophile. Substitution on 2-chloropropane is S_N2 and reaction is faster with the better nucleophile diethylamine.

10-29

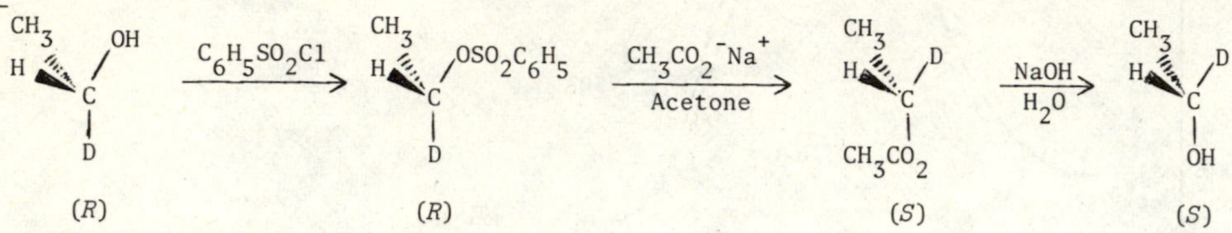

10-30 The adjacent methylene groups on either side of the ether oxygen atom are held back in the cyclic THF structure. The oxygen atom is more free to function as a Lewis base. THF is also slightly more polar than diethyl ether and this will make a contribution to its solvation.

10-31 A phenyl substituent favors S_N1 reactivity by stabilizing an adjacent carbocation through resonance. This dominates any electron withdrawing inductive effect.

10-32 Departure of acetic acid from protonated A forms a tertiary carbocation. That planar intermediate leads to a racemic alcohol product on reaction with water.

10-33

a) As the nucleophile becomes less solvated it is freer for reaction. When reactions are carried out under similar conditions, substitution in aprotic HMPT proceeds more effectively.

b) Inversion of configuration provides strong evidence for substitution by the S_N2 mechanism. Apparently the enhanced nucleophilicity in HMPT overcomes the severe steric inhibition to back sided attack on the primary carbon atom of the neopentyl group.

10-34

a) Reaction of each of the tertiary haloalkanes proceeds by an S_N1 mechanism. They form the same carbocation in their rate controlling steps, thus give the same mixture of alcohol and ether products on subsequent substitution by water and methanol.

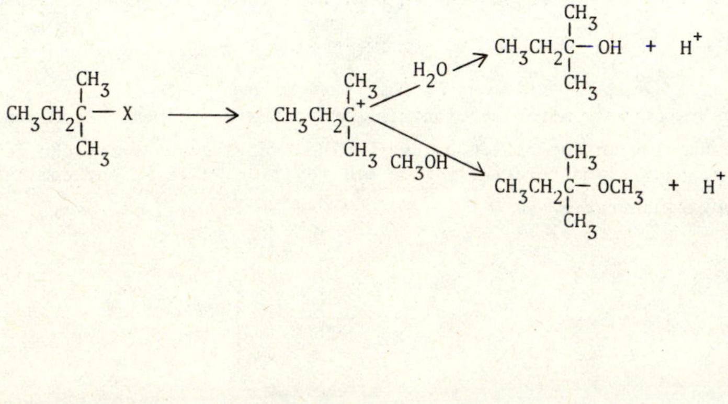

Contd...

10-34 Contd...

b) The rate controlling step depends on relative leaving group abilities of the halide. It is found that I > Br > Cl.

10-35

A ⟶ B S_N2 - inversion		
B ⟶ C No effect on asymmetric carbon atom - retention		Overall retention
C ⟶ D S_N2 - inversion		
D ⟶ E No effect on asymmetric carbon atom - retention		

10-36 Ring opening occurs readily with ethylene oxide because the protonation makes the oxygen atom a good leaving group. No such catalysis is possible for the saturated hydrocarbon cyclopropane.

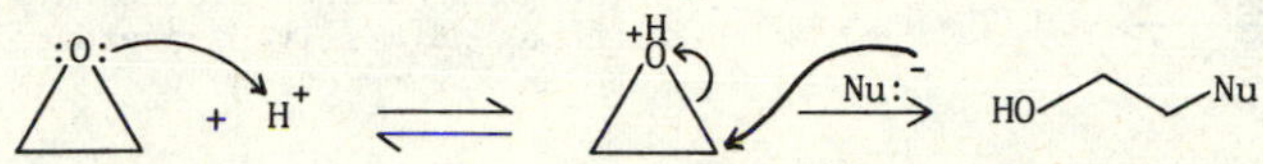

10-37 The rate of reaction is the rate of carbocation formation and is independent of the nucleophile. However the product mixture will depend on the nucleophilicity of Br^- and Cl^- even though the addition occurs after the slow step of the overall reaction. Br^- is a better nucleophile than Cl^- in this reaction.

10-38

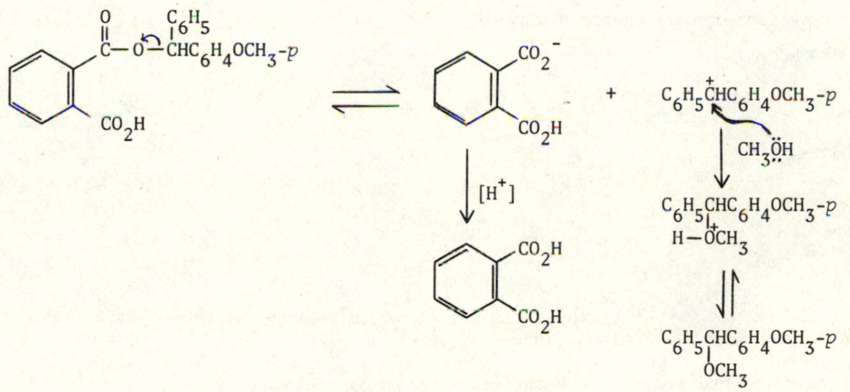

10-39

a) In the highly polar solvent formic acid, the benzylic alcohol solvolyzes by an S_N1 pathway with racemization stereochemistry.

b) In this case bromide replaces bromide in an S_N2 reaction. Each substitution gives an inverted molecule which cancels the rotation of an unreacted molecule.

Contd...

10-39 Contd...

c) In azide the negative charge is more delocalized than in the amide anion. Charge is spread over three atoms.

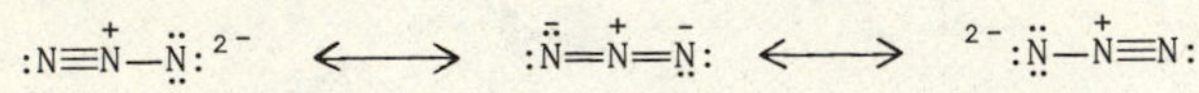

d)

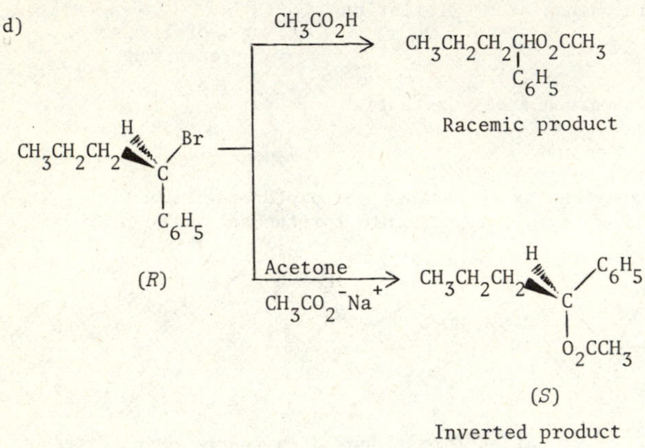

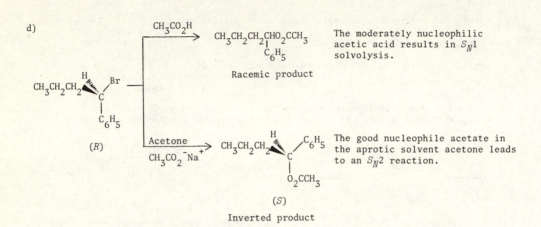

The moderately nucleophilic acetic acid results in S_N1 solvolysis.

The good nucleophile acetate in the aprotic solvent acetone leads to an S_N2 reaction.

e) The structure of the rigid bicyclic molecule inhibits backsided attack of a nucleophile by an S_N2 pathway. Formation of a carbocation in an S_N1 process is inhibited because the bridgehead carbon atom cannot become planar.

f) Only the trans isomer can undergo a backsided S_N2 cyclization.

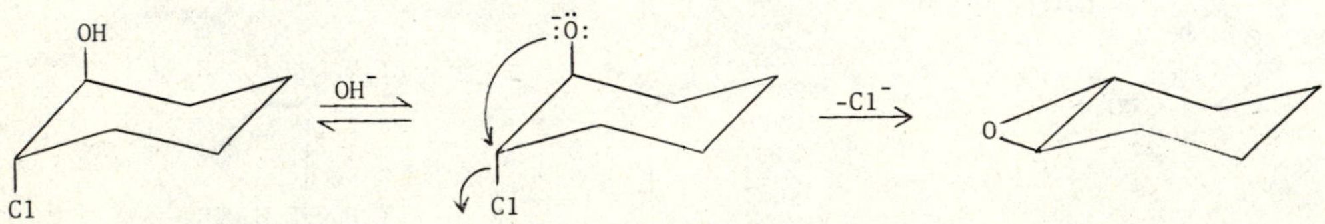

10-40

a) Chlorocyclopentane is faster because less strain is introduced in the substitution transition state for the cyclopentane.

b) These S_N2 reactions proceed faster with the primary substrate.

c) The moderate nucleophile, H_2O, in the polar solvent H_2O favors an S_N1 mechanism. The tertiary substrate is faster.

d) The phosphorus atom is the better nucleophile, thus reacts more rapidly in this S_N2 reaction.

e) The substrate with the better leaving group, iodide, reacts faster.

f) Methyl at the beta position of the branched substrate slows this S_N2 reaction by steric inhibition to back sided approach of the nucleophile.

11 The Scope of Nucleophilic Substitution

11-1 The $ZnCl_2$, a Lewis acid, enhances the leaving ability of the hydroxy group. Reaction favors an S_N1 pathway thus shows a reactivity order of tert > sec > prim.

11-2

a) $C_6H_5CH_2OH \xrightarrow{PBr_3} C_6H_5CH_2Br$

b) (R)-$C_6H_5CH(OH)CH_3 \xrightarrow[\text{or } SOCl_2/\text{pyridine}]{PCl_3}$ (S)-$C_6H_5CH(Cl)CH_3$

c) $CH_3CH_2OH \xrightarrow[(P\ +\ I_2)]{PI_3} CH_3CH_2I$

d) (R)-$CH_3CH(OH)CH_2CH_3 \xrightarrow[Et_2O]{SOCl_2}$ (R)-$CH_3CH(Cl)CH_2CH_3$

11-3 A skeletal rearrangement produces the same relatively stable tertiary carbocation in each case. Migration of hydride occurs with 3-methyl-2-butanol and migration of methyl takes place with 2,2-dimethyl-1-propanol.

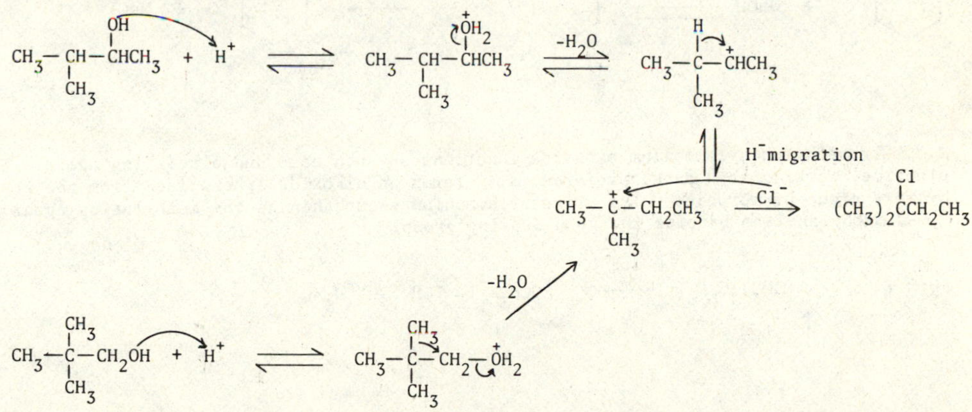

11-4 In aqueous base these allylic halides are expected to react by an S_N1 mechanism. The nucleophile can add to the common allylic carbocation at either of two carbon atoms.

$$CH_3CH{=}CH{-}CH_2Cl \text{ and } CH_3CH(Cl){-}CH{=}CH_2 \longrightarrow [CH_3CH{=}CH{-}\overset{+}{C}H_2 \longleftrightarrow CH_3\overset{+}{C}H{-}CH{=}CH_2]\ Cl^-$$

$$\xrightarrow[(-H^+)]{H_2O} CH_3CH{=}CHCH_2OH + CH_3CH(OH)CH{=}CH_2$$

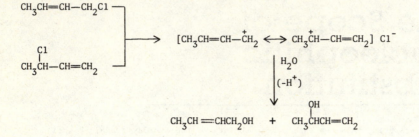

11-5 Formation of a haloalkane from an alcohol (the reverse of the hydrolysis reaction) requires that hydroxy be converted to a good leaving group. Acid produced in the neutral hydrolysis can be the catalyst for a reverse reaction. Basic hydrolysis produces alcohol and the halide anion so that no acid catalyst is available for the reverse reaction.

11-6 Iodide is a better nucleophile than water and a better leaving group than chloride. Iodide substitutes for chloride, then water displaces iodide. Only a catalytic amount of iodide is added since it is regenerated as the alcohol product forms.

$$CH_3CH_2CH_2Cl + :\ddot{I}:^- \longrightarrow CH_3CH_2CH_2I + :\ddot{C}l:^- \quad \text{fast}$$

$$CH_3CH_2CH_2I + H_2\ddot{O} \longrightarrow CH_3CH_2CH_2OH + :\ddot{I}:^- + (H_3O^+) \quad \text{fast}$$

11-7

i) $$(CH_3)_3C\ddot{O}:^-Na^+ + CH_3{-}OSO_3CH_3 \longrightarrow (CH_3)_3COCH_3 + CH_3OSO_3^-Na^+$$

ii)

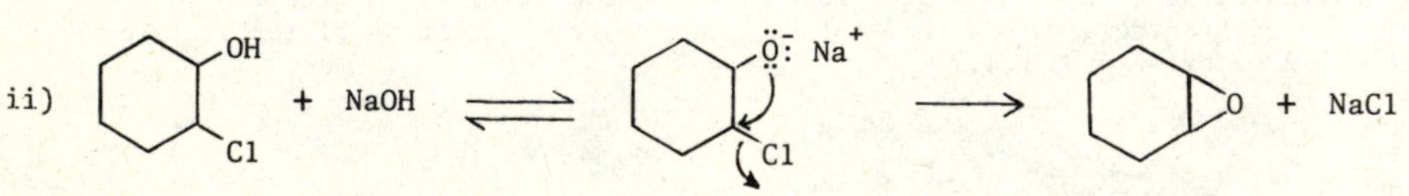

11-8 *N,N,N*-Trimethylphenylammonium ethoxide functions as both base and alkylating agent. The ethoxide, working as a base, preferentially forms an alkoxide (phenoxide) from the phenolic hydroxy group of morphine. The trimethylammonium group then is the methylating agent since *N,N*-dimethylaniline becomes the good leaving group.

$$R\ddot{O}:^- + CH_3{-}\overset{+}{N}(CH_3)_2C_6H_5 \longrightarrow ROCH_3 + (CH_3)_2\ddot{N}C_6H_5$$

11-9 Alcohol is protonated by the acid in either mechanism so that the initial leaving group is unchanged. Since bisulfate is the better nucleophile, we would expect the alkyl sulfate to form more readily and then lead to the ether.

11-10 The ether cleaves under acidic conditions so as to give the more stable carbocation or the more favorable carbocation-like transition state. In this case, the benzylic carbocation is favored.

$$C_6H_5CH_2OCH_3 \xrightarrow{HI} C_6H_5CH_2I + CH_3OH$$

11-11 Although the alcohol initially forms on ether cleavage, excess HX rapidly converts alcohol to an haloalkane.

11-12 In acid, the protonated epoxide first opens (S_N1) to give the more favorable secondary carbocation, then the alcohol adds. In base, an S_N2 reaction takes place as alkoxide attacks at the less substituted carbon atom of the epoxide.

11-13

a) $CH_3CH_2CH_2OCH_3$

b) 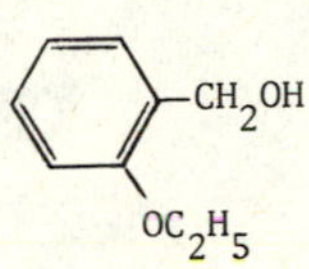

c) $(CH_3)_2CHI + C_6H_5OH$

d) $C_2H_5OCH(CH_3)—CH(OH)CH_3$

e) 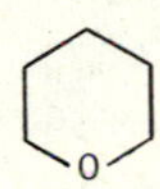

f) $(CH_3)_3CI + CH_3OH$ (or CH_3I)

11-14 If alkyl-oxygen cleavage takes place, the oxygen atom of the alcohol product must come from the water.

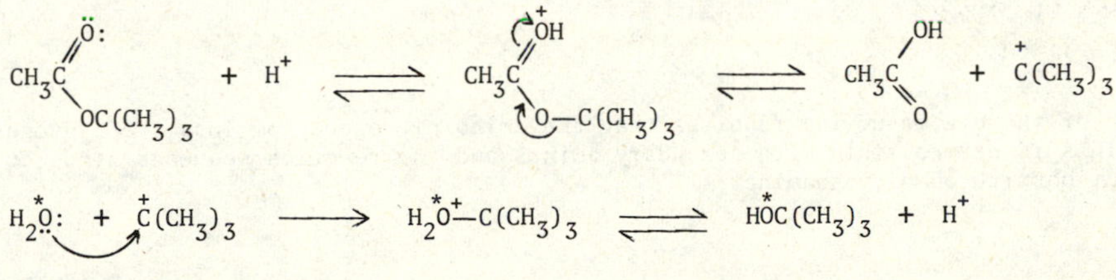

11-15 The ethanethiol is a weak acid ($pK_a \simeq 11$). In the presence of ethoxide, thiolate anion forms and functions as a nucleophile toward unreacted bromoethane.

$$C_2H_5Br + HS^-Na^+ \longrightarrow C_2H_5SH + NaBr$$

$$C_2H_5SH + NaOEt \longrightarrow C_2H_5S^-Na^+ + EtOH$$

$$C_2H_5S^-Na^+ + C_2H_5Br \longrightarrow C_2H_5SC_2H_5 + NaBr$$

11-16

a) Trimethyloxonium fluoroborate is an excellent methylating agent.

b) Fluoroborate is not a nucleophilic anion (see problem solution 10-14). A better nucleophile such as the halides would immediately undergo substitution on the cation and destroy the salt.

$$X^- + CH_3{-}\overset{+}{O}(CH_3)_2 \longrightarrow CH_3X + O(CH_3)_2$$

11-17

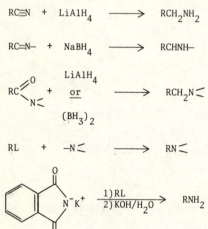

$$RC{\equiv}N + LiAlH_4 \longrightarrow RCH_2NH_2$$

$$RC{=}N{-} + NaBH_4 \longrightarrow RCHNH{-}$$

$$RC(=O)N< + LiAlH_4 \text{ or } (BH_3)_2 \longrightarrow RCH_2N<$$

$$RL + {-}N< \longrightarrow RN<$$

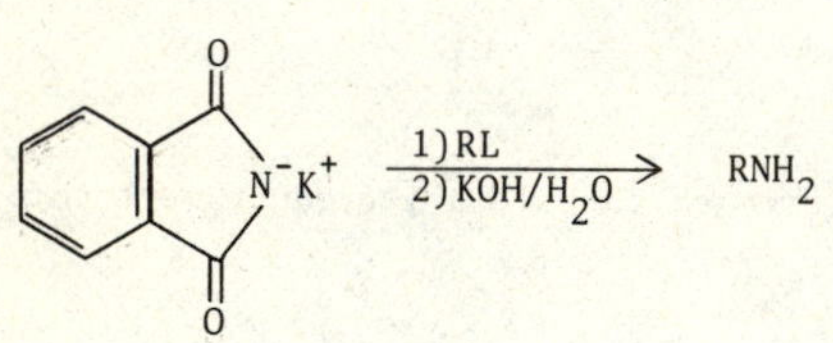

11-18 Formation of the diazonium ion requires that the amino nitrogen atom lose two hydrogen atoms. This is not possible with secondary amines and the reaction sequence stops at the *N*,*N*-disubstituted nitrosoamine.

11-19

a) The critical distance between the two quaternary ammonium groups is the same as in decamethonium dibromide (and curare).

b) Succinyl choline is a diester. Many methods of metabolic degradation of esters are available in biological systems.

11-20

$$HOOH \quad H^+ \rightleftharpoons HO{-}\overset{+}{O}H_2$$

$$(CH_3)_3\ddot{N} \quad H{-}O{-}\overset{+}{O}H_2 \rightleftharpoons (CH_3)_3\overset{+}{N}{-}O{-}H \quad H_2O \rightleftharpoons (CH_3)_3\overset{+}{N}{-}\ddot{O}\!:^- \quad H_3O^+$$

11-21 The first methyl group increases the electron density at the nitrogen atom to which it is attached. That nitrogen atom is the better nucleophile in the second alkylation step.

11-22

a) $CH_3CH_2CH_2CH_2OH + HCl \longrightarrow CH_3CH_2CH_2CH_2Cl \xrightarrow[\text{or } NH_3\text{(excess)}]{\text{1) Potassium phthalimide; 2) }KOH/H_2O/\Delta} CH_3CH_2CH_2CH_2NH_2$

b) $CH_3CH_2CH_2Br + (CH_3)_2NH \xrightarrow{Na_2CO_3} CH_3CH_2CH_2N(CH_3)_2$

c) $H_2NNH_2 + (CH_3)_2CHCl \xrightarrow{Na_2CO_3} H_2NNHCH(CH_3)_2 \xrightarrow[Na_2CO_3]{CH_3CH_2Br} CH_3CH_2NHNHCH(CH_3)_2$

(The more bulky isopropyl group inhibits substitution of the second alkyl group on the same nitrogen atom.)

d) $C_6H_5CH_2Cl + (CH_3)_3N \longrightarrow C_6H_5CH_2\overset{+}{N}(CH_3)_3 \; Cl^-$

e) $(CH_3)_2NH + C_2H_5I \xrightarrow{Na_2CO_3} (CH_3)_2NC_2H_5 \xrightarrow{H_2O_2} (CH_3)_2\overset{+}{N}(O^-)C_2H_5$

11-23

$$\underset{(R)}{[CH_3CH_2CH(CH_3)O]_3P\!:} + CH_3{-}I \longrightarrow \underset{(R)}{[CH_3CH_2CH(CH_3)O]_3\overset{+}{P}CH_3} \; I^-$$

$$\underset{(R)}{[CH_3CH_2CH(CH_3)O]_2}\overset{+}{P}(CH_3){-}O{-}\underset{(R)}{CH(CH_3)CH_2CH_3} + I^- \xrightarrow{(S_N2\text{ inversion})}$$

$$\underset{(R)}{[CH_3CH_2CH(CH_3)O]_2P(CH_3){=}O} + \underset{(S)}{CH_3CH_2CH(CH_3)I}$$

11-24

a) $(CH_3)_2CHCH(Cl)CO_2H + NH_3\text{(excess)} \xrightarrow{H_2O} (CH_3)_2CHCH(NH_2)CO_2H$

2-Amino-3-methylbutanoic acid

(Valine)

Contd...

11-24 Contd...

b) $ClCH_2CH_2OH + (C_2H_5)_2NH \longrightarrow (C_2H_5)_2NCH_2CH_2OH$

2-(*N*,*N*-Diethylamino)ethanol

c) $C_6H_5CH_2Cl + C_6H_5NH_2 \longrightarrow C_6H_5CH_2NHC_6H_5$

N-Phenylbenzylamine

d) $3\ CH_3I + C_6H_5CH_2NH_2 \longrightarrow C_6H_5CH_2\overset{+}{N}(CH_3)_3\ I^-$

Benzyl-*N*,*N*,*N*-trimethylammonium iodide

e) $C_2H_5Br + (C_6H_5CH_2)_3P \longrightarrow (C_6H_5CH_2)_3\overset{+}{P}C_2H_5\ Br^-$

Ethyltribenzylphosphonium bromide

f) $C_6H_5CH_2CH_2Br + (CH_3)_2S \longrightarrow C_6H_5CH_2CH_2\overset{+}{S}(CH_3)_2Br^-$

Dimethyl-2-phenylethylsulfonium bromide

11-25 Neighboring group participation by the sulfur is considerably better than by oxygen. One reason is that the sulfur atom is a better nucleophile than oxygen. A second factor is that sulfur is a larger atom with longer bonds. The three ring intermediate formed by the sulfur atom is less strained (easier to form in the rate-controlling step) than that of the oxygen atom.

11-26 In concentrated base the large excess of hydroxide substitutes for bromide with typical S_N2 inversion. In the dilute base the intramolecular substitution by carboxylate is the favored first step.

11-27

a) 2-(*S*), 3-(*R*), 3-Chloro-2-butanol

b)

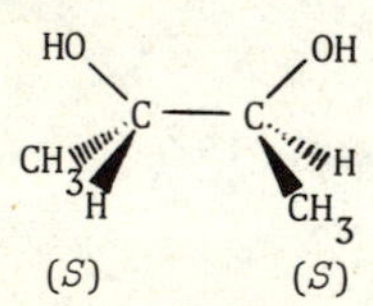

Contd...

11-27 Contd...

c)

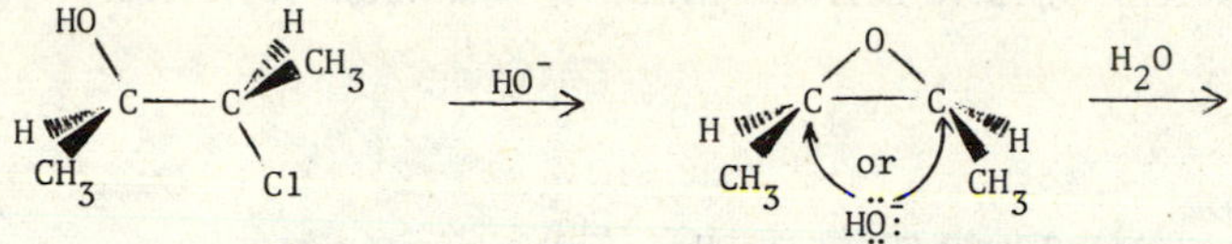

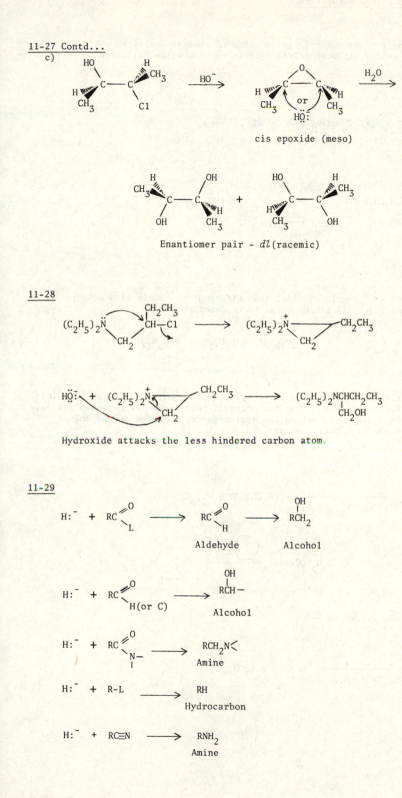

cis epoxide (meso)

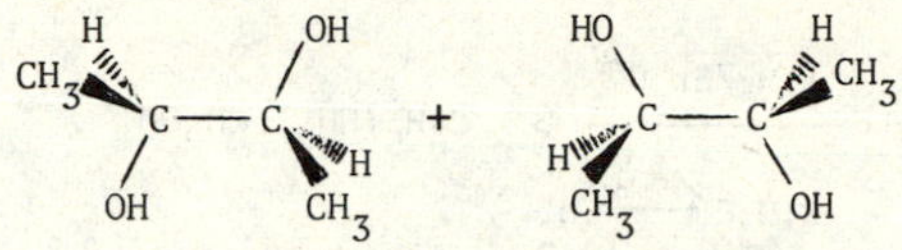

Enantiomer pair - *dl*(racemic)

11-28

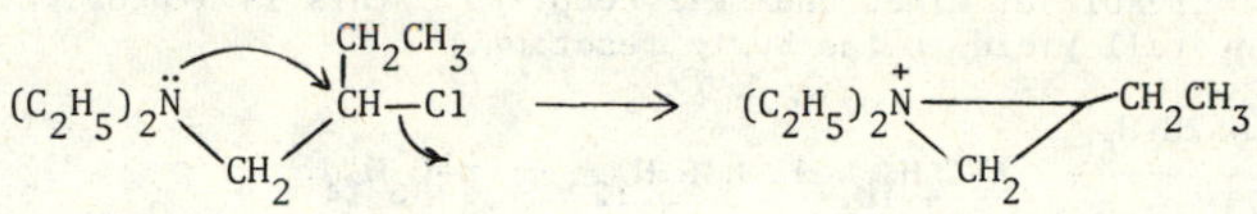

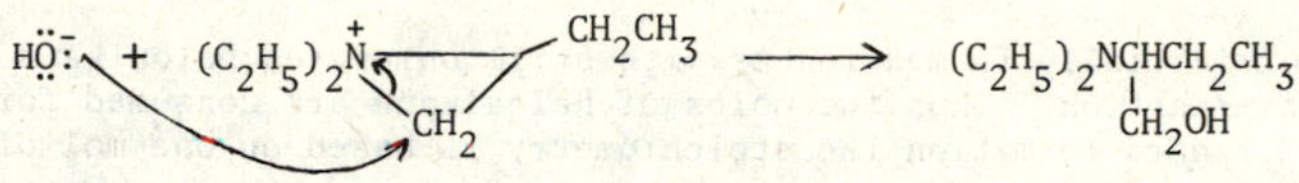

Hydroxide attacks the less hindered carbon atom.

11-29

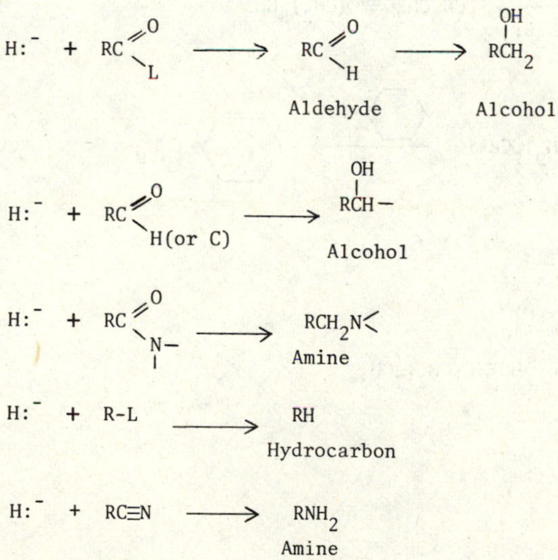

11-30 Solvation of the cyanide ion through hydrogen bonding to ethanol decreases the effective nucleophilicity of the anion. Aprotic solvents minimize such anion solvation.

11-31

$$C_6H_5CH_2CH_2Br \xrightarrow[\substack{2)\,H_2C\overset{O}{\frown}CH_2 \\ 3)\,H_3O^+}]{1)\,Mg/Et_2O} C_6H_5CH_2CH_2CH_2CH_2OH \xrightarrow{HBr}$$

$$C_6H_5CH_2CH_2CH_2CH_2Br \xrightarrow[\substack{2)\,H_2C\overset{O}{\frown}CH_2 \\ 3)\,H_3O^+}]{1)\,Mg/Et_2O} C_6H_5(CH_2)_5CH_2OH$$

11-32 Three products will form as a result of mixed and self coupling. This is a problem in addition to a usually poor overall yield in the Wurtz reaction.

$$CH_3CH_2Br + CH_3CH_2CH_2Br \xrightarrow[\Delta]{Na/C_6H_6} n\text{-}C_4H_{10} + n\text{-}C_5H_{12} + n\text{-}C_6H_{14}$$

11-33 Some of the Grignard reagent initially formed reacts with still unreacted haloalkane in a process resembling the Wurtz reaction. Thus two moles of haloalkane are consumed for only one mol of magnesium. In Grignard formation the stoichiometry is based on one mol of magnesium per one mol of the organohalogen compound.

11-34

a) $$CH_3CH_2C{\equiv}CH + NaNH_2 \xrightarrow[2)\,CH_3(CH_2)_4CH_2Br]{1)\,NH_3/-33°} CH_3CH_2C{\equiv}C(CH_2)_5CH_3$$

b) 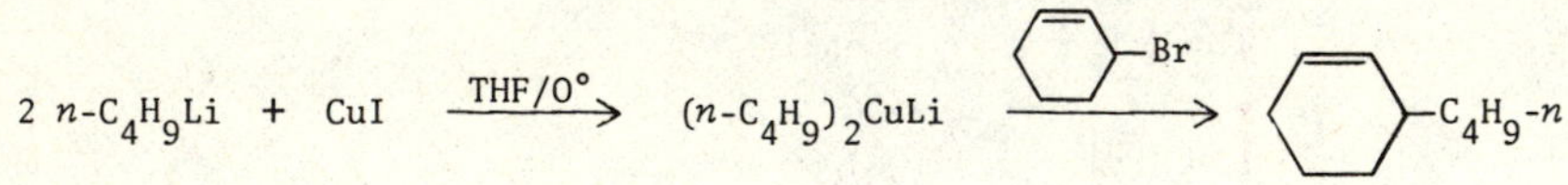

c) $$CH_3Br + NaCN \xrightarrow{DMSO} CH_3CN$$

d) $$HC{\equiv}CH + NaNH_2 \xrightarrow[2)\,CH_3(CH_2)_6CH_2Br]{1)\,NH_3/-33°} HC{\equiv}C(CH_2)_7CH_3$$

11-35

a) $(CH_3)_2CHCH_2Br$

b) 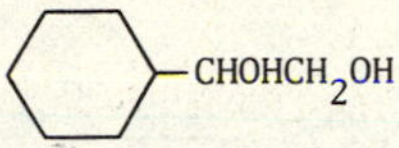

c) 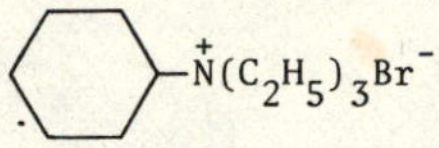

d) $CH_3\overset{\overset{Br}{|}}{C}HCH_3$

e) $(CH_3)_2CHI$

f) $CH_3C\equiv CCH_2CH_2CH_2Cl$

g) 

h) $CH_3CH_2CH_2I$

i) CH_3OCH_3

j) $CH_3CH_2CH_2NH_2$

k) $(CH_3)_3CI$ + CH_3CH_2OH
(or CH_3CH_2I)

l) $C_6H_5CH_2Cl$

11-36

a) A = $HC\equiv C\colon^- Na^+$

B = $HC\equiv CCH_2CH_2CH_2OH$

C = $SOCl_2$

D = $HC\equiv CCH_2CH_2CH_2CN$

b) E = CH_3CH_2Br

F = $CH_3CH_2NH_2$

c) G = $CH_2{=}CHCH_2O^-Na^+$

H = 

I = PBr_3

J = $CH_2{=}CHCH_2OCH_2CH_3$

d) K = 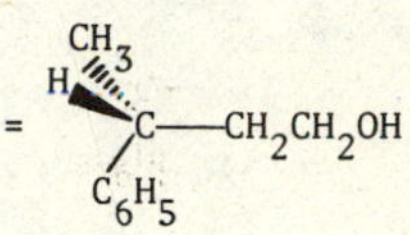

L = 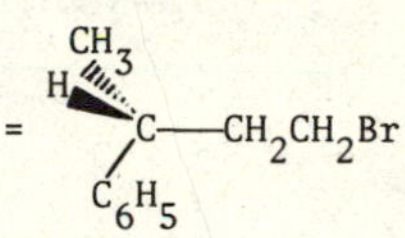

M = $LiAlH_4/Et_2O$

11-37

a) With silver nitrate, tertiary haloalkanes ionize to form a tertiary carbocation and a precipitate forms rapily. The speed of ionization and hence precipitate formation decreases in the order: tertiary > secondary > primary, The reverse is true for S_N2 substitution of I^- for Br^- where primary > secondary > tertiary. If the unknown reacts rapidly with $AgNO_3$/EtOH and not at all with NaI/acetone it is tertiary; if a precipitate forms slowly with both tests then the unknown is secondary. If the unknown reacts rapidly with NaI/acetone and not with $AgNO_3$/EtOH it is primary.

b)

CH_3CH_2Br	$AgNO_3$/EtOH - no reaction	NaI/Acetone - fast
$(CH_3)_2CHBr$	$AgNO_3$ - slow	NaI/Acetone - slow
$(CH_3)_3CCl$	$AgNO_3$/EtOH - fast	NaI/Acetone - no reaction

11-38

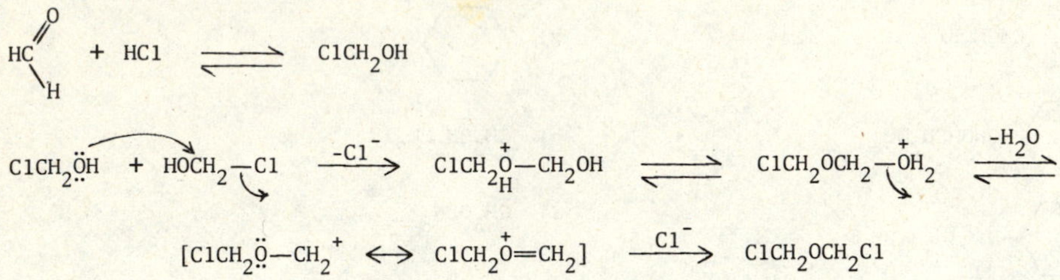

11-39 The product has a plane of symmetry.

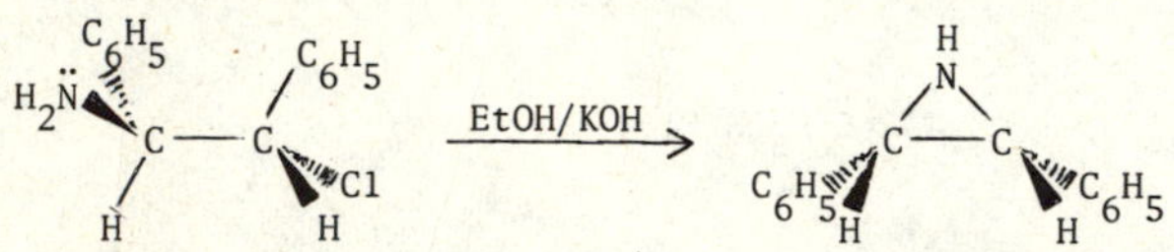

11-40

a) 

b) 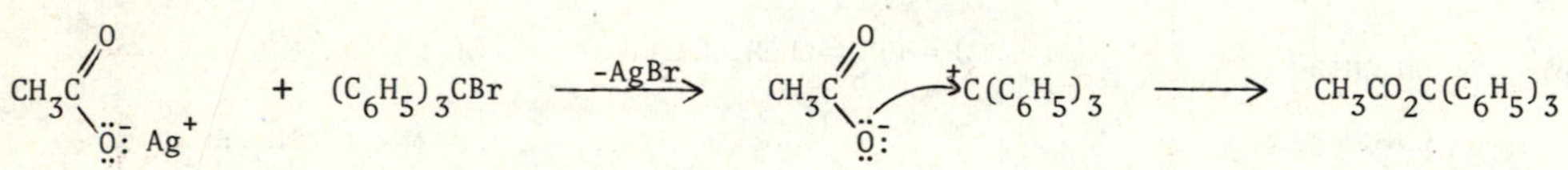

c) The nitrogen atom acts as a neighboring group to displace Br^-. The cyclic product can be recovered, or re-opened to give substitution.

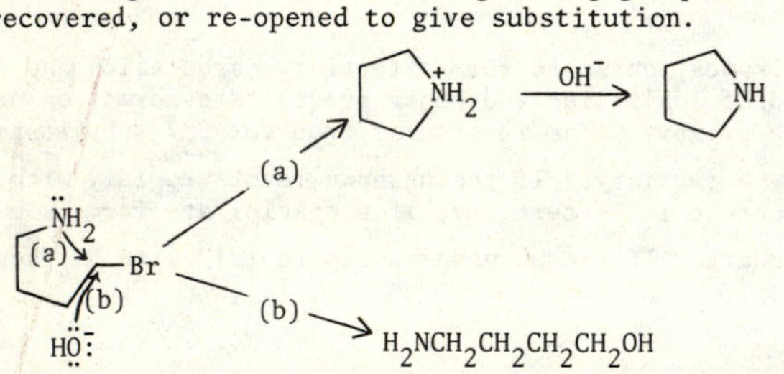

d) 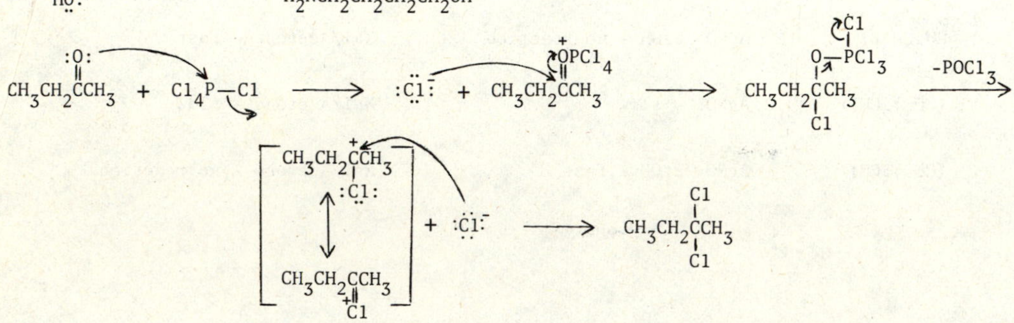

11-41

a) $(CH_3CH_2)_3COH$

b) 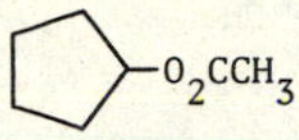

c) *m*-$CH_3C_6H_4CH_2CN$

d) 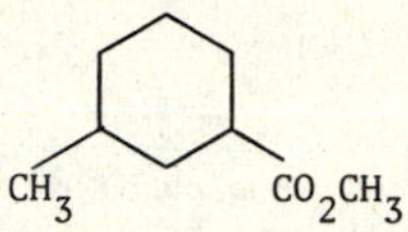

e) $CH_3\overset{NH_2}{C}HCH_2CH_3$

f) $C_6H_5CH_2CH_2OCH_2CH_3$

g) 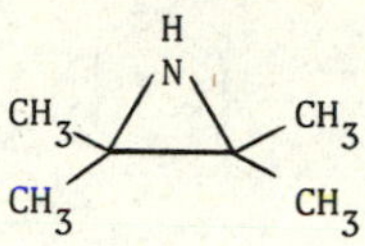

h) $CH_3(CH_2)_4CH_3$

i) 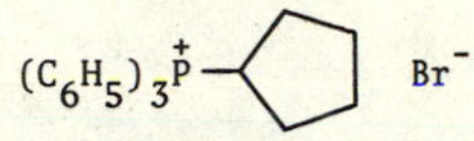

j) 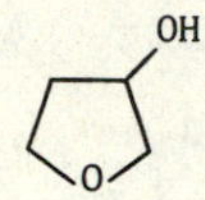

k) (S)-$CH_3CH_2\overset{CN}{C}HCH_3$

l) 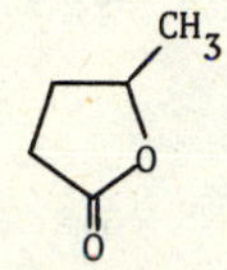

m) $CH_3CH_2S{-}SCH_2CH_3$

11-42

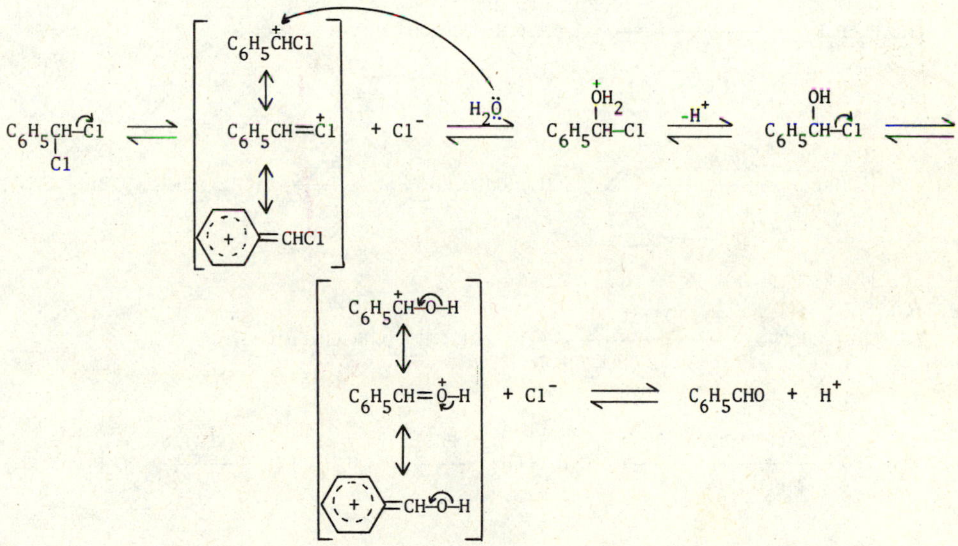

11-43

a) In the trans isomer the carbonyl oxygen atom can assist in displacing the bromide by a neighboring group effect. This is not possible in the cis isomer.

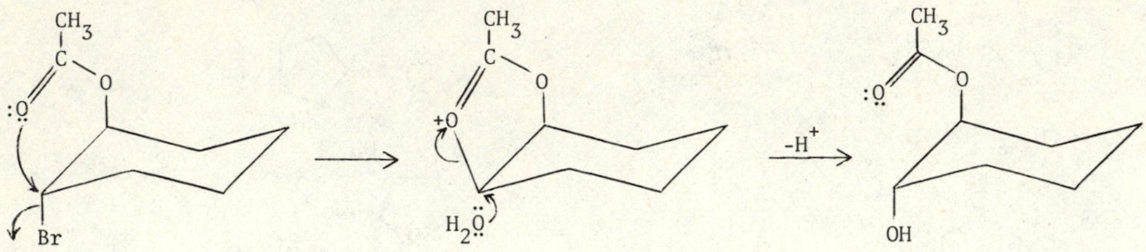

b) The carbonyl carbon atom is too hindered for nucleophilic addition by an alcohol or the reagents required for acyl halide formation. However, reaction with base readily removes the less hindered proton from the carboxy hydroxy group. Alkylation of the carboxylate anion can then take place at that much less hindered oxygen anion.

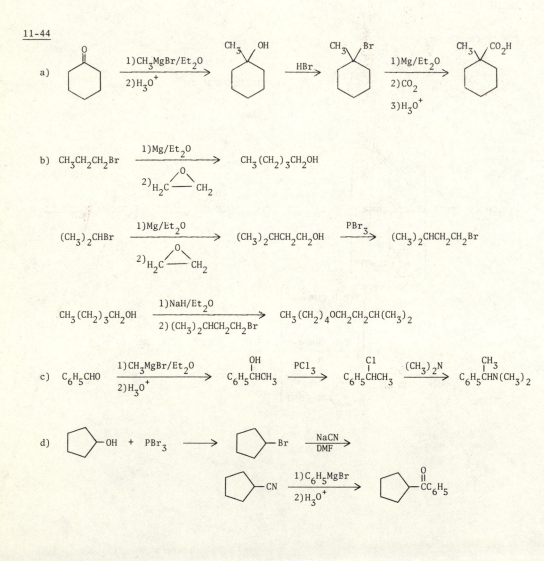

11-44

a) 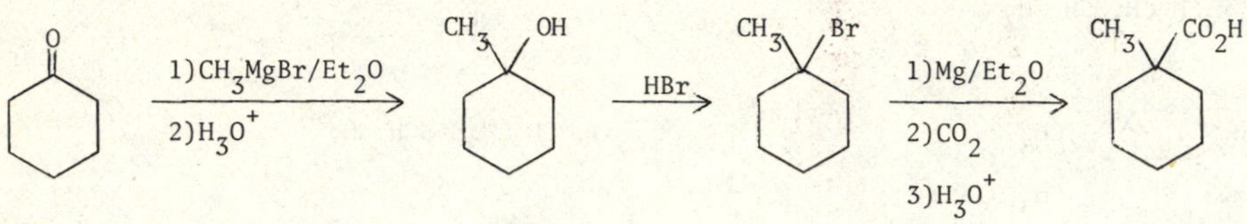

b) 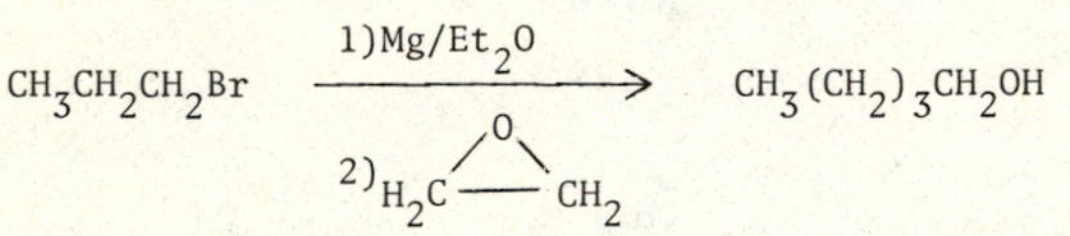

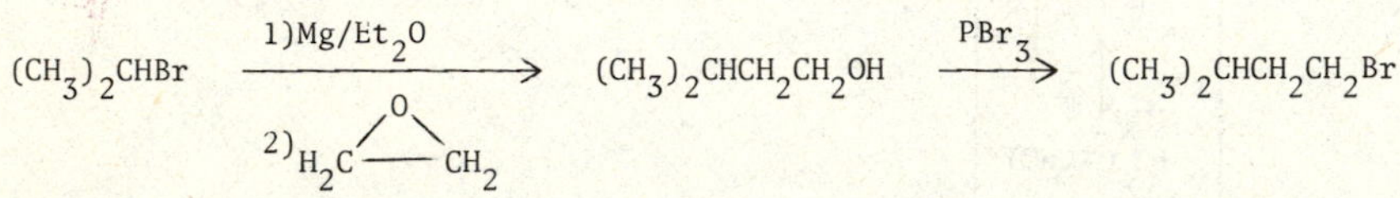

$$CH_3(CH_2)_3CH_2OH \xrightarrow[2)\,(CH_3)_2CHCH_2CH_2Br]{1)\,NaH/Et_2O} CH_3(CH_2)_4OCH_2CH_2CH(CH_3)_2$$

c) $$C_6H_5CHO \xrightarrow[2)\,H_3O^+]{1)\,CH_3MgBr/Et_2O} C_6H_5CH(OH)CH_3 \xrightarrow{PCl_3} C_6H_5CH(Cl)CH_3 \xrightarrow{(CH_3)_2N} C_6H_5CH(CH_3)N(CH_3)_2$$

d) 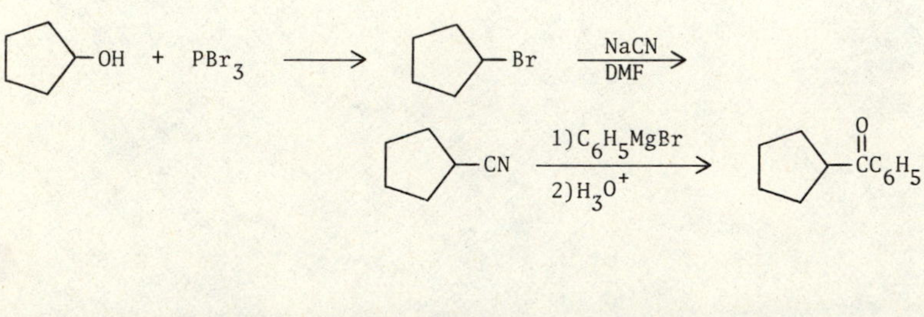

11-45

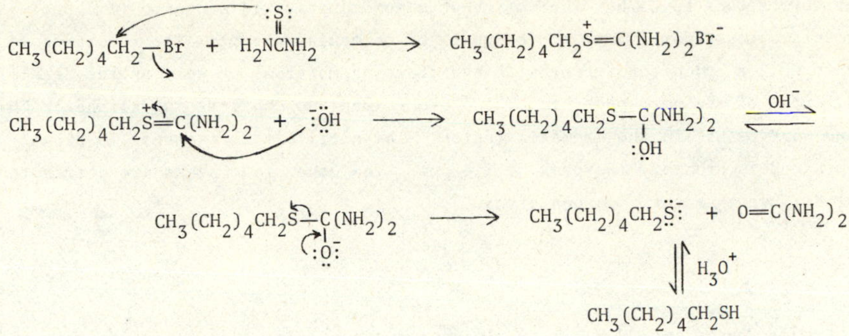

11-46

a) Elemental analysis provides an empirical formula of $C_6H_{13}NO_2$ and the mass spectrum confirms this as the molecular formula. The IR spectrum suggests some type of carbonyl group. The mode of formation of A suggests that dimethylamine has displaced a halogen in a substitution reaction, thus compound B must also have the carbonyl group. The nmr spectrum is consistent with a dimethylamino group (singlet at 2.3 ppm) and the ethyl group of an ester (triplet-quartet at 1.2 and 4.2 ppm). The singlet at 3.2 ppm must be a $-CH_2-$ deshielded by a carbonyl and another group.

b) $$(CH_3)_2NH + \underset{B}{XCH_2CO_2C_2H_5} \xrightarrow{Na_2CO_3} \underset{A}{(CH_3)_2NCH_2CO_2C_2H_5}$$

(X = halogen)

11-47

B = 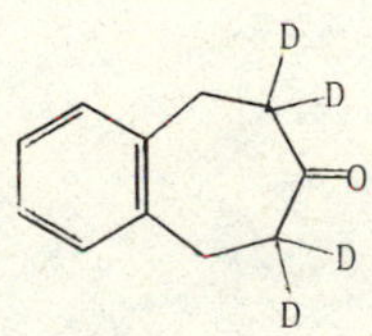

D = 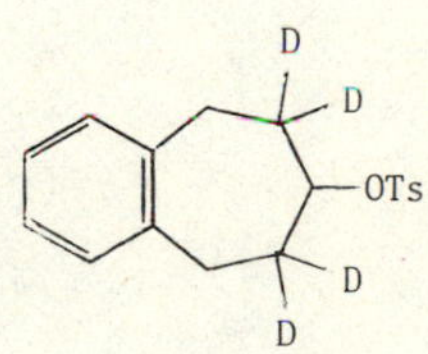

C = 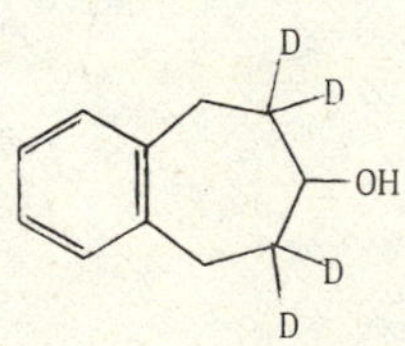

E = 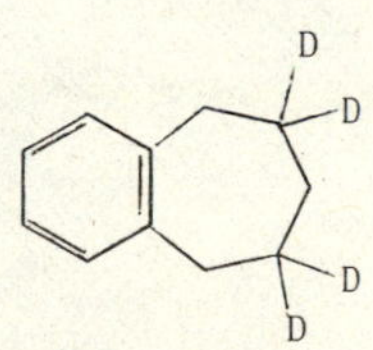

The IR peak at 2160 cm^{-1} is due to C—D stretching.

11-48 The IR spectrum suggests a triple bond (2240 cm^{-1}). The odd molecular weight suggests that the IR peak corresponds to C≡N. The nmr shows a monosubstituted aromatic (7.3 ppm). These fragments account for 103 (77 + 26) units of molecular weight. The remaining 14 units correspond to CH_2, a group consistent with the starting material as well as the deshielded singlet (3.7 ppm) of the nmr spectrum. The ^{13}C-nmr spectrum confirms the aliphatic CH_2 (23 ppm) and five peaks in the aromatic region. The nitrile ^{13}C resonance is usually about 120 ppm and is probably the peak at 118 ppm. The other four peaks are attributed to the four different aromatic carbon atoms.

$$C_6H_5CH_2C{\equiv}N$$

11-49 The IR spectrum suggests an —OH and a triple bonded group. The even molecular weight (M^+ = 98) indicates zero or an even number of nitrogen atoms. The ^1H-nmr confirms an —OH (or NH) at 2.8 ppm which exchanges with D_2O. When combined with the IR information that peak must be an OH. The ^{13}C-nmr spectrum shows six different kinds of carbon atoms. These, plus the total of ten hydrogen atoms from the ^1H-nmr and the hydroxy oxygen suggests a molecular formula of $C_6H_{10}O$ which fits the mass spectral molecular weight. An IHD =2 is consistent with the presence of —C≡C—. With 2-butanone as starting material, the structure is identified.

$$HC{\equiv}CH + NaNH_2 \xrightarrow{NH_3(l)} Na^{+\,-}C{\equiv}CH$$

$$CH_3\overset{O}{\overset{\|}{C}}CH_2CH_3 + Na^{+\,-}C{\equiv}CH \xrightarrow[2)\,H_3O^+]{1)\,NH_3(l)} CH_3\underset{\underset{C\equiv CH}{|}}{\overset{\overset{OH}{|}}{C}}CH_2CH_3$$

12 The Alpha-Carbanion—Nucleophilic-Electrophilic Reactivity of Carbonyl Compounds

12-1

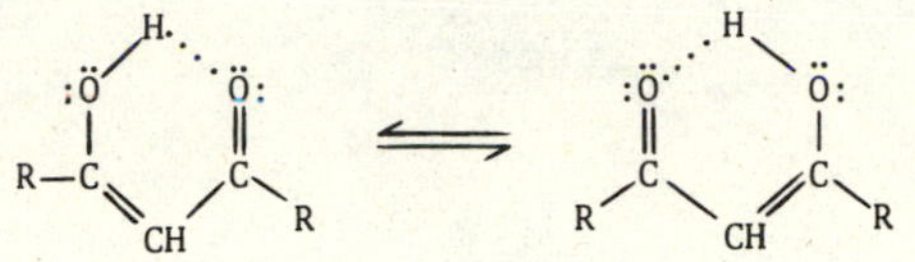

1,3-Dicarbonyl compounds are stabilized by intramolecular H-bonding in a six-membered cyclic structure.

12-2 Such an intramolecular proton shift is not consistent with the catalytic role of added acid or base.

12-3 The structure with the charge on oxygen is expected to make the greater contribution to the resonance hybrid since oxygen is more electronegative and can accommodate the negative charge better than carbon can.

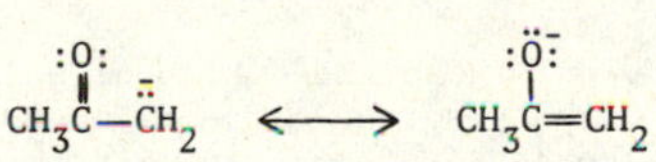

12-4 Those data confirm that formation of the carbon-halogen bond is not rate controlling. No information is provided about the actual rate of the halogenation step in each of the reactions, except that all are faster than enolization.

12-5

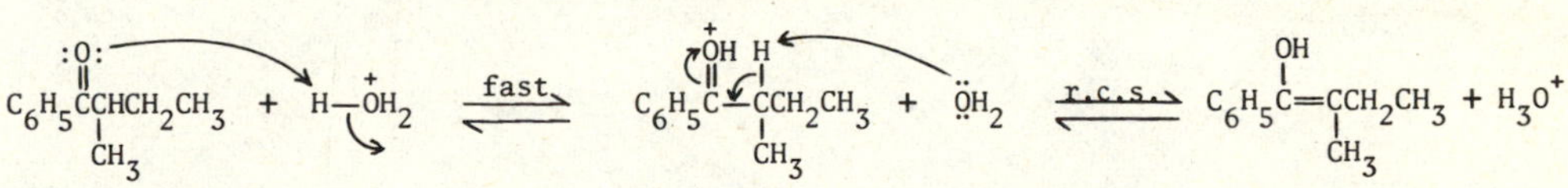

Optically active

Achiral

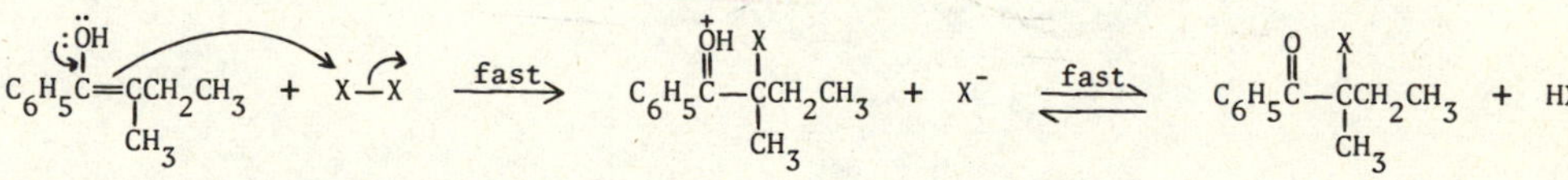

Racemic

12-6

$$CH_3CH_2CHO \ + \ CH_3CH_2CH_2CHO \ \xrightarrow{\text{base}}$$

$$\underset{\underset{CH_3}{|}}{CH_3CH_2\overset{\overset{OH}{|}}{CH}CHCHO} \ + \ \underset{\underset{C_2H_5}{|}}{CH_3CH_2CH_2CH\!-\!\overset{\overset{OH}{|}}{C}HCHO} \ + \ \underset{\underset{CH_3}{|}}{CH_3CH_2CH_2\overset{\overset{OH}{|}}{C}HCHCHO} \ + \ \underset{\underset{C_2H_5}{|}}{CH_3CH_2\overset{\overset{OH}{|}}{C}HCHCHO}$$

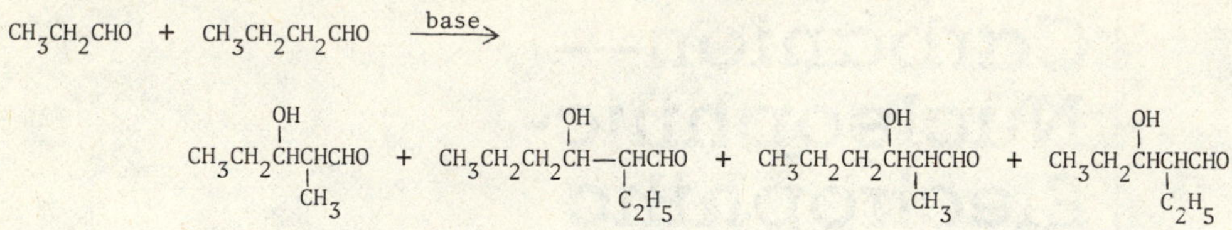

12-7

a) Formaldehyde has no alpha hydrogen atoms, thus cannot function as the nucleophile. Also, the branched aldehyde is crowded so that self condensation is minimized.

b)

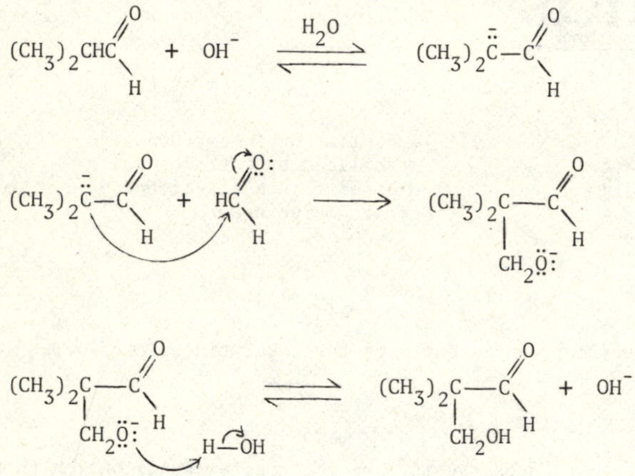

12-8 To obtain an α,β-unsaturated carbonyl product the compound acting as the nucleophile in the aldol reaction must have two hydrogen atoms alpha to the carbonyl group. One is lost during formation of the enolate anion and the second is lost in the dehydration step. In this example, the product from self condensation of 2-methylpropanal, does not have that second proton.

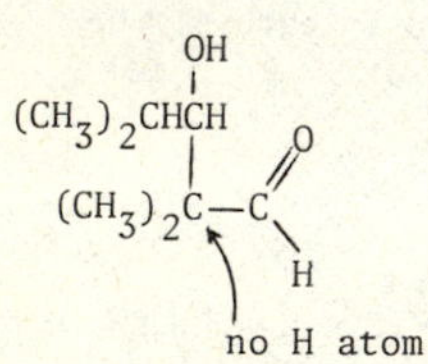

12-9 Two enolate anions can form from the unsymmetrical ketone,

$$\overset{-}{:}\!CH_2\overset{\overset{O}{\|}}{C}CH_2CH_3 \quad \text{and} \quad CH_3\overset{\overset{O}{\|}}{\underset{}{C}}\overset{..}{C}HCH_3 .$$

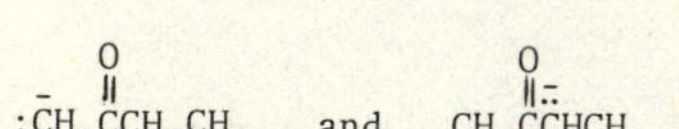

Contd...

12-9 Contd...

The aldol products are--

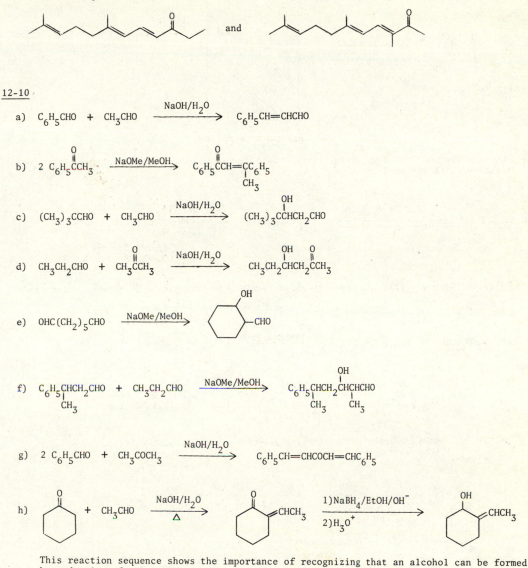

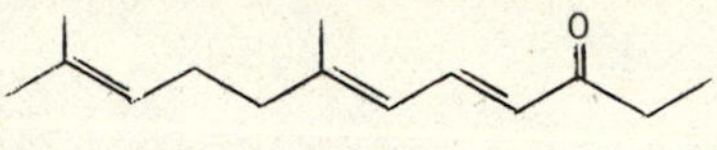

and

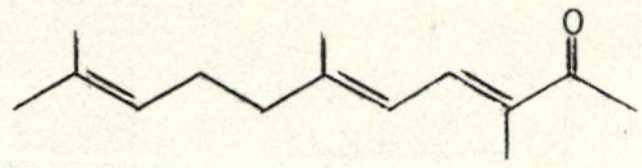

12-10

a) $C_6H_5CHO + CH_3CHO \xrightarrow{NaOH/H_2O} C_6H_5CH{=}CHCHO$

b) $2\ C_6H_5\overset{O}{\overset{\|}{C}}CH_3 \xrightarrow{NaOMe/MeOH} C_6H_5\overset{O}{\overset{\|}{C}}CH{=}\underset{CH_3}{\underset{|}{C}}C_6H_5$

c) $(CH_3)_3CCHO + CH_3CHO \xrightarrow{NaOH/H_2O} (CH_3)_3C\overset{OH}{\overset{|}{C}H}CH_2CHO$

d) $CH_3CH_2CHO + CH_3\overset{O}{\overset{\|}{C}}CH_3 \xrightarrow{NaOH/H_2O} CH_3CH_2\overset{OH}{\overset{|}{C}H}CH_2\overset{O}{\overset{\|}{C}}CH_3$

e) 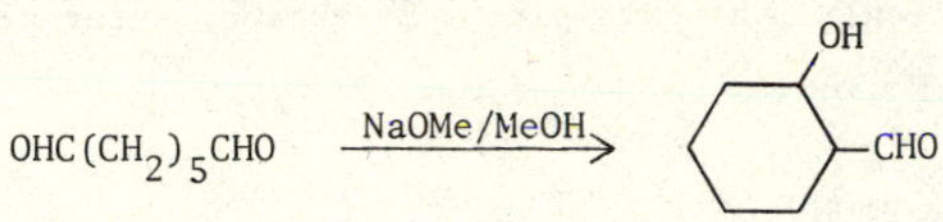

f) $C_6H_5\underset{CH_3}{\underset{|}{C}H}CH_2CHO + CH_3CH_2CHO \xrightarrow{NaOMe/MeOH} C_6H_5\underset{CH_3}{\underset{|}{C}H}CH_2\overset{OH}{\overset{|}{C}H}\underset{CH_3}{\underset{|}{C}H}CHO$

g) $2\ C_6H_5CHO + CH_3COCH_3 \xrightarrow{NaOH/H_2O} C_6H_5CH{=}CHCOCH{=}CHC_6H_5$

h) 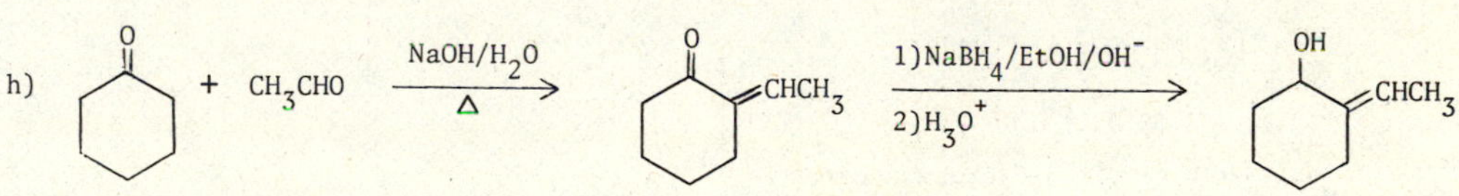

This reaction sequence shows the importance of recognizing that an alcohol can be formed by reduction of a ketone.

12-11 Both experiments are consistent with initial formation of the ester enolate anion.

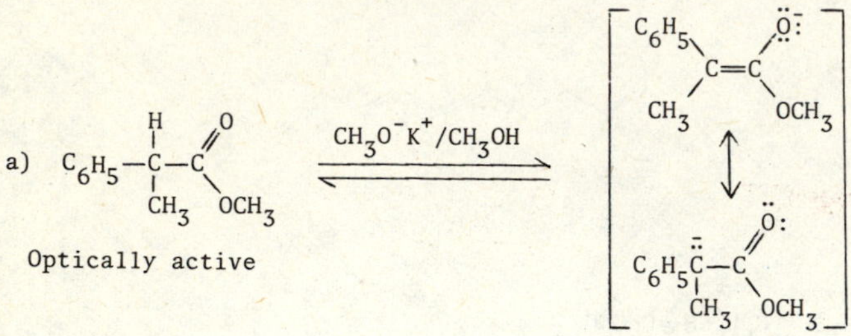

Optically active

Achiral intermediate

(Condensation may not occur with this compound since the equilibrium cannot be shifted by formation of a new enolate anion.)

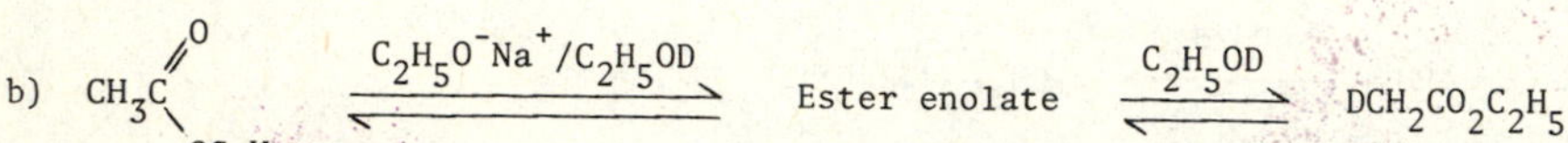

Partially exchanged ester

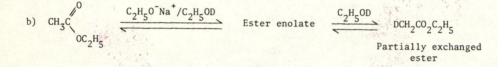

12-12 To avoid the possibility of transesterification which would give a mixture of ester products.

12-13 $CH_3CH_2CO_2C_2H_5 + CH_3CH_2CH_2CO_2C_2H_5 \xrightarrow[2)H_3O^+]{1)NaOEt/EtOH}$

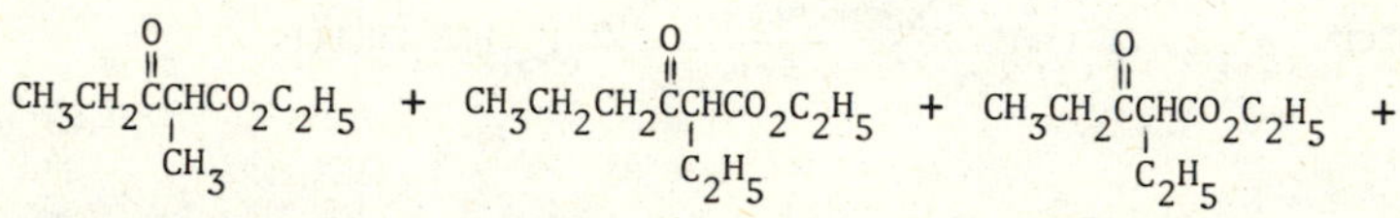

Ethyl 2-methyl-3-oxopentanoate + Ethyl 2-ethyl-3-oxohexanoate + Ethyl 2-ethyl-3-oxopentanoate +

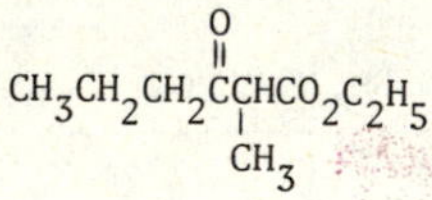

Ethyl 2-methyl-3-oxohexanoate

12-14 The strong base converts all of the ester to its enolate anion. (An alkoxide base gives an equilibrium mixture of ester and alkoxide.) The enolate anion is not electrophilic so that self condensation does not occur. Then the second ester is added and enolate addition takes place.

12-15

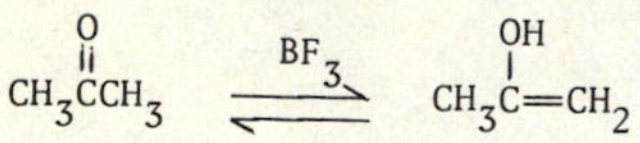

Acid-catalyzed enolization

Contd...

12-15 Contd...

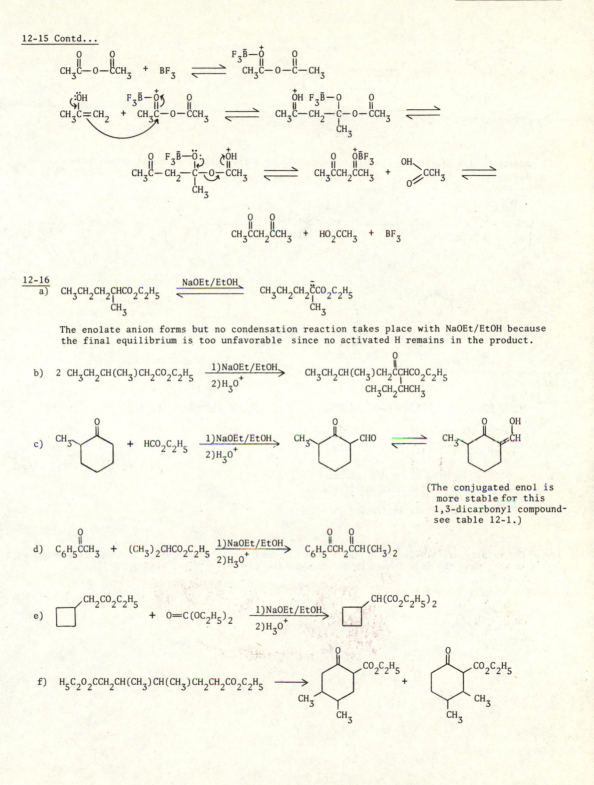

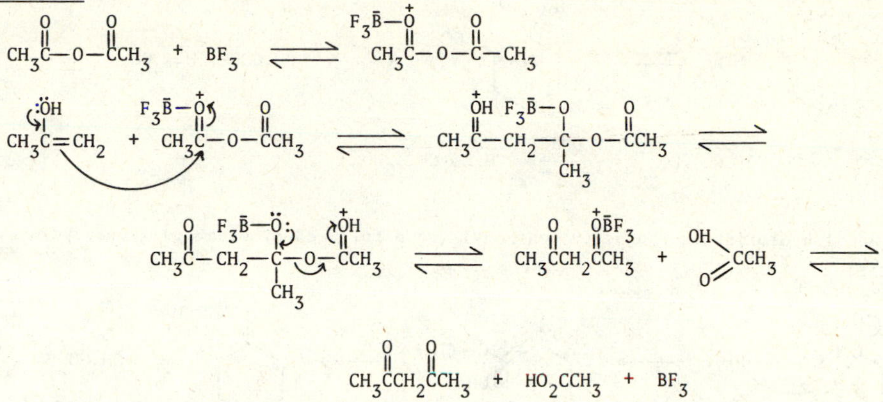

12-16

a) 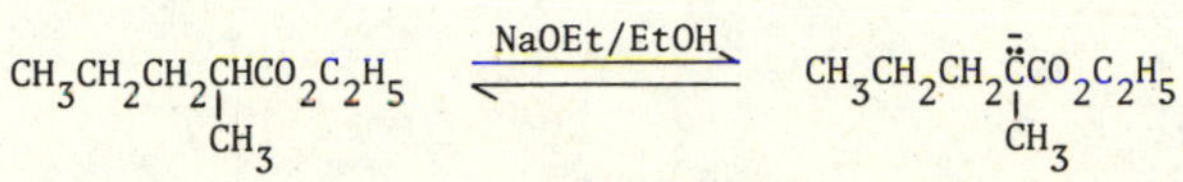

The enolate anion forms but no condensation reaction takes place with NaOEt/EtOH because the final equilibrium is too unfavorable since no activated H remains in the product.

b) 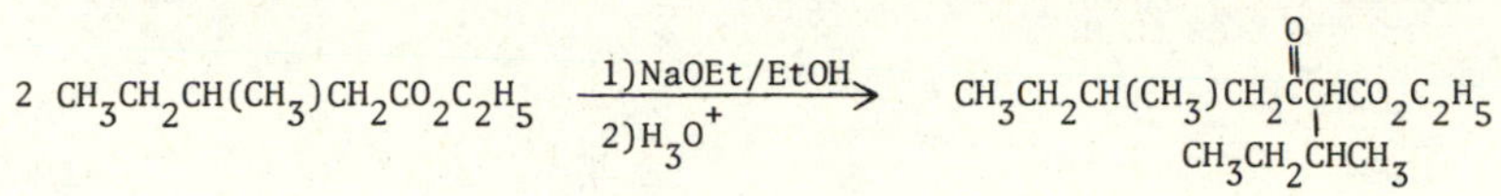

c) 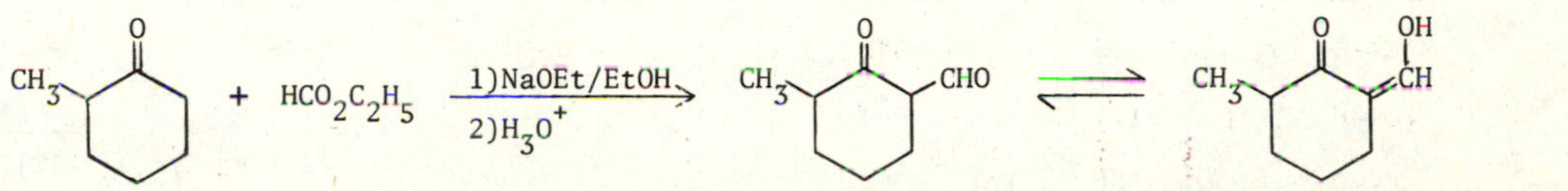

(The conjugated enol is more stable for this 1,3-dicarbonyl compound- see table 12-1.)

d) 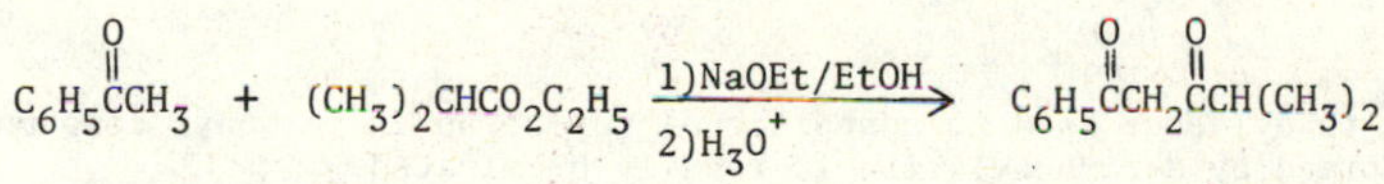

e) 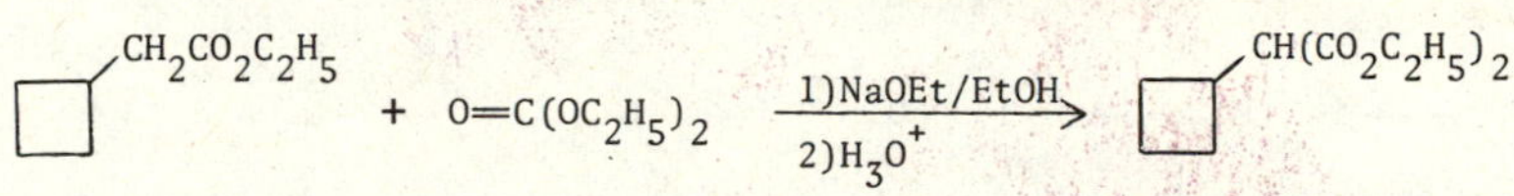

f) 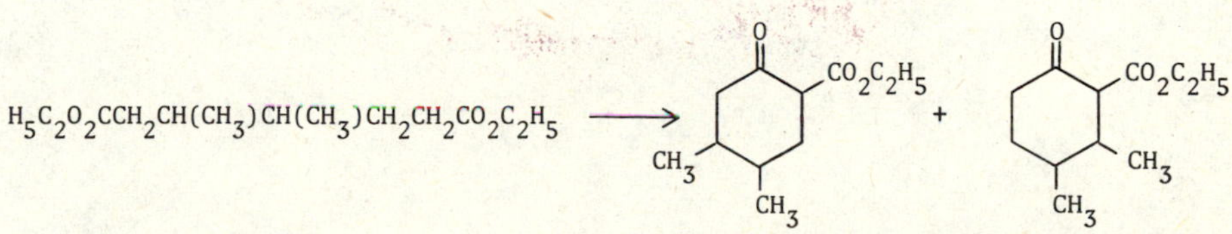

12-17

a) 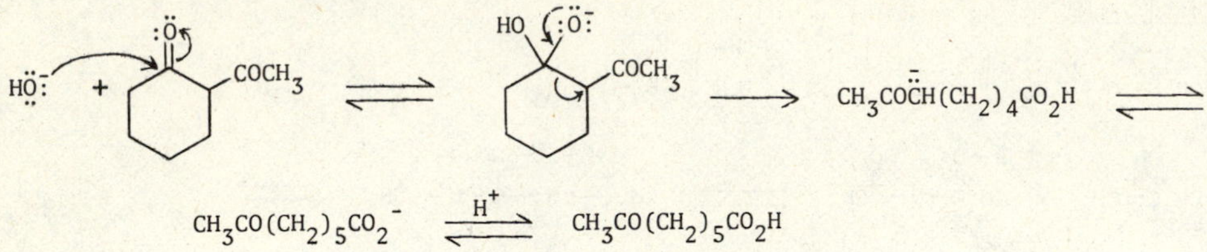

b) Addition of hydroxide to the acetyl carbonyl group followed by cleavage gives cyclohexanone and acetic acid.

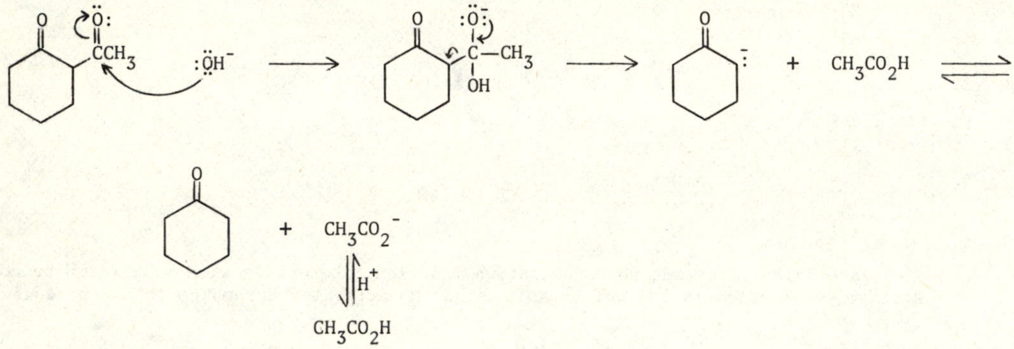

12-18

Bonds broken	kcal/mol	kJ/mol	Bonds formed	kcal/mol	kJ/mol
C—C	83	347	C—H	-99	-414
O—H	111	464	C=O	-192	-803
C—O	86	359		-291 kcal/mol	-1217 kJ/mol
	280 kcal/mol	1170 kJ/mol			

$\Delta H° = 280 - 291 = -11$ kcal/mol (-46 kJ/mol)

12-19 The results are similar to evidence used to support enol formation in carbonyl compounds. In this case the enol formed by decarboxylation is readily brominated.

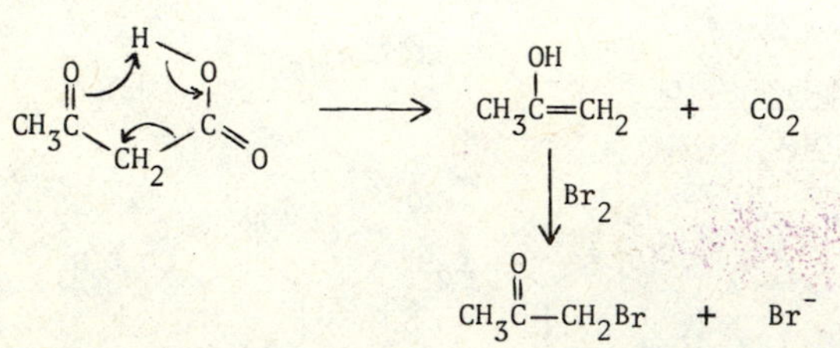

12-20 β-Keto esters form the enol of a ketone which is more stable than the enol of the carboxylic acid which forms in the *gem*-diacid decarboxylation.

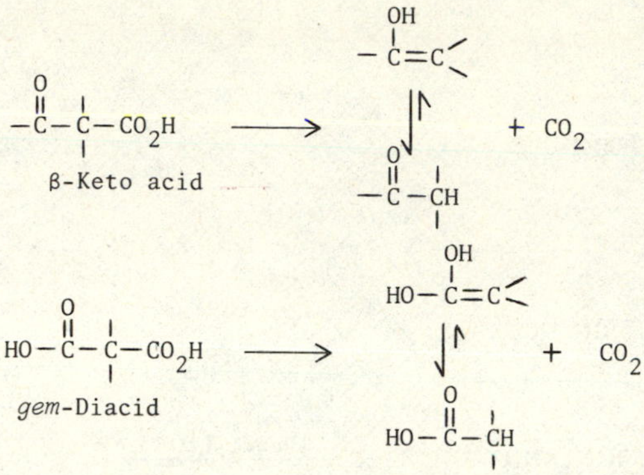

12-21

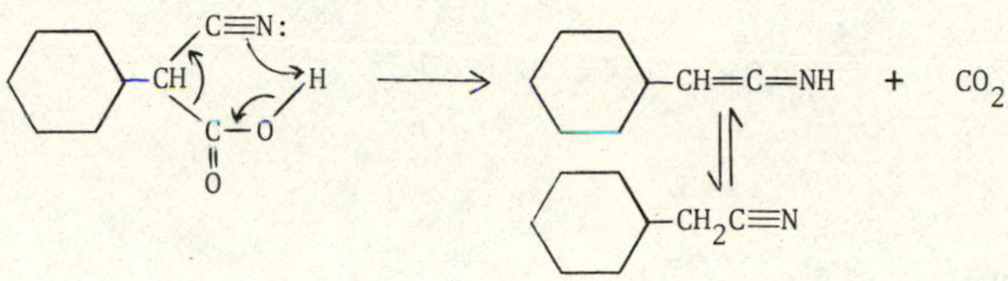

12-22

a) $CH_3CH{=}CHCH{=}CHCH_2CO_2H$

3,5-Heptadienoic acid

b) $C_6H_5CH_2CH(CH_3)CO(CH_2)_3CO_2CH_3$

Methyl 6-methyl-5-oxo-7-phenylheptanoate

c) $C_6H_5COCH(CH_3)CH_2CH_3$

2-Methyl-1-phenyl-1-butanone

d) O_2NCH_3

Nitromethane

e) 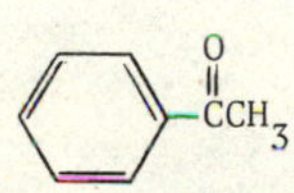

Acetophenone
(1-Phenylethanone)

f) $K^+\ ^-O_2CCH_2CH(CH_3)CH_2CH_2CO_2^-\ K^+$

Dipotassium 3-methylhexanedioate
(Dipotassium 3-methyladipate)

12-23

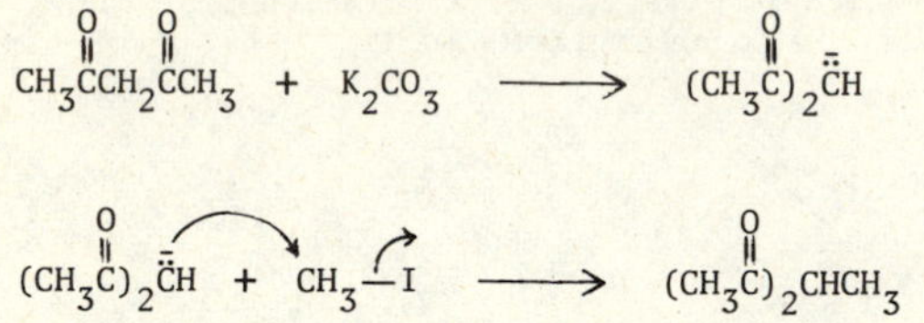

12-24

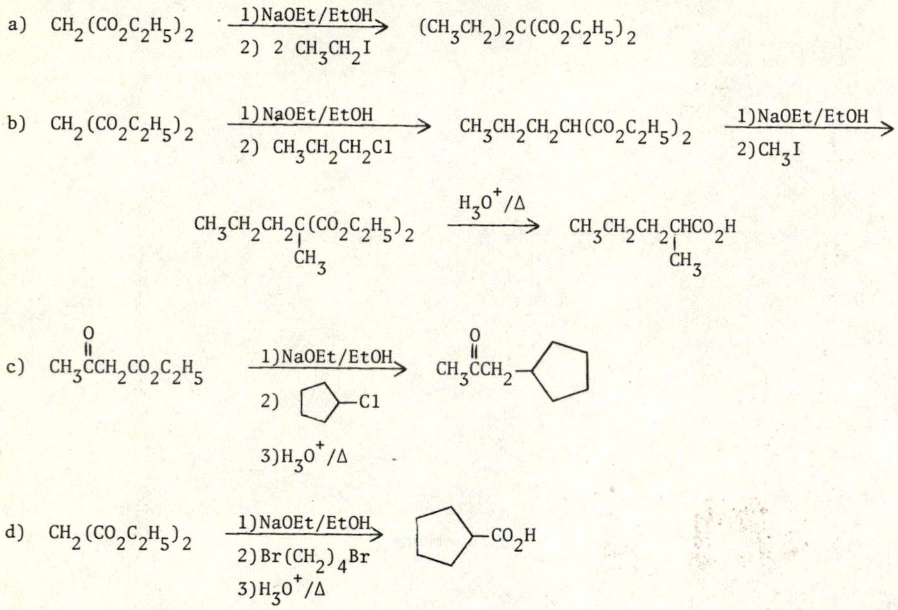

12-25 The S_N2 reaction depends on a good nucleophile. Carbon is less electronegative than an oxygen atom so that the nonbonding electrons of the carbanion are more available to function as a nucleophile than those on the oxyanion. In the S_N1 reaction a reactive carbocationic intermediate is involved. Reaction is less selective and the relative importance of alkylation increases at the more electron rich oxygen atom.

12-26

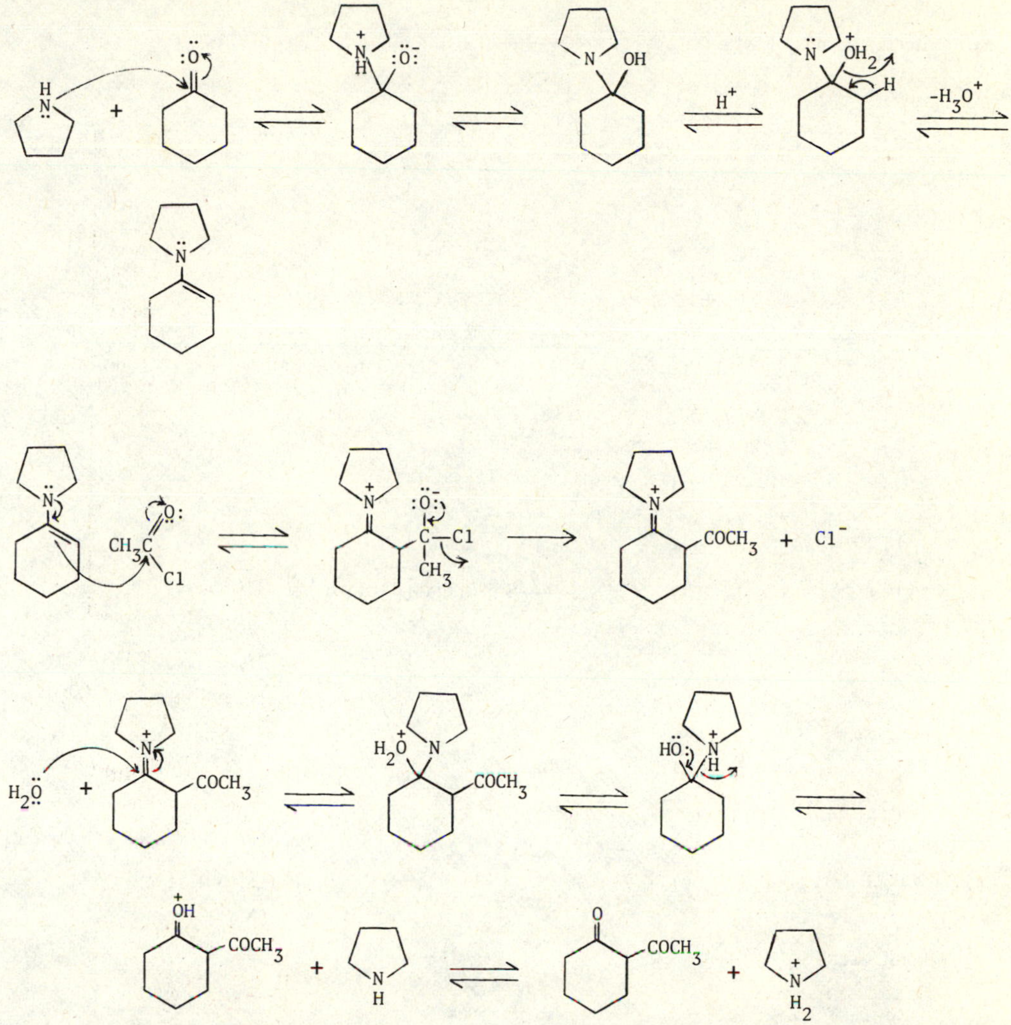

12-27

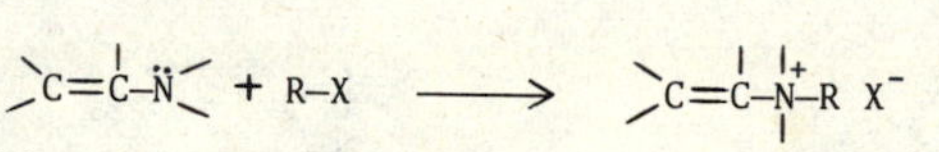

This quaternary ammonium salt could be removed by aqueous extraction from the organic product.

12-28

a) $C_6H_5CH_2CN + Li^+\bar{N}[CH(CH_3)_2]_2 \longrightarrow C_6H_5\ddot{C}HCNLi^+ + HN[CH(CH_3)_2]_2$

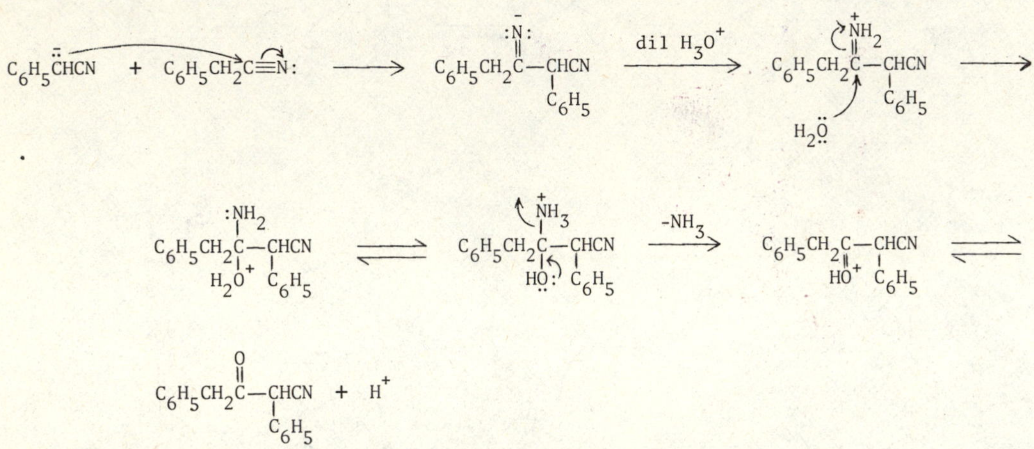

b) 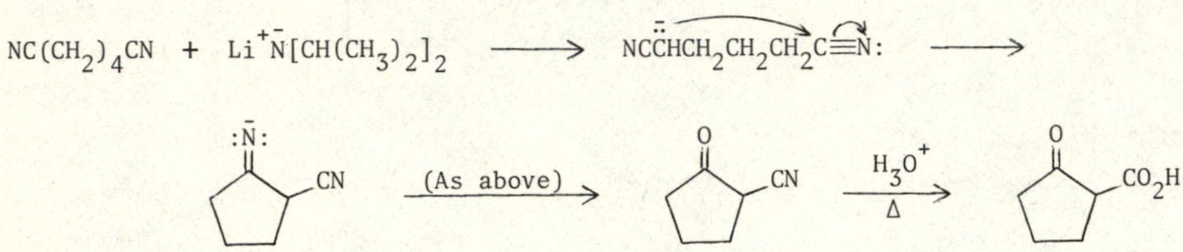

12-29

a) $HOCH_2CH_2NO_2$

b) $(HOCH_2)_3CNO_2$

c) 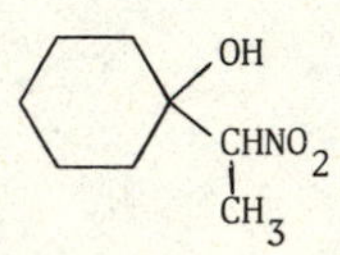

d) $C_6H_5CH{=}\underset{\displaystyle CN}{C}CH(CH_3)_2$

12-30 Nitrogen, a second row element, cannot normally accommodate more than eight electrons in its outer shell.

12-31

a) $(C_6H_5)_3\overset{+}{P}-\bar{C}HC_6H_5 + HCHO \longrightarrow C_6H_5CH{=}CH_2 + (C_6H_5)_3P{=}O$

b) 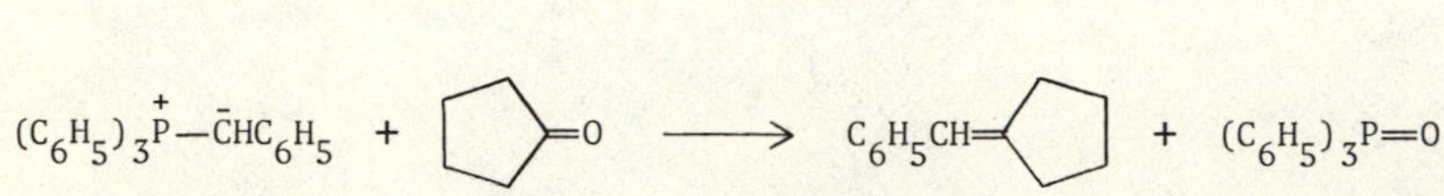

c) $(C_6H_5)_3\overset{+}{P}-\bar{C}HC_6H_5 + C_6H_5CHO \longrightarrow C_6H_5CH{=}CHC_6H_5 + (C_6H_5)_3P{=}O$

Z and *E*

12-32

a) $CH_3CH_2CH_2CHO \xrightarrow{NaOH/H_2O} CH_3CH_2CH(CHO){-}CH(OH)CH_2CH_2CH_3$

b) 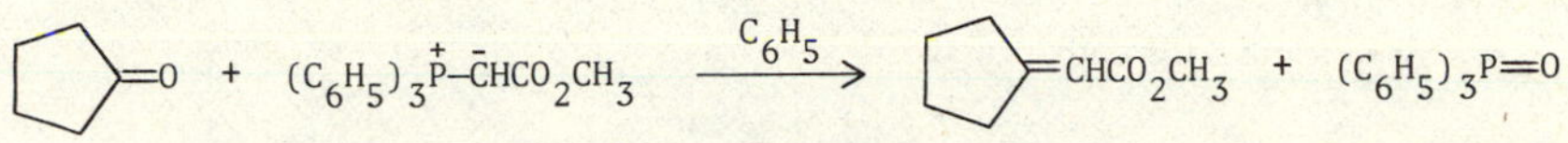

c) $CH_3CH_2COCH_2COCH_3 \xrightarrow[2)\,C_2H_5Br]{1)\,NaOEt/EtOH} CH_3CH_2COCH(C_2H_5)COCH_3$

d) $C_2H_5O_2CCO_2C_2H_5 + 2CH_3CO_2C_2H_5 \xrightarrow{NaOEt/EtOH} C_2H_5O_2CCH_2CO{-}COCH_2CO_2C_2H_5$

e) $CH_2(CN)_2 + CH_3CH_2CO_2C_2H_5 \xrightarrow{NaOEt/EtOH} CH_3CH_2COCH(CN)_2$

f) 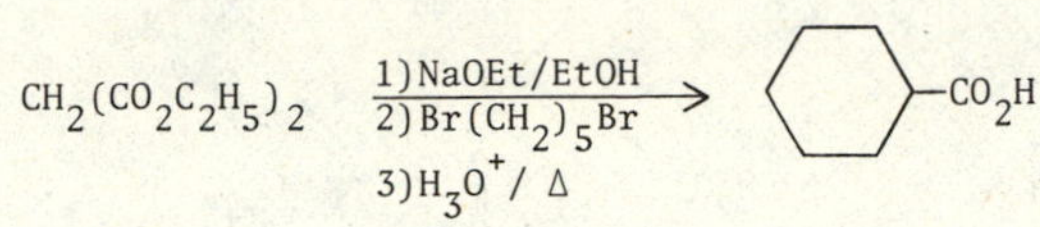

g) 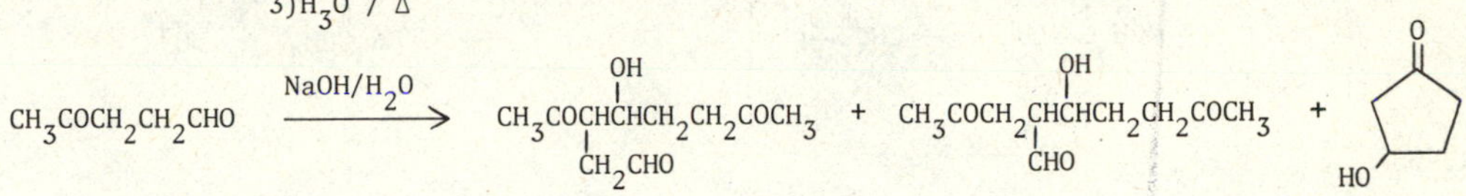

h) $C_6H_5CH_2CO_2CH_3 + HCO_2CH_3 \xrightarrow{NaOMe/MeOH} C_6H_5CH(CHO)CO_2CH_3$

i) $(CH_3)_2CHCOCH_3 \xrightarrow[2)\,CH_3CH_2I]{1)\,LDA/DME} (CH_3)_2CHCOCH_2CH_2CH_3$

j) $C_6H_5COCOC_6H_5 + C_6H_5CH_2COCH_2C_6H_5 \xrightarrow{KOH/EtOH} C_6H_5COC(C_6H_5){=}C(C_6H_5)COCH_2C_6H_5$

k) 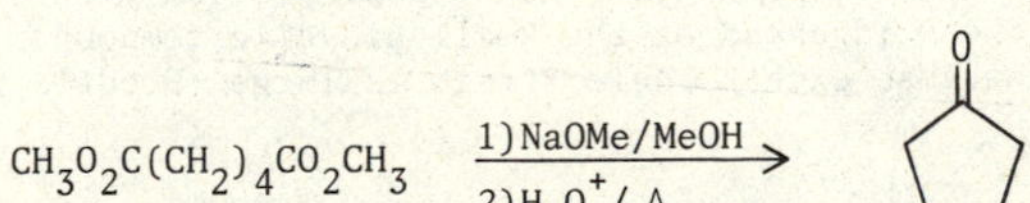

l) $CH_3COCH_2CO_2C_2H_5 \xrightarrow[2)\,ICH_2CO_2C_2H_5]{1)\,NaOEt/EtOH} CH_3COCH(CO_2C_2H_5)CH_2CO_2C_2H_5$

Contd...

12-32 Contd...

m) $C_6H_5COCH_3 + C_6H_5CO_2C_2H_5 \xrightarrow[2)\,H_3O^+]{1)\,NaOEt/EtOH} C_6H_5\overset{O}{\overset{\|}{C}}CH_2\overset{O}{\overset{\|}{C}}C_6H_5$

n) $(CN)_2CH_2 \xrightarrow[2)\,C_6H_5CH_2Cl]{1)\,NaH/Et_2O} (CN)_2CHCH_2C_6H_5$

12-33 Both $\overset{+}{N}$ and $\overset{+}{P}$ can stabilize the adjacent carbanion by an electrostatic inductive effect. In addition the $\overset{+}{P}$ can make use of empty *d*-orbitals to delocalize the negative charge.

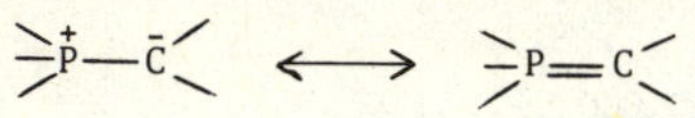

12-34

$$CH_3\overset{O}{\overset{\|}{C}}CH_2CO_2H + \underset{}{\overset{NaOH/H_2O}{\rightleftharpoons}} CH_3\overset{O}{\overset{\|}{C}}-CH_2-C\overset{O}{\underset{:\ddot{O}:^-}{}} \longrightarrow$$

$$\left[CH_3\overset{:\ddot{O}:^-}{\overset{|}{C}}=CH_2 \longleftrightarrow CH_3\overset{:O:}{\overset{\|}{C}}-\ddot{C}H_2^-\right] + CO_2 \xrightarrow{H_2O} CH_3\overset{O}{\overset{\|}{C}}CH_3 + H\ddot{O}:^-$$

12-35

a) A $\equiv C_6H_5CHO + CH_3CH_2\overset{O}{\overset{\|}{C}}CH_2CH_3$

b) B $\equiv$ 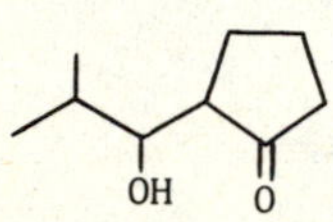

c) C $\equiv (C_2H_5)_2C{=}O + NCCH_2CO_2C_2H_5$

d) D $\equiv C_6H_5\underset{NO_2}{\underset{|}{C}}{=}C(CH_3)_2$

e) E $\equiv (CH_3)_2C{=}CHCH_2CH_2CH{=}C(CH_3)_2$

12-36 A cyclic, six-membered ring transition state proposed for the decarboxylation reaction would require formation of a double bond at the bridgehead of the small bicyclic compound. Such bridgehead double bonds are forbidden because excess angle strain is large (Bredt's rule).

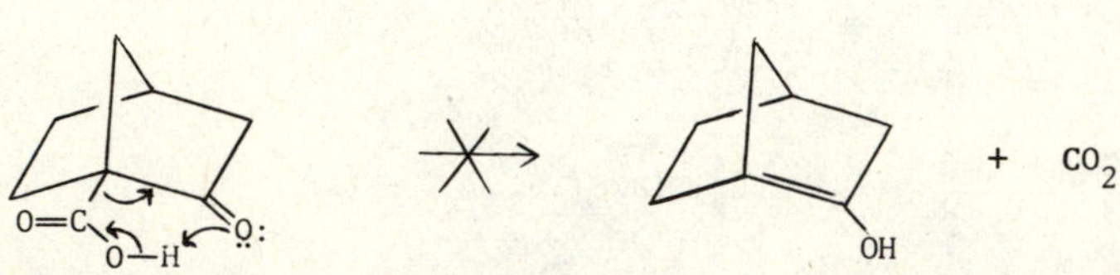

12-37

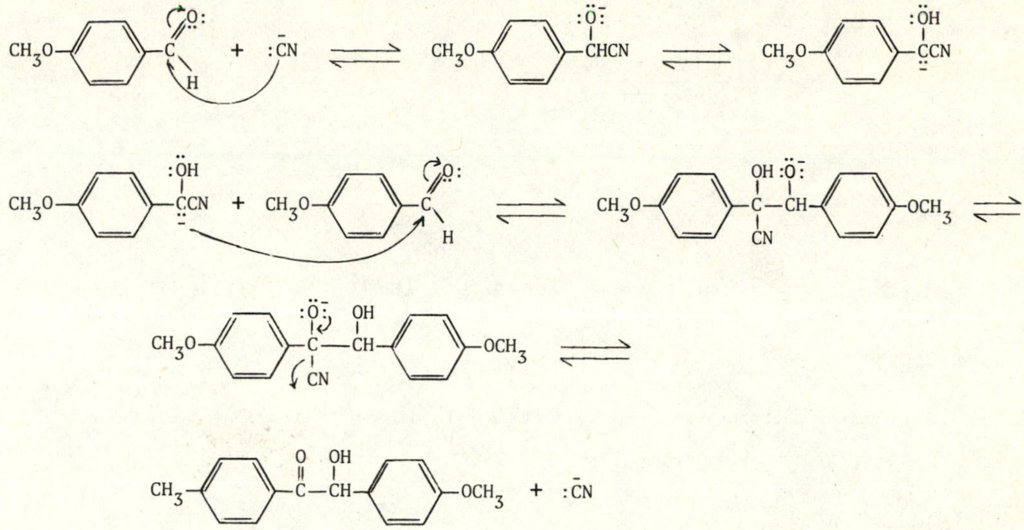

12-38

a)

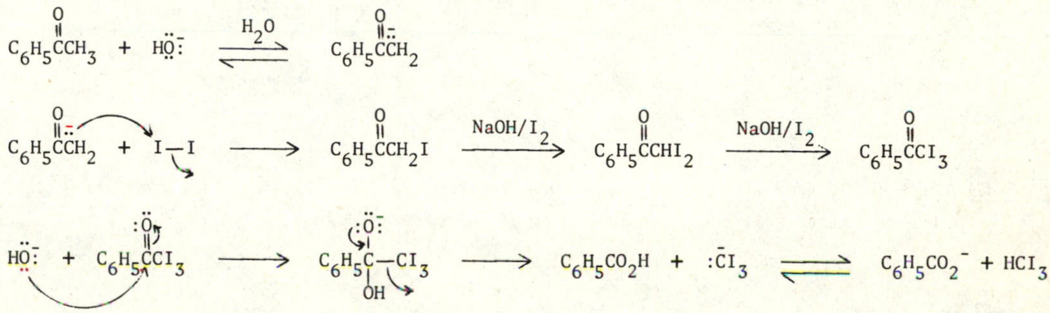

b) Yes; it is a methyl carbonyl compound.

c) $CH_3CH_2CH(OH)CH_3 \xrightarrow[\text{(Oxidation)}]{NaOI} CH_3CH_2C(=O)CH_3 \xrightarrow[I_2]{NaOH/H_2O} CH_3CH_2C(=O)CI_3 \longrightarrow$

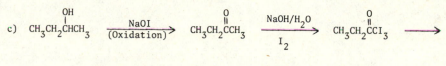

$CH_3CH_2CO_2^-Na + HCI_3$

12-39

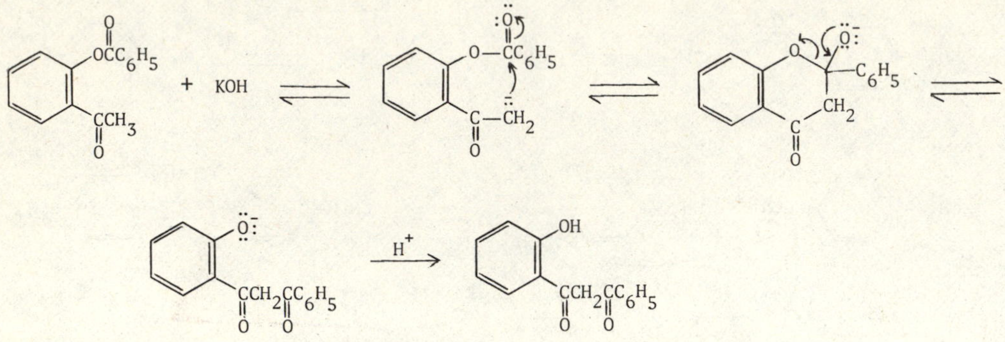

12-40 A Cannizzaro reaction takes place with excess formaldehyde functioning as the reducing agent.

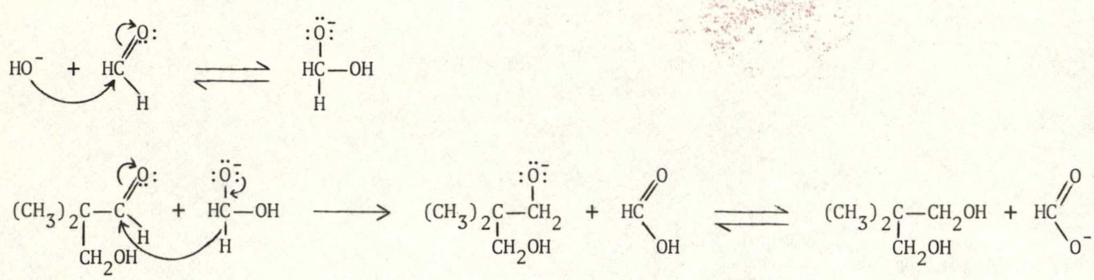

(From prob. 12-7)

12-41 Loss of CO_2 from the carbamic acid involves the departure of an amine. Decarboxylation of the carbamate salt would require loss of the very poor leaving group, $R_2\ddot{N}^-$.

12-42

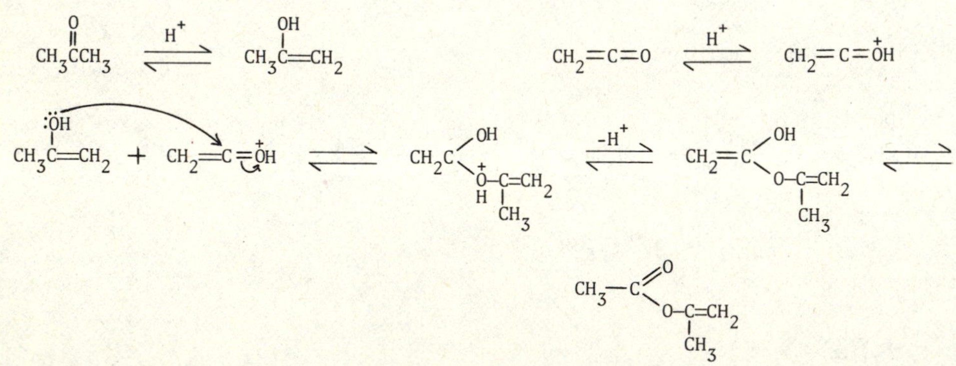

12-43

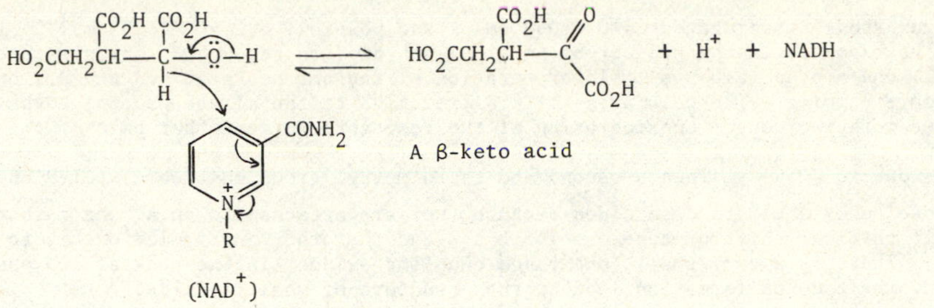

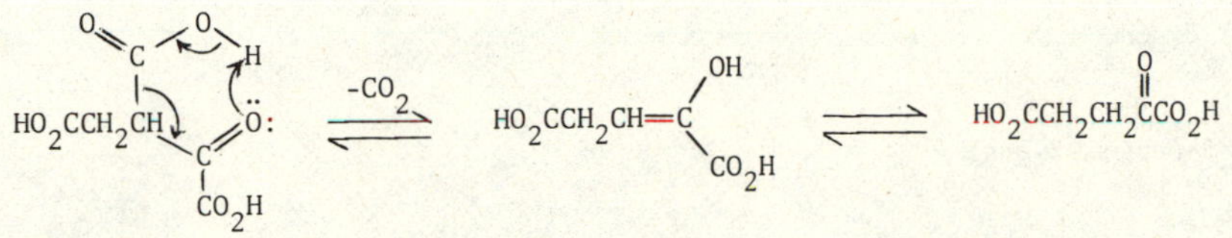

12-44

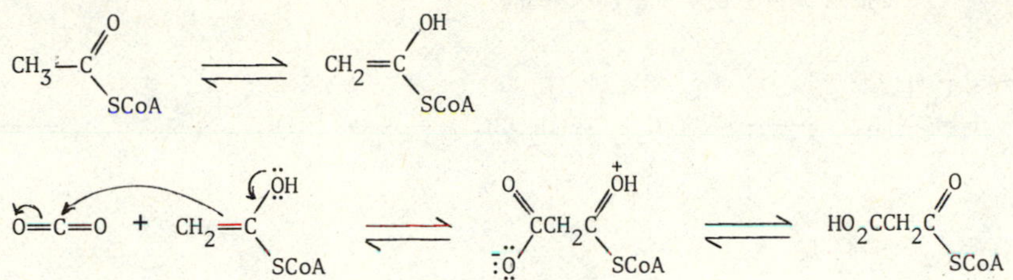

12-45 One equivalent of base forms the more stable enolate anion which is acylated by ethyl benzoate. Two equivalents of strong base produce a dianion.

$$:\bar{C}H_2-\overset{\overset{\displaystyle O}{\|}}{C}-\bar{C}H-CHO$$

Acylation takes place at the less stable (more reactive) terminal carbanion.

12-46 The ^{1}H-nmr shows hydrogen atoms in the aromatic region, but gives little more information about the structure. The two peaks at 680 cm^{-1} and 740 cm^{-1} in the IR spectrum are consistent with a monosubstituted aromatic, and the peak at 1660 cm^{-1} could be a carbonyl shifted by conjugation, or a terminal alkene. Major information for the structure is derived from the expected reaction, an aldol addition-dehydration.

$$C_6H_5CHO + CH_3\overset{\overset{\displaystyle O}{\|}}{C}C_6H_5 \xrightarrow{NaOH/H_2O} C_6H_5CH{=}CH\overset{\overset{\displaystyle O}{\|}}{C}C_6H_5 + H_2O$$

Chalcone
(1,3-Diphenyl-2-propen-1-one)

The IR peak at 1660 cm^{-1} is thus assigned to the conjugated carbonyl. (The carbon-carbon double bond stretching absorption would be very weak for this internal alkene.) The nmr spectrum shows that the alkene hydrogens are shifted into the aromatic region by conjugation.

12-47

a) The IR spectrum shows a conjugated C=O (1690 cm^{-1}) and possibly a C=C (1620 cm^{-1}). Those groups would be expected in the product of an aldol-dehydration sequence. The nmr spectrum confirms an alkene proton (6.0 ppm). Interpretation of the nmr peaks at 1.9 and 2.1 ppm is a little more complex. The peak areas of 6:3 (relative to the alkene proton) suggest a total of three methyl groups. Consideration of the reactants suggests that part of the peak at 2.1 ppm is due to $CH_3\overset{O}{\overset{\|}{C}}$—. Then a second and third methyl group must come at 1.9 and 2.1 ppm. Those peaks are both deshielded because they are attached to an alkene carbon but differ in chemical shift because one is in a *Z* and the other an *E* relationship to the acetyl group. There is a very small long range coupling evident in the peak at 1.9 ppm. These data and the mode of formation lead to the structure of mesityl oxide.

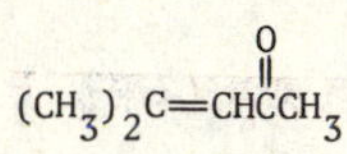

Mesityl oxide

(4-Methyl-3-penten-2-one)

b)

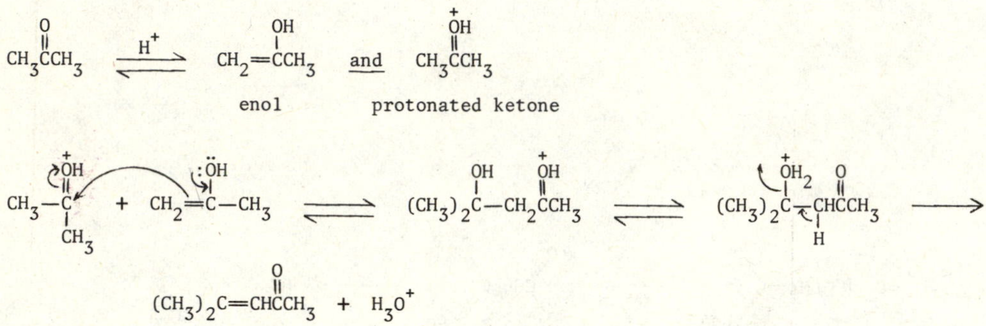

12-48 2,4-Pentanedione is a 1,3-dicarbonyl compound, thus exists in equilibrium with its enol. Both forms are evident in the spectra.

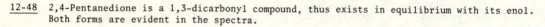

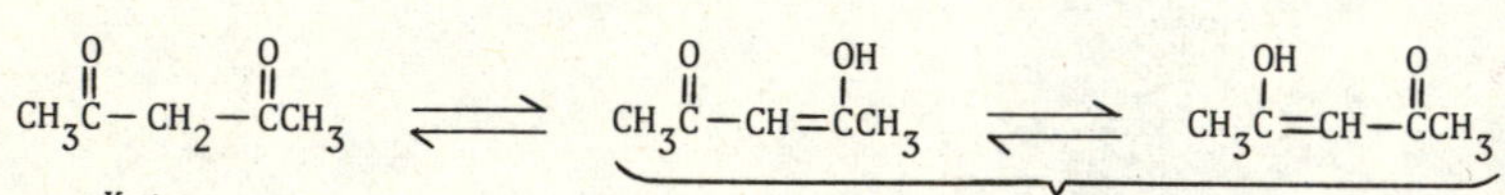

The keto form is expected to show an ^1H-nmr peak at about 2 ppm (2.2 ppm; CH_3CO) and a deshielded methylene $-COCH_2CO-$ at about 3.6 ppm. The equivalent enol forms will show the terminal methyls (2.0 ppm) and two hydrogen atoms (5.5 ppm) which equilibrate between the hydroxy and alkene positions.

In the ^{13}C-nmr spectrum the keto form shows three different carbon atoms CH_3 (30 ppm), CH_2 (58 ppm), and CO (202 ppm). The enol form has CH_3 (24 ppm), =CH (100 ppm), and the equivalent carbonyl-carbinol carbon atoms (191 ppm).

The spectra support the assumption that 1,3-dicarbonyl compounds favor their enol form.

12-49 The IR spectrum suggests —OH and C=O, structural features consistent with formation of the product via a base promoted aldol reaction. The ^{13}C-nmr spectrum suggest five different kinds of carbon atoms and the ^{1}H-nmr integral areas suggest twelve hydrogens. If we assume a molecular formula of $C_5H_{12}O_2$ we are 12 units short of the molecular weight (M^+ = 116). A molecular formula of $C_6H_{12}O_2$ is suggested. This is consistent with two identical CH_3 groups, which is why there are only five carbon resonance peaks. The ^{1}H-nmr singlet at 1.3 ppm and an area of six confirms this. The formula has IHD = 1 consistent with the carbonyl group. The D_2O experiment with the ^{1}H-nmr confirms that the peak at 4.0 ppm is —OH, and the fact that all proton resonance peaks are singlets indicates none of their carbon atoms are next to each other. The compound and its preparation are -

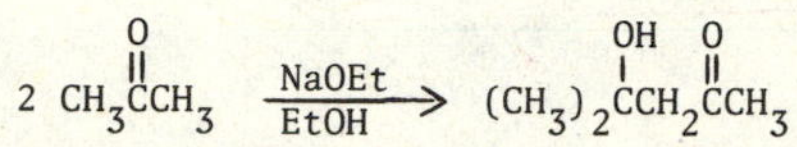

The ^{13}C-nmr coupling data (carbon atoms split by attached hydrogens) can be used to confirm chemical shift assignments: CH_3's - 29.4 and 31.8 ppm; CH_2 - 54.7 ppm; COH - 69.6 ppm; C=O - 210 ppm.

13 Nucleophilic Additions and Substitutions in Synthesis

13-1 Refer to previous chapters in textbook for specific examples.

13-2 The yield of a multistep synthesis is the mathematical product of the yields of each individual step.

Yield = .93 x .85 x .87 (x 100) = 69%

13-3 Refer to problem 13-1 and textbook.

13-4 The following are representative of the types of compounds that can be prepared from methanol or its derivatives.

C_1 compounds:

$$CH_3OH + \left\{\begin{array}{l} HX \\ \text{or} \\ PX_3 \\ \text{or} \\ SOCl_2 \end{array}\right. \longrightarrow CH_3X \qquad X = \text{halogen}$$

$$CH_3X + NH_3 \longrightarrow CH_3NH_2$$

$$CH_3X \xrightarrow[2)H_3O^+]{1)Mg/Et_2O} CH_4$$

$$CH_3OH + \begin{array}{c} NaOCl \\ \text{or} \\ CrO_3\cdot\text{pyridine} \end{array} \longrightarrow HCHO \xrightarrow[\text{Oxidation}]{\text{Further}} HCO_2H$$

$$HCO_2H + NH_3 \xrightarrow{\Delta} HC(=O)NH_2$$

$$CH_3X + H_2S \longrightarrow CH_3SH$$

Contd...

13-4 contd...

C_2 compounds:

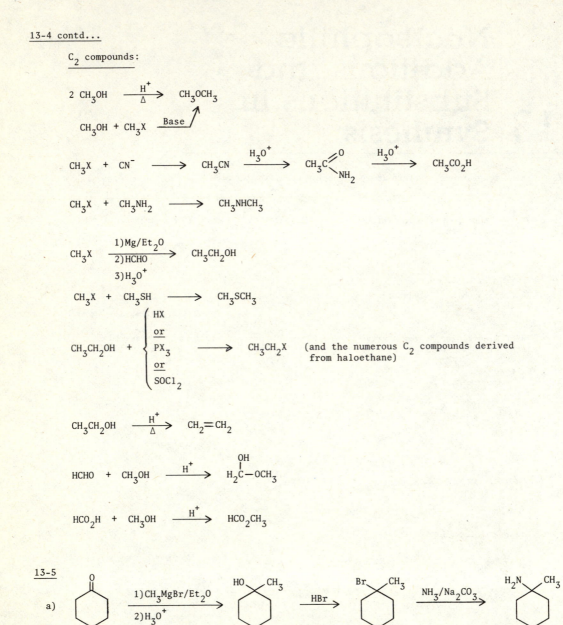

$$2\ CH_3OH \xrightarrow[\Delta]{H^+} CH_3OCH_3$$

$$CH_3OH + CH_3X \xrightarrow{Base} CH_3OCH_3$$

$$CH_3X + CN^- \longrightarrow CH_3CN \xrightarrow{H_3O^+} CH_3C(=O)NH_2 \xrightarrow{H_3O^+} CH_3CO_2H$$

$$CH_3X + CH_3NH_2 \longrightarrow CH_3NHCH_3$$

$$CH_3X \xrightarrow[\substack{2)\,HCHO \\ 3)\,H_3O^+}]{1)\,Mg/Et_2O} CH_3CH_2OH$$

$$CH_3X + CH_3SH \longrightarrow CH_3SCH_3$$

$$CH_3CH_2OH + \left\{\begin{array}{l} HX \\ \text{or} \\ PX_3 \\ \text{or} \\ SOCl_2 \end{array}\right. \longrightarrow CH_3CH_2X$$

(and the numerous C_2 compounds derived from haloethane)

$$CH_3CH_2OH \xrightarrow[\Delta]{H^+} CH_2{=}CH_2$$

$$HCHO + CH_3OH \xrightarrow{H^+} H_2C(OH){-}OCH_3$$

$$HCO_2H + CH_3OH \xrightarrow{H^+} HCO_2CH_3$$

13-5

a)

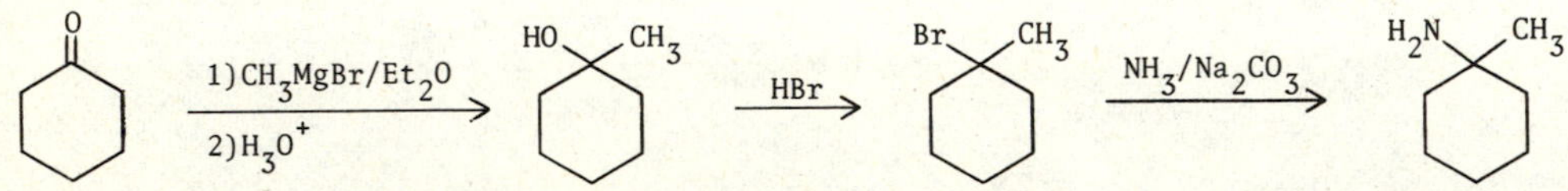

b) By comparing the target to the starting material, we see that a construction (carbon-carbon bond formation) must be accomplished at the ring carbon atom that originally was a carbonyl. An amino group must also be introduced at that same position. Amino can be introduced by a substitution so that a possible last synthetic step is established. That precursor must have a reasonable leaving group. We chose bromide (a halogen) because it can be readily introduced by substitution for an alcohol hydroxy. The hydroxy is the transition between the unsaturated carbonyl and the tetracoordinate atom of the target.

13-6

a) This strategy is based on the fact that an anion can be formed on either side of the ketone carbonyl. With excess strong base a dianion is formed in which the less stable terminal anion is the more reactive. That enables a nucleophilic substitution to be used to add the necessary isopropyl end group.

b) Major competing reactions would be anion addition or substitution at the carbonyls and elimination of the haloalkane.

13-7

a) $C_6H_5CH_2CN + CH_3CH_2MgBr \xrightarrow[2)H_3O^+]{1)Et_2O} C_6H_5CH_2\overset{O}{\overset{\|}{C}}CH_2CH_3$

b) $2\ CH_3CHO \xrightarrow[\Delta]{NaOH/H_2O} CH_3CH{=}CHCHO \xrightarrow[2)H_3O^+]{1)C_2H_5MgBr/Et_2O} CH_3CH{=}CH\overset{OH}{\overset{|}{C}}HCH_2CH_3$

c) $C_6H_5CH_2CN \xrightarrow[2)H_3O^+]{1)LiAlH_4/Et_2O} C_6H_5CH_2CH_2NH_2$

d) $2\ CH_3CHO \xrightarrow[\Delta]{NaOH/H_2O} CH_3CH{=}CHCHO \xrightarrow[2)H_3O^+]{1)NaBH_4/THF} CH_3CH{=}CHCH_2OH$

e) $2\ CH_3\overset{O}{\overset{\|}{C}}CH_3 \xrightarrow[\Delta]{NaOEt/EtOH} (CH_3)_2C{=}CH\overset{O}{\overset{\|}{C}}CH_3 \xrightarrow[2)CH_3COCH_3\ \ 3)\ \Delta]{1)NaOEt/EtOH}$

Not isolated

$(CH_3)_2C{=}CH\overset{O}{\overset{\|}{C}}CH{=}C(CH_3)_2 \xrightarrow[2)H_3O^+]{1)NaBH_4/THF} (CH_3)_2C{=}CH\overset{OH}{\overset{|}{C}}HCH{=}C(CH_3)_2$

f) $C_6H_5CHO + CH_3CHO \xrightarrow{NaOH/H_2O} C_6H_5CH{=}CHCHO \xrightarrow{CH_3CHO/NaOH/H_2O}$

Not isolated

$C_6H_5CH{=}CHCH{=}CHCHO \xrightarrow[2)H_3O^+]{1)NaBH_4/THF} C_6H_5CH{=}CHCH{=}CHCH_2OH$

g) $CH_3CHO + 2\ C_2H_5SH \xrightarrow[(-H_2O)]{H^+} CH_3CH(SC_2H_5)_2$

13-8

a) $C_6H_5CO_2CH_3 + 2\ CH_3MgBr \xrightarrow[2)H_2O/NH_4Cl]{1)Et_2O} C_6H_5C(OH)(CH_3)_2$

b)

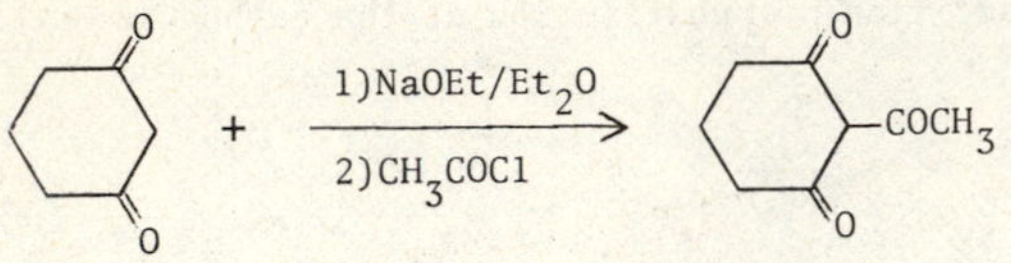

c) $CH_3CO_2H + C_2H_5OH \xrightarrow{H^+} CH_3CO_2C_2H_5$

$2\ CH_3CO_2C_2H_5 \xrightarrow{NaOEt/EtOH} CH_3COCH_2CO_2C_2H_5 \xrightarrow[2)H_3O^+]{1)LiAlH_4/Et_2O} CH_3CHOHCH_2CH_2OH$

d) $(CH_3)_2CHCH_2CO_2H \xrightarrow[2)CH_3NH_2]{1)SOCl_2} (CH_3)_2CHCH_2C(=O)NHCH_3$

e) $HO_2CCO_2H + C_2H_5OH(excess) \xrightarrow{H^+} C_2H_5O_2CCO_2C_2H_5$

$CH_3CH_2CH_2CO_2C_2H_5 \xrightarrow[2)C_2H_5O_2CCO_2C_2H_5]{1)\ LDA/THF} CH_3CH_2CH(CO_2C_2H_5)COCO_2C_2H_5$

f)

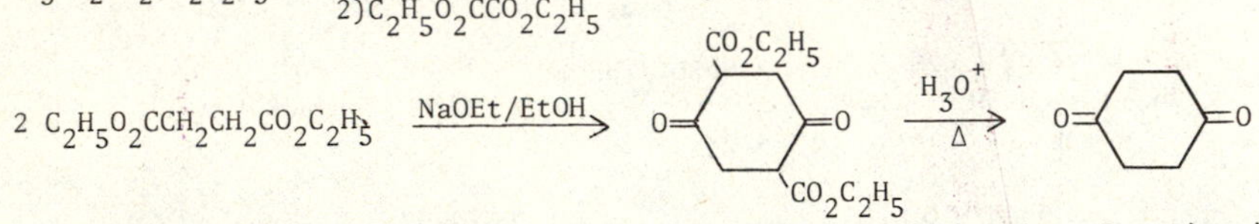

Cyclization to form a six-membered ring is relatively favorable. Reaction is carried out in very dilute solution so that the initial substitution is followed by intra rather than intermolecular reaction.

13-9

a) In the Knoevenagel condensation the α-hydrogen atoms are activated by two ester groups. An enolate anion is formed under very mild basic conditions so that self condensation of the aldehyde is minimal. Heat promotes dehydration without isolation of the initial adduct.

b) In the Perkin reaction the carboxylate salt functions as the base which abstracts an α-hydrogen atom from the anhydride. Hydrolysis provides a cinnamic acid (a 3-phenyl-2-propenoic acid) in the most common examples of the Perkin reaction.

c) In the Stobbe condensation the anion formed from diethyl succinate adds to a ketone. The intermediate undergoes an intramolecular substitution on the more remote ester group to give a five-membered lactone. Reaction of the lactone with base generates a new carbanion which leads to a relatively stable carboxylate anion on ring opening. This type of mechanistic sequence is required to account for hydrolysis of only one ester group.

$C_2H_5O_2CH_2CH_2CO_2C_2H_5 + (CH_3)_3CO^- \rightleftharpoons C_2H_5O_2CH_2\ddot{C}HCO_2C_2H_5 + (CH_3)_3COH$

Contd...

13-9 Contd...

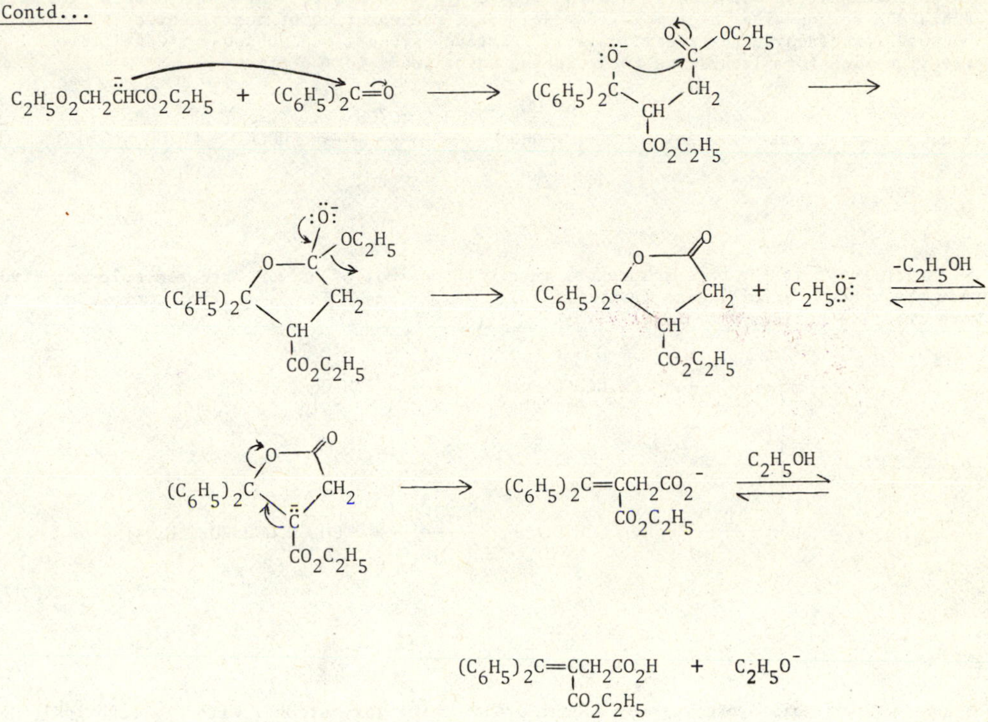

13-10

a)

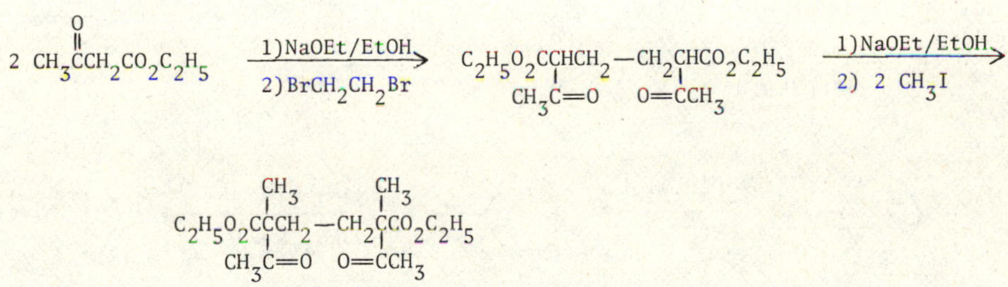

A

b)

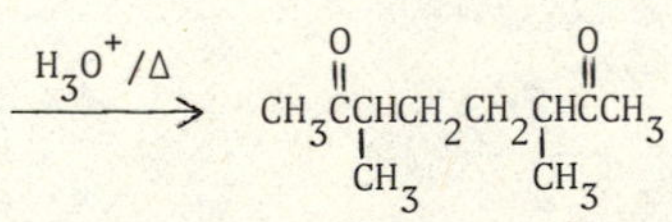

B

13-11 Three successive aldol reactions with formaldehyde replace the three α-hydrogen atoms of acetaldehyde and give the triol-aldehyde. Self condensation of acetaldehyde is minimal because formaldehyde has a particularly reactive carbonyl. A subsequent Cannizzaro reaction with formaldehyde as the reducing agent leads to the product.

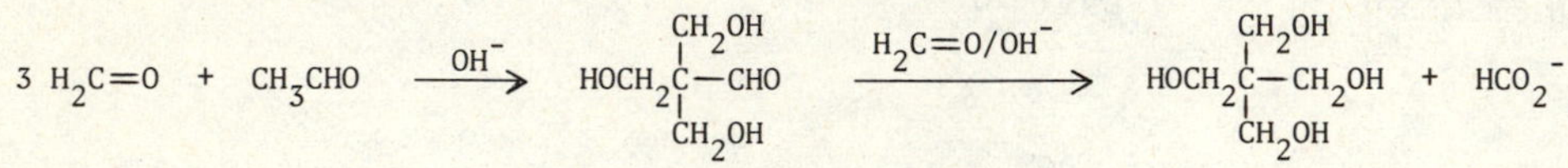

13-12 The anion forms at the more acidic position with one mol of base. When two moles of strong base are used, a second, less favorable anion also forms. The less stable carbanion is the more reactive nucleophile and is alkylated first.

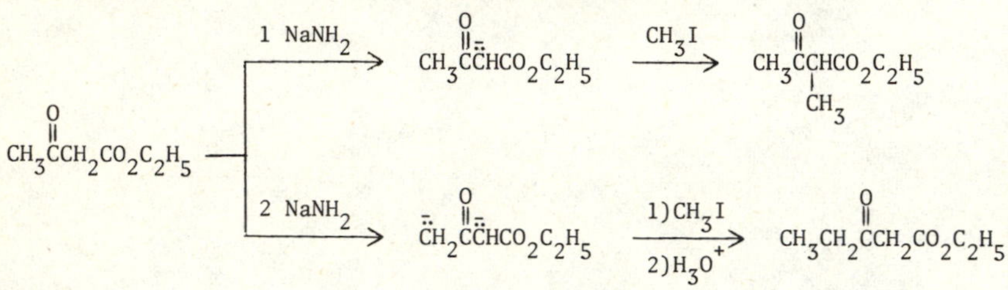

13-13

a) In approaching this synthesis we recognize that a tertiary alcohol with two identical alkyl groups can be prepared from a Grignard reagent plus an ester. The ester, methyl cyclohexylcarboxylate, is a substituted acetate. It can be prepared from diethylmalonate.

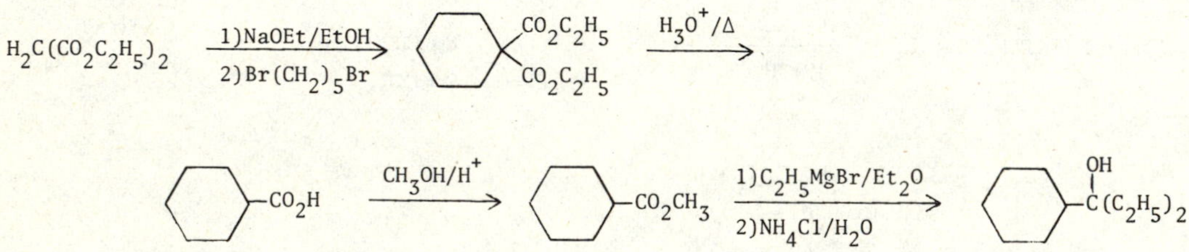

b) We recognize that the requisite tertiary alcohol with three different groups could come from Grignard addition to a ketone. The ketone precursor to the alcohol product is a substituted acetone.

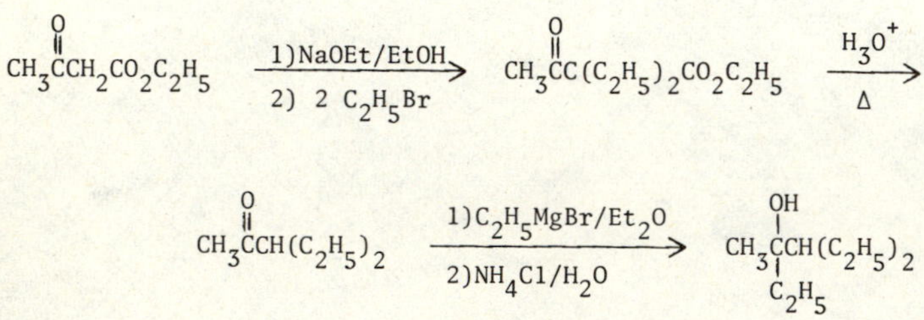

Contd...

13-13 Contd...

c) To minimize steric problems, add the larger group (CH_3CH_2) first, then the smaller group (CH_3).

$$CH_3\overset{O}{\overset{\|}{C}}CH_2CO_2C_2H_5 \xrightarrow[\text{2)}CH_3CH_2Br]{\text{1)NaOEt/EtOH}} CH_3\overset{O}{\overset{\|}{C}}-\overset{CH_2CH_3}{\underset{}{C}}HCO_2C_2H_5 \xrightarrow[\text{2)}CH_3I]{\text{1)NaOEt/EtOH}}$$

$$CH_3\overset{O}{\overset{\|}{C}}-\overset{CH_2CH_3}{\underset{CH_3}{C}}CO_2C_2H_5 \xrightarrow[\Delta]{H_3O^+} CH_3\overset{O}{\overset{\|}{C}}\overset{}{\underset{CH_3}{C}}HCH_2CH_3$$

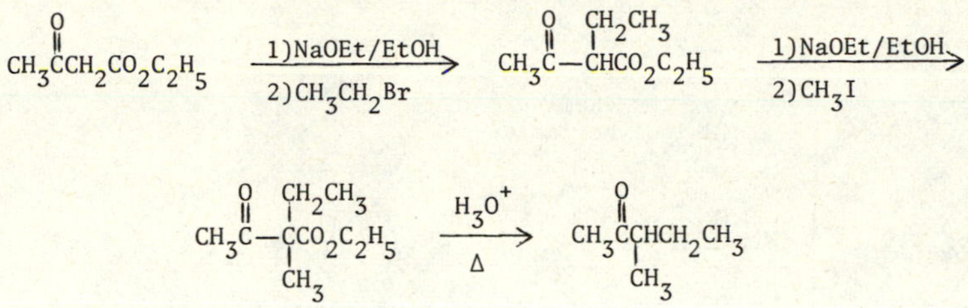

d) $CH_2(CO_2C_2H_5)_2 \xrightarrow[2)\ 2\ C_2H_5Br]{1)NaOEt/EtOH} (C_2H_5)_2C(CO_2C_2H_5)_2 \xrightarrow[\Delta]{H_3O^+} (C_2H_5)_2CHCO_2H$

e) $CH_2(CO_2C_2H_5)_2 \xrightarrow[2)C_6H_5CH_2Cl]{1)NaOEt/EtOH} C_6H_5CH_2CH(CO_2C_2H_5)_2 \xrightarrow{H_3O^+/\Delta}$

$$C_6H_5CH_2CH_2CO_2H \xrightarrow{PBr_3} C_6H_5CH_2\overset{Br}{\overset{|}{C}}HCO_2H$$

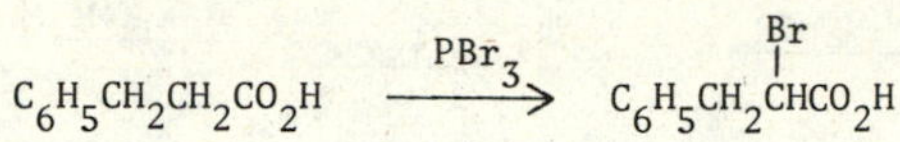

13-14

$$(C_2H_5O)_3P + C_6H_5CH_2Cl \longrightarrow (C_2H_5O)_3\overset{+}{P}CH_2C_6H_5\ Cl^- \longrightarrow (C_2H_5O)_2\overset{O}{\overset{\|}{P}}CH_2C_6H_5 + C_2H_5Cl$$
$$ A B$$

$$B \xrightarrow[DMF]{NaOMe} (C_2H_5O)_2\overset{O}{\overset{\|}{P}}-\overset{..}{C}HC_6H_5 \xrightarrow{C_6H_5CHO} C_6H_5CH{=}CHC_6H_5 + (C_2H_5O)_2\overset{O}{\overset{\|}{P}}-O^-Na^+$$
$$ C$$

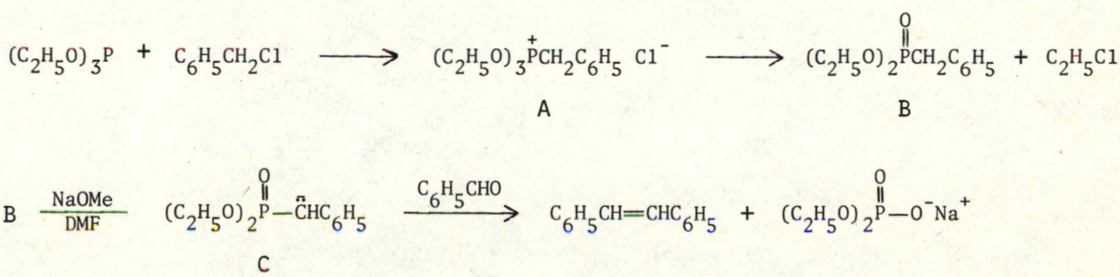

The phosphonium salt A is unstable and forms B by the Arbusov reaction. Addition of base forms the phosphonate ylid which undergoes a Wittig type reaction with benzaldehyde.

13-15

a) $C_6H_5CHO + \begin{matrix} NaBH_4 \\ \text{or} \\ LiAlH_4 \end{matrix} \xrightarrow[\text{2)}H_3O^+]{\text{1)}Et_2O} C_6H_5CH_2OH \xrightarrow{SOCl_2} C_6H_5CH_2Cl \xrightarrow{Mg/Et_2O}$

$$C_6H_5CH_2MgCl \xrightarrow[\text{2)}H_3O^+]{\text{1)}H_2C\overset{O}{\overset{\diagup\diagdown}{\longrightarrow}}CH_2/Et_2O} C_6H_5CH_2CH_2CH_2OH$$

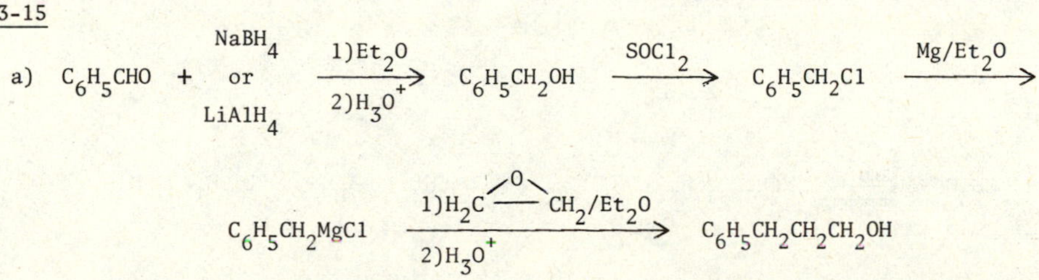

Contd...

13-15 Contd...

b) 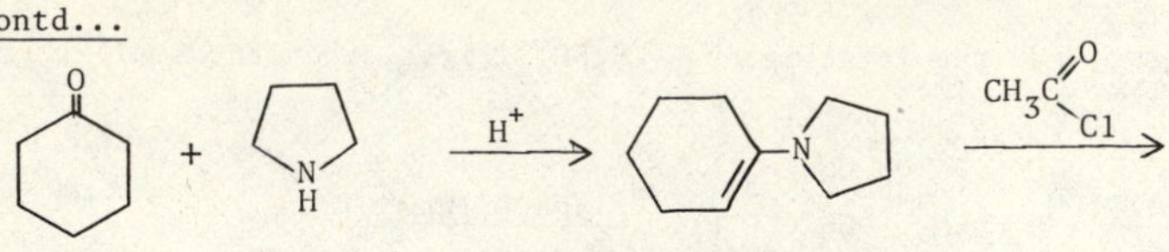

(Use of the enamine avoids polyacylation)

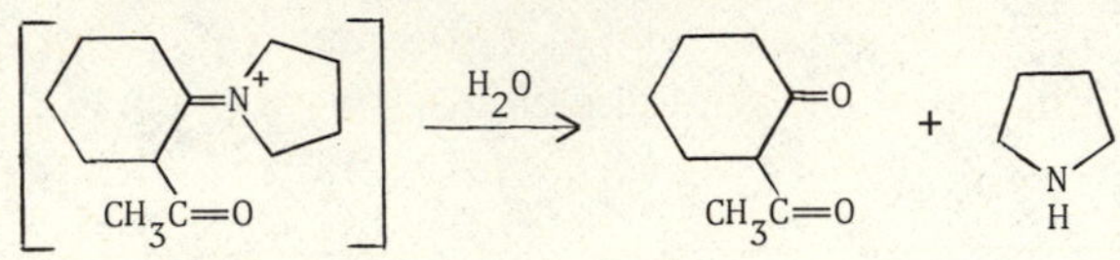

c) 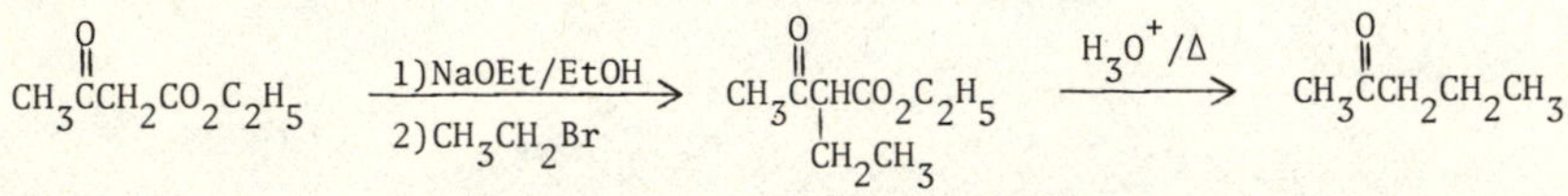

d) 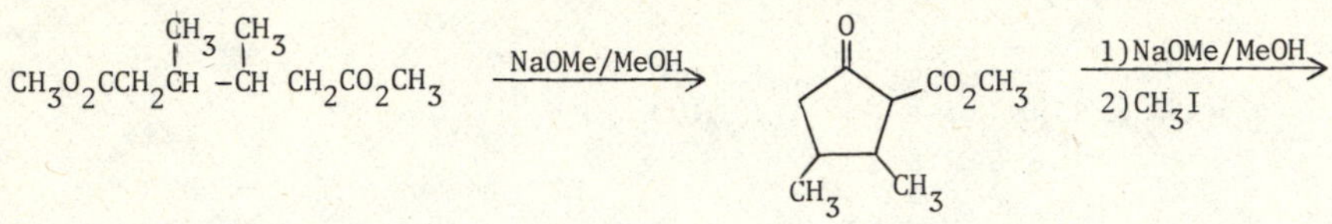

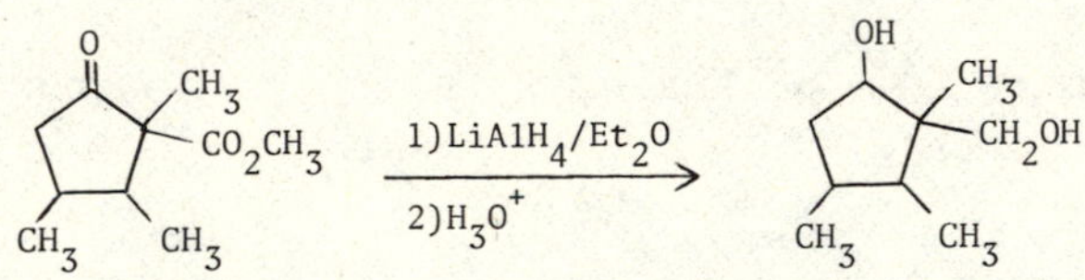

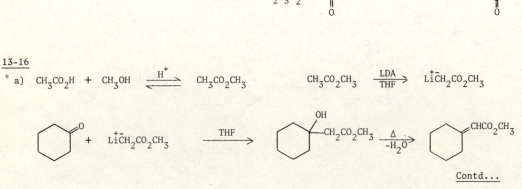

e) $HO_2CCH_2CH_2CO_2H + C_2H_5OH(\text{excess}) \xrightarrow{H_3O^+} C_2H_5O_2CCH_2CH_2CO_2C_2H_5$

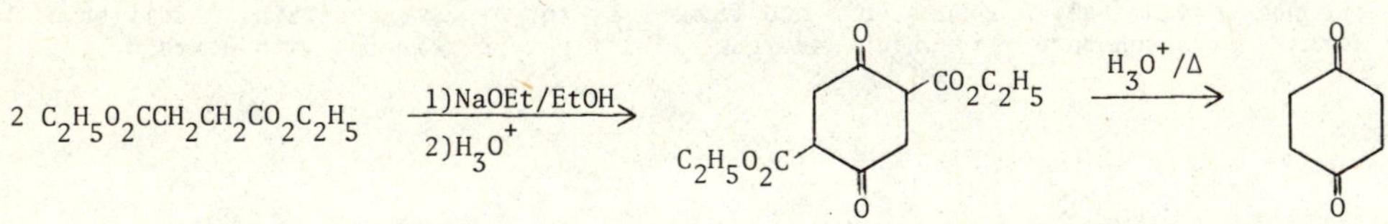

13-16

a) $CH_3CO_2H + CH_3OH \underset{}{\overset{H^+}{\rightleftharpoons}} CH_3CO_2CH_3$ $\qquad CH_3CO_2CH_3 \xrightarrow[\text{THF}]{\text{LDA}} \overset{+}{Li}\overset{-}{C}H_2CO_2CH_3$

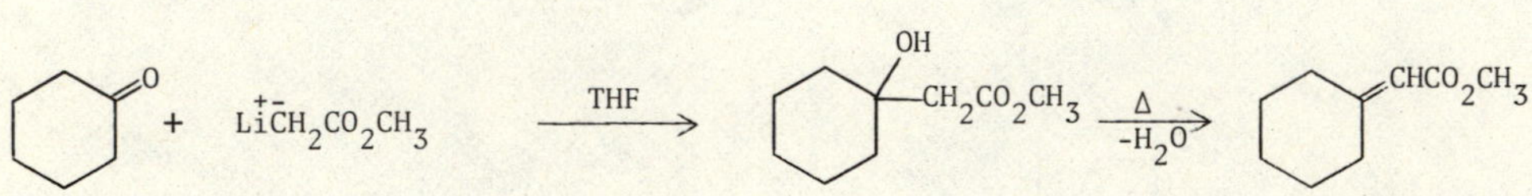

Contd...

13-16 contd...

b) $HO_2CCO_2H + 2\ CH_3OH \xrightarrow{H^+} CH_3O_2CCO_2CH_3$

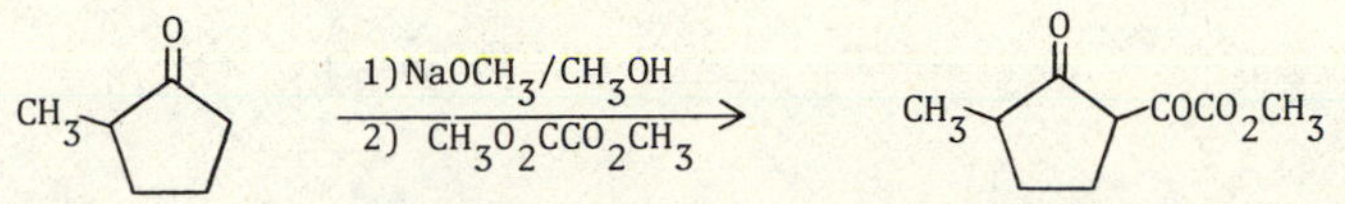

Steric factors favor formation of this isomer. Also, this 1,3-dicarbonyl activated α-H shifts the equilibrium whereas the other isomer has no similar opportunities.

c) $CH_3CO_2Cl + C_2H_5OH \longrightarrow CH_3CO_2C_2H_5$

$CH_3CO_2C_2H_5 + CH_3\overset{O}{\overset{\|}{C}}CH_3 \xrightarrow{NaOET/EtOH} CH_3\overset{O}{\overset{\|}{C}}CH_2\overset{O}{\overset{\|}{C}}CH_3$

$CH_3\overset{O}{\overset{\|}{C}}CH_2\overset{O}{\overset{\|}{C}}CH_3 + CH_3CO_2C_2H_5 \xrightarrow{NaOEt/EtoOH} (CH_3CO)_3CH$

d) $CH_3CH_2Br \xrightarrow[Et_2O]{Mg} CH_3CH_2MgBr$

$(CH_3CH_2)_2CO + C_6H_5CHO \xrightarrow{NaOEt/EtOH} CH_3CH_2\overset{O}{\overset{\|}{C}}\underset{CH_3}{\underset{|}{C}}{=}CHC_6H_5$

$CH_3CH_2MgBr + Li \xrightarrow{Et_2O} CH_3CH_2Li$

$CH_3CH_2Li + CH_3CH_2\overset{O}{\overset{\|}{C}}\underset{CH_3}{\underset{|}{C}}{=}CHC_6H_5 \xrightarrow[2)\ NH_4Cl/H_2O]{1)\ Et_2O} CH_3CH_2\overset{OH}{\overset{|}{C}}CH_2CH_3$ with $CH_3C{=}CHC_6H_5$ attached below the carbinol carbon

e) $2\ CH_3CH_2CHO \xrightarrow[\Delta]{NaOH/H_2O} CH_3CCHO$ with $=CHCH_2CH_3$ below ($CH_3\overset{\|}{C}CHO$ over CH_3CH_2CH)

$CH_3\underset{\|}{C}CHO$ (CH_3CH_2CH below) $+ HCN \longrightarrow CH_3\underset{\|}{C}{-}\overset{OH}{\overset{|}{C}}HCN$ (CH_3CH_2CH below) $\xrightarrow{H_3O^+} CH_3\underset{\|}{C}\overset{OH}{\overset{|}{C}}HCO_2H$ (CH_3CH_2CH below)

f) $2\ CH_3CO_2C_2H_5 \xrightarrow{NaOEt/EtOH} CH_3\overset{O}{\overset{\|}{C}}CH_2CO_2C_2H_5 \xrightarrow[2)H_3O^+]{1)LiAlH_4/Et_2O} CH_3\overset{OH}{\overset{|}{C}}HCH_2CH_2OH$

13-17

a)

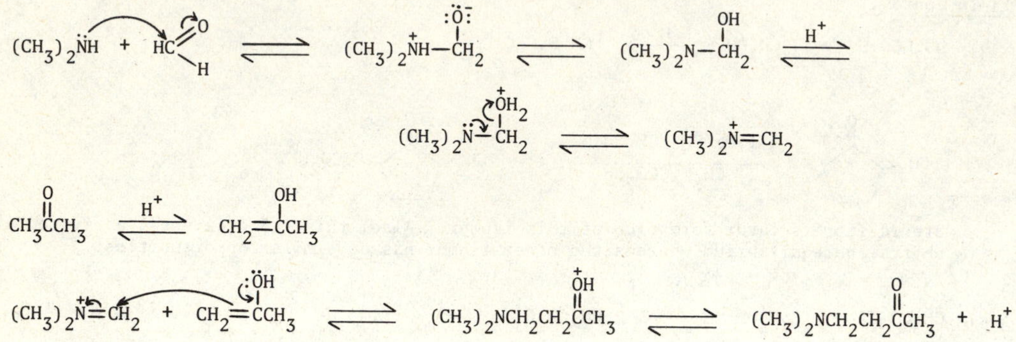

b) The reactants fit togethter in the following way:

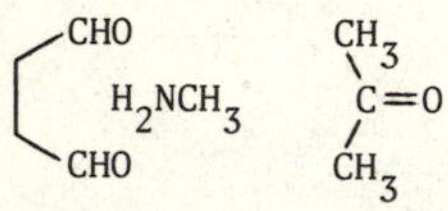

c)

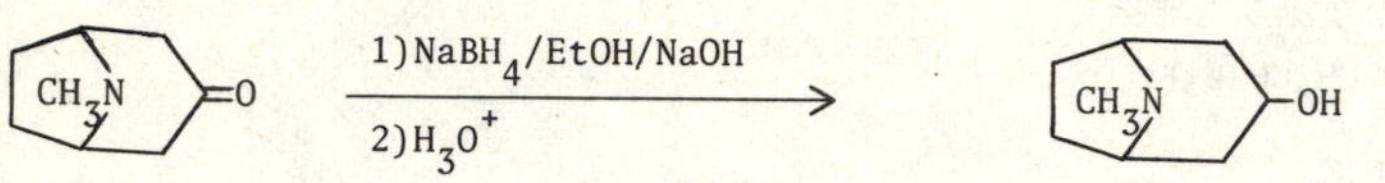

13-18

a)

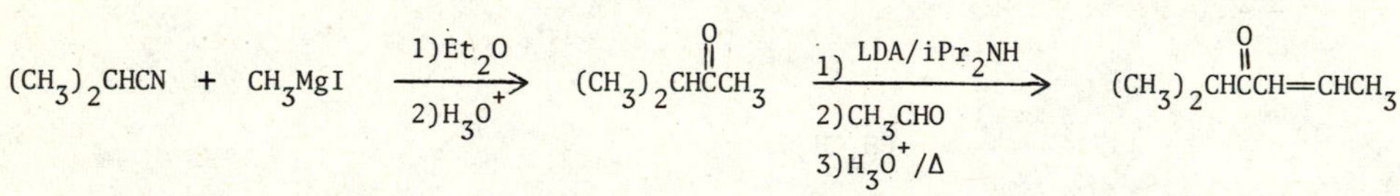

b)

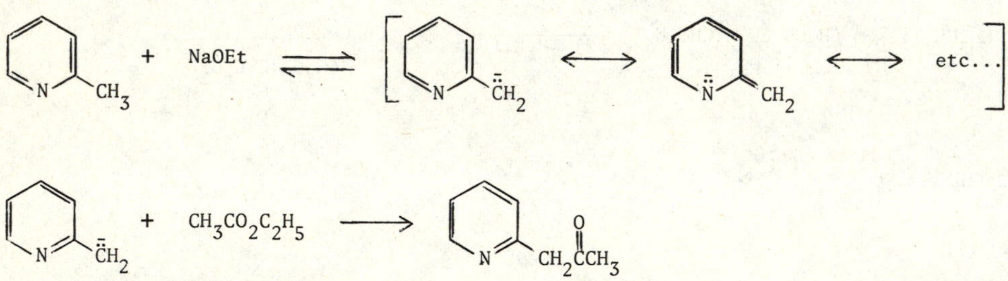

c)

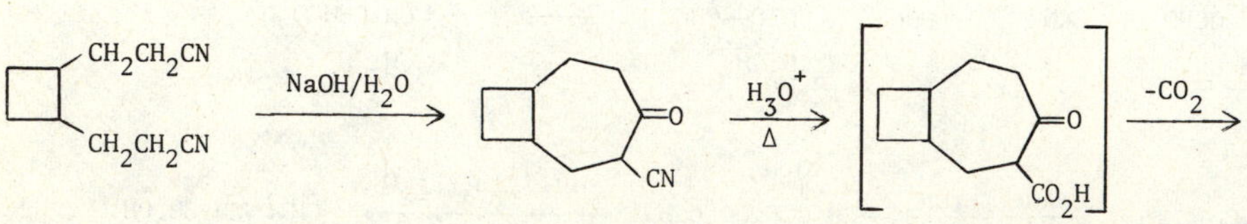

not isolated

Contd...

13-18 contd...

c) contd...

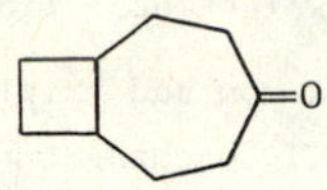

d)

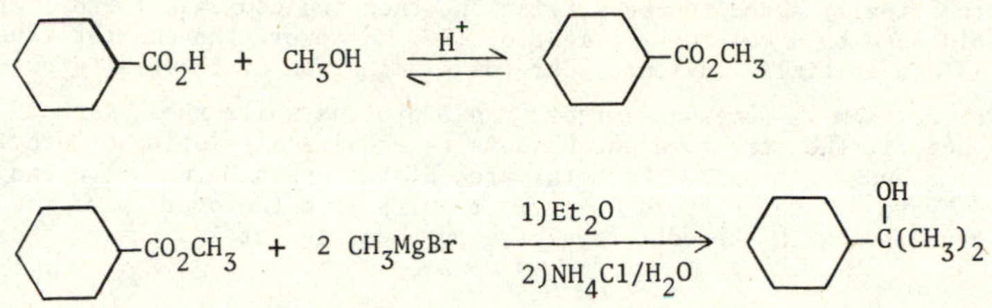

Note that the weakly acidic workup is necessary to avoid elimination of water from this tertiary alcohol.

e)

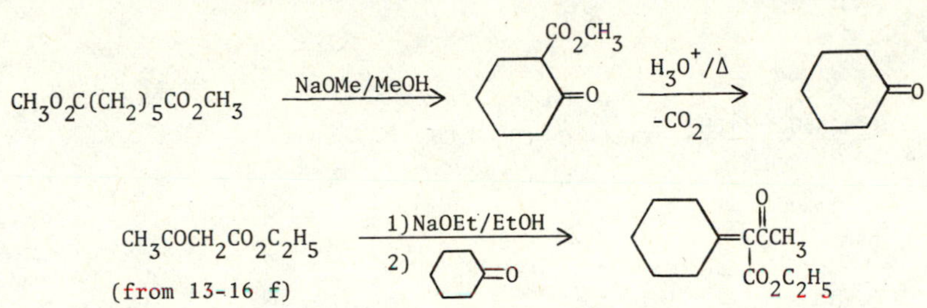

(from 13-16 f)

f) $$CH_3CH_2CH_2Br \xrightarrow[2)\,CO_2\quad 3)\,H_3O^+]{1)\,Mg/Et_2O} CH_3CH_2CH_2CO_2H \xrightarrow{SOCl_2} CH_3CH_2CH_2C(=O)Cl$$

$$2\ (CH_3)_2CHBr \xrightarrow[2)\,CdCl_2]{1)\,Mg/Et_2O} [(CH_3)_2CH]_2Cd$$

$$C_6H_5Br \xrightarrow{Mg/Et_2O} C_6H_5MgBr$$

$$CH_3CH_2CH_2C(=O)Cl + [(CH_3)_2CH]_2Cd \xrightarrow[2)\,H_3O^+]{1)\,Et_2O} CH_3CH_2CH_2\overset{O}{\overset{\|}{C}}CH(CH_3)_2 \xrightarrow[2)\,H_2O/NH_4Cl]{1)\,C_6H_5MgBr/Et_2O}$$

$$CH_3CH_2CH_2C(C_6H_5)(OH)CH(CH_3)_2$$

13-19 The molecular weight of ethyl acetoacetate is 130. The addition of four carbon atoms and associated hydrogens would be required to give a molecular weight near 186. The nmr singlet at 2.1 ppm could account for $CH_3C(=O)—$ while the quartet at 4.2 ppm and triplet at 1.3 are due to the $CH_3CH_2O—$ group. Both of these groups are essentially unchanged from ethyl acetoacetate. Alkylation must have taken place at the methylene next to the carbonyl. The triplet at 0.8 ppm with integral area of 6 is consistent with two equivalent methyl groups. If four carbon atoms were actually added in the alkylation, then two equivalent ethyl groups are indicated. That would lead to a molecular weight of 186. However, the quartet expected for the $—CH_2—$ groups is not initially obvious. Careful examination of integral areas shows that the singlet at 2.1 ppm is somewhat larger than 3 protons while the *apparent triplet* at 1.9 ppm does not fit the structure which seems to require alkylation of ethyl acetoacetate by two ethyl groups. Actually the total area of the peaks between 1.6 and 2.1 ppm corresponds to 7 protons. The pattern observed results from the overlap of a singlet at 2.1 ppm and a quartet with skewed multiplet areas centered at 1.9 ppm.

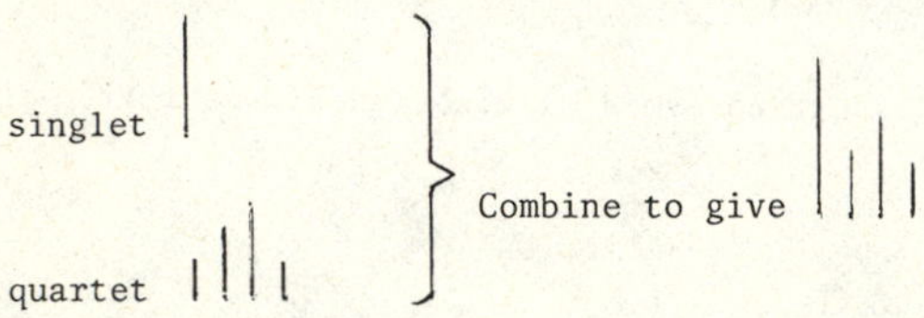

b) $CH_3C(=O)CH_2CO_2CH_2CH_3 \xrightarrow[\text{2) 2 } CH_3CH_2Br]{\text{1) NaOEt/EtOH}} CH_3C(=O)—C(CH_2CH_3)_2—CO_2CH_2CH_3$

13-20 The IR spectrum shows a C=O and the ^{1}H-nmr indicates a monosubtituted aromatic. The 3:2 aromatic multiplet is typical of $C_6H_5CO—$ where the ortho hydrogen atoms are more deshielded than the meta and para hydrogens. The ^{13}C-nmr spectrum also shows the four different aromatic carbons and carbonyl carbon (200 ppm) typical of this grouping. It accounts for 105 units of molecular weight leaving 43 (148-105). The ^{13}C-nmr shows three additional carbon atoms and the ^{1}H-nmr has seven additional hydrogen atoms. These account for the remaining molecular weight. The ^{1}H-nmr splitting patterns for this C_3H_7 fragment lead to the structure.

$C_6H_5C(=O)CH_2CO_2H \xrightarrow{CH_3OH/H^+} C_6H_5C(=O)CH_2CO_2CH_3 \xrightarrow[\text{2) } CH_3CH_2I \text{; 3) } H_3O^+/\Delta]{\text{1) NaOMe/MeOH}} C_6H_5C(=O)CH_2CH_2CH_3$

14 Elimination Reactions—Alkenes and Alkynes

14-1

a) 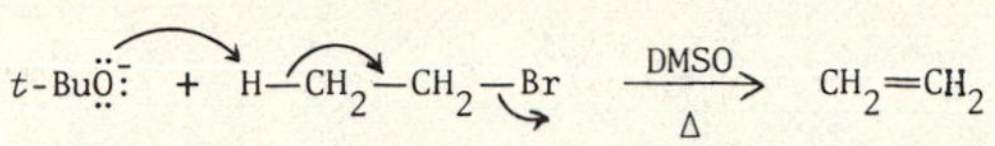

b) 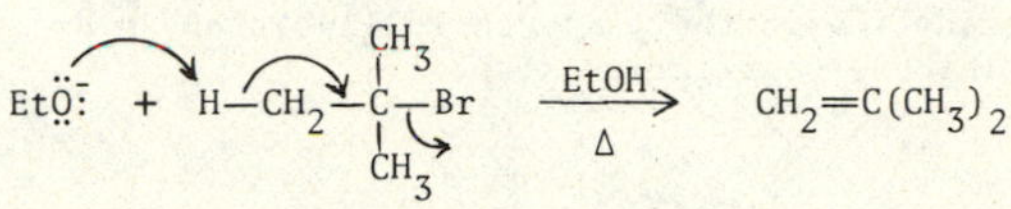

c) 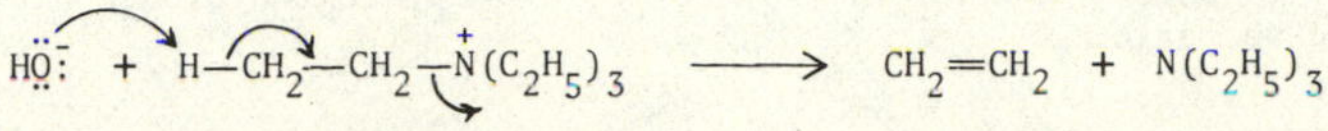

d) 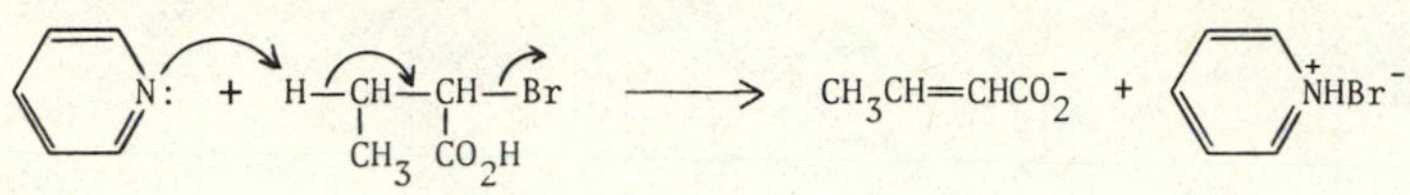

14-2

a) 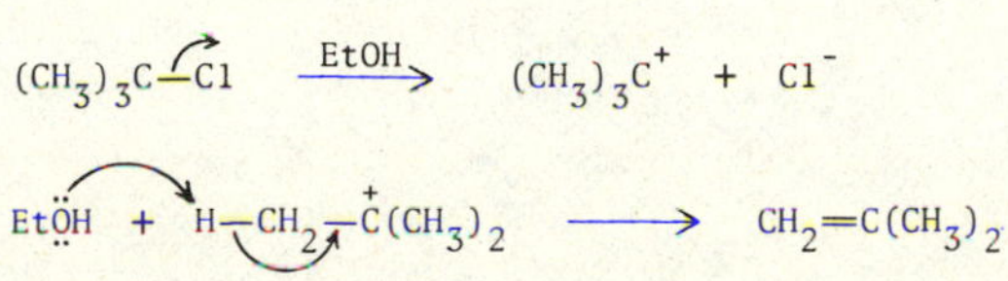

b) 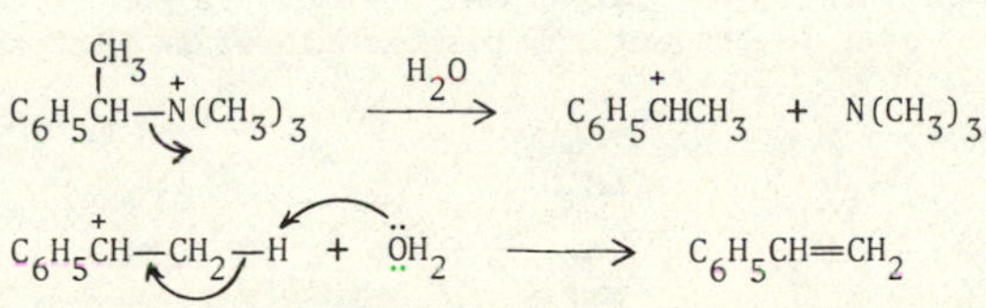

c) 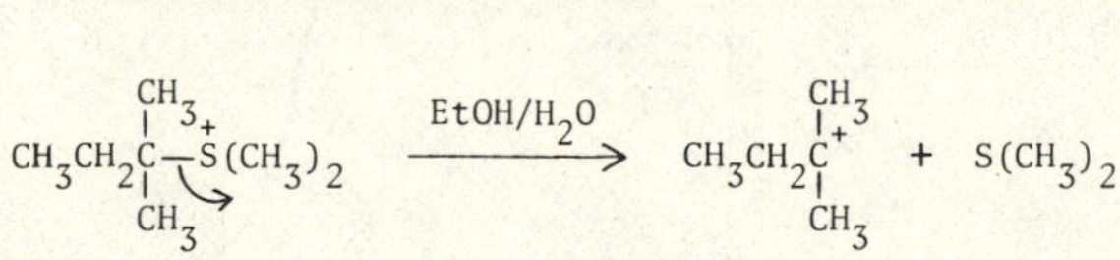

Contd...

14-2 Contd...

c) Contd..

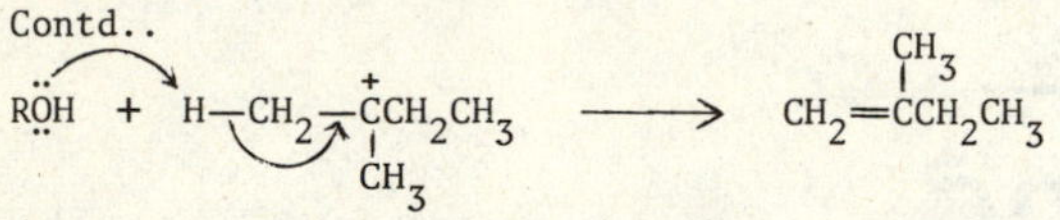

and

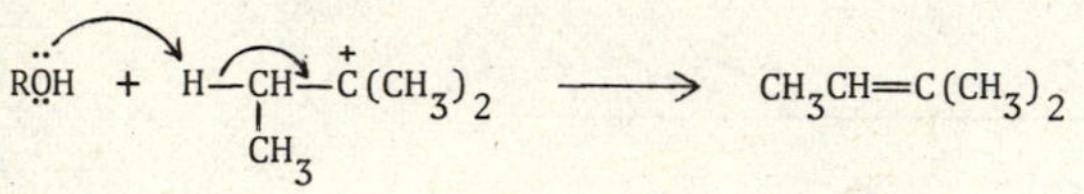

14-3 The exchange of deuterium for a proton takes place via the initially formed carbanion intermediate. Since departure of a fluoride ion is the rate-controlling step, hydrogen-deuterium exchange occurs during an initial pre-equilibrium step.

14-4 The carbonyl grcup and chlorine atom enhance the acidity of the adjacent hydrogen atom. Carbanion formation should be rapid.

14-5

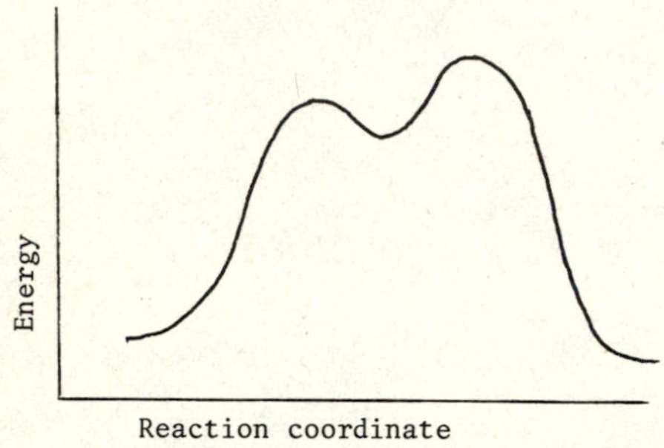

*E*1cB

To account for the hydrogen-deuterium exchange, the carbanion intermediate must return to starting material more rapidly than it goes to product in this example of an *E*1cB reaction.

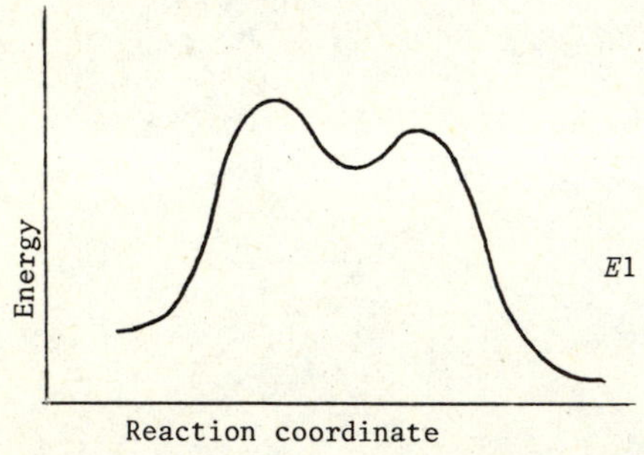

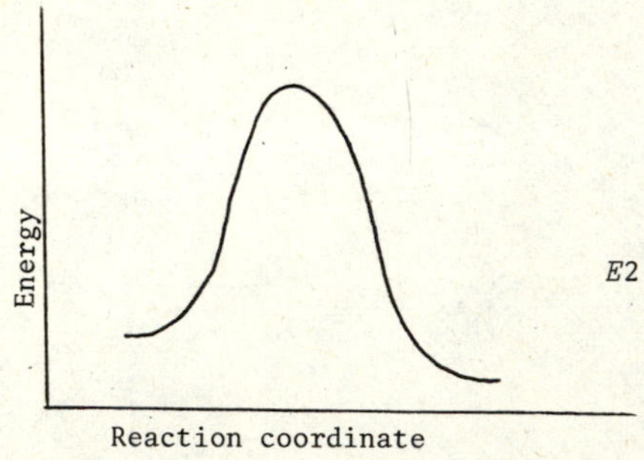

14-6 The smaller nitrogen kinetic isotope effect for the 2-phenylethylammonium salt demonstrates that C—$\overset{+}{N}$ bond breaking is not as important in forming the transition state. This is expected for an "*E*1cB like" *E*2 process where cleavage of the β-hydrogen bond is kinetically more important. The phenyl group enhances the acidity of the β-hydrogen atom relative to that of the ethylammonium salt making C—H bond cleavage occur earlier.

14-7 These anions are the conjugate bases of reasonably strong acids. They are weak bases, but not good nucleophiles.

14-8 Acetate is a stronger base than chloride. (Acetate is the conjugate base of a weaker acid.) Elimination is favored relative to substitution with the stronger base.

14-9 The reactivity order is the order of alkene stability. The results suggest that product stability is reflected in transition state stability.

14-10 Alkene stability will depend, in part, on the number of C-H bonds adjacent to the double bond if stabilization of the alkene by hyperconjugation is considered. Product A has eight C-H bonds (two methyl and one ethyl group) adjacent to the double bond while B has only seven.

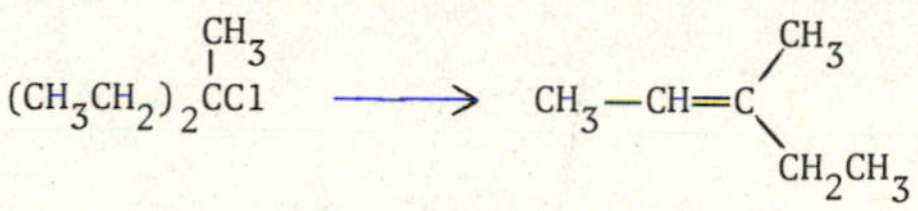

A

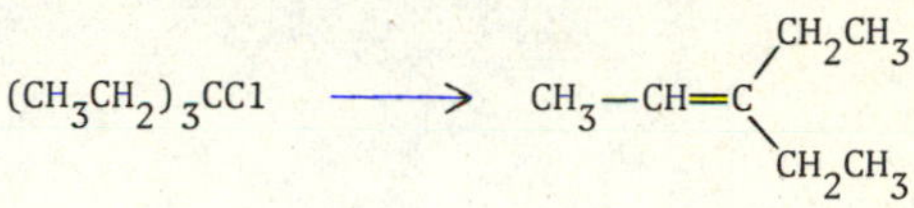

B

14-11

a) Solvent will have no effect on the S_N1:*E*1 rate ratio since both mechanisms have the same rate-controlling step; formation of a carbocation. Partition between elimination or substitution products occurs after the slow step.

b)

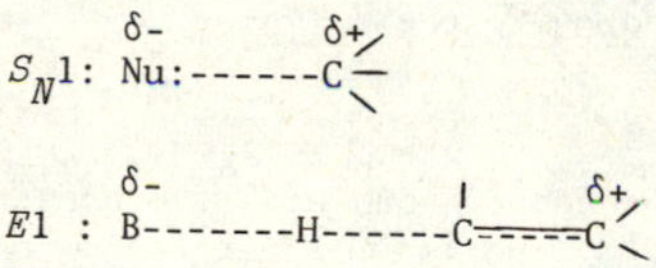

14-12 In aqueous alcohol media ethers from substitution by the alcohol and alcohols from substitution by water are common products.

14-13

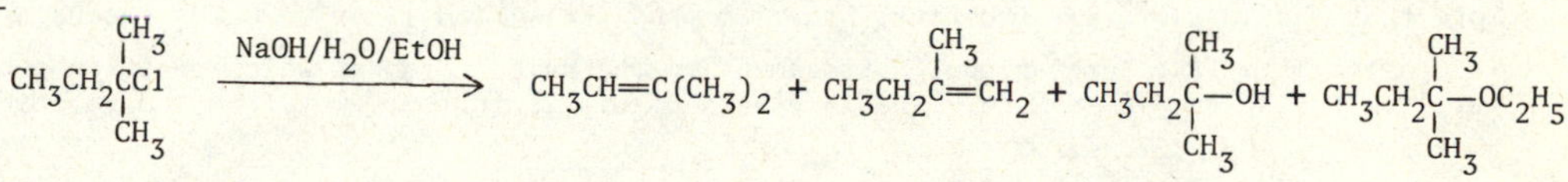

14-14 A larger number of atoms and associated bond charges are involved in elimination relative to substitution. The greater thermal energy associated with increasing temperature thus has a relatively greater influence on favoring the elimination process.

14-15 Formation of the 2-pentenes proceeds preferentially through the less crowded transition state to give the more stable *E* isomer. The 1-pentene, though the less substituted alkene, forms through a less crowded pathway than the *Z*-2-pentene.

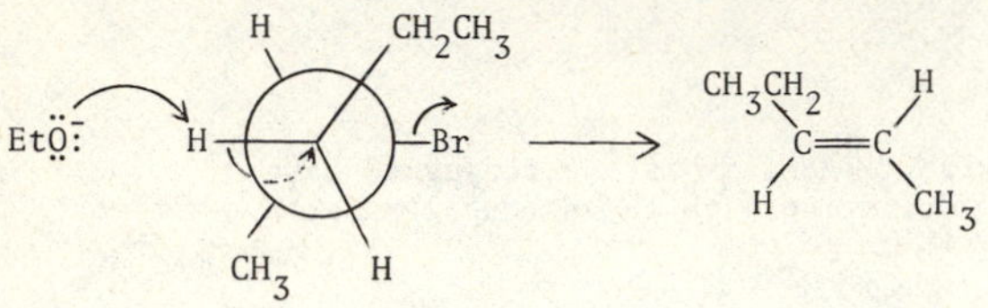

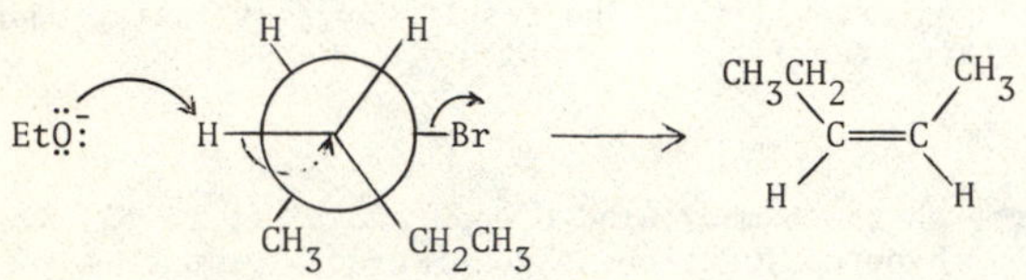

(Three groups in a gauche orientation.)

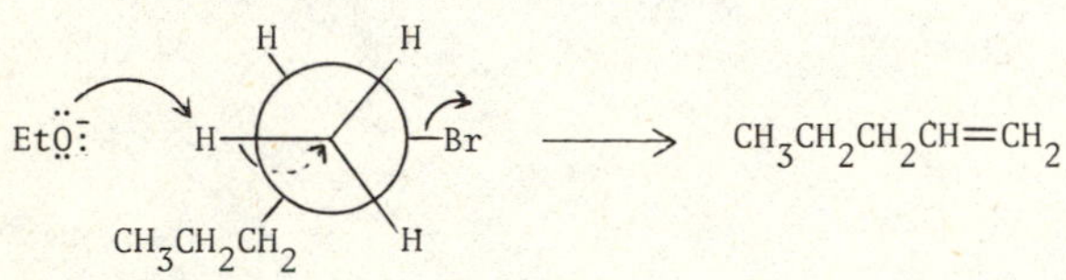

14-16 An *E*1 reaction is expected to occur because of the good leaving group on a secondary carbon atom, crowding at the β-carbon, and the nonbasic solvolytic conditions. The more highly substituted products predominate with the *E*-2-alkene being favored over the *Z* isomer because of steric crowding in the transition state leading to the *Z* isomer.

14-17 The 1-phenylpropene is stabilized by conjugation of the double bond with the benzene ring and this stabilization is reflected in the energy of the transition state.

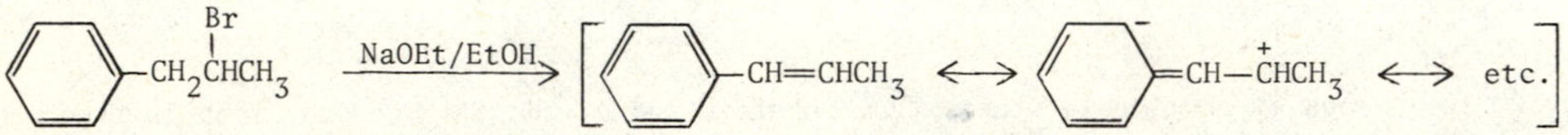

14-18 As the size of the leaving group increases, more product of Hofmann orientation is observed. Note that the tosylate is effectively smaller than the $—\overset{+}{S}(CH_3)_2$ or $—SO_2CH_3$ groups.

Though tosylate is a large group, attachment to the carbon atom at which reaction occurs involves a single C—O—S bond. The largeness of OTs is further away from the reaction center.

14-19 The 1-alkene product increases as the size of the base increases because approach by base to the hydrogen at the number three carbon (to give the 2-alkene) becomes more crowded.

14-20

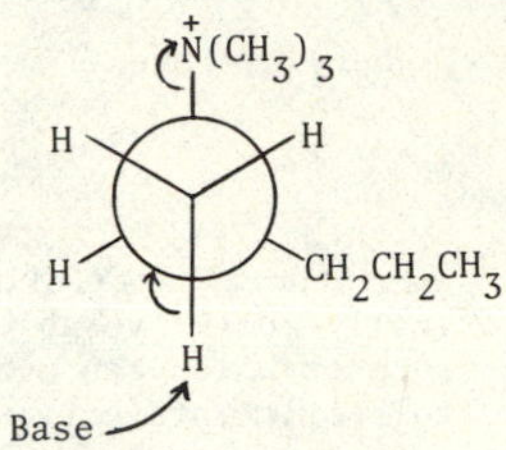

Conformation leading to the less-substituted alkene.

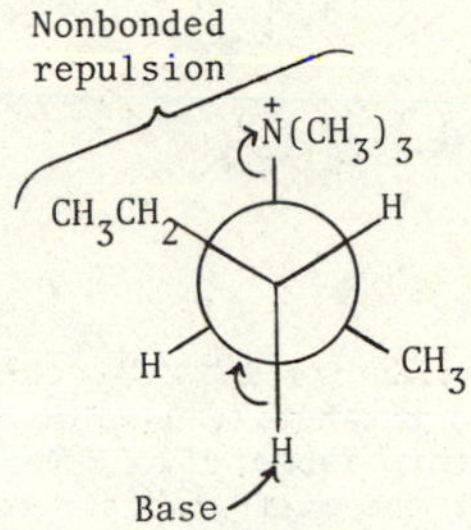

Conformation leading to the more-substituted alkene

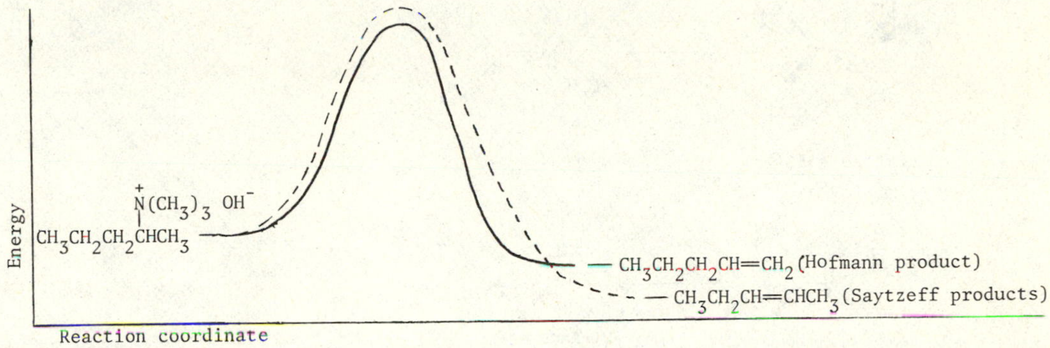

14-21

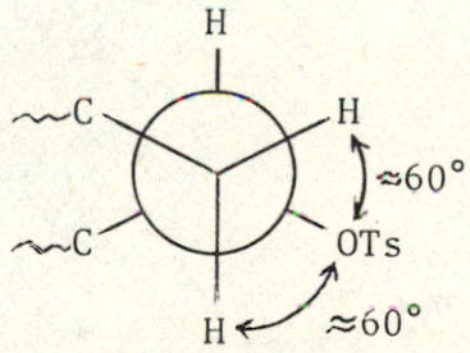

14-22 We expect loss of the benzylic β-hydrogen atom to be most favorable. Only the cis isomer has the benzylic H atom and OTs group antiperiplanar. The trans isomer must change to an unfavorable diaxial conformation for *anti-E2* elimination to occur, and then a less acidic β-hydrogen atom is removed. An *E*1 process is more likely to take place from the trans diequatorial conformation.

Contd...

14-22 contd...

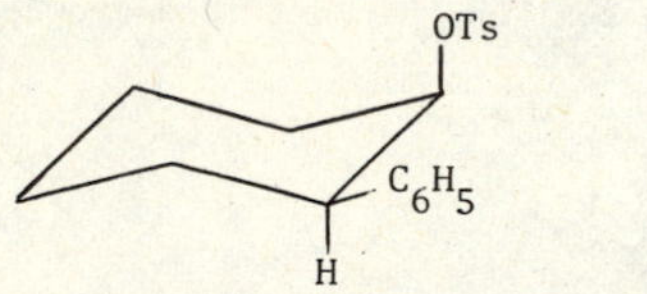

cis-2-Phenylcyclohexyl tosylate

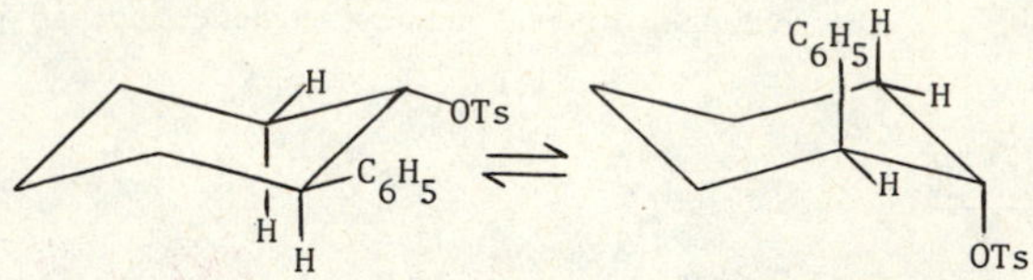

trans-2-Phenylcyclohexyl tosylate

14-23 (Refer to fig. 14-2)
For the threo isomer 1*S*, 2*R* (or 1*R*, 2*S*), the rotamer which can undergo anti elimination (A) is also the most stable rotamer, thus provides the energetically most favorable pathway for reaction. For the erythro isomer 1*R*, 2*R* (or 1*S*, 2*S*), the rotamer which can undergo anti elimination (A') is not the most stable rotamer in regard to steric interaction. Anti elimination is therefore slower for the erythro isomer.

14-24

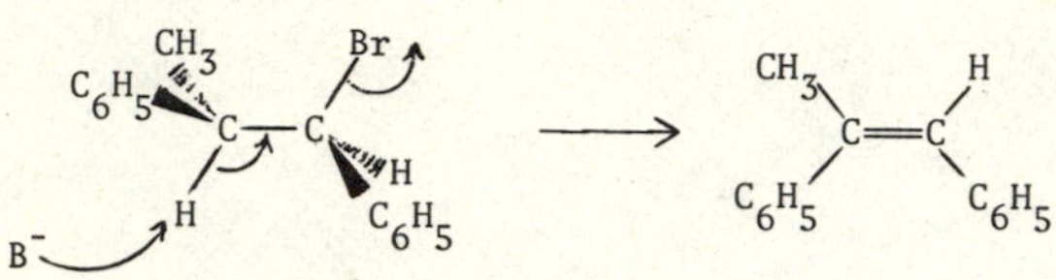

Erythro (1*R*, 2*R*)

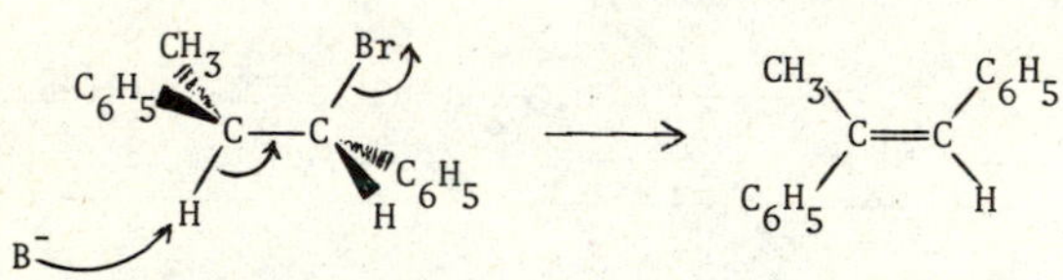

Threo (1*S*, 2*R*)

14-25

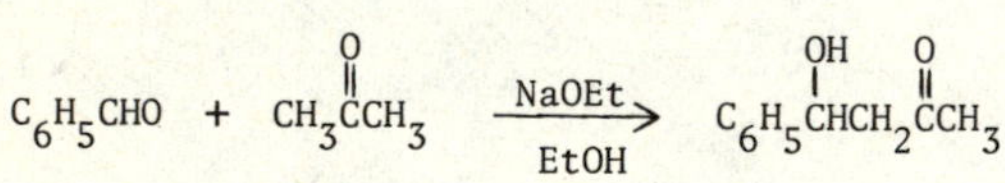

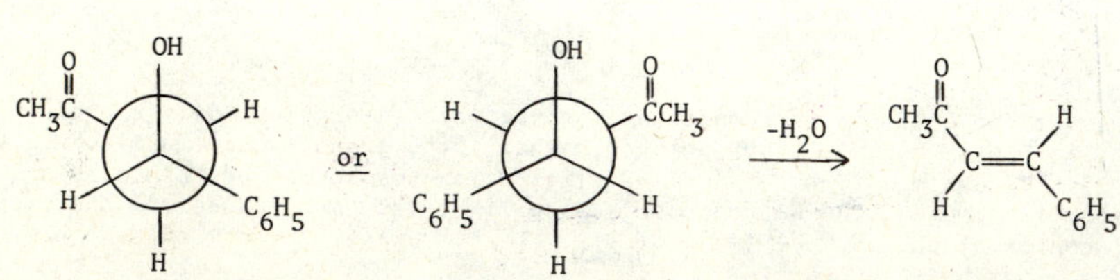

Anti elimination of H-OH from the most favorable rotamers of the enantiomeric aldol adducts leads to the *E* stereoisomer.

14-26

a) Because of the large size of the *t*-butyl group, flipping the ring of *trans*-4-*t*-butylcyclohexyl tosylate to all axial substitutents is much less favorable than a simila- ring flip by menthyl chloride.

b) If an *E*1 mechanism were involved, the elimination could have occured from a β-hydrogen atom on either side of the carbocation intermediate to give a mixutre of 2- and 3-menthenes. (See sec. 14-3 for further discussion of alkene stability.)

14-27 One isomer has no H and Cl atoms anti to each other in either ring conformation. *E*2 elimination is very unfavorable.

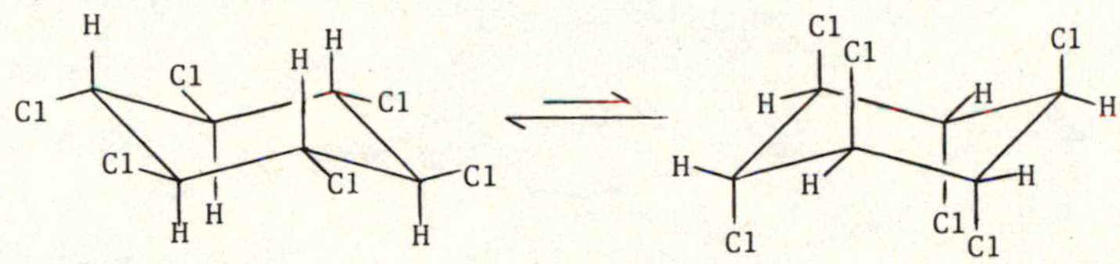

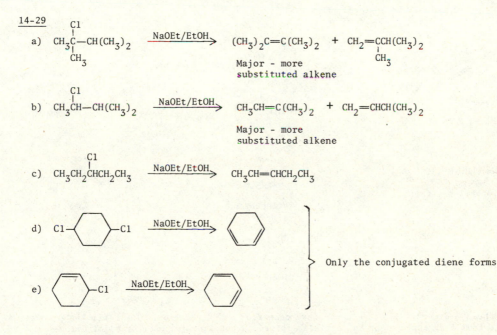

14-28 The H and Cl atoms are coplanar in A, but in B their dihedral angle is about 60°. The *E*2 process is favored in the coplanar system, A.

14-29

a) $CH_3C(Cl)(CH_3)—CH(CH_3)_2 \xrightarrow{NaOEt/EtOH} (CH_3)_2C{=}C(CH_3)_2 + CH_2{=}C(CH_3)CH(CH_3)_2$

Major - more substituted alkene

b) $CH_3CH(Cl)—CH(CH_3)_2 \xrightarrow{NaOEt/EtOH} CH_3CH{=}C(CH_3)_2 + CH_2{=}CHCH(CH_3)_2$

Major - more substituted alkene

c) $CH_3CH_2CH(Cl)CH_2CH_3 \xrightarrow{NaOEt/EtOH} CH_3CH{=}CHCH_2CH_3$

d) 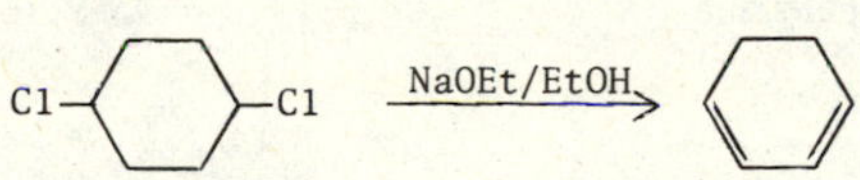

e) 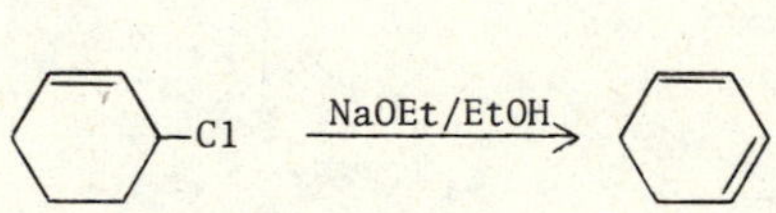

} Only the conjugated diene forms (d and e)

14-30 The trans diaxial isomer has a more favorable molecular dipole moment because the moments of each C-Br bond are in opposite directions and cancel each other. The bond dipole moments of the diequatorial isomer add together to give the molecule a significant dipole moment.

Contd...

14-30 contd...

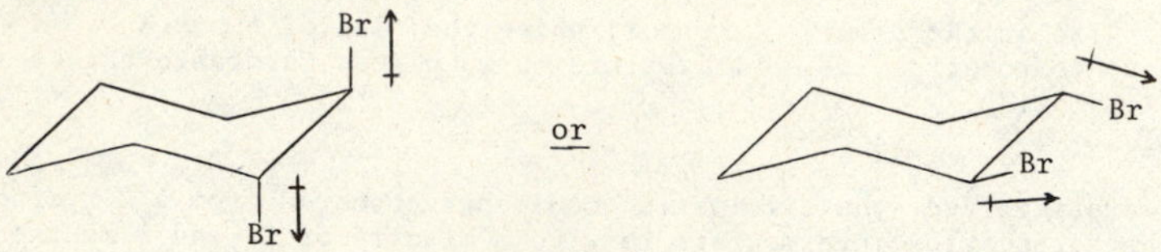

14-31 The bromine atoms of *cis*-1,2-dibromocyclohexane cannot attain the *anti*-periplanar arrangement preferred for elimination. However S_N2 displacement of one bromide by iodide gives a trans product from which *anti*-elimination can occur after a ring flip.

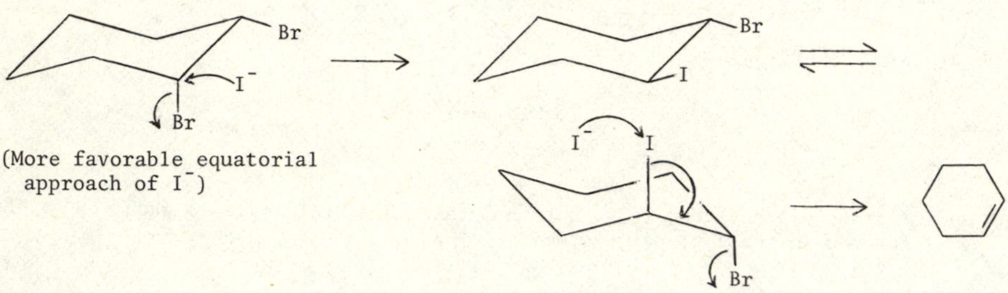

(More favorable equatorial approach of I^-)

14-32

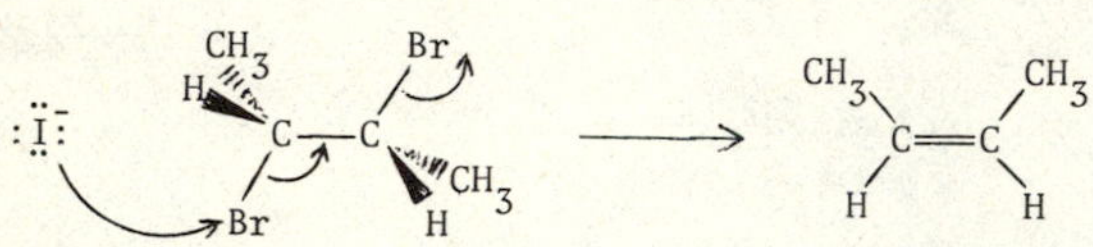

Z-2-Butene

14-33

a) $C_6H_5CH{=}CHCH(CH_3)C_6H_5$

1,3-Diphenyl-1-butene

b) $CH_3CH{=}C{=}CH_2$

1,2-Butadiene

(Methylallene)

c) 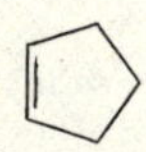

Cyclopentene

d) $(CH_3)_2C{=}C{=}O$

2-Methyl-1-oxopropene

(Dimethylketene)

14-34 Only the 3,3-dimethyl-1-butene could have come from an *E*2 reaction. The other two products result from rearrangement of an initially formed carbocation followed by elimination. An *E*1 pathway can account for all three products.

Contd...

14-34 contd...

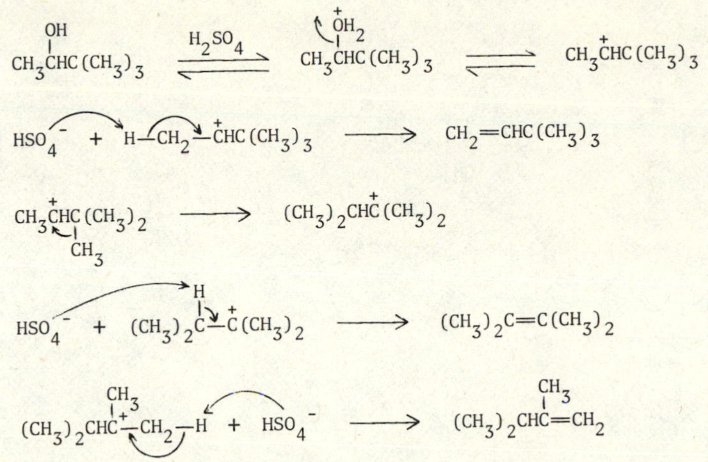

14-35

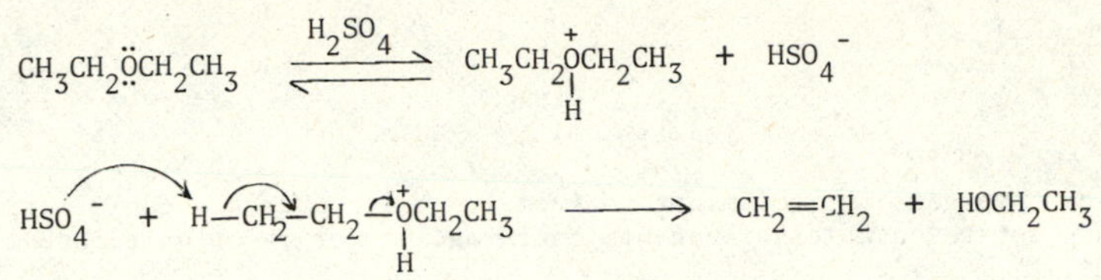

14-36

a) $CH_2{=}C(CH_3)CH{=}CH_2$

b) $CH_2{=}C(CH_3){-}C(CH_3){=}CH_2$

c)

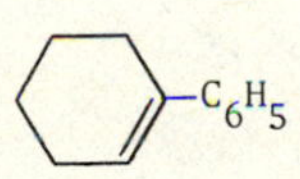

d) $C_6H_5CH{=}CHC_6H_4NO_2$-*p*

e)

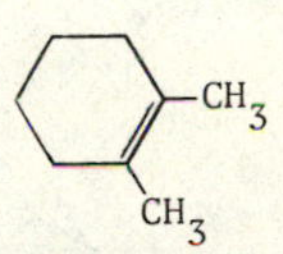

(Methyl rearrangement gives the more stable carbocation and then the more stable alkene.)

14-37 The trans isomer has no β-hydrogen atoms coplanar with the quaternary nitrogen atom. Ring flip would force two large groups into axial positions. Substitution by hydroxide on an *N*-methyl group is the preferred reaction. The cis isomer can undergo anti-periplanar elimination.

14-38

a) The reactions are both anti *E*2 eliminations.

Contd...

14-38 Contd...

a) Contd...

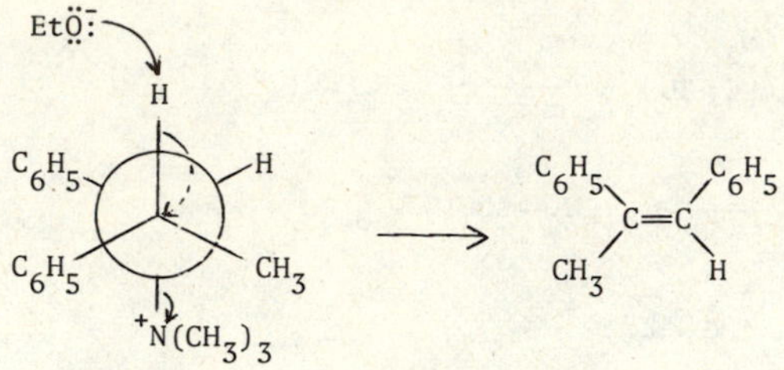

Erythro Z

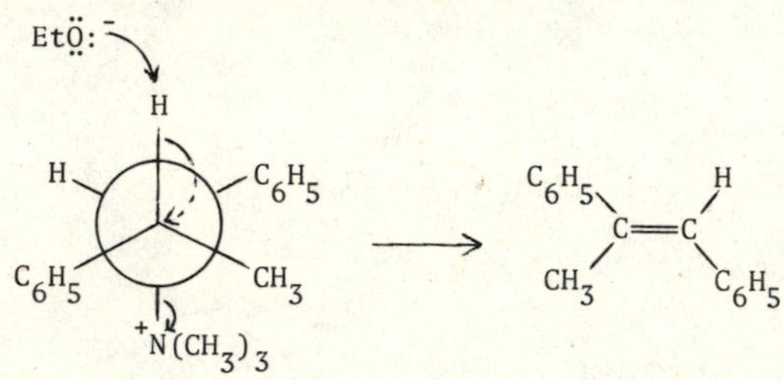

Threo *E*

b) The threo configuration undergoes anti elimination from its most favorable conformation. The erythro isomer must rotate to a less favorable conformation for reaction to occur.

14-39

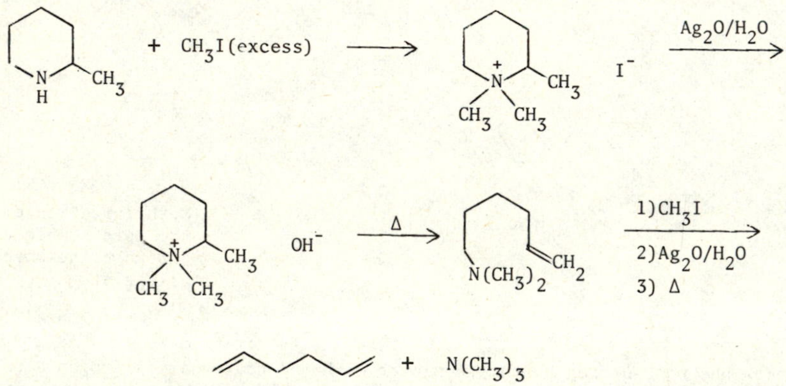

14-40

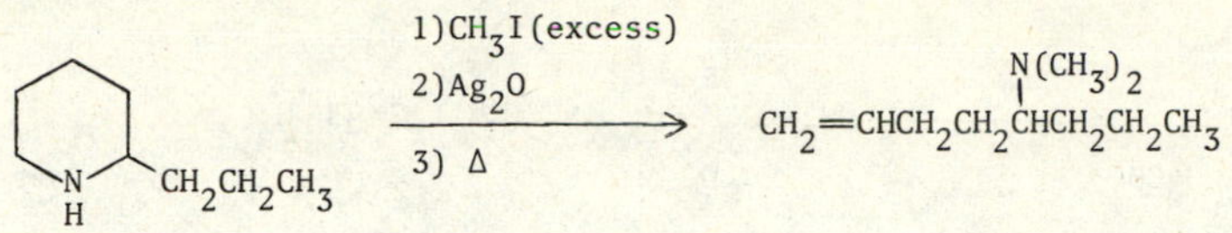

The alternate directions of elimination would produce the optically inactive products $(CH_3)_2N(CH_2)_3CH{=}CHCH_2CH_2CH_3$ and $(CH_3)_2N(CH_2)_4CH{=}CHCH_2CH_3$. As expected the less substituted alkene product is formed.

14-41

a) and b)

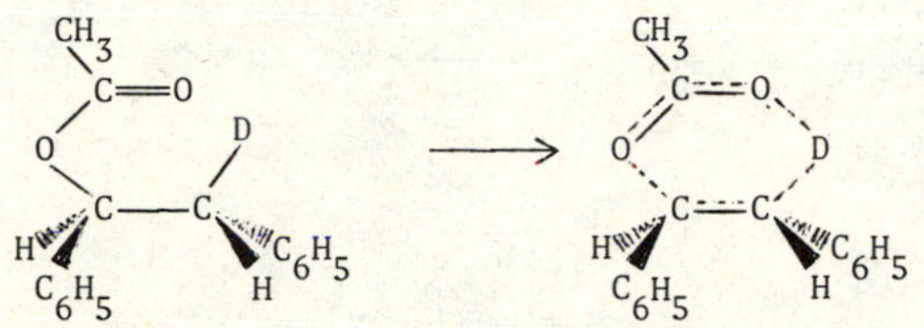

Threo (1*S*, 2*R*)

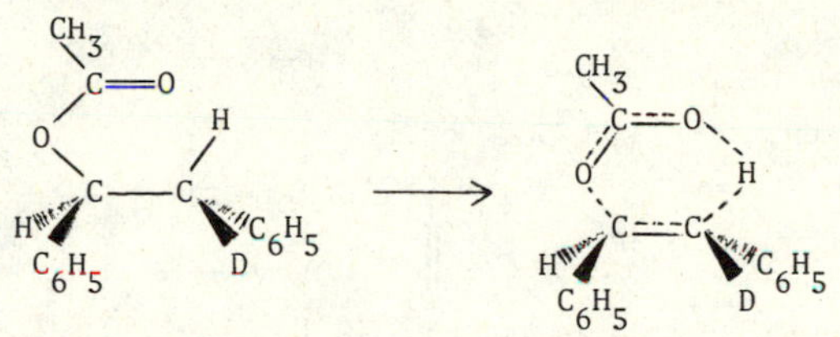

Erythro (1*S*, 2*S*)

c) The transition state leading to the *E* isomer is of lower energy in both cases because the phenyl groups are further apart. Syn coplanarity can be attained from erythro or threo and lead to *E* products.

14-42

a) The high temperatures at which pyrolysis is usually carried out results in very little selectivity in the choice of β-hydrogen atoms.

b)

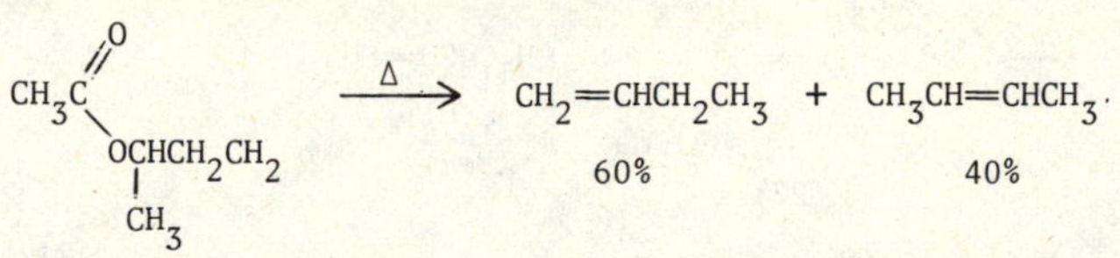

Two internal H's
Three terminal H's

Contd...

14-42 contd...

b) Contd...

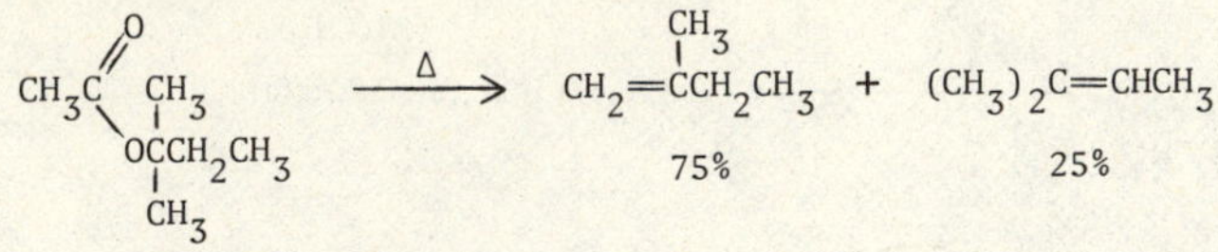

Two internal H's
Six terminal H's

14-43

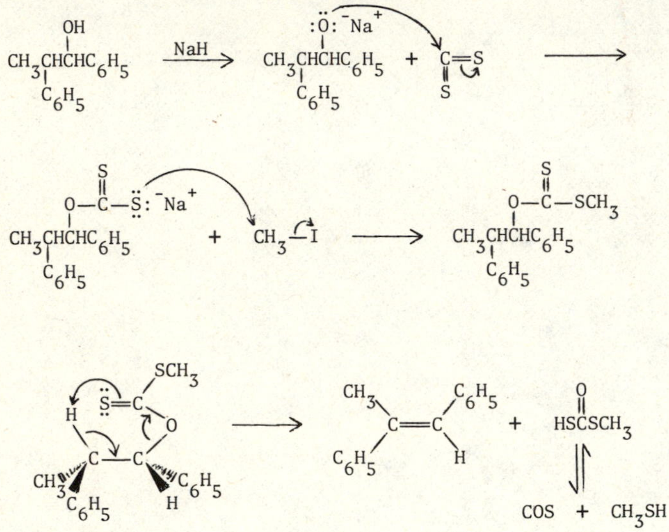

Erythro

14-44 The amine oxide pyrolysis proceeds by a syn elimination to give *Z* product. The Hofmann elimination is an anti elimination giving predominately *E* product.

14-45

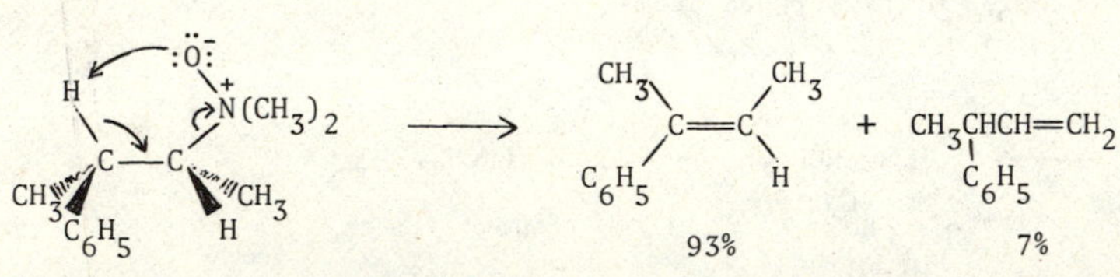

The *E* alkene is readily produced via a syn elimination. A syn elimination can also produce the less stable terminal alkene. Formation of the *Z* alkene would require an anti-elimination which cannot occur with the amine oxide.

14-46 Only the *Z* isomer has the H and Cl atoms in an anti-periplanar configuration favorable for *E*2 elimination.

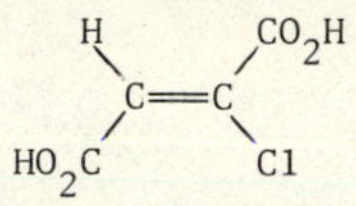

14-47

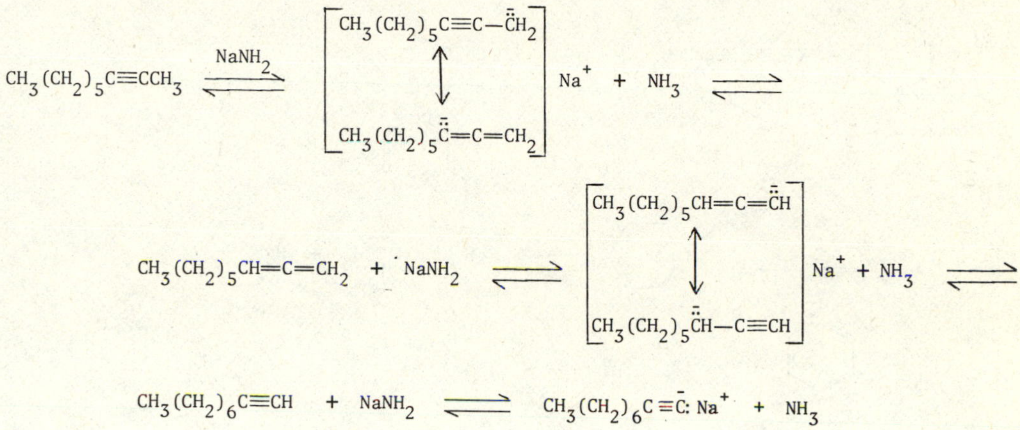

14-48

a) $n\text{-}C_5H_{11}C{\equiv}CH$

1-Heptyne

b) $C_6H_5C{\equiv}CH$

Phenylacetylene

(Phenylethyne)

c) 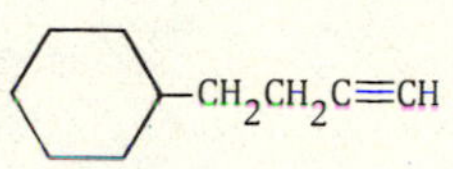

4-Cyclohexyl-1-butyne

d) $CH_3C{\equiv}CCH_3$

2-Butyne

14-49 The isotope effect indicates a slower rate for the reaction of deuterio compared to protio alcohol. This shows that the C-H (or C-D) bond must break in the rate-controlling step of the oxidation. An *E*2 process is consistent with this result.

14-50 The process is analogous to elimination via acetate esters.

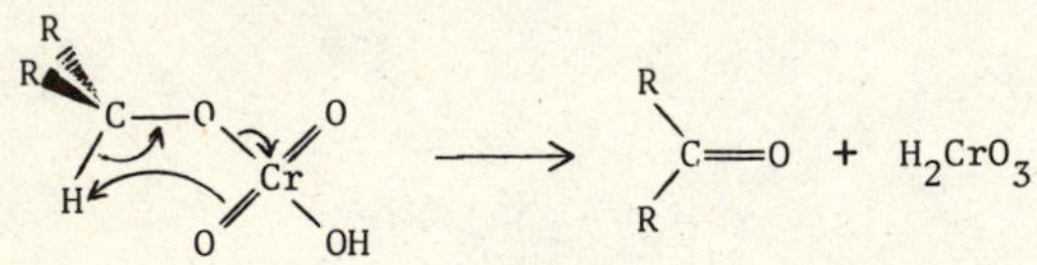

14-51

a)

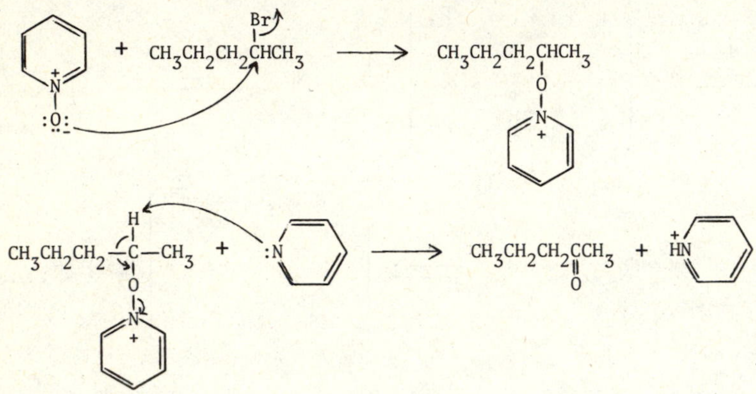

b)

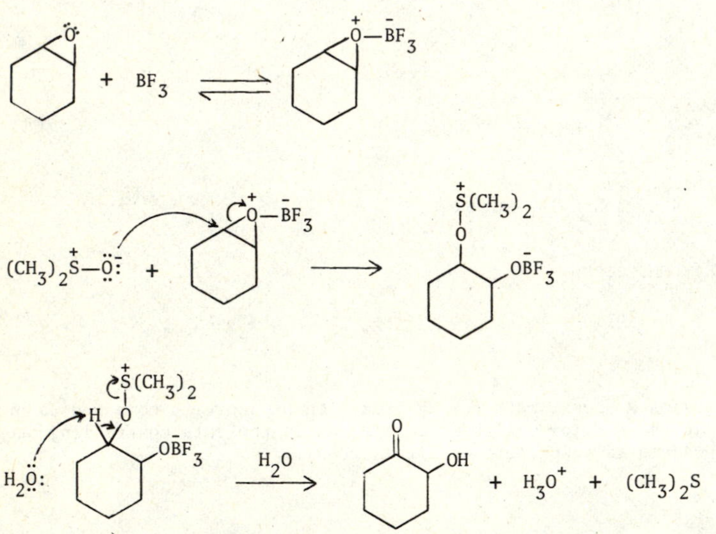

14-52 The cyclic iodate ester proposed as the oxidation intermediate requires a cis arrangement of the two hydroxy groups. A trans isomer must isomerize to the cis isomer for oxidation to occur.

14-53

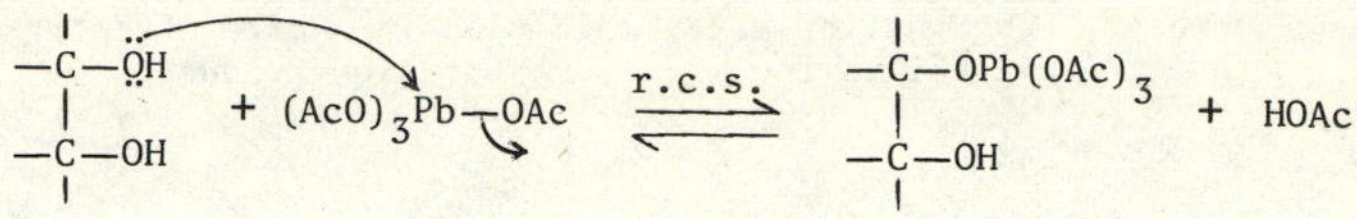

This step is second order. Additional HOAc would reverse the reaction by a mass action effect.

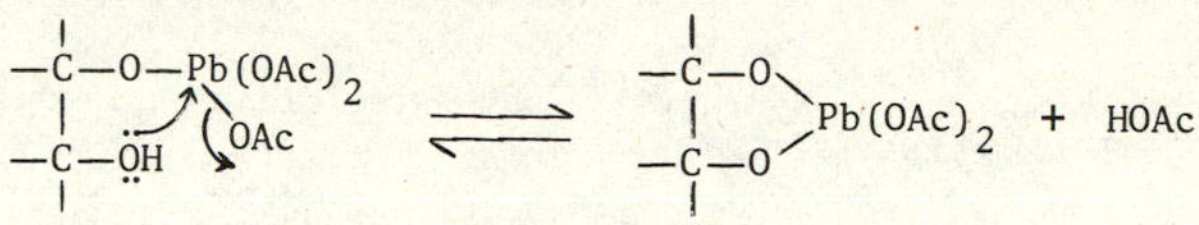

Cis glycol is required to form the cyclic ester.

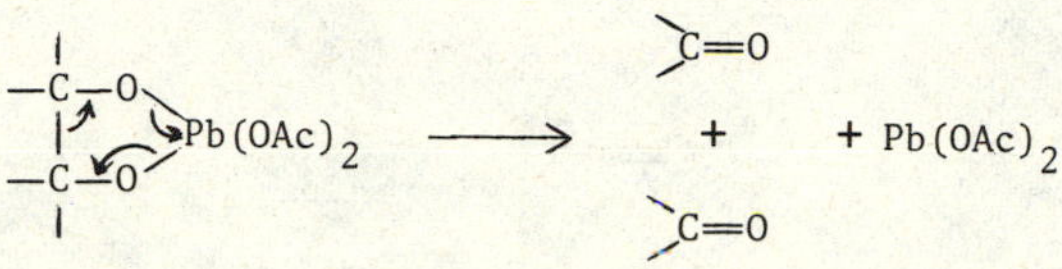

14-54

a) 

b) $(CH_3)_2C{=}CH_2$

c) 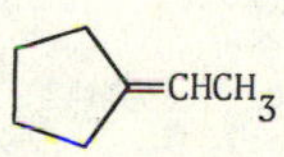

d) $(CH_3)_2CHCH{=}C(CH_3)_2$ + $(CH_3)_2CHCH_2C(CH_3){=}CH_2$

e) 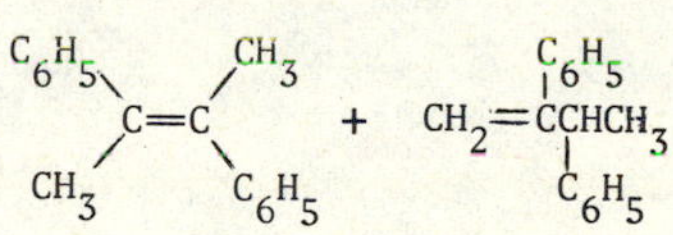

f) $CH_2{=}C(CH_3)CH_2CH_3$ + $(CH_3)_2C{=}CHCH_3$ + CH_3CO_2H

g) 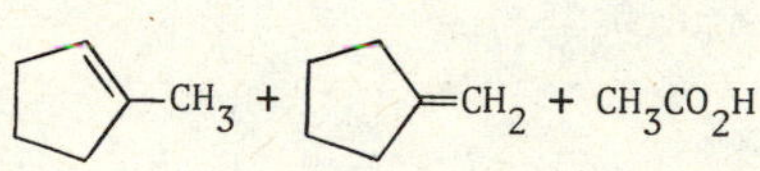

h) 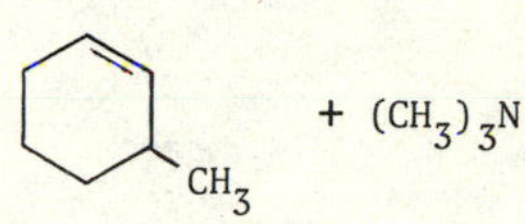

i) $OHC(CH_2)_4CHO$

j) $CH_3CH_2CH(CH_3)CH{=}CH_2$ + $(CH_3)_3N$

k) $CH_2{=}CHCH_2CH_3$ + $CH_3CH{=}CHCH_3$ + CH_3CO_2H

l) $HC{\equiv}CH$ + 2 $(CH_3)_3N$

m) *p*-$ClC_6H_4C{\equiv}CC_6H_4Cl$-*p*

14-55 In the polar neutral medium, *E*1 and S_N1 solvolysis processes occur. Substitution predominates in the absence of base. The smaller water moleculae is a more effective nucleophile than ethanol so that the alcohol is the major substitution product.

14-56

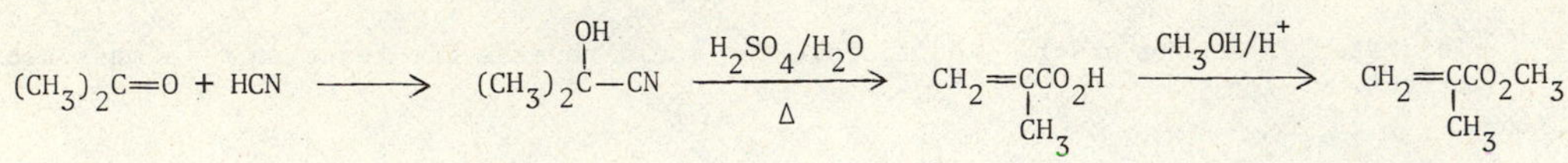

(The commercial process accomplishes elimination, hydrolysis, and esterification in one step by treating the cyanohydrin with acidic methanol.)

14-57 Because of free rotation about the single bond either of the two hydrogen atoms can be abstracted from the number 3 carbon atom.

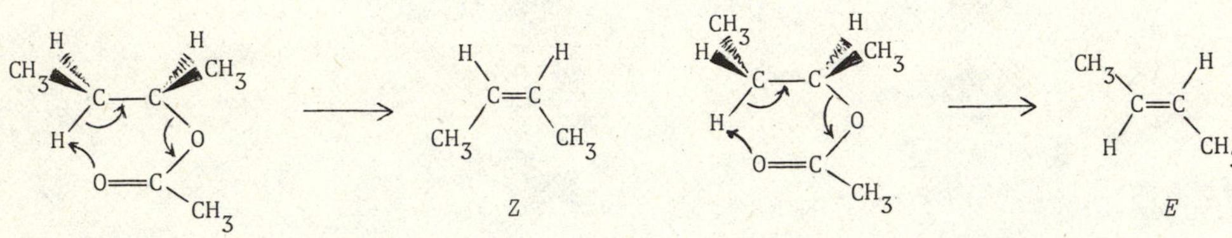

14-58

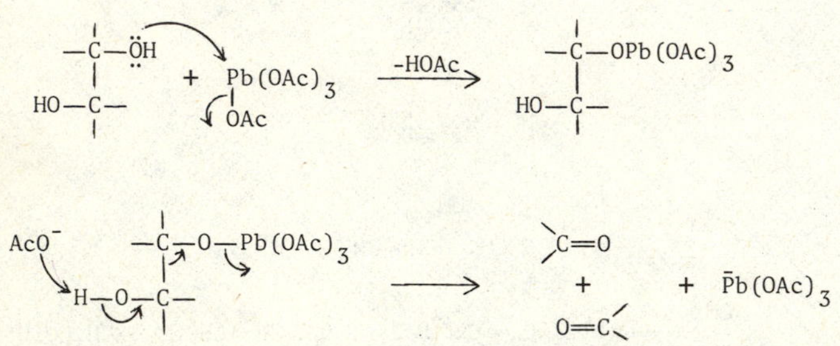

14-59

a) $C_6H_5CH_2CHO$

b)

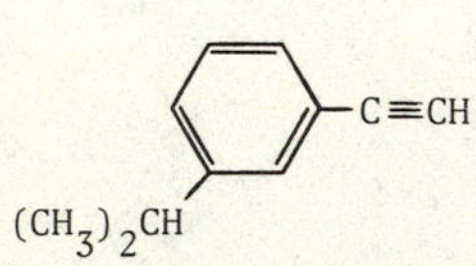

c)

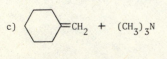

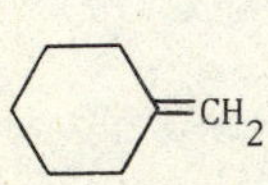 + $(CH_3)_3N$

d)

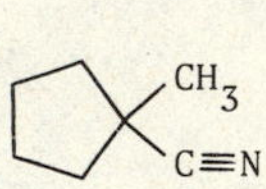

e) $CH_3CH{=}CHCO_2CH_3$

f)

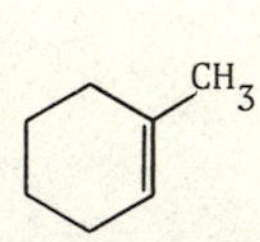

g)

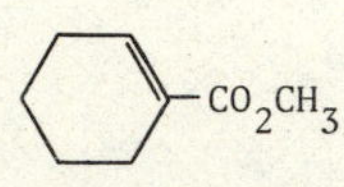

h) $CH_3CH_2CH{=}CH_2$

Contd...

14-59 contd...

i) 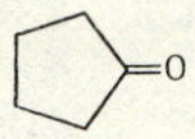

j) 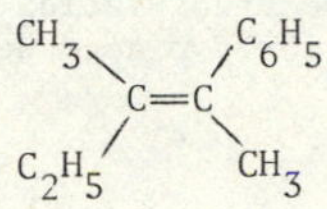

k) $CH_2{=}CHC(CH_3)_3$

l) 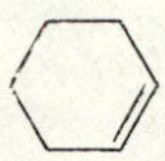

m) $CH_2{=}CH{-}CH{=}CH_2$

14-60 The negative oxygen atom of the amine oxide is solvated strongly by hydrogen bonding to water molecules. The oxyanion must lose much of this water of solvation before it can effectively act as a base. Under anhydrous conditions with an aprotic solvent, the oxygen atom is relatively free.

14-61

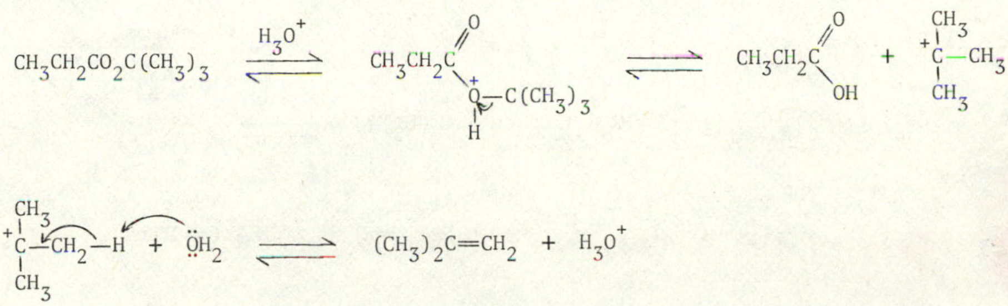

14-62 The amino group in A can act as an internal base to remove the β-hydrogen atom. A six-membered cyclic transition state results in formation of the 1-alkene. Compound B reacts via a normal bimolecular *E*2 process in which steric factors favor, but do not require, removal of the less hindered terminal β-hydrogen atom.

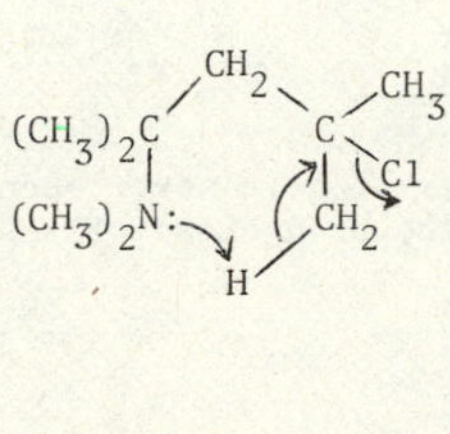

14-63

a) 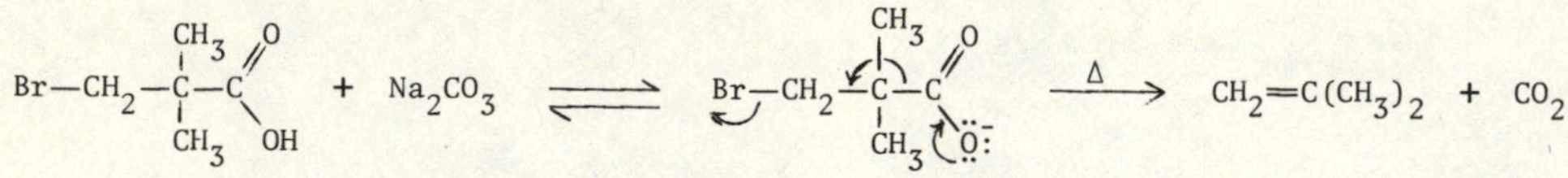

b) In this reaction the driving force is formation of a stable alkene. Decarboxylation of β-keto acids depends on formation of an enolate anion which has a similarly placed double bond.

14-64

a) Elimination takes place as one carbon-halogen bond is converted to the organolithium.

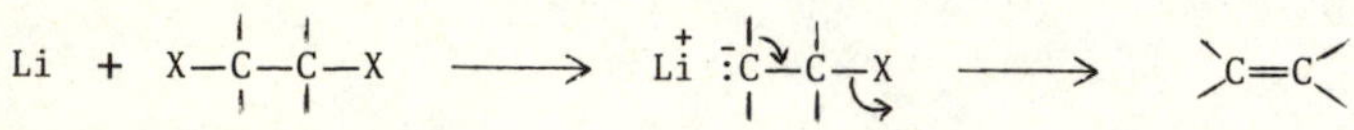

b) Loss of H_2O in the fragmentation proceeds in the direction expected for formation of the more stable benzylic cation.

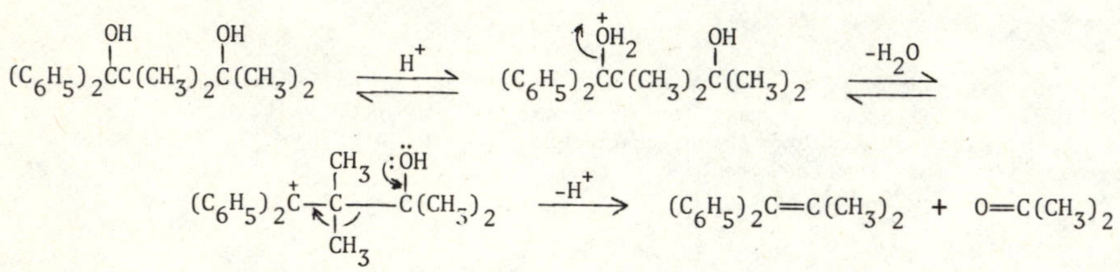

rather than

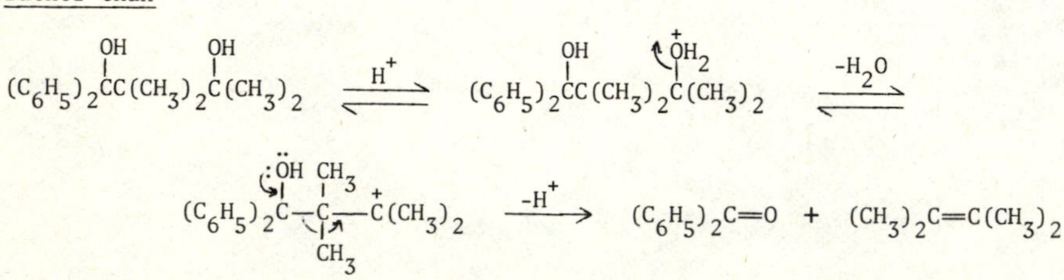

c) The more acidic β-hydrogen atom is adjacent to the phenyl ring. Acidity factors govern the the direction of elimination in this example. The product is also a conjugated alkene.

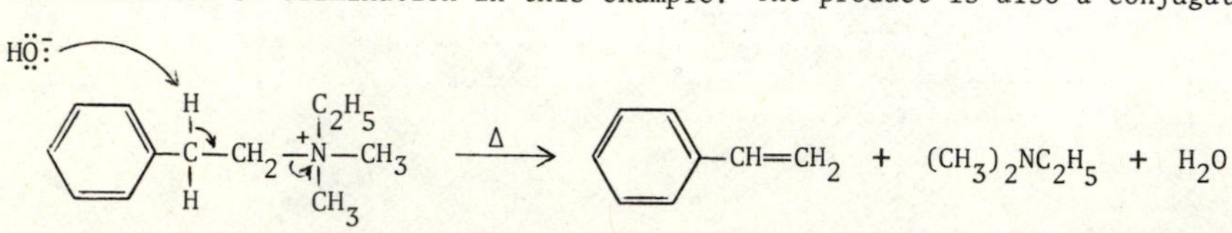

d) Small bicyclic molecules with bridgehead double bonds such as G have never been prepared. Bredt's rule states that introduction of a double bond at the bridgehead of small bicyclic systems is prohibited.

14-65

a) $CH_3CH_2CH_2Br \xrightarrow{Mg/Et_2O} CH_3CH_2CH_2MgBr \xrightarrow[2)\ H_3O^+]{1)\ (C_6H_5)_2C{=}O/Et_2O}$

$CH_3CH_2CH_2C(OH)(C_6H_5)_2 \xrightarrow{-H_2O} CH_3CH_2CH{=}C(C_6H_5)_2$

The tertiary benzylic alcohol readily dehydrates during acid workup.

b) $CH_3CHO + H_2NOH \longrightarrow CH_3CH{=}NOH \xrightarrow[(-H_2O)]{SOCl_2} CH_3C{\equiv}N$

$CH_3C{\equiv}N \xrightarrow[2)\ CH_3CHO]{1)\ LDA/THF} CH_3CH(OH)CH_2CN \xrightarrow{H_3O^+ \text{ or } HO^-/H_2O} CH_3CH{=}CHCN$

c) $HO{\cdot}CH_2CH_2CH_2Br \xrightarrow{H_2SO_4/\Delta} CH_2{=}CHCH_2Br$

$2\ CH_2{=}CHCH_2Br \xrightarrow[2)\ HCO_2Et;\ 3)\ H_3O^+]{1)\ Mg/Et_2O} CH_2{=}CHCH_2CH(OH)CH_2CH{=}CH_2 \xrightarrow[2)\ LiAlH_4THF;\ 3)\ H_3O^+]{1)\ p\text{-}TsCl/pyridine}$

$CH_2{=}CHCH_2CH_2CH_2CH{=}CH_2$

d) 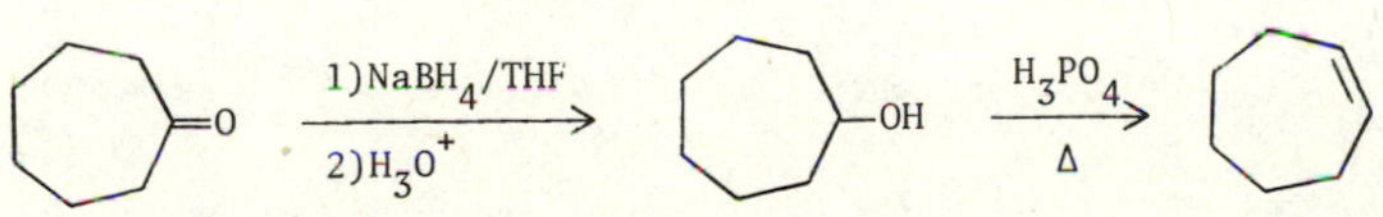

e) $C_6H_5C(=O)CH_3 \xrightarrow[2)\ H_3O^+]{1)\ LiAlH_4/Et_2O} C_6H_5CH(OH)CH_3 \xrightarrow{PBr_3} C_6H_5CH(Br)CH_3$

$HC{\equiv}CH + NaNH_2 \longrightarrow NaC{\equiv}CH$

$C_6H_5CH(Br)CH_3 + NaC{\equiv}CH \xrightarrow[0°C]{THF} C_6H_5CH(CH_3)C{\equiv}CH$ (Low temperature minimizes competing elimination.)

14-66 The IR absorption of A suggests a ketone. By making the reasonable assumption that the product should have some similarity to the starting material we deduce that the integral areas of the nmr must correspond to 6:4:4. Again relying on the starting material structure, we propose that there are two methyl groups (0.9 ppm), two methylene groups adjacent to a carbonyl (2.3 ppm) and two other similar methylene groups (1.2 - 1.7 ppm). We conclude that A is formed by an elimination reaction followed by tautomerization of the initially formed enol to a more stable ketone.

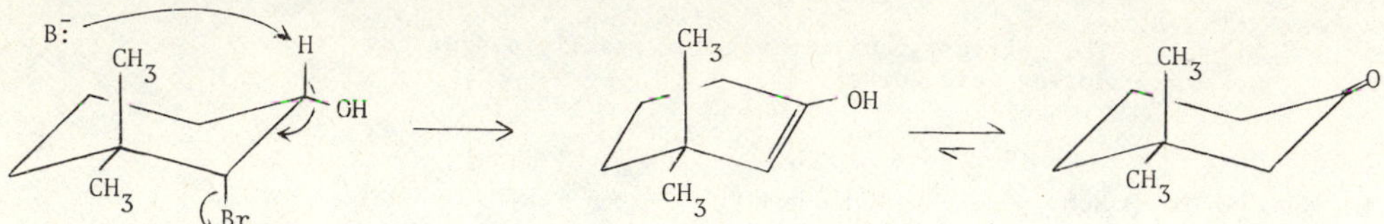

A

The notable feature about the nmr spectrum of B is that there are two protons (using the same deductions as above about the actual numbers of protons) deshielded as if they were adjacent to an oxygen atom. The absence of functional group IR absorption suggests an ether, in this case an epoxide. The trans isomer has Br and OH anti and can readily form an epoxide through intramolecular displacement of bromide.

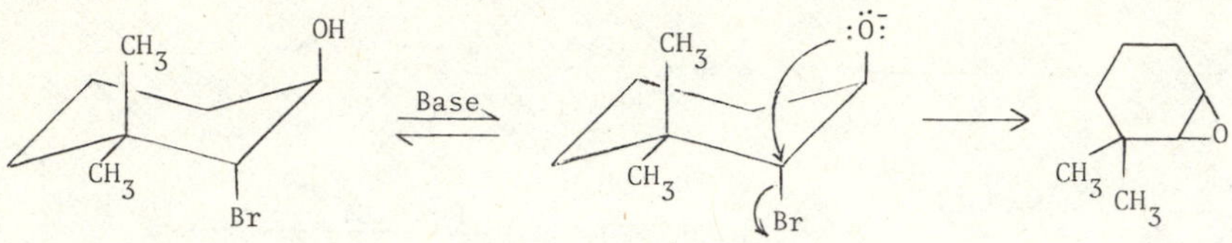

B

14-67 *IHD* = 8 for A. The IR spectrum suggests that A is an alcohol and the ^{1}H-nmr spectrum, when adjusted to the integral areas corresponding to the molecular formula, indicates two methyl groups (1.2ppm), two phenyl groups, and two deshielded protons. The structure is symmetrical and can only fit the glycol shown below.

Periodic acid accomplishes oxidative cleavage of glycols. Compound B has a carbonyl (IR 1690 cm^{-1}) and a monosubstituted aromatic and deshielded methyl based on its ^{1}H-nmr. It must be acetophenone, a conclusion supported by its ^{13}C-nmr spectrum which shows an alkyl carbon and four different aromatic carbons; the ortho, meta, and para carbons and the (very small) substituted carbon. The ketone carbon is small and deshielded at 198ppm.

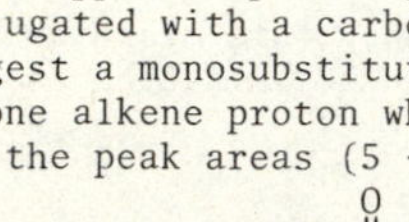

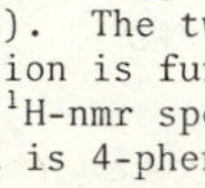

A B

14-68 *IHD* = 6. The ^{1}H-nmr spectrum indicates an aromatic ring, an alkene proton deshielded to 6.6-6.8ppm and possibly a methyl ketone. The IR suggests a double bond ($\approx$1600 cm^{-1}) conjugated with a carbonyl group ($\approx$1680 cm^{-1}). The two peaks at 680 cm^{-1} and 740 cm^{-1} suggest a monosubstituted aromatic. Conjugation is further suggested by the presence of one alkene proton which must be under the ^{1}H-nmr spectral aromatic resonance to account for the peak areas (5 + 1 = 6). The compound is 4-phenyl-3-buten-2-one (benzalacetone).

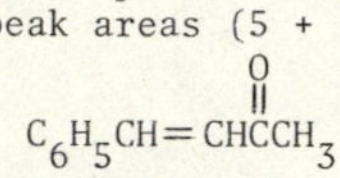

14-69 Elemental analysis and the mass spectrum lead to a molecular formula of $C_{10}H_{14}N_2$.

Nicotine must be an amine and possess a pyridine ring with a substituent at the C-3 carbon atom. Formation of a dihydrochloride indicates two basic nitrogen atoms are present. The nmr spectrum shows aromatic protons (the peaks deshielded to 8.6ppm are a part of this), some deshielded aliphatic protons and a possible deshielded methyl singlet. The Hofmann degradation sequence is most helpful since it indicates that the substituent on the pyridine ring of nicotine has at least four carbon atoms. Furthermore, the fact that one Hofmann elimination still retains the nitrogen atom shows that the substituent is a cyclic amine containing four carbon atoms. The structure of nicotine is...

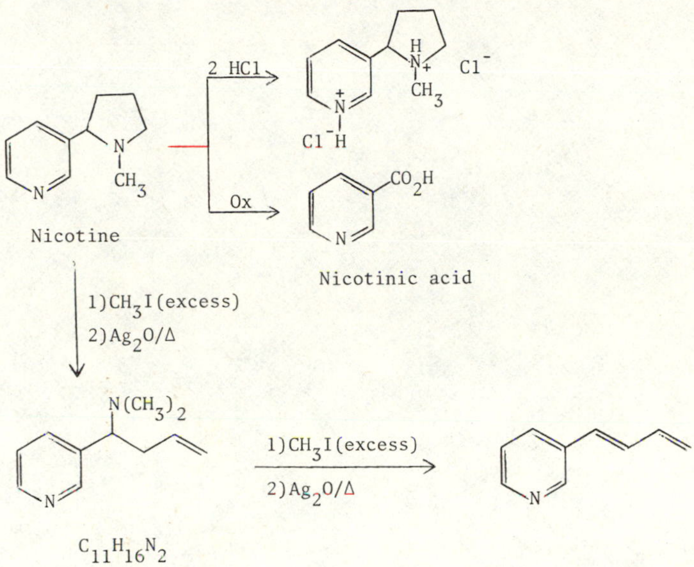

14-70 *IHD* = 2.

The IR spectrum shows a carbonyl (1760 cm^{-1}) and an unsymmetrical alkene (1640 cm^{-1}). There is no OH in the IR spectrum so the oxygen atoms of the molecular formula cannot be alcohol or a carboxylic acid. The ^{13}C-nmr spectrum shows a normal aliphatic carbon atom and three deshielded carbon atoms, one of which must be the carbonyl carbon. The other two could be from the alkene, but deshielded by the oxygen atoms. We have accounted for:

$C{=}O$ and $C{=}C$

The molecular formula precludes an aromatic so that the peaks at 7.2-7.5 ppm in the ^1H-nmr spectrum must be due to a strongly deshielded alkene hydrogen such as $O{-}CH{=}$ and the singlet at 2.1 ppm is probably $CH_3\overset{|}{C}{=}O$. The only reasonable structure is -

$CH_3CO_2CH{=}CH_2$

The complex splitting in the ^1H-nmr spectrum is due to coupling of the terminal alkene hydrogens (4.4-5.0 ppm) with each other as well as with the highly deshielded (by oxygen) internal alkene hydrogen (7.1-7.5 ppm).

15 Electrophilic Additions to Unsaturated Carbon

15-1 If the initial protonation step were a rapid acid-base reaction as is observed in most carbonyl addition reactions, H and D would exchange faster than hydration occurs. The slow electrophilic addition of H^+ must be followed by a rapid product-forming step.

15-2 Electron donating groups enhance reactivity while electron withdrawing groups decrease reactivity at the double bond.

$p\text{-}CH_3OC_6H_4CH{=}CHCO_2H > C_6H_5CH{=}CHCO_2H > p\text{-}ClC_6H_4CH{=}CHCO_2H > p\text{-}NO_2C_6H_4CH{=}CHCO_2H$

15-3 The more stable intermediate dication formed on addition of H^+ has positive charges the furthest from each other.

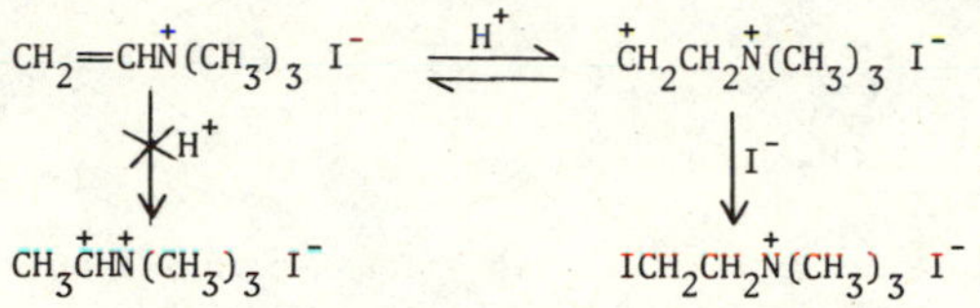

15-4

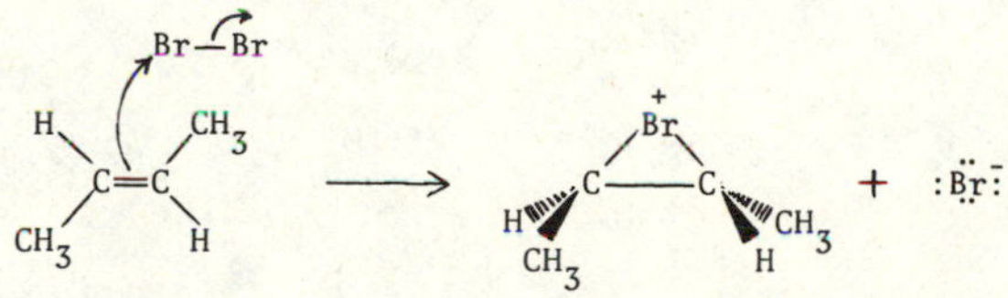

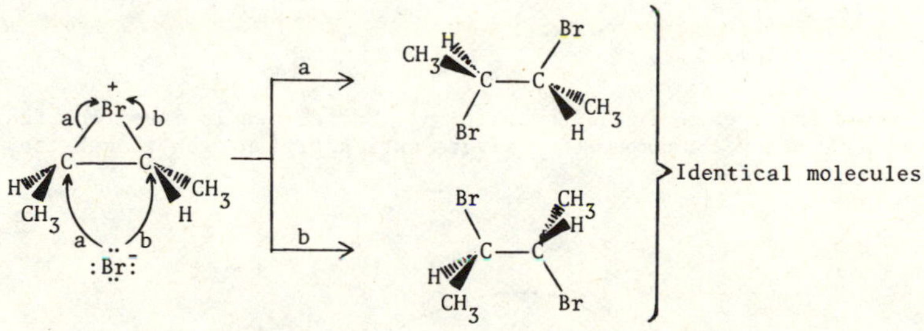

15-5 The phenyl group stabilizes the carbocation by resonance delocalization of the positive charge.

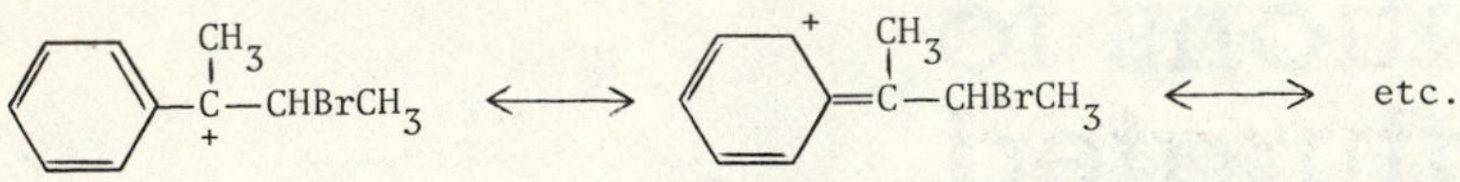

15-6

a)

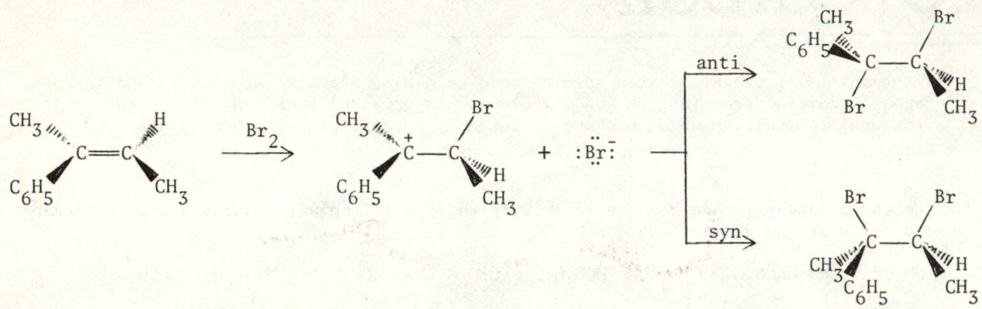

Each stereochemical pathway leads to a mixture of enantiomers. Only one enantiomer is shown in each sequence above.

b)

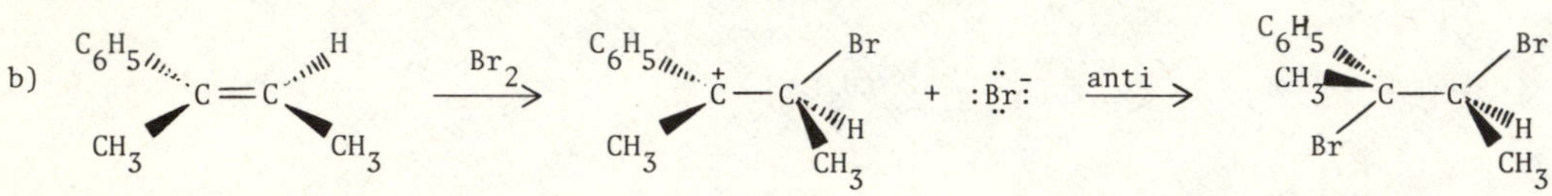

This is the same stereoisomer that is obtained by syn addition to *Z*-2-phenyl-2-butene.

c) Anti addition to the *Z* isomer produces the stereoisomer with the sterically more favorable arrangement of large, medium, and small groups, whereas anti addition to the *E* isomer produces the less favorable isomer.

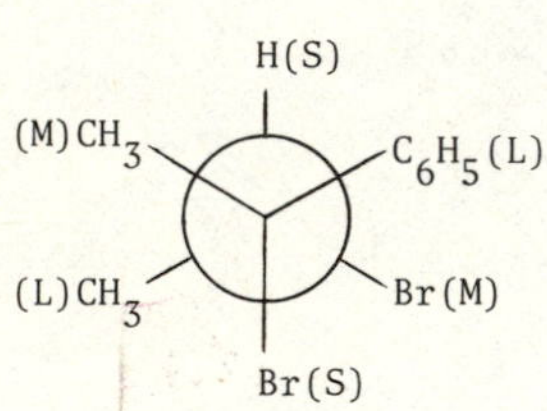

(More stable enantiomer)

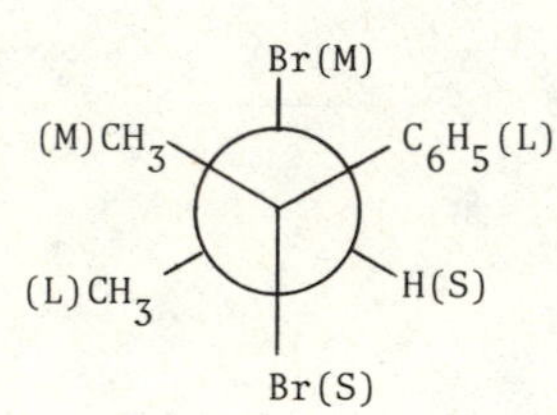

It has been suggested that, even when a carbocation intermediate is involved, there is some degree of bromonium ion character that favors anti addition even though the less favorable isomer may be formed.

15-7

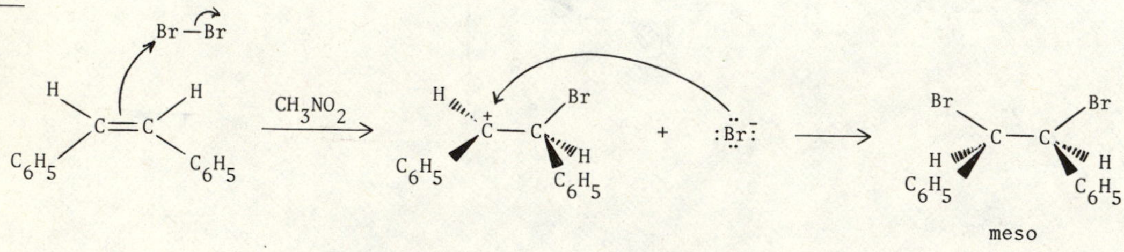

meso

15-8

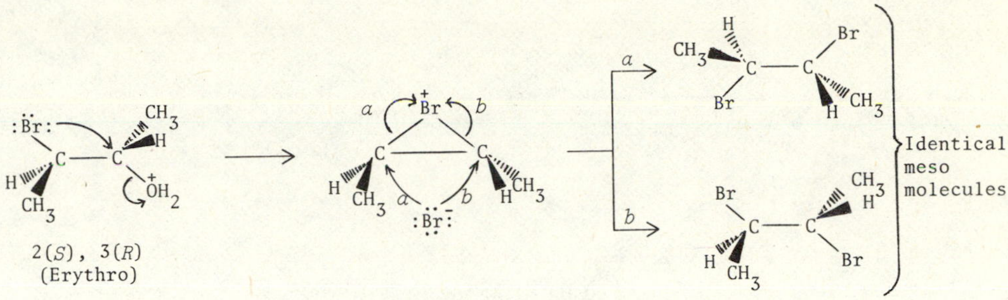

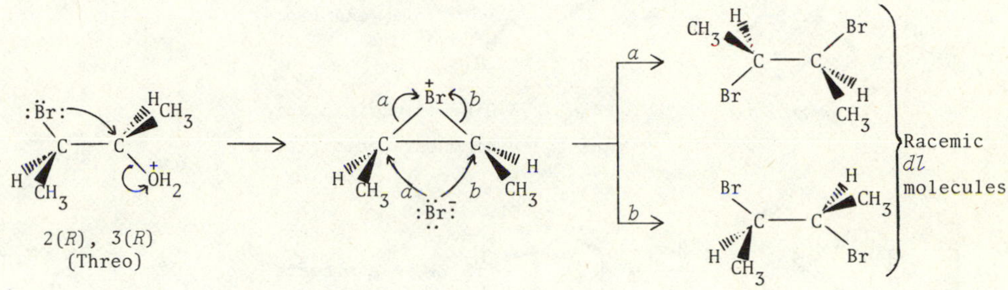

15-9 The phenyl group provides stability to a noncyclic cation intermediate so that stereospecificity is lost. The sterically more favored approach of chloride leads to a predominance of the less crowded erythro product.

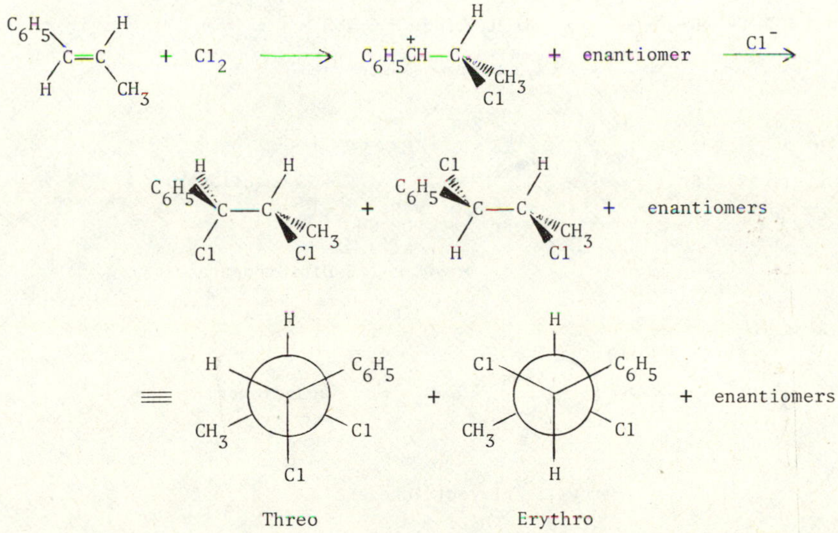

15-10 The intermediate bromonium ion reacts with bromide formed in the reaction or with the added chloride ion.

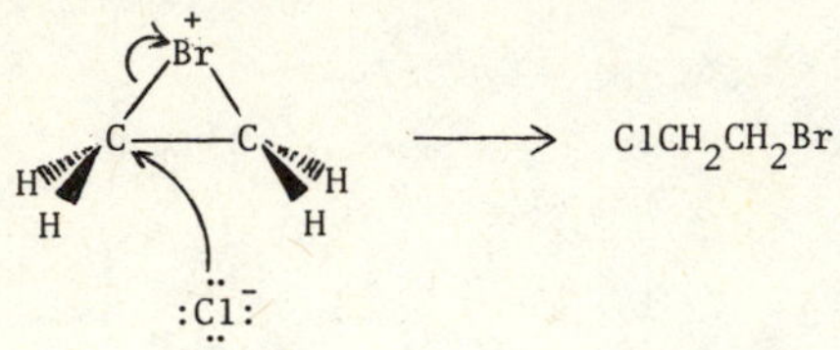

15-11

a) 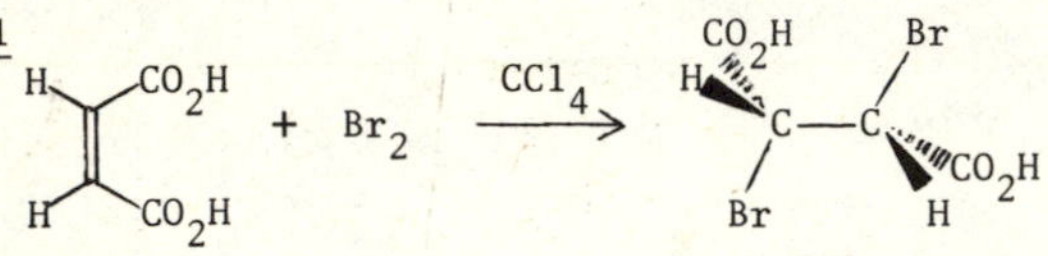 + enantiomer

dl-2,3-Dibromosuccinic acid
(*dl*-2,3-Dibromobutanedioic acid)

b) 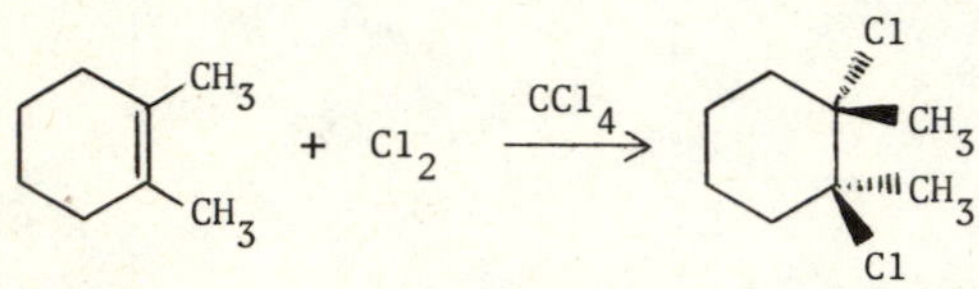 + enantiomers

trans-1,2-Dichloro-1,2-dimethylcyclohexane

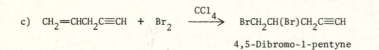

c) $CH_2{=}CHCH_2C{\equiv}CH + Br_2 \xrightarrow{CCl_4} BrCH_2CH(Br)CH_2C{\equiv}CH$

4,5-Dibromo-1-pentyne

d) 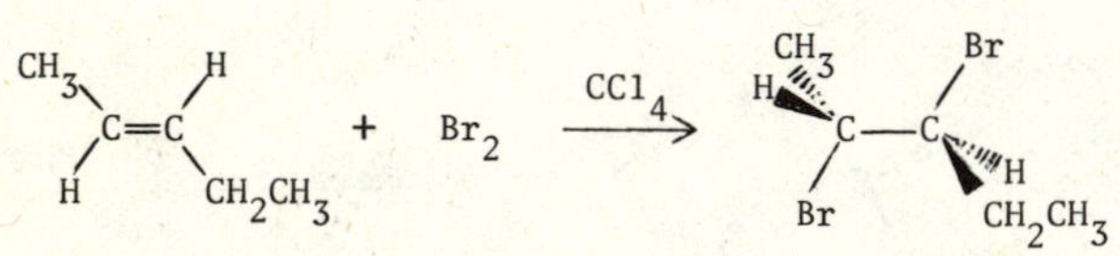 + enantiomer

erythro-2,3-Dibromopentane

e) 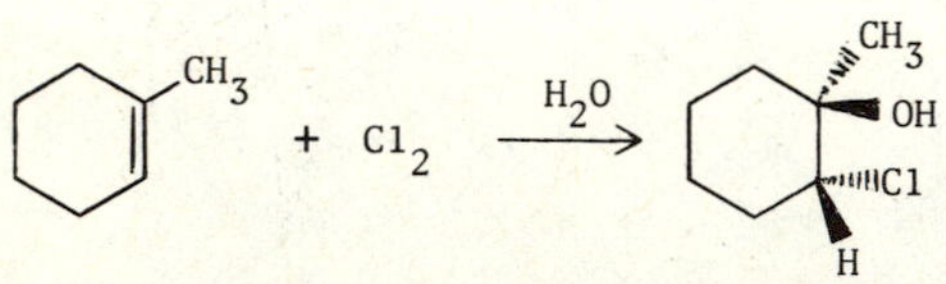 + enantiomer

1-Methyl-2-chlorocyclohexanol

f) 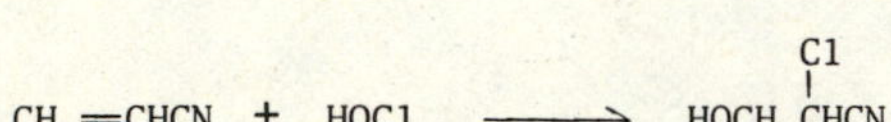

2-Chloro-3-hydroxypropanenitrile

15-12 Acetate from the solvent competes with bromide as the potential nucleophile that adds to the initially formed cation.

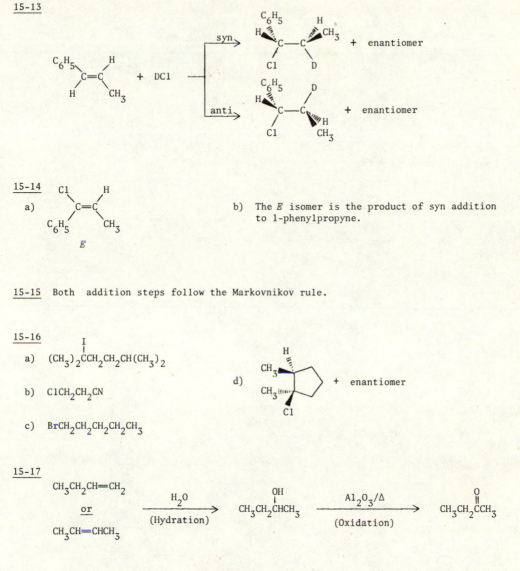

15-13

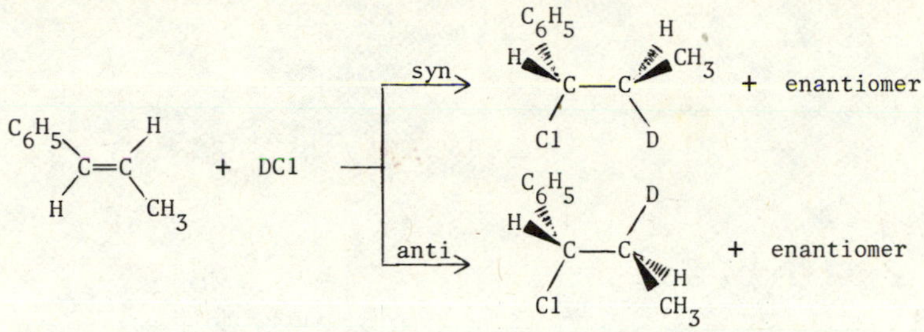

15-14

a) 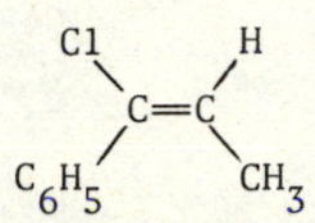

E

b) The *E* isomer is the product of syn addition to 1-phenylpropyne.

15-15 Both addition steps follow the Markovnikov rule.

15-16

a) $(CH_3)_2CI(CH_2CH_2CH(CH_3)_2$ — i.e. $(CH_3)_2\overset{I}{C}CH_2CH_2CH(CH_3)_2$

b) $ClCH_2CH_2CN$

c) $BrCH_2CH_2CH_2CH_2CH_3$

d) 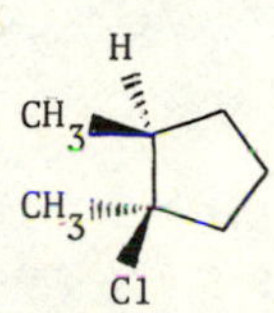 + enantiomer

15-17

$CH_3CH_2CH{=}CH_2$ or $CH_3CH{=}CHCH_3$ $\xrightarrow[\text{(Hydration)}]{H_2O}$ $CH_3CH_2\overset{OH}{C}HCH_3$ $\xrightarrow[\text{(Oxidation)}]{Al_2O_3/\Delta}$ $CH_3CH_2\overset{O}{\overset{\|}{C}}CH_3$

15-18 The addition of H^+ to the double bond is more favorable when a tertiary rather than a secondary carbocation is formed.

15-19

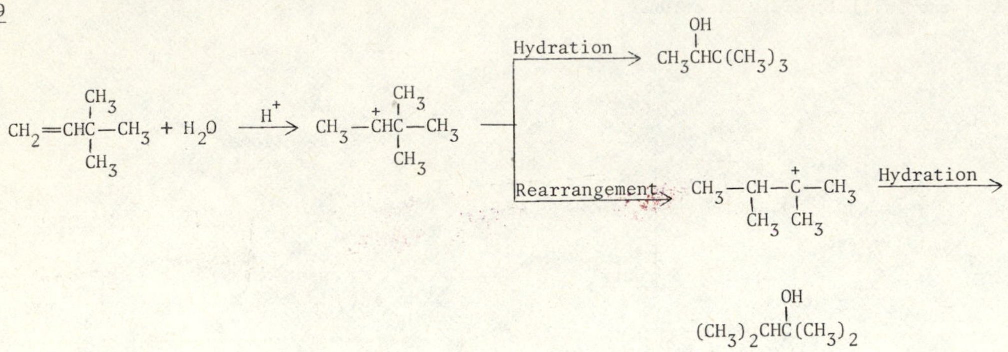

Rearrangement provides a more stable intermediate carbocation.

15-20

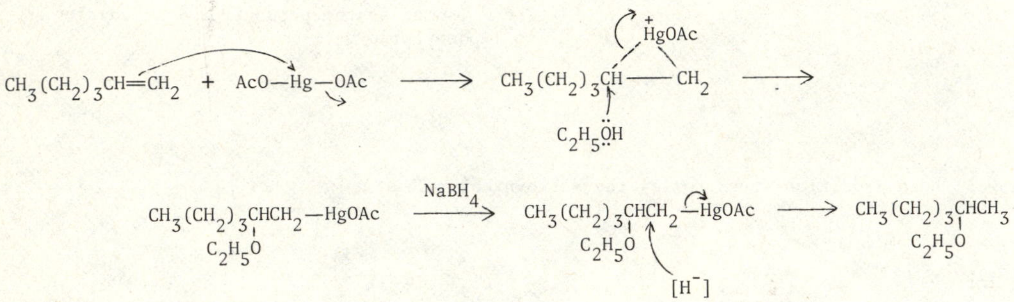

15-21

a)

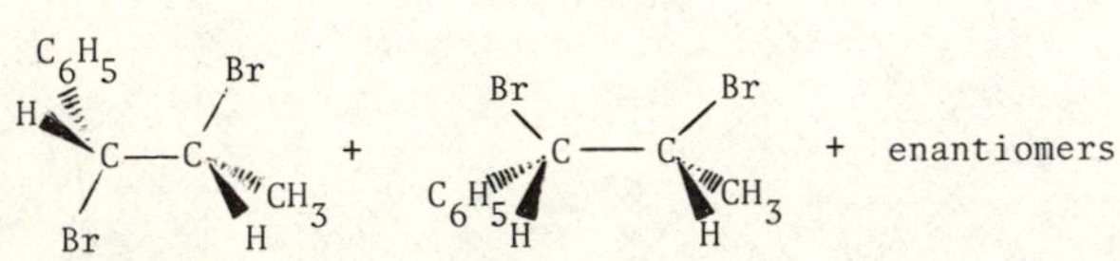

1,2-Dibromo-1-phenylpropane

b)

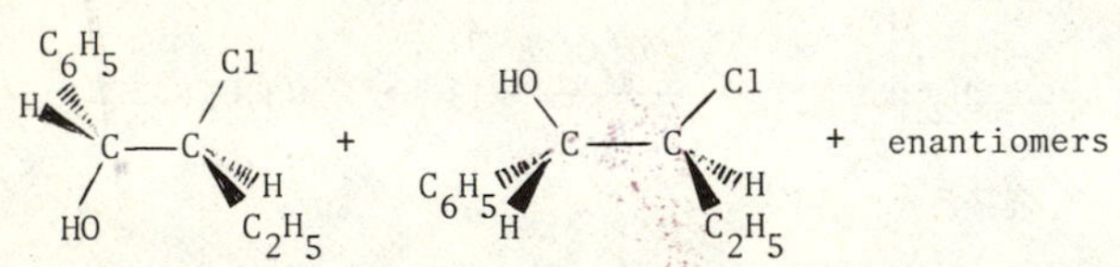

2-Chloro-1-phenyl-1-butanol

c)

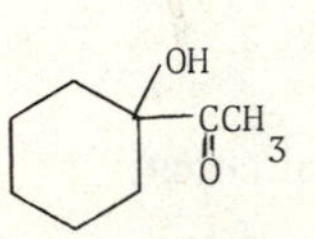

1-Acetylcyclohexanol

d)

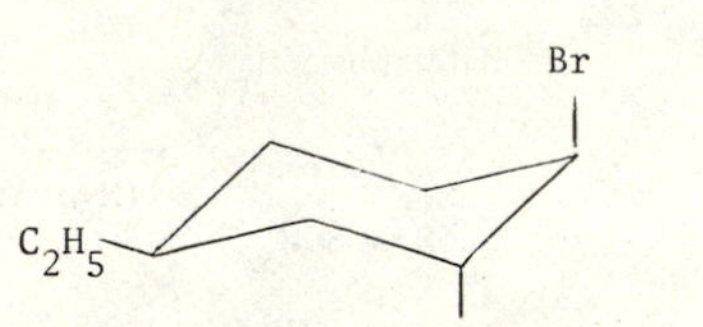

1,2-Dibromo-4-ethylcyclohexane

Contd...

15-21 contd...

e) 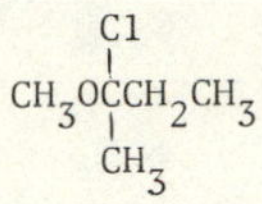

2-Chloro-2-methoxybutane

g) 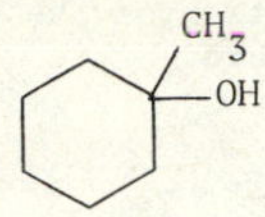

1-Methylcyclohexanol

f) $(CH_3)_2CHCH_2CH_2CH_2Br$

1-Bromo-4-methylpentane

15-22

a) $CH_3CH_2CH{=}CH_2 \xrightarrow[2)\,H_2O_2/H_2O/NaOH]{1)\,(BH_3)_2/THF} CH_3CH_2CH_2CH_2OH$

b) 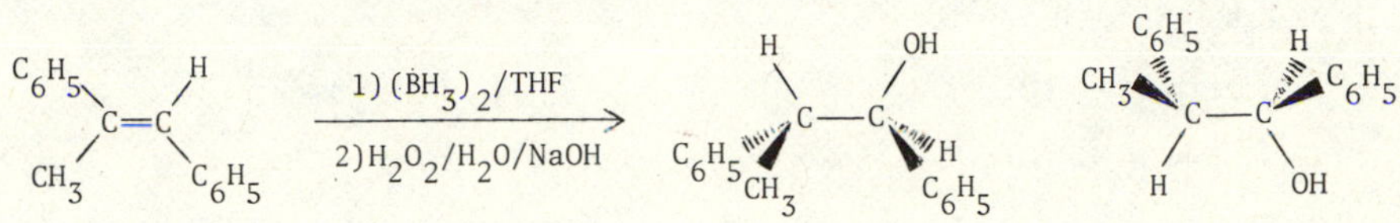

(*R*,*R*) (*S*,*S*)

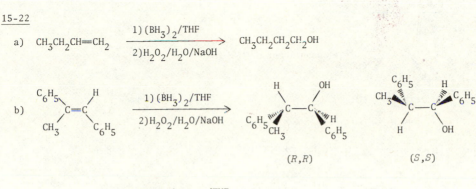

c) $CH_3CH_2C{\equiv}CH \xrightarrow[2)\,H_2O_2/H_2O/NaOH]{1)\,\text{Disiamylborane/THF}} CH_3CH_2CH_2CHO$

15-23

a) The meso product is formed by permanganate oxidation; a syn hydroxylation. The *dl* product is formed from hydrolysis of the epoxide; an anti hydroxylation.

b) 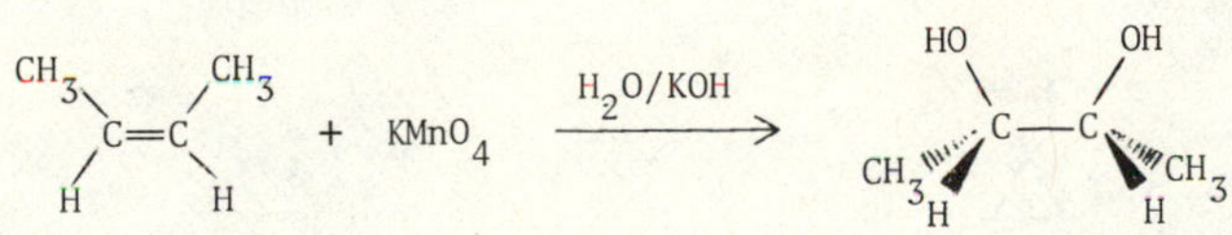

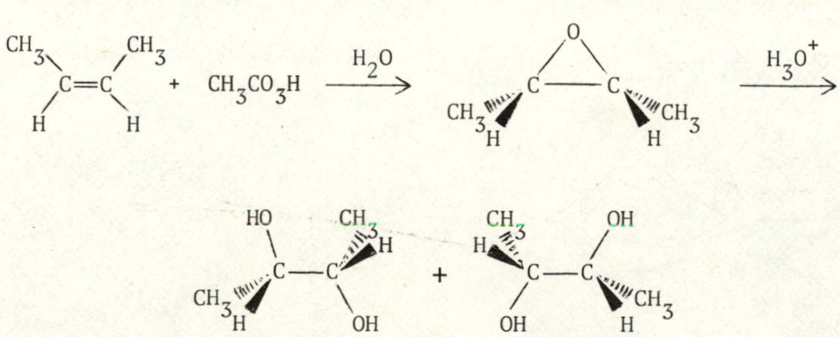

15-24 The aldehyde group can be converted to an acetal which will be stable under the basic conditions. After completion of the oxidation, the acetal protecting group is readily removed by using dilute acid.

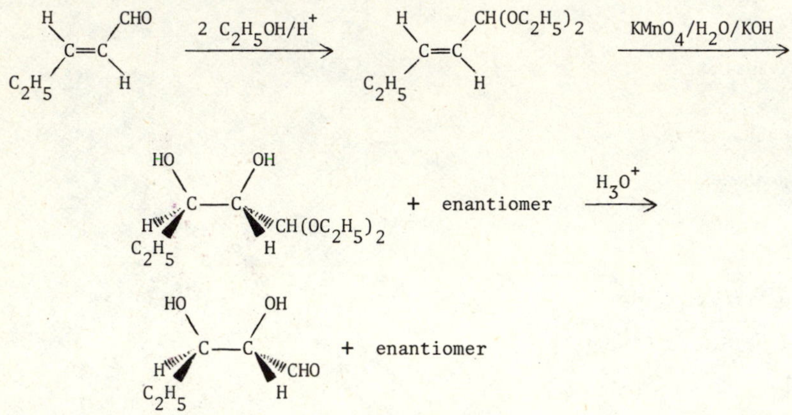

15-25

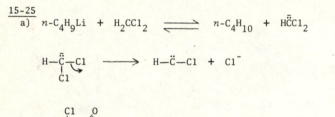

a) $n\text{-}C_4H_9Li + H_2CCl_2 \rightleftharpoons n\text{-}C_4H_{10} + H\ddot{\bar{C}}Cl_2$

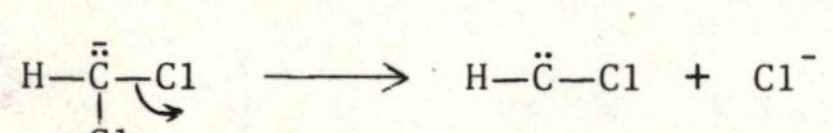

b) 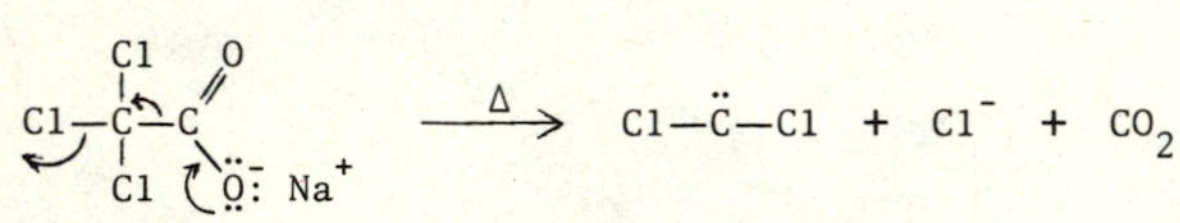

15-26

a) 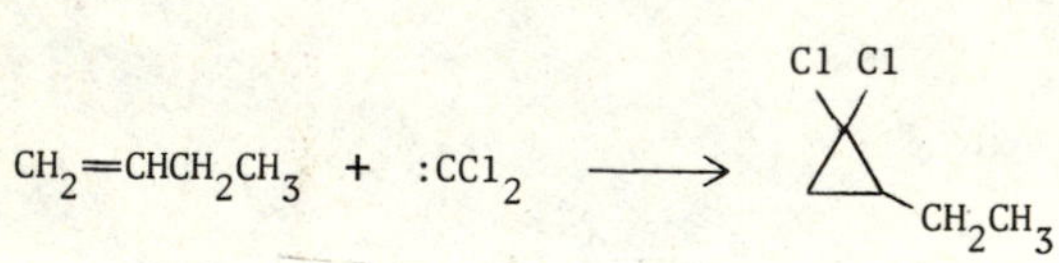

b) 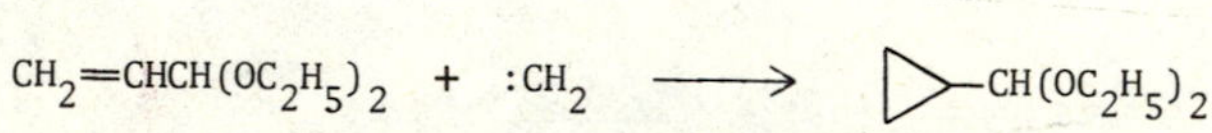

c) 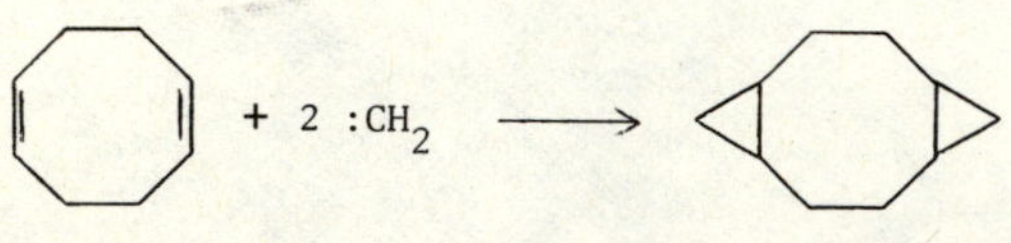

Contd...

– 15-26 contd...

d) 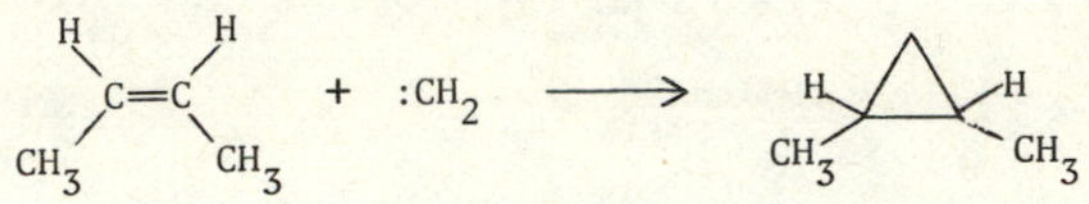

e) 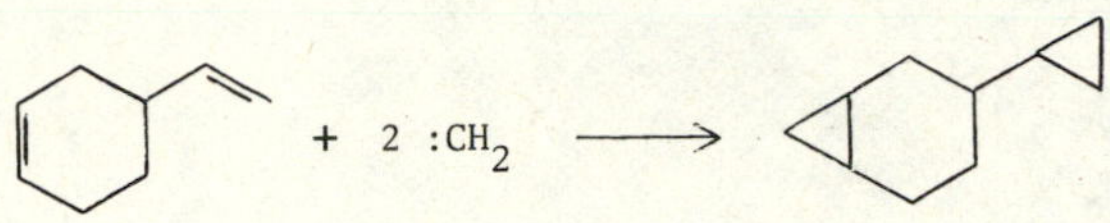

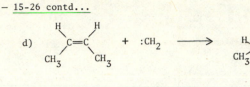

15-27 Hydrogenation indicates that only one double bond is present, thus A must have a cyclic structure in addition to the carboxy group to account for an *IHD* = 3. Since A is optically active, two sets of reasonable structures for A and B are -

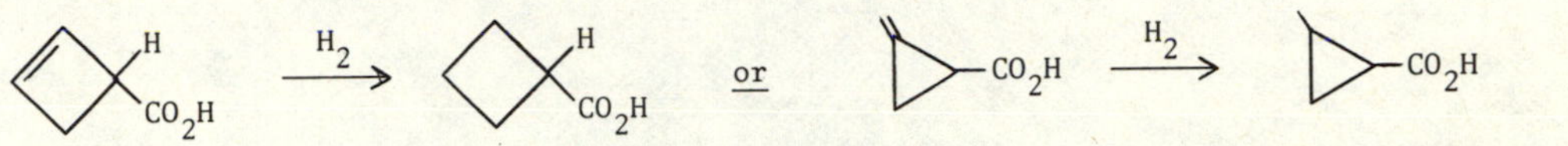

A B A B

15-28 The lead "poisons the catalyst and destroys the effectiveness of that kind of exhaust system to reduce air polluting contaminants. Furthermore, reduction of the emission of lead compounds will reduce potential health risks from such toxic materials.

15-29

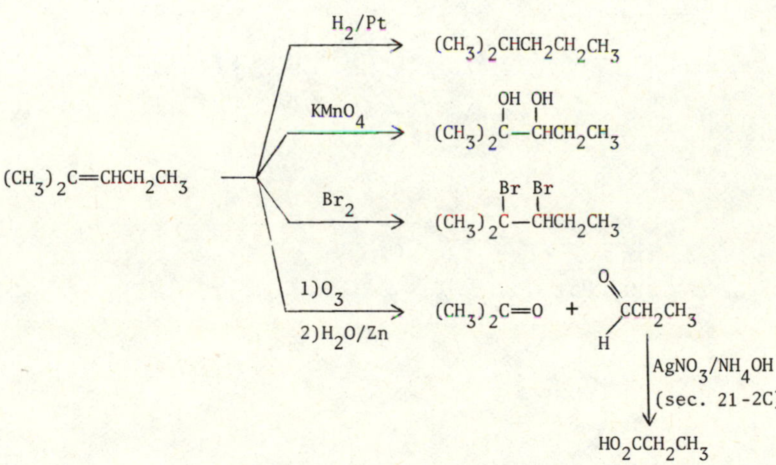

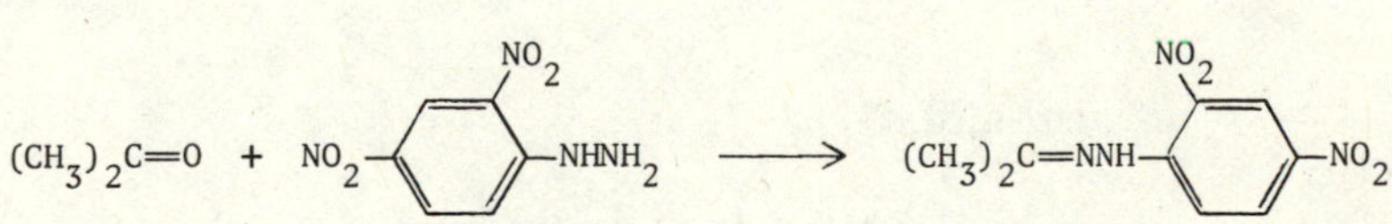

Contd...

15-29 contd...

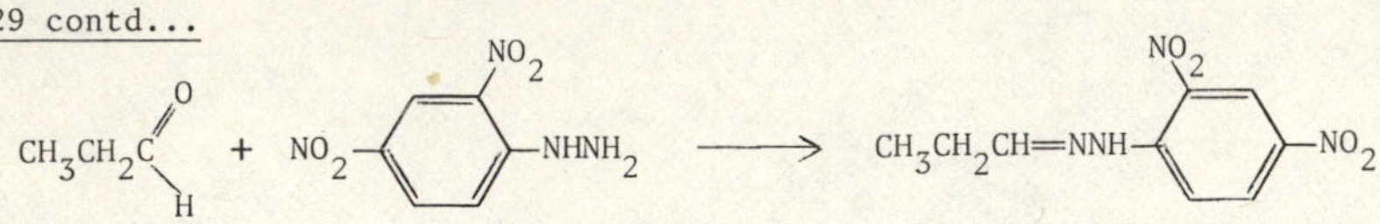

15-30

a) 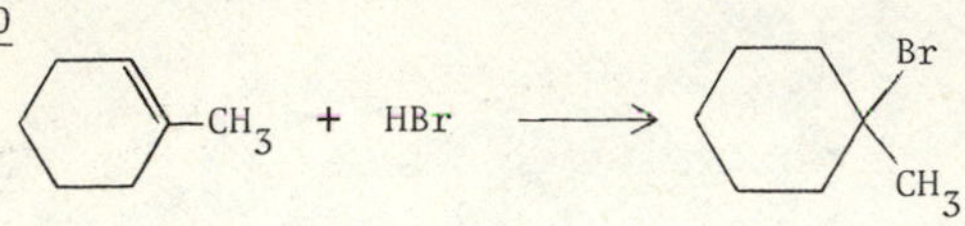

b) 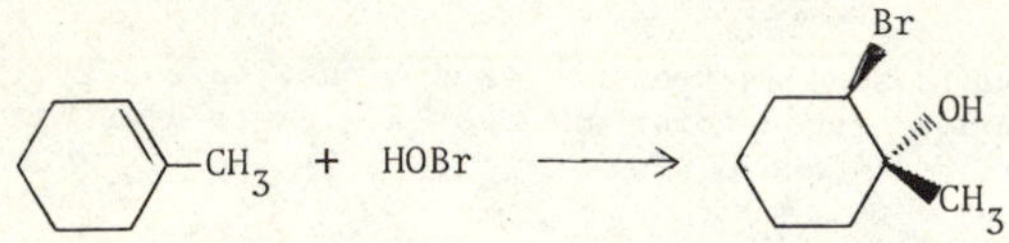

c) $(CH_3)_2C{=}CHCH_3 + HBr \xrightarrow{\text{(Peroxide)}} (CH_3)_2CHCH(Br)CH_3$

d) 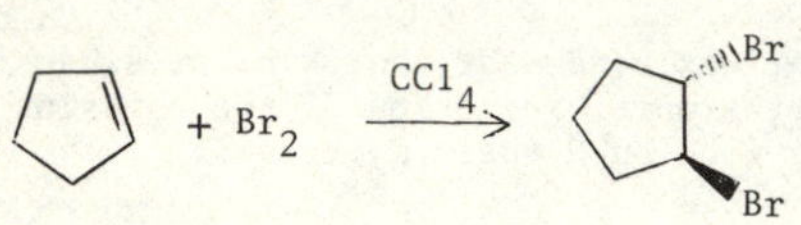

e) $CH_3CH{=}CH_2 + H_2O \xrightarrow{H_2SO_4} CH_3CH(OH)CH_3$

f) 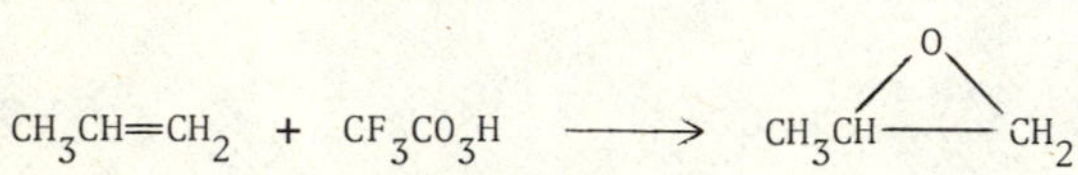

g) 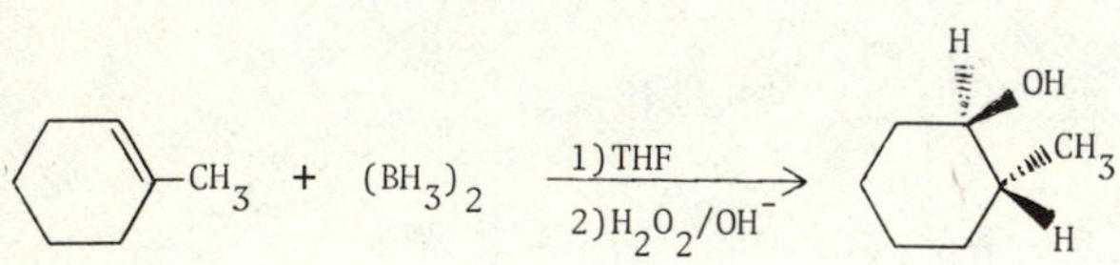

h) 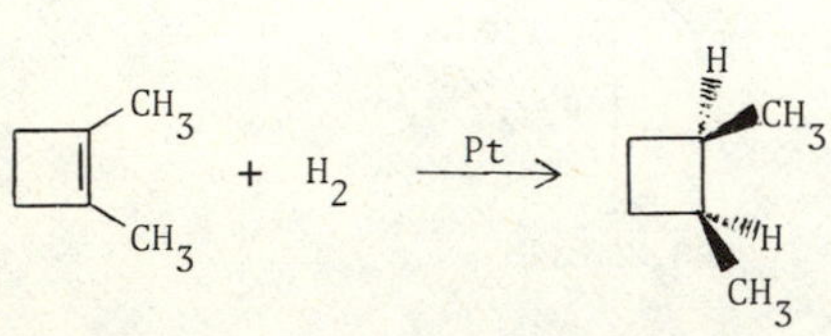

i) 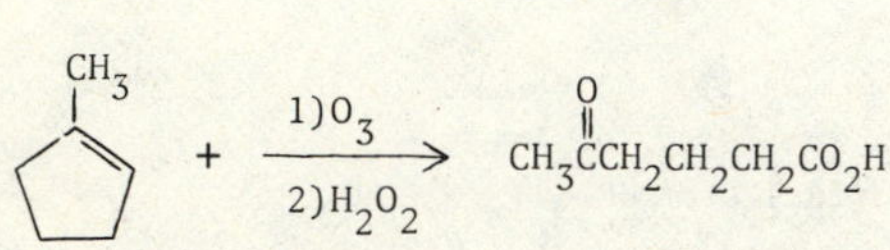

Contd...

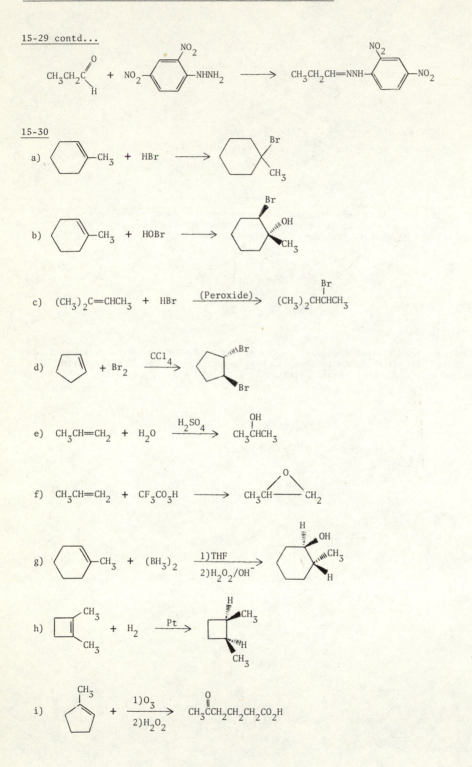

15-30 contd...

j) 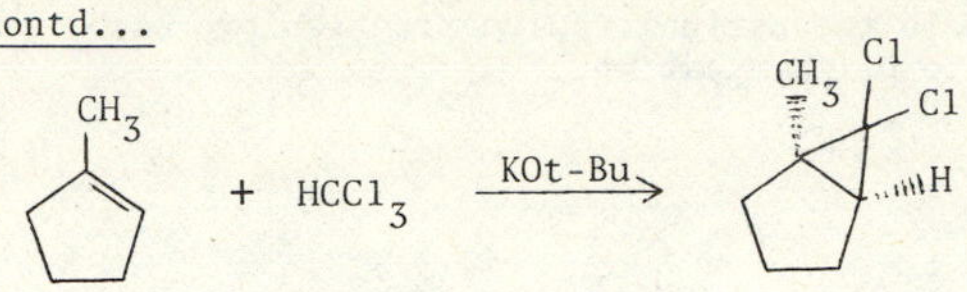

15-31

a) Molecular formula = C_4H_8 ; *IHD* = 1

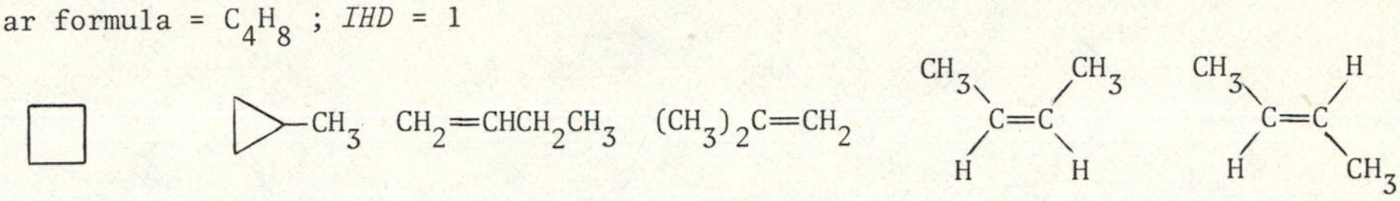

(a) (b) (c) (d) (e) (f)

b) c,d,e, and f

c) c

15-32 *IHD* = 3 for A and *IHD* = 2 for B and C. Since B and C are diketones, this fact accounts for their *IHD* value. The diketones were formed by ozonolysis so that A must possess two double bonds and a cyclic structure. Furthermore, since ketones, but not aldehydes, are formed on ozonolysis, each of the carbon-carbon double bonds must be tetrasubstituted. Five structures are possible. Structures (a), (b) and (c) are reasonable. Cyclobutadienes such as (d) and (e) have never been isolated under normal conditions.

(a) 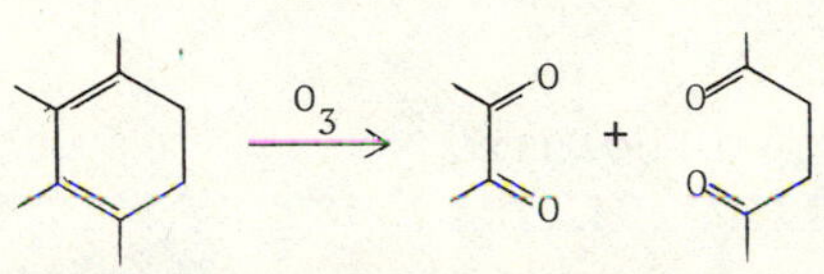

(b) 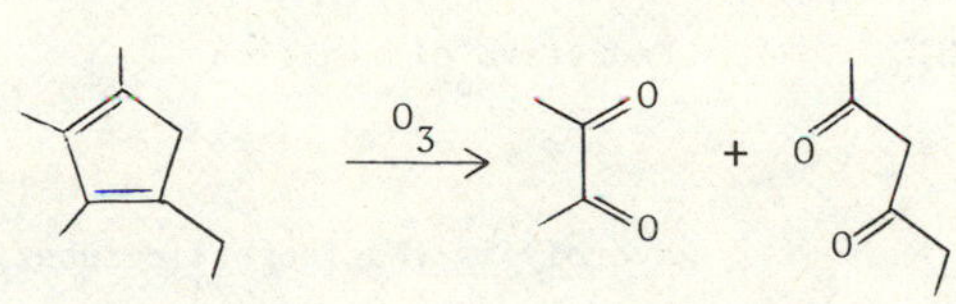

(c) 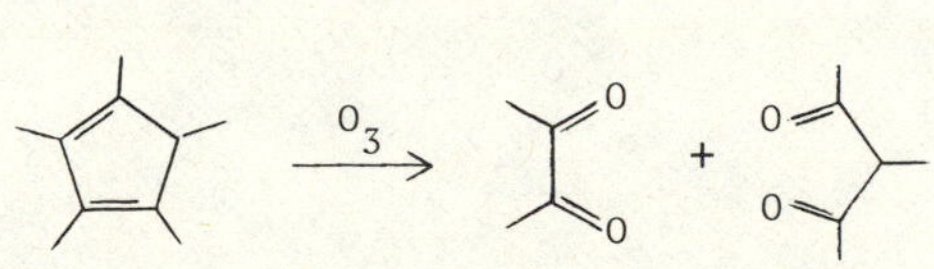

(d) 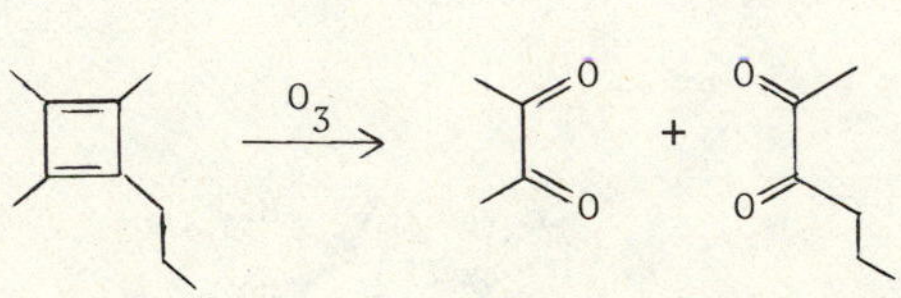

(e) 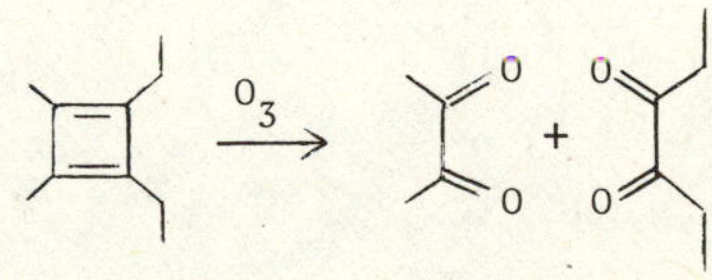

15-33 Formation of a carbocation is a key step in the cationic polymerization process. Ethylene would produce an energetically unfavorable primary cation.

15-34

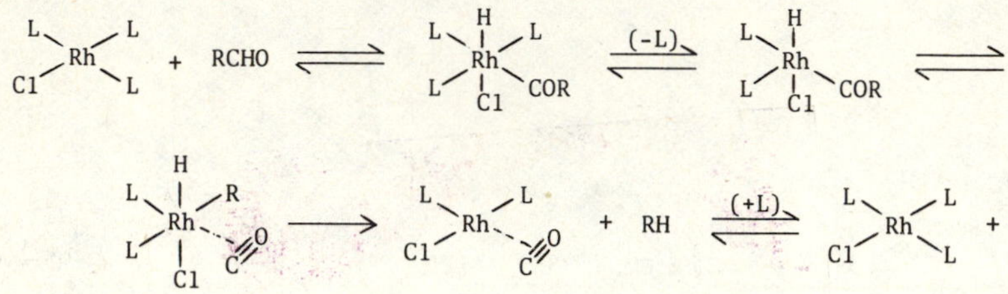

This decarbonylation sequence differs from the reverse of the analogous hydrogenation in that the order of loss of CO (analogous to the alkene) and of RH (analogous to H_2) are reversed.

15-35

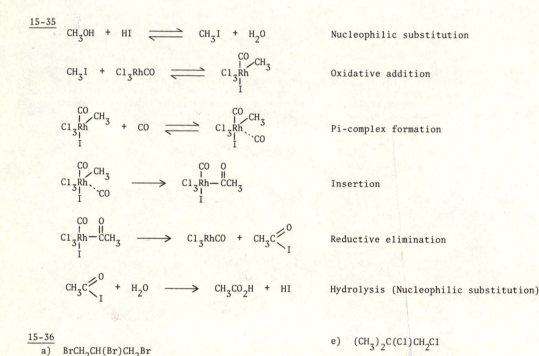

$CH_3OH + HI \rightleftharpoons CH_3I + H_2O$ Nucleophilic substitution

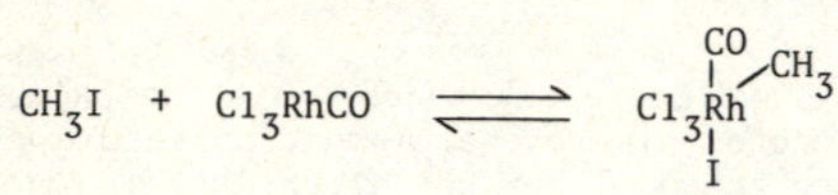

Oxidative addition

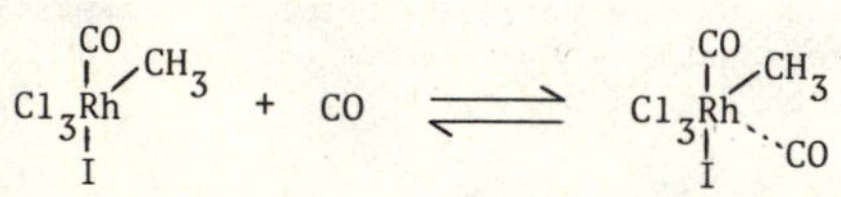

Pi-complex formation

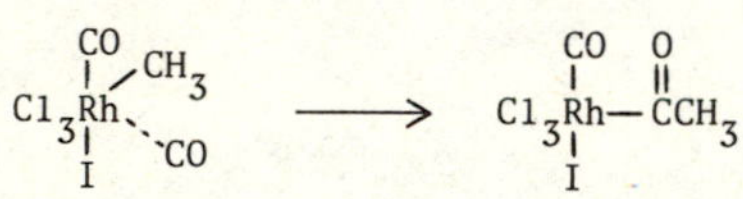

Insertion

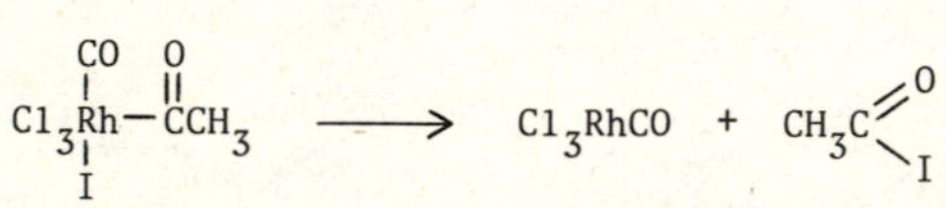

Reductive elimination

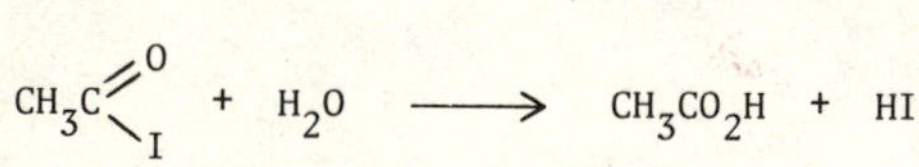

Hydrolysis (Nucleophilic substitution)

15-36

a) $BrCH_2CH(Br)CH_2Br$

b) $ClCH_2CH(Cl)CH_2CH_2CH{=}CHCO_2CH_3$

c) $Cl_3CCO_2C(CH_3)_3$

d) $CH_3(CH_2)_5\overset{O}{\overset{\|}{C}}CH_3$

e) $(CH_3)_2C(Cl)CH_2Cl$

f)

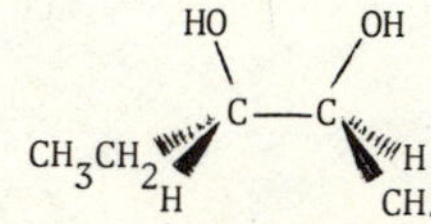

g)

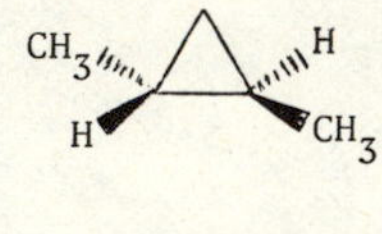

Contd...

15-36 contd...

h) $(CH_3)_3CCH_2CH_2CH(OH)CH_3$

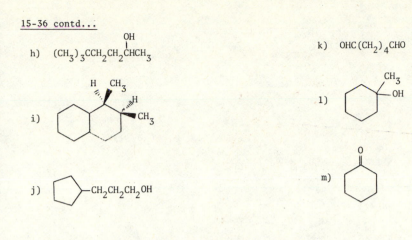

i) 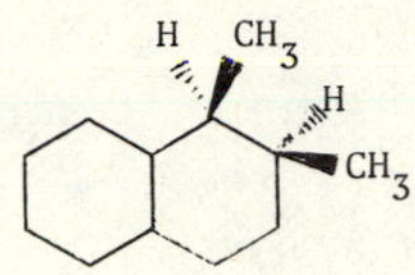

j) 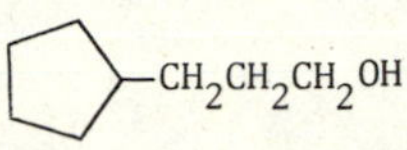

k) $OHC(CH_2)_4CHO$

l) 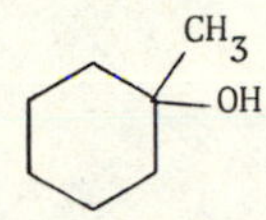

m) 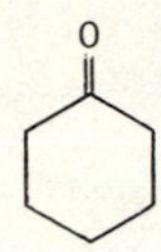

15-37

$$(CH_3)_2C{=}\overset{*}{C}H_2 + Cl_2 \longrightarrow (CH_3)_2\overset{+}{C}{-}\overset{*}{C}H_2Cl + Cl^-$$

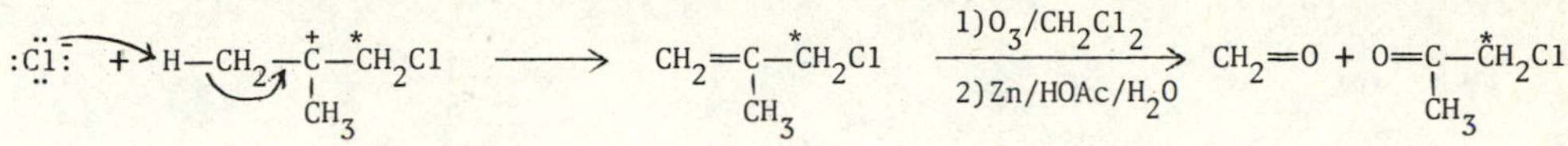

A

Addition of Cl^+ gives a relatively stable tertiary carbocation. Chloride-promoted elimination leads to A.

15-38 The carbocation intermediate that forms from either alkene is essentially identical. If we make the reasonable assumption that the transition states leading to that common intermediate are very similar, then the rate difference reflects the fact that 1-butene, the thermodynamically less stable isomer, begins the reaction energetically closer to the transition state than does 2-butene.

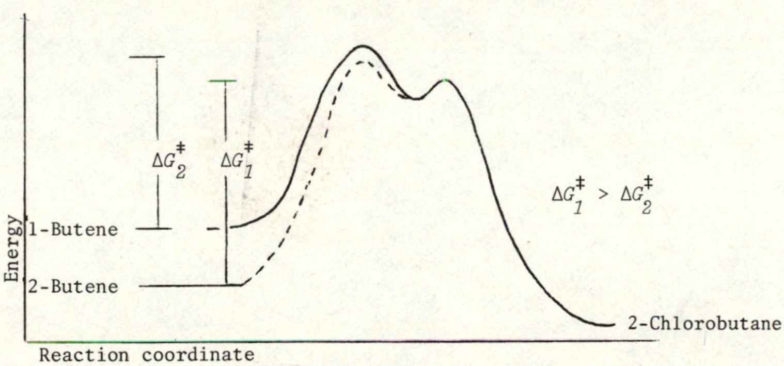

15-39

a)

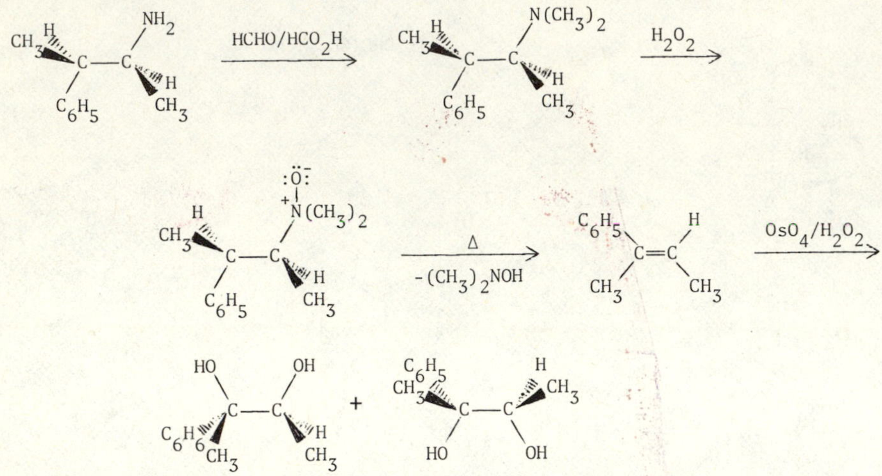
HCHO/HCO2H
H2O2
Δ
-(CH3)2NOH
OsO4/H2O2
+

b)

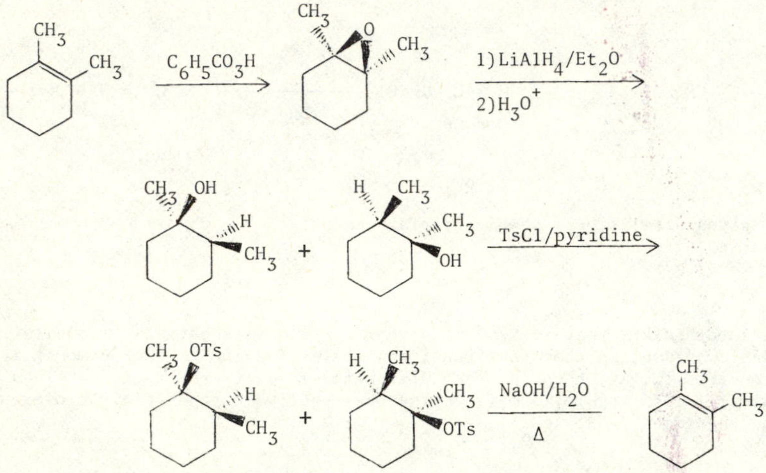
C6H5CO3H
1) LiAlH4/Et2O
2) H3O+
TsCl/pyridine
NaOH/H2O
Δ

c)

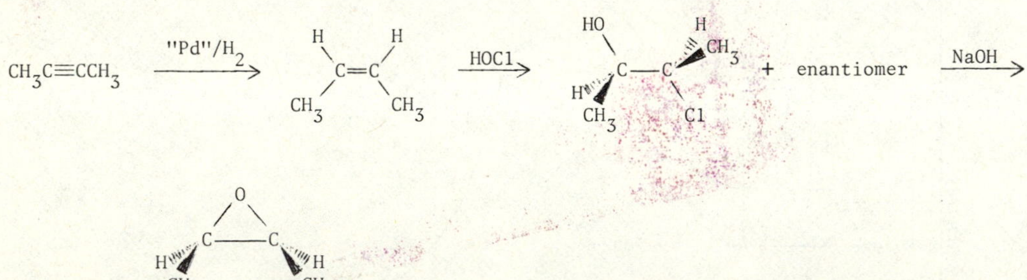
"Pd"/H2
HOCl
+ enantiomer
NaOH

15-40

a) $(CH_3)_2CHCH(OH)CH_3$

b) 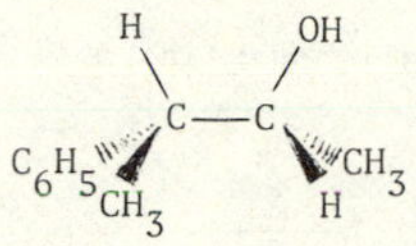 + enantiomer

c) $(CH_3)_3CCH_2CI(CH_3)_2$

d) 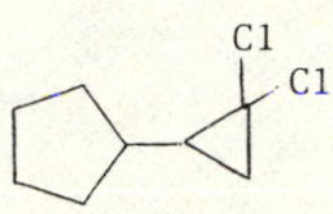

e) $ClCH_2CH(OH)Cl \longrightarrow ClCH_2CHO$

Unstable - this intermediate is an HCl adduct of an aldehyde

f) p-$CH_3C_6H_4CH(Cl)CH_3$

g) 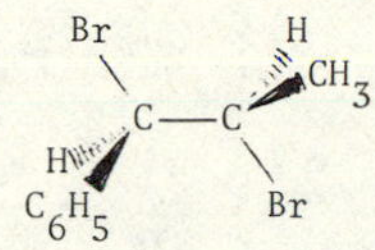 + enantiomer

threo

h) $CH_3CH_2C(OH)(CH_3)C(=O)CH_3$

i) 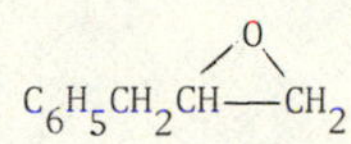

j) 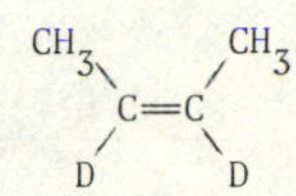

15-41

a) 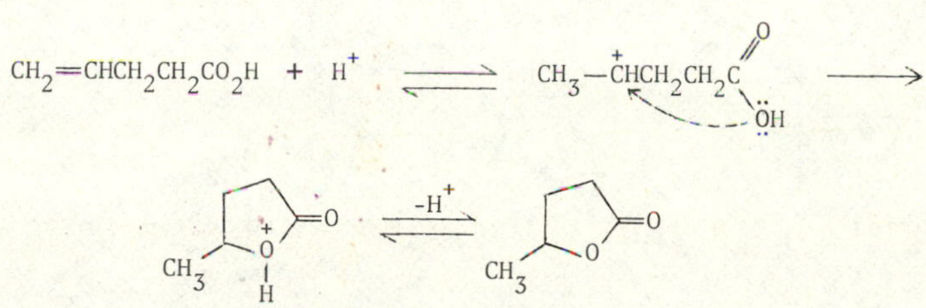

15-42

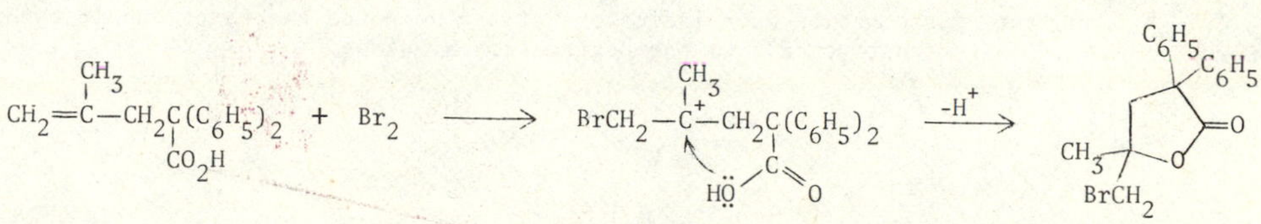

15-43

$$2\ CH_3CHO \xrightarrow{OH^-} CH_3CH(OH)CH_2CHO \xrightarrow[-H_2O]{\Delta} CH_3CH{=}CHCHO$$

Contd...

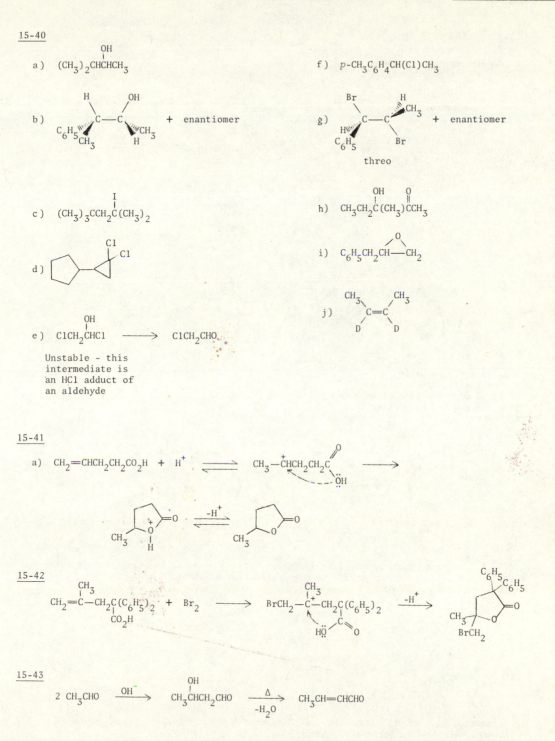

15-43 contd...

$$CH_3CH{=}CHCHO \xrightarrow[\text{Cu catalyst}]{H_2} CH_3CH_2CH_2CH_2OH$$

1-Butanol

(In the laboratory reduction of the carbonyl group would be accomplished using a metal hydride.)

$$CH_3CH{=}CHCHO \xrightarrow{H_2/Pt} CH_3CH_2CH_2CHO \xrightarrow{NaOH} CH_3CH_2CH_2CH(OH)CH(CH_2CH_3)CHO \xrightarrow[-H_2O]{\Delta}$$

$$CH_3CH_2CH_2CH{=}C(CH_2CH_3)CHO \xrightarrow{H_2/Pt} CH_3CH_2CH_2CH_2CH(CH_2CH_3)CH_2OH$$

2-Ethyl-1-hexanol

15-44

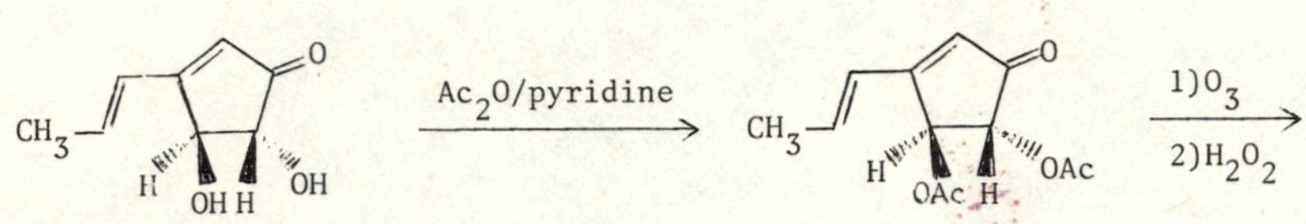

Terrein

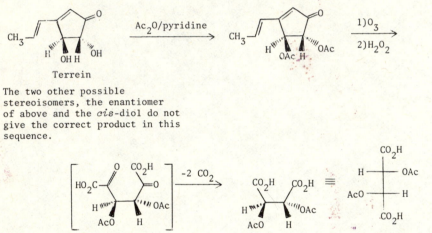

The two other possible stereoisomers, the enantiomer of above and the *cis*-diol do not give the correct product in this sequence.

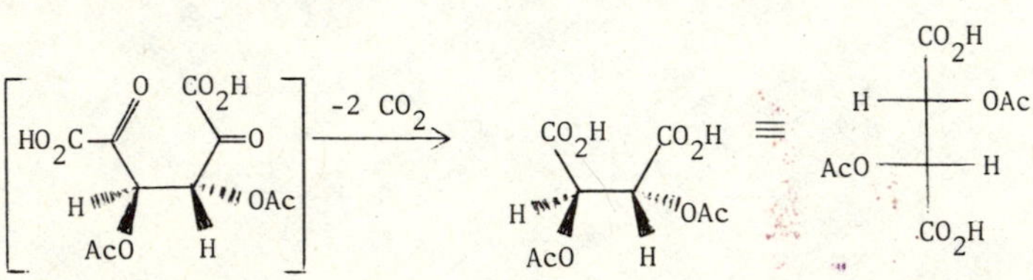

not isolated

Diacetate of (+)-tartaric acid

15-45 *IHD* = 3 and the hydrogenation data indicate that the compound has three double bonds. Four isomers can be constructed from the degradation by ozone.

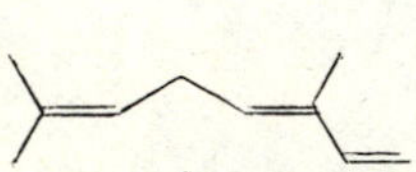

A

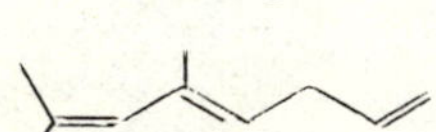

B

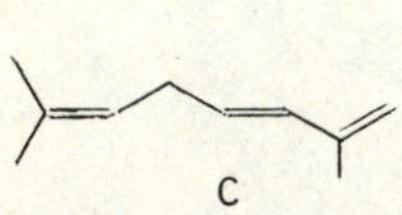

C

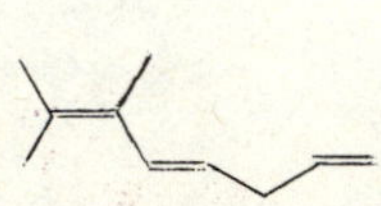

D

In section 23-2 we will learn that only compound A has the head to tail isoprenoid structure typical of terpenes.

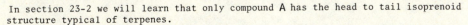

15-46 The terminal carbon-carbon double bond is less hindered, so that the oxymercuration step can be selectively carried out.

15-47

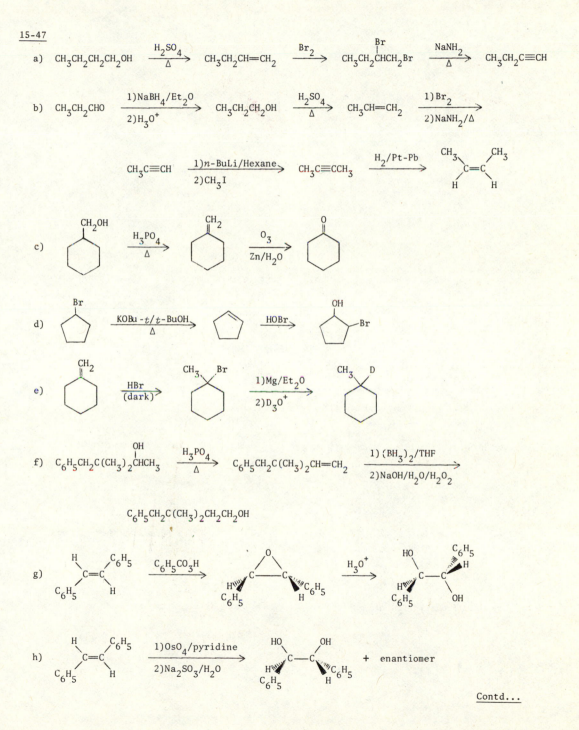

a) $CH_3CH_2CH_2CH_2OH \xrightarrow[\Delta]{H_2SO_4} CH_3CH_2CH{=}CH_2 \xrightarrow{Br_2} CH_3CH_2CHBrCH_2Br \xrightarrow[\Delta]{NaNH_2} CH_3CH_2C{\equiv}CH$

b) $CH_3CH_2CHO \xrightarrow[2)\,H_3O^+]{1)\,NaBH_4/Et_2O} CH_3CH_2CH_2OH \xrightarrow[\Delta]{H_2SO_4} CH_3CH{=}CH_2 \xrightarrow[2)\,NaNH_2/\Delta]{1)\,Br_2}$

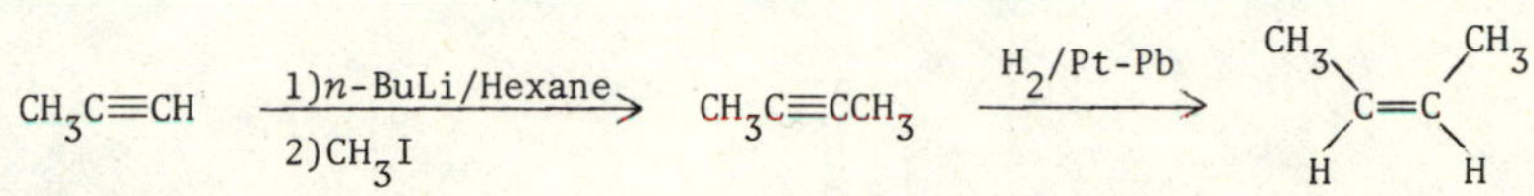

c) 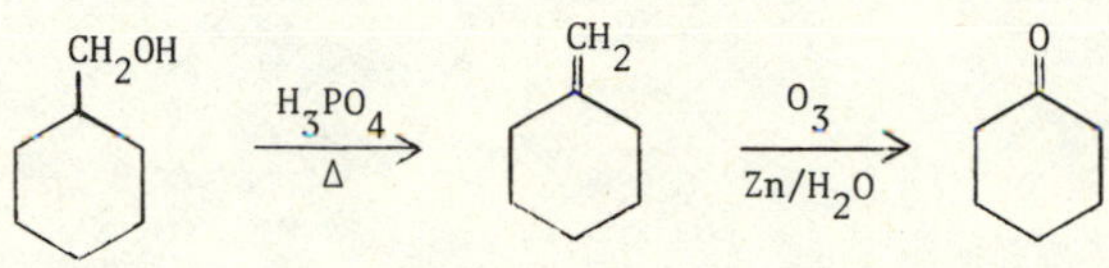

d) 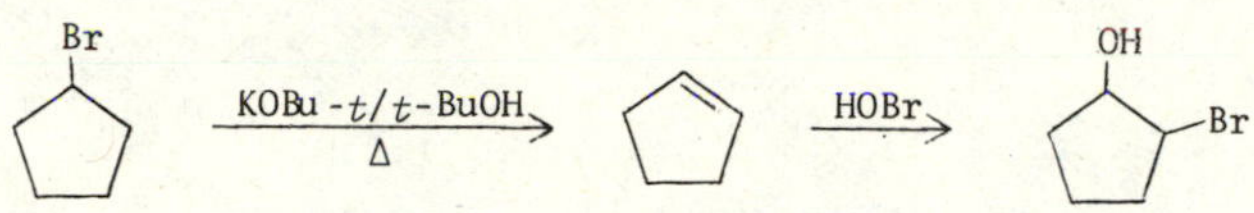

e) 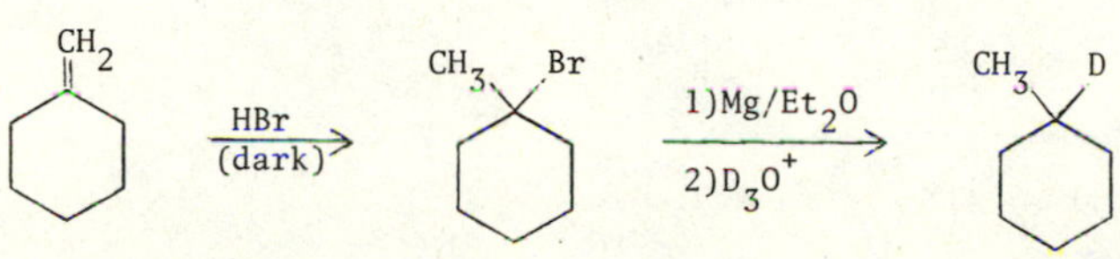

f) $C_6H_5CH_2C(CH_3)_2CH(OH)CH_3 \xrightarrow[\Delta]{H_3PO_4} C_6H_5CH_2C(CH_3)_2CH{=}CH_2 \xrightarrow[2)\,NaOH/H_2O/H_2O_2]{1)\,(BH_3)_2/THF}$

$C_6H_5CH_2C(CH_3)_2CH_2CH_2OH$

g) 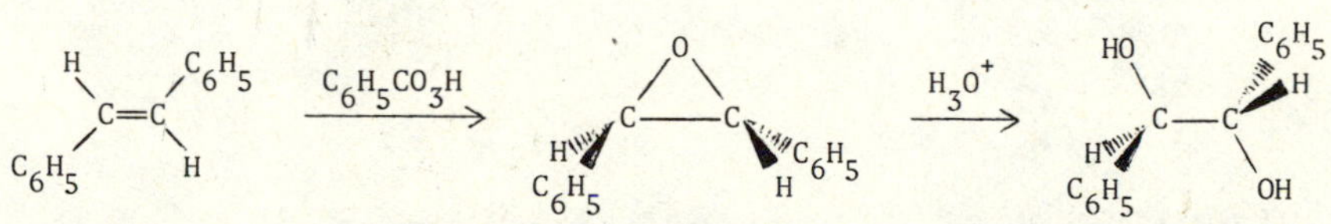

h) 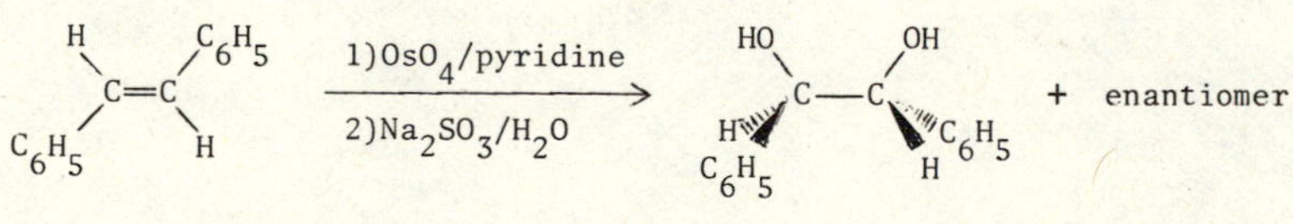

Contd...

15-47 contd...

i) 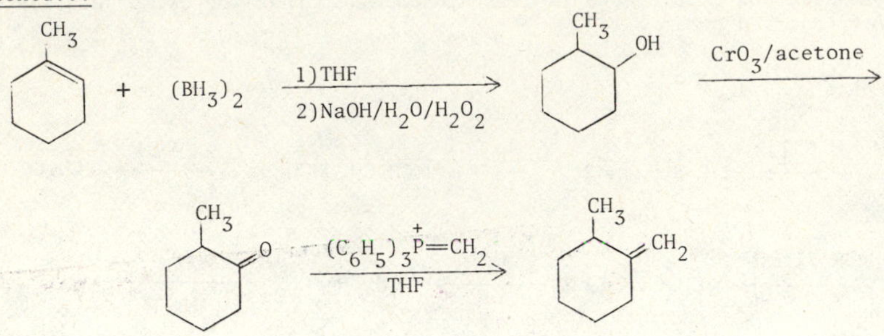

j) 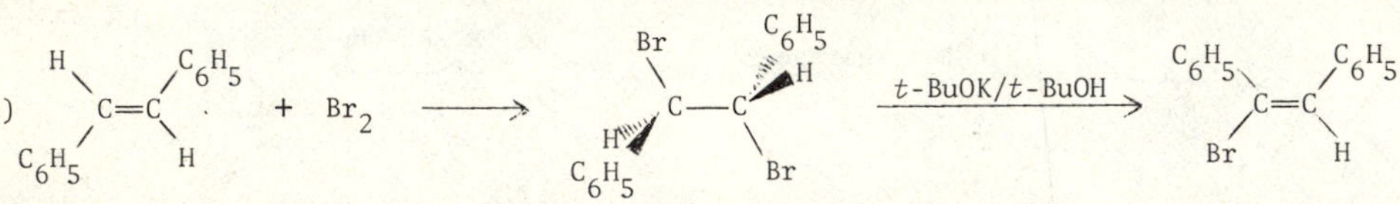

15-48 Compound A : *IHD* = 2; Compounds B and C: *IHD* = 1. With an *IHD* = 2, ozonolysis suggests a carbon-carbon double bond and the $LiAlH_4$ reduction is consistent with a carbonyl group.

$$CH_3\overset{O}{\overset{\|}{C}}CH_2CH{=}CHCH_2\overset{OH}{\overset{|}{C}}HCH_3 \xrightarrow{LiAlH_4} CH_3\overset{OH}{\overset{|}{C}}HCH_2CH{=}CHCH_2\overset{OH}{\overset{|}{C}}HCH_3 \xrightarrow{H_2/Pt} CH_3\overset{OH}{\overset{|}{C}}H(CH_2)_4\overset{OH}{\overset{|}{C}}HCH_3$$

A — (A must have at least one chiral center.)

B (opt. active); C (meso) — (A new chiral center is produced in the reduction step.)

D (opt. active); E (meso)

$$A \xrightarrow{NaOI} 2\ CHI_3 + HO_2CCH_2CH{=}CHCH_2CO_2H$$

F

(The haloform reaction shows the presence of a methyl ketone or methyl carbinol.)

$$A \xrightarrow[2)\,Zn/H_2O]{1)\,O_3/HOAc} OHCCH_2\overset{OH}{\overset{|}{C}}HCH_3 + CH_3\overset{O}{\overset{\|}{C}}CH_2CHO$$

G H

(Ozonolysis cleaves a double bond and two carbonyl groups are formed at the original alkene carbon atoms.)

$$G \xrightarrow{NaOI} HO_2CCH_2CO_2H + CHI_3 \xrightarrow{\Delta} CH_3CO_2H + CO_2$$

(The haloform reaction cleaves the bond to methyl and oxidizes the aldehyde group.)

Contd...

15-48 contd...

$$H \xrightarrow{[Ox]} CH_3\overset{O}{\overset{\|}{C}}CH_2CO_2H \xrightarrow{\Delta} CH_3\overset{O}{\overset{\|}{C}}CH_3 + CO_2$$

(Oxidation converts the aldehyde group to a carboxylic acid which readily decarboxylates. H must be a β-ketoacid or related compound.)

The sequence recorded above shows how all of the data fits together. Solving of the problem commonly begins at the end of the chemical elucidation and works backwards to construct each of the pieces in the puzzle.

15-49 *IHD*= 3 for A and B. The molecular formula indicates that the nmr integral areas actually correspond to double the number of H atoms. The 3:2 triplet-quartets must correspond to two equivalent ethyl groups in each compound probably attached to an oxygen atom to account for the deshielded CH_2. Catalytic hydrogenation, the presence of four oxygen atoms, and the nmr data suggest that A and B are diesters containing one double bond. The H's that can be attributed to alkenyl grouping in A and B have been converted to two equivalent CH_2 groups in C deshielded by C=O.

$$C_2H_5O_2CCH{=}CHCO_2C_2H_5 \xrightarrow{H_2/Pt} C_2H_5O_2CCH_2CH_2CO_2C_2H_5$$

(*Z*) and (*E*)

A + B C

15-50 *IHD* = 2 and the hydrogenation data indicates that the compound must be a cycloalkene. The nmr spectrum suggests a methyl group, slightly deshielded, one alkene proton, and a multiplet, probably due to three CH_2 groups.

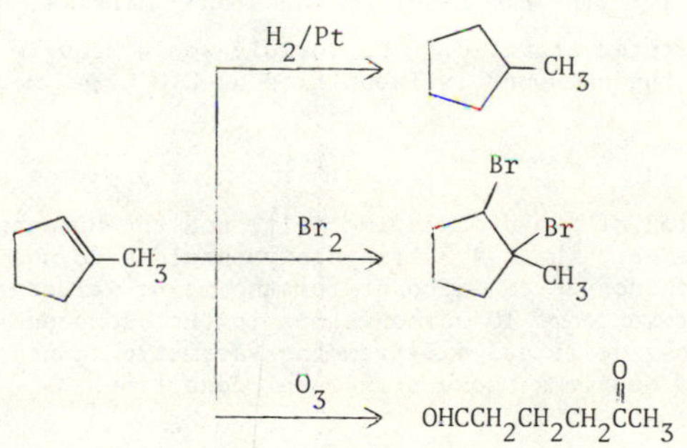

15-51 *IHD* = 6 and the hydrogenation results suggest that two triple bonds are present. The nmr peak areas actually correspond to 6:2 protons indicative of two equivalent methyl groups and a CH_2 or two equivalent C-H groups. The position of the CH_3 resonance suggests that there are two methyl ketones, a conclusion supported by the formation of oximes and the molecular formula. Only one structure (A) is consistent with these data.

Contd...

15-51 contd...

$$\underset{\textbf{A}}{CH_3\overset{\overset{O}{\|}}{C}C\equiv CCH_2C\equiv C\overset{\overset{O}{\|}}{C}CH_3} \xrightarrow{H_2} \underset{\textbf{B}}{CH_3\overset{\overset{O}{\|}}{C}CH{=}CHCH_2CH{=}CH\overset{\overset{O}{\|}}{C}CH_3} \xrightarrow{H_2}$$

$$\textbf{A} \xrightarrow{H_2} \underset{\textbf{C}}{CH_3\overset{\overset{O}{\|}}{C}(CH_2)_5\overset{\overset{O}{\|}}{C}CH_3}$$

$$A \xrightarrow{H_2NOH} CH_3\overset{\overset{NOH}{\|}}{C}C\equiv CCH_2C\equiv C\overset{\overset{NOH}{\|}}{C}CH_3$$

$$B \xrightarrow{H_2NOH} CH_3\overset{\overset{NOH}{\|}}{C}CH{=}CHCH_2CH{=}CH\overset{\overset{NOH}{\|}}{C}CH_3$$

$$C \xrightarrow{H_2NOH} CH_3\overset{\overset{NOH}{\|}}{C}(CH_2)_5\overset{\overset{NOH}{\|}}{C}CH_3$$

$$B \xrightarrow[2)\ Zn/HOAc]{1)\ O_3} 2\ CH_3\overset{\overset{O}{\|}}{C}CHO + OHCCH_2CHO$$

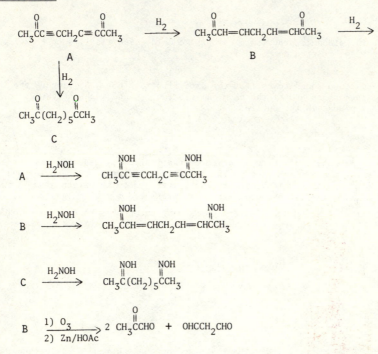

15-52 Combustion analysis provides an empirical formula of C_7H_5 and the mass spectrum indicates $C_{14}H_{10}$ is the molecular formula; *IHD* = 10. The nmr spectrum shows only aromatic H's and the IR spectrum suggests mono-substituted aromatic. The formula would require two aromatic rings plus two carbons so that the compound is identified as $C_6H_5C\equiv CC_6H_5$.

15-53 The presence of a nitrogen atom, formation of a hydrochloride salt, and the Hofmann degradation show that mescaline is an amine. The 3,4,5-trihydroxybenzaldehyde and its presumed precursor, 3,4,5-trimethoxybenzaldehyde can account for the major structural component of mescaline. In fact, the presence of 10 carbon atoms in that compound and *IHD* = 4 for mescaline suggests that mescaline is a 3,4,5-trimethoxyaromatic connected to a two carbon alkylamine. The nmr spectrum confirms those structural deductions.

Contd...

15-53 contd...

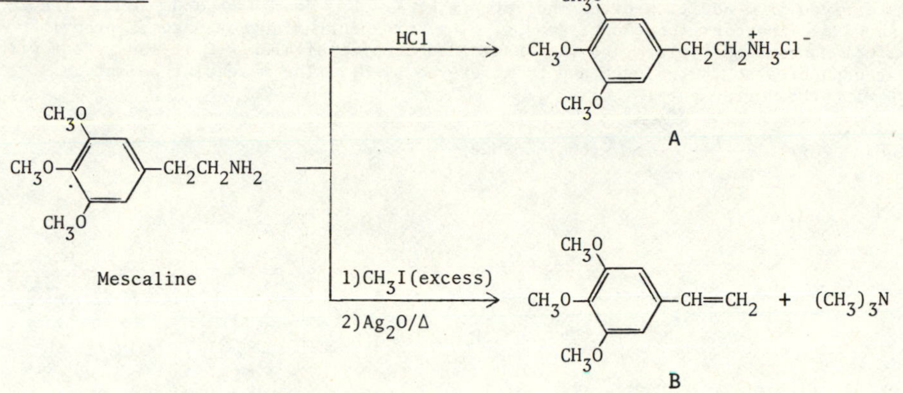

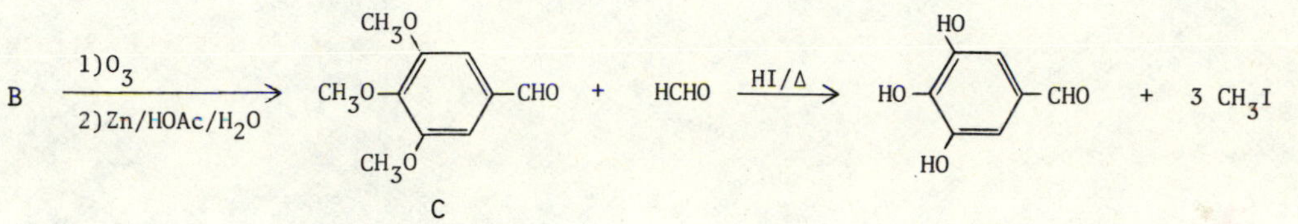

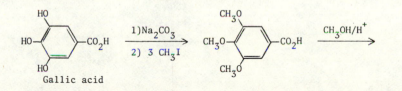

The peak areas in the nmr spectrum of gallic acid actually account for 4:2 hydrogen atoms. Two are typically aromatic and the other four could be accounted for by carboxy and hydroxy protons which rapidly exchange, thus give a single peak. The spectral data does not provide a unique structure for gallic acid but our knowledge of the mescaline structure suggests 3,4,5-trihydroxybenzoic acid.

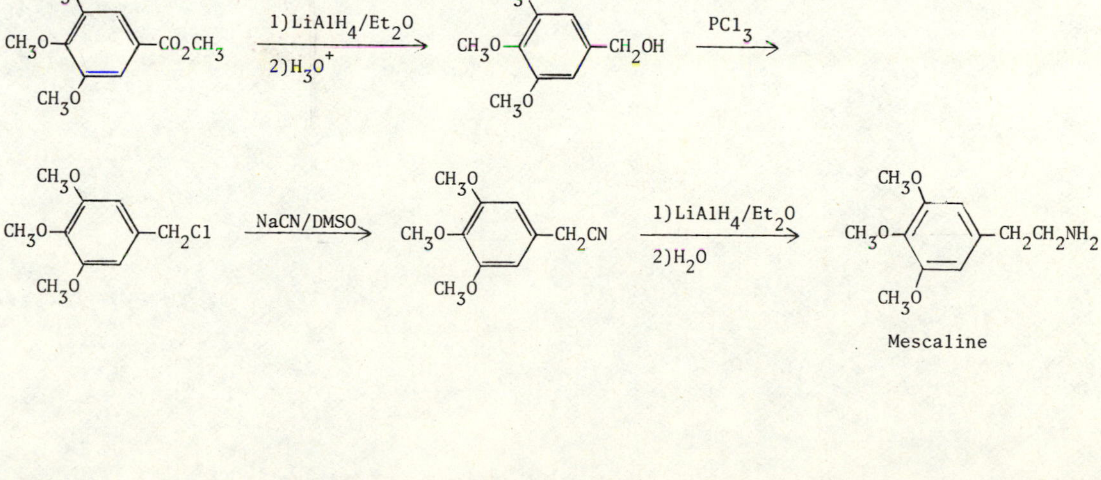

15-54 The two approximately equal [M^+] peaks in the mass spectrum and the information that the product is derived from addition of HX indicates that X = Br (from ^{79}Br and ^{81}Br). The IR spectrum shows no important functional groups. The nmr spectrum suggests an isopropyl group (the 6:1 doublet-multiplet) and the deshielded doublet of area = 2 is consistent with a methylene connected to the halogen and to a methyne (C—H). The molecular weight is consistent with those fragments.

$$(CH_3)_2C{=}CH_2 + HBr \xrightarrow{\text{Peroxide}} (CH_3)_2CHCH_2Br$$

16 Additions to Conjugated Compounds

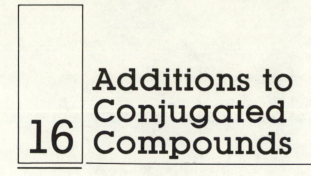

16-1

a)

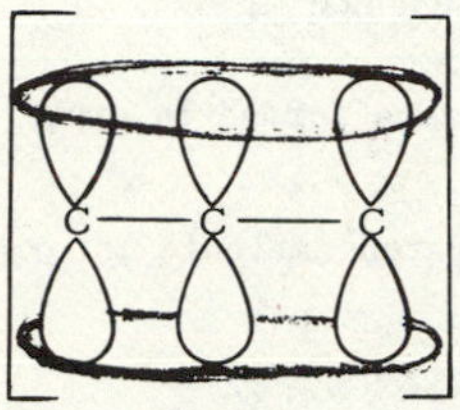

b) $CH_2{=}CH{-}CH{=}CH_2 + H^+ \rightleftharpoons CH_2{=}CH{-}CH_2{-}\overset{+}{C}H_2$

This potential cation is primary and nonstabilized. It would be considerably less stable than the allylic cation resonance structures depicted in the text.

16-2

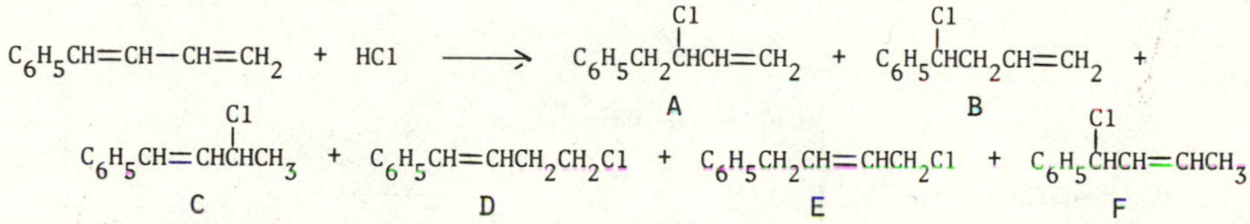

Compound D is unlikely since its carbocation precursor is primary and nonstabilized. Compound C and F come from the most favorable carbocations.

16-3

a)

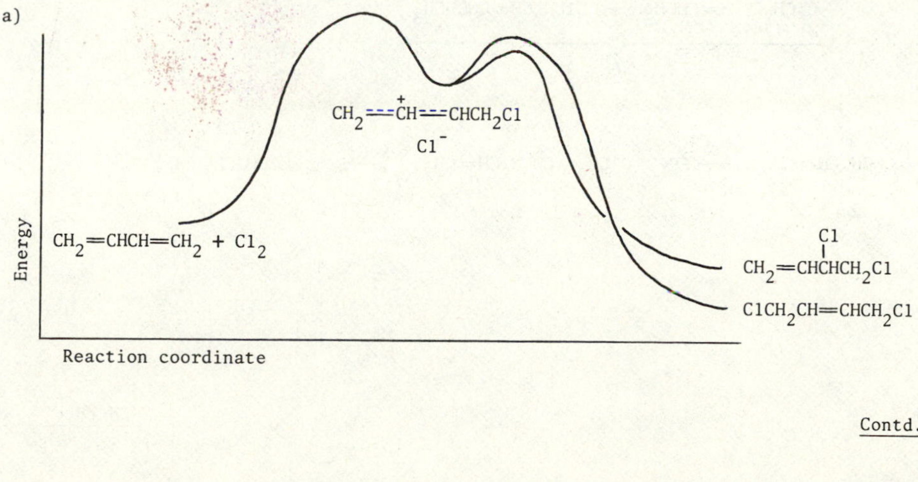

Contd...

16-3 contd...

b) 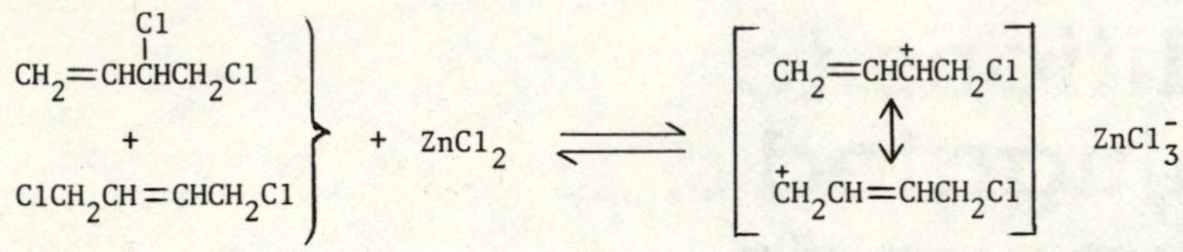

ZnCl$_2$ is a Lewis acid catalyst for the isomerization.

16-4

a) $C_6H_5CH(Cl)\text{—}CH(Cl)CH{=}CH_2$ $\qquad C_6H_5CH(Cl)CH{=}CHCH_2Cl$ $\qquad C_6H_5CH{=}CHCH(Cl)CH_2Cl$

(Note that the symmetrical reagent leads to less products than were formed in prob. 16-2.

b) Conjugation with the benzene ring favors the most highly conjugated cationic intermediate which then leads to the most highly conjugated product.

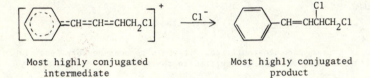

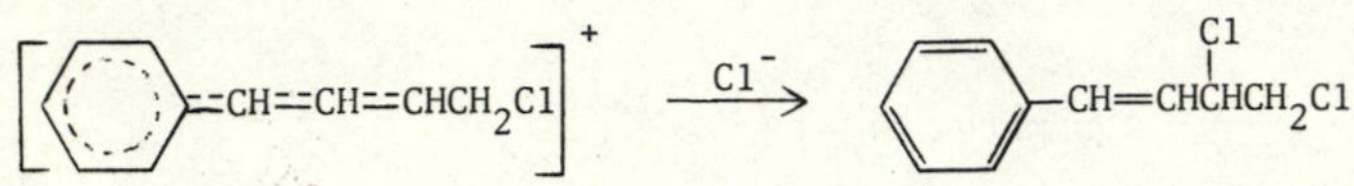

Most highly conjugated intermediate $\qquad$ Most highly conjugated product

16-5

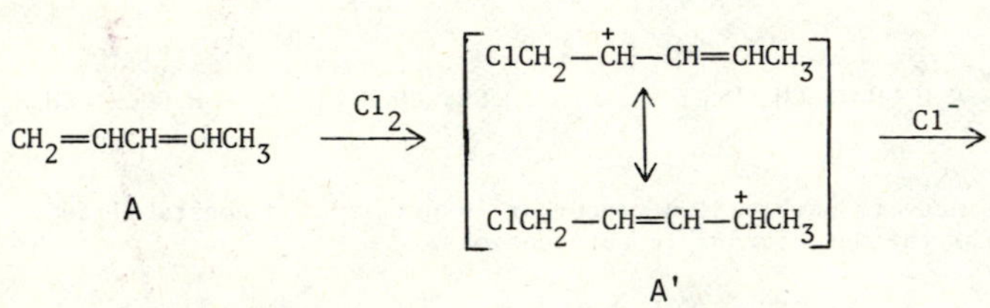

$ClCH_2CH(Cl)CH{=}CHCH_3 + ClCH_2CH{=}CHCH(Cl)CH_3$

A''

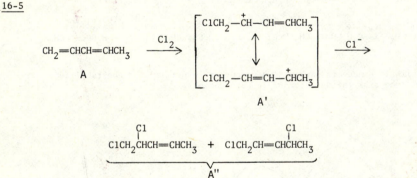

$CH_2{=}CHCH_2CH{=}CH_2 \xrightarrow{Cl_2} ClCH_2\text{—}\overset{+}{C}HCH_2CH{=}CH_2 \xrightarrow{Cl^-} ClCH_2CH(Cl)CH_2CH{=}CH_2$

B $\qquad$ B' $\qquad$ B''

Contd...

16-5 contd...

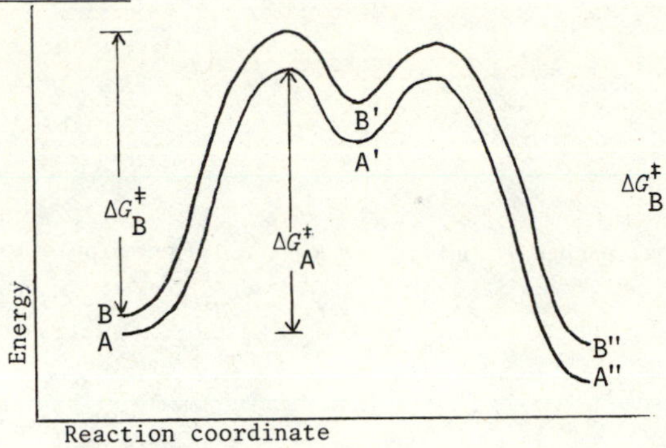

$\Delta G^{\ddagger}_{B} > \Delta G^{\ddagger}_{A}$

The conjugated diene A is not only more stable than B but the resonance-stabilized intermediate A′ and the transition state leading to it are relatively more stabilized than is B′. Addition of Cl_2 to A is therefore more rapid than similar addition to B.

16-6

$$CH_2{=}CHCH{=}CHCH{=}CH_2 + H^+ \rightleftharpoons$$

$$[CH_3{-}\overset{+}{C}HCH{=}CH{=}CH_2 \longleftrightarrow CH_3CH{=}CH\overset{+}{C}HCH{=}CH_2 \longleftrightarrow CH_3CH{=}CHCH{=}CH\overset{+}{C}H_2]$$

Delocalization of the positive charge over a six atom conjugated system is particularly favorable. Resonance stabilization of this cationic intermediate is more important than stabilization of the starting material because the latter would require structures in which charge is generated and separated.

16-7

a) $$CH_2{=}\overset{CH_3}{\overset{|}{C}}CH{=}CH_2 + Br_2 \xrightarrow[\Delta]{HOAc} BrCH_2\overset{CH_3}{\overset{|}{C}}{=}CHCH_2Br$$

b) $$CH_2{=}\overset{CH_3}{\overset{|}{C}}{-}\overset{CH_3}{\overset{|}{C}}{=}CH_2 + HBr \longrightarrow CH_3\overset{CH_3}{\overset{|}{C}}{=}\overset{CH_3}{\overset{|}{C}}CH_2Br$$

c) $$CH_2{=}CH{-}CH{=}CH{-}CH{=}CH_2 + Br_2 \xrightarrow{HOAc} BrCH_2{-}\overset{Br}{\overset{|}{C}}HCH{=}CHCH{=}CH_2 + BrCH_2CH{=}CHCH{=}CHCH_2Br$$

(3,6-Dibromo-1,4-hexadiene is non-conjugated and thus less likely to form in this reaction.)

d) 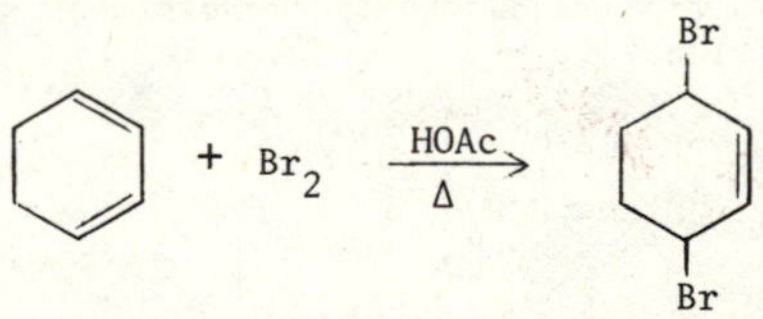

e) $$CH_2{=}CH{-}CH{=}CH_2 + CH_3OH + Cl_2 \longrightarrow ClCH_2\overset{OCH_3}{\overset{|}{C}}HCH{=}CH_2 + ClCH_2CH{=}CHCH_2OCH_3$$

16-8

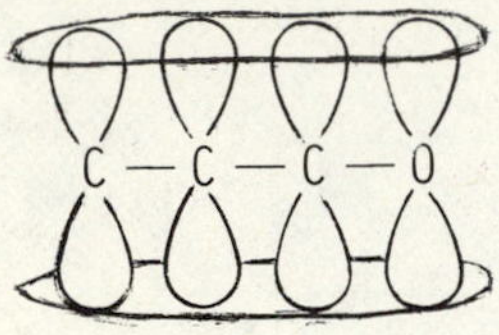

16-9 A charge separated resonance structure for 2-propenal contributes to the larger dipole moment of that compound.

16-10

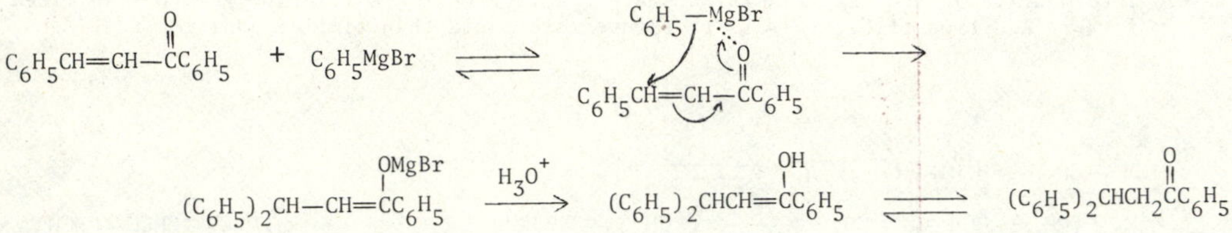

16-11 The terminal methyl group is vinylogous with a carbon atom alpha to the carbonyl group. The methyl hydrogen atoms possess an acidity comparable to those of α-hydrogen atoms because of resonance stabilization of the conjugate base.

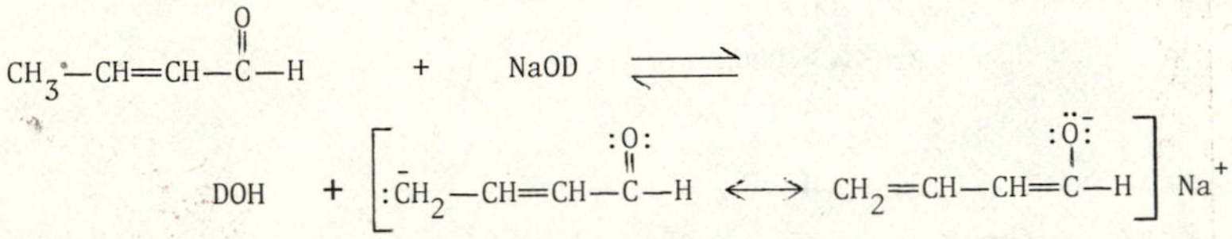

Exchange at the α-carbon atom would require formation of a carbanion which is not easily stabilized by resonance.

16-12 Reaction proceeds via a 1,4-addition. The proton initially adds to the more basic carbonyl oxygen atom and promotes conjugate addition by chloride.

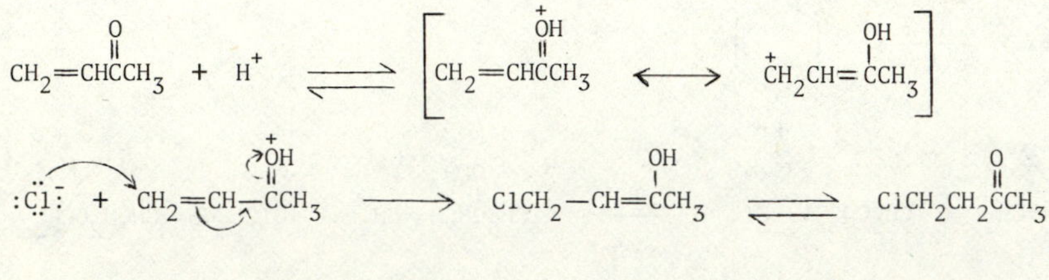

16-13 Reaction between C and D is preferred because the enolate (of diethyl malonate) is formed more easily under less basic conditions. Furthermore, the conjugated ketones such as C, are usually more reactive electrophiles than conjugated esters.

16-14

a) $$CH_2(CO_2C_2H_5)_2 + NaOC_2H_5 \xrightleftharpoons{C_2H_5OH} Na^+ :\overset{-}{C}H(CO_2C_2H_5)_2 + C_2H_5OH$$

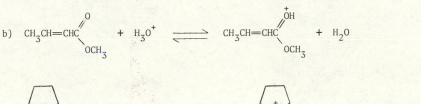

$$(H_5C_2O_2C)_2\ddot{\overset{-}{C}}H + CH_2{=}CH{-}C{\equiv}N: \rightleftharpoons (H_5C_2O_2C)_2CHCH_2CH{=}C{=}\ddot{N}^-$$

$$(H_5C_2O_2C)_2CHCH_2CH{=}C{=}\ddot{N}^- + C_2H_5\ddot{O}{-}H \longrightarrow (H_5C_2O_2C)_2CHCH_2CH_2C{\equiv}N + C_2H_5O^-$$

b) $$CH_3CH{=}CHC(=O)OCH_3 + H_3O^+ \rightleftharpoons CH_3CH{=}CHC(={\overset{+}{O}}H)OCH_3 + H_2O$$

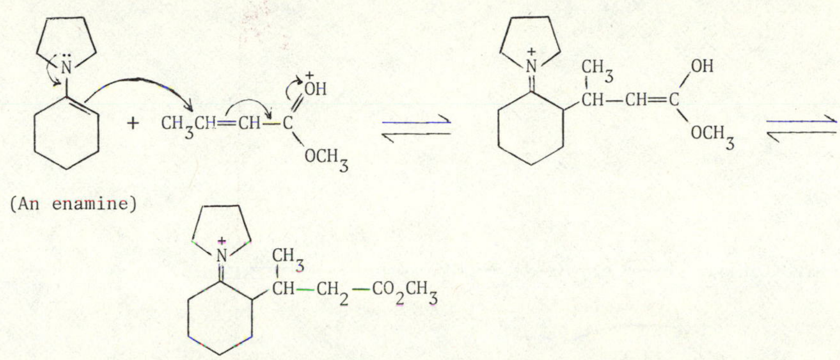

(An enamine)

Contd...

16-14 contd...

b) Contd...

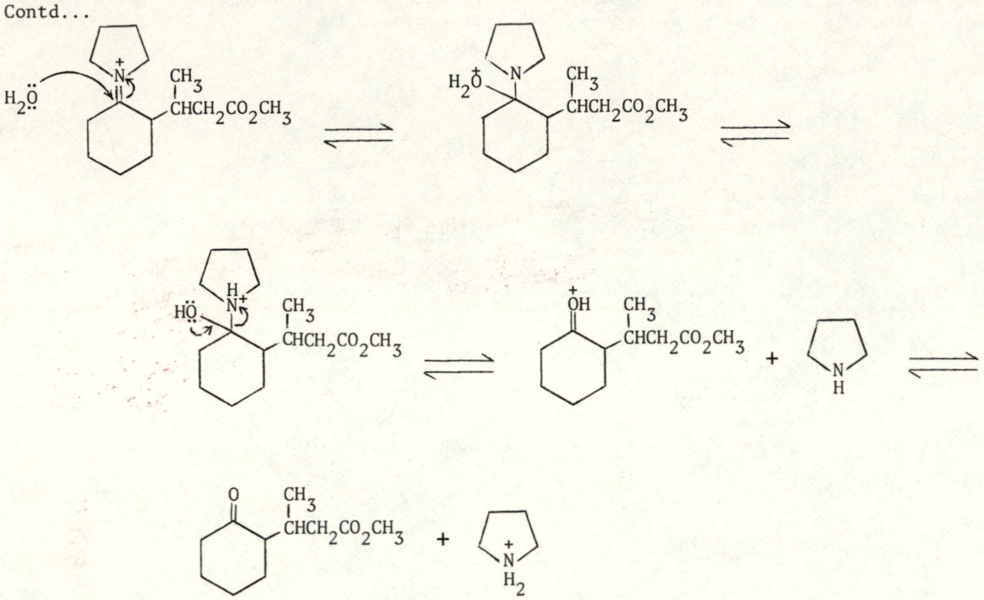

c) $CH_2(CO_2C_2H_5)_2 \xrightarrow[C_2H_5OH]{NaOC_2H_5} Na^+\bar{C}H(CO_2C_2H_5)_2$

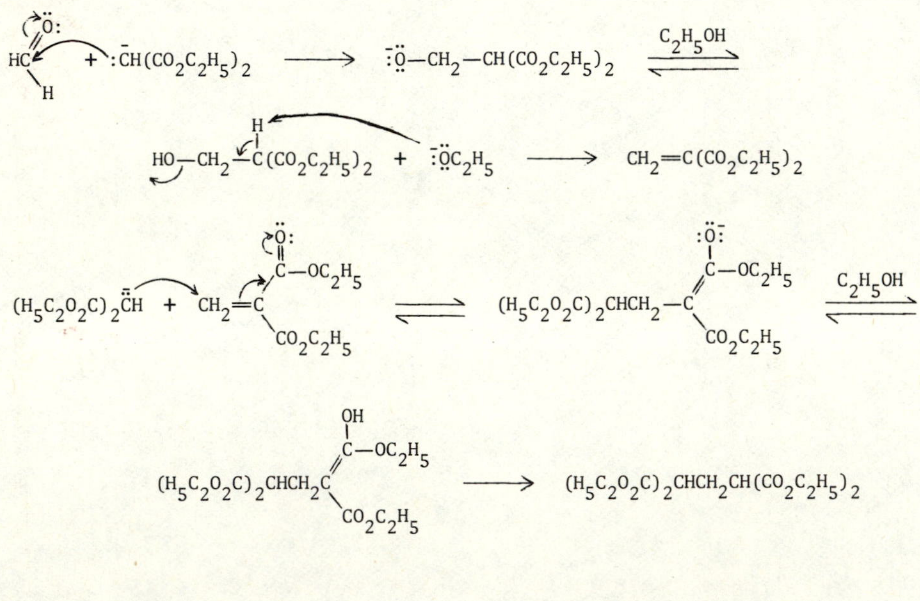

16-14 contd...

d) 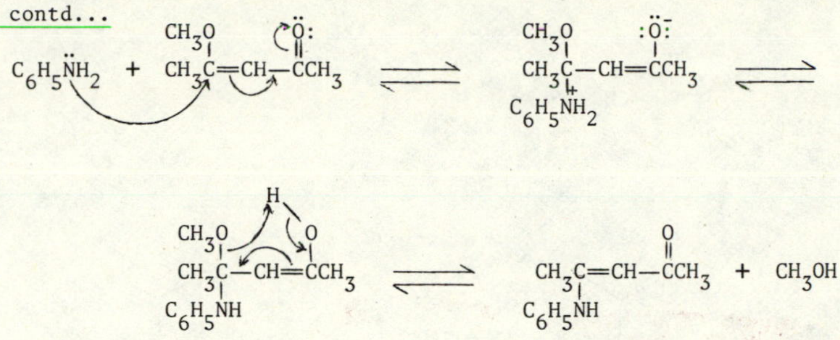

e) 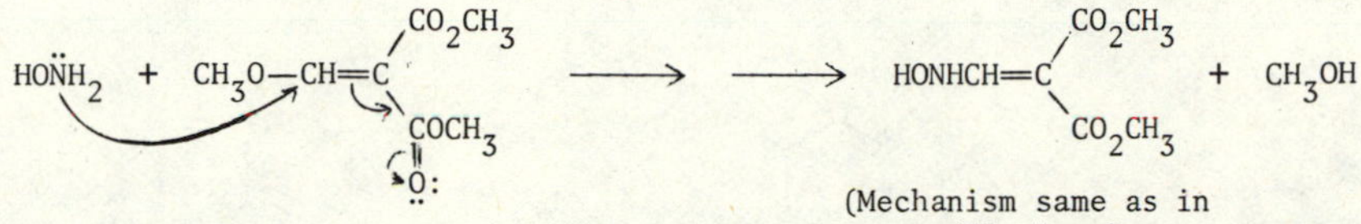

(Mechanism same as in part (d) above)

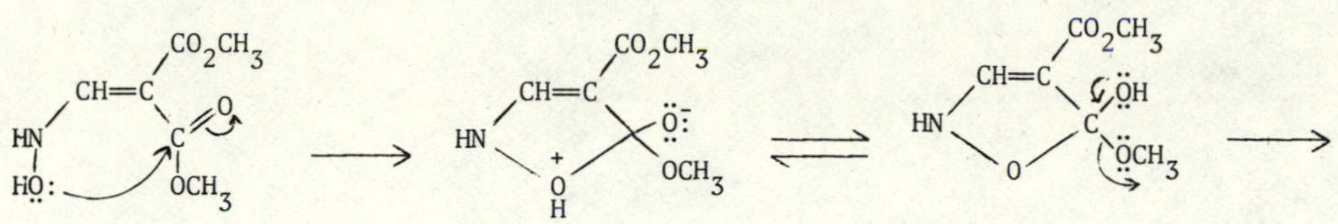

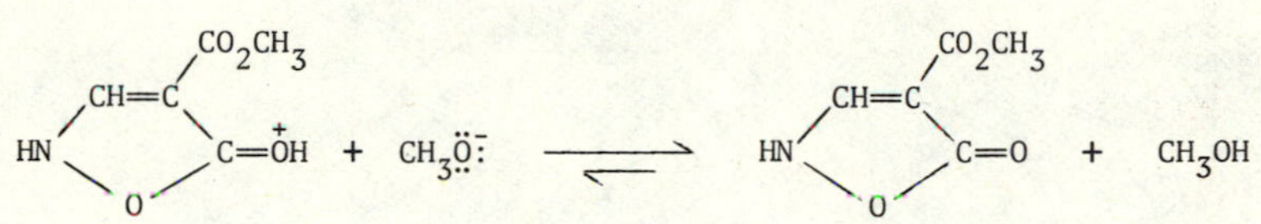

16-15

a) 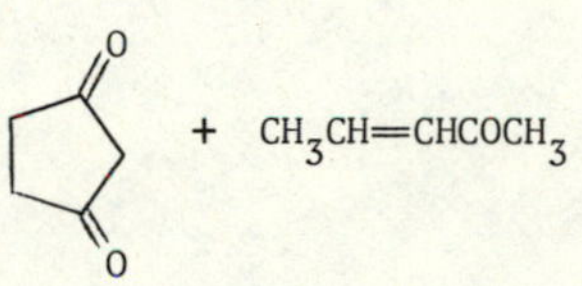

b) $(CH_3)_2NH$ + $CH_2{=}CHCN$

c) $(CH_3O_2C)_2CH_2$ + $CH_2{=}CHCO_2CH_3$

d) 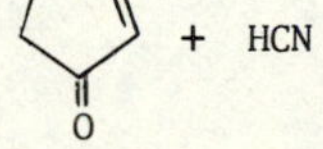

e) 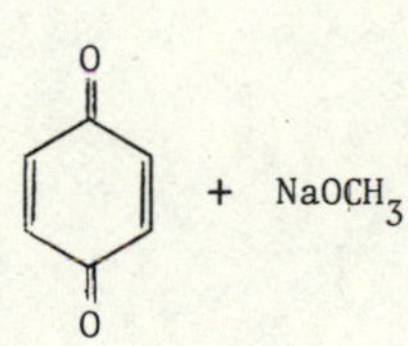

f) 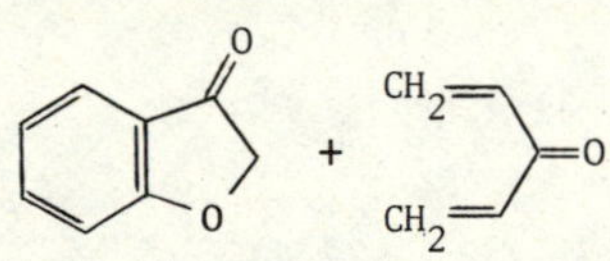

16-16

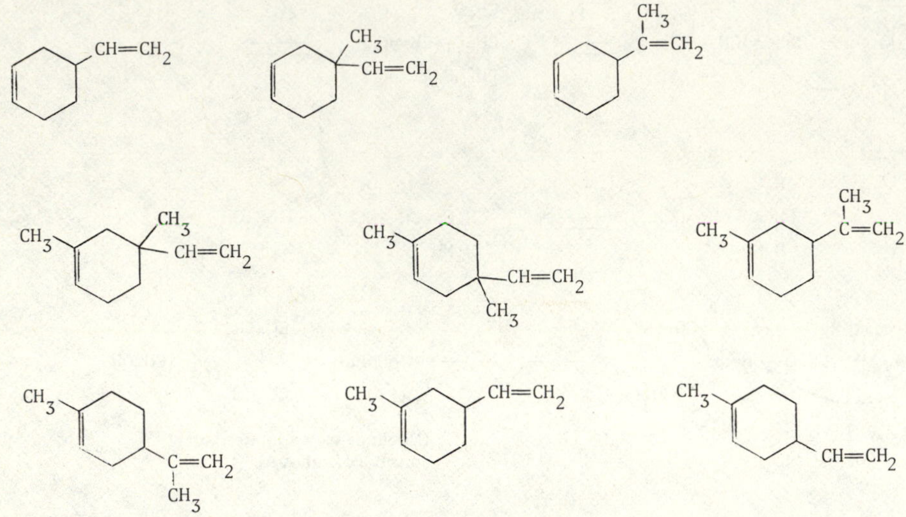

16-17

a) 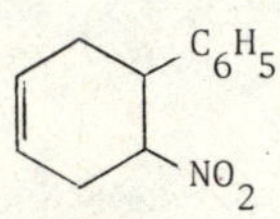

b) 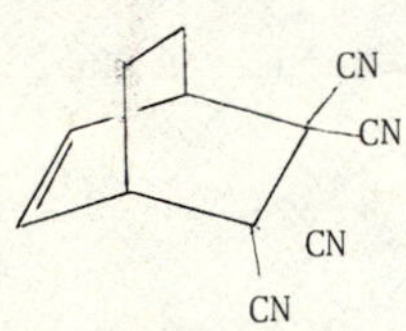

c) 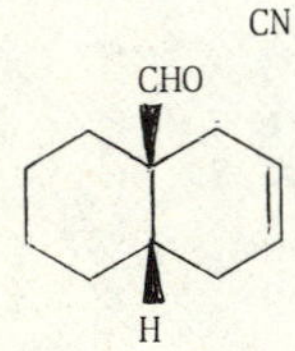

d) 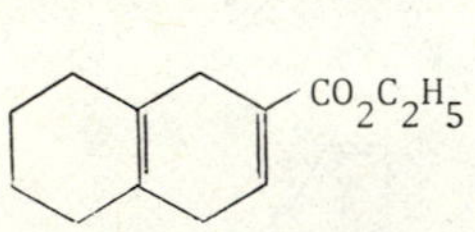

e) 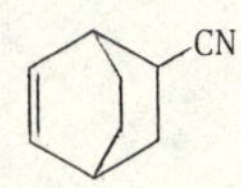

f) 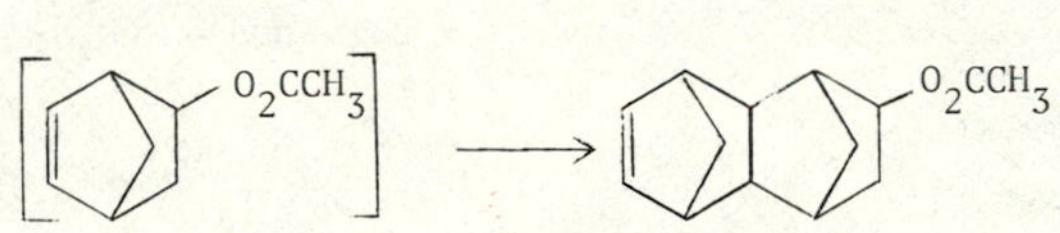

g) 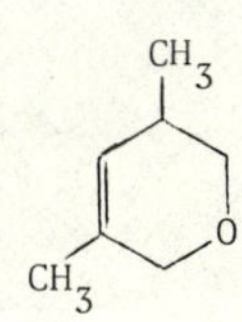

16-18

a) 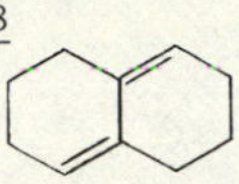 has a fixed *s*-trans configuration, thus cannot function as a Diels-Alder diene.

b) $CH_2{=}CH{-}CH{=}C(CH_3)_2$ is crowded at one end of the diene and thus cannot readily become *s*-cis.

c) Only one set of pi electrons from the alkyne are involved. A double bond remains.

16-19

a) 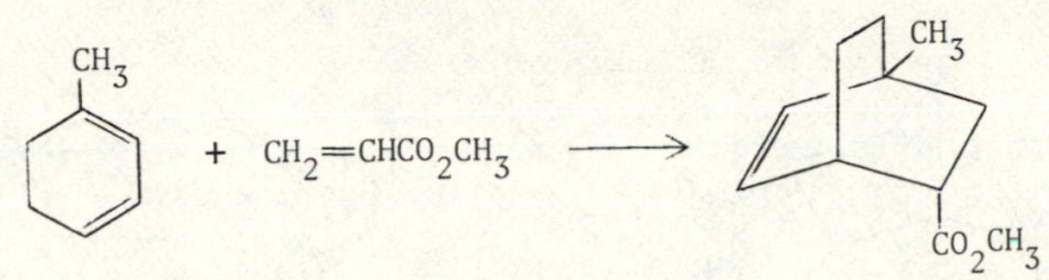

The less crowded transition state leads to this regiochemistry.

b) 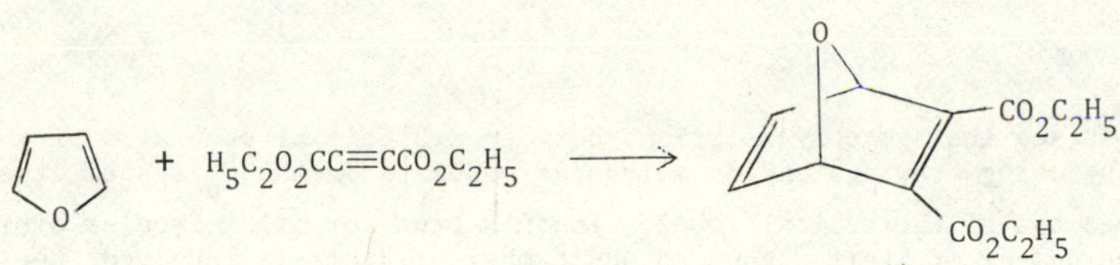

c) 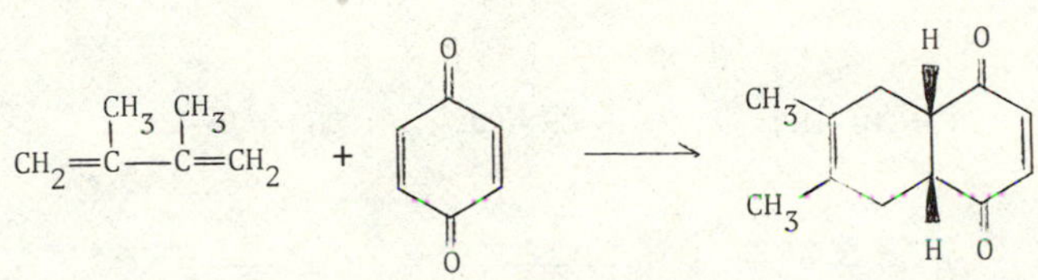

d) 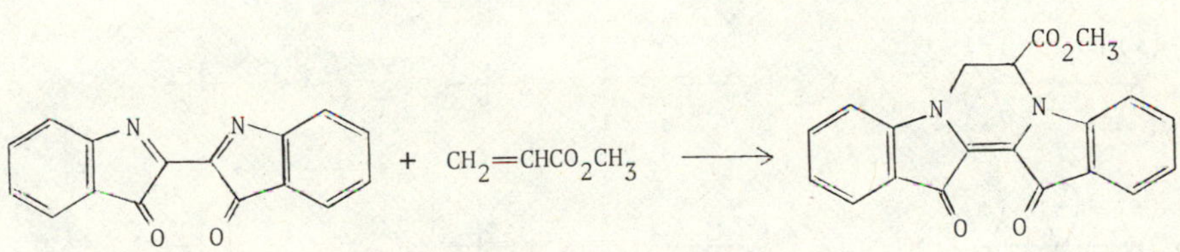

e) 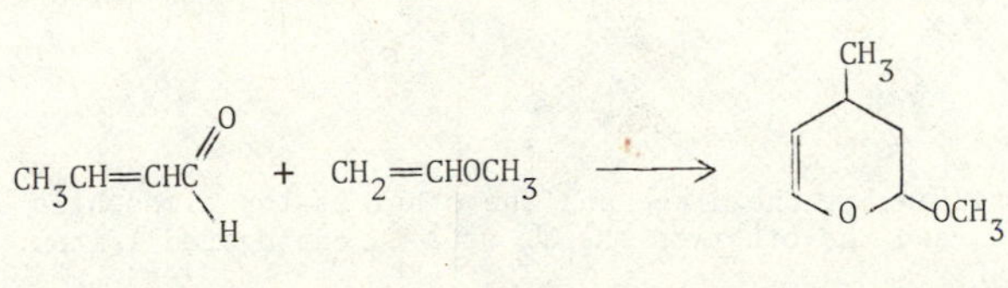

Contd...

16-19 contd...

f) $CH_2{=}CHCH{=}CH_2$ + $CH_2{=}C(CN)_2$ ⟶ 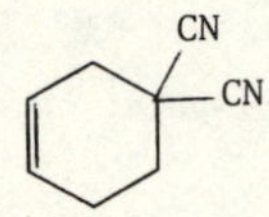

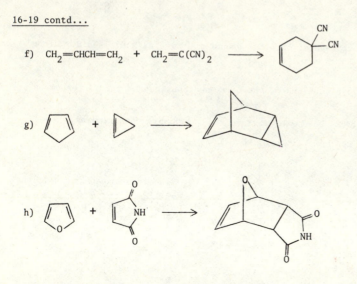

g) 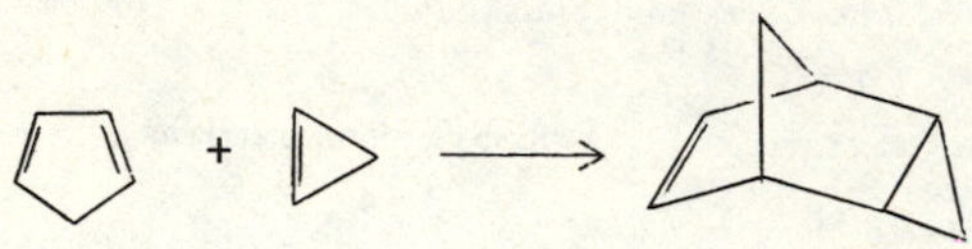

h) 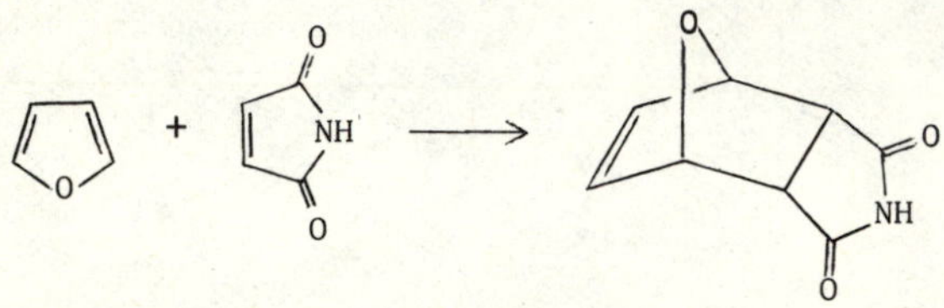

16-20 The student can best see the symmetry relations between orbitals at each atom of a conjugated system by using drawings and/or molecular models. In the C_6 system, the plane and axis lie between two of the orbital lobes. That is true for all molecular orbitals formed with an even number of atoms. When an odd number of atoms is involved, the plane and axis is located at the center orbital lobe.

16-21

$CH_2{=}CH{-}\overset{+}{C}H_2$ $CH_2{=}CH{-}\dot{C}H_2$ $CH_2{=}CH{-}\overset{..}{C}H_2$

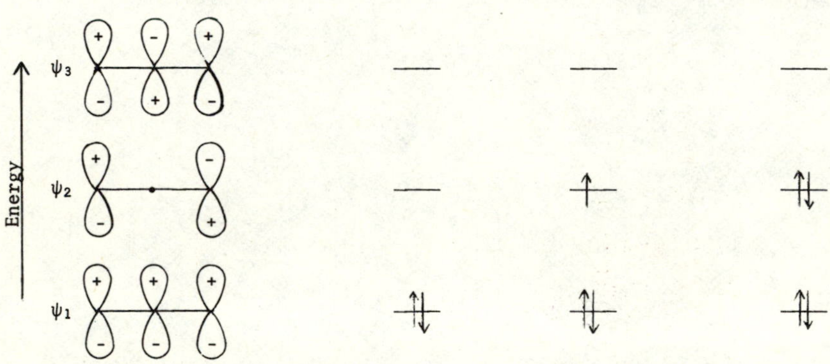

16-22 Treat one molecule of cyclopentadiene as the diene and the other as the dienophile. One will thus be based on the ψ_2 and the other on the ψ_3 of a C_4 conjugated system.

Contd...

16-22 contd...

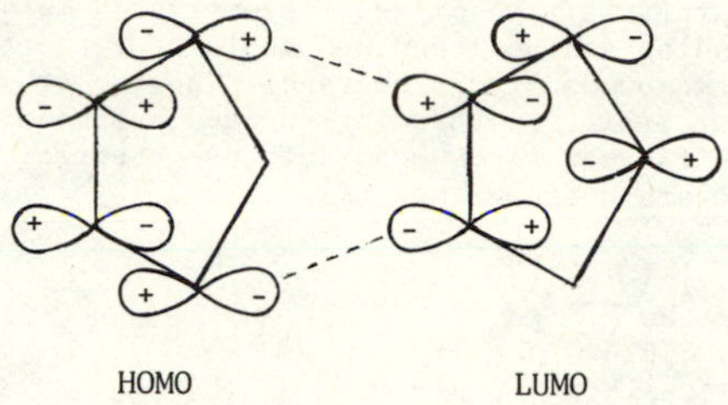

16-23

a) An allowed [4 + 2] cycloaddition in which one molecule of the α,β-unsaturated aldehyde is the diene and another is the dienophile.

b) An allowed [4 + 2] cycloaddition in which one double bond of allene is the dienophile.

c) A forbidden [2 + 2] cycloaddition.

d) An allowed [6 + 4] cycloaddition.

e) A forbidden [4 + 4] cycloaddition.

f) In problem 16-23 (d) the product represents a [6 + 4] cycloaddition. A suprafacial-suprafacial process between a butadiene and hexatriene system is symmetry allowed.

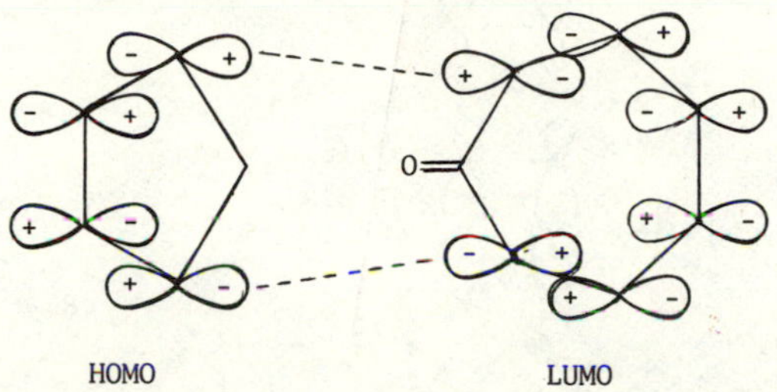

In problem 16-23 (e) the product represents a [4 + 4] cycloaddition. A suprafacial-suprafacial process between two butadiene systems is not symmetry allowed. An allowed suprafacial-antarafacial reaction would be geometrically difficult to attain in this case.

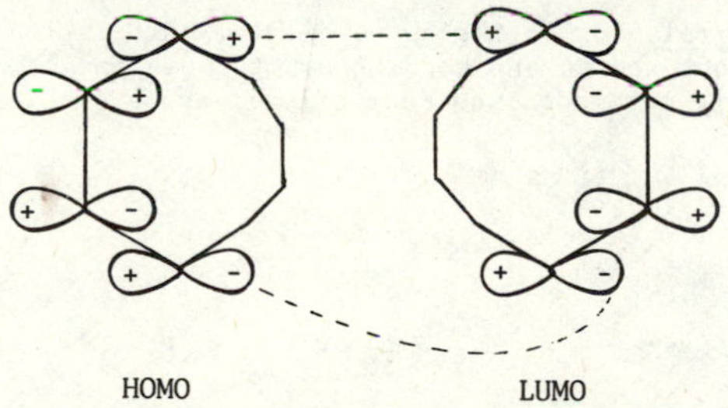

16-24 In order for the two ethylene molecules to combine in a geometrically feasible manner, the [2 + 2] cycloaddition must be suprafacial-suprafacial. A plane of symmetry is maintained. Thus we consider a symmetry plane for the bonding and antibonding orbitals of the two ethylene molecules as well as for the new sigma bonds of cyclobutane. The two ethylene molecules provide a total of four molecular orbitals, two bonding and two antibonding. Four molecular orbitals are developed for the two new sigma bonds of the potential cyclobutane product. Two are bonding and two are antibonding.

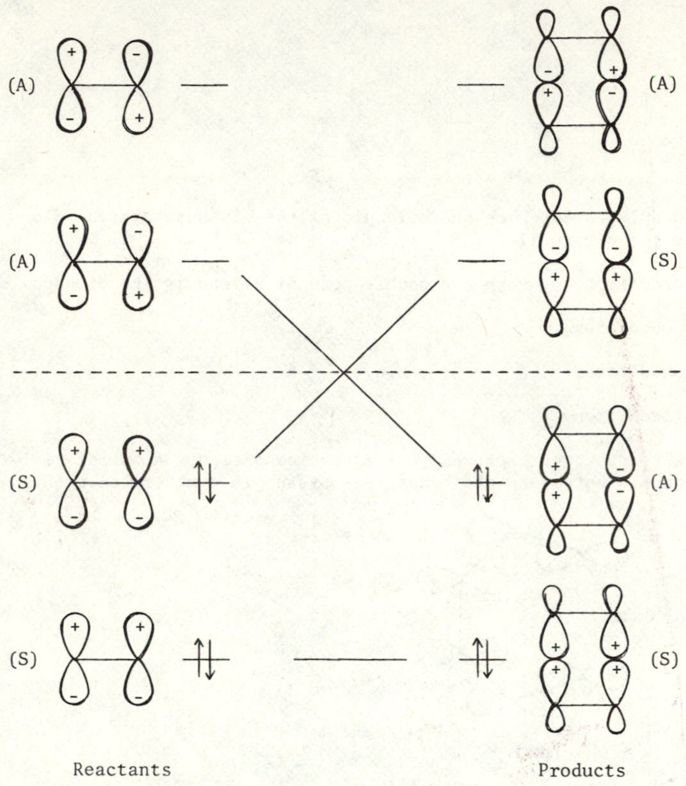

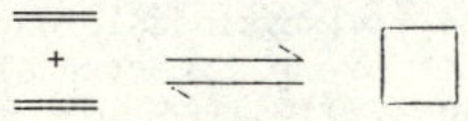

In order to maintain orbital symmetry correlation in the above diagram, correlation must occur between a bonding orbital of reactant and an antibonding orbital of product. The [2 + 2] cycloaddition is therefore symmetry forbidden and energetically unfavorable.

16-25

a) 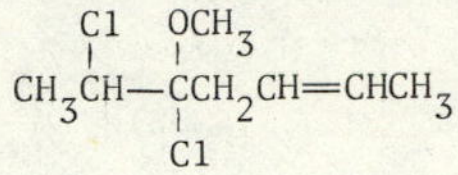

b) 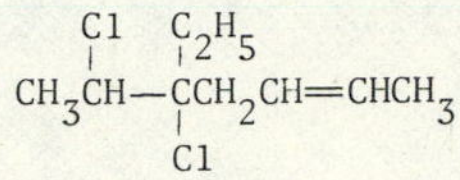

c) 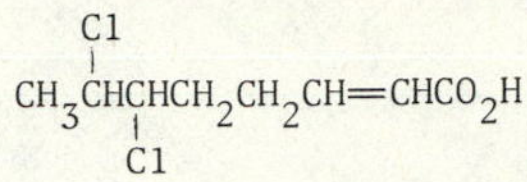

d) 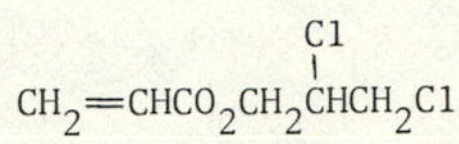

e) 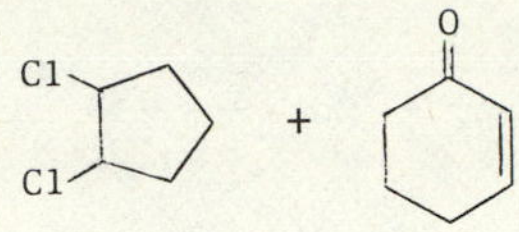

f) 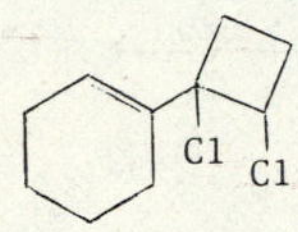

[The more strained double bond is the more reactive. In this example angle strain is relieved in going from sp^2 (120°) to sp^3 (109°)]

g) 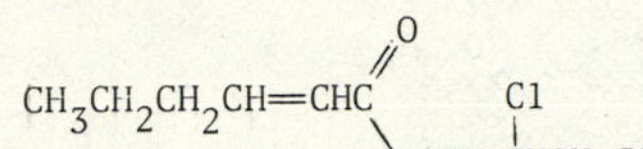

16-26

a) 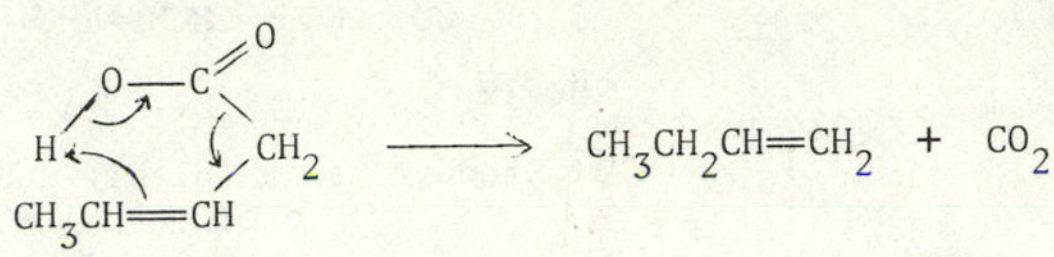

b) 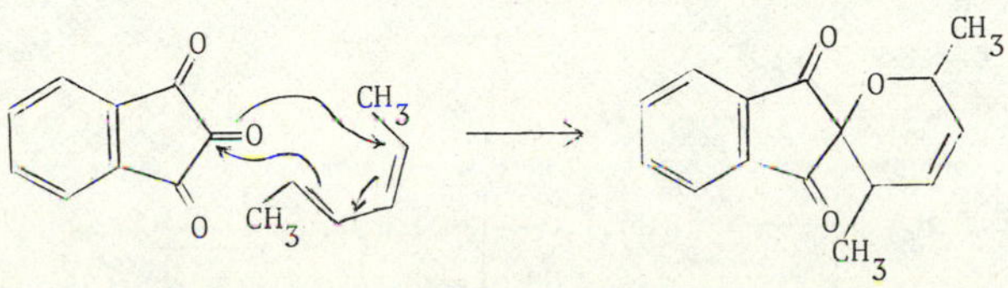

c) 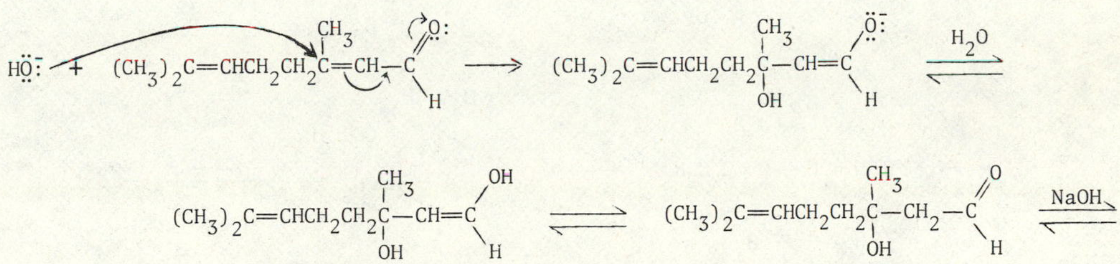

(This is an aldol)

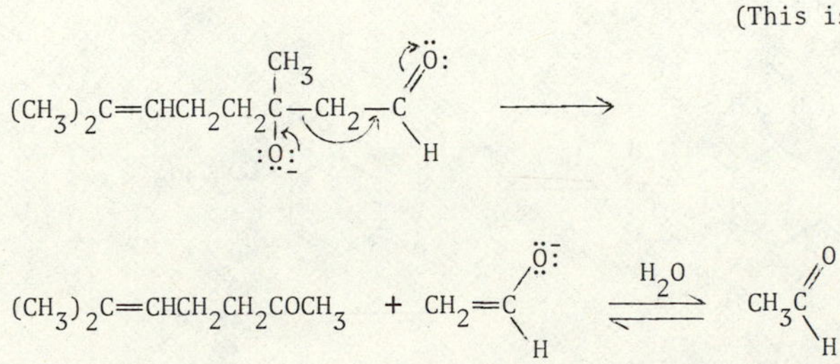

(The final step is a reverse aldol reaction.)

Contd...

16-26 contd...

d) 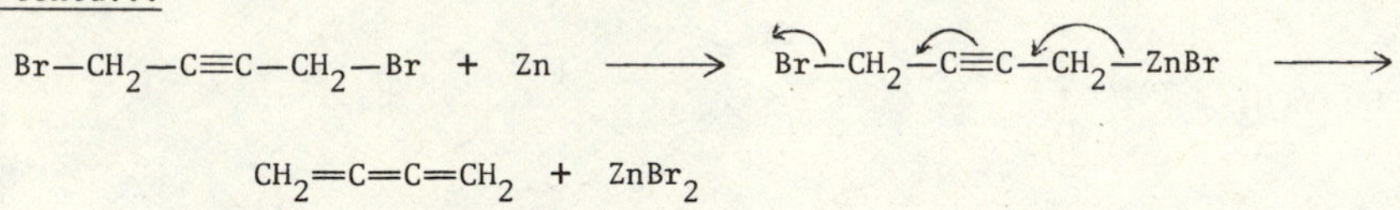

e) 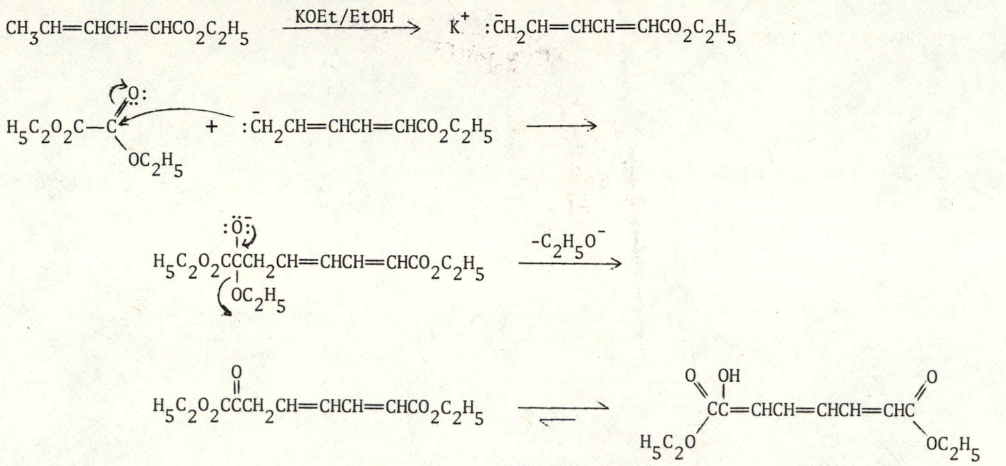

(The product is the highly conjugated enol.)

f) 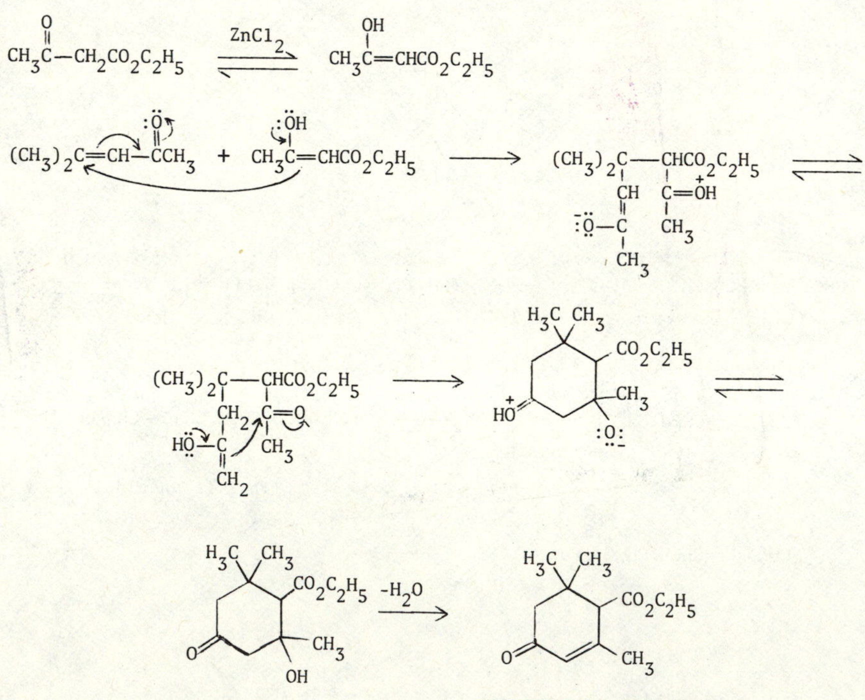

Contd...

16-26 contd...

The Lewis acid promotes decarboxylation of the vinylogous β-keto ester.

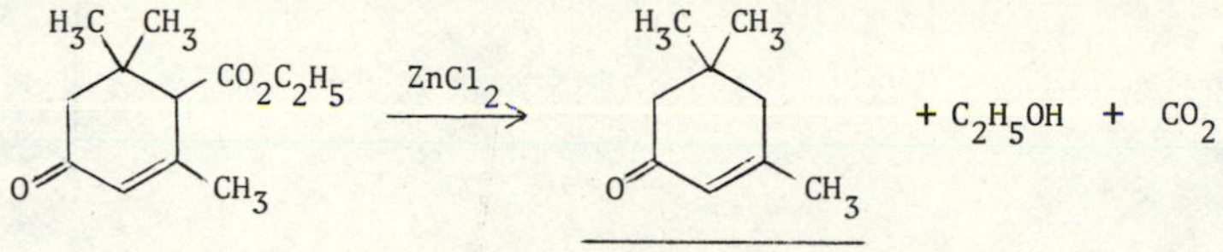

Cyclization in the opposite direction produces the third product.

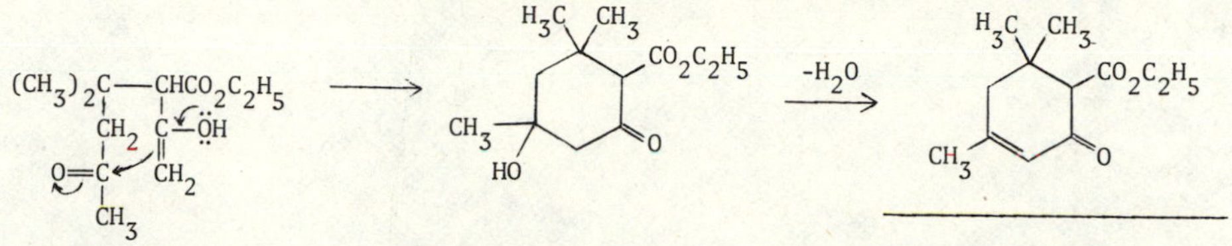

Cyclization to the ester carbonyl group produces the fourth product.

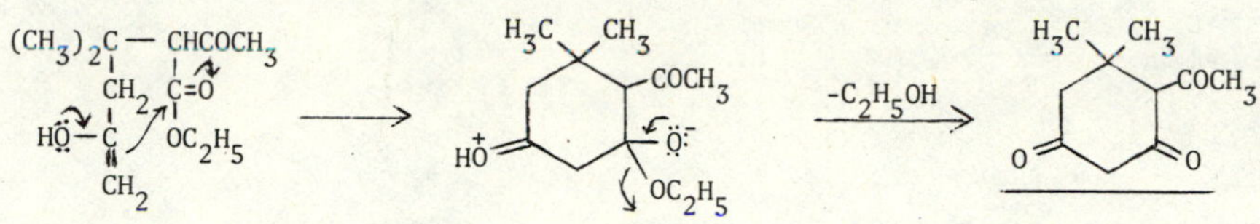

g)

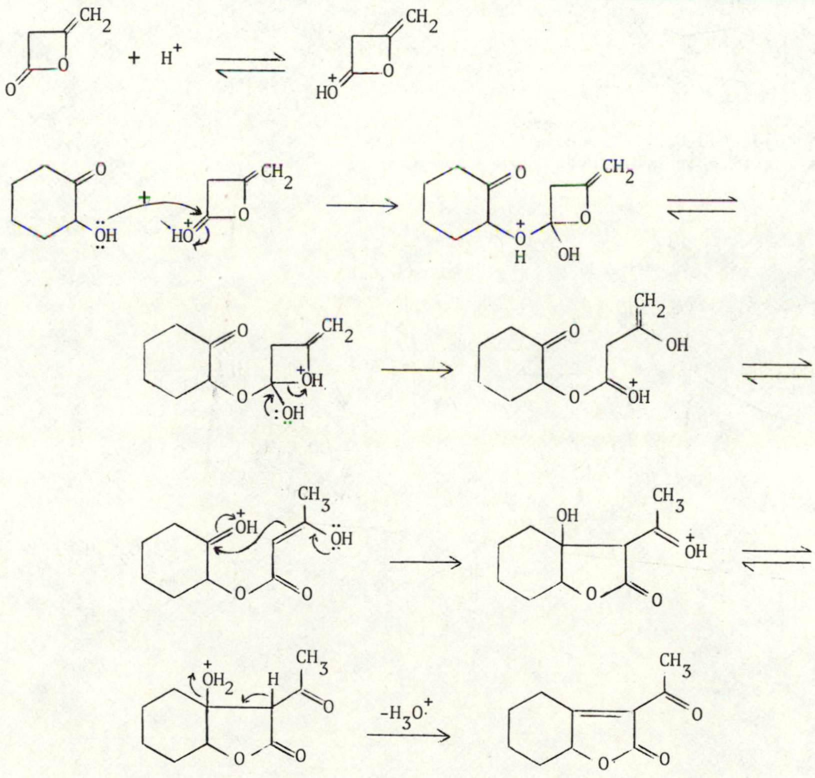

16-27 Protonation of the carbonyl oxygen atom enhances the electrophilicity of the carbon-carbon double bond in the α,β-unsaturated carbonyl compound. Similar enhancement is not available to the alcohol.

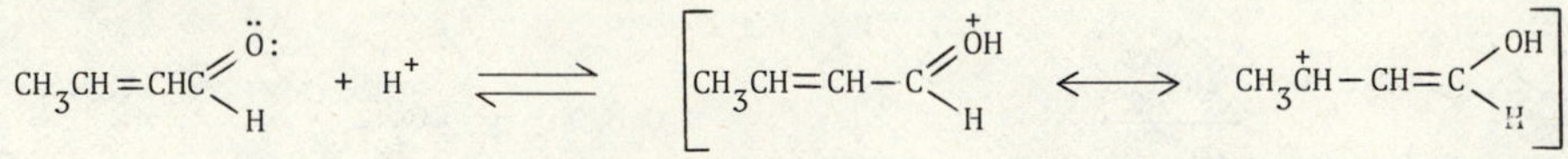

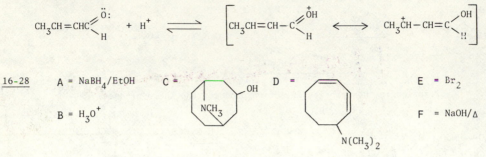

16-28 A = $NaBH_4$/EtOH

B = H_3O^+

C = 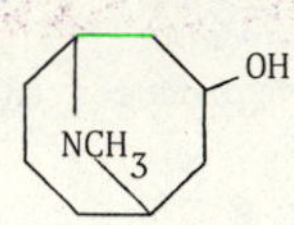

D = 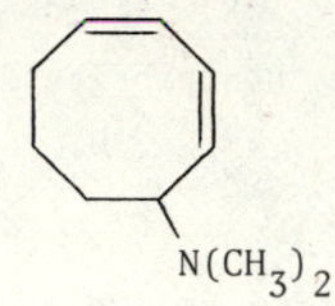

E = Br_2

F = NaOH/Δ

16-29

a) 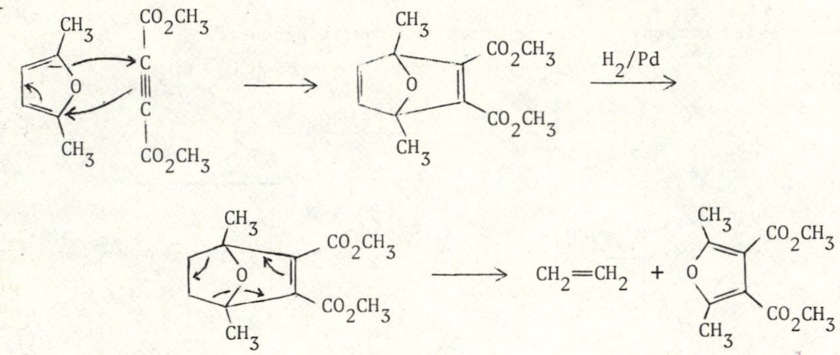

(The conjugated double bond is less reactive in hydrogenation.)

b) 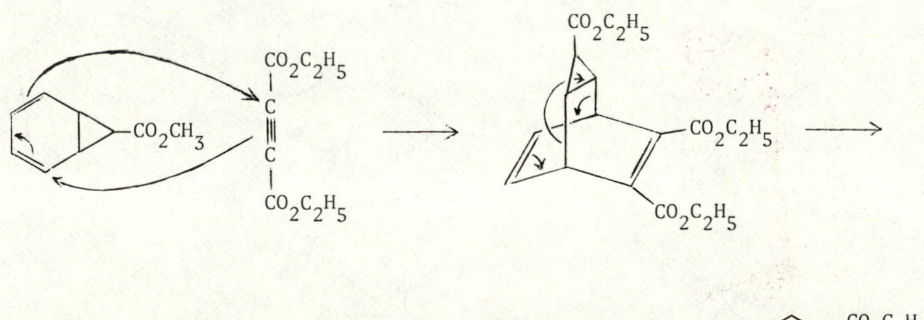

16-30

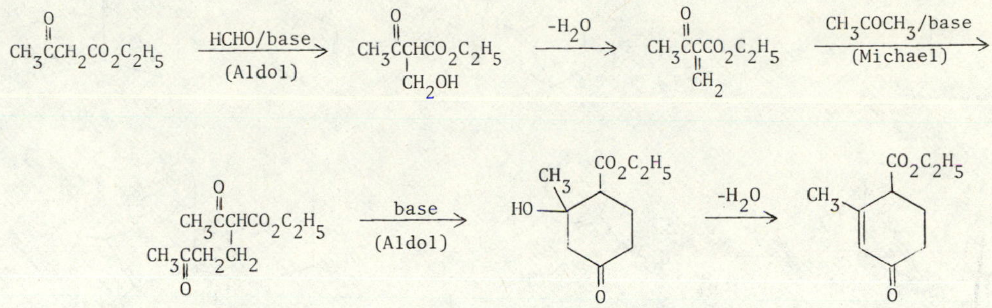

16-31

a) The problem is treated as a [6 + 2] cycloaddition. In the HOMO-LUMO method Ψ_1 of the C_2 portion can be the HOMO and Ψ_4 of the C_6 portion can be the LUMO. The cycloaddition would be a symmetry allowed suprafacial-antarafacial process. The actual reaction is forbidden since the geometry would require a suprafacial-suprafacial process.

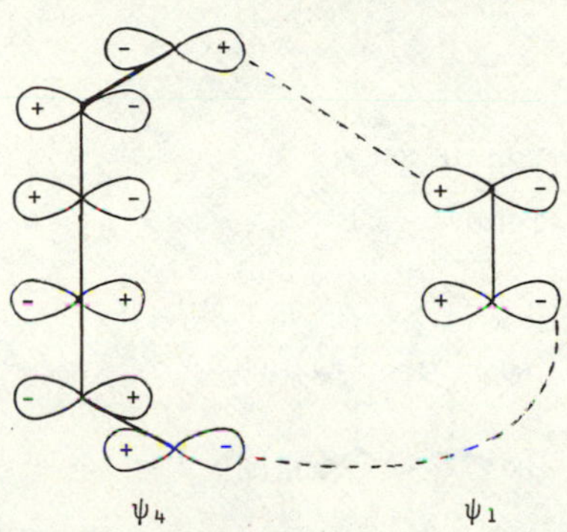

b) The azide group can be treated as a C_3 system with four conjugated electrons ($-\ddot{\bar{N}}-\ddot{N}=\ddot{N}^+$), that is, as an allyl carbanion ($>C=C-\ddot{\bar{C}}-$). We can thus use the allyl ψ_3 as our LUMO and ψ_1 of a C_2 system as the HOMO. The reaction is an allowed suprafacial-suprafacial process.

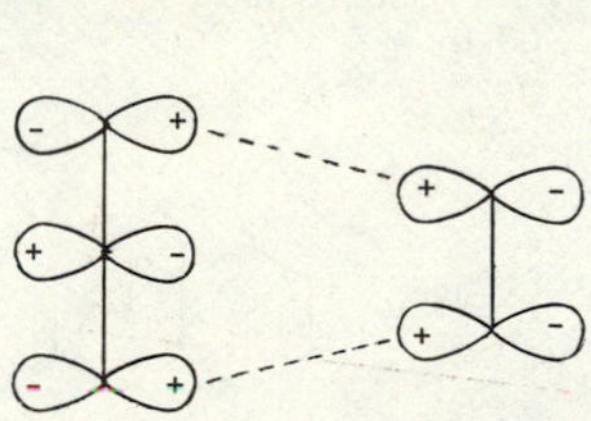

16-32

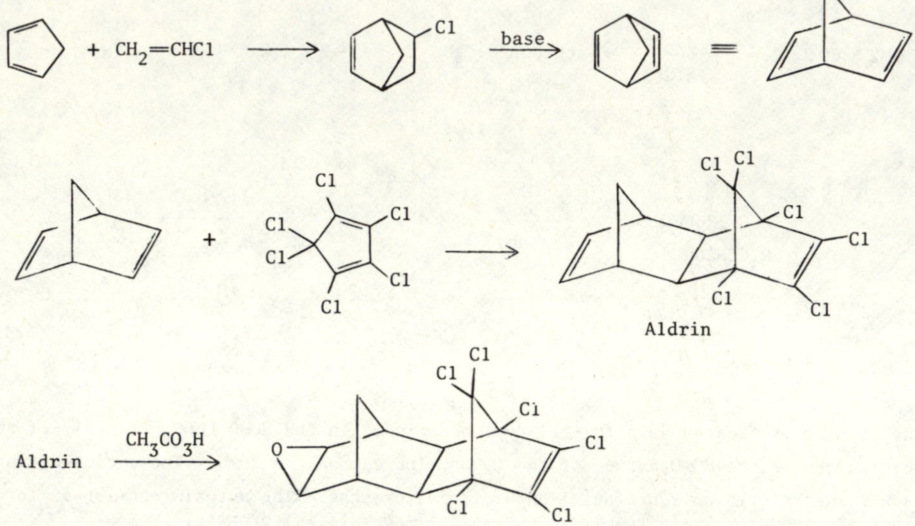

16-33

a)

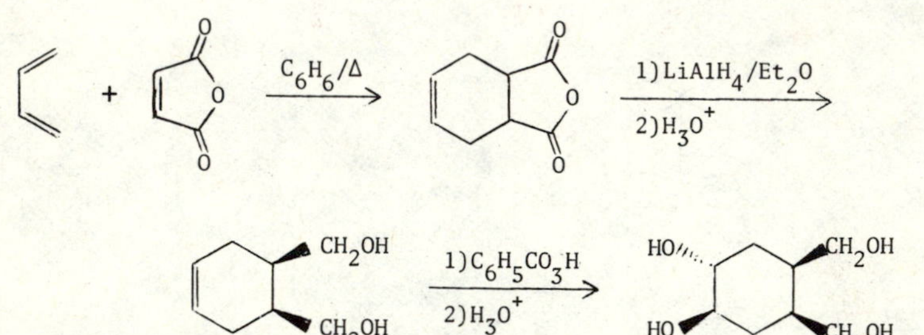

b)

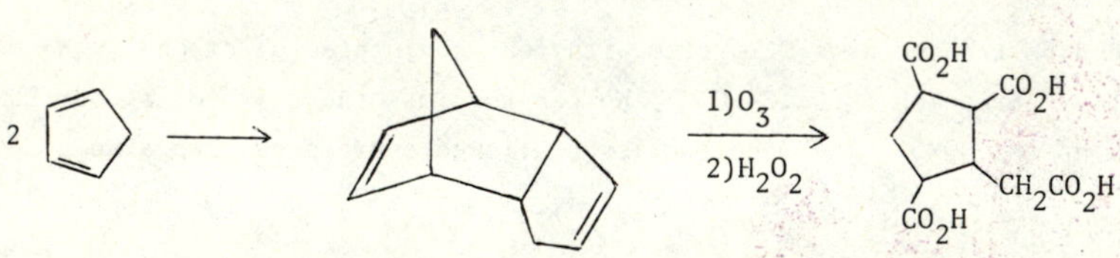

c)

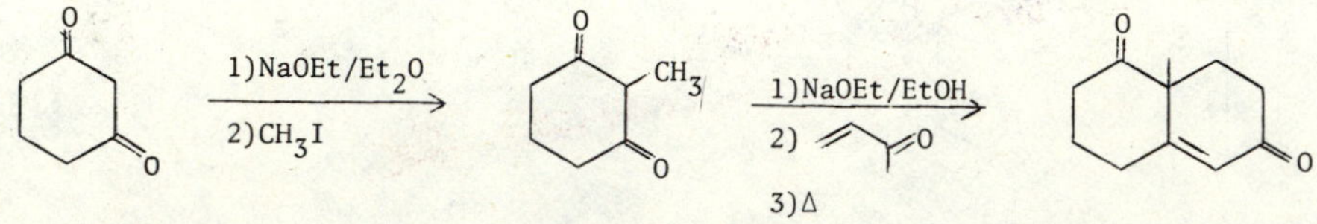

Contd...

16-33 contd...

d) 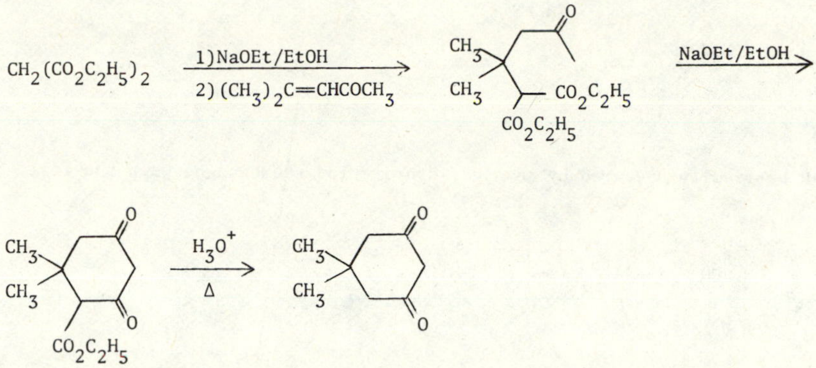

16-34 *IHD* = 3. Catalytic hydrogenation indicates only one carbon-carbon double bond. The IR spectrum suggests an ether (C-O-C stretching at 1100 cm^{-1}) and the peaks at 1667 and 1620 cm^{-1} are indicative of a conjugated carbonyl and double bond since they are shifted to lower frequencies. The UV absorption is consistent with an α,β-unsaturated carbonyl compound that has a substituent on the α and β positions (calc. λ_{max} 237 nm). Those data would account for the two oxygen atoms and two of the *IHD*. The nmr spectrum shows a singlet (3) at 2.4 ppm typical of $CH_3\overset{O}{\overset{\|}{C}}-$ and only a single proton in the alkene region (6.2 ppm).

Those data suggest $\overset{H}{\diagdown}C{=}C{-}\overset{O}{\overset{\|}{C}}CH_3$ as a structural fragment. The triplet (1.0 ppm) and quartet (3.2 ppm) are surely an ethyl group attached to an oxygen atom; $-OCH_2CH_3$. The multiplet (1-2 ppm) would fit a chain of three CH_2 groups. The one proton triplet deshielded to 4.8 ppm can be due to $CH_2-\overset{|}{C}H-O-$. All of these fragments can be combined to form a cyclic structure (that accounts for the third *IHD*) consistent with all of the data.

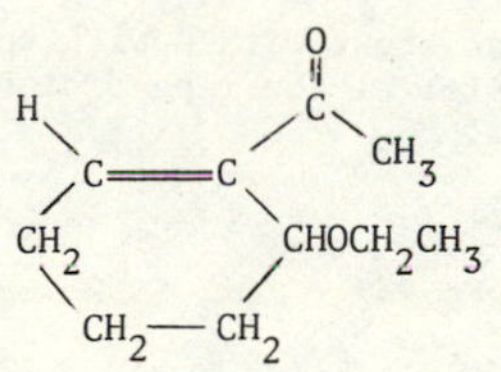

16-35

a) The IR spectrum of C is consistent with an α,β-unsaturated carbonyl compound and the UV spectrum suggests that one substituent is located on that chromophore. The nmr spectrum shows two alkene protons and possibly a methyl ketone (2.2 ppm, s, 3). The *IHD* of 3 suggests a cyclic structure in addition to the α,β-unsaturated carbonyl. The broad nmr peak at 1.7 ppm would fit $-(CH_2)_4-$ and the single proton multiplet at 2.5 ppm would be consistent with a ring position at which a substituent group is located. A cyclopentane with a C_4-α,β-unsaturated carbonyl side chain will fit the data.

Contd...

16-35 contd...

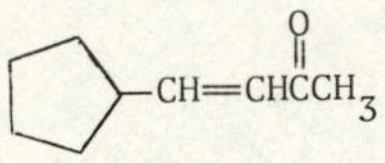

C

b) Compound C can be readily prepared by an aldol-dehydration sequence between the following reactants.

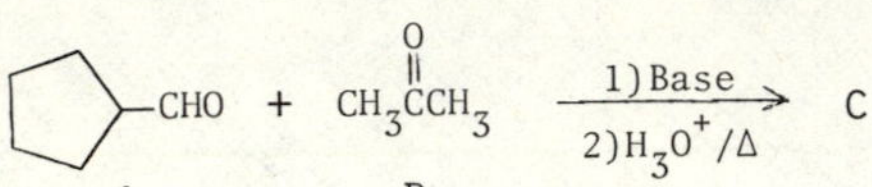

A B

16-36 The following sequence is expected for the reactions indicated.

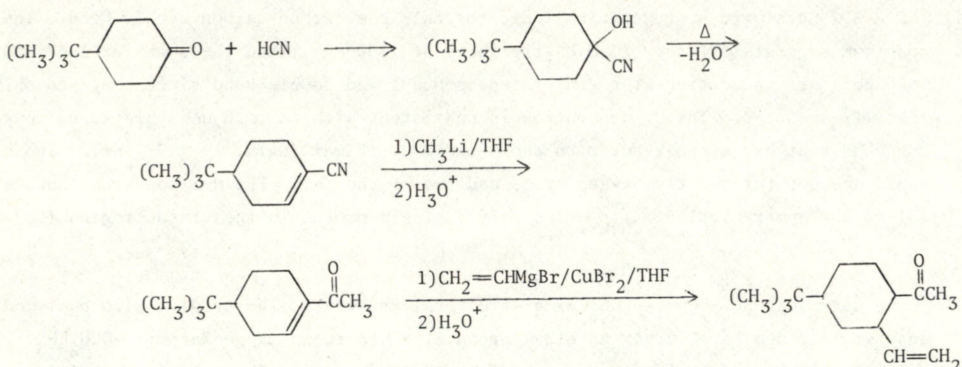

A

A has MW = 208. The two major peaks in the mass spectrum are due to $CH_3\overset{O}{\overset{\|}{C}}-$(43) and $(CH_3)_3C-$(57). The IR absorption is normal for a ketone as is the unsymmetrical alkene absorption. The nmr chemical shift values are consistent with A with the methyl and *tert*-butyl peaks both overlapping the broad multiplet of the ring protons.

16-37 Elemental analysis gives an empirical formula of $C_{10}H_{16}O$ and mass spectral analysis confirms this as the molecular formula. The nmr multiplet at 4.9 - 6.2 ppm for 4 protons suggests that at least two double bonds are present. The IR spectrum is consistent with an α,β-unsaturated carbonyl structure. Two double bonds and a carbonyl group would account for the *IHD* of 3. No aldehydic proton is present in the nmr spectrum so the carbonyl group must be a ketone. The groups of singlets in the nmr spectrum of 6,3, and 3 protons indicates two equivalent and two different methyl groups. The two equivalent methyl groups are not deshielded while the two nonequivalent methyl groups are consistent with a methyl ketone and a methyl attached to a carbon-carbon double bond. Two reasonable structural formulas, A and B, should be considered (no differentiation is made between geometrical isomers).

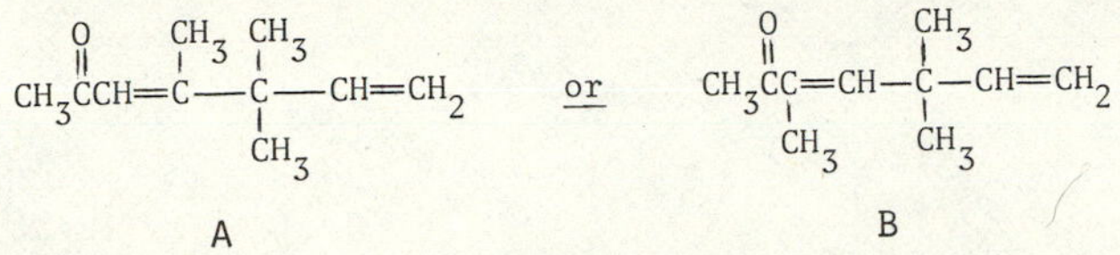

Structure A is the more probable since the alkene C-H beta to the carbonyl group on an α,β-unsaturated carbonyl compound such as structure B is typically deshielded into the aromatic region (~ 7 ppm) in the ^1H nmr spectrum.

17 The Mechanism of Electrophilic Aromatic Substitution

17-1

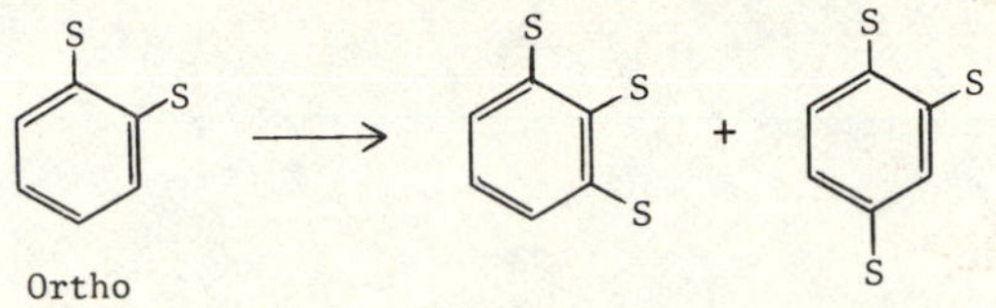

Ortho

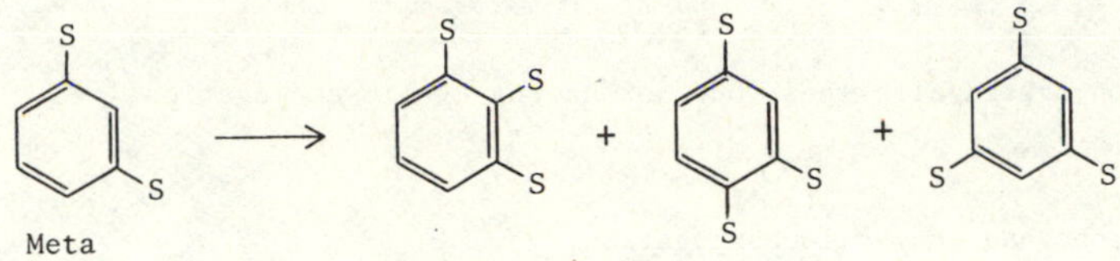

Meta

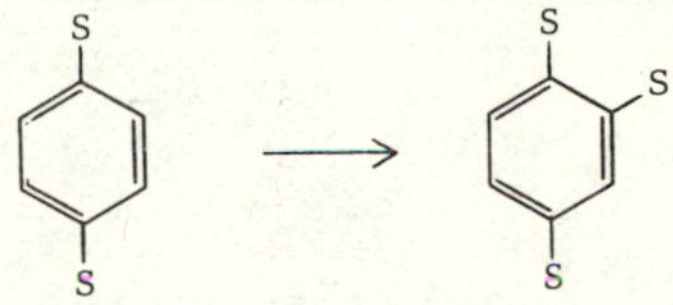

Para

17-2

Substituents adjacaent on a triangle face

Substituents adjacent on a square face

Substituents nonadjacent (only possible on a square face)

17-3

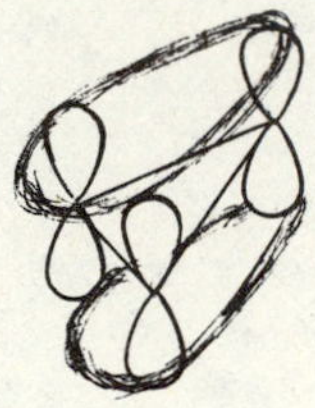

Cyclopropenyl cation

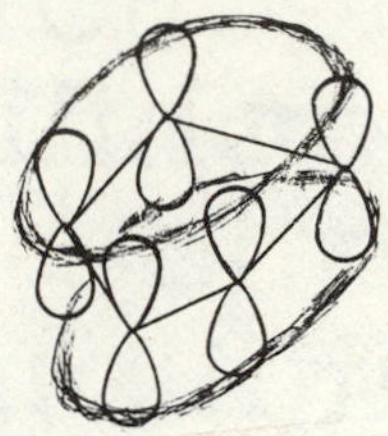

Cyclopentadienyl cation

17-4 Conjugated cyclic compounds with $4n + 2$ electrons,
∴ Aromatic: a, c, d, f, g

Conjugated cyclic compounds with $4n$ electrons,
∴ Nonaromatic: b, h

Compounds with $4n + 2$ conjugated electrons, but not having cyclic conjugation,
∴ Nonaromatic: e, j

Compounds with $4n$ electrons and noncyclic conjugation,
∴ Nonaromatic: i

17-5

If we assume that four cyclooctene double bonds are equivalent to a hypothetical nonstabilized cyclooctatetraene, then 4 x (-23 kcal/mol) = -92 kcal/mol [4 x (-97 kJ/mol) = -388 kJ/mol] is the energy released on hydrogenation of that species. Since -101 kcal/mol (-422 kJ/mol) is released on hydrogenation of cyclooctatetraene, the actual compound is less stable by 9 kcal/mol (34 kJ/mol) than the hypothetical model.

17-6

a) If cyclooctatetraene were planar, the adjacent p-orbitals of the double bonds would be expected to interact. This would give an unfavorable $4n$ π-electron system.

b) The dianion is a $4n + 2$ electron system, thus favors a planar aromatic configuration.

c) Cyclooctatetraene has typical single and double bonds. The dianion is aromatic and all of the carbon-carbon bonds are equivalent.

17-7

a) Azulene is a cyclic conjugated compound with $4n + 2$ ($n = 2$) electrons.

b) 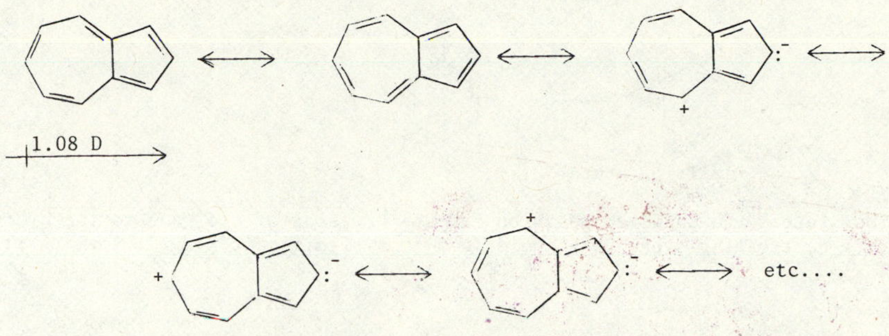

The dipole is directed toward the five membered ring. Structures with a delocalized negative charge in a five membered ring and a delocalized positive charge in a seven membered ring resemble the cyclopentadiene anion and tropylium cation, both of which show some aromatic stability.

17-8 The [14]-annulene is a $4n + 2$ electron system and is found to be aromatic. [16]-Annulene is a $4n$ electron system and is nonaromatic.

17-9 The difference, 32 kcal/mol (134 kJ/mol), is approximately equal to the loss of resonance energy going from benzene to a 1,3-cyclohexadiene.

17-10

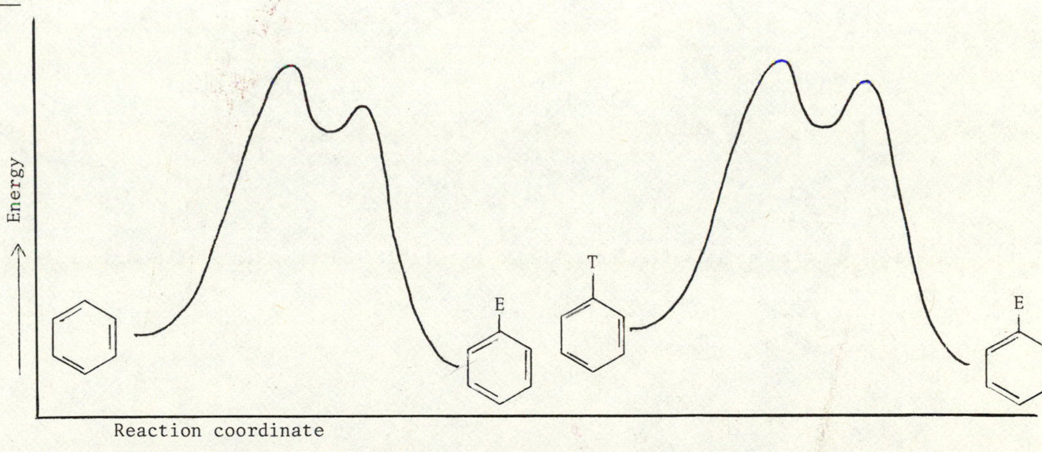

The first rate-controlling steps are essentially the same. The second is slower (higher energy) for tritiated benzene, but neither second step affects the rates of reaction.

17-11

a) Rate = $k(E^{+})(ArH)$

b) 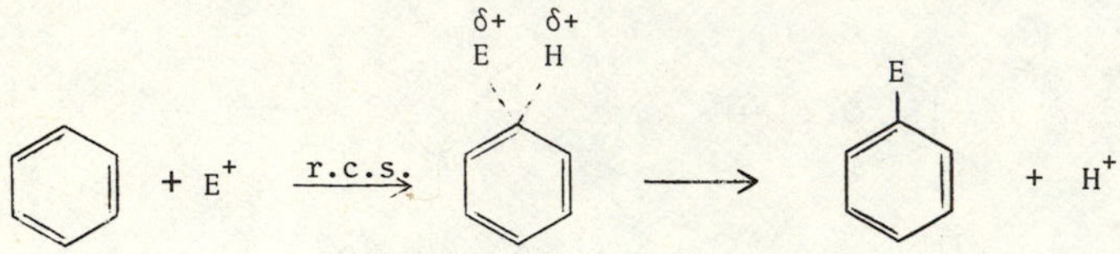

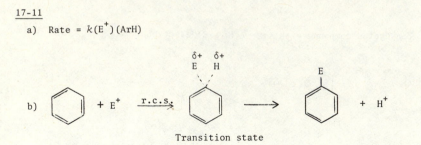

Transition state

The transtition state for a one-step reaction depicts breaking of the bond to the hydrogen atom in the rate-controlling step, thus would lead to a significant kinetic isotope effect.

17-12

a) The intermediates leading to ortho and para substitution have more resonance structures and are energetically more favorable than the intermediate of meta substitution.

Ortho: 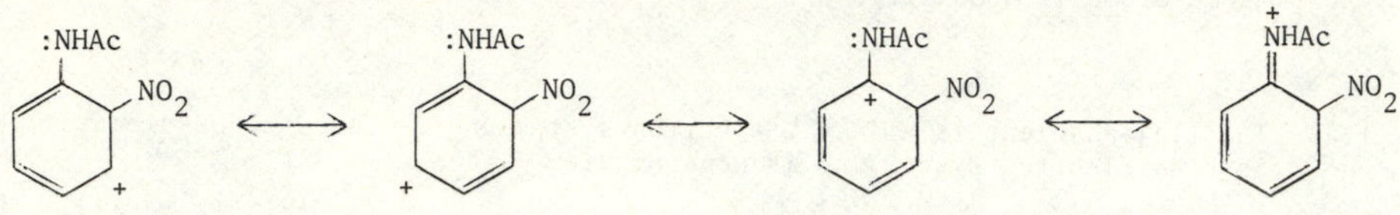

Para: 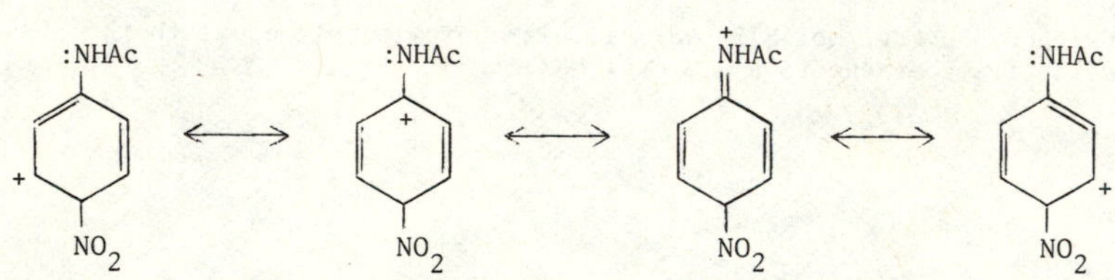

Meta: 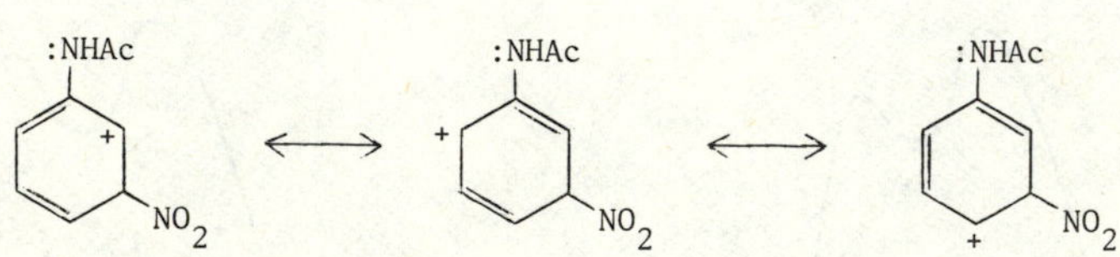

b) The amide substituent is sufficiently large to inhibit substitution at the ortho position.

17-13

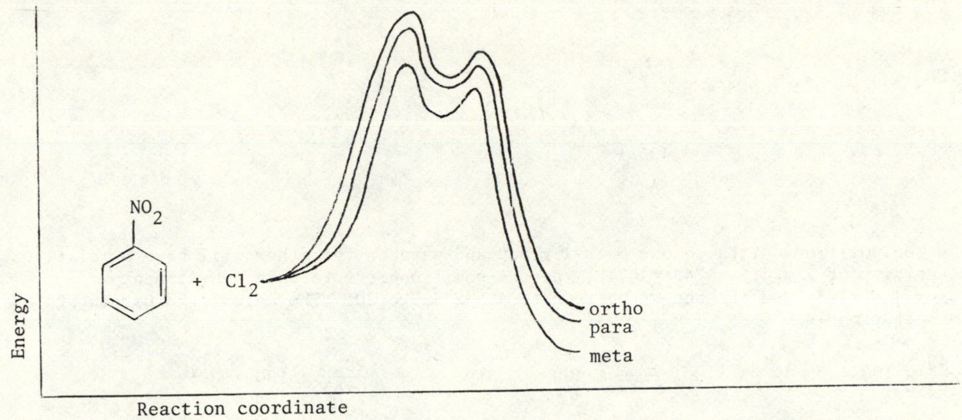

17-14 Toluene is 61 times more reactive than benzene in this reaction. $\frac{98.4}{1.6} = 61$

17-15 The electron donating resonance effect by the nitrogen and oxygen containing substituents is much greater than the electron withdrawing inductive effect, even in the ground state of the molecules.

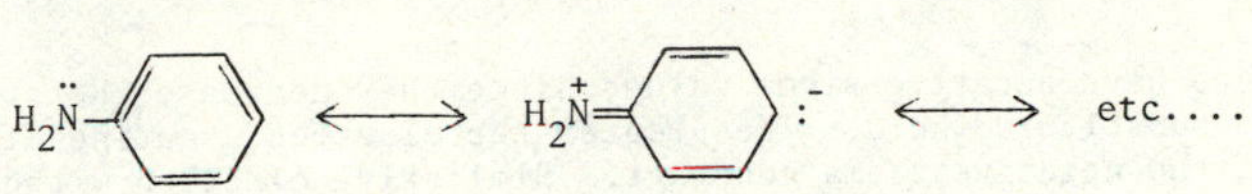

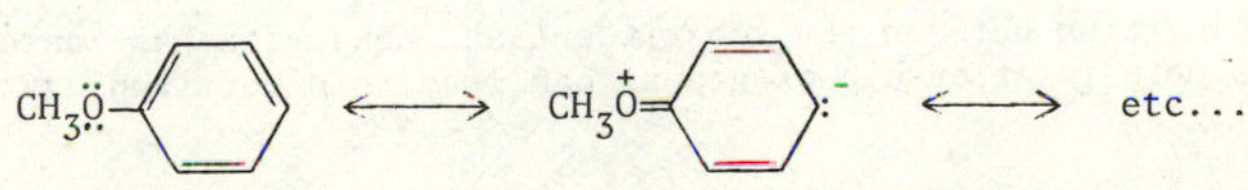

17-16

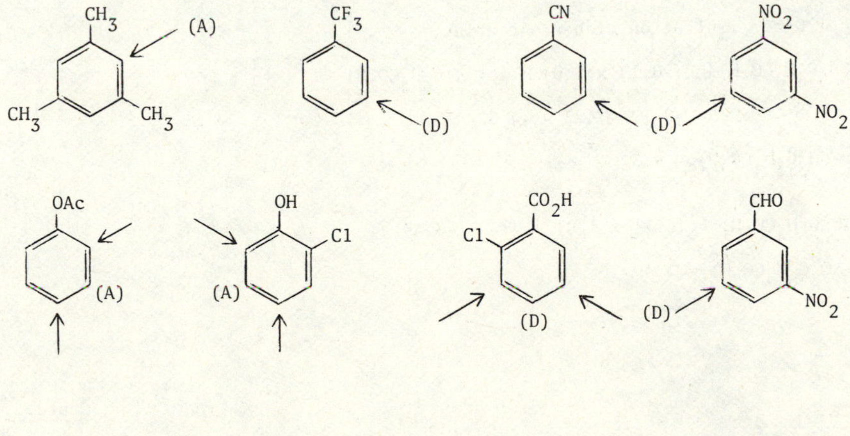

17-17

$o_f^{CH_3} = 3 \times .329 \times 605 = 597$

$m_f^{CH_3} = 3 \times .003 \times 605 = 5.4$

$p_f^{CH_3} = 6 \times .668 \times 605 = 2,425$

17-18 The two positions ortho to one methyl group and para to the other will have a relative reactivity of 2 x 4.5 x 749 = 6740, the one position meta to each substituent 4.8 x 4.8 = 23, and the one ortho to both 4.5 x 4.5 = 20. The calculated percentage of substitution is:

$\frac{6740}{6783} \times 100 = 99.4\%$ $\frac{23}{6783} \times 100 = 0.3\%$ $\frac{20}{6783} \times 100 = 0.3\%$

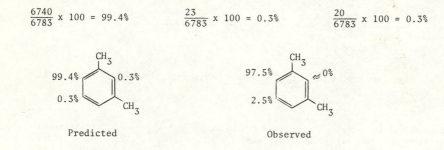

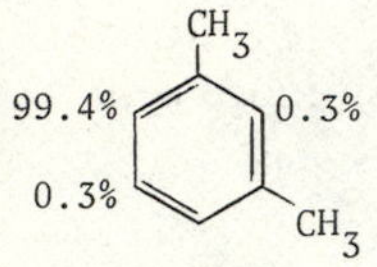

Predicted

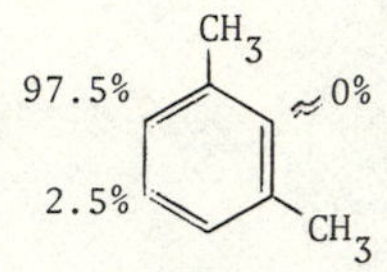

Observed

17-19 Electron donating substituents have negative sigma values since they decrease the stability of the conjugate base of the substituted acid. The greater the electron donating effect, the greater the magnitude of the negative sigma constant. Similarly, electron withdrawing substituents stabilize the negative charge on the conjugate base and enhance acidity, thus they have positive values for sigma. The σ_m and σ_p values are expected to differ significantly when direct conjugation between the substituent and conjugate base anion can occur. The para, but not the meta relationship of groups can lead to direct conjugation.

17-20

a) The pK_a value for $C_6H_5CO_2H$ is 4.2 .

Since $\rho = 1.0$ for ionization of benzoic acids;

$\log K(p\text{-}CH_3C_6H_4CO_2H) = (-0.17 \times 1.0) + \log K(C_6H_5CO_2H)$

$= -0.17 - 4.2 = -4.37$

$\therefore pK_a\ (p\text{-}CH_3C_6H_4CO_2H) = 4.37.$

b) $\log K(m\text{-}NO_2C_6H_4CO_2H) = (0.71 \times 1.0) - 4.2 = -3.49$

$\therefore pK_a(m\text{-}NO_2C_6H_4CO_2H) = 3.49.$

Contd...

17-20 contd...

c) The Hammett linear free energy relationship can be extended to reaction rates by using the sigma substituent values determined from acidities.

$$\log k(p\text{-}ClC_6H_4CO_2C_2H_5) = \rho\sigma + \log 82$$

$$= (2.43 \times 0.23) + 1.91$$

$$= 0.56 + 1.91 = 2.47$$

$$\therefore k(p\text{-}ClC_6H_4CO_2C_2H_5) = 295 \text{ L/mol-sec}$$

d) A positive value of ρ indicates that reaction is favored by electron withdrawal by a substituent. This is consistent with addition of a nucleophile in the rate-controlling step.

17-21

a) Aromatic with $4n + 2$ electrons

b) A $4n$ electron system; not aromatic

c) One of the resonance structures illustrates some degree of aromatic character.

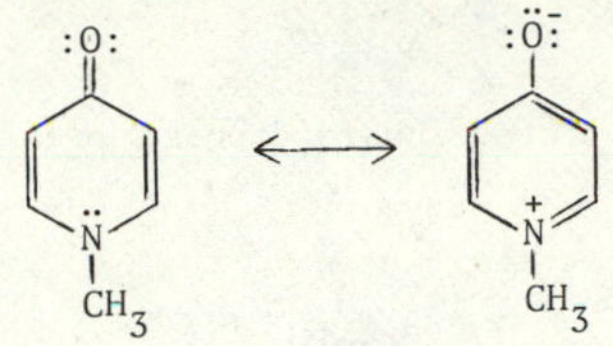

d) A resonance structure shows that each ring independently has some aromatic character.

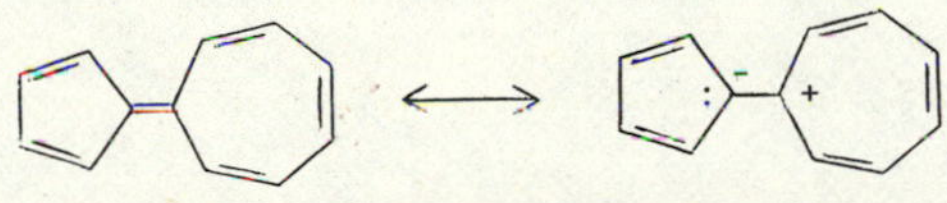

e) The structure is seen to have $4n + 2$ electrons ($n = 2$) when only alternate multiple bonds are involved in the cyclic conjugated system.

f) Six electrons are in conjugation but not in a cyclic array, thus the compound is not aromatic.

g) Cyclic conjugation of 6 electrons through the empty orbital on each boron atom provides an aromatic system.

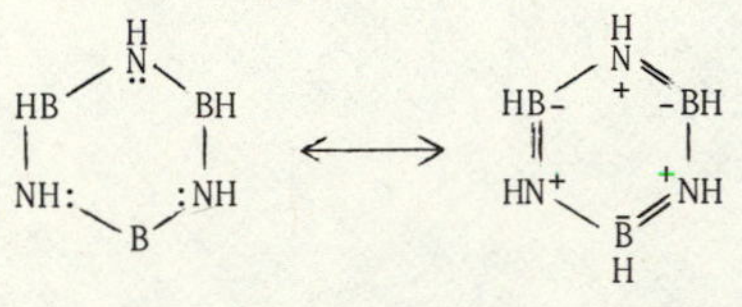

h) An electron pair on oxygen participates in conjugation for this aromatic heterocycle.

17-22

a)

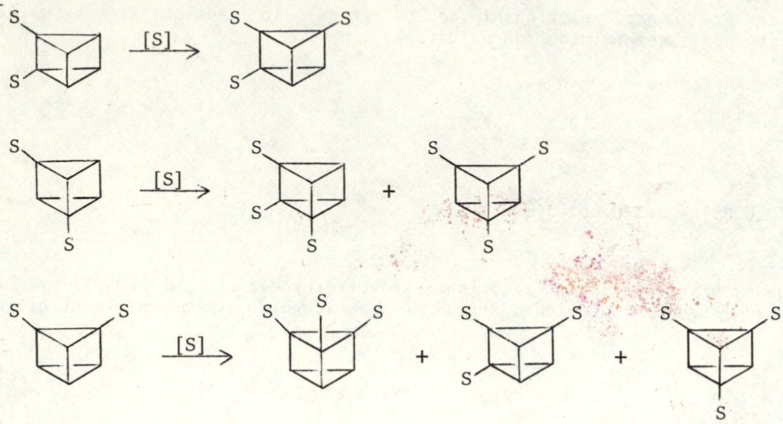

b) The Kekulé structure that gives two trisubstituted products is ortho disubstituted and leads to a 1,2-disubstituted cyclohexane. The disubstituted Ladenberg isomer that gives two trisubstituted products does not have the substituents on adjacent carbon atoms, thus does not lead to a 1,2-disubstituted cyclohexane.

17-23 Nonbonded interactions between the four hydrogen atoms pointed inward forces the molecule out of planarity.

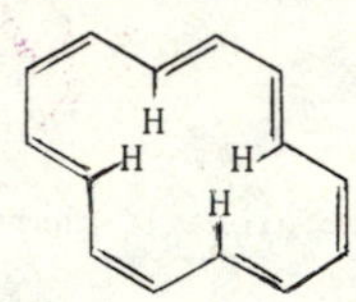

17-24

a) $ICl \rightleftharpoons I^+ + Cl^-$

b) $HOBr + H^+ \rightleftharpoons HOH + Br^+$

c) $I_2 + SbCl_5 \rightleftharpoons I^+ + ISbCl_5^-$

17-25 The nitrogen (and oxygen) atoms are electronegative and withdraw electron density, thus nitroso is deactivating. However, the nonbonding electrons on nitrogen favor ortho-para orientation.

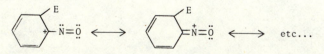

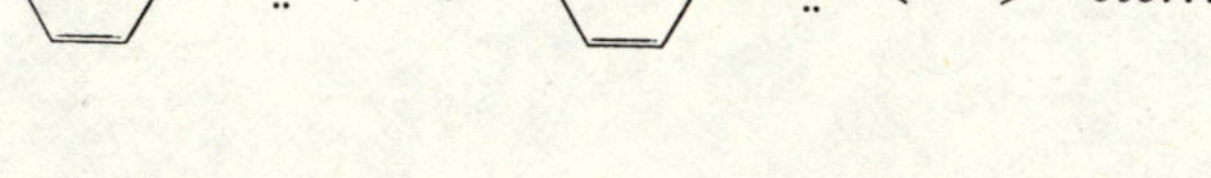

17-26

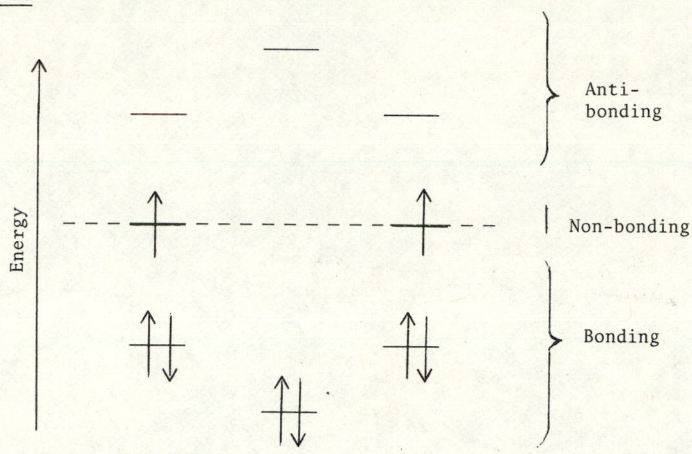

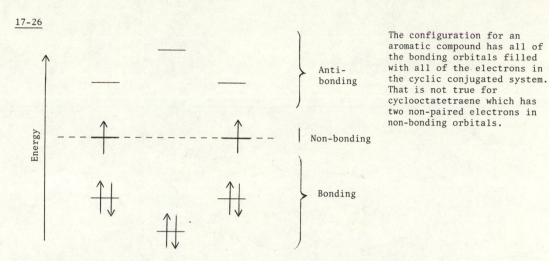

The configuration for an aromatic compound has all of the bonding orbitals filled with all of the electrons in the cyclic conjugated system. That is not true for cyclooctatetraene which has two non-paired electrons in non-bonding orbitals.

(Also see considerations of cyclooctatetraene in problems 17-5 and 17-6.)

17-27 Compound A reacts as an alkyl substituted aromatic, thus favors ortho-para orientation. In compound B the electron withdrawing (meta-directing) cyano group is conjugated with the aromatic ring.

17-28 One ring can provide electron density to the cationic intermediate of the other through conjugation.

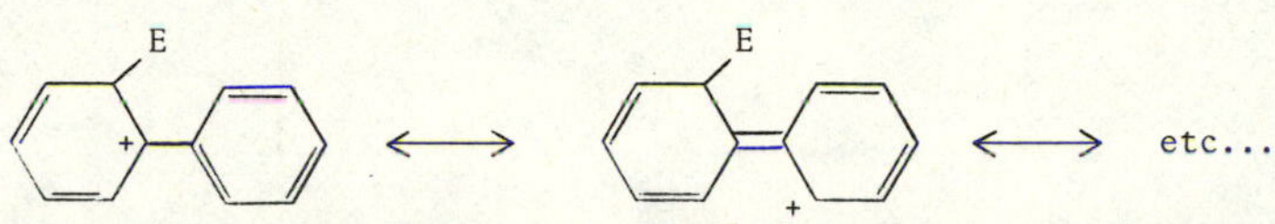

17-29

a) Normal position for alkene protons.

b) Normal position for alkene protons. Resonance between adjacent double bonds is minimal since the molecule is nonplanar.

c) The two methyl groups are in the center of an aromatic ring current and are strongly shielded.

d) The indicated CH_2 groups lie over the central part of the aromatic ring and are slightly shielded by the aromatic ring current.

18 The Scope of Aromatic Substitutions

18-1 Mononitration is readily accomplished on the activated phenol. But each nitro group deactivates the ring so that polynitration requires more extreme conditions.

18-2

a)

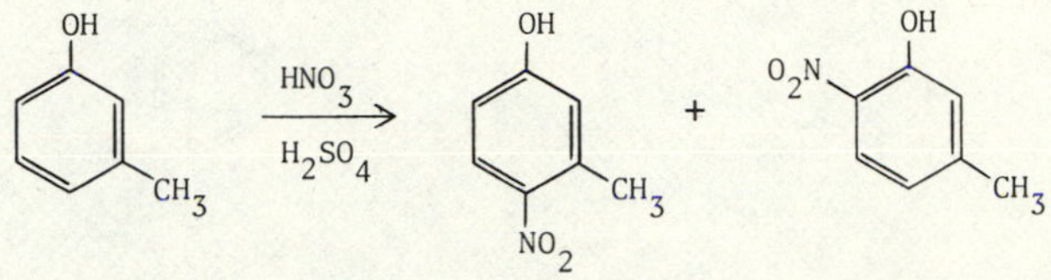

Ortho and para to the directing groups. Ortho substitution takes place at the less hindered position.

b)

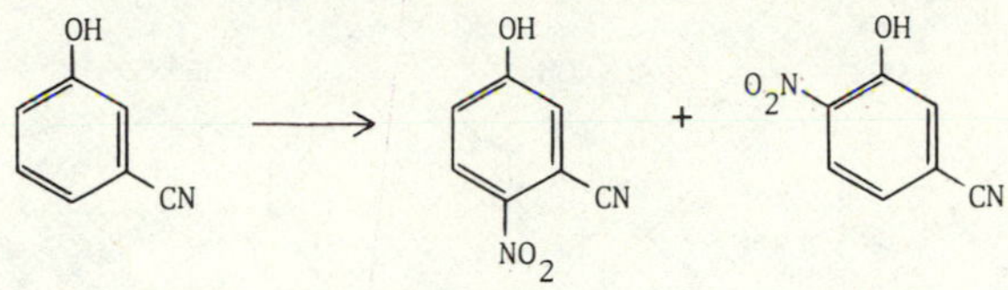

Ortho and para to the directing OH group. Ortho substitution takes place at the less hindered position.

c)

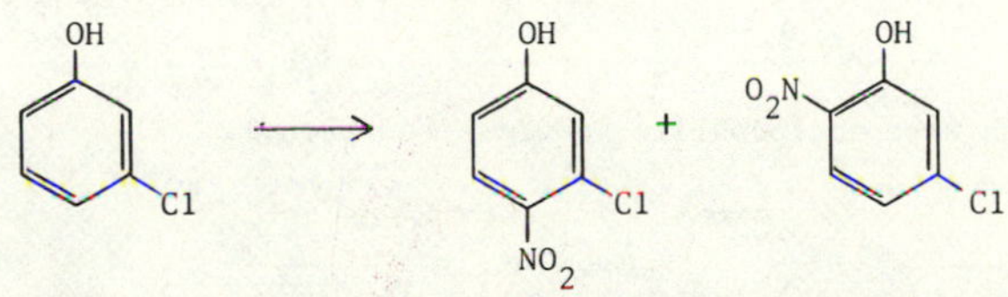

Both groups direct to the same positions as in (a).

d)

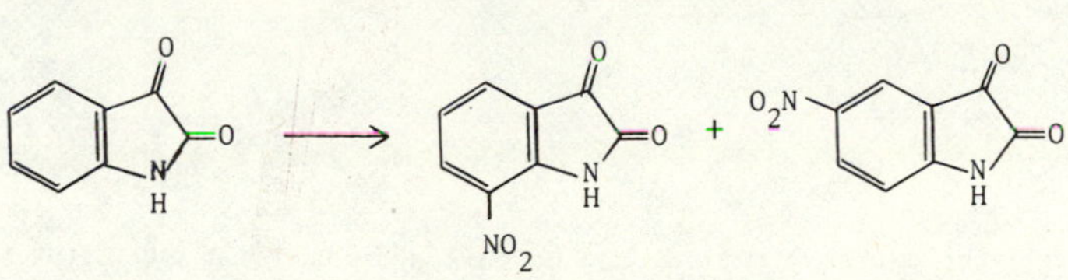

The amide nitrogen atom directs ortho and para and the carbonyl directs meta. Both direct to the same positions.

e)

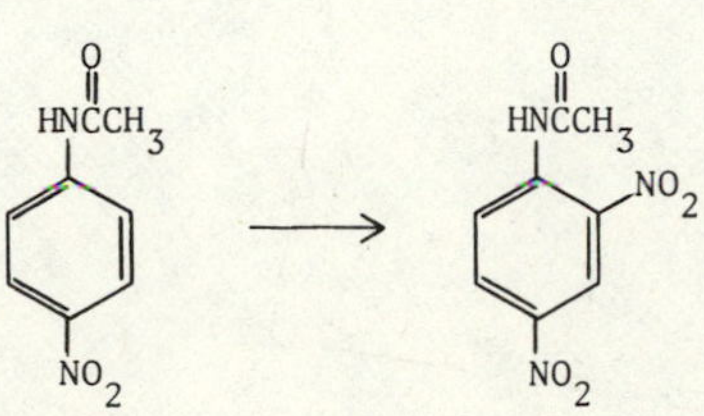

Ortho to the amide and meta to the nitro group.

18-3 In aqueous base the very reactive phenoxide anion is formed. The electron withdrawing effect as each bromine atom substitutes favors formation of the anion even more. The partially substituted material reacts more readily than phenol so that only tribromophenol is recovered.

18-4 The second Br_2 molecule acts as a catalyst to enhance the displacement of Br^-.

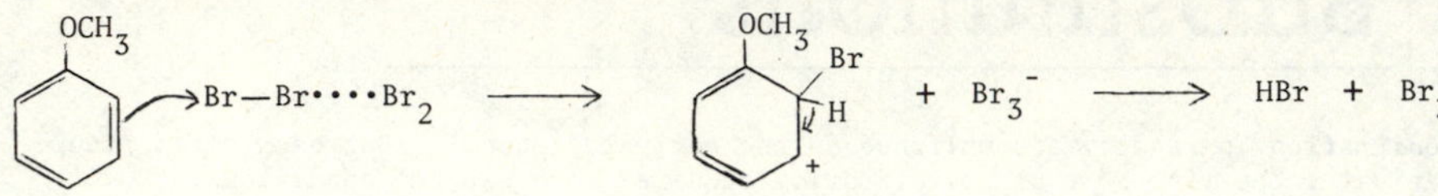

18-5

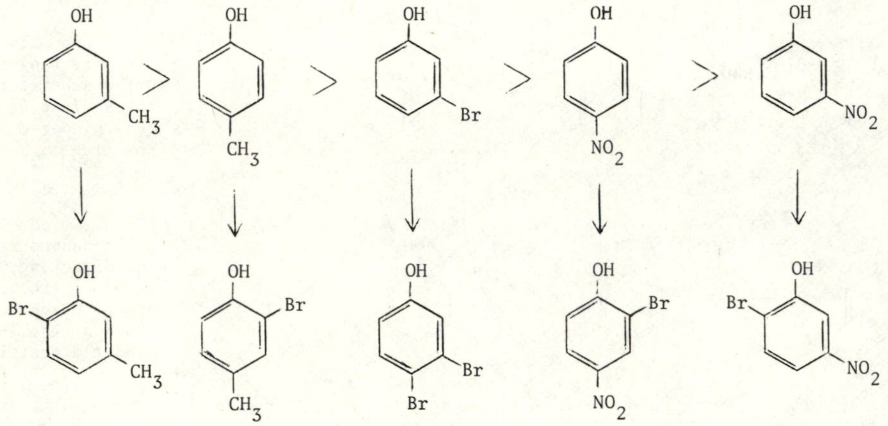

18-6 Base forms the phenoxide, a much more nucleophilic species than phenol.

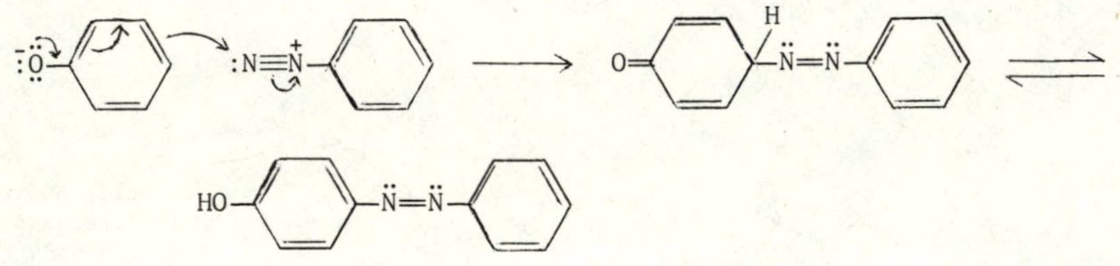

18-7 A cationic leaving group, the 2-propyl cation, can be formed in the ipso substitution on 1-isopropyl-4-methylbenzene. The same process with 1,4-dimethylbenzene would require formation of a very unstable methyl cation.

18-8

a) 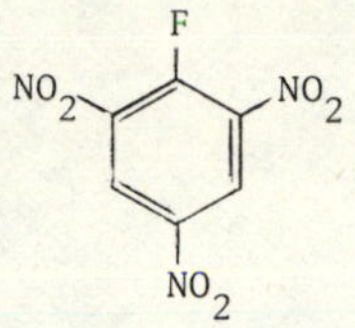

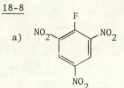

1-Fluoro-2,4,6-trinitrobenzene

b) 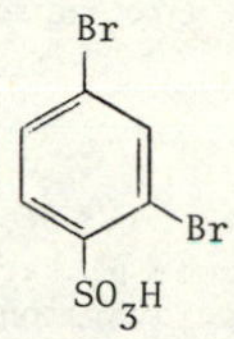

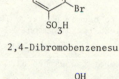

2,4-Dibromobenzenesulfonic acid

c) 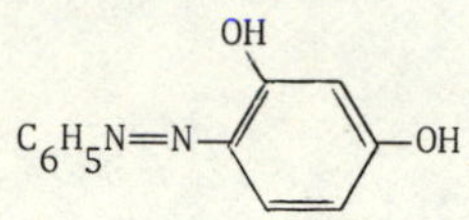

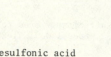

2,4-Dihydroxyazobenzene

d) 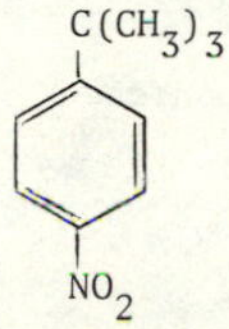

p-*t*-Butylnitrobenzene

e) 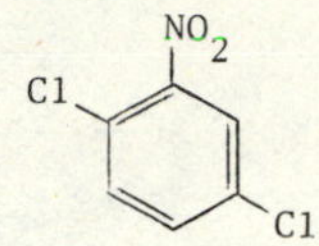 + 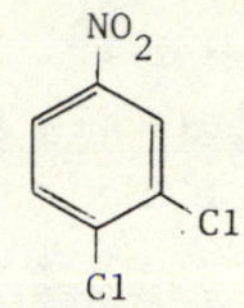

2,5-Dichloronitrobenzene 3,4-Dichloronitrobenzene

f) 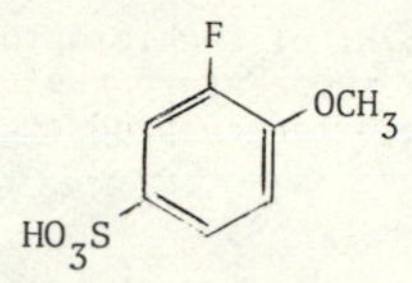 + 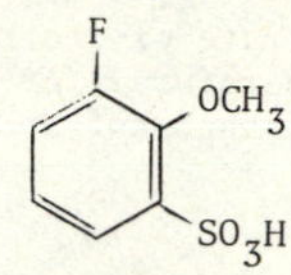

3-Fluoro-4-methoxybenzenesulfonic acid 3-Fluoro-2-methoxybenzenesulfonic acid

g) 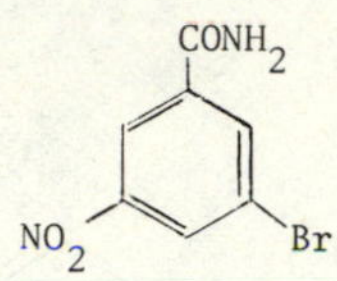

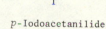

5-Bromo-3-nitrobenzamide

h) 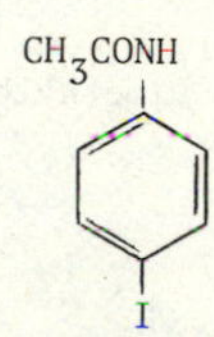

p-Iodoacetanilide

18-9 A *tert*-butyl cation is formed in each case.

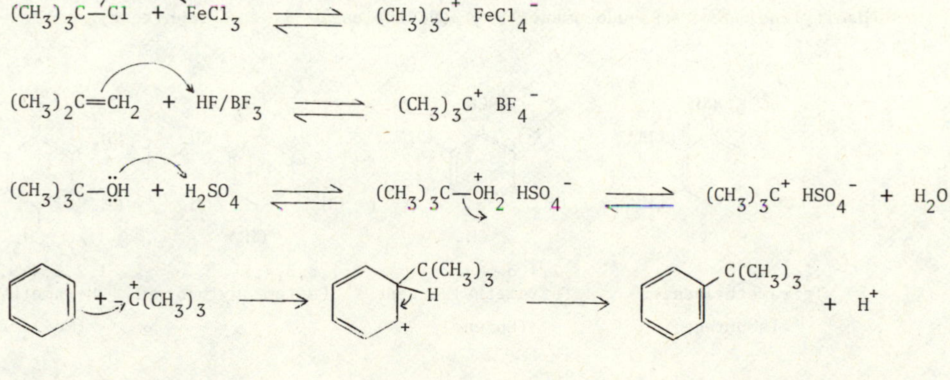

18-10

$$CH_3CH_2CH_2Br + AlBr_3 \rightleftharpoons CH_3\overset{H}{C}H\text{-}CH_2\text{-}Br\cdots AlBr_3 \longrightarrow CH_3CH{=}CH_2 + \overset{+}{H}\cdots Br\text{-}\overset{-}{Al}Br_3 \rightleftharpoons$$

$$(CH_3)_2\overset{+}{C}H \quad Br\overset{-}{Al}Br_3 \longrightarrow (CH_3)_2CHBr + AlBr_3$$

Formation of a secondary carbocation favors rearrangement.

18-11 The rate of rearrangement of 1-chloropropane to 2-chloropropane depends on the reaction temperature. At lower temperature the Friedel-Crafts reaction is slightly more rapid than rearrangement and nonrearranged product predominates.

18-12

a) Racemization demonstrates that the intermediate is sufficiently free to become planar. This is consistent with a carbocation intermediate. An S_N2 process would lead to inversion stereochemistry.

b) Trideuteriomethyl is a different substituent than methyl, thus the number two carbon atom is chiral.

18-13

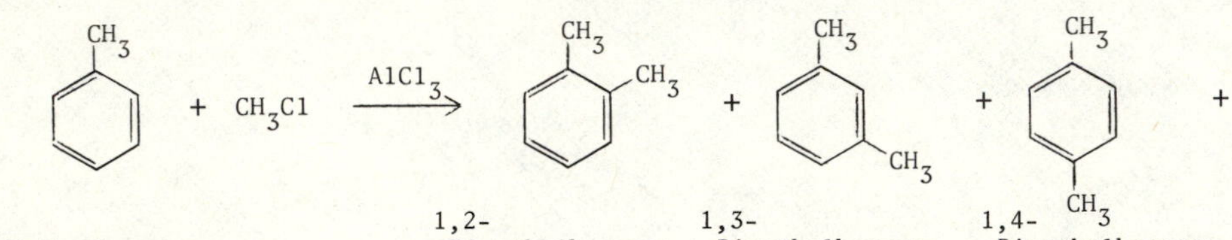

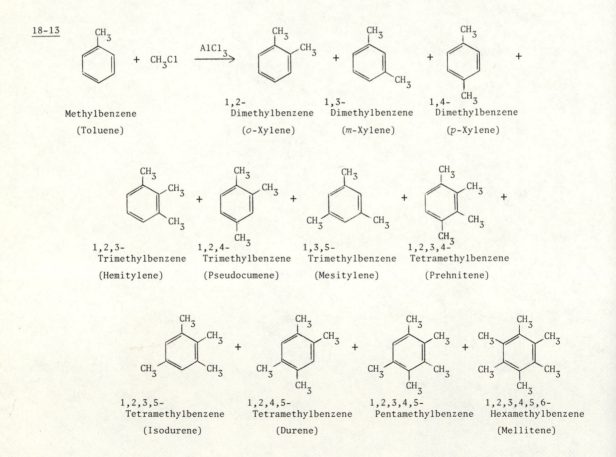

Methylbenzene (Toluene) + CH_3Cl $\xrightarrow{AlCl_3}$ 1,2-Dimethylbenzene (*o*-Xylene) + 1,3-Dimethylbenzene (*m*-Xylene) + 1,4-Dimethylbenzene (*p*-Xylene) +

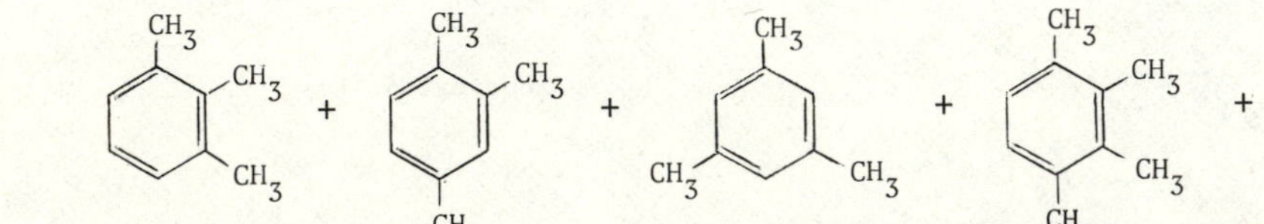

1,2,3-Trimethylbenzene (Hemitylene) + 1,2,4-Trimethylbenzene (Pseudocumene) + 1,3,5-Trimethylbenzene (Mesitylene) + 1,2,3,4-Tetramethylbenzene (Prehnitene) +

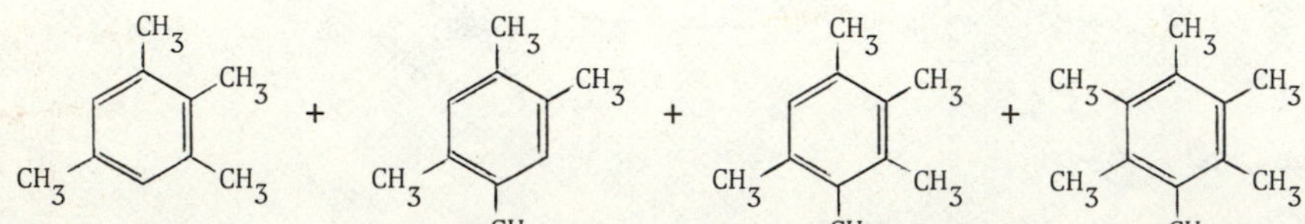

1,2,3,5-Tetramethylbenzene (Isodurene) + 1,2,4,5-Tetramethylbenzene (Durene) + 1,2,3,4,5-Pentamethylbenzene + 1,2,3,4,5,6-Hexamethylbenzene (Mellitene)

18-14 The result is consistent with the cationic character of the migrating group since a tertiary carbocation is formed more readily than a primary carbocation.

18-15

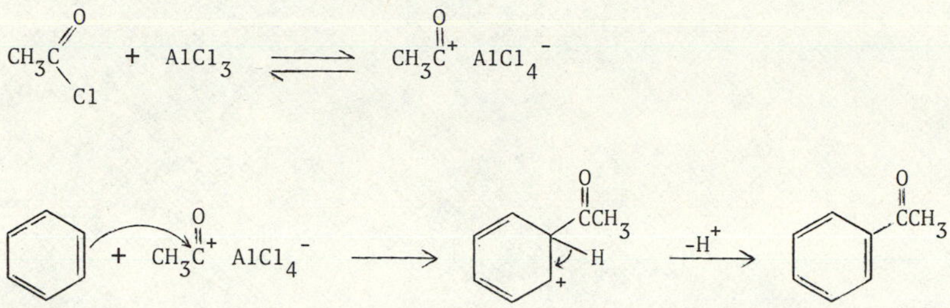

18-16 The exchange of chlorine atoms is consistent with formation of an acyl cation and supports the suggestion that such an intermediate is also involved in the Friedel-Crafts acylation.

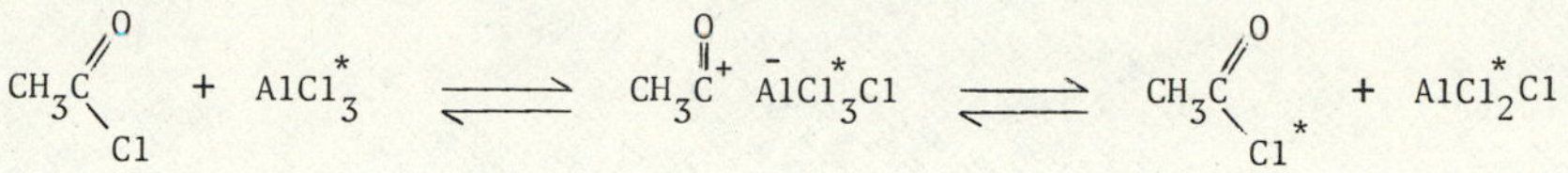

18-17 The two carbonyl groups of the anhydride complex the catalyst.

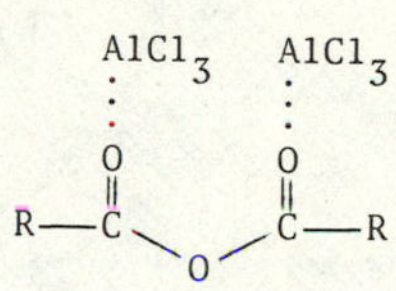

18-18

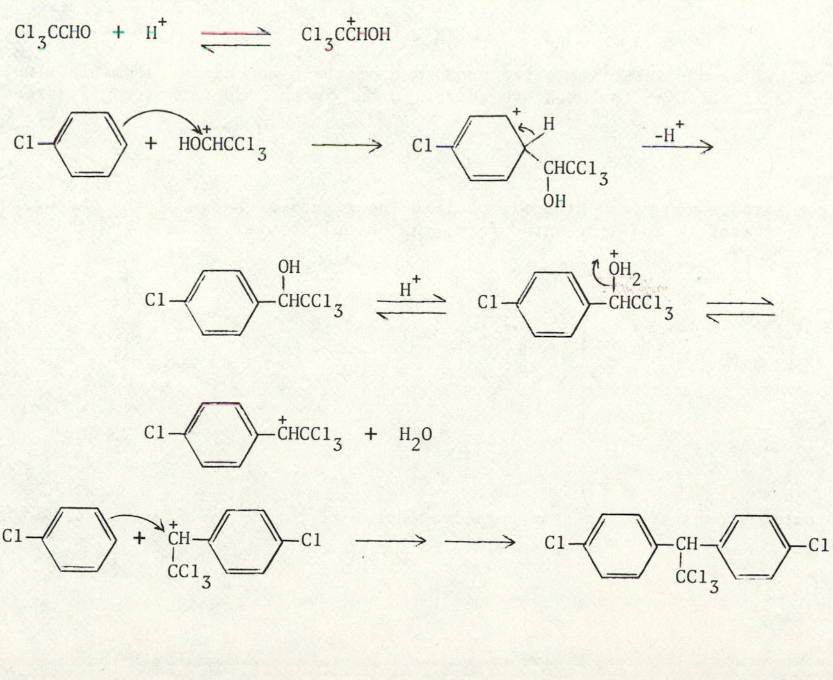

18-19

a)

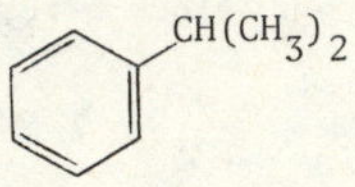

Isopropylbenzene

b)

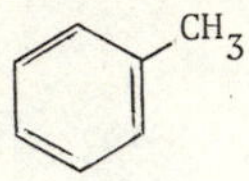

Toluene

c)

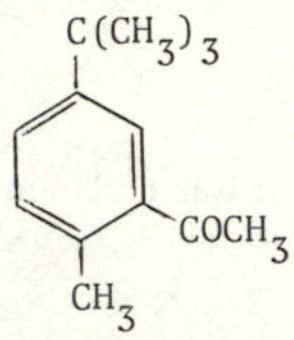

5-*tert*-Butyl-2-methylacetophenone

d)

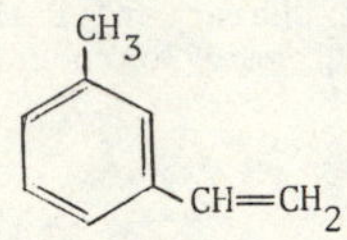

3-Methylstyrene

e)

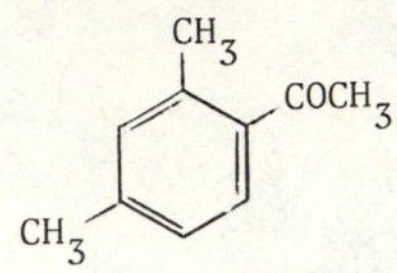

2,4-Dimethylacetophenone

f)

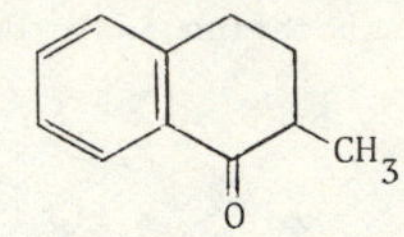

2-Methyl-1-tetralone

g)

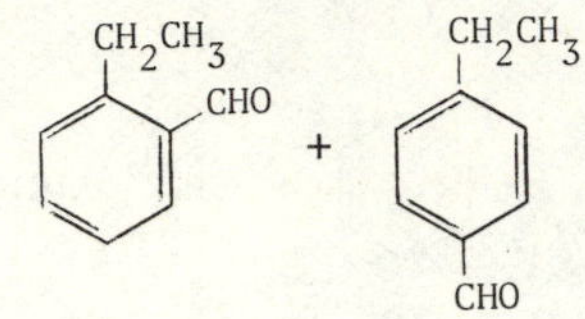

o-Ethylbenzaldehyde *p*-Ethylbenzaldehyde

18-20

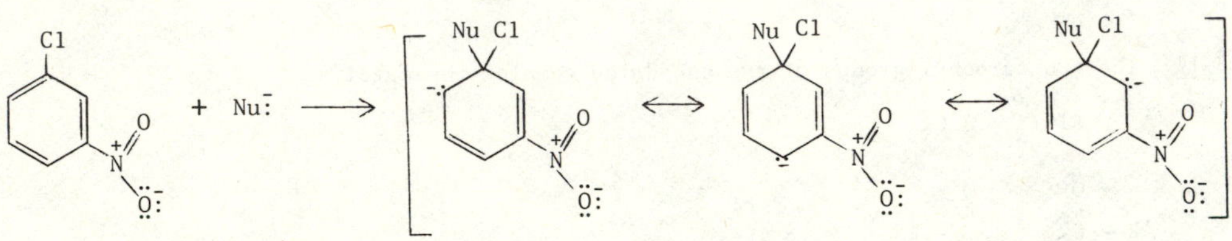

No direct conjugation between the electron pair and the nitro group is possible so as to stabilize the negative charge of the anionic intermediate. Anion stabilization is almost always required for nucleophilic aromatic substitution to take place.

18-21 The 2,6-dimethyl groups force the nitro group out of the plane of the aromatic ring. Resonance stabilization of the negative charge formed in the addition step is markedly decreased.

18-22 The electron withdrawing nitro groups stabilize the negative charge of the conjugate base by resonance, therefore the protonated form (the phenol) is very acidic ($pK_a \approx 0$).

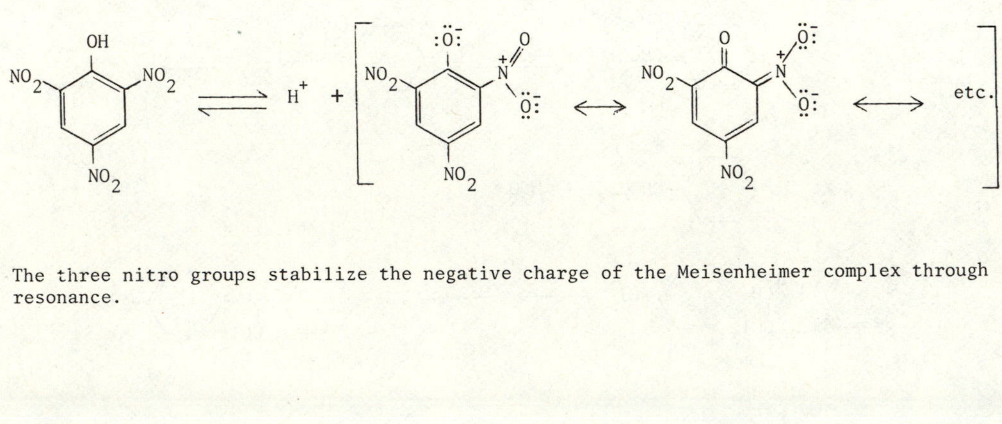

18-23 The three nitro groups stabilize the negative charge of the Meisenheimer complex through resonance.

18-24 If an S_N2 mechanism were involved in nucleophilic aromatic substitution, the nucleophile would approach the reaction center from the back side; the side opposite the leaving group. This would require approach through the plane of the aromatic ring, a rather unlikely process. In the actual addition step of nucleophilic aromatic substitution, it is believed that the nucleophile approaches the reaction center perpendicular to one side of the aromatic ring as the pi-electrons move away on the opposite side.

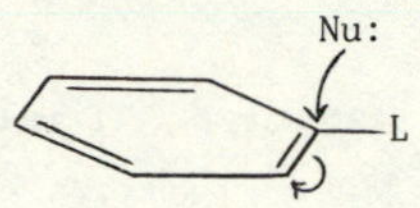

18-25 The data suggest that halogen does not leave in the rate controlling step but does have an influence on this step. That is, two steps are involved. The more electronegative fluorine enhances attack of the nucleophile by inductive stabilization of the forming negative charge, thus increasing the rate of the rate controlling step.

18-26 Nucleophilic substitution on a haloalkane is considerably more favorable than on a haloaromatic since aromaticity is disrupted in the addition step with the latter substrates.

18-27 The addition-elimination pathway normally occurs only when electron withdrawing groups are available to stabilize the intermediate anion (Meisenheimer complex).

18-28

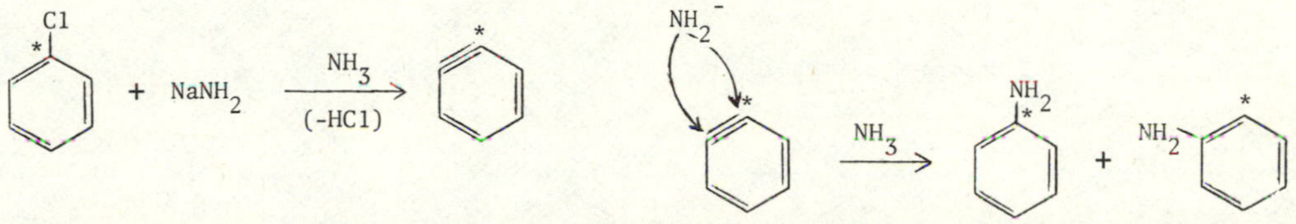

18-29 Reaction of phenoxide (formed from the phenol product plus base) with unreacted chlorobenzene produces the ether. At the high temperature of this process both addition-elimination and benzyne pathways are probably involved.

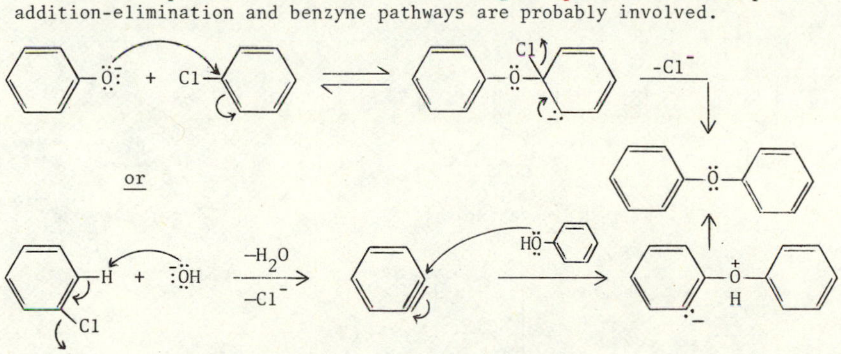

18-30 The isomerization is consistent with (though doesn't necessarily require) a phenyl cation intermediate.

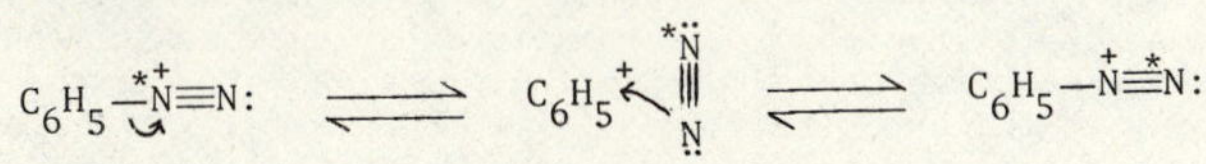

18-31 The phenyl cation ion is very unstable relative to *tert*-butyl. It reacts very quickly, thus is much less selective than *tert*-butyl.

18-32

a)

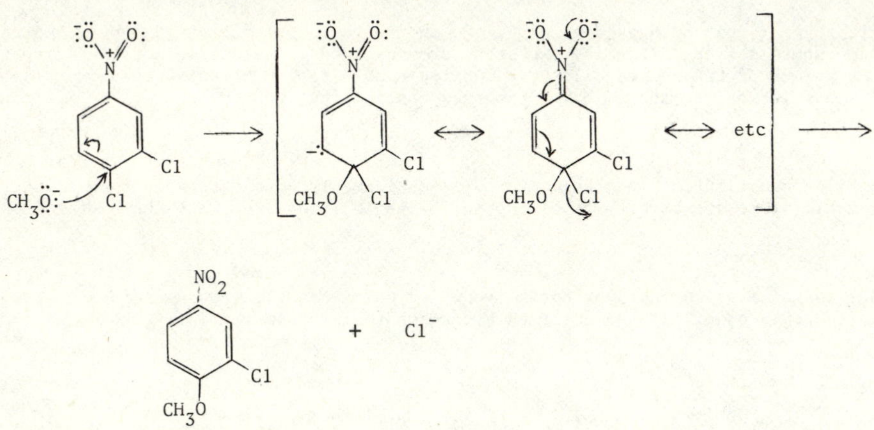

2-Chloro-4-nitroanisole

b)

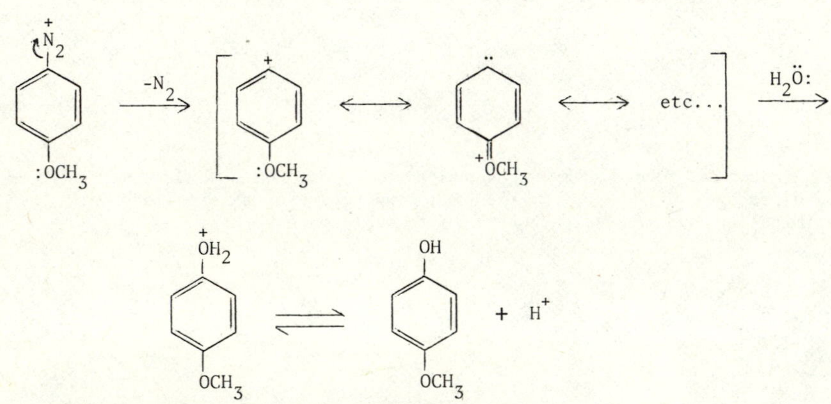

p-Hydroxyanisole

c)

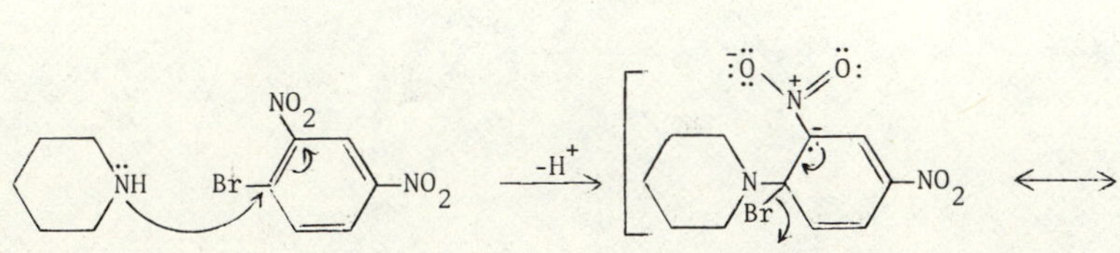

Contd...

18-32 contd...

c) Contd...

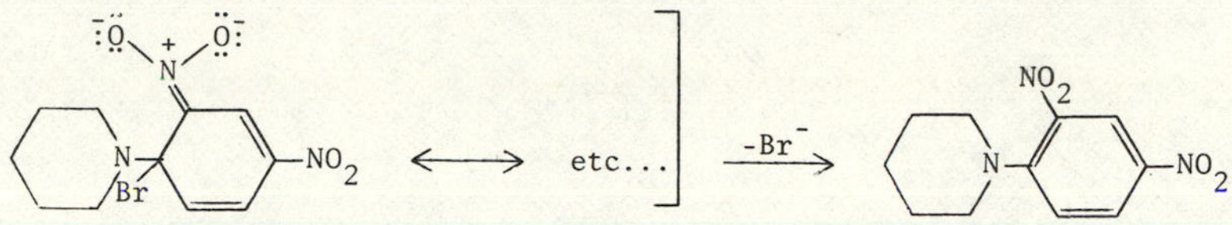

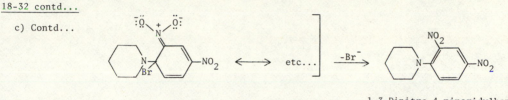

1,3-Dinitro-4-piperidylbenzene

d)

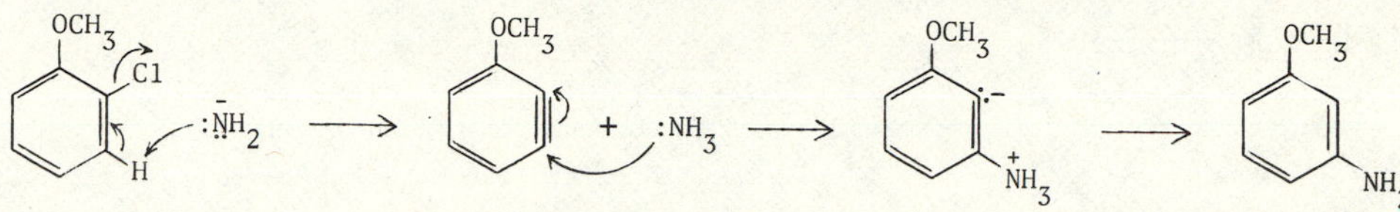

m-Methoxyaniline

(The ortho position is sufficiently hindered so that meta substitution predominates.)

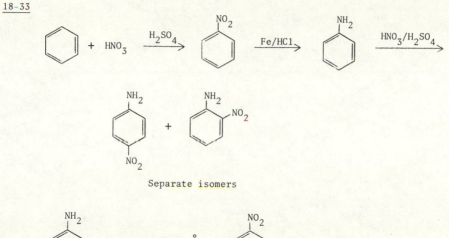

18-33

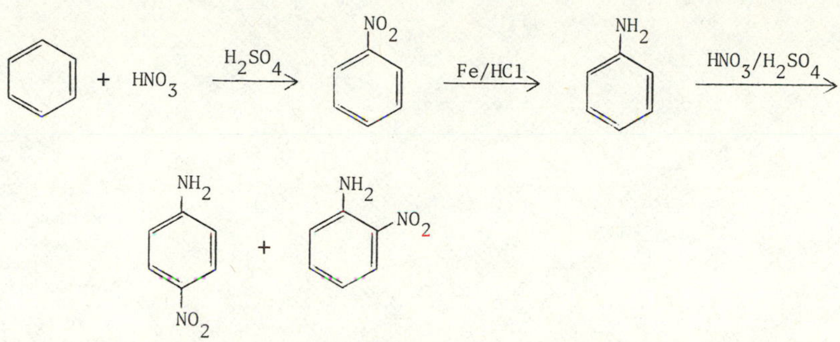

Separate isomers

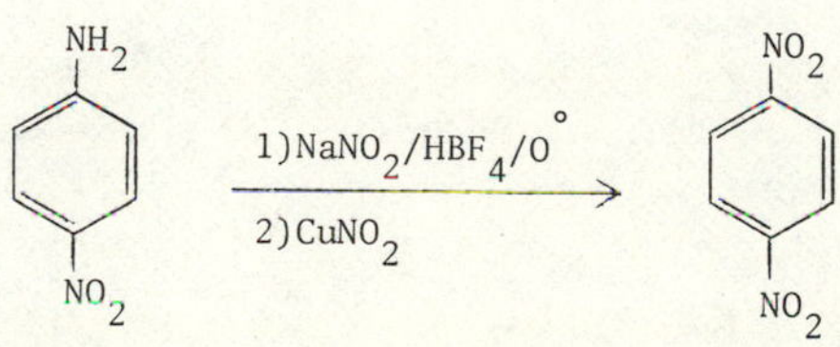

18-34 Delocalization of the positive charge on the nitrogen atom by the aromatic ring enhances stability relative to that observed with an alkyl substrate.

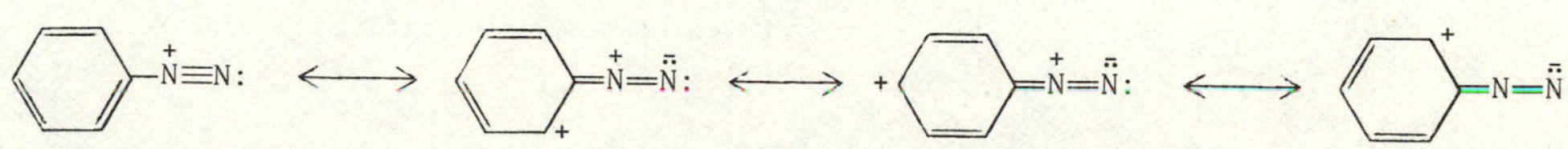

18-35

a)

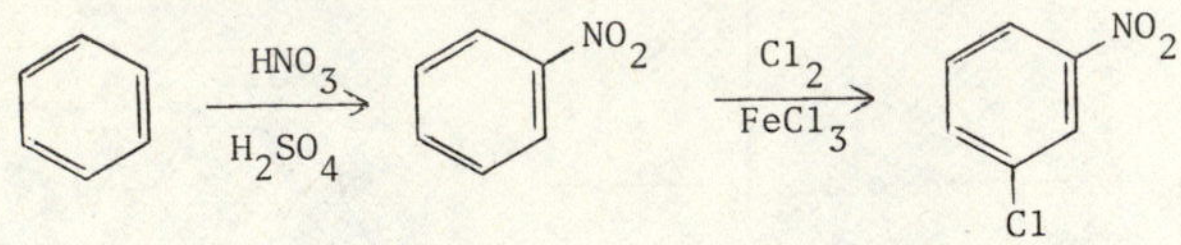

Substitute NO_2 group first to direct Cl meta.

b)

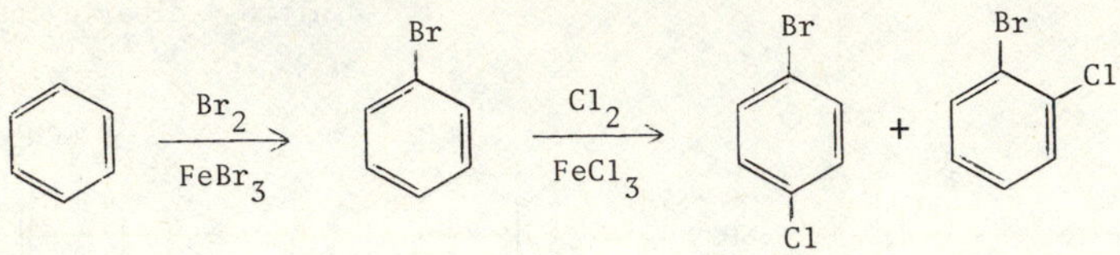

Separate isomers

c)

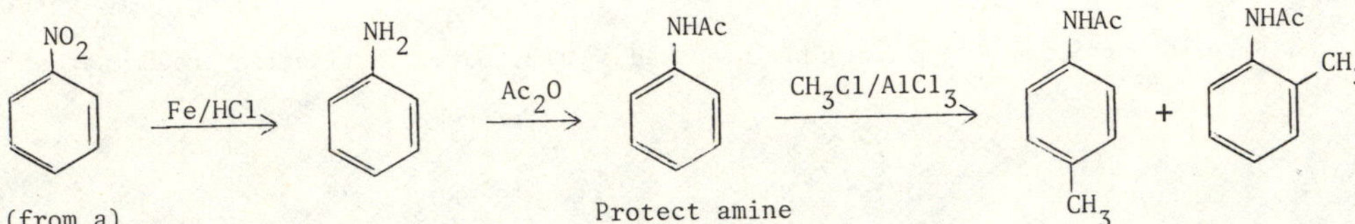

(from a)

Protect amine group from acid catalyst of next two steps.

Separate isomers

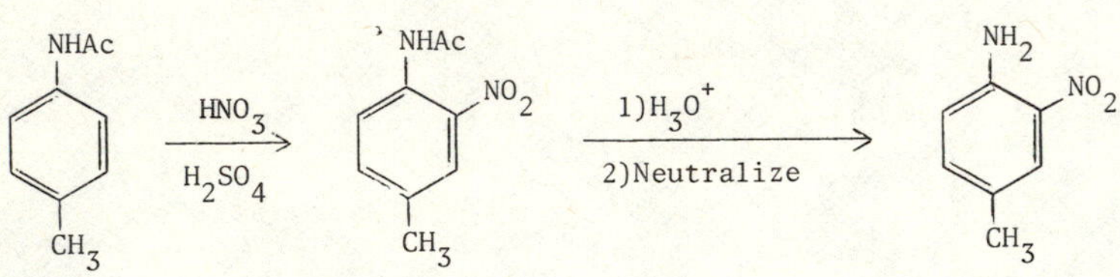

Friedel-Crafts reaction does not take place on a deactivated ring. The NO_2 must go on after the CH_3.

d)

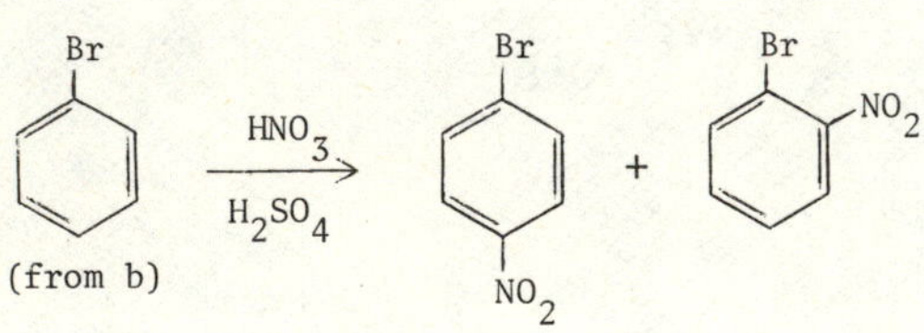

(from b)

Separate isomers

Substitute Br first to direct NO_2 para.

e)

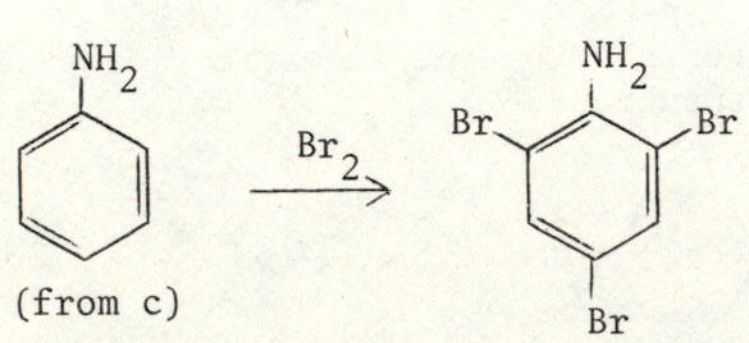

(from c)

No catalyst needed with activated aromatic.

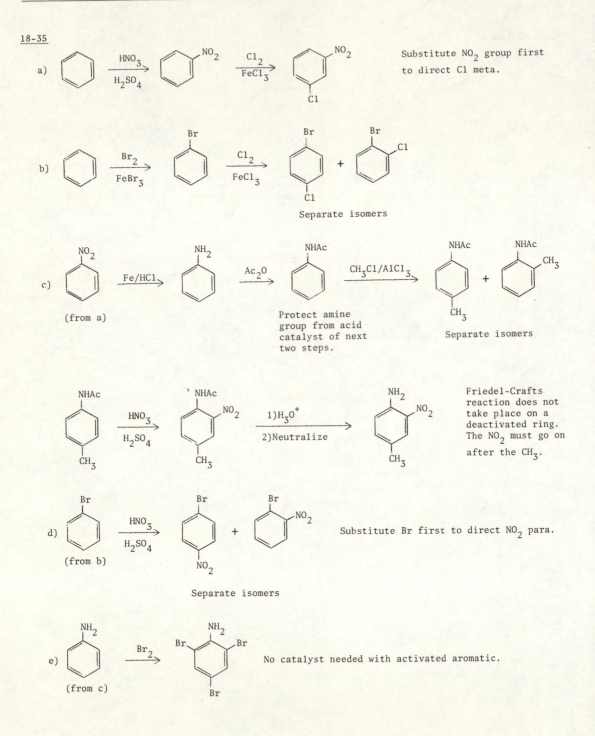

18-35 contd...

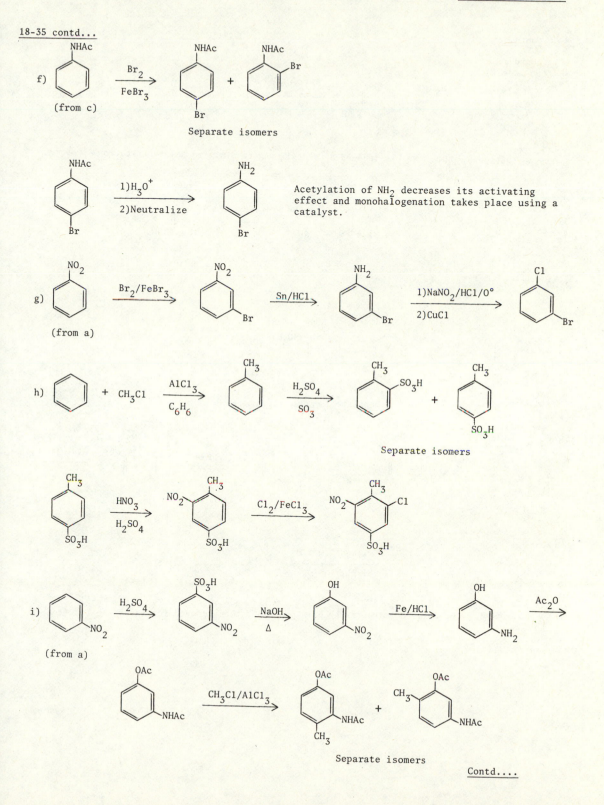

f) 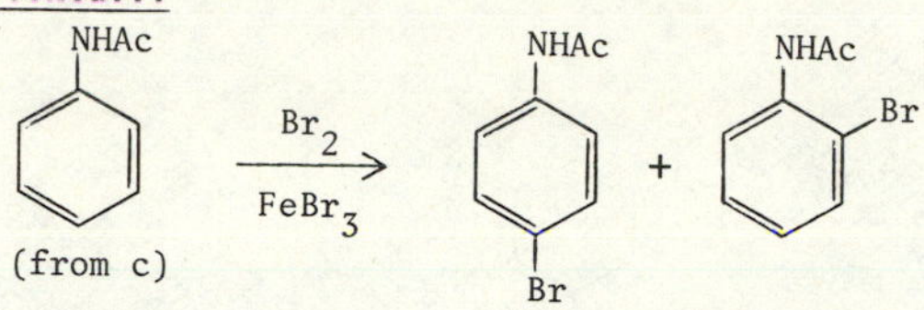

(from c)

Separate isomers

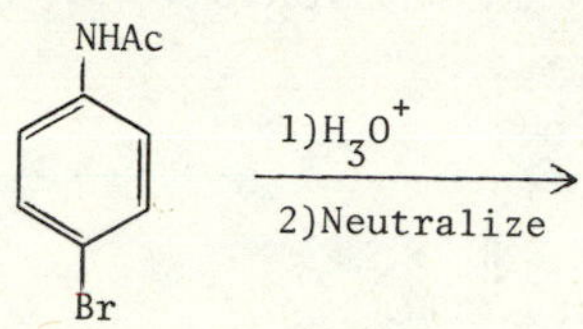

Acetylation of NH_2 decreases its activating effect and monohalogenation takes place using a catalyst.

g) 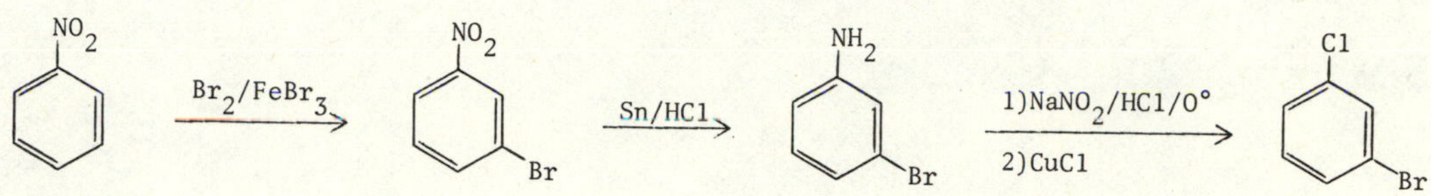

(from a)

h) 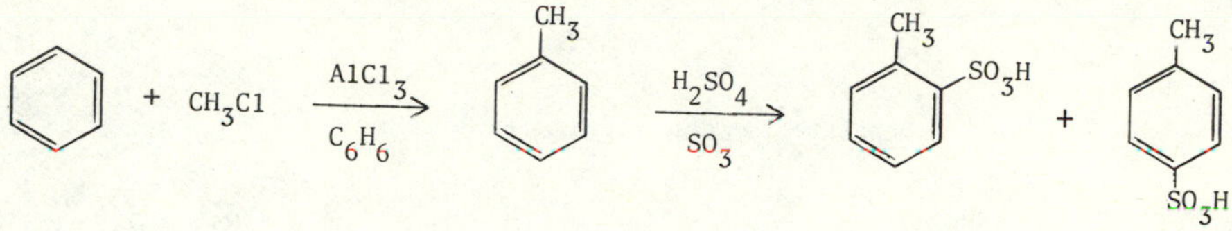

Separate isomers

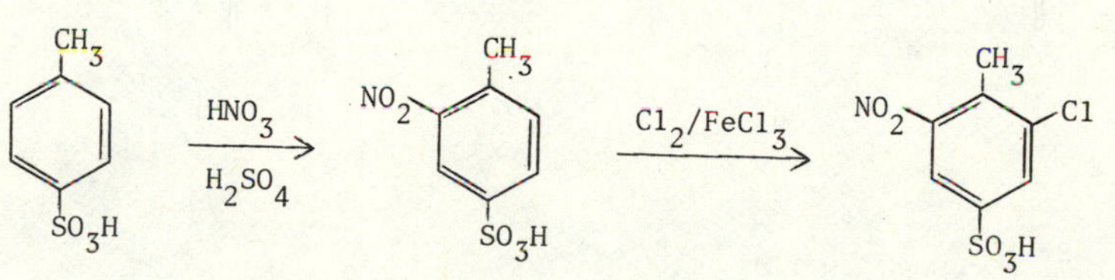

i) 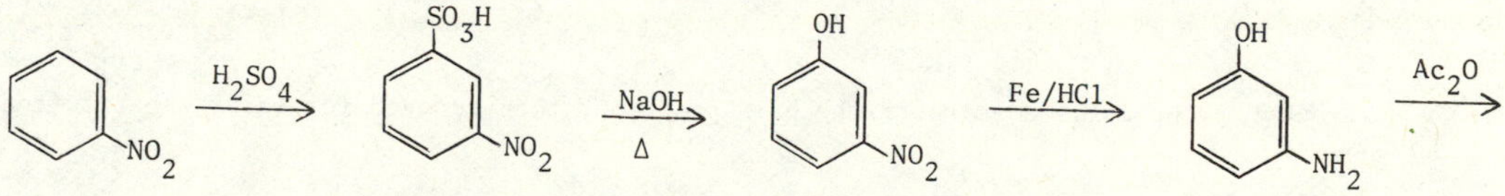

(from a)

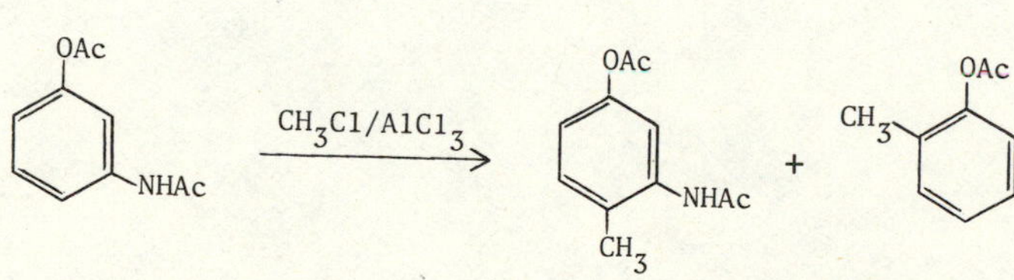

Separate isomers

Contd....

18-35 contd...

i) contd...

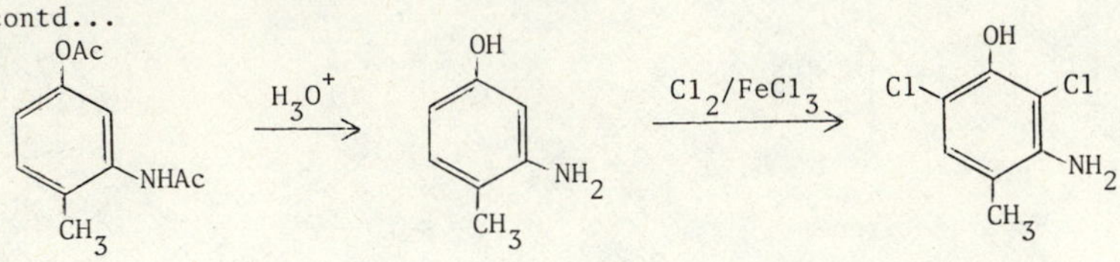

Activation by OH and NH_2 is reduced by acetyl group to minimize polysubstitution.

j)

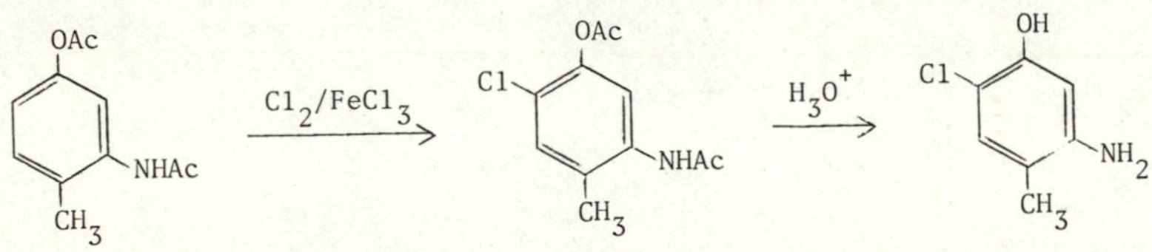

(from i)

k)

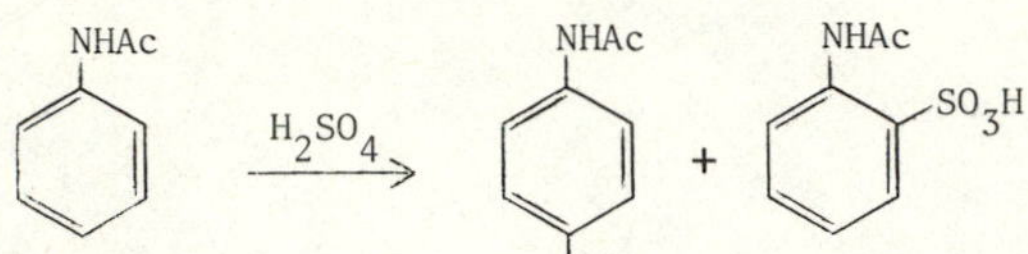

(from c)

Separate isomers

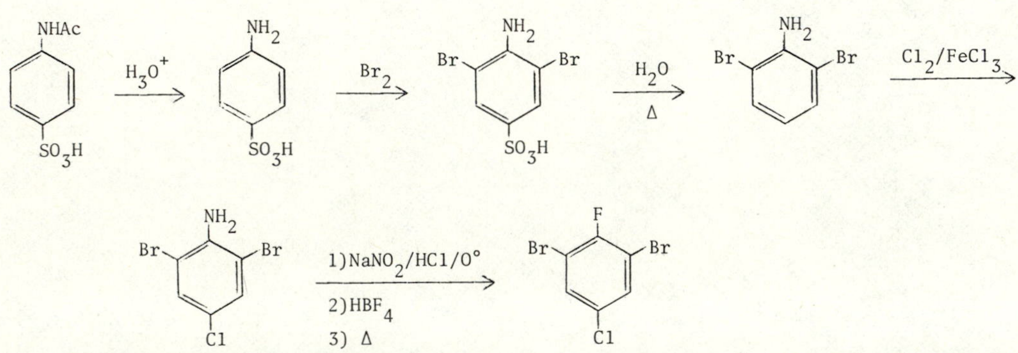

The SO_3H group is used to block the para position, then removed for later substitution.

18-36

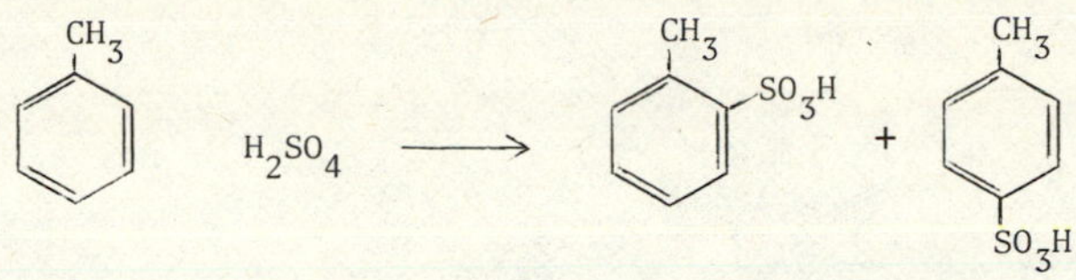

Separate isomers

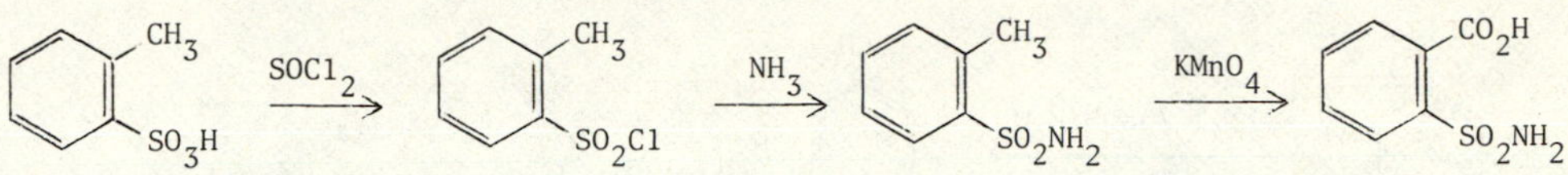

18-37

a)

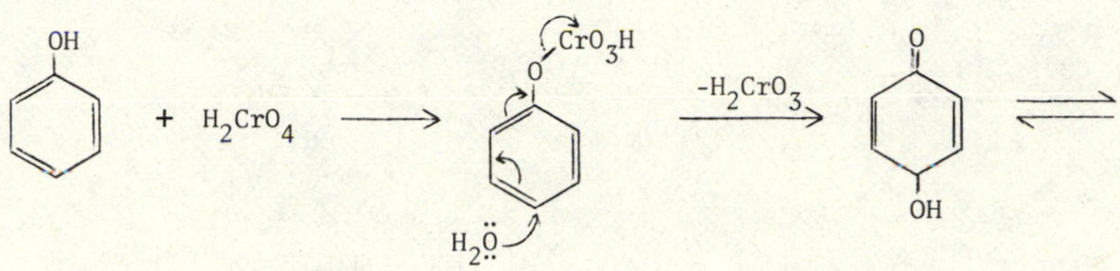

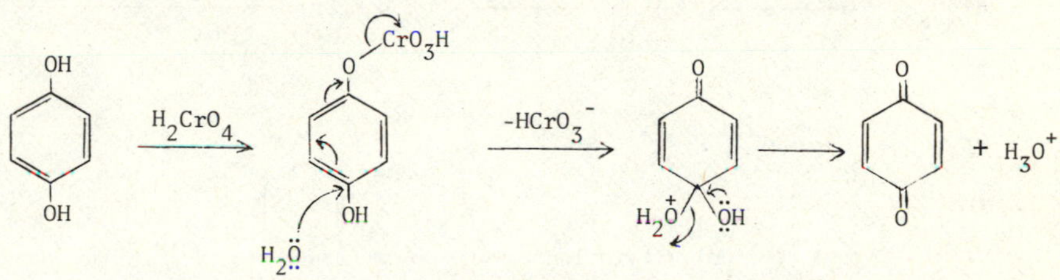

b) A *meta*-quinone does not exist. The aromaticity would be lost and there is no possibility for conjugation between the two carbonyl groups of the potential product.

18-38 Benzene is much more resistant to catalytic reduction than is cyclohexadiene or cyclohexene. Once a molecule of benzene begins the reduction sequence it reacts much more rapidly than an unchanged molecule.

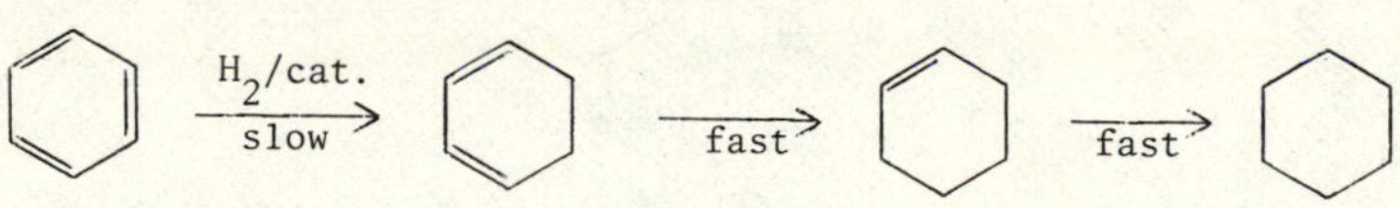

18-39

a)

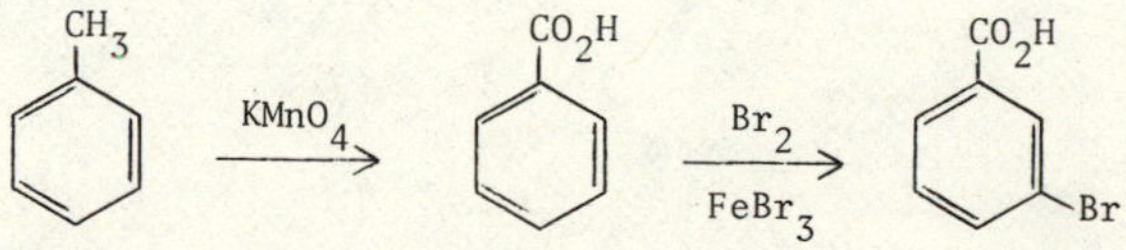

Contd....

18-39 contd...

b) 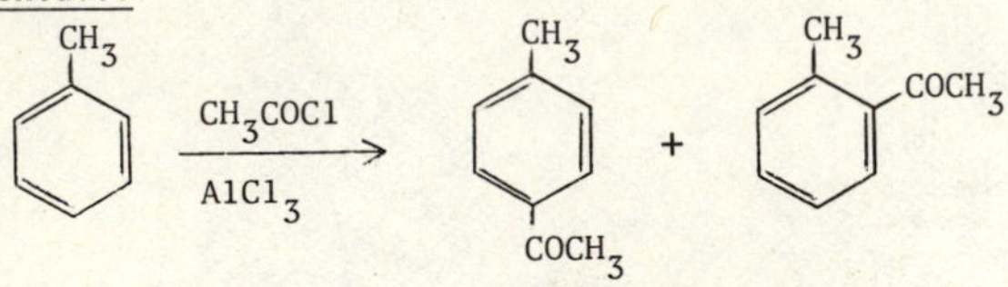

Separate isomers

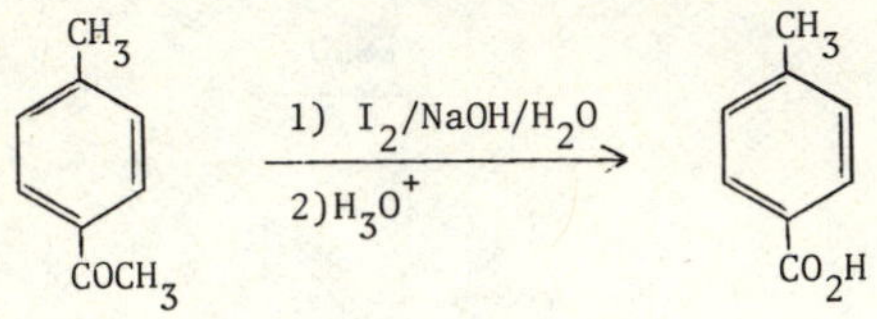

c) $C_6H_6 \xrightarrow{CH_3COCl/AlCl_3} C_6H_5COCH_3 \xrightarrow[2)H_3O^+]{1)LiAlD_4/Et_2O} C_6H_5-C(D)(OH)-CH_3$

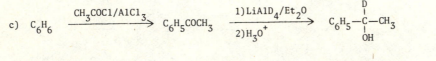

d) $C_6H_6 \xrightarrow{Br_2/FeBr_3} C_6H_5Br \xrightarrow{Mg/Et_2O} C_6H_5MgBr \xrightarrow[2)H_3O^+]{1)CH_2-CH_2(O)/Et_2O} C_6H_5CH_2CH_2OH$

18-40

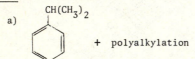

a) 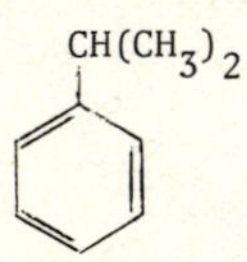 + polyalkylation

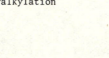

b) 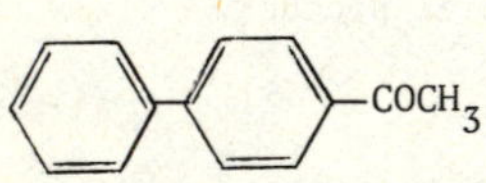

c) 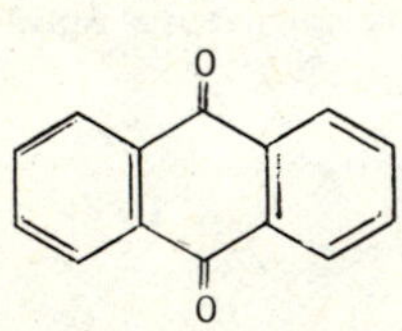

d) 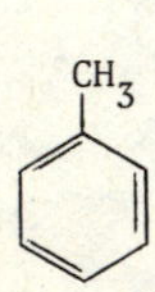

e) 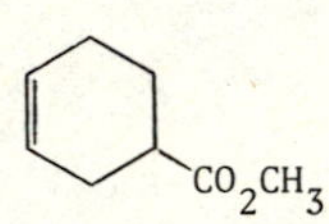

f) 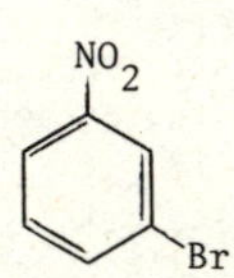

g) 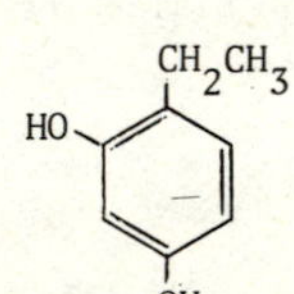

h) 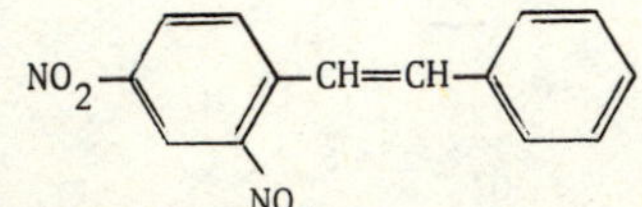

i) 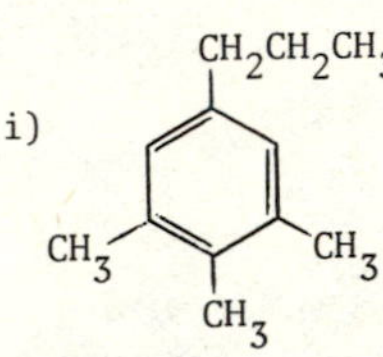

18-41 Substitution of an electronegative halogen atom onto the aromatic ring decreases the basicity of aniline. HX formed in the substitution protonates aniline more effectively than it protonates the haloanilines. The haloanilines are more reactive than the anilinium ion, thus react further.

18-42

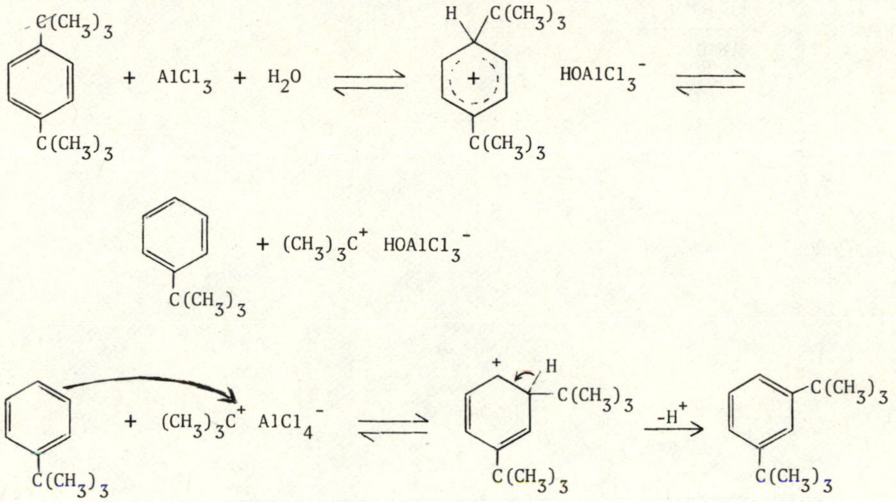

The *o*-di-*tert*-butylbenzene would be very crowded.

18-43

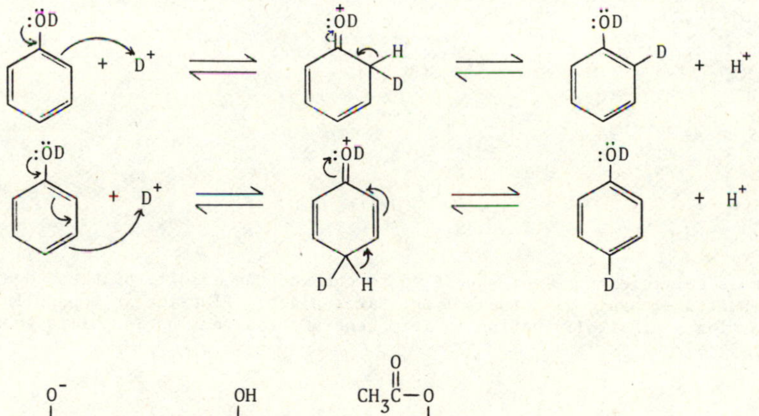

Note that the phenolic hydrogen rapidly exchanges with the deuterated medium.

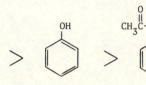

18-44

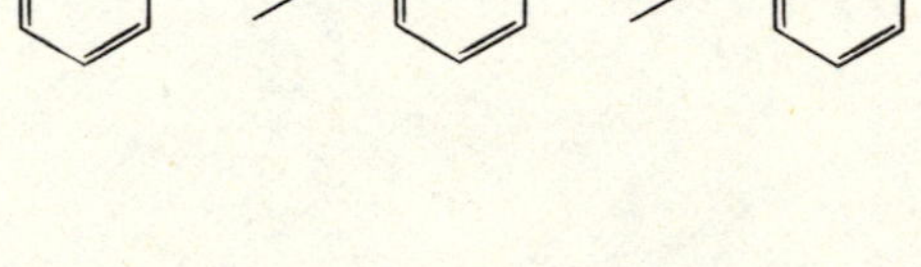

18-45 *IHD* = 5. Reaction with phenylhydrazine suggests an aldehyde or ketone and the positive haloform test indicates a methyl ketone. Compound B is a carboxylic acid produced from the haloform reaction. Compound C is an alcohol derived from the ketone A and can also undergo the haloform reaction. The acid D with mp of 121-122° is benzoic acid, an observation consistent with the *IHD* of 5 for A.

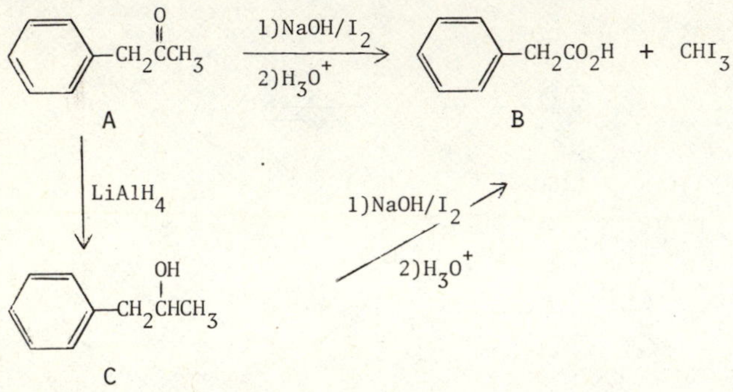

A or B or C 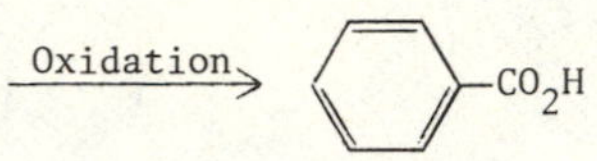

D

18-46 The unprotonated aniline is considerably more reactive in electrophilic substitution than is the protonated form. As long as at least a small equilibrium concentration of the unprotonated form is present, it will nitrate in the ortho and para positions. Rapid proton transfer reestablishes the equilibrium to provide additional aniline.

18-47

a) 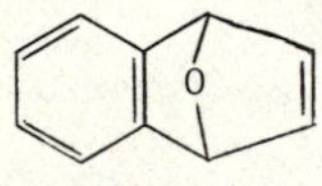

b) 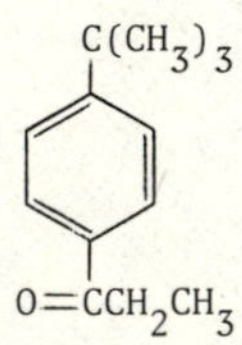

18-48 The fluorine atom is sufficiently electronegative to enhance the acidity of the α-deuterium atoms and base-promoted exchange with the solvent takes place. Fluoride is a poor leaving group in this reaction so that elimination to form benzyne or direct nucleophilic substitution does not occur.

18-49 Each alcohol leads to the same tertiary carbocation which is the electrophile in these Friedel-Crafts alkylations.

Contd...

18-49 contd...

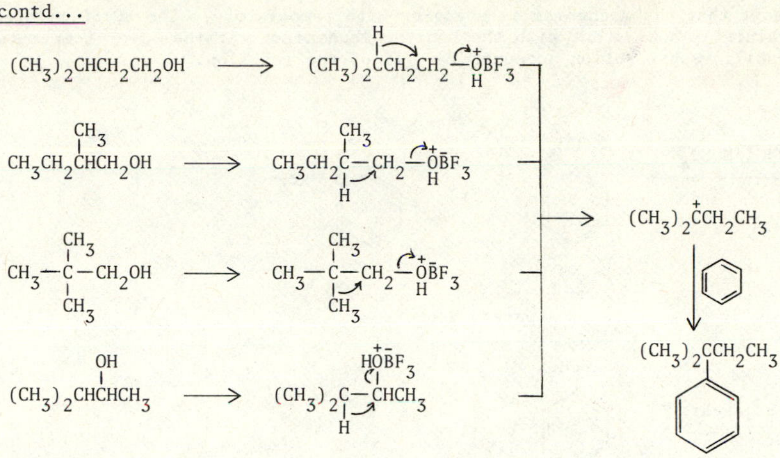

18-50

a)

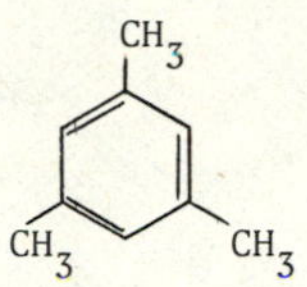

b)

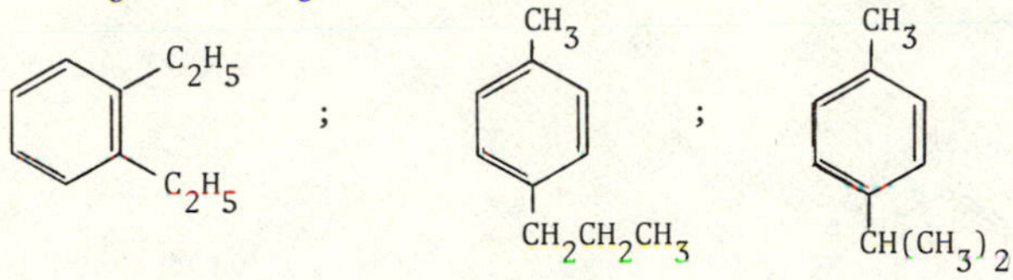

c)

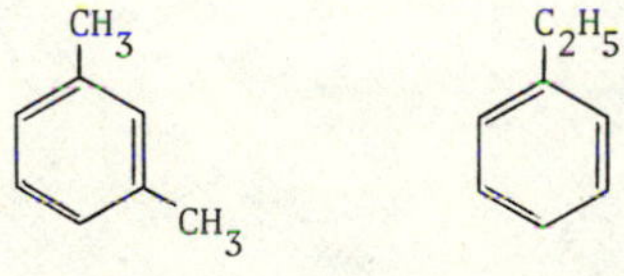

18-51

a) $A \equiv (CH_3CO)_2O$

$B \equiv$

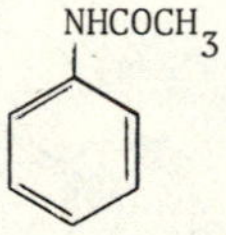

$C \equiv$

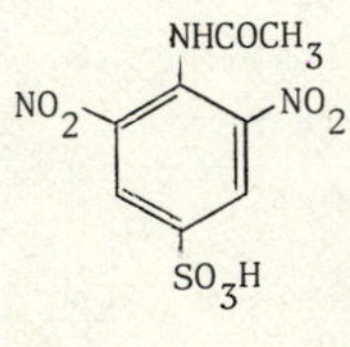

b) $D \equiv$

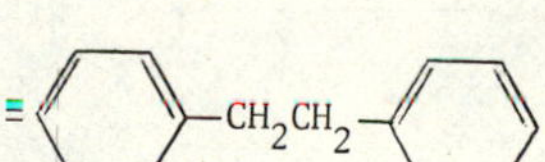

$E \equiv$

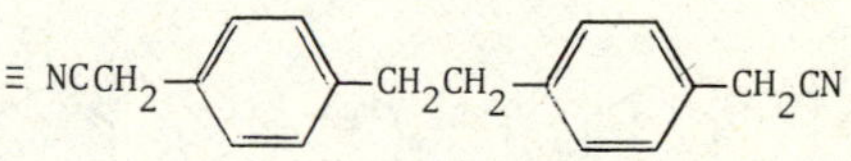

$F \equiv$

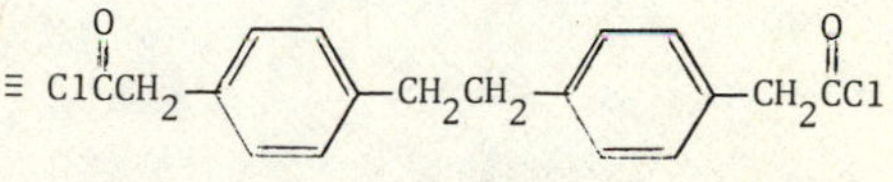

18-52 The data suggest that the mechanism is changing with temperature. The 50:50 mixture at higher temperature is consistent with the benzyne mechanism. At the lower temperature direct nucleophilic substitution accounts for 71% of the reaction.

18-53

a)

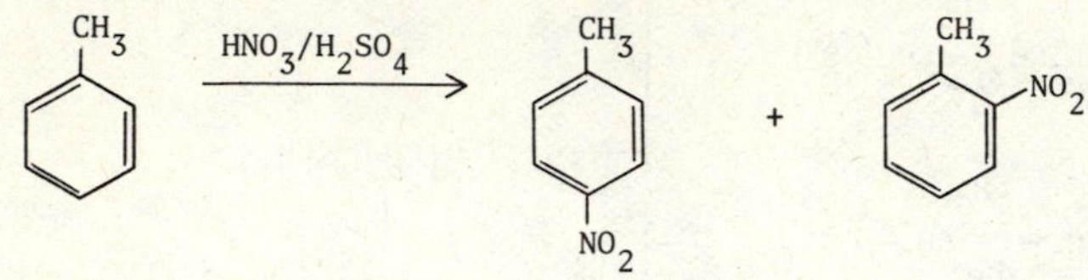

Separate isomers

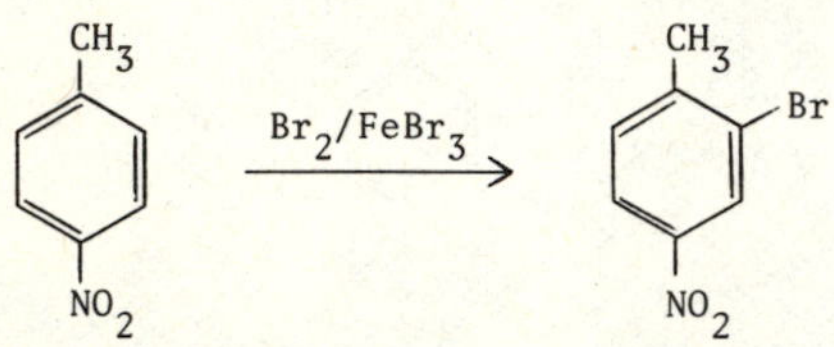

b)

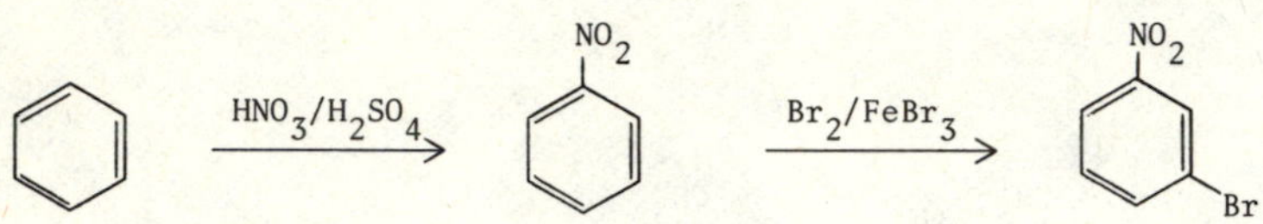

c)

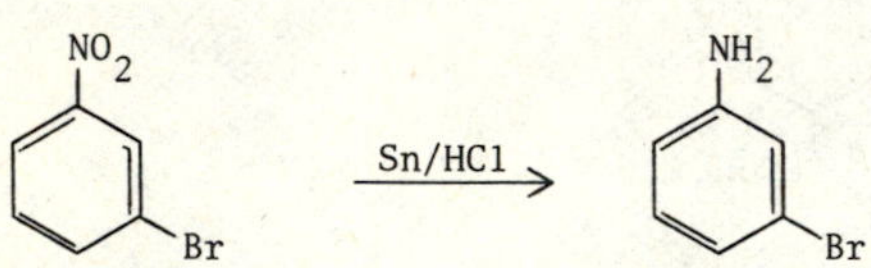

(from b)

d)

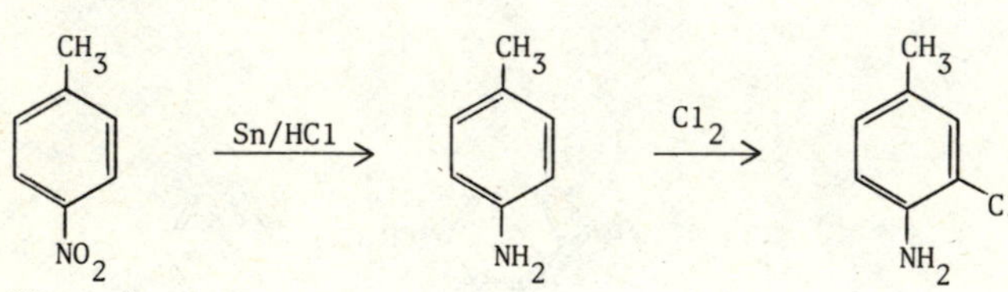

(from a)

e)

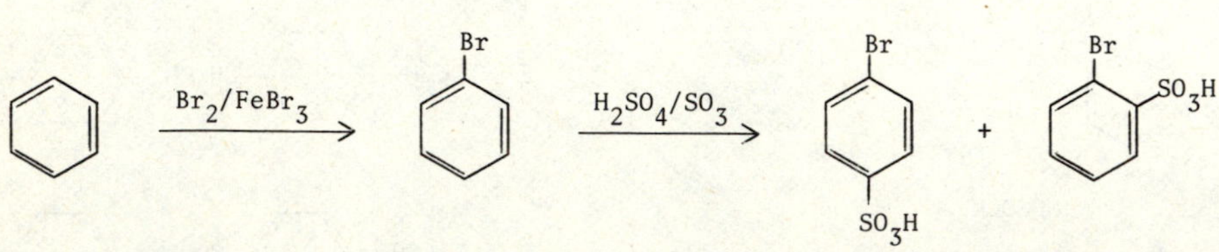

Separate isomers

Contd...

18-53 contd...

f) 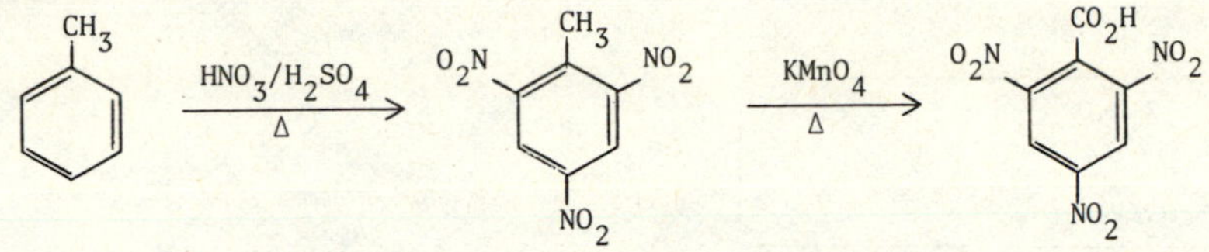

g) 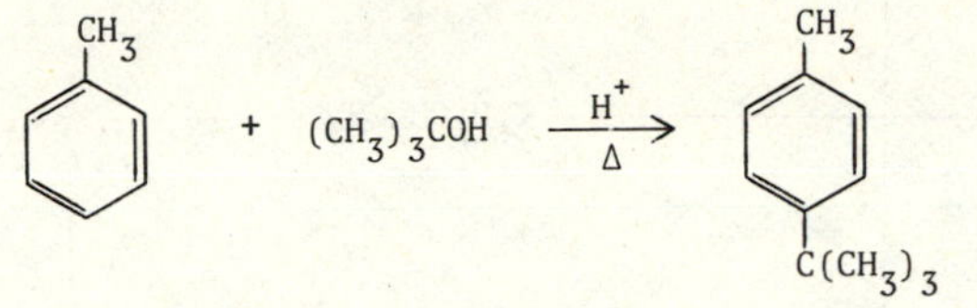

18-54

Acidic conditions:

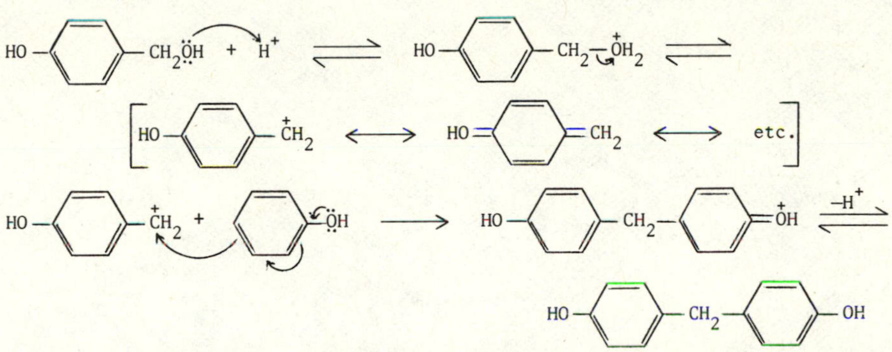

Basic conditions:

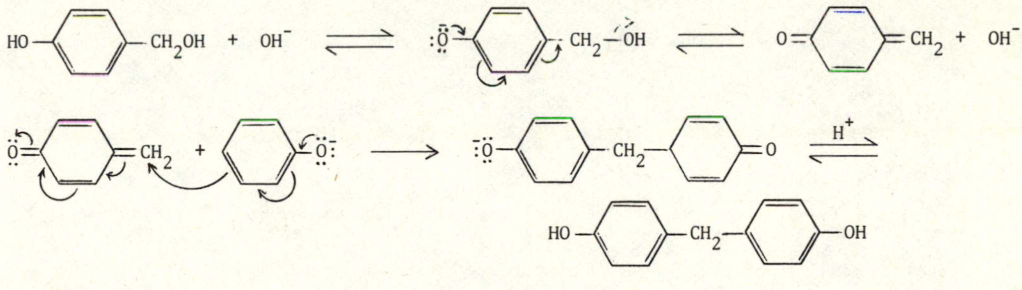

18-55

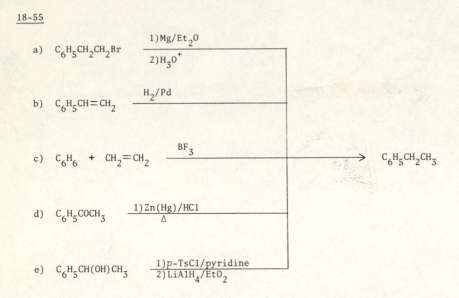

a) $C_6H_5CH_2CH_2Br \xrightarrow[2)\,H_3O^+]{1)\,Mg/Et_2O} C_6H_5CH_2CH_3$

b) $C_6H_5CH{=}CH_2 \xrightarrow{H_2/Pd} C_6H_5CH_2CH_3$

c) $C_6H_6 + CH_2{=}CH_2 \xrightarrow{BF_3} C_6H_5CH_2CH_3$

d) $C_6H_5COCH_3 \xrightarrow[\Delta]{1)\,Zn(Hg)/HCl} C_6H_5CH_2CH_3$

e) $C_6H_5CH(OH)CH_3 \xrightarrow[2)\,LiAlH_4/EtO_2]{1)\,p\text{-}TsCl/pyridine} C_6H_5CH_2CH_3$

18-56

a)

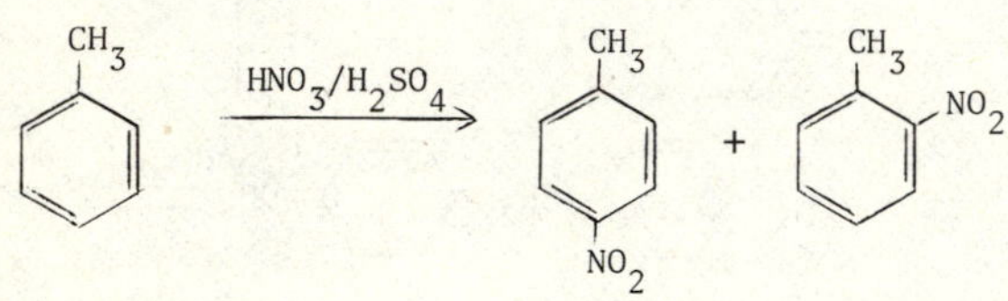

Separate isomers

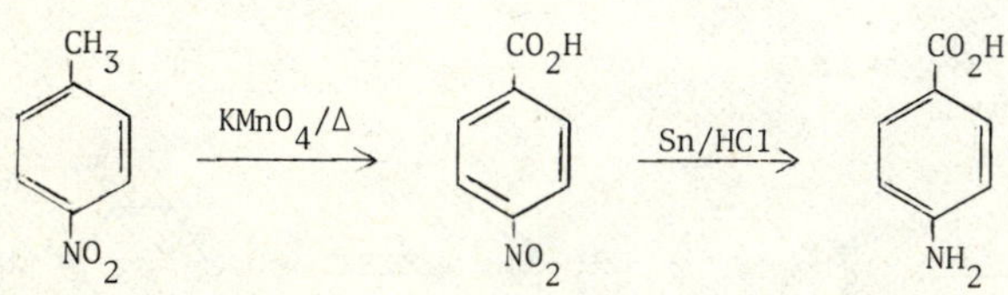

b)

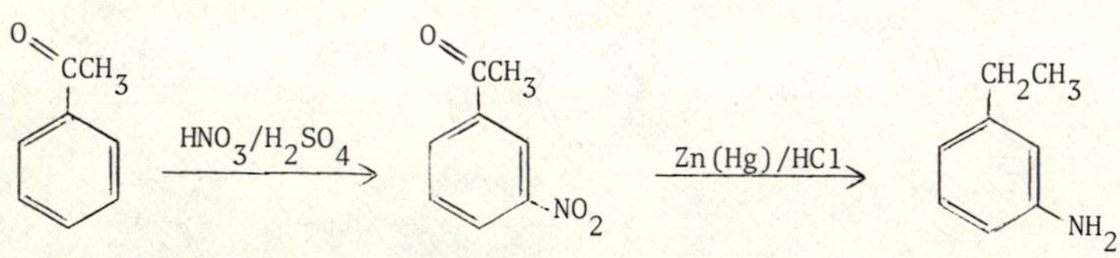

Contd...

18-56 contd...

c)

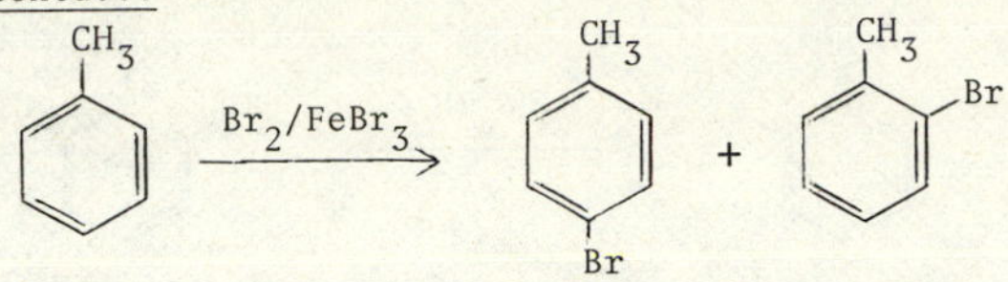

Separate isomers

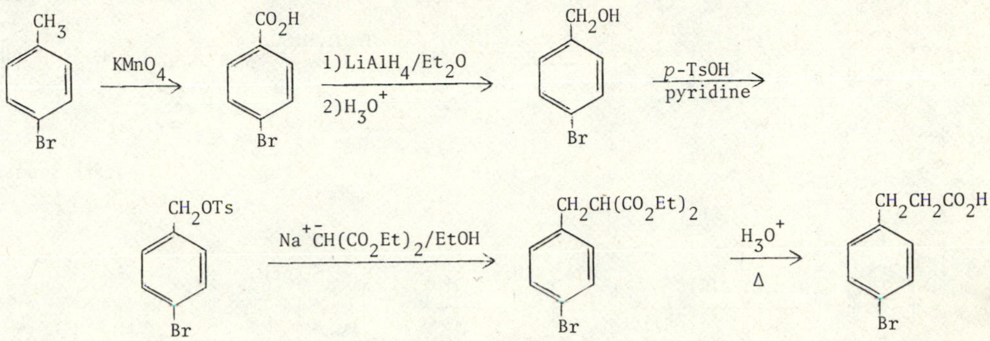

d)

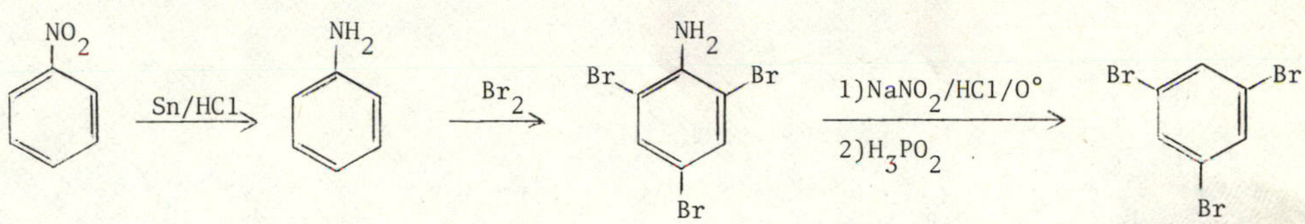

e)

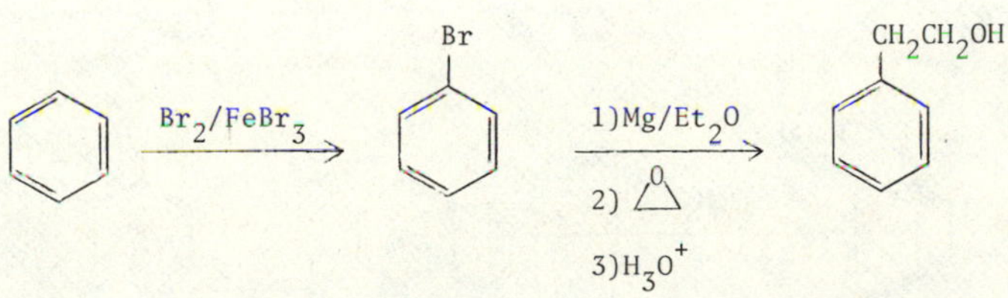

f)

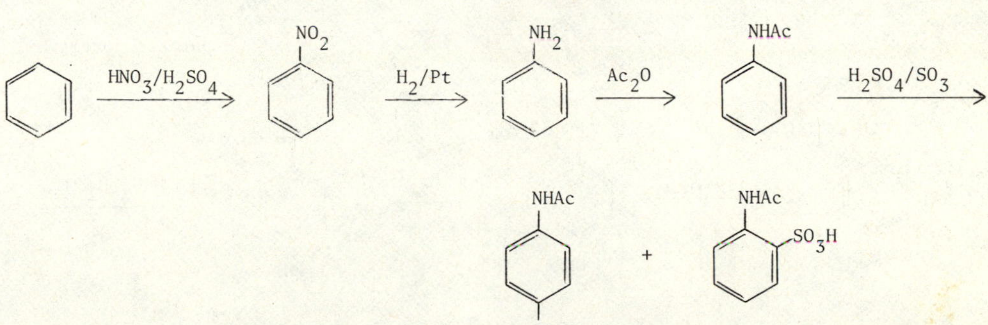

Separate isomers

Contd...

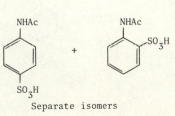

18-56 contd...

f) contd...

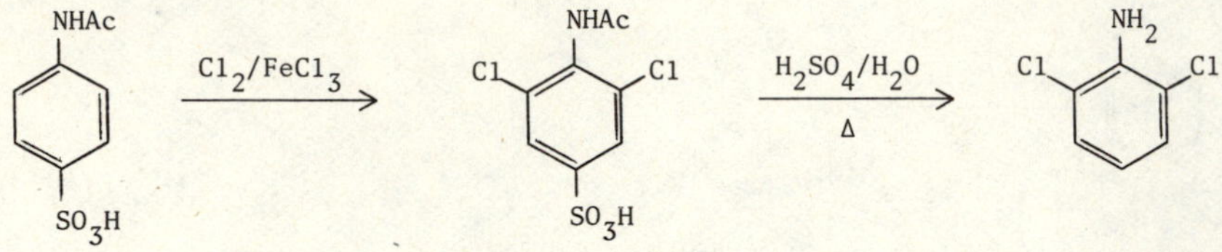

g)

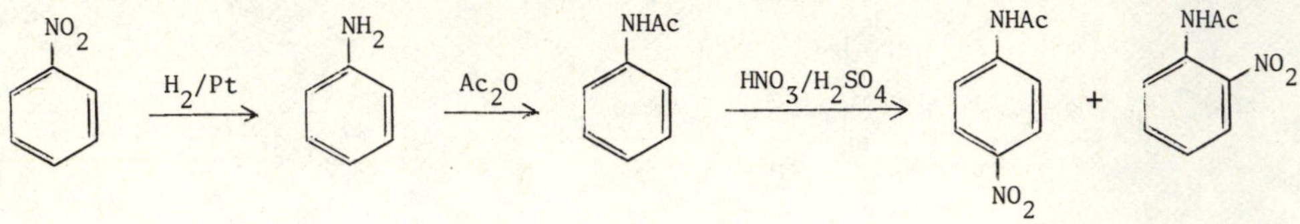

Separate isomers

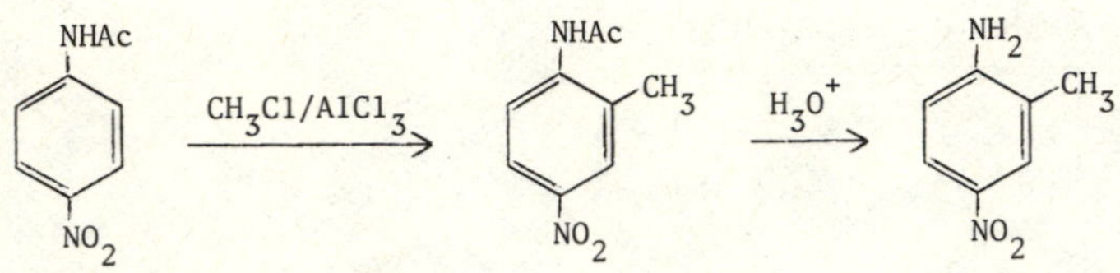

h)

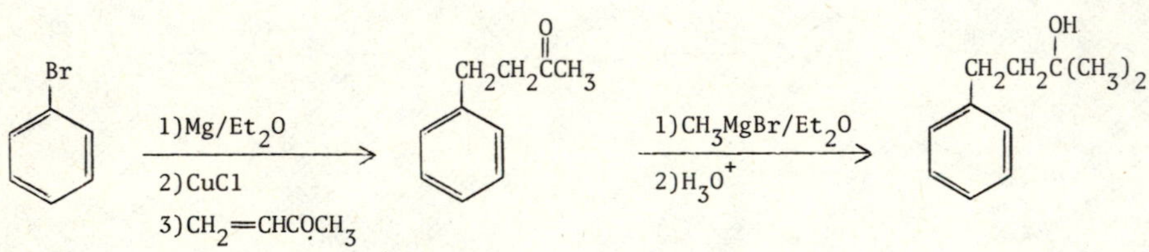

i)

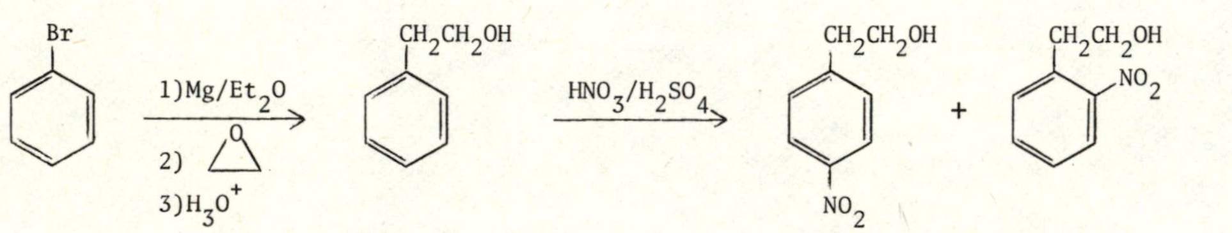

Separate isomers

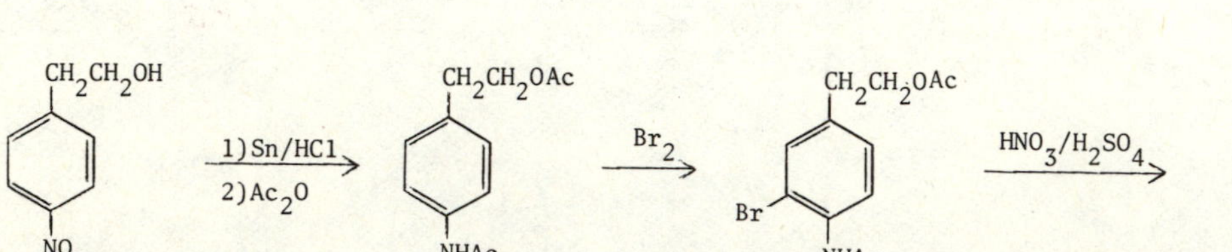

Contd...

18-56 contd...

i) contd...

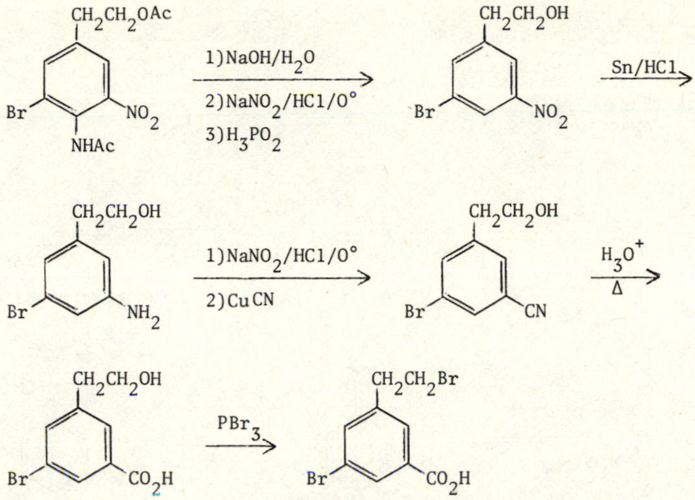

j)

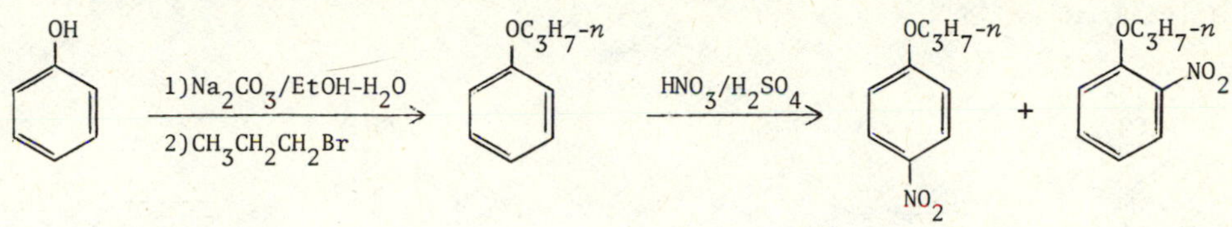

Separate isomers

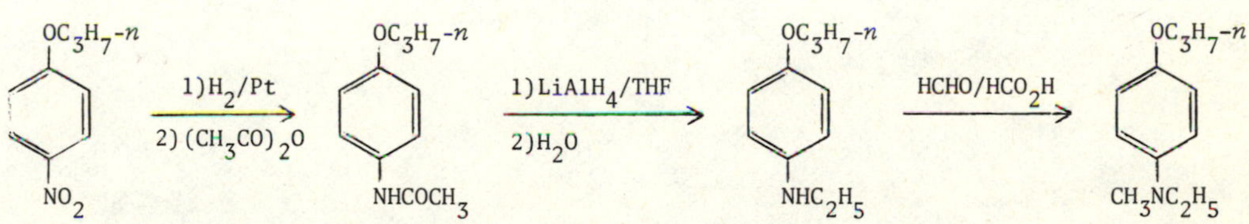

k)

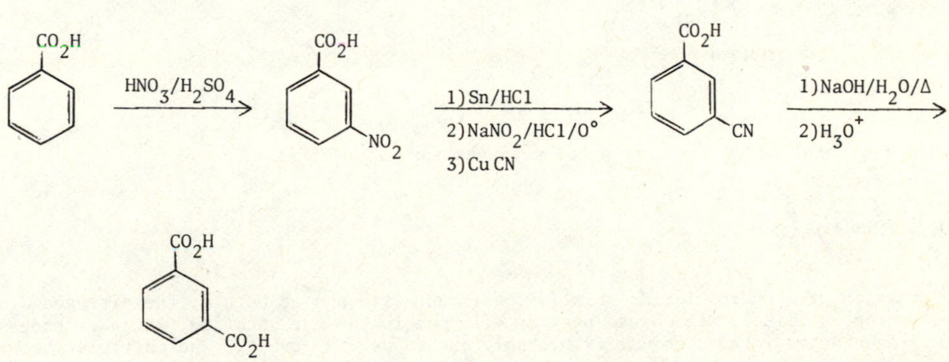

18-57

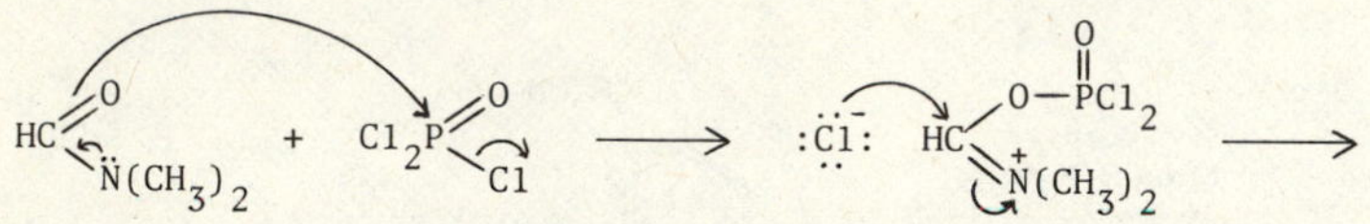

N,*N*,-Dimethylformamide Phosphorus oxychloride

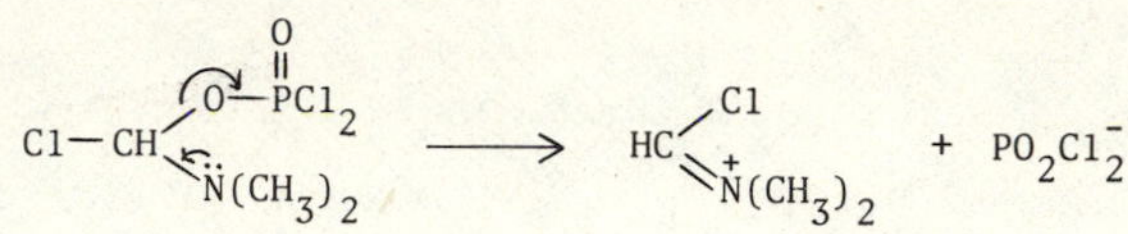

The electrophile

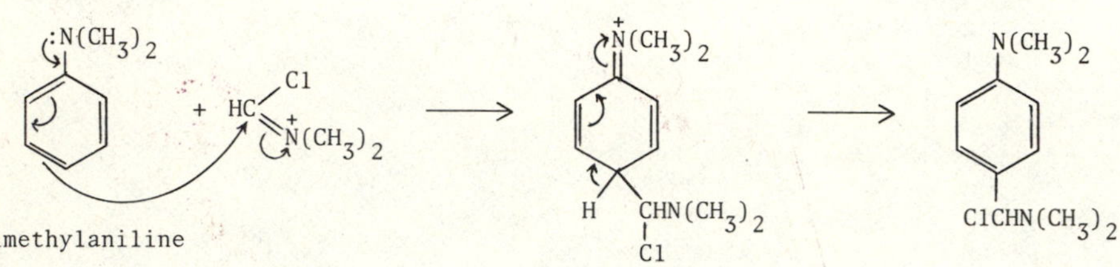

N,*N*-Dimethylaniline

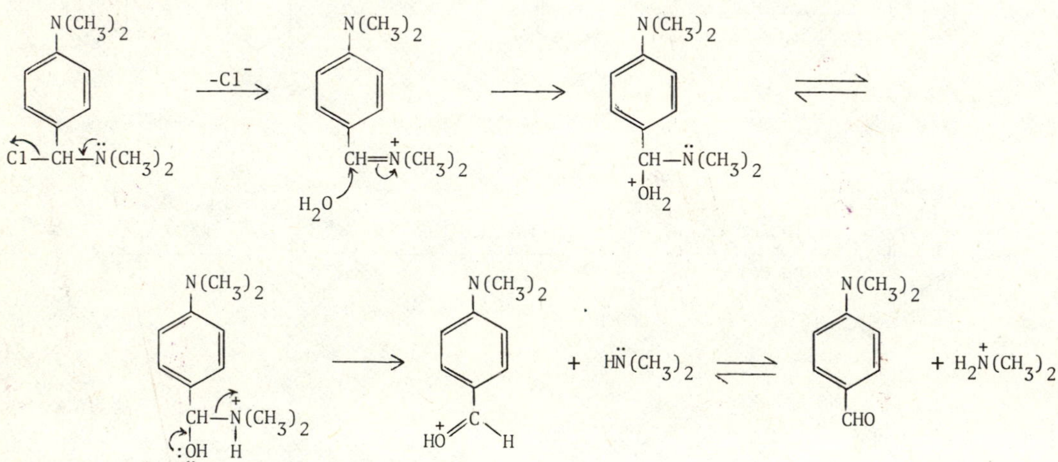

18-58 *IHD* = 5, and the UV spectral model accounts for all of this.

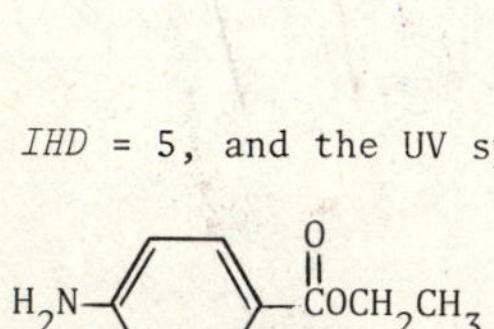

Formation of a hydrochloride salt is consistent with one or both of the nitrogen atoms existing as amines. The broad peak at 4.15 ppm in the nmr spectrum, which exchanges two hydrogen atoms for deuterium, is probably due to an $-NH_2$ group. The multiplet between 6.5

Contd...

18-58 contd...

and 8.0 is due to the *para*-disubstituted aromatic ring. The data to this point suggests that the remaining atoms of novocaine are connected to the ethyl group of the ester in the UV model. The nmr absorption at 4.3 ppm is probably the $-OCH_2-$ group and must be adjacent to a CH_2 group to account for the triplet. These data suggest that the rest of the molecule, $C_4H_{10}N$, is attached to that end atom of the model. The nmr triplet at 1.0 suggests two identical methyl groups split by adjacent CH_2 groups. That is surely two ethyl groups and the requisite CH_2 quartets are deshielded to 2.6. The five peak pattern can originate from the overlap of that quartet and the triplet for the CH_2 group beta to the ester oxygen atom and deshielded by attachment to a nitrogen atom.

triplet at 2.8 ppm 

quartet at 2.6 ppm 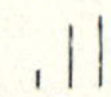

Combine to give these five peaks. 

The structure of novocaine is:

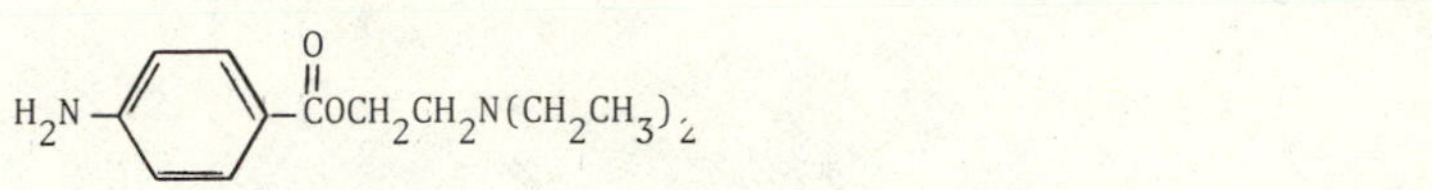

18-59 The IR spectrum suggests the presence of a carbonyl group (1670 cm^{-1}), possibly an amide, and the ^1H-nmr spectrum shows an aromatic group, probably *para*-disubstituted. Those account for the *IHD* of 5. The ^1H-nmr spectrum splitting patterns show an ethyl group (1.4 and 4.0 ppm) with the CH_2 group deshielded, probably because of attachment to an oxygen atom. (Attachment to a nitrogen atom or an aromatic ring would not deshield that much.) The ^1H-nmr peak at 2.1 ppm is consistent with a methyl next to the carbonyl group. The disappearance, on treatment with D_2O, of the nmr peak at 8 ppm and the IR peak at 3350 cm^{-1} is typical of an N–H or O–H group. (The N–D or O–D stretching frequencies after H–D exchange are near 2400 cm^{-1} in the IR.) We have evidence for the following groups:

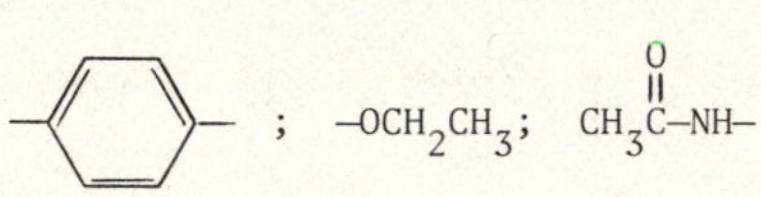

These fragments account for all of the atoms of phenacetin and lead to:

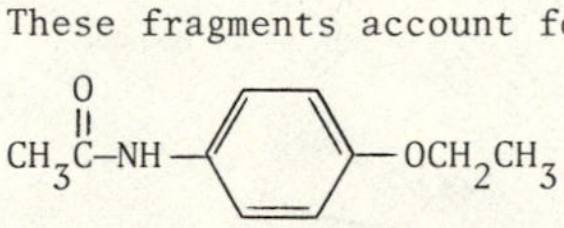

Phenacetin

The ^{13}C-nmr is consistent with this and shows δppm 15(CH_3); 24($\underline{C}H_3CO$); 64(CH_2); 115 and 122 (aromatic CH's); 131(Aryl-N); 156(Aryl-O); 169(CO).

19 Polycyclic and Heterocyclic Aromatic Compounds

19-1

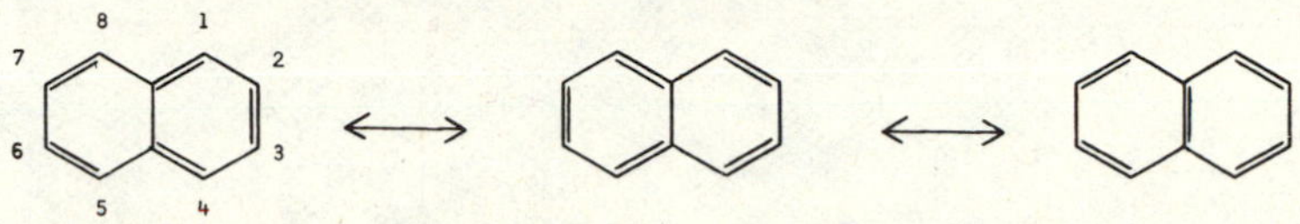

We can account for the trend in bond lengths by considering the amount of double or single bond character as represented by the resonance structures. Thus the 2-3 and identical 6-7 bonds are designated twice as double bonds and four times as single bonds. They therefore have more single bond character and are relatively longer. A similar analysis can be applied to the other bonds.

19-2

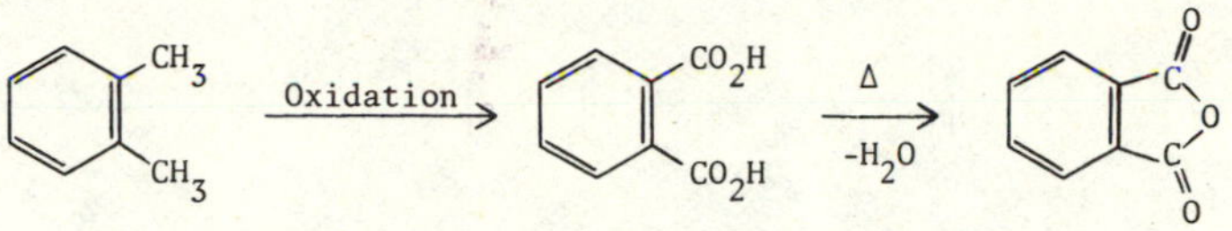

19-3 An α-substituent is close to the hydrogen atom on the 8-position (a *peri* relationship). There is less steric hindrance between the β-substituent and the adjacent hydrogen atoms since the groups point away from each other.

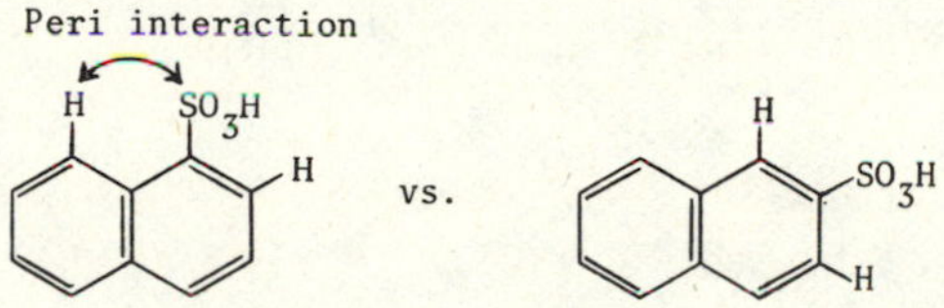

19-4 Reaction at the 9,10-positions disrupts aromaticity the least. The two benzenoid structures of the intermediates possess more resonance stabilization than one naphthalene structure.

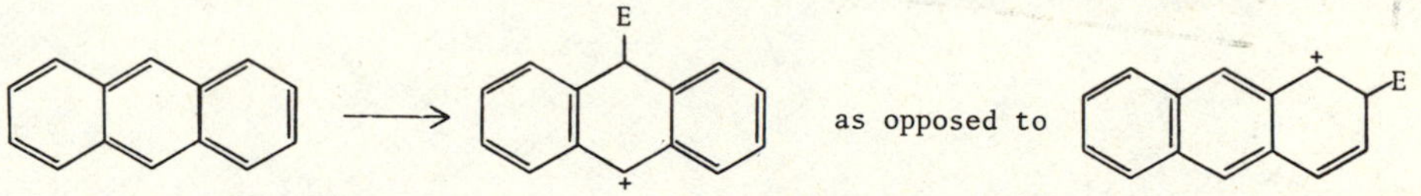

19-5 With a substituent at position 1, the intermediates from substitution at 2 or 4 benefit from conjugation with the substituent without disrupting aromaticity of the second ring.

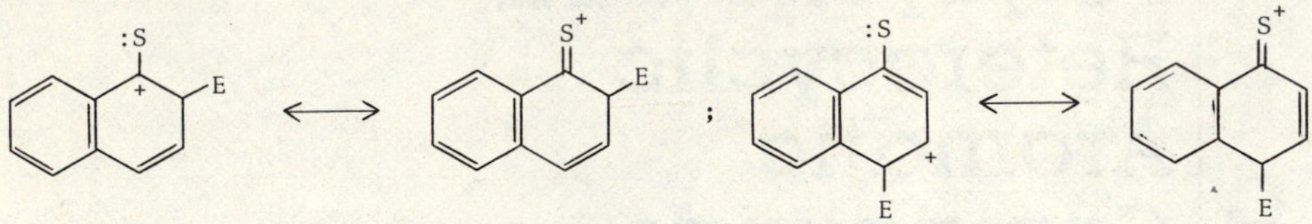

With a substituent at position 2, the intermediate from substitution at 1, but not at 3 or 4, benefits from conjugation with the substituent without disrupting aromaticity of the second ring.

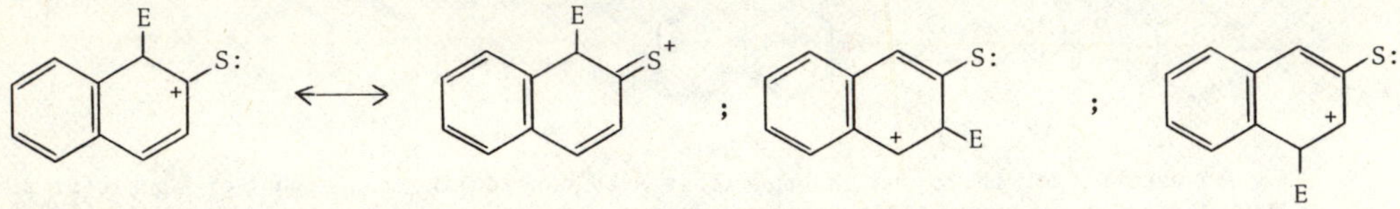

19-6

a) 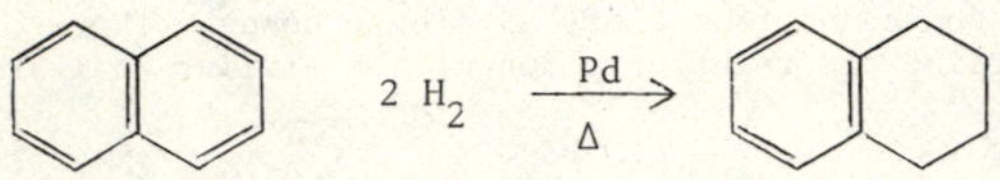

b) 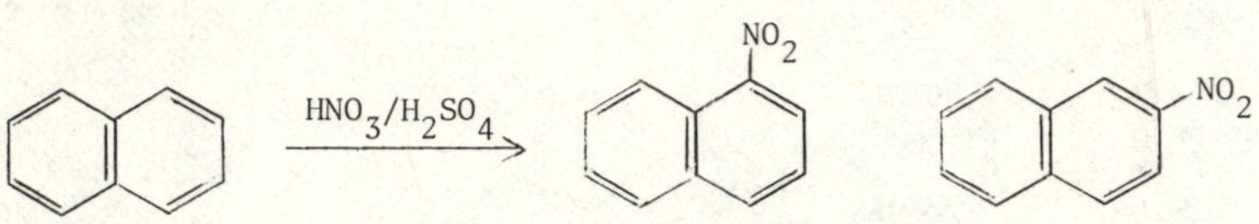

Separate isomers

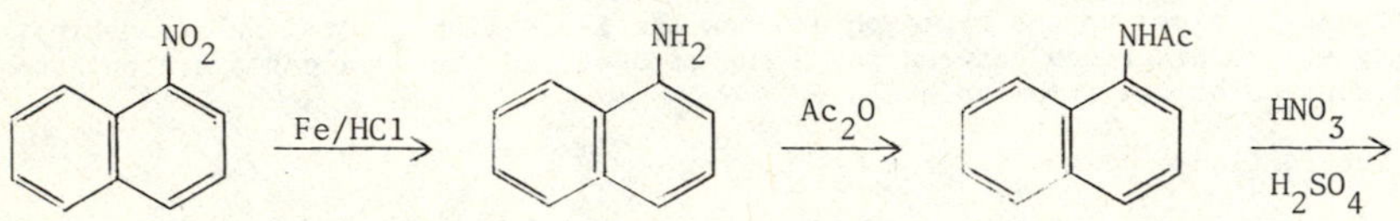

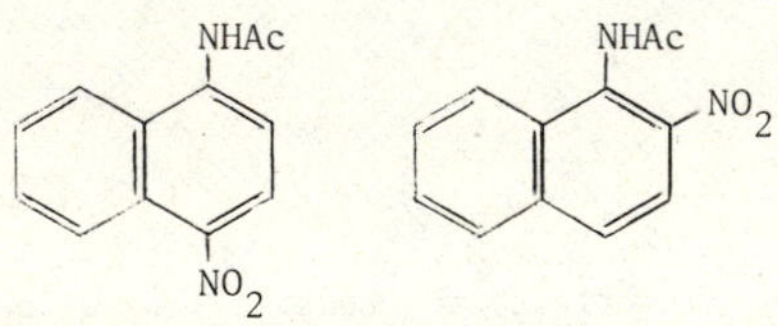

Separate isomers

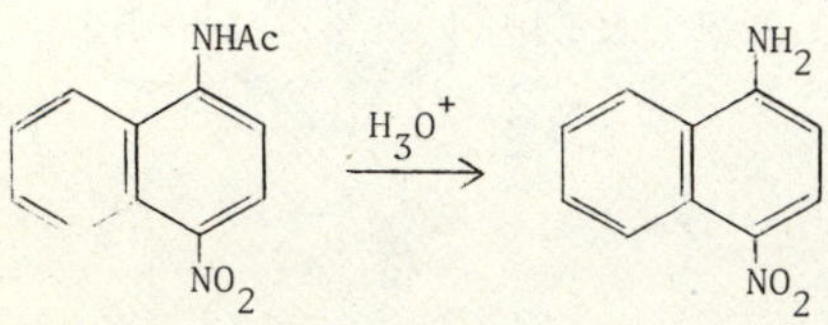

Contd...

19-6 contd...

c) 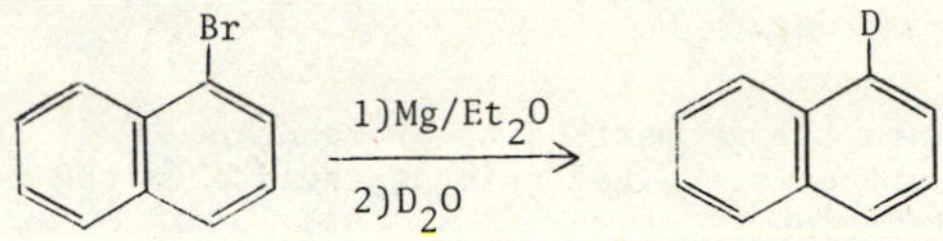

d) 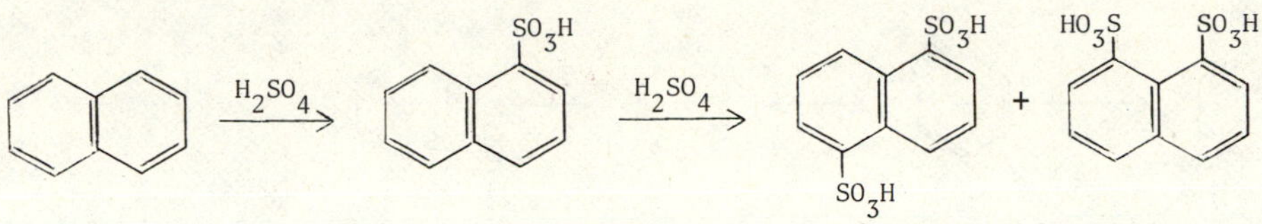

Separate isomers

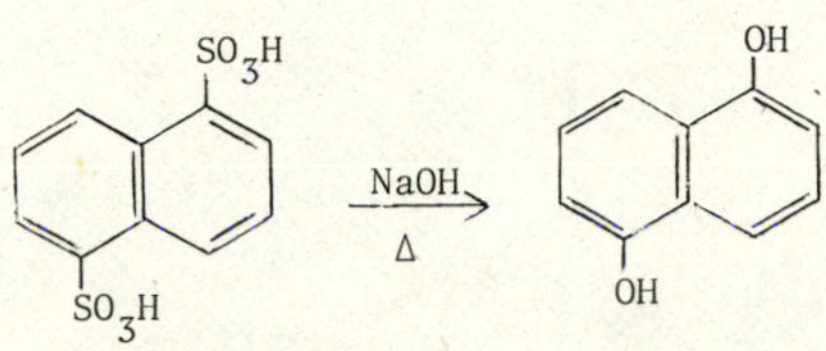

e) 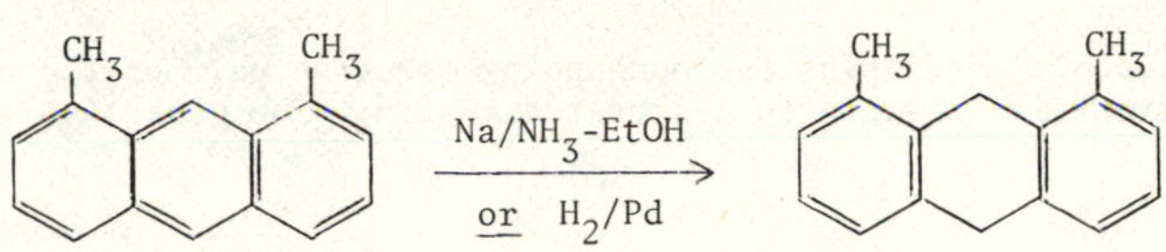

f) 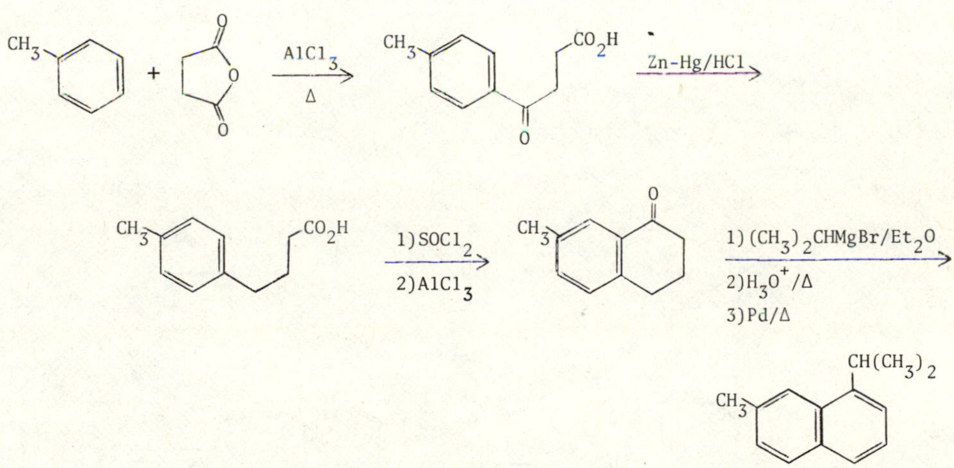

g) 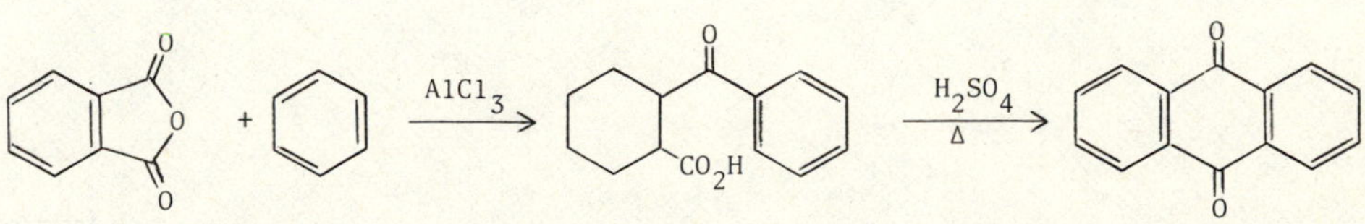

19-7 The dipole moment of piperidine is due to the electronegativity difference between the nitrogen and attached carbon atoms. Pyridine has additional charge separated resonance structures which account for a greater dipole moment.

19-8 The electron pair of aniline is delocalized into the aromatic ring by resonance. It is much less available to function as a base. Furthermore, the conjugate acid obtained from protonation of aniline loses resonance stabilization.

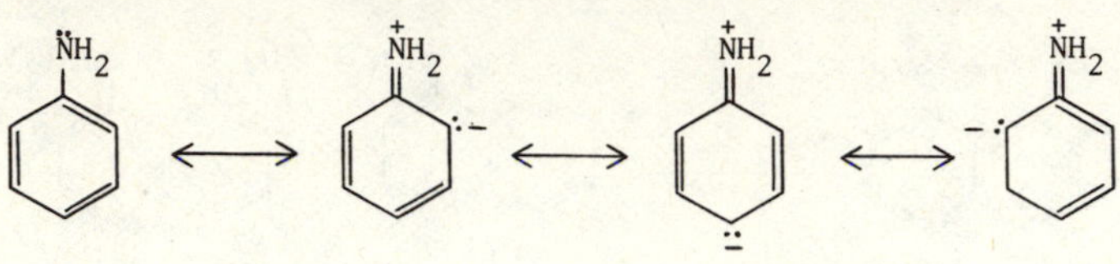

19-9

a)

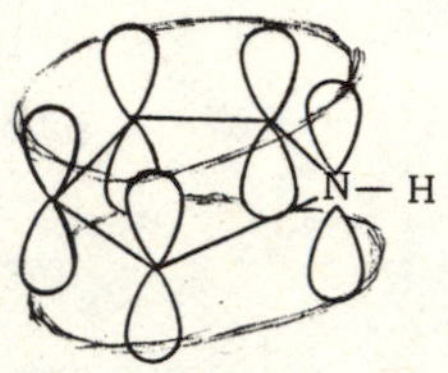

b) The four electrons of the two double bonds plus the nonbonding electron pair of the nitrogen atom make a total of six electrons in this cyclic conjugated system. The Hückel rule is followed with $n = 1$.

c) Resonance structures with a negative charge on nitrogen would require that the nitrogen atom possesses ten electrons.

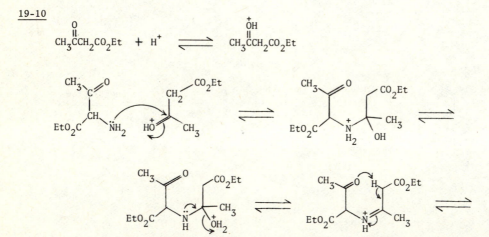

19-10

$$CH_3\overset{O}{\overset{\|}{C}}CH_2CO_2Et + H^+ \rightleftharpoons CH_3\overset{\overset{+}{O}H}{\overset{\|}{C}}CH_2CO_2Et$$

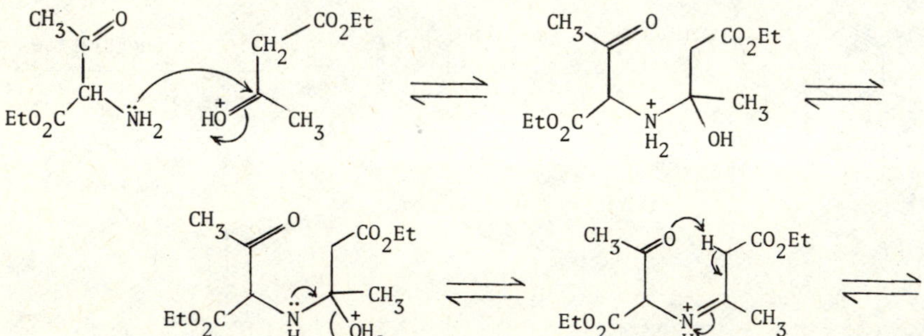

Contd...

19-10 contd...

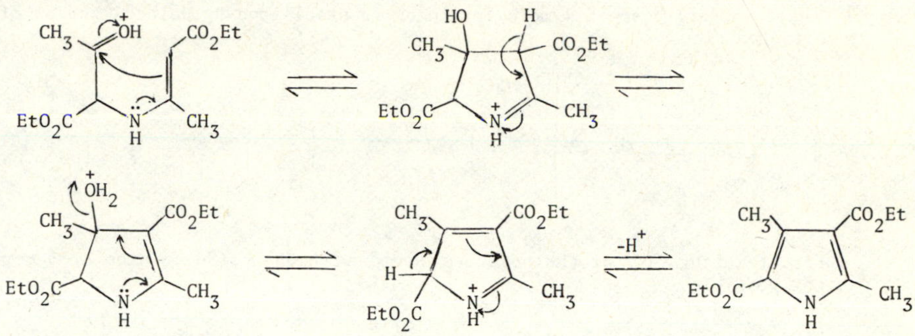

19-11

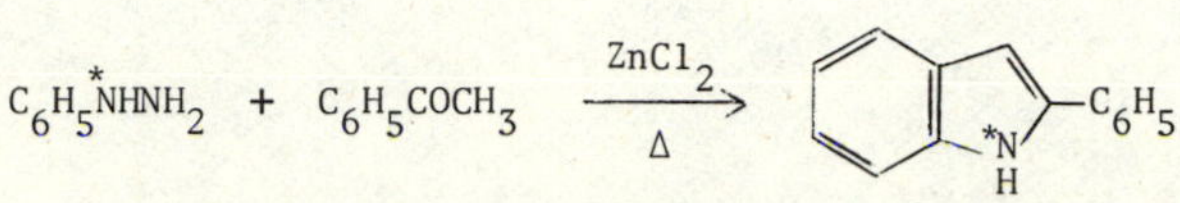

The experiment confirmed that the β-nitrogen atom of phenylhydrazine is lost as ammonia.

19-12 The cyclohexanone structure is retained and leads to a tricyclic heterocycle known as a carbazole.

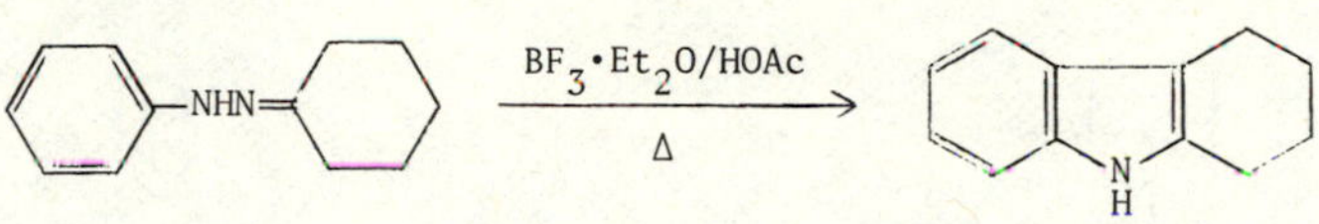

1,2,3,4-Tetrahydrocarbazole

19-13

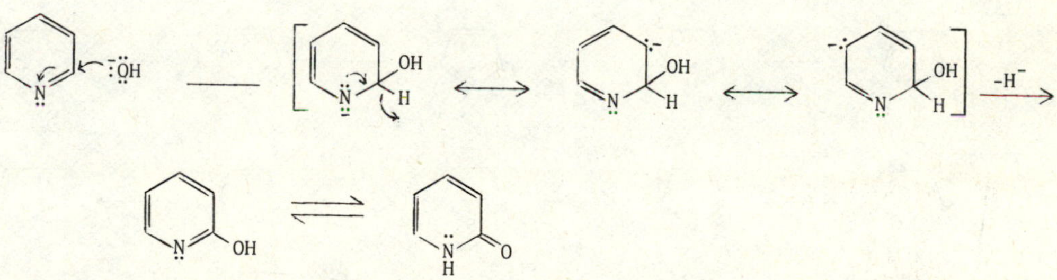

2-Pyridone is a resonance stabilized tautomer of 2-hydroxypyridine.

19-14

a) The conjugate base of 4-methylpyridine is stabilized by delocalization of charge by the nitrogen atom.

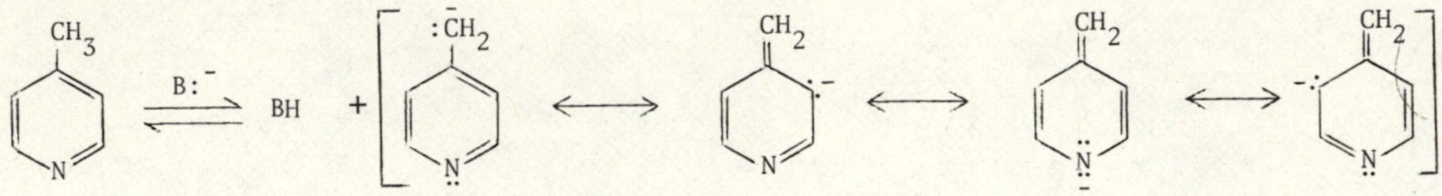

Direct conjugation of the negative charge with the nitrogen atom of 3-methylpyridine is not possible.

b) The decarboxylation of A leads by a favorable cyclic pathway, to a conjugated intermediate. Decarboxylation of B leads to an unstable aryl anion.

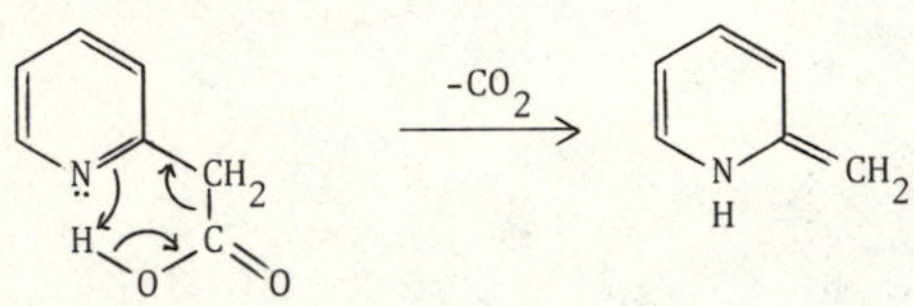

A

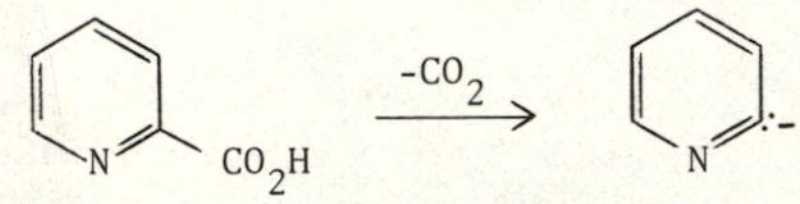

B

19-15 Pyridine oxide is more reactive than pyridine in both electrophilic and nucleophilic substitution. The oxygen anion can function as an electron donor while the positive nitrogen is an electron acceptor.

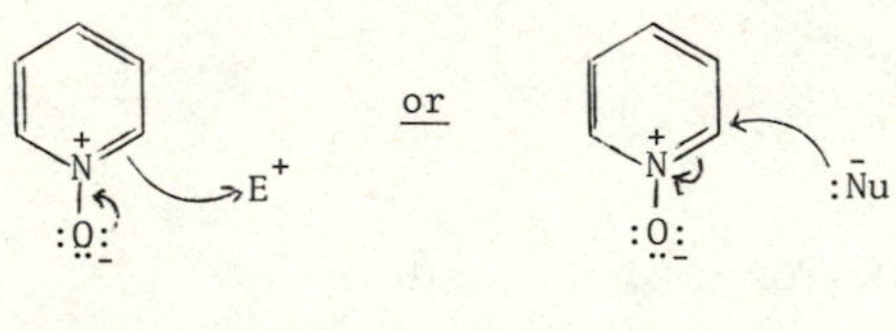

19-16

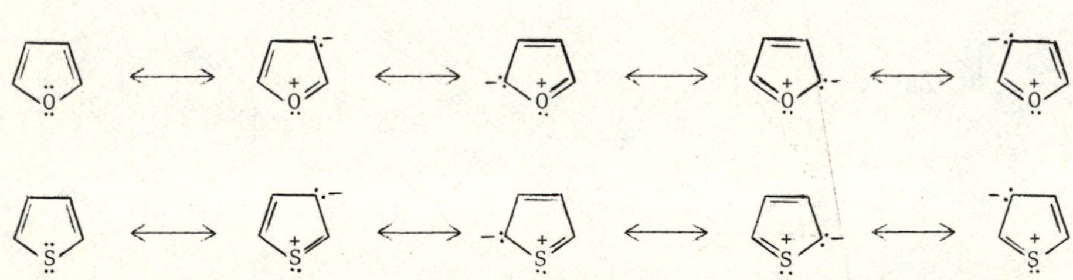

19-17

a)

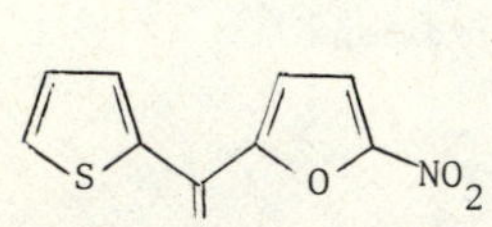

b)

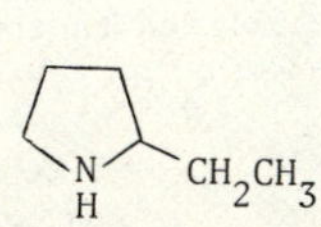

c)

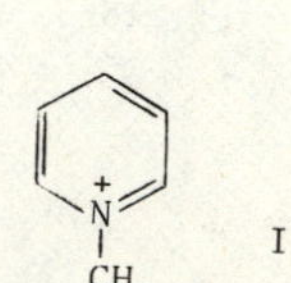

Contd...

19-17 contd...

d) 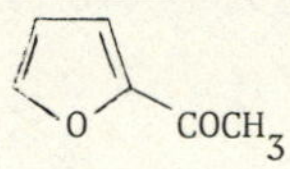

e) 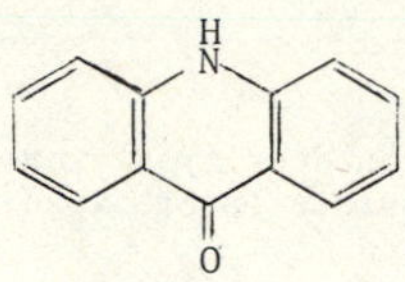

f) 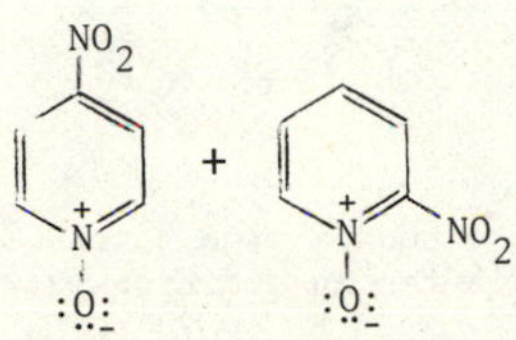

Major Minor

g) 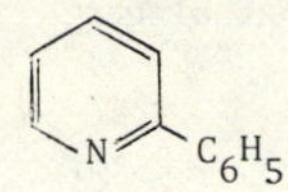

h) 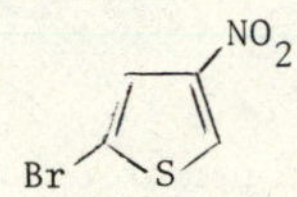

i) 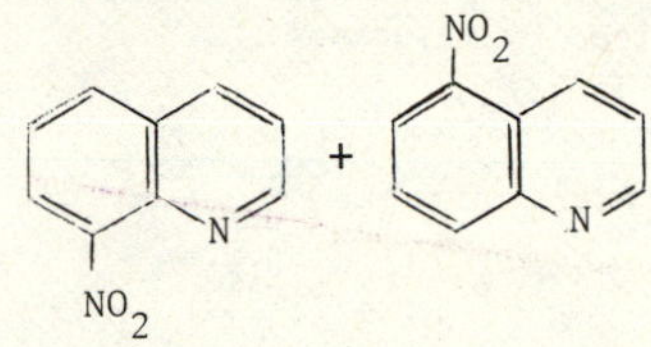

j) 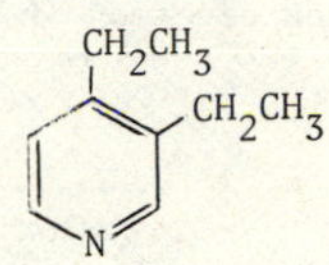

19-18

a) 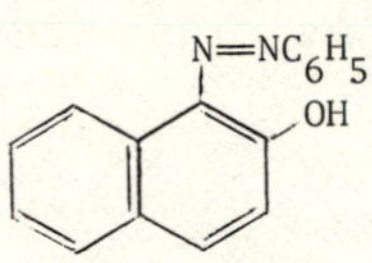

b) 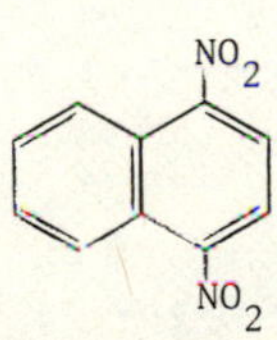

c) 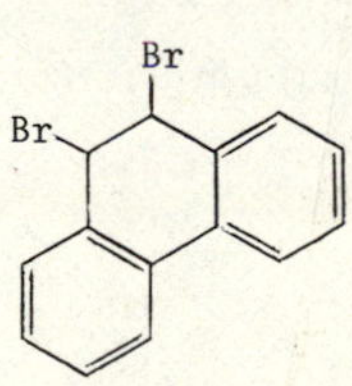

d) 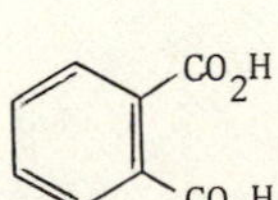

e) 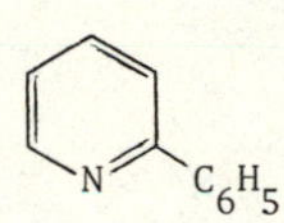

f) 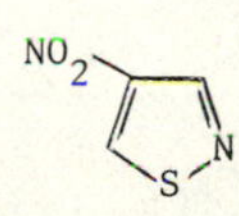

g) 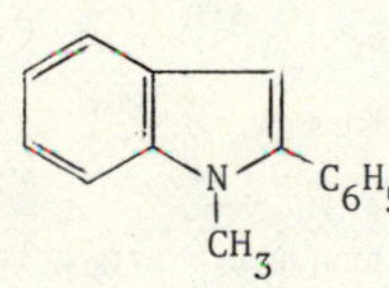

h) 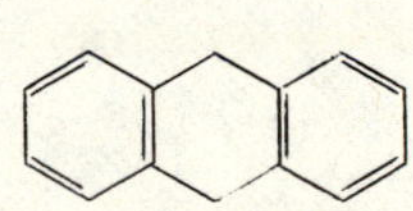

i) 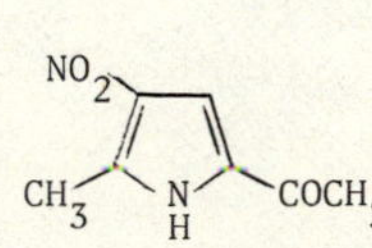

j) 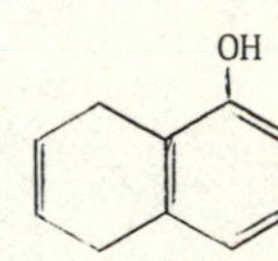

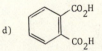

(The less stabilized ring is reduced.)

k) 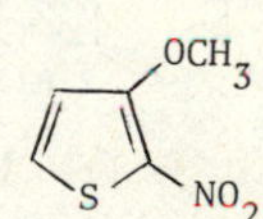

19-19 There are two mononitronaphthalenes.

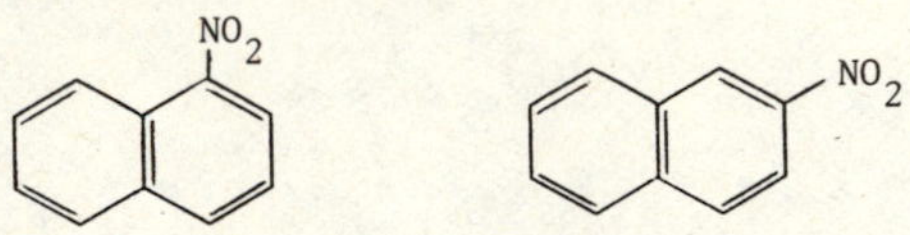

19-20 Oxazole is more basic then pyrrole. The oxygen can enhance electron density at the nitrogen atom by resonance. Furthermore, the conjugate acid of oxazole (protonated form) is stabilized by resonance.

19-21 The Hückel $4n+2$ rule (n =1) can be accounted for if each double bonded atom, including one nitrogen, contributes one electron and the N–H nitrogen contributes an electron pair.

19-22

a)

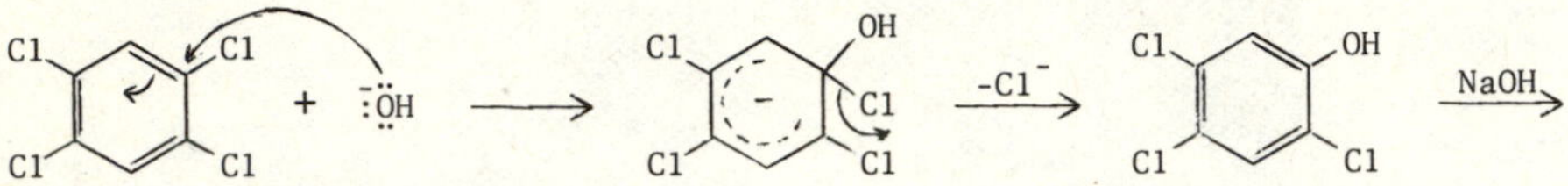

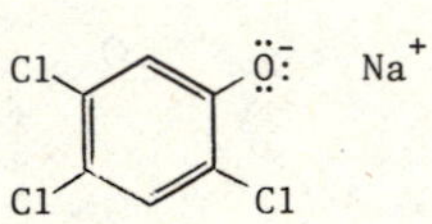

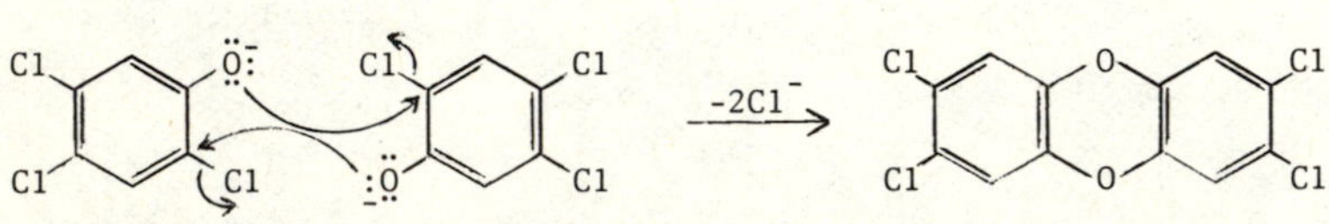

This is probably a two step process

b) The reaction is a double electrophilic aromatic substitution by formaldehyde. Formation of the methylene bridge probably takes place one step at a time.

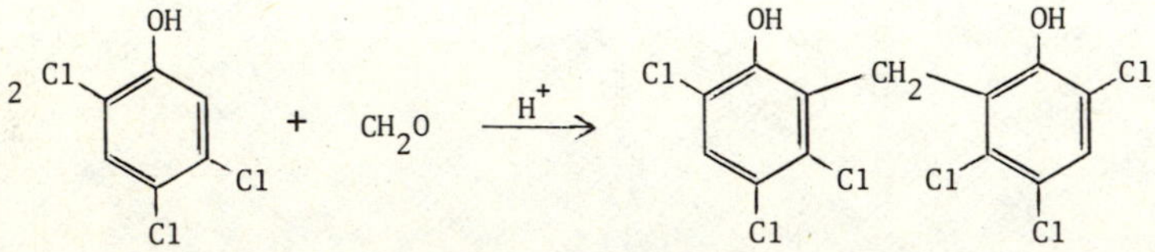

19-23

a) Indole has a cyclic array of ten interacting electrons ($4n+2$ = 10; n=2) which includes eight pi electrons and two electrons from the nitrogen atom.

b) The intermediate cation formed by electrophilic addition at carbon 3 can be stabilized by the nitrogen atom without disrupting aromaticity in the six membered ring.

Contd...

19-23 contd...

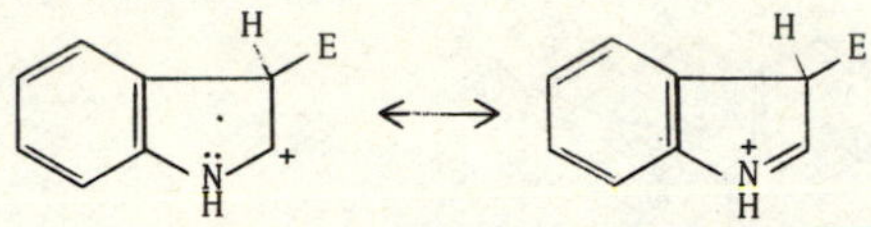

19-24 The amino group is an electron donating substituent and thus it enhances electrophilic substitution. Positions 3 and 5 are ortho-para to the amino substituent and benefit from stabilization of the intermediate carbocation.

19-25

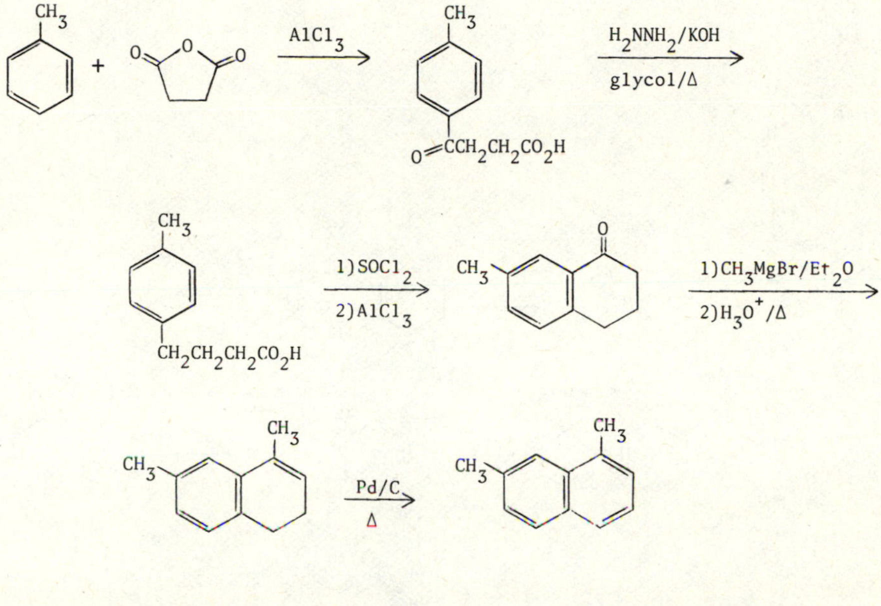

19-26 Benzyne is the dienophile which adds across the 9,10 position of anthracene (the diene).

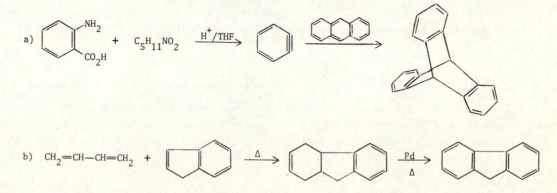

a)

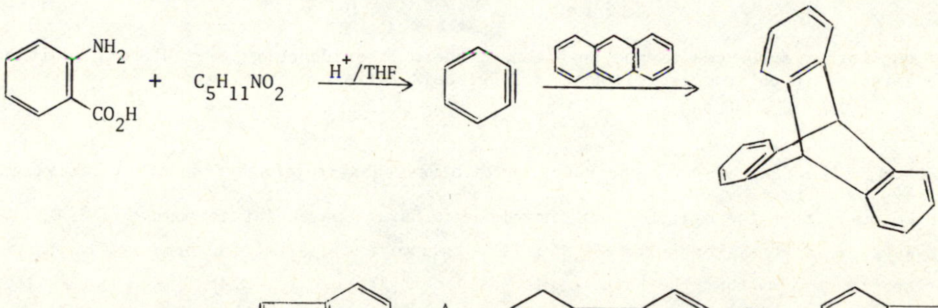

b)

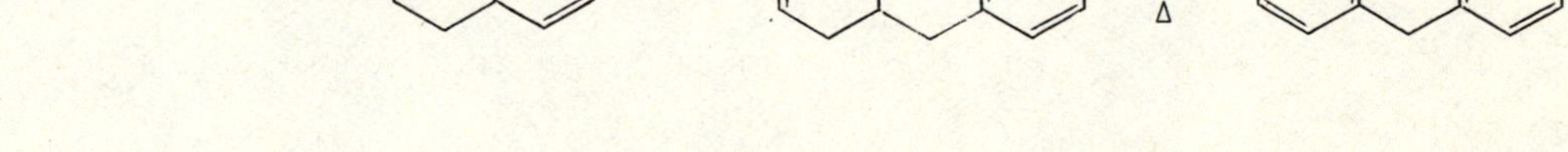

19-27

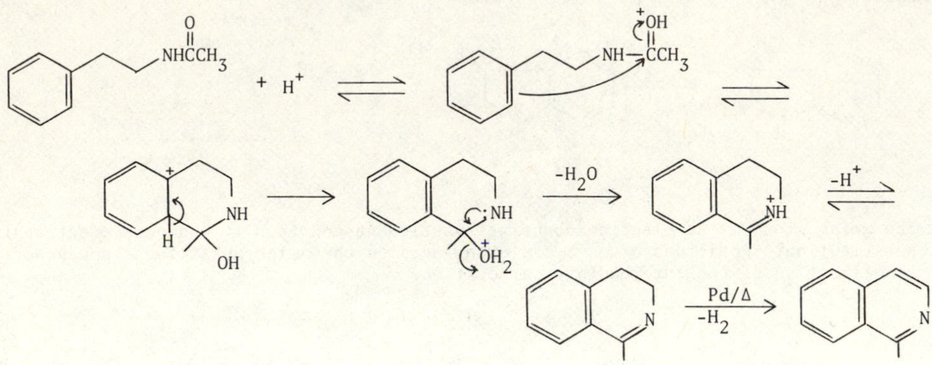

19-28

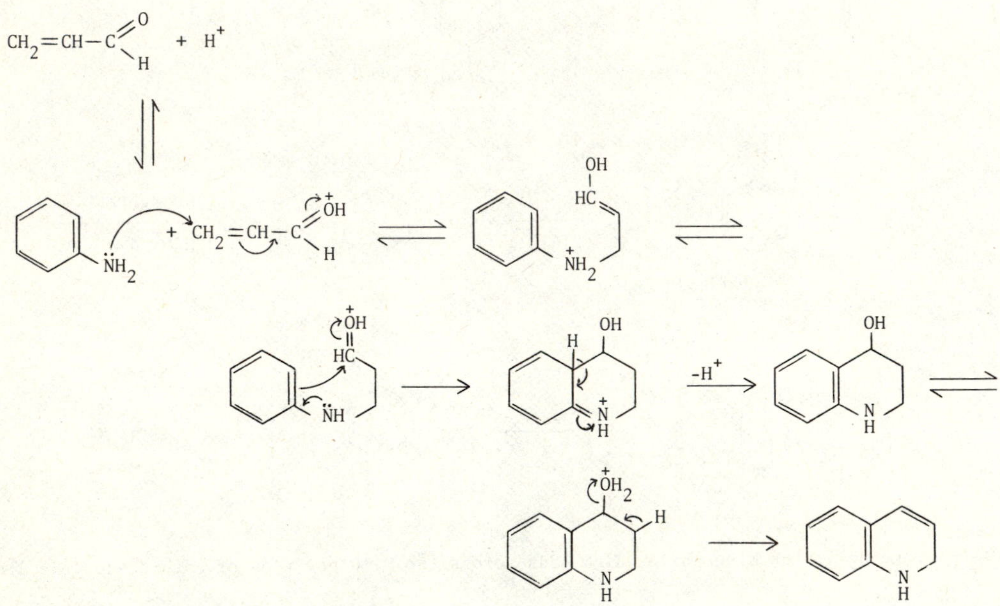

Although other sequences can be envisioned, there is evidence that a Michael addition by the amino nitrogen atom is followed by a Friedel-Crafts cyclization.

19-29 *IHD* = 4. If we assume that the reaction involves substitution by the ethyl formyl group ($CH_3CH_2O\overset{O}{\overset{\|}{C}}-$), then the original heterocycle must have a molecular formula of C_5H_6O. An ethyl ester is consistent with the triplet (1.2ppm) and quartet (3.8ppm) in the nmr. The nmr spectrum shows that the heterocycle has a methyl group (s, 2.4ppm) and two single protons (6.1 and 7.1ppm) on adjacent carbon atoms (J = 5 Hz). The only compound consistent with these data is

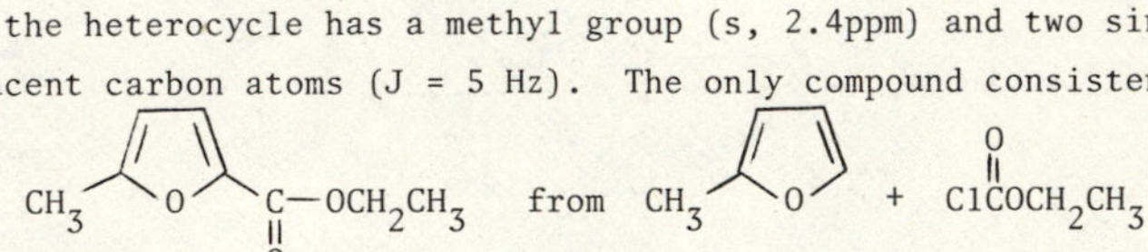

19-30 The 3-hydroxythiophene exists as a mixture of the keto and enol tautomers. The carbonyl group is shifted to a low frequency because it is conjugated with the double bond and sulfur atom.

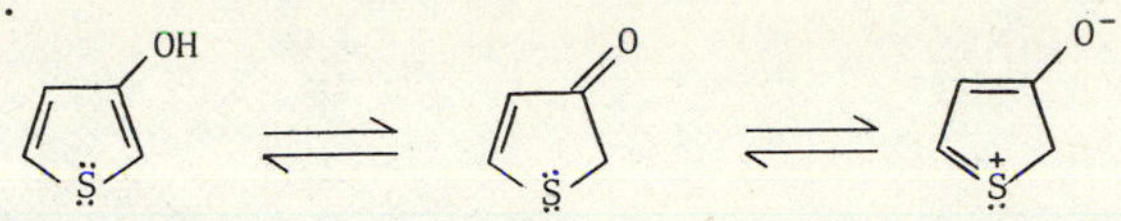

19-31 The UV model compound is

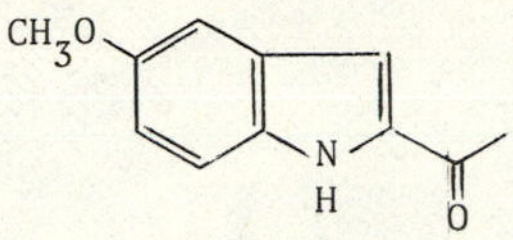

The *IHD* = 8 requires that the unknown have one more ring or double bond than the model and the molecular formula requires two additional carbon atoms compared to the model. The nmr spectrum shows three aromatic protons and the methoxy singlet (3.8 ppm). The two triplets indicate adjacent, deshielded methylene groups ($—CH_2—CH_2—$). The singlet at 2.3 ppm is a methyl group attached to a nitrogen atom or an aromatic ring carbon atom.
Possible structures are

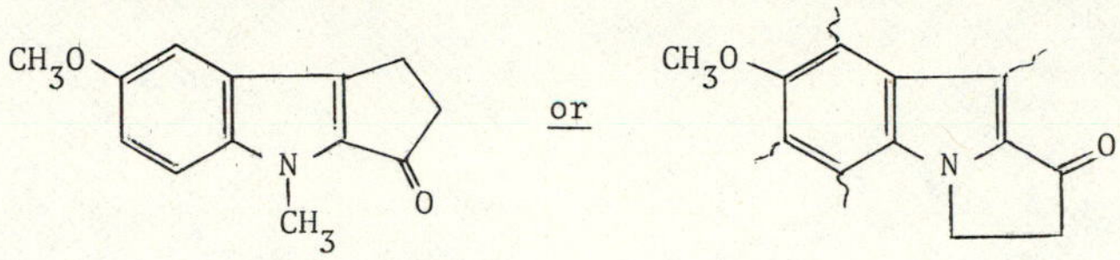

(With a CH_3 group at one of the wavy bond lines.)

20 Organic Synthesis

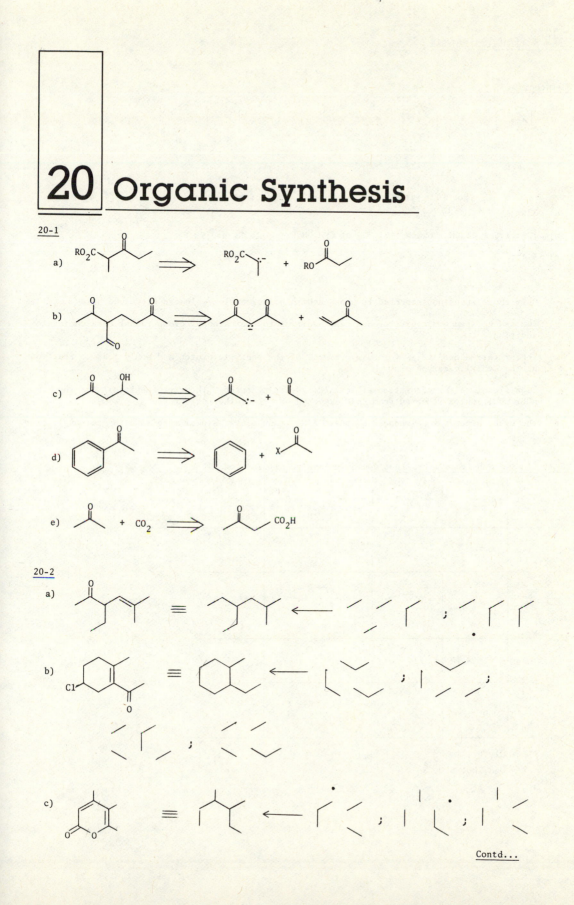

20-1

a) 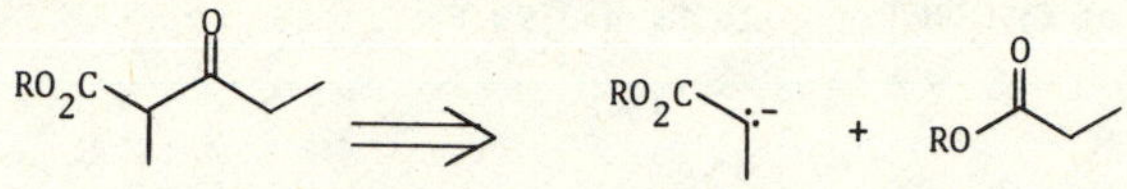

b) 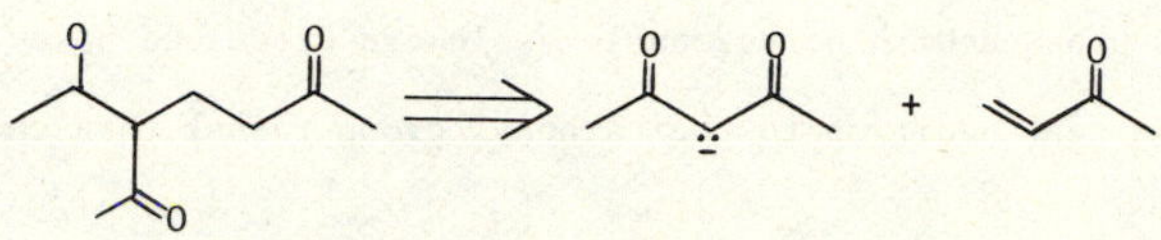

c) 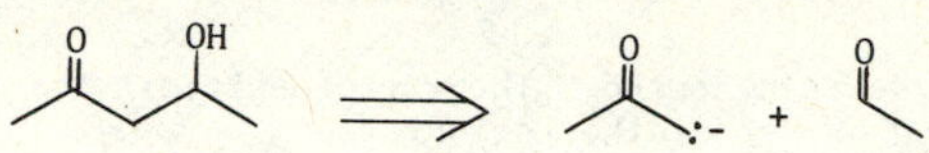

d) 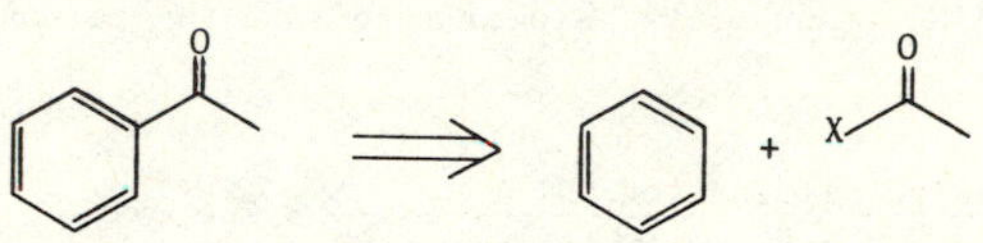

e) 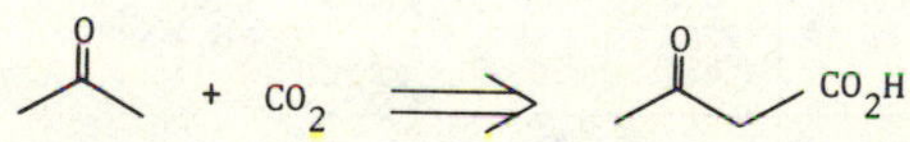

20-2

a) 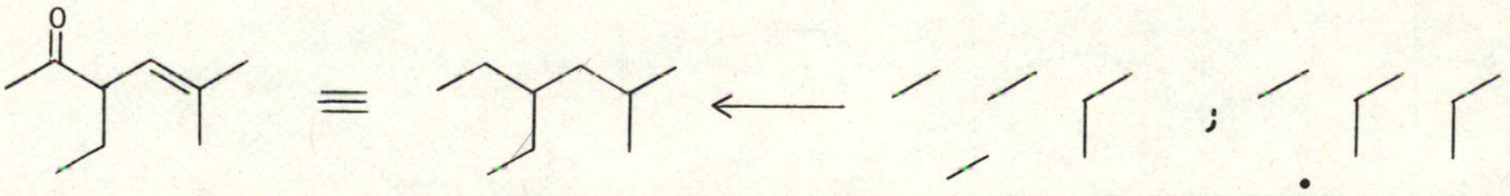

b) 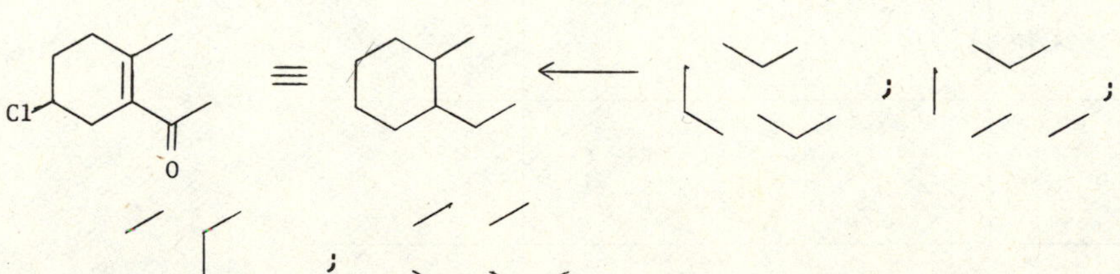

c) 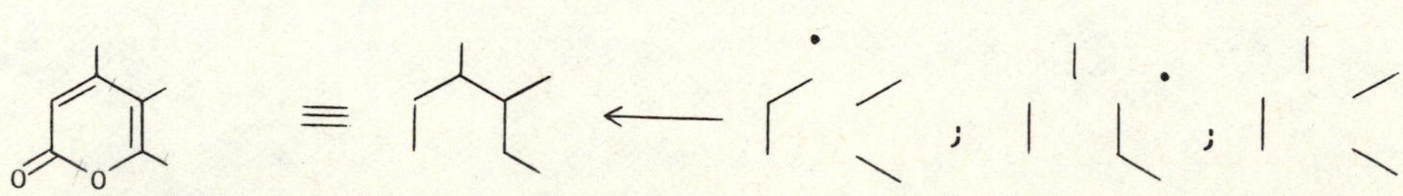

Contd...

20-2 contd...

d) 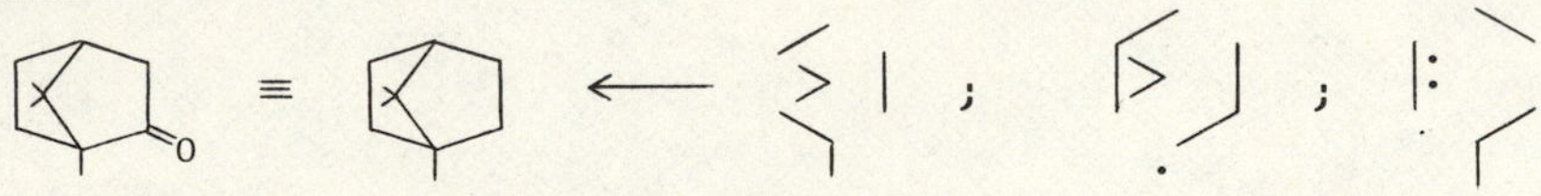

20-3

	(For 100% yield)	(For calculated yield)
a) 32%;	2.6g	8.2g
b) 12%; 1.2 g of $CH_2(CO_2C_2H_5)$; 1.3 g of $C_6H_5CH_2Br$		10.0g and 10.8g
c) 3.4%;	0.47g	13.8g

20-4 Only those steps not expected to proceed with a reasonable yield are discussed below.

a) The second step would give predominant addition to the carbonyl group rather than conjugate addition.

b) The preferred enolate is formed by removal of the benzylic proton and would lead to reaction at the benzylic carbon atom.

c) Significant self condensation of acetaldehyde is expected. The regiospecificity of the HOBr addition is reversed from that expected by Markovnikov orientation.

d) The cyano group may be reduced by the $LiAlH_4$, though at a slower rate than the carbonyl.

20-5 Substrate types that commonly lead to elimination

Reagents	OH H –C–C–	*X H –C–C–	H *X *X H –C–C–C–C– H H	H Y –C–C–
Acid	>C=C<			
Base		>C=C<	>C=C–C=C< or –C≡C–	>C=C< (Y = $\overset{+}{N}$<)
Heat				>C=C<
I_2 or Zn			–C–C=C–C– H H	

*X = Cl, Br, I

Y = OAc, NO, SeO, $\overset{+}{N}$<

20-6 The epoxide plus nucleophile would correspond to a span of three for product formation.

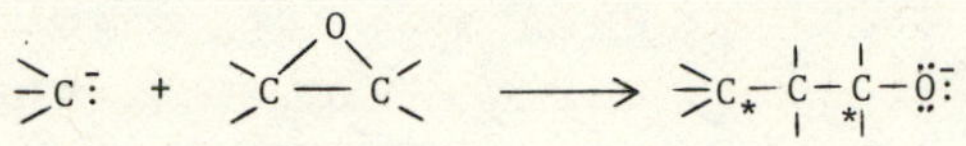

20-7

Target | Span | Potential precursors

a)
b)
c)

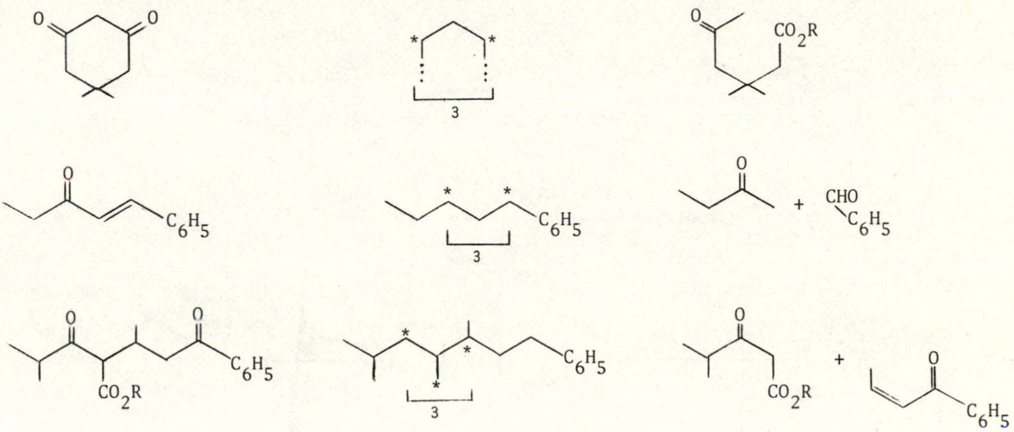

20-8

a) $C_6H_5COCH_2CH_3 \xrightarrow[2)\,H_3O^+]{1)\,NaBH_4/THF} C_6H_5CH(OH)CH_2CH_3 \xrightarrow[\Delta]{H_3PO_4} C_6H_5CH{=}CHCH_3$

$C_6H_5CH{=}CHCH_3 + H_2O \xrightarrow{H_2SO_4} C_6H_5CH(OH)CH_2CH_3 \xrightarrow{CrO_3/H_2SO_4} C_6H_5COCH_2CH_3$

b) As in part (a). Conjugated double bond forms.

$C_6H_5CH{=}CHCH_3 \xrightarrow[2)\,H_2O_2/H_2O/OH^-]{1)\,(BH_3)_2/THF} C_6H_5CH_2CH(OH)CH_3 \xrightarrow[2)\,\Delta]{1)\,CrO_3/pyridine} C_6H_5CH_2COCH_3$

c) $C_6H_5CH_2COCH_3 \xrightarrow{NH_3} C_6H_5CH_2C({=}NH)CH_3 \xrightarrow[2)\,H_2O]{1)\,NaBH_4/THF} C_6H_5CH_2CH(NH_2)CH_3$

d)

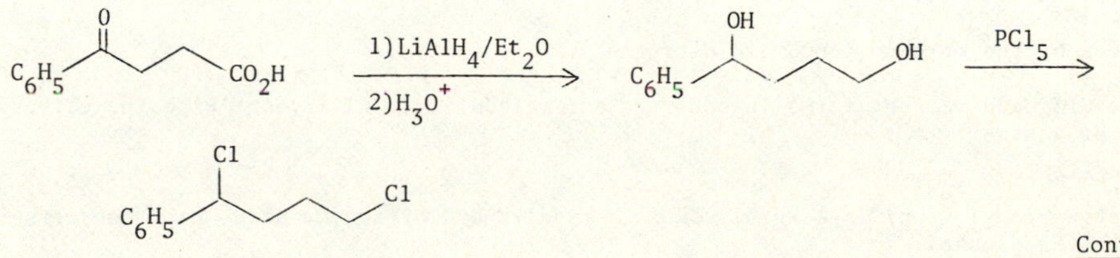

Contd...

20-8 contd...
d) contd...

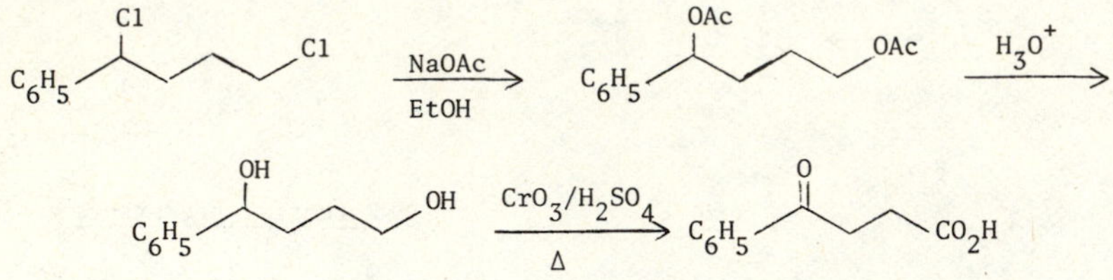

e)

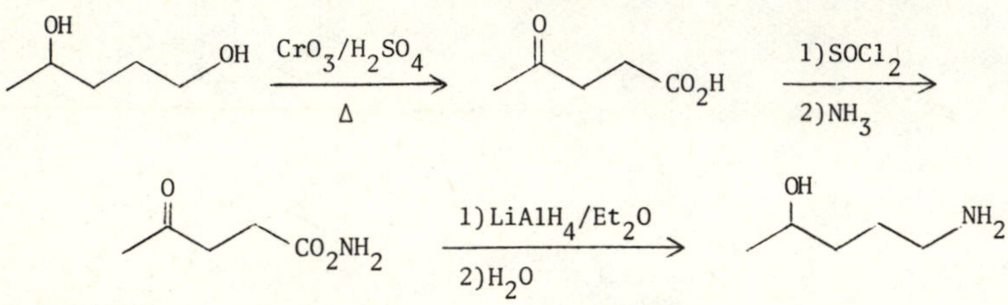

f)

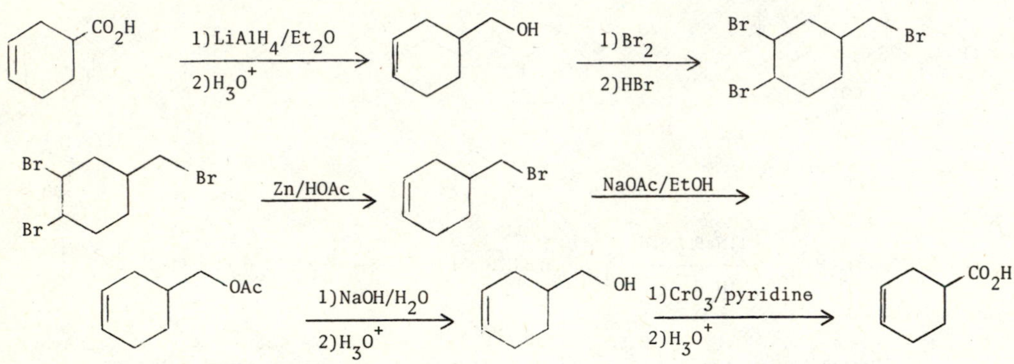

20-9 Only those steps not expected to proceed with a reasonable yield are discussed below.

a) Intramolecular cyclization (lactone formation) is more favorable.

b) Feasible as written.

c) Addition of HCN to the aldehyde is more rapid.

d) Feasible as written.

e) Substitution is more rapid at the primary carbon atom.

f) Feasible as written.

g) Reaction by the more nucleophilic nitrogen atom is faster.

h) As the Grignard reagent forms in one reactant molecule, it will react with the carbonyl group of another.

20-10 The tetrapyranyl ether is a ketal, thus is readily hydrolyzed by acid which converts it back to the alcohol.

20-11

a) Protection of the hydroxy group as a tetrahydropyranyl ether or as a ketal is possible, but would then require a nonacidic esterification such as $SOCl_2/NaOH/CH_3OH$. A simpler approach to the desired esterification would be to form the carboxylate salt, then add a methylating agent such as CH_3I.

b) Satisfactory as written.

c) Protect the aldehyde as an actal, then add HCN under neutral or slightly basic conditions.

d) Satisfactory as written.

e) Substitution at the primary carbon atom by AcO^- would protect that position. After cyanide substitution at the secondary carbon, hydrolysis to an alcohol, under mild conditions so as not to hydrolyze the nitrile, then conversion to the chloride accomplishes the synthesis.

f) Satisfactory as written.

g) In chapter 22 we will learn that many methods are available for protecting amino groups in the presence of most other functional groups.

h) Protect the carbonyl group as ketal.

20-12

a) $HC{\equiv}C^-Na^+ + CH_3CH_2CH_3Br \xrightarrow{THF} CH_3CH_2CH_2C{\equiv}CH \xrightarrow[2)\,CH_3CH_2CH_2Br]{1)\,NaNH_2/THF}$

$CH_3CH_2CH_2C{\equiv}CCH_2CH_2CH_3 \xrightarrow[HOAc]{H_2/Pt} CH_3(CH_2)_6CH_3$

b) $3\ C_2H_5MgBr + (MeO)_2C{=}O \xrightarrow[2)\,H_3O^+/\Delta]{1)\,Et_2O} [(C_2H_5)_3COH] \xrightarrow{H^+}$

$(C_2H_5)_2C{=}CHCH_3 \xrightarrow[HOAc]{H_2/Pt} (C_2H_5)_3CH$

c) $2\ CH_3CO_2CH_3 \xrightarrow{NaOMe/MeOH} CH_3COCH_2COCH_3 \xrightarrow[2)\,2\ CH_3I]{1)\,NaOMe/MeOH}$

$CH_3COC(CH_3)_2COCH_3 \xrightarrow[DMSO]{H_2NNH_2/NaOH} (CH_3CH_2)_2C(CH_3)_2$

d) $(CH_3)_2CHCuCl + CH_2{=}CHCN \xrightarrow{Et_2O} (CH_3)_2CHCH_2CH_2CN \xrightarrow[2)\,H_3O^+]{1)\,(CH_3)_2CHMgBr/Et_2O}$

$(CH_3)_2CHCH_2CH_2COCH(CH_3)_2 \xrightarrow[DMSO]{H_2NNH_2/\underline{t}\text{-}BuOK} (CH_3)_2CH(CH_2)_3CH(CH_3)_2$

20-13

i) 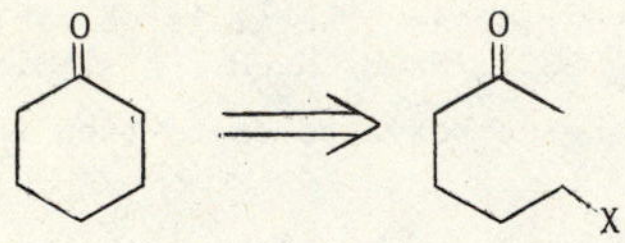

iii) 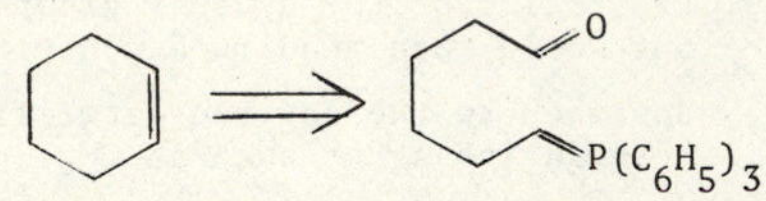

ii) 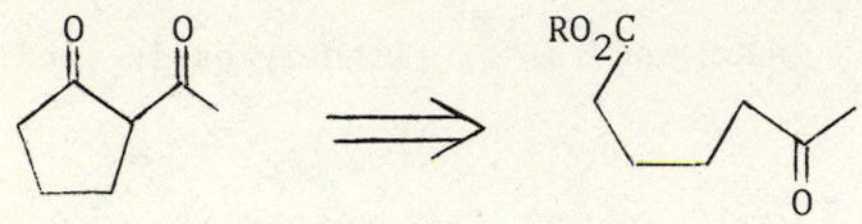

iv) 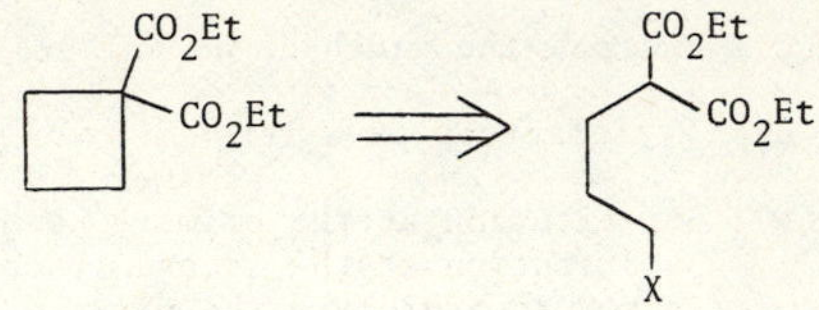

20-14

a) 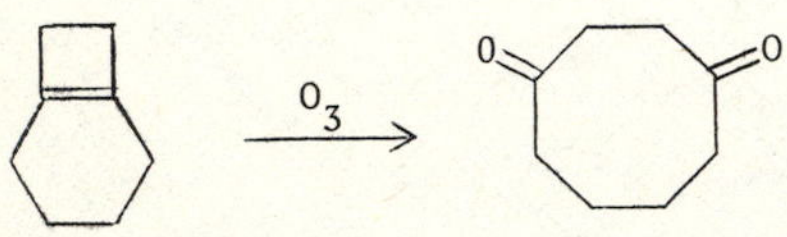

b) 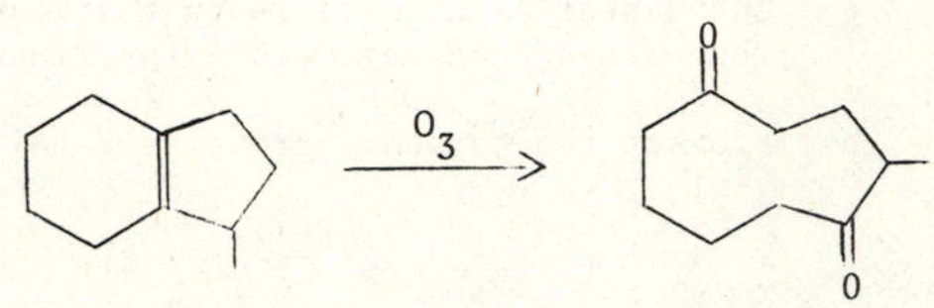

20-15 In outlining the following construction sequences only the major steps for altering the carbon skeleton are included.

a) 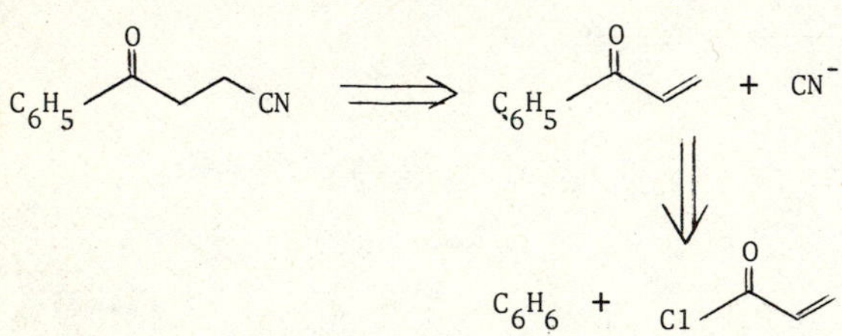

b) 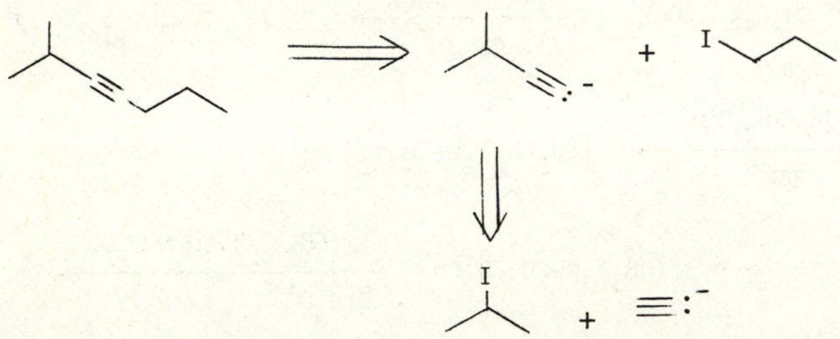

Contd...

20-15 contd...

c)

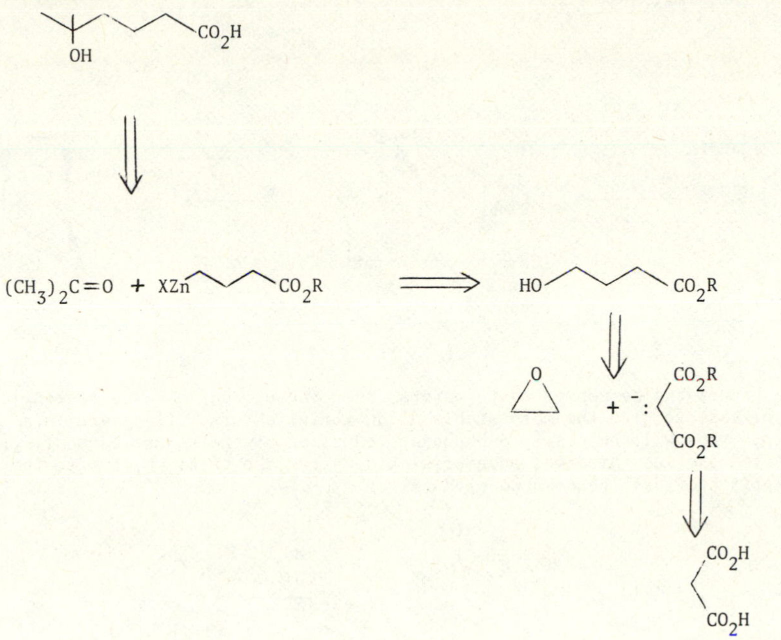

d)

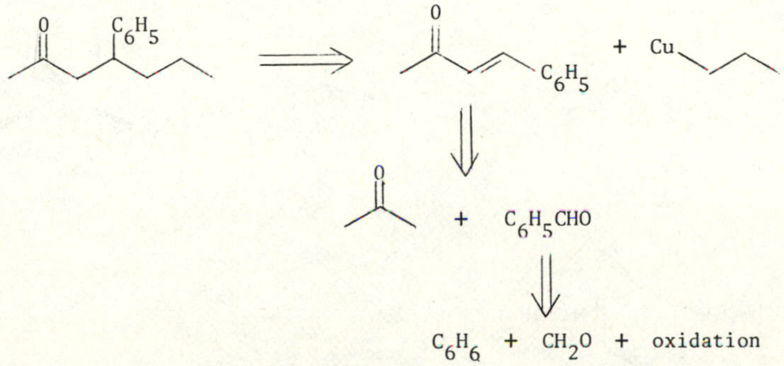

e)

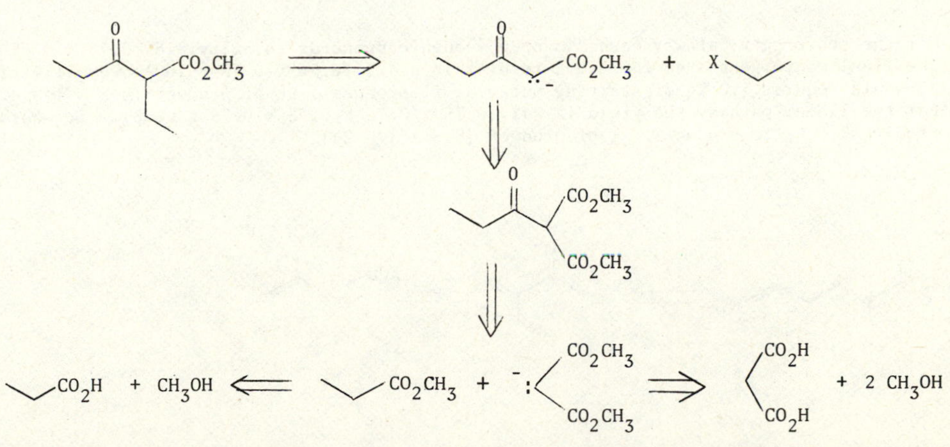

20-16 The reaction proceeds through an achiral enolate to give equal amounts of the two enantiomers, a racemic mixture.

20-17

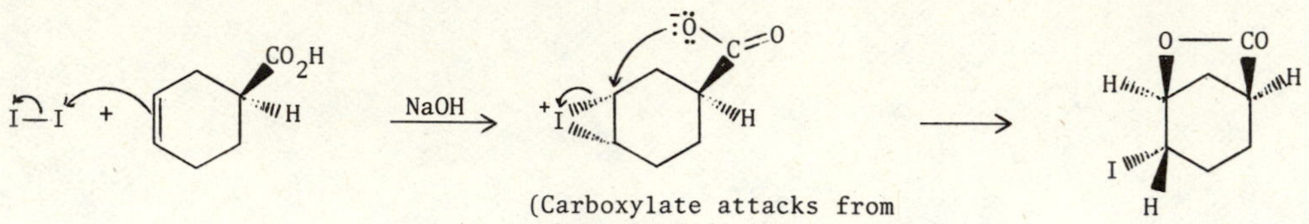

(Carboxylate attacks from above as I departs to below.)

20-18 The carboxylate groups can be converted to esters, then the product allowed to equilibrate in dilute acid or base to give the more stable diequatorial esters. (They are in a 1,3-relationship, thus will be cis.) Subsequent reduction of the ketone carbonyl group by approach of H^- from the less hindered equatorial side gives the axial OH, cis to the ester groups. Hydrolysis provides the desired product.

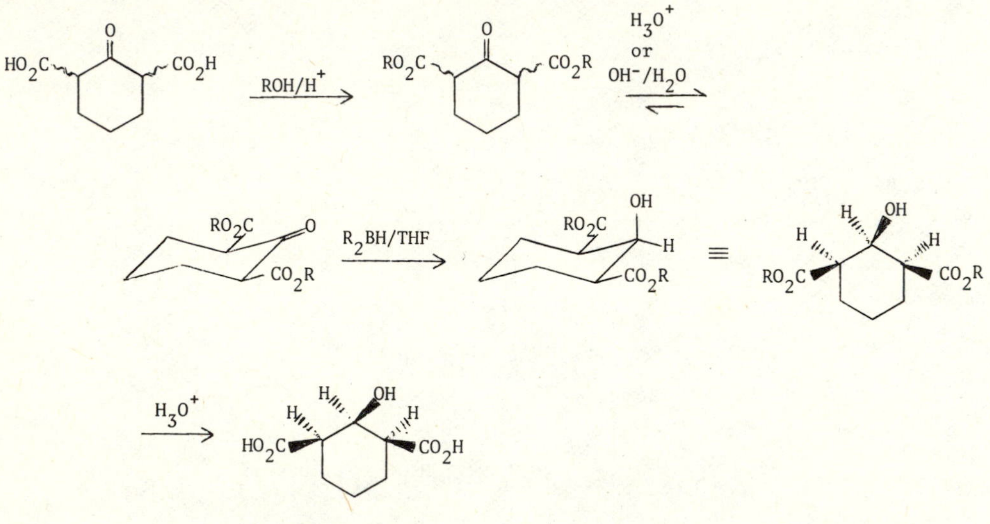

20-19 For the convergent pathway each two step sequence proceeds in 56% yield (0.75 x 0.75 x 100). The final convergent step therefore results in a 42% (0.56 x 0.75 x 100) overall yield. It would require 1.2 kg of starting material to produce 1 kg of product (0.5 x 1/0.42). For the linear pathway the yield is 24% (0.75 x 0.75 x 0.75 x 0.75 x 0.75). It would require 2.1 kg to produce 1 kg of product (0.5 x 1/0.24).

20-20

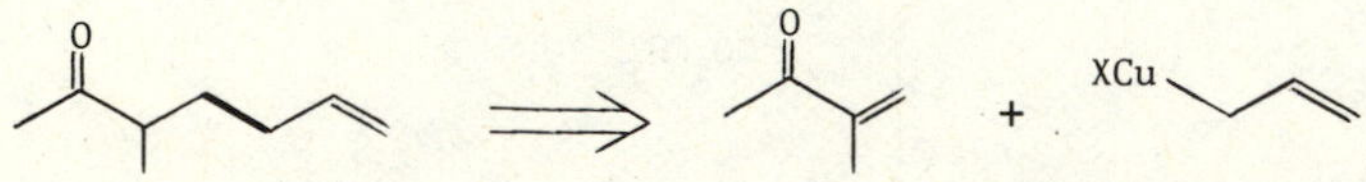

20-21 The final step involves a conjugate addition and removal of the ester group which is present to ensure regiospecificity in the addition. All of the other construction steps involve small available synthons and relatively simple reactions. However as a route to 3-methyl-6-hepten-2-one this approach would not be preferred since selective reduction of the 2,6-diketone would be quite difficult.

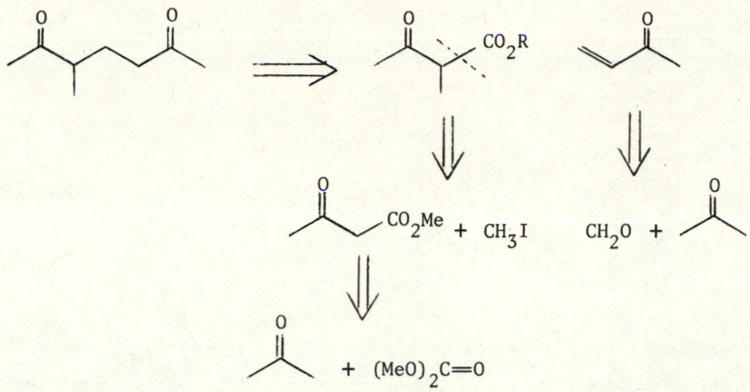

20-22

a) Chirality at that carbon atom is destroyed in the next step so it isn't important.

b)

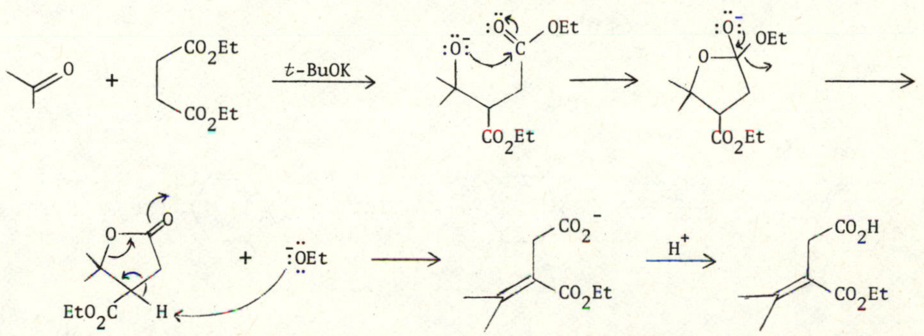

c) See fifth entry, table 20-3.

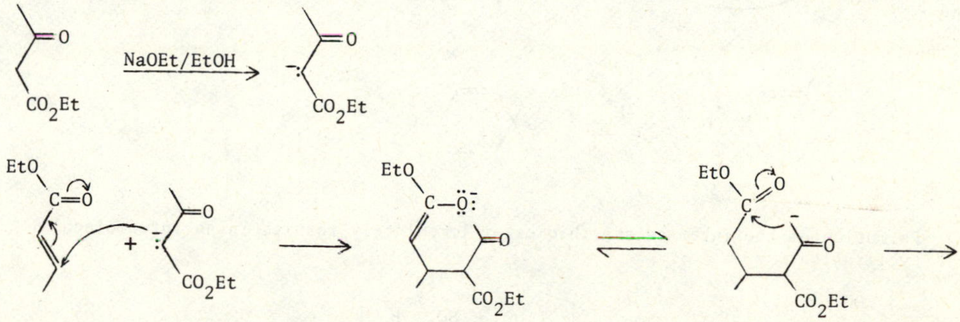

Contd...

20-22 contd...

c) contd...

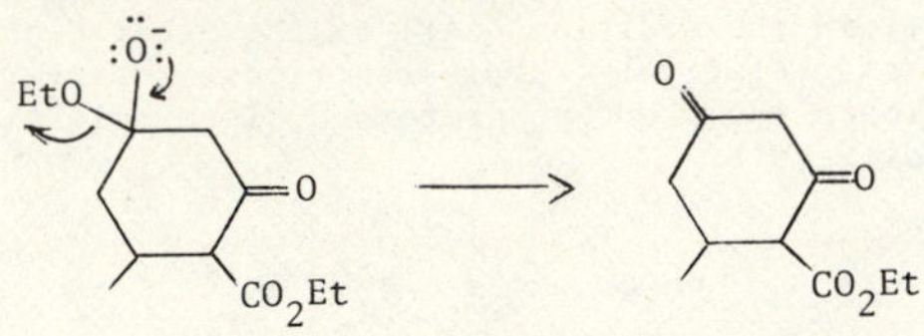

d)

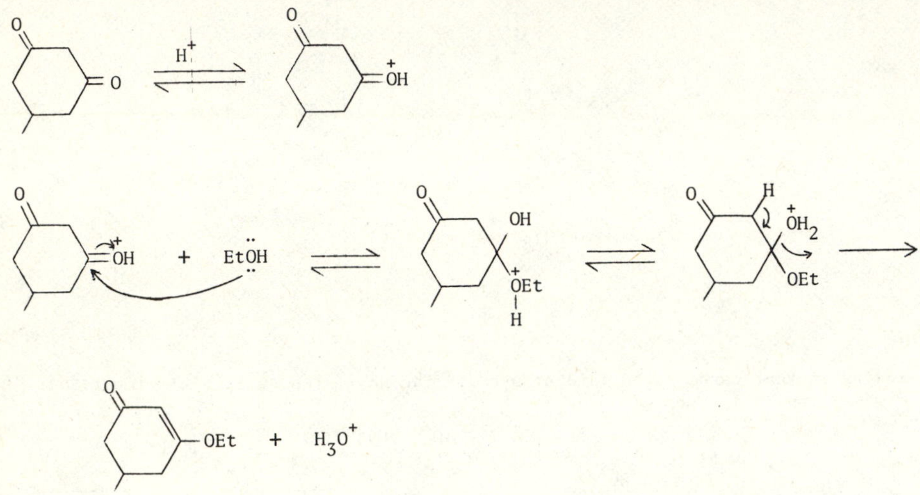

e)

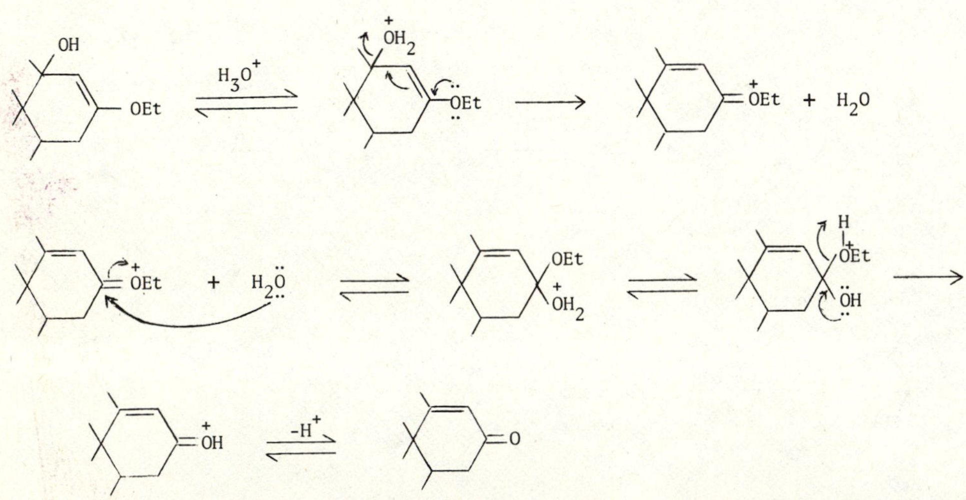

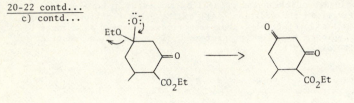

f) Formation of the intermediate thio ester immediately results in an intramolecular cyclization.

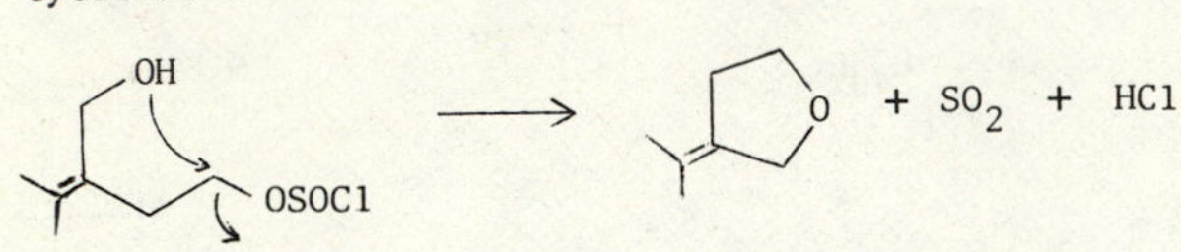

Contd...

20-22 contd...

f) contd...

However, conversion of the diol to a dianion with methyllithium keeps the two ends apart and inhibits ring formation as a sulfonate diester rapidly forms. Then substitution by chloride gives the desired product.

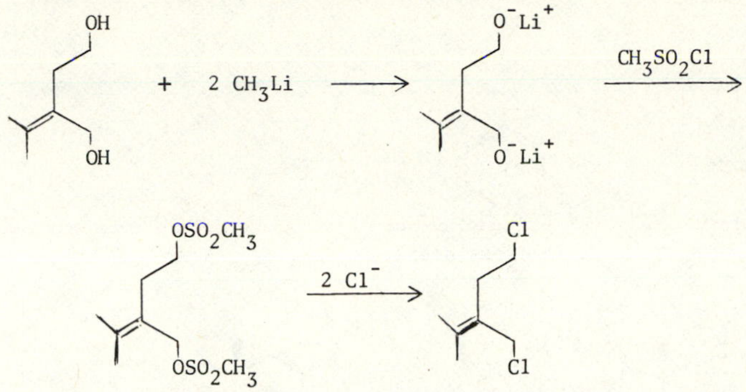

20-23

a)

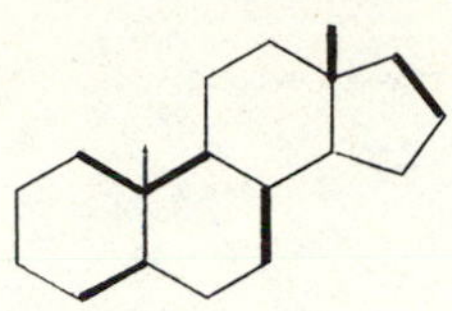

b)

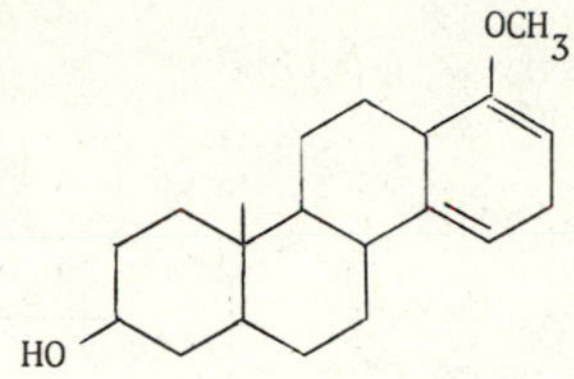

c)

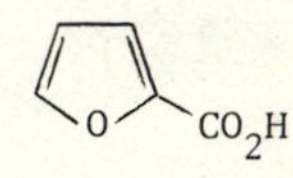

d)

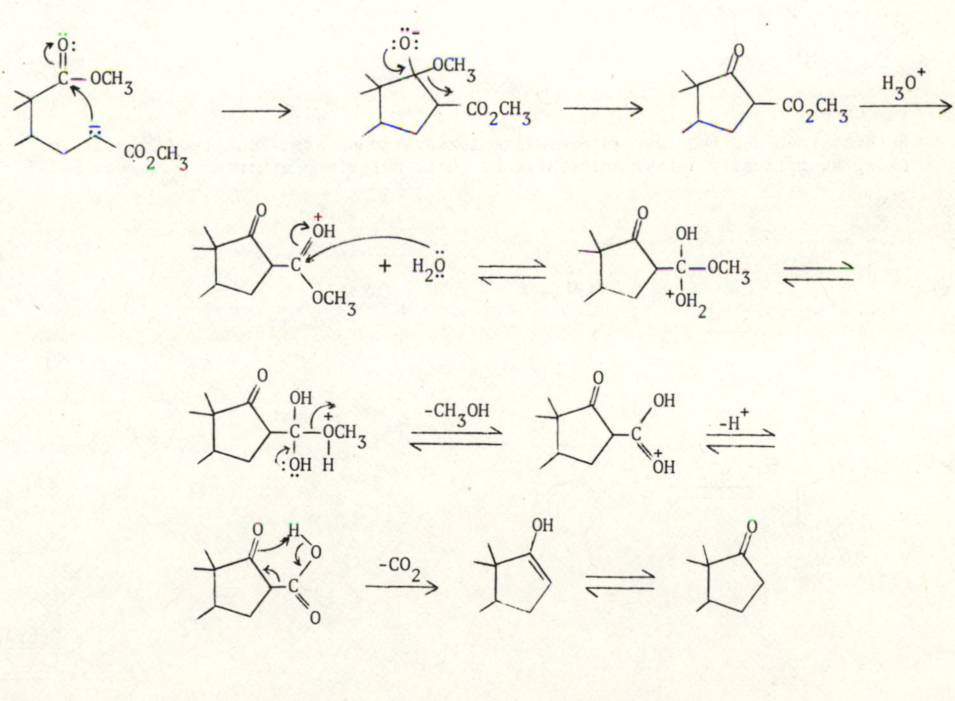

20-24

a) 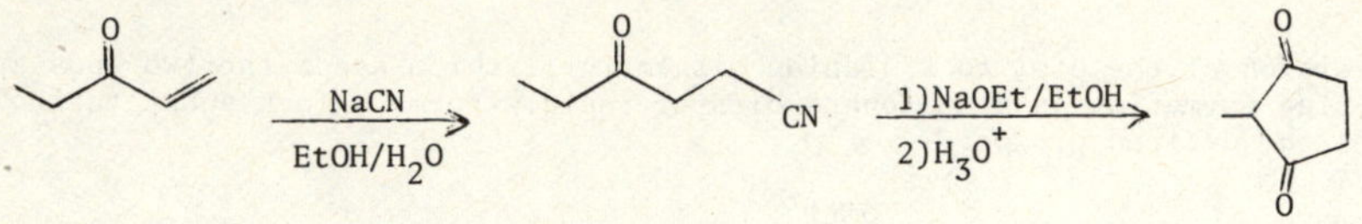

b) 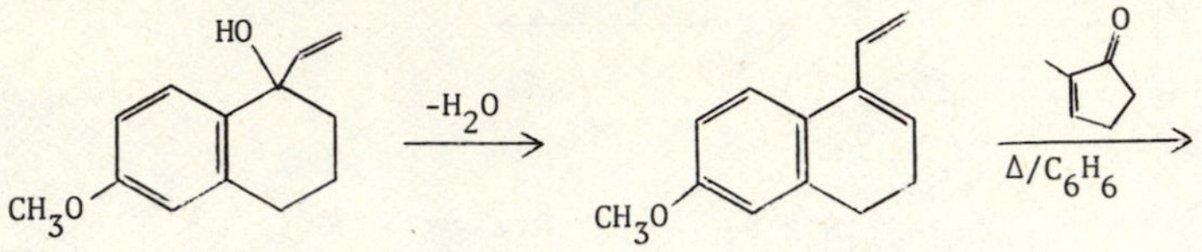

(This is an *s*-cis diene.)

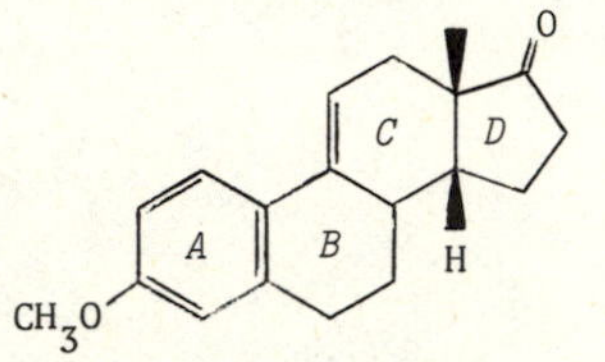

Stereochemistry at *C-D* ring junction is wrong

or

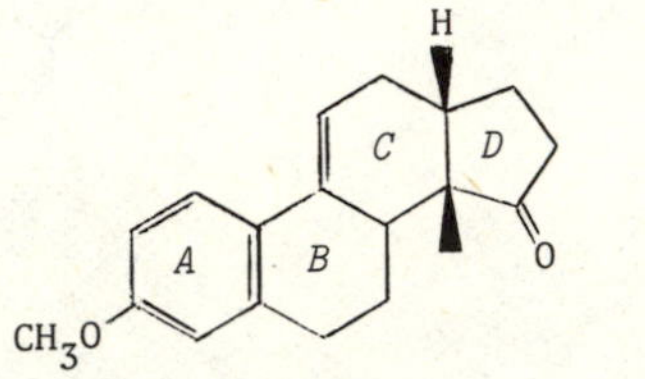

Regiochemistry is wrong

c) 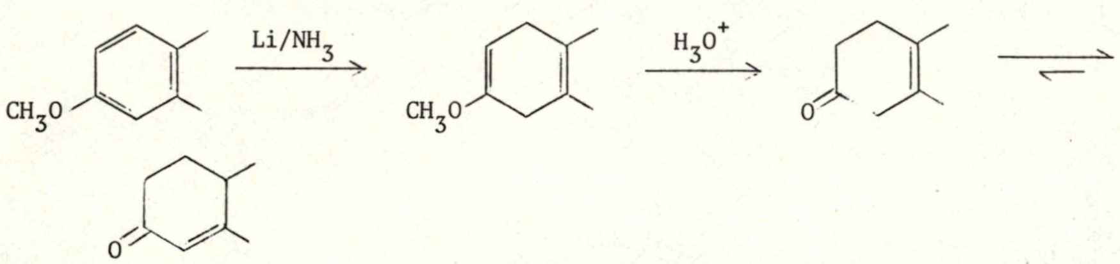

d) Both estrone and the last intermediate to β-vetivone are alcohols. They can be resolved using an optically active amine salt of their phthalate esters. (See sec. 9-7)

e) 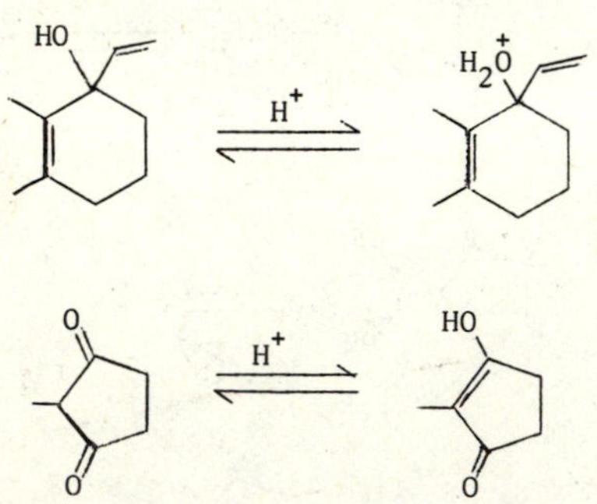

Contd...

20-24 contd...

e) contd...

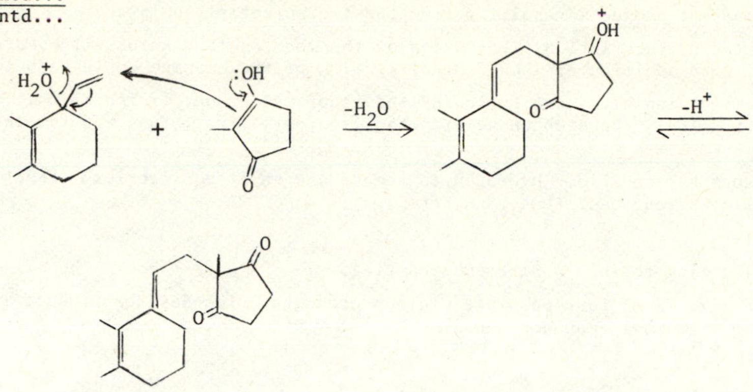

20-25

a) Following the sequence from citral to intermediate A we have: an aldol, an acid catalyzed cyclization, an addition and substitution, and a decarboxylative elimination. From 3-buten-2-one to intermediate B we have: an addition, a rearrangement, and an acid-base reaction. From A plus B to vitamin A we have: an addition, a reduction, a protecting esterification, an elimination, and a deprotecting reduction.

b) The Grignard reagent functions as a base forming potential anions at O and the acetylenic carbon.

c)

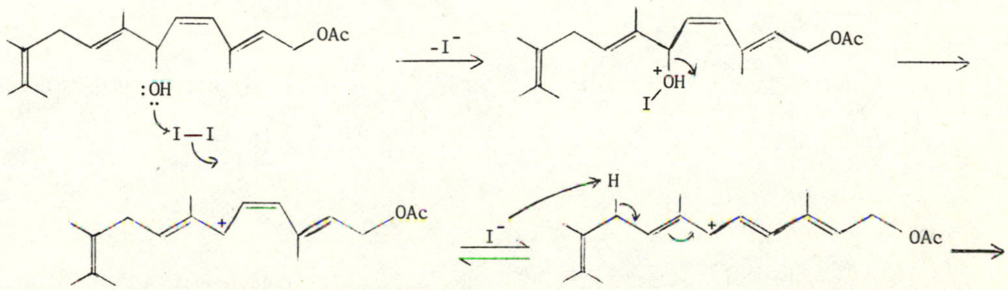

Isomerization of the Z double bond readily occurs in this conjugated system.

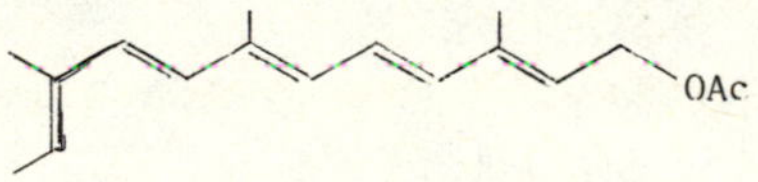

d) Acetylation is more rapid at the less hindered primary hydroxy group.

20-26

a) Would give a mixture of products resulting from the two potential enolates.

b) Would give mono plus polyalkylation. Formation of the mono enolate anion with a strong base such as LDA, then addition of 1 CH_3I or alkylation of the enamine would be better.

c) Steric hindrance might inhibit substitution by the isopropyl group. If feasible it would have been better to alkylate acetophenone with the isopropyl first, then add the two methyl groups.

d) This Claisen product has no acidic hydrogen to remove and shift equilibrium. Reaction must be carried out using strong base (LDA, NaH, etc.).

e) Feasible as written.

f) A would give cyclization to form a six-membered ring.

g) We would expect a mixture of four possible Claisen products. The desired product is best made by alkylation of methyl acetoacetate.

20-27

a) $CH_3CH_2COCH(CH_3)CO_2CH_3 \Longrightarrow CH_3CH_2COCH_2CO_2CH_3 + CH_3X$

b) 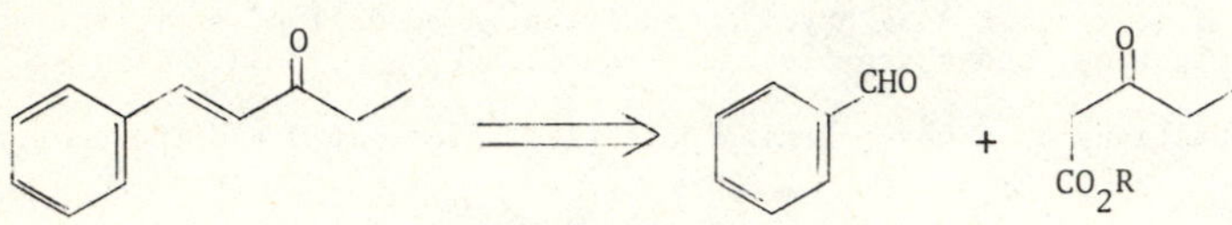

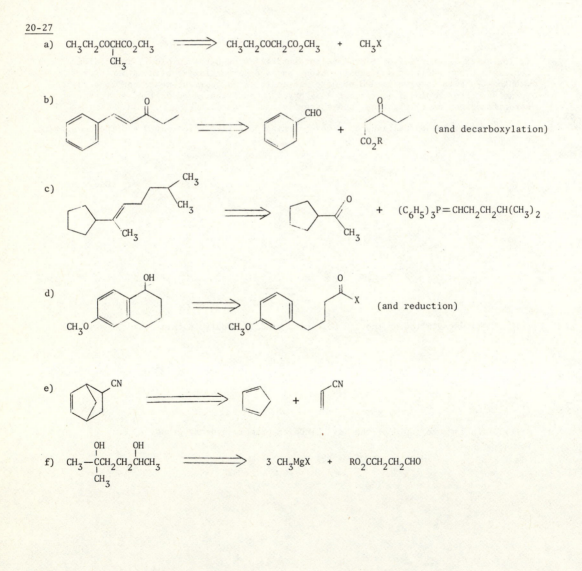

(and decarboxylation)

c) 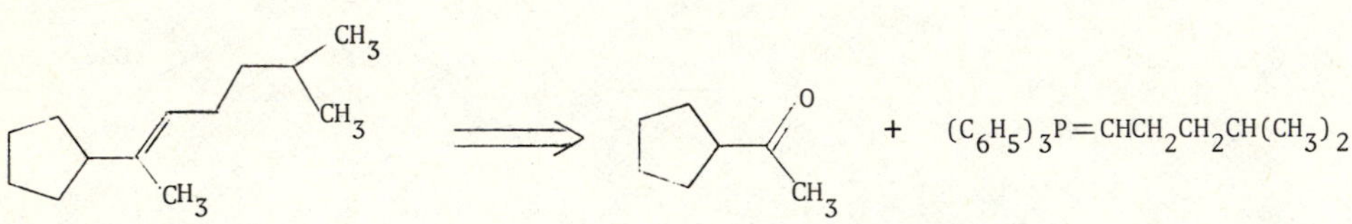

d) 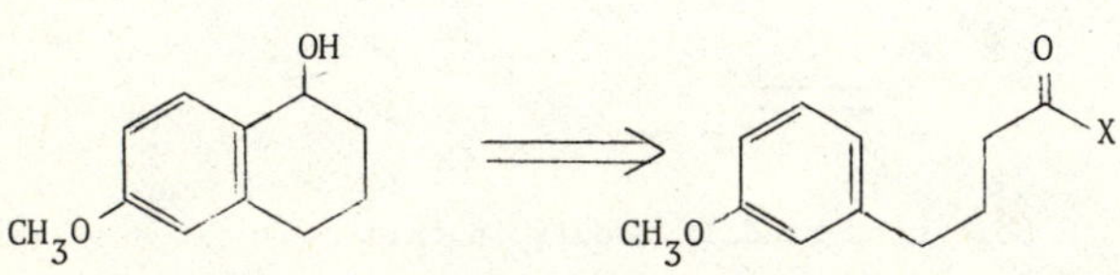

(and reduction)

e) 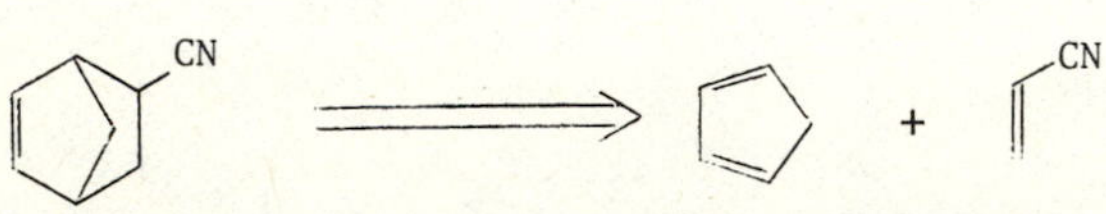

f) $CH_3-C(OH)(CH_3)CH_2CH_2CH(OH)CH_3 \Longrightarrow 3\ CH_3MgX + RO_2CCH_2CH_2CHO$

20-28

a) No; both carbonyl groups would reduce. One must protect the aldehyde carbonyl as an acetal.

b) No; would give mixture of potential Claisen products. No direct method for using the given starting materials is obvious.

c) Unlikely; would result principally in elimination of HCl and epoxide formation. It is best to substitute OAC for Cl, then to hydrolyze the ester.

d) Not as written; two moles of Grignard reagent are required. The first would form a carboxylate salt and the second might add to the carbonyl, though that sequence is usually not successful. Subsequent formation of the lactone would require acid catalysis.

e) No; catalytic hydrogenation would also reduce the carbon-carbon double bond. The use of $LiAlH_4$ would be selective for reduction of the cyano group.

20-29

a) $R_2C{=}O + XZnCH_2CO_2CH_3 \longrightarrow R_2C(OH)CH_2CO_2CH_3$

b) $ROH \xrightarrow{PBr_3} RBr \xrightarrow[2)\,CH_2O]{1)\,Mg/Et_2O} RCH_2OH$

c) $RCHO + (C_6H_5)_3P{=}CH_2 \longrightarrow RCH{=}CH_2 \xrightarrow[2)\,H_2O/H_2O_2/NaOH]{1)\,(BH_3)_2/THF}$

$RCH_2CH_2OH \xrightarrow[\text{Pyridine}]{CrO_3} RCH_2CHO$

d) $ROH \xrightarrow{PBr_3} RBr \xrightarrow[2)\ \ ]{1)\,Mg/Et_2O} RCH_2CH_2OH$

e) $ArBr \xrightarrow[2)\,CH_2O]{1)\,Mg/Et_2O} ArCH_2OH \xrightarrow{PBr_3} ArCH_2Br \xrightarrow[DMSO]{NaCN} ArCH_2CN \xrightarrow[\Delta]{H_3O^+} ArCH_2CO_2H$

f) $ROH \xrightarrow[2)\,Mg/Et_2O]{1)\,PBr_3} RMgBr \xrightarrow[CuCl_2]{CH_2{=}CHCO_2R} RCH_2CH_2CO_2R \xrightarrow{LiAlH_4} RCH_2CH_2CH_2OH$

g) $RCOCH_3 \xrightarrow[2)\,BrCH_2CO_2R']{1)\,LDA/THF} RCOCH_2CH_2CO_2R' \xrightarrow{H_3O^+/\Delta} RCOCH_2CH_2CO_2H$

h) $RCH_2OH \xrightarrow[2)\,NaCN/DMSO]{1)\,PBr_3} RCH_2CN \xrightarrow{LiAlH_4} RCH_2CH_2NH_2$

Contd...

20-29 contd...

i) $R_2CHOH \xrightarrow[H_2SO_4]{CrO_3} R_2C{=}O \xrightarrow{(C_6H_5)_3P{=}CHCH_3} R_2C{=}CHCH_3$

j) $RCO_2H \xrightarrow[2)\,(CH_3)_2NH]{1)\,SOCl_2} RC(=O)N(CH_3)_2 \xrightarrow{LiAlH_4} RCH_2N(CH_3)_2$

20-30

a) 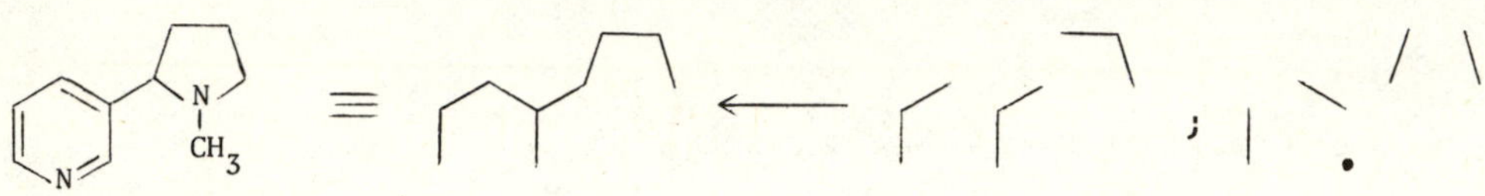

b) 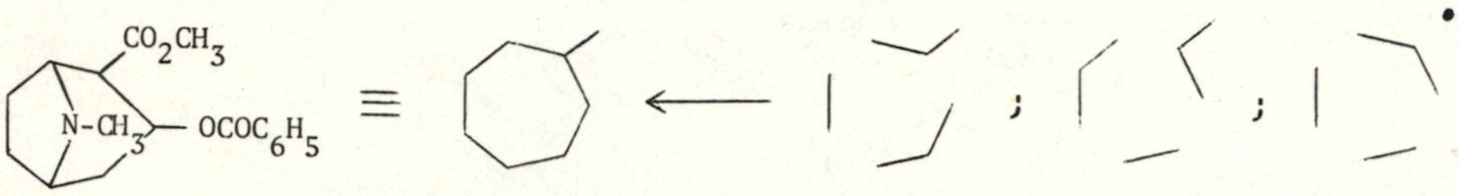

20-31

a) i) $CH_3CO_2H \xrightarrow[2)\,NaOC_2H_5/C_2H_5OH]{1)\,C_2H_5OH/H^+} CH_3COCH_2CO_2C_2H_5 \xrightarrow[\Delta]{H_3O^+} CH_3COCH_3 + CO_2 + C_2H_5OH$

ii) $CH_3CN + CH_3MgI \xrightarrow{Et_2O} CH_3C(=NMgI)CH_3 \xrightarrow[HCl]{H_2O} CH_3COCH_3$

iii) $CH_3CHO + CH_3MgI \xrightarrow[2)\,H_3O^+]{1)\,Et_2O} CH_3CH(OH)CH_3 \xrightarrow[H_2SO_4]{CrO_3} CH_3COCH_3$

b) i) $CH_3Li + CH_2O \xrightarrow[2)\,H_3O^+]{1)\,Hexane} CH_3CH_2OH \xrightarrow{PBr_3} CH_3CH_2Br$

$CH_3CH_2OH \xrightarrow{CrO_3\cdot pyridine} CH_3CHO$ (Distill volatile product)

$CH_3CH_2Br \xrightarrow{Mg/Et_2O} CH_3CH_2MgBr \xrightarrow[2)\,H_3O^+]{1)\,CH_3CHO} CH_3CH_2CH(OH)CH_3 \xrightarrow{CrO_3/H_2SO_4} CH_3CH_2COCH_3$

<u>or</u>

$CH_3I \xrightarrow{NaCN/DMSO} CH_3CN \xrightarrow[2)\,H_3O^+]{1)\,{}^*CH_3CH_2MgBr/Et_2O} CH_3CH_2COCH_3$

Contd...

20-31 contd...

b) contd...

ii) $^*CH_3CHO \xrightarrow[2)\,H_3O^+]{1)\,CH_3Li/Hexane} CH_3CH(OH)CH_3 \xrightarrow{CrO_3/H_2SO_4}$

$CH_3COCH_3 \xrightarrow[2)\,H_3O^+]{1)\,CH_3MgBr/Et_2O} (CH_3)_3COH$

or

$^*CH_3CH_2OH \xrightarrow{CrO_3/H_2SO_4} CH_3CO_2H \xrightarrow{CH_3OH/H^+} CH_3CO_2CH_3 \xrightarrow[2)\,H_3O^+]{1)\,2\ CH_3MgBr/Et_2O}$

$(CH_3)_3COH$

iii) $^*CH_3CH_2Br \xrightarrow[2)\,CH_2O\ \ 3)\,H_3O^+]{1)\,Mg/Et_2O} CH_3CH_2CH_2OH \xrightarrow{HI} CH_3CH_2CH_2I$

*Prepared in an earlier part of problem 20-31.

20-32

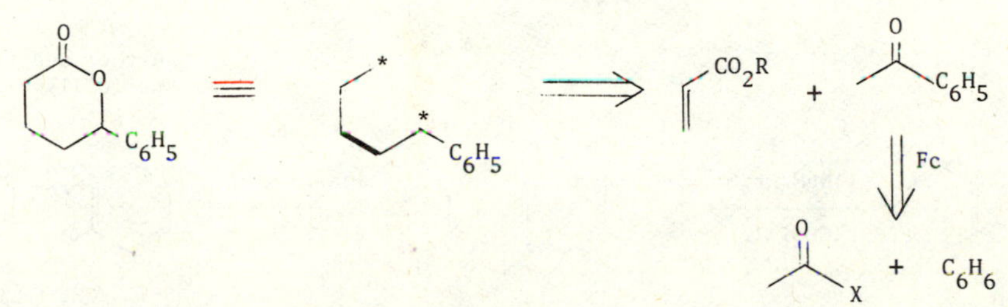

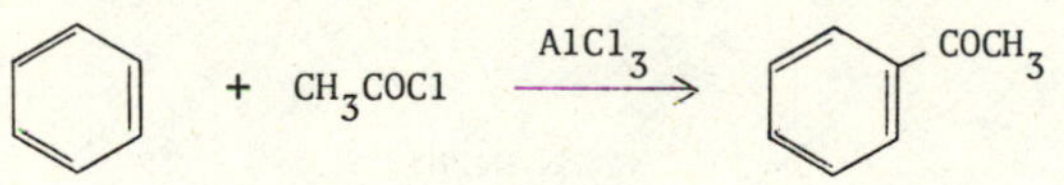

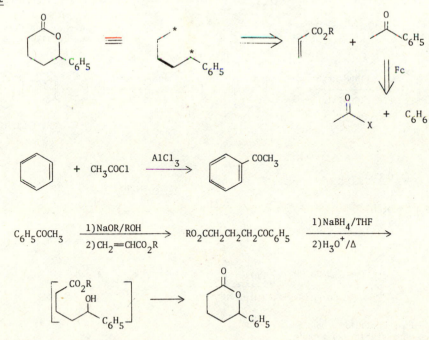

$C_6H_5COCH_3 \xrightarrow[2)\,CH_2{=}CHCO_2R]{1)\,NaOR/ROH} RO_2CCH_2CH_2CH_2COC_6H_5 \xrightarrow[2)\,H_3O^+/\Delta]{1)\,NaBH_4/THF}$

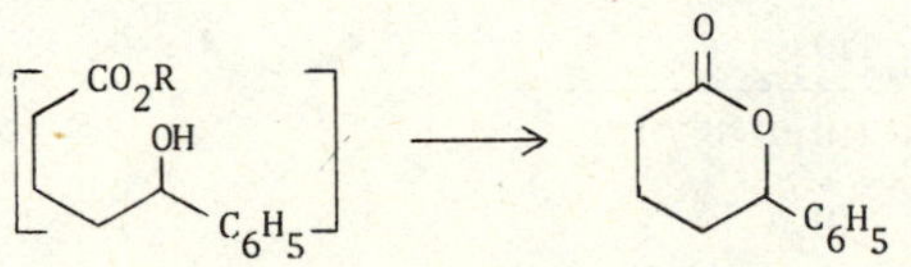

20-33

a) 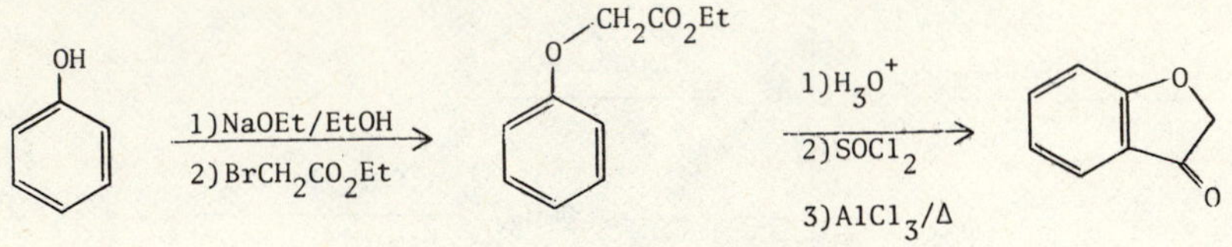

b) 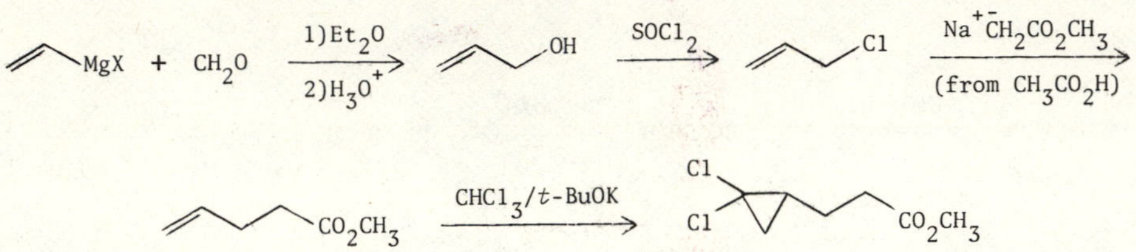

c)

$$CH_3CO_2CH_3 \quad 2\ CH_3MgX \xrightarrow[2)\,H_2O/NH_4Cl]{1)\,Et_2O} (CH_3)_3COH \xrightarrow{HCl} (CH_3)_3CCl \xrightarrow[2)\,CO_2 \quad 3)\,H_3O^+]{1)\,Mg/Et_2O}$$

$$(CH_3)_3CCO_2H \xrightarrow{SOCl_2} (CH_3)_3CCOCl \xrightarrow[2)\,H_3O^+]{1)\,(CH_3)_2CHMgBr/Et_2O} (CH_3)_3C\overset{O}{\overset{\|}{C}}CH(CH_3)_2$$

Further addition is sterically hindered

d) 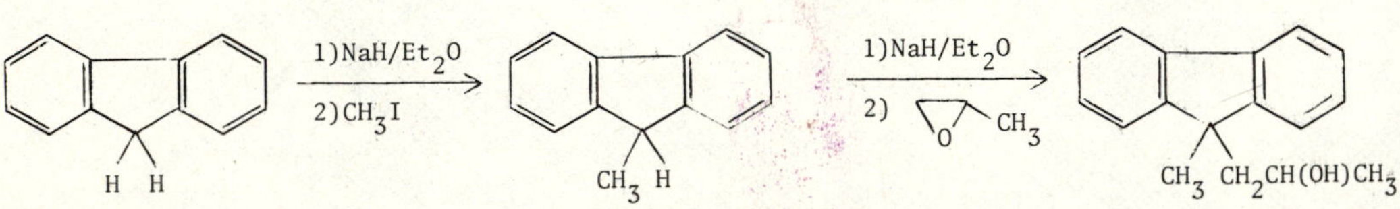

e) 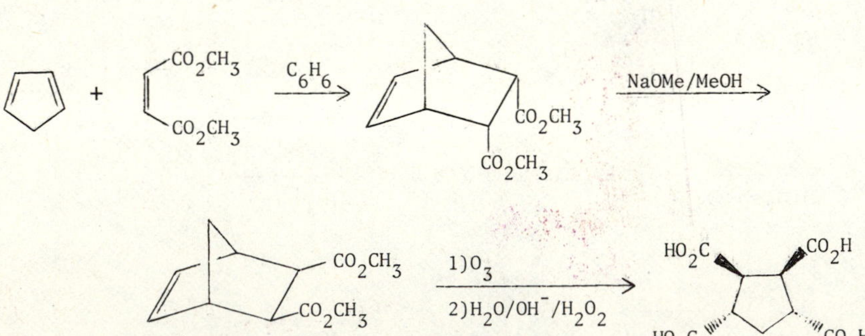

Base equilibration to more stable exo isomer

Contd...

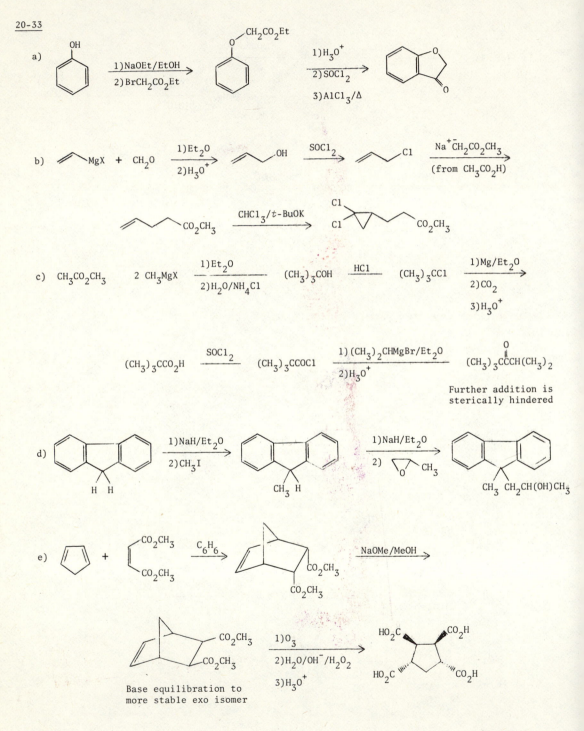

20-33 contd...

f) 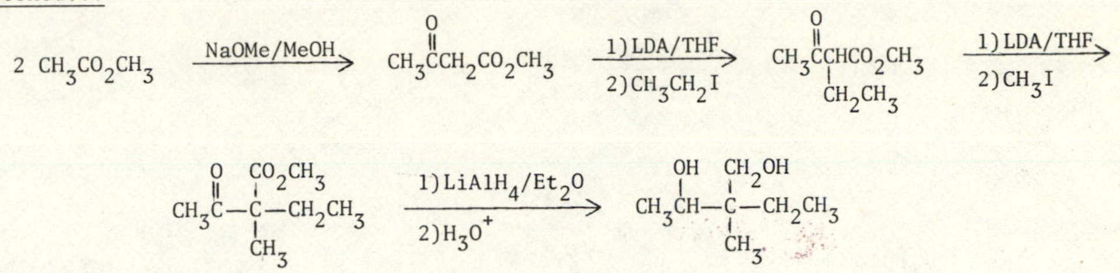

20-34

A ≡ 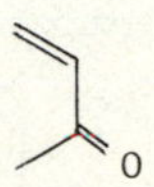

B ≡ NaOEt/EtOH

C ≡ 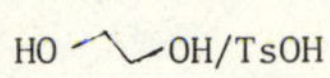

D ≡ $(BH_3)_2$/THF

E ≡ H_2O/NaOH/H_2O_2

F ≡ 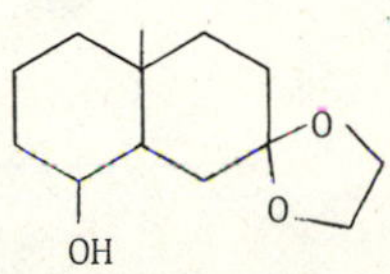

(mixture of regioisomers)

G ≡ CrO_3/Acetone

H ≡ $(C_6H_5)_3P{=}CH_2$

I ≡ 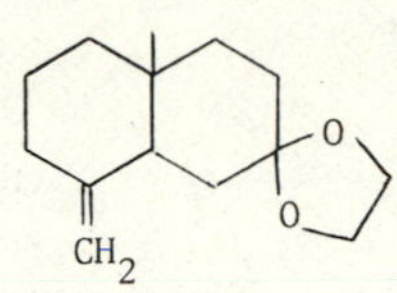

J ≡ 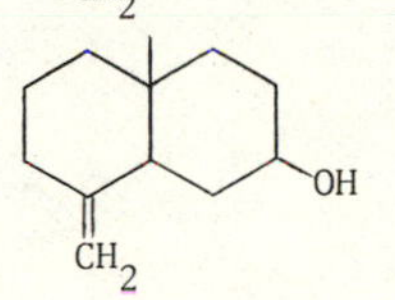

K ≡ PBr_3

L ≡ Mg/Et_2O

M ≡ CO_2

N ≡ CH_2N_2

O ≡ $2CH_3MgBr/Et_2O$

P ≡ H_3O^+

20-35

a) 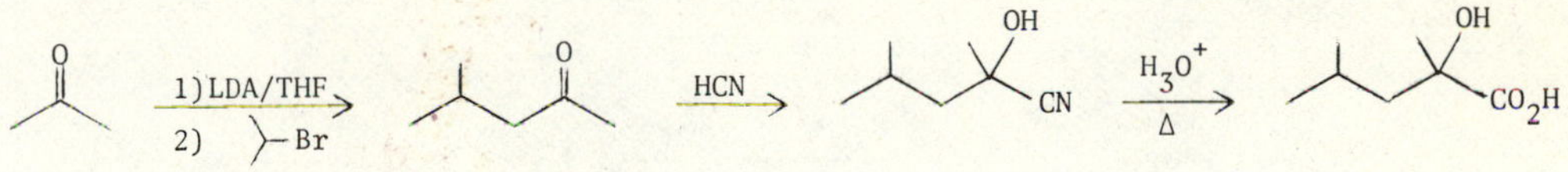

b) 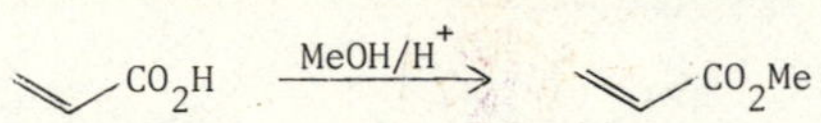

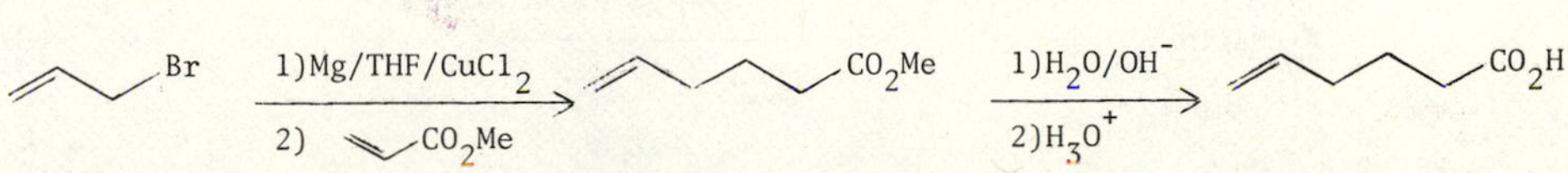

c) $CH_3CH_2CHO + XZnCH_2CO_2Me \xrightarrow[2)\,H_3O^+]{1)\,Et_2O} CH_3CH_2CH(OH)CH_2CO_2Me \xrightarrow[\text{Pyridine}]{CrO_3}$

Contd...

20-35 contd...

c) Contd...

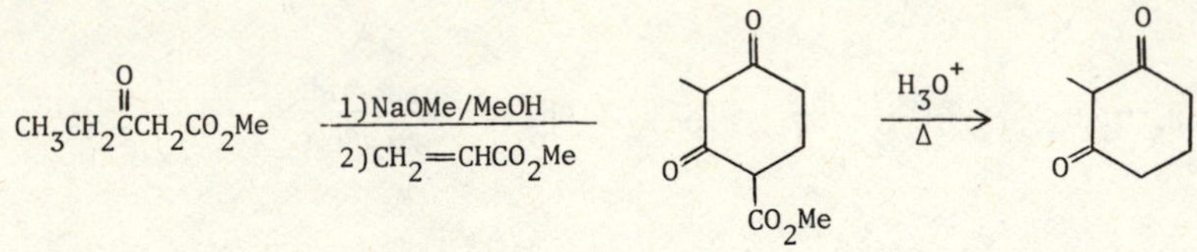

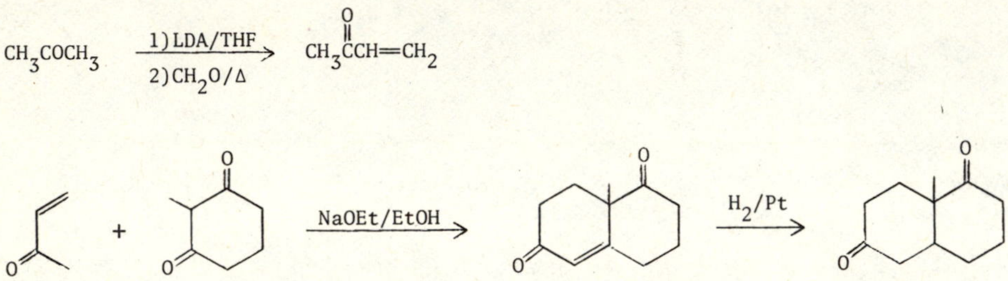

d)

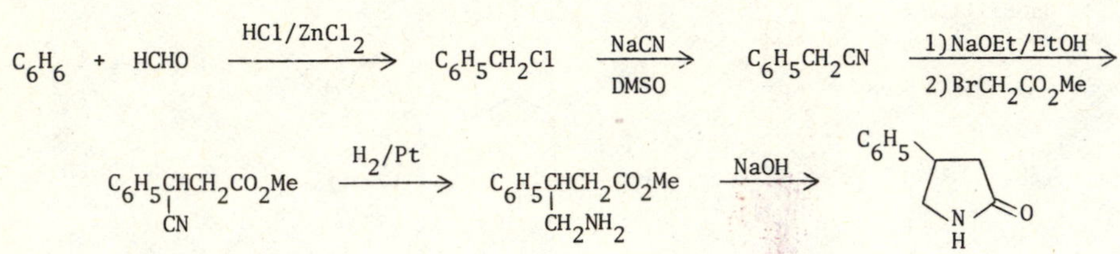

e)

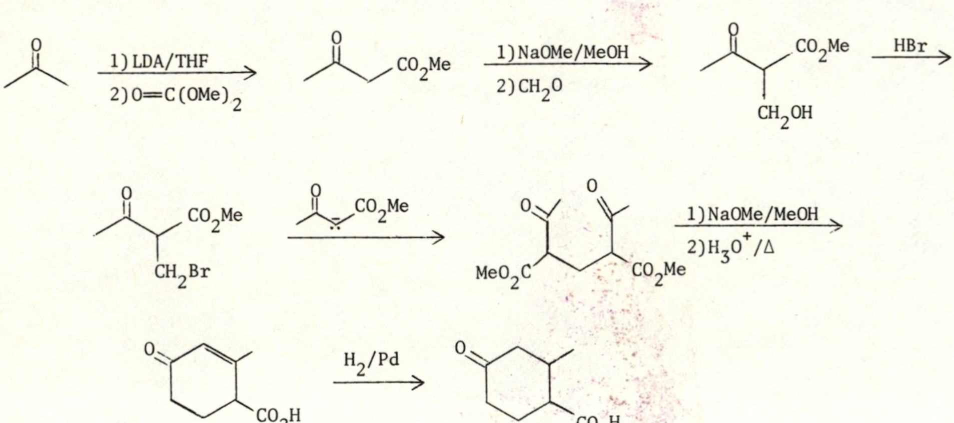

f)

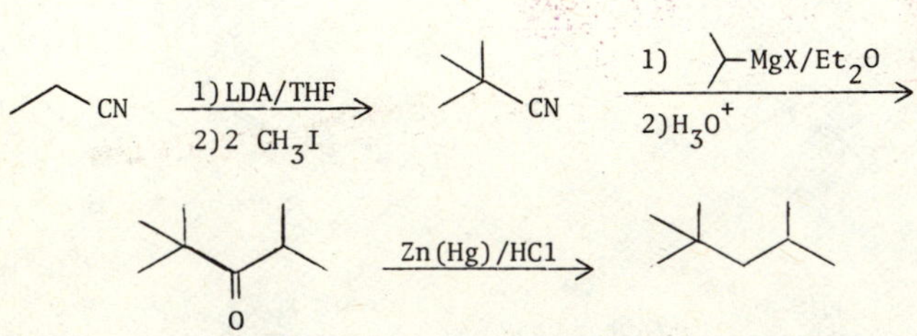

20-35 contd...

g) $CH_2(CO_2H)_2 \xrightarrow{MeOH/H^+} CH_2(CO_2Me)_2 \xrightarrow[2)\ 2\ CH_2{=}CHCOCH_3]{1)\ LDA/THF}$

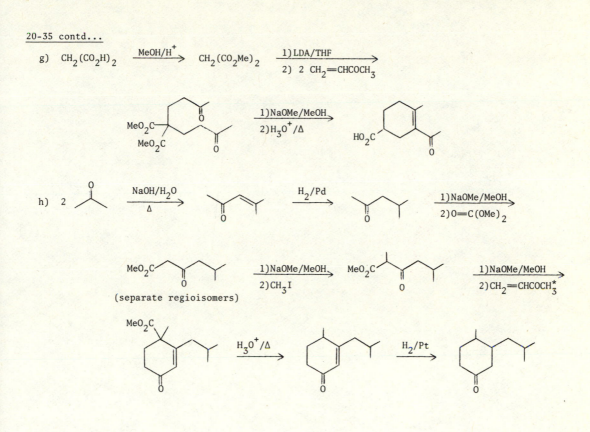

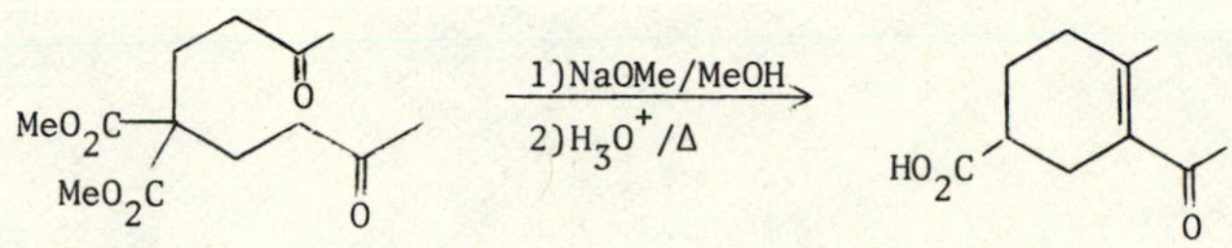

h)

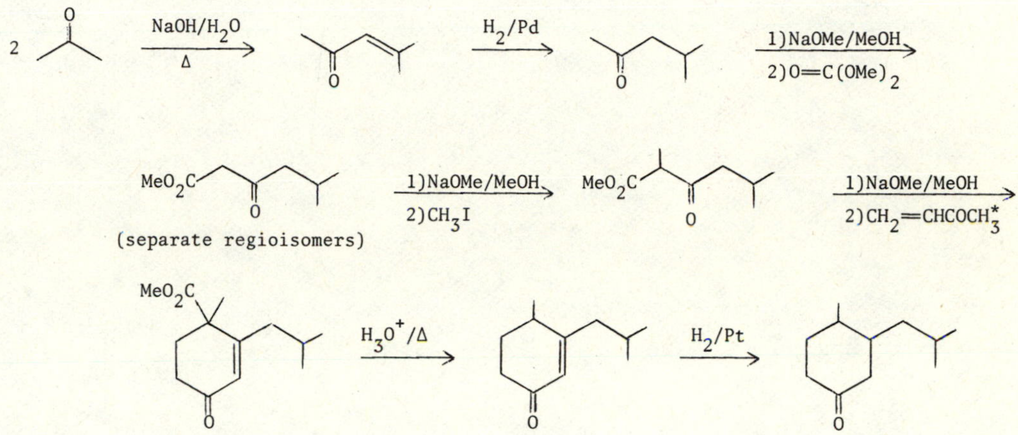

21 Carbohydrates and Nucleosides

21-1

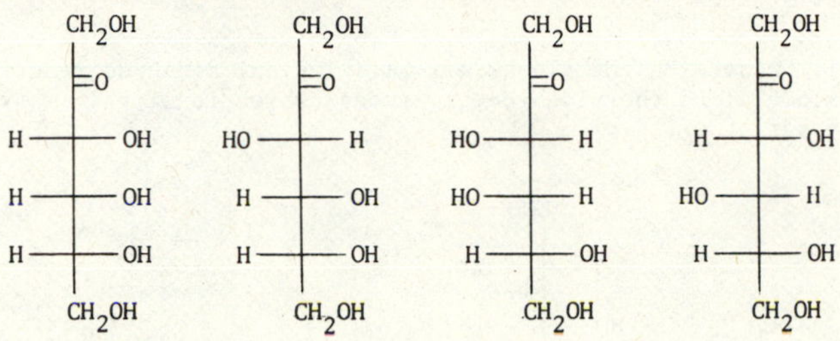

21-2 D-Glucose ≡ 2*R*,3*S*,4*R*,5*R*-2,3,4,5,6-Pentahydroxyhexanal
L-Glucose ≡ 2*S*,3*R*,4*S*,5*S*-2,3,4,5,6-Pentahydroxyhexanal

21-3 Common names are included. Students may find the plane or axis of symmetry in each meso form more obvious by using Haworth structures (sec. 21-2B).

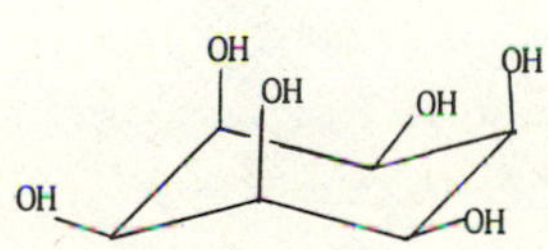

cis-Inositol
meso

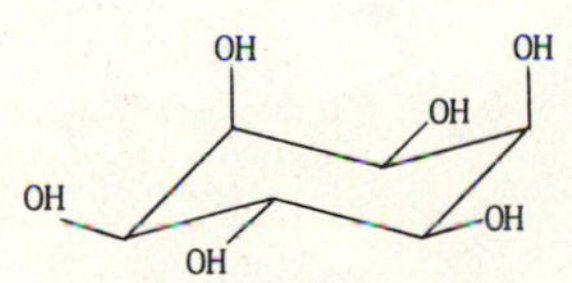

epi-Inositol
meso

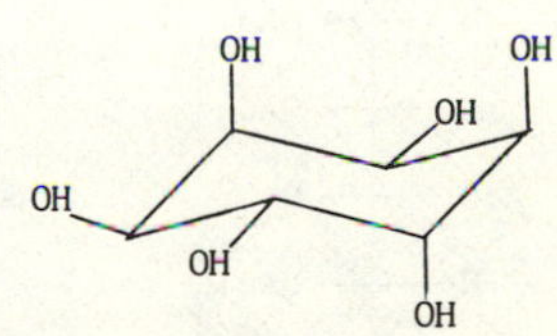

allo-Inositol
meso

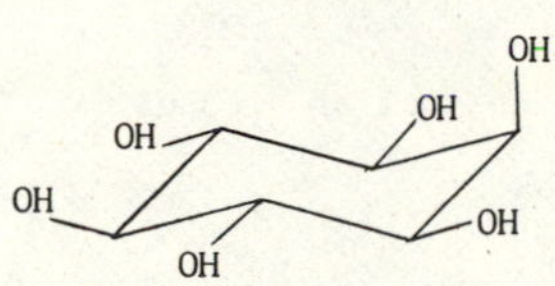

myo-Inositol
meso

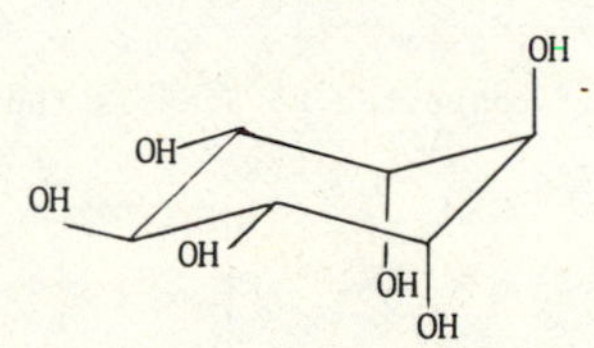

muco-Inositol
meso

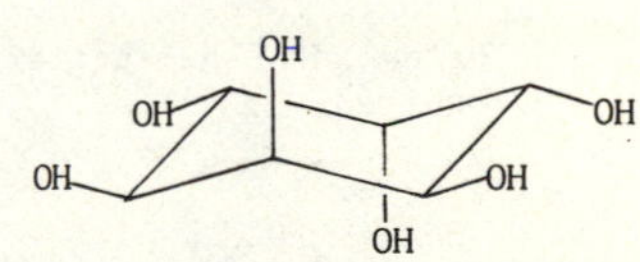

neo-Inositol
meso

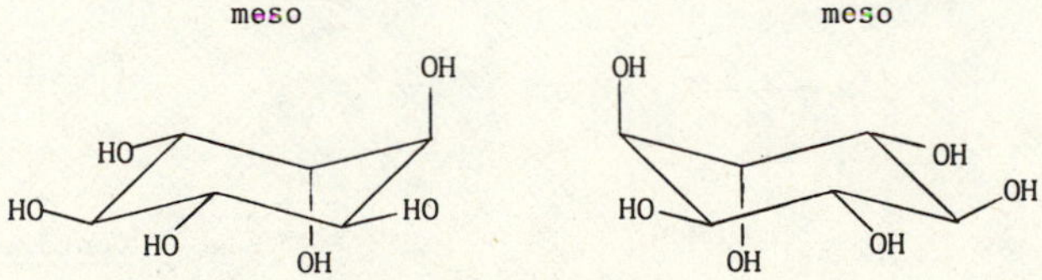

chiro-Inositols
dl-pair

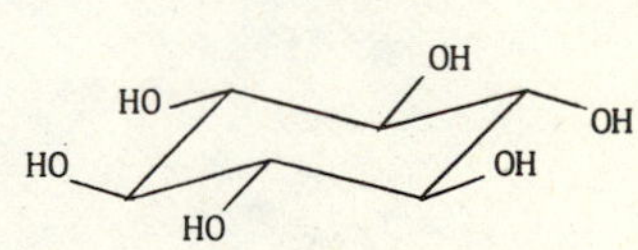

scyllo-Inositol
meso

21-4

1) The six carbon atoms of the *n*-hexane were derived from the six carbons of glucose. The cyclic hemiacetal structure readily opens under the reaction conditions. (The original reaction sequence involved conversion of glucose to its hexaiodo derivative, then reduction.)

2) Typical reactions of aldehydes and ketones.

3) Mild oxidation to a carboxylic acid is typical of aldehydes.

4) Reduction is consistent with an aldehyde and acetylation confirms six (5 + 1) hydroxy groups after reduction of the carbonyl group.

21-5

1) The hemiacetal exists in equilibrium with the aldehyde form. Typical aldehyde reactions require conversion of the hemiacetal to the aldehyde, a process which is slow in many cases. Bisulfite addition does not occur.

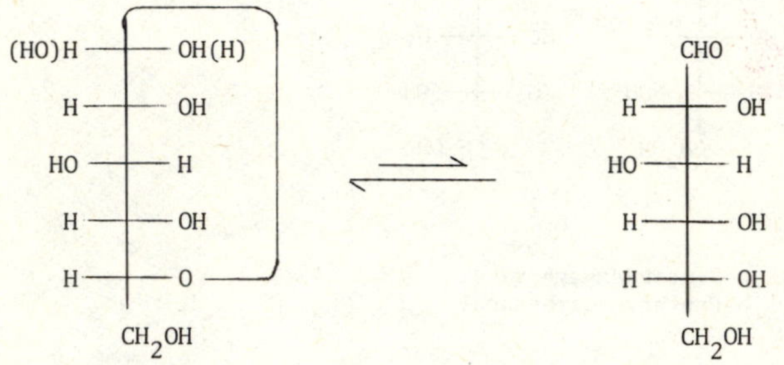

2) The pentacetates are epimers and differ in configuration at the anomeric carbon atom.

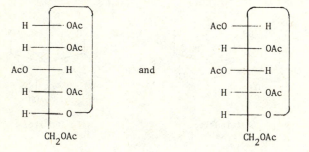

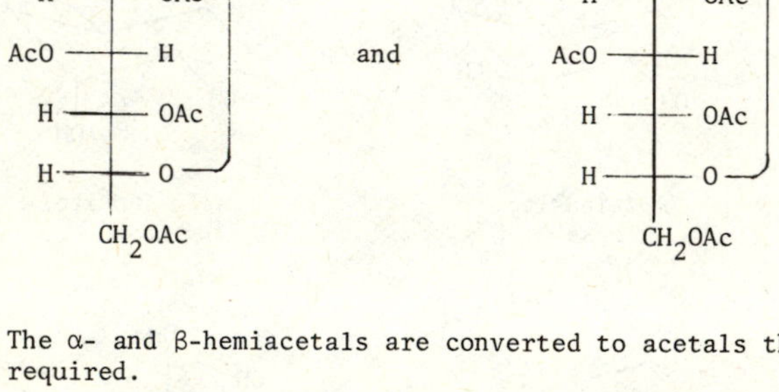

3) The α- and β-hemiacetals are converted to acetals thus only one mol of alcohol is required.

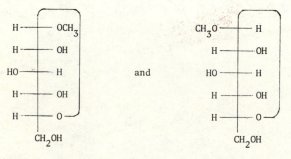

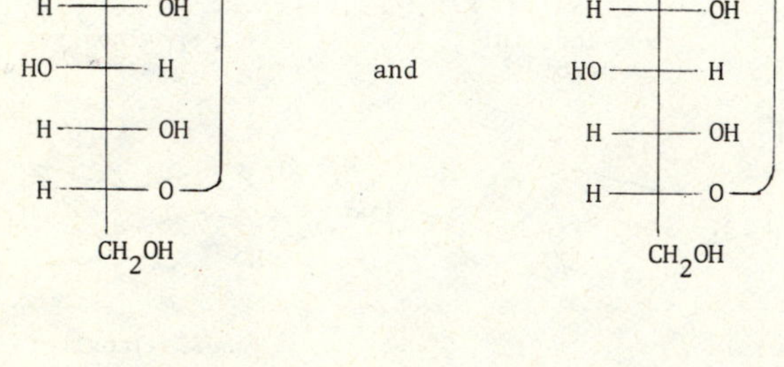

Contd...

21-5 contd...

4) These are the two hemiacetals.

5) Both hemiacetals interconvert through the open chain structure. The same position of equilibrium is reached starting from either anomer.

21-6

a)

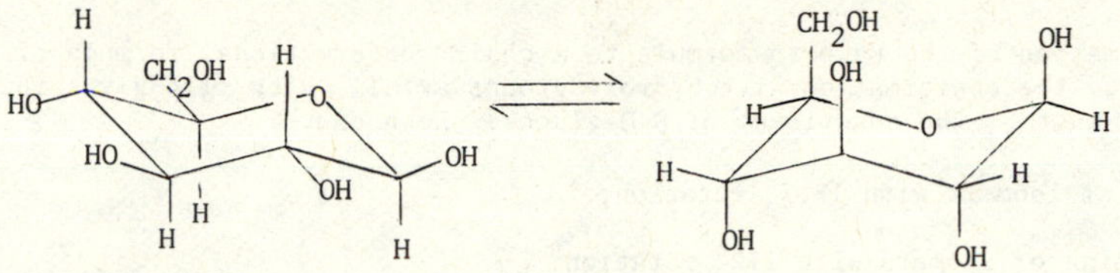

A B

b) Conformation A is expected to represent glucose because all large groups are equatorial.

21-7

a) β-D-Glucose

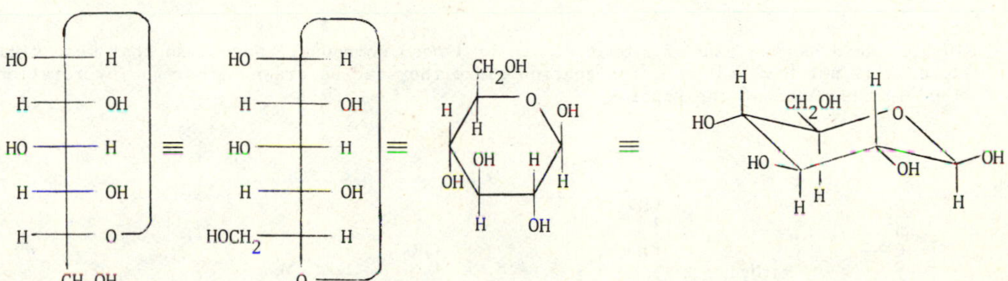

Fischer Haworth Chair

β-L-Glucose

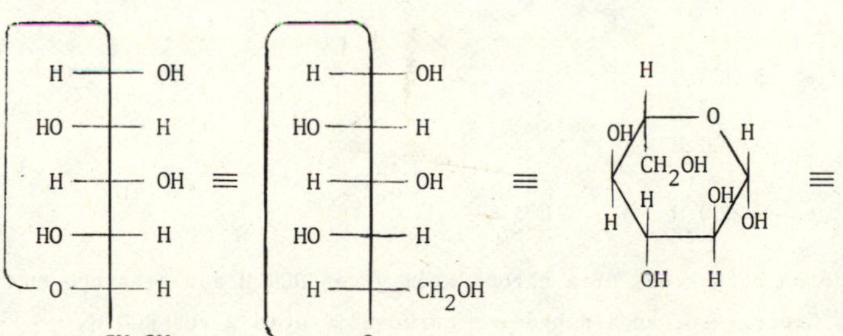

Fischer Haworth

Contd...

21-7 contd...
a) contd...

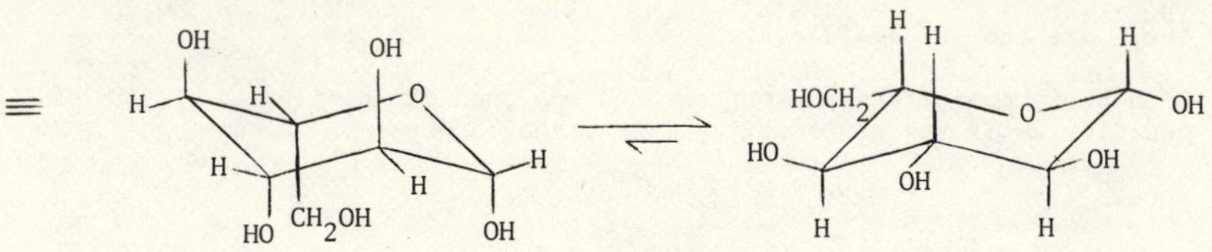

Chair

Note that transformation of the Haworth formula to a chair conformational formula of β-L-glucose leads to the conformation with hydroxy groups axial. Ring flip gives the expected conformation which must be the enantiomer of β-D-glucose shown above.

b) Let X = fraction of isomers with 18.7° rotation;

then 1 - X = fraction of isomers with 112° rotation.

$$18.7°X + 112°(1 - X) = 52.7°$$

$$X = 0.636$$

$$1 - X = 0.364$$

The more abundant isomer (63.6%) is the β-anomer with all large groups equatorial in the chair form. The α-isomer (36.4%) has the anomeric hydroxy group axial.

21-8 Glutaric acid has a plane of symmetry. It is a meso compound. Note also that meso aldaric acids will not have a D or L designation since they can be drawn either way by rotation of 180° in the plane of the paper.

21-9

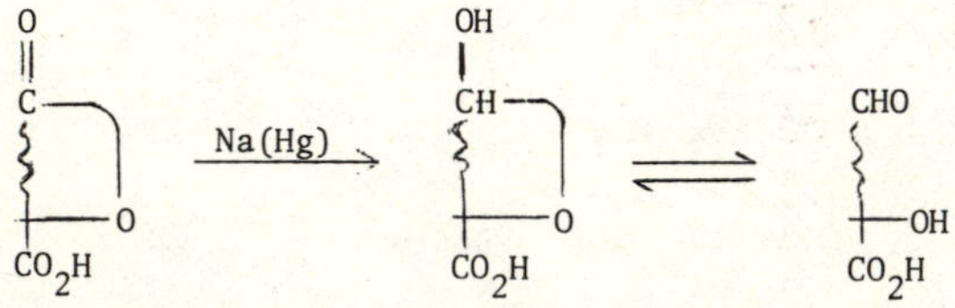

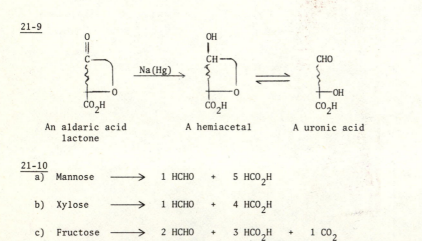

An aldaric acid lactone　　　A hemiacetal　　　A uronic acid

21-10
a) Mannose ⟶ 1 HCHO + 5 HCO_2H

b) Xylose ⟶ 1 HCHO + 4 HCO_2H

c) Fructose ⟶ 2 HCHO + 3 HCO_2H + 1 CO_2

Note that oxidative cleavage on both sides of a carbon atom gives HCO_2H and cleavage on only one side gives R–CHO. Cleavage of an aldehyde or carboxylic acid gives RCO_2H. Cleavage on both sides of the ketone carbonyl of fructose gives CO_2.

21-11 Products differ in configuration at the original anomeric carbon atom.

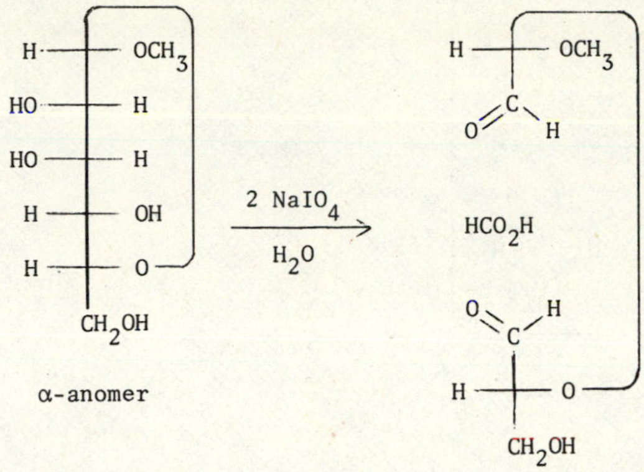

α-anomer

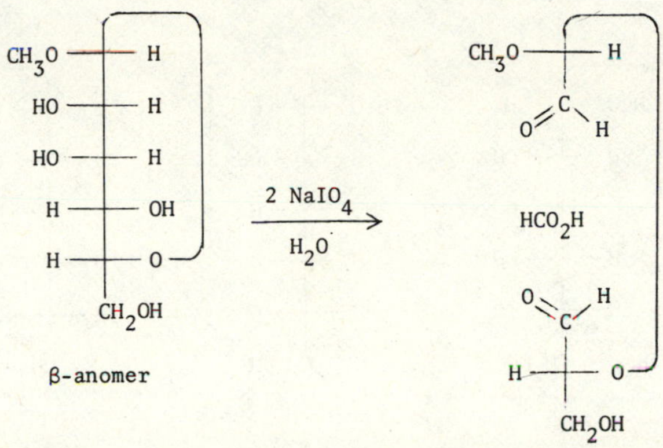

β-anomer

21-12

a)

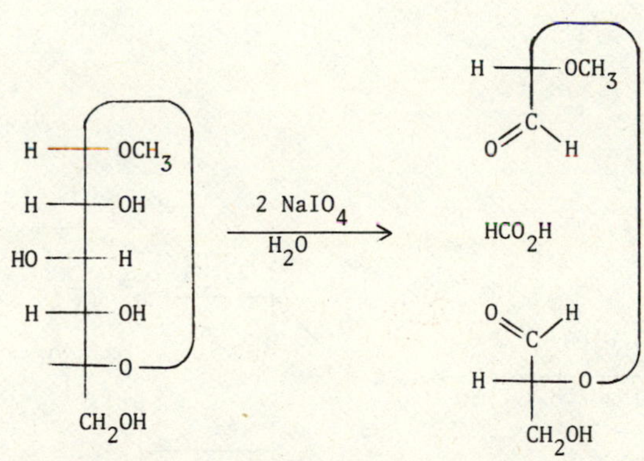

α-anomer

Contd...

21-12 contd...

a) contd...

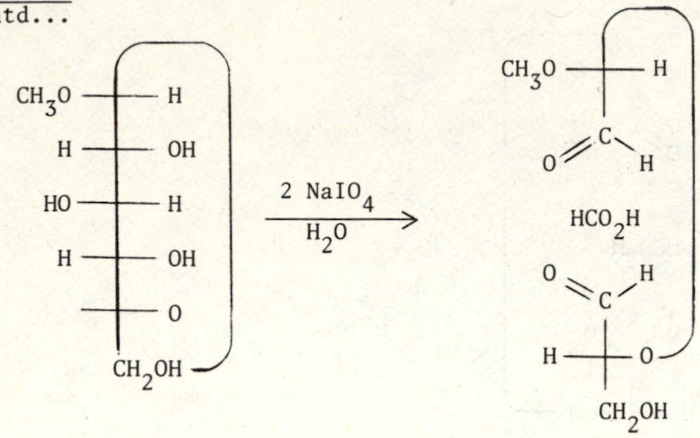

b) Glucose and mannose give the same product since the point of difference, carbon #2, loses its stereochemical identity in the cleavage.

21-13

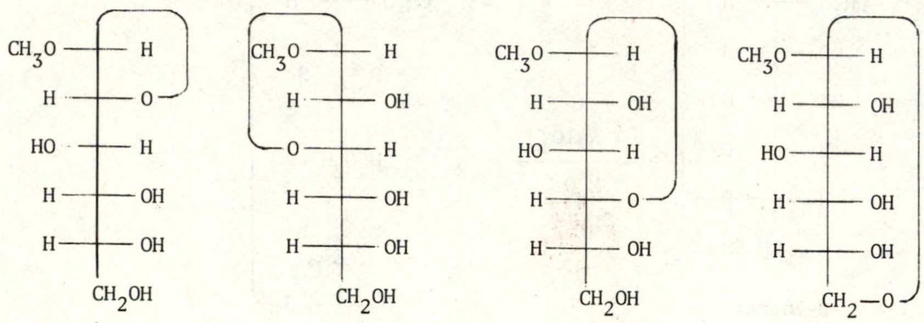

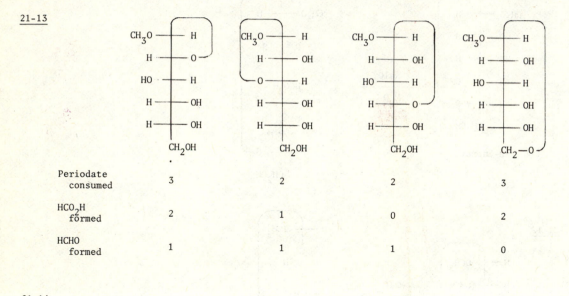

Periodate consumed	3	2	2	3
HCO_2H formed	2	1	0	2
HCHO formed	1	1	1	0

21-14

a)

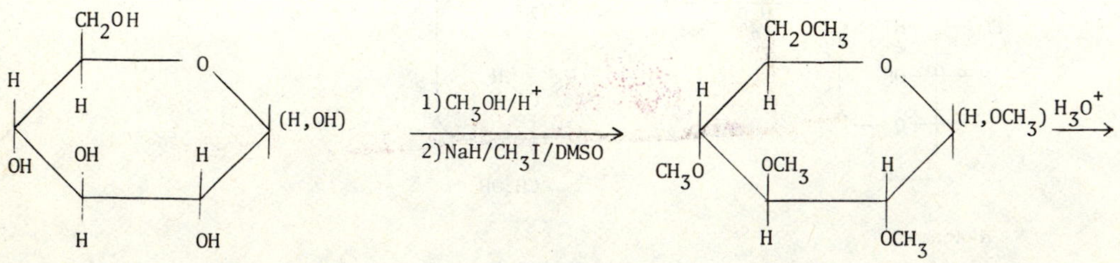

Contd...

21-14 contd...

a) contd...

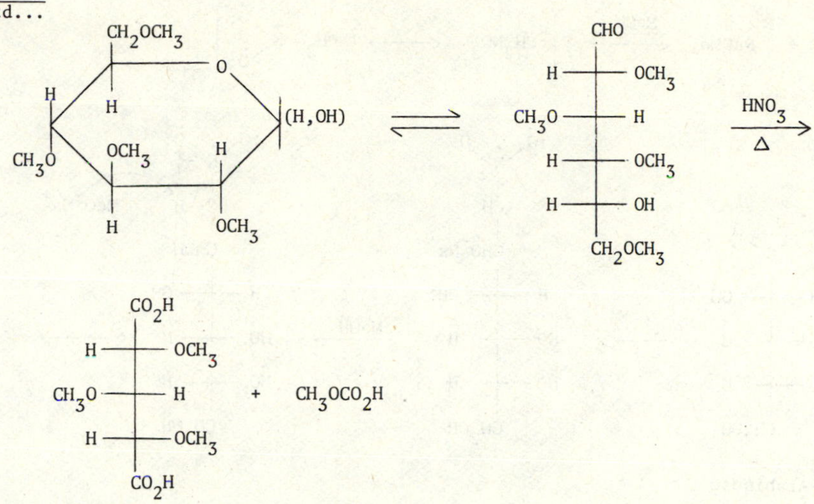

b)

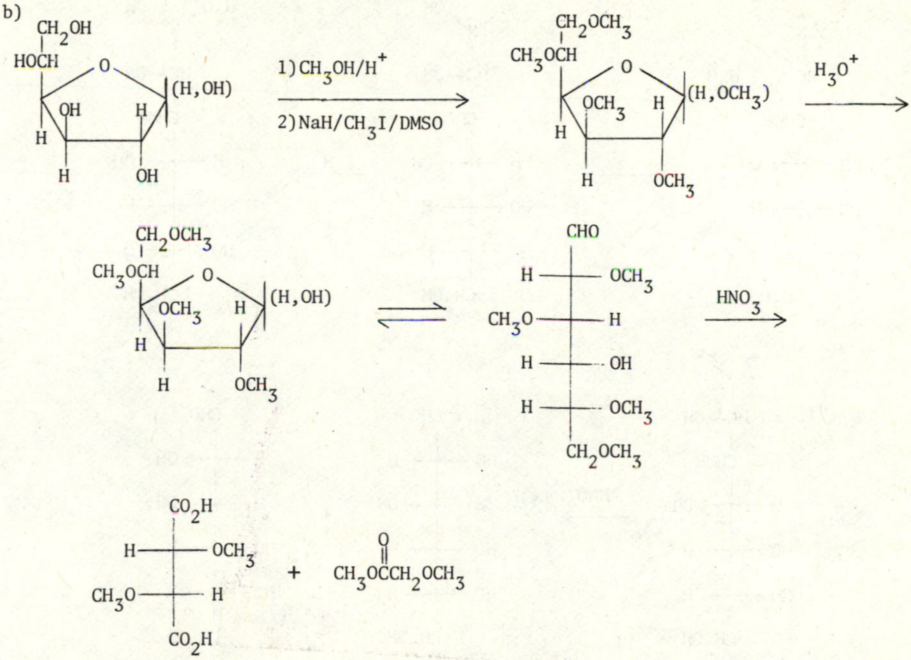

(Largest product fragment)

21-15 The products are diastereomers, thus differ in physical properties and in the thermodynamics for their formation. In this example one epimer has greater nonbonded repulsion between the C-2 and C-4 hydroxy groups than does the other.

21-16

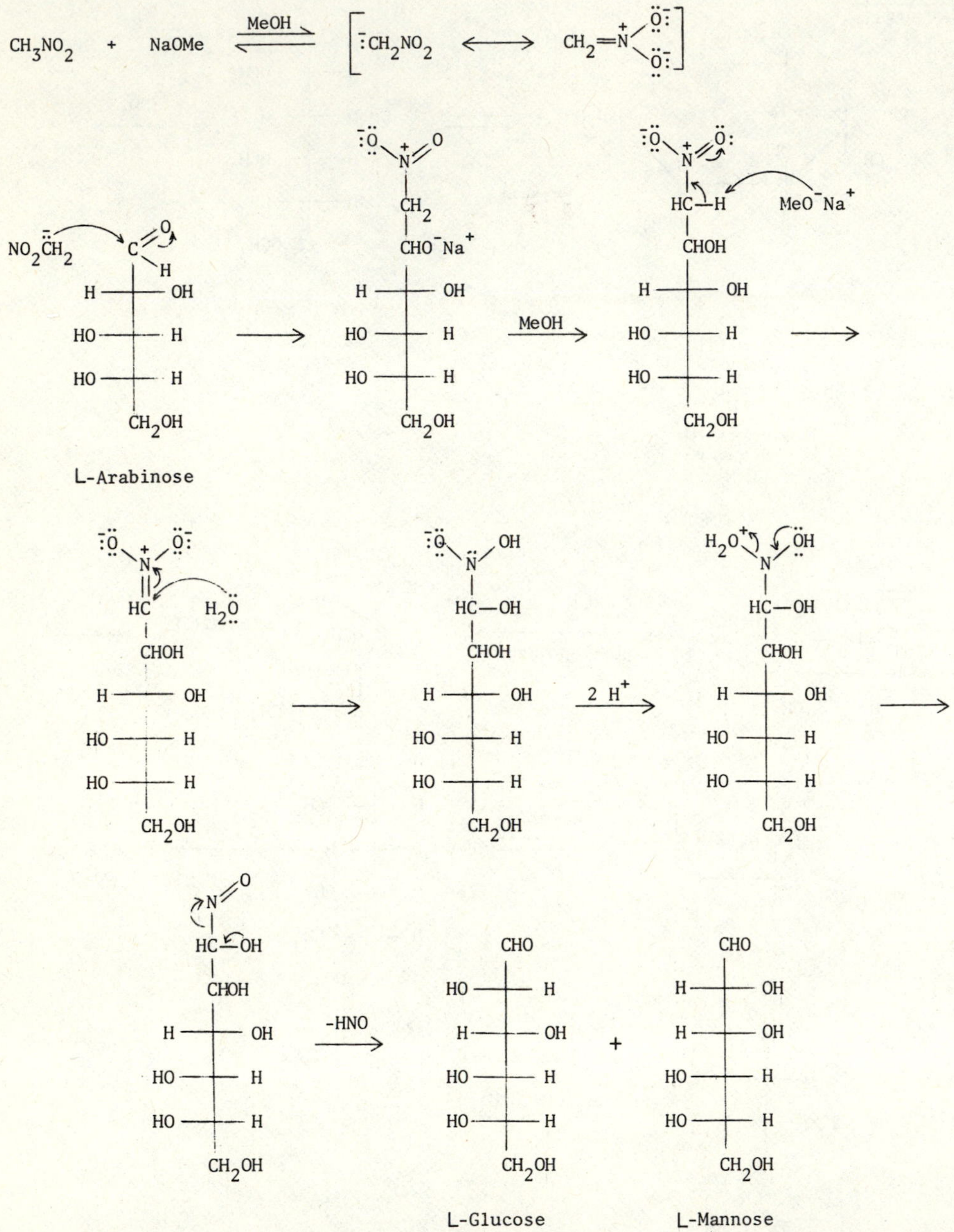

21-17 Allose and galactose give meso aldaric acids, thus are eliminated as possibilities. Since glucose and mannose are known to differ only at carbon 2, altrose and talose, the sugars with identical configurations to allose and galactose at carbon atoms 3,4, and 5, must also be rejected as possibilities.

21-18 Both aldaric acids are meso.

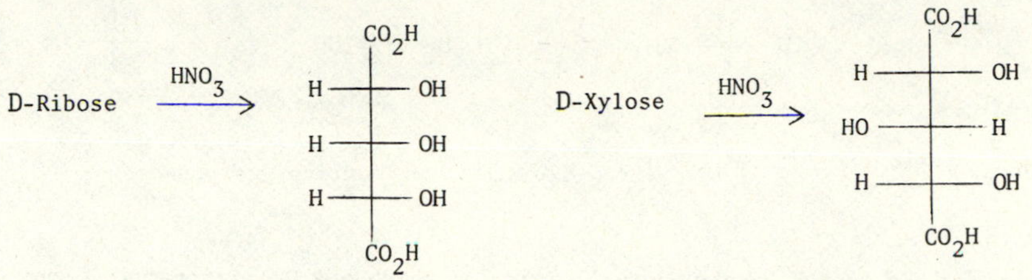

21-19

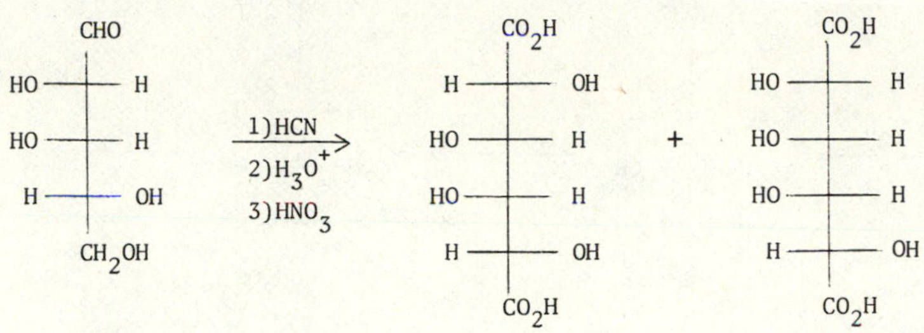

D-Lyxose Meso Optically active

The D-Lyxose structure would not lead to two optically active C_6 aldaric acids.

21-20

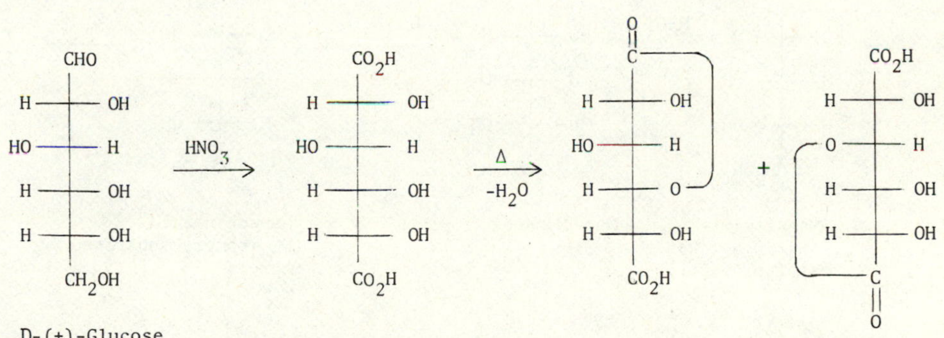

D-(+)-Glucose

Different γ-lactones

Contd...

21-20 Contd...

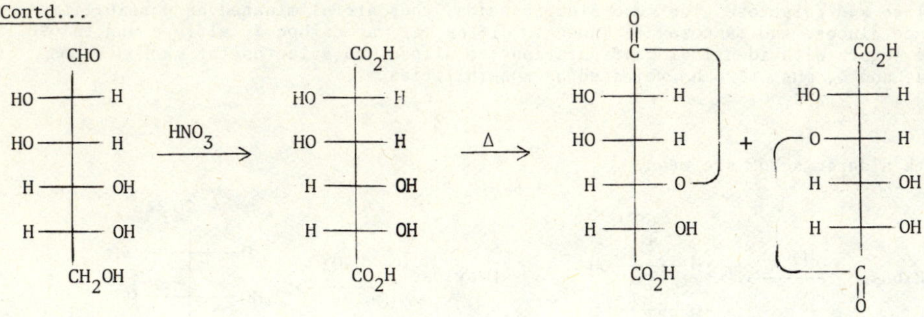

D-(+)-Mannose

Identical compounds

21-21 Only D isomer shown though Fischer actually carried through a DL mixture until the mannonic acid was resolved. The reagents used by Fischer are shown.

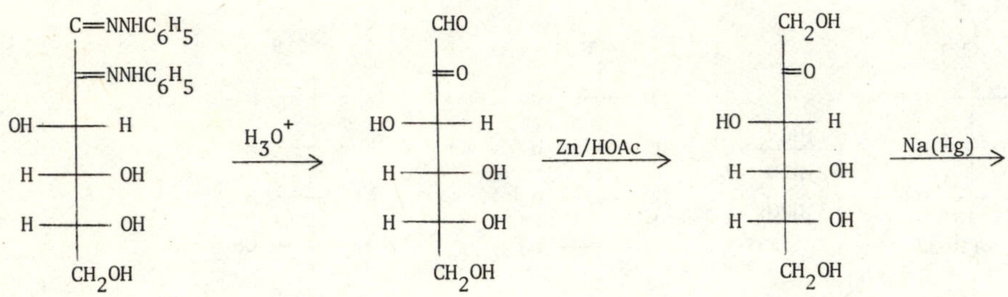

Osazone

Fructose

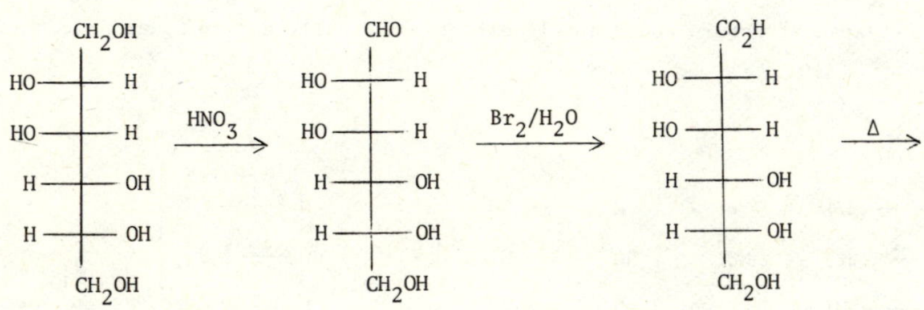

Mannitol (This epimer was major product.)

Mannose

Mannonic acid (DL isomers separated here)

Contd...

21-21 Contd...

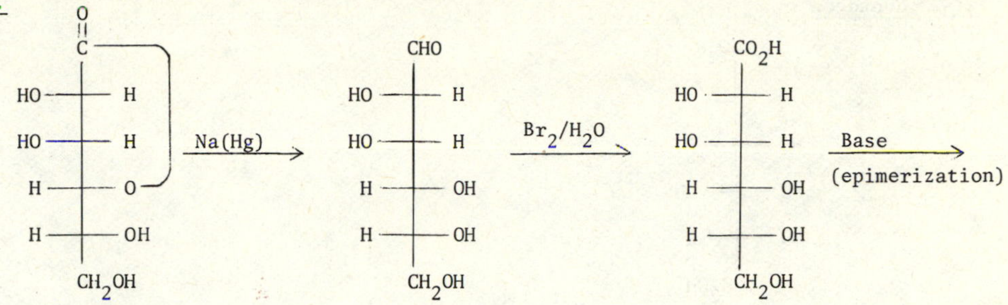

γ-D-Lactone D-Mannose D-Mannonic acid

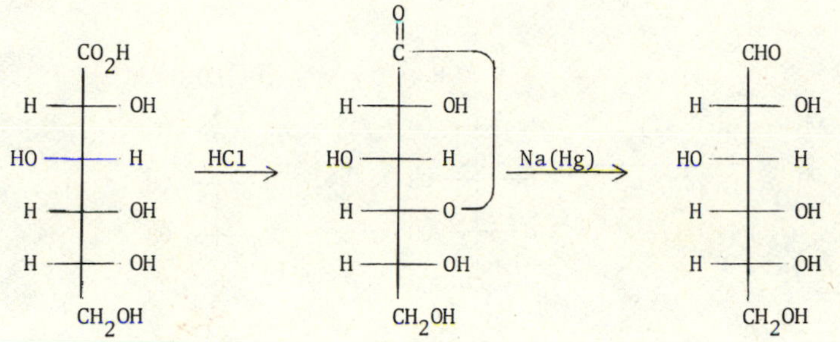

D-Gluconic acid γ-D-Lactone D-Glucose

21-22

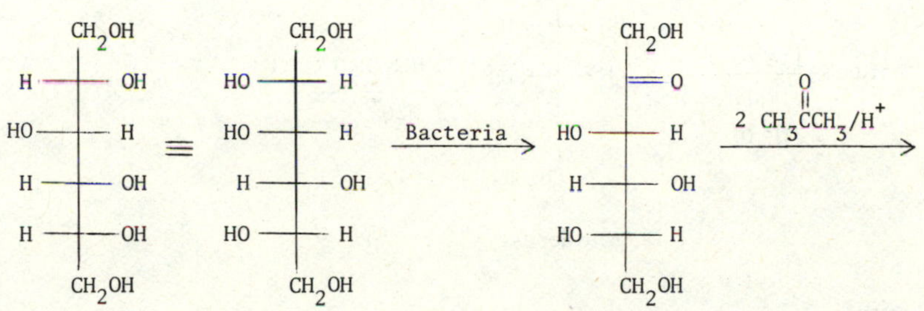

D-Glucitol L-Sorbose

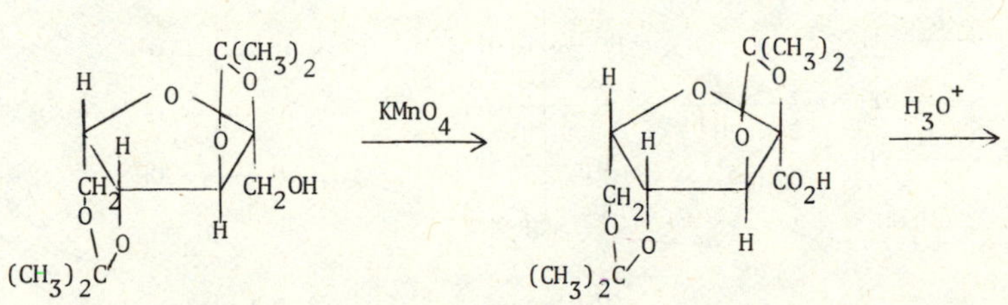

Contd...

21-22 Contd...

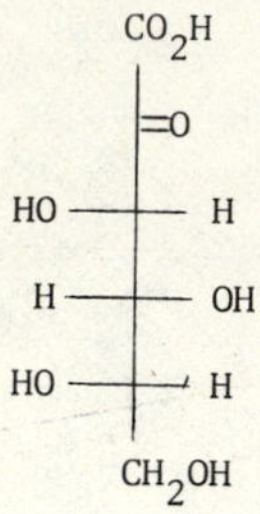

2-Keto-L-Gulonic acid

21-23

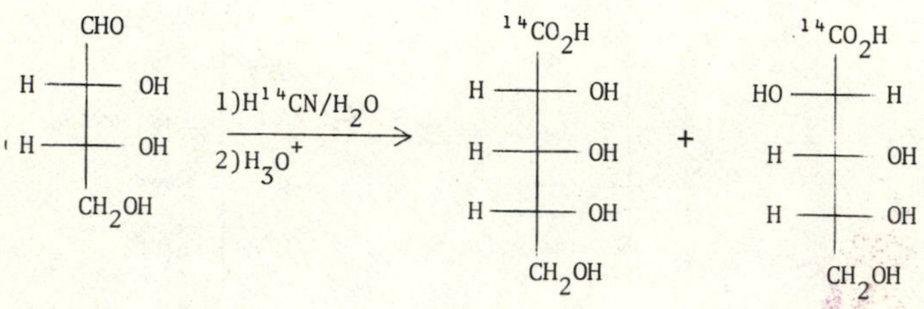

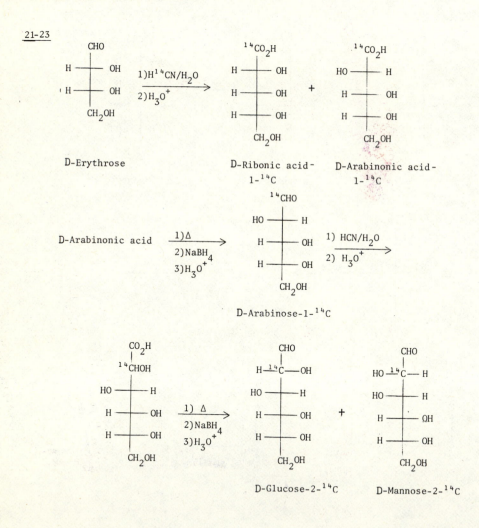

D-Erythrose D-Ribonic acid-1-^{14}C D-Arabinonic acid-1-^{14}C

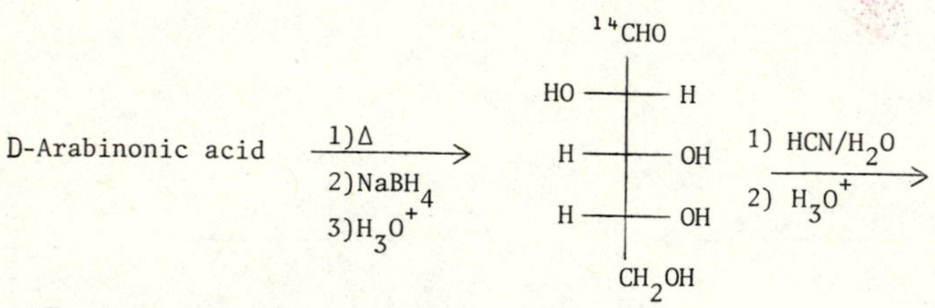

D-Arabinose-1-^{14}C

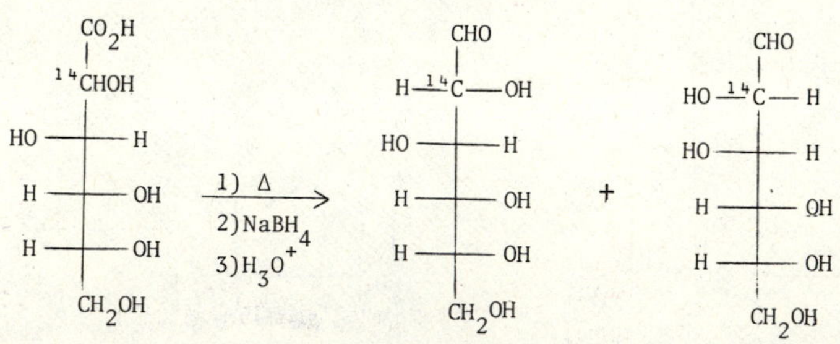

D-Glucose-2-^{14}C D-Mannose-2-^{14}C

21-24

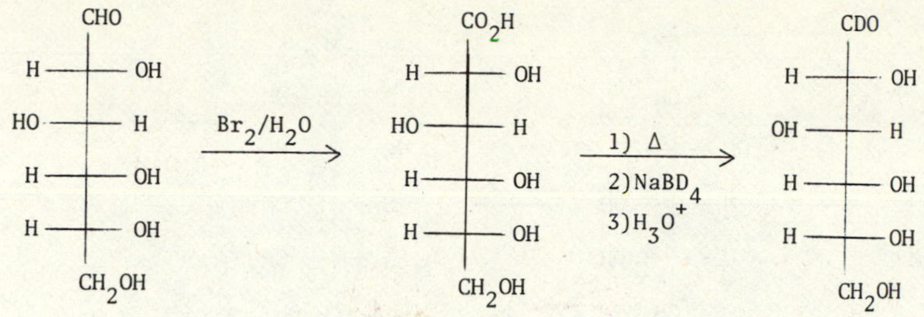

21-25

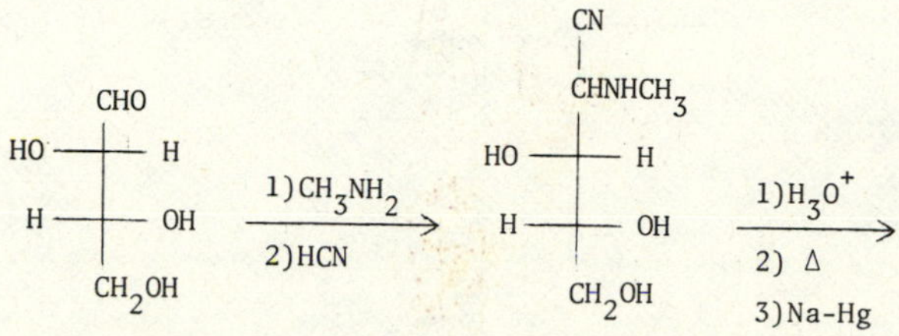

D-Threose

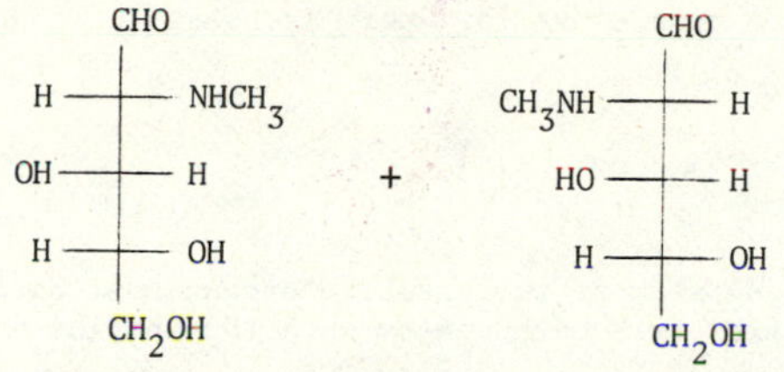

N-Methyl-D-xylosamine *N*-Methyl-D-lyxosamine

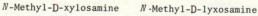

Separate epimers

21-26
a)

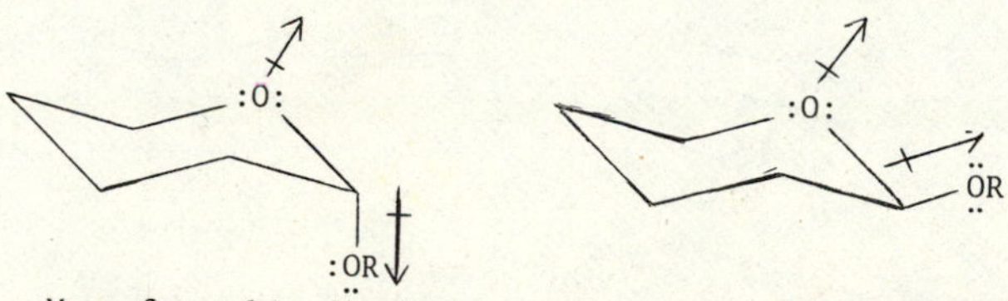

More favorable dipole interactions

b) More favorable solvation of the less hindered equatorial methoxy group stabilizes the β-anomer.

21-27

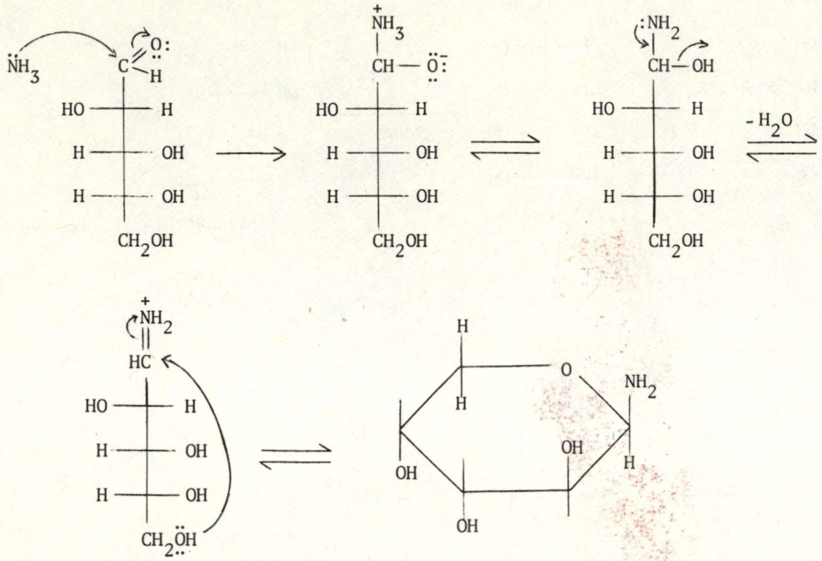

21-28 Interaction between the cyclic oxygen atom and C—Cl bond favors an axial chlorine atom so as to move negative ends of the dipoles as far apart as possible (the anomeric effect; prob. 21-26).

21-29

a) The result demonstrates that the oxygen atom at the anomeric carbon must have come from the water, thus the aglycone departs from this carbon atom with the initial anomeric oxygen.

b) The relatively stable *tert*-butyl cation departs and produces t-$Bu^{18}OH$. Both of the results above are similar to those observed during ester hydrolysis.

21-30

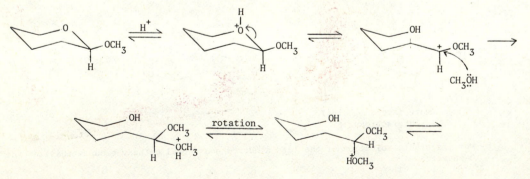

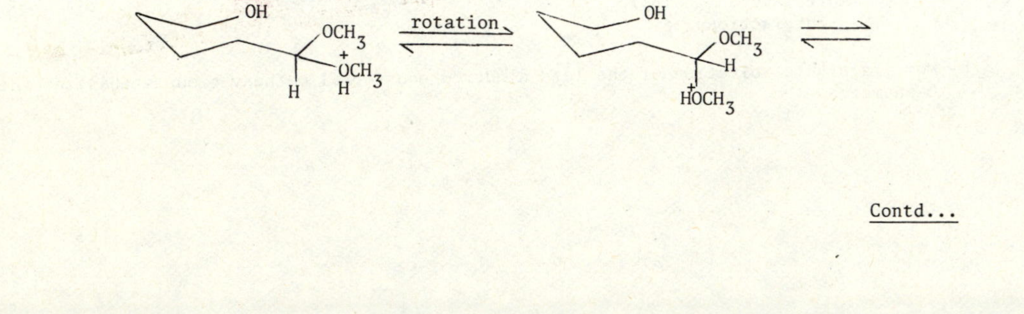

Contd...

21-30 contd...

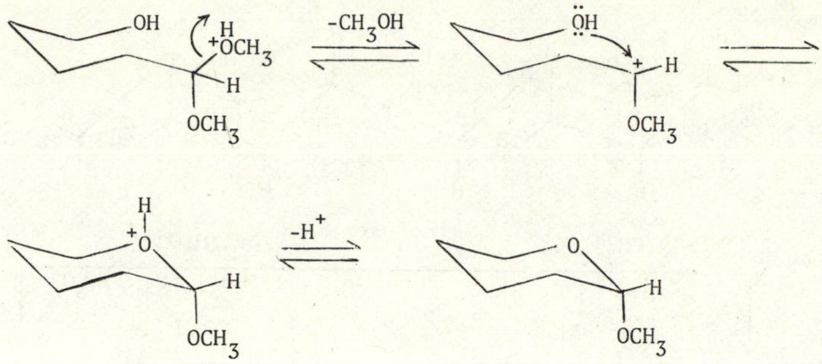

21-31

a) Invert sugar is an equal mixture of D-fructose and D-glucose. It will have an optical rotation of

$$\frac{-92.4° + 52.7°}{2} = -19.9°$$

b) The absence of mutarotation indicates that sucrose does not easily form a mixture of anomers. It is often easier to obtain crystals from a solution of one component rather than from a mixture of closely related isomers because identical molecules can form a more ordered crystal.

21-32

a) D-Glucose and D-galactose, the components of lactose, are connected by a β-glycoside linkage.

b) Either glucose or galactose must have a hemiacetal group in the lactose structure.

c) The tetramethylgalactose fragment shows that galactose is present as a pyranose ring. Furthermore, the only remaining hydroxy group is at the anomeric carbon atom. Since this is the only hydroxy available to form a glycoside linkage to the glucose, galactose cannot be the reducing portion of lactose. The trimethylglucose fragment does not differentiate between a furanose and pyranose structure. Since glucose must be the reducing portion of lactose the anomeric hydroxy must be free. Thus the hydroxy groups at 4 and 5 will form the cyclic structure and the glycoside linkage.

d) The tetramethylgluconic acid fragment shows that the 4-hydroxy group of glucose forms the glycoside bond to galactose. The 5-hydroxy group must therefore be involved in a pyranose ring. The chemistry of this structural elucidation is:

Contd...

21-32 contd...

d) contd...

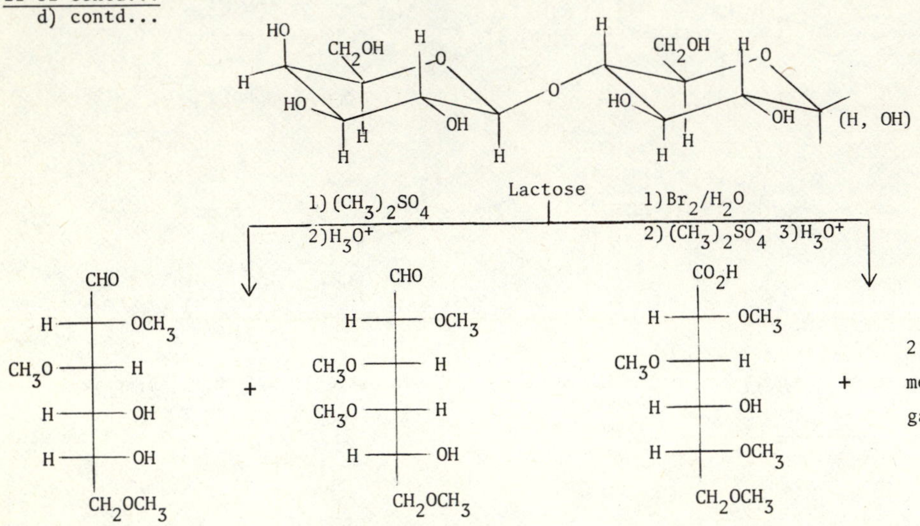

2,3,4,6-Tetra-*O*-methyl-D-galactose

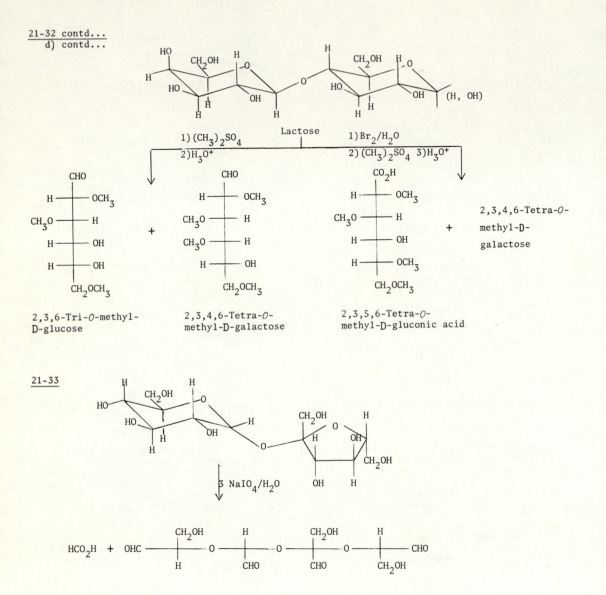

2,3,6-Tri-*O*-methyl-D-glucose + 2,3,4,6-Tetra-*O*-methyl-D-galactose

2,3,5,6-Tetra-*O*-methyl-D-gluconic acid

21-33

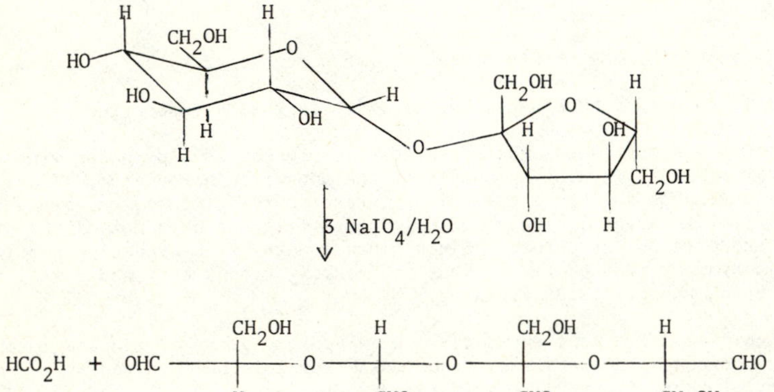

21-34 Maltose andcellobiose are reducing sugars. α,α-Trehalose and raffinose do not have hemiacetal or hemiketal groups, therefore are nonreducing.

21-35

a) The 2,3,6-trimethyl derivative is characteristic of the 1,4-linked polyglucose.

The 2,3-dimethyl derivative is derived from glucose fragments 1,4-linked in a straight chain but also containing a branch at carbon 6.

The 2,3,4,6-tetramethyl derivatives are from the ends of the polysaccharide chains.

b) Amylose has no branches thus would give principally the 2,3,6-trimethyl derivative along with a small quantity of the 2,3,4,6-tetramethyl derivative from the terminal groups.

21-36 The tight linear helix of amylose favors intramolecular hydrogen bonding and inhibits solvation. The more open branched structure of amylopectin more readily interacts with solvent.

21-37 Pulping is carried out in strongly basic media. The glucoside (acetal) bonds are relatively stable toward base.

21-38

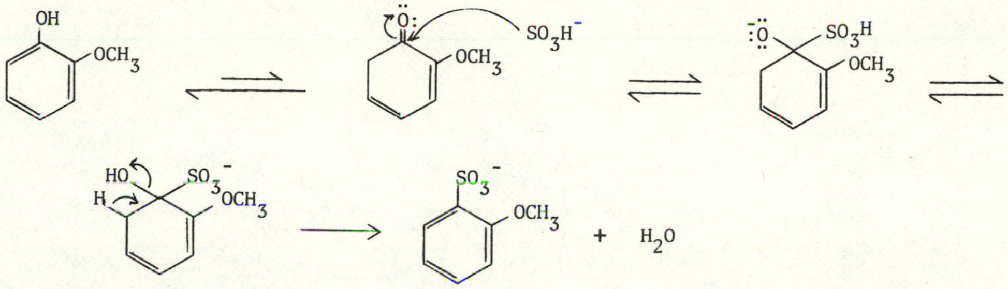

21-39

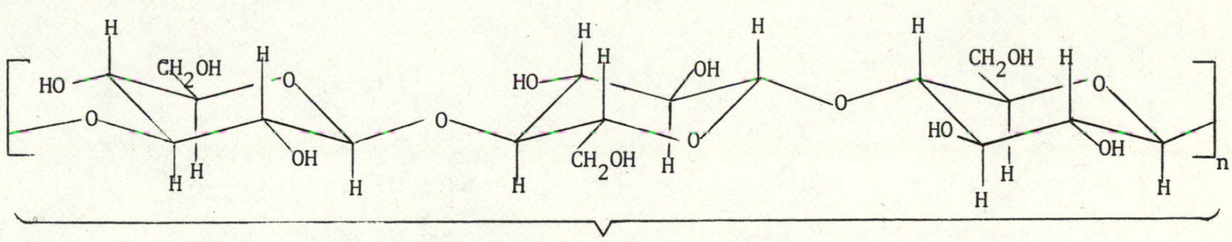

Cellotriose

21-40 They exist as mono- or dianions (or an equilibrium mixture of the two) since the pK_a values of phosphoric acid are:

$$pK_{a1} = 2.12 \; ; \; pK_{a2} = 7.21.$$

21-41

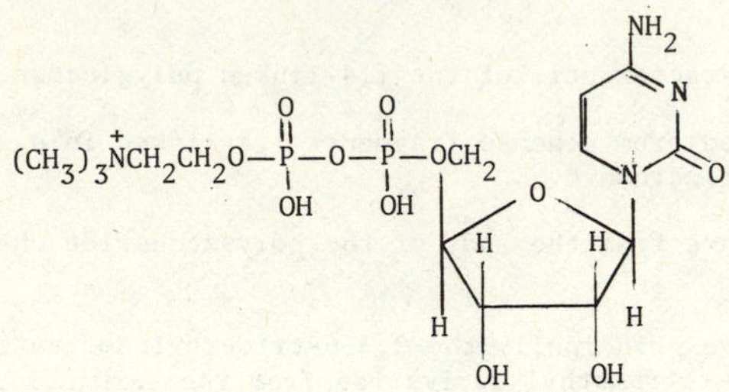

21-42 Nucleosides and nucleotides refer to the purine and pyrimidine glycosides derived from nucleic acids. Riboflavin does not fit that definition nor does it contain the ribose carbohydrate units characteristic of nucleosides and nucleotides.

21-43

a)

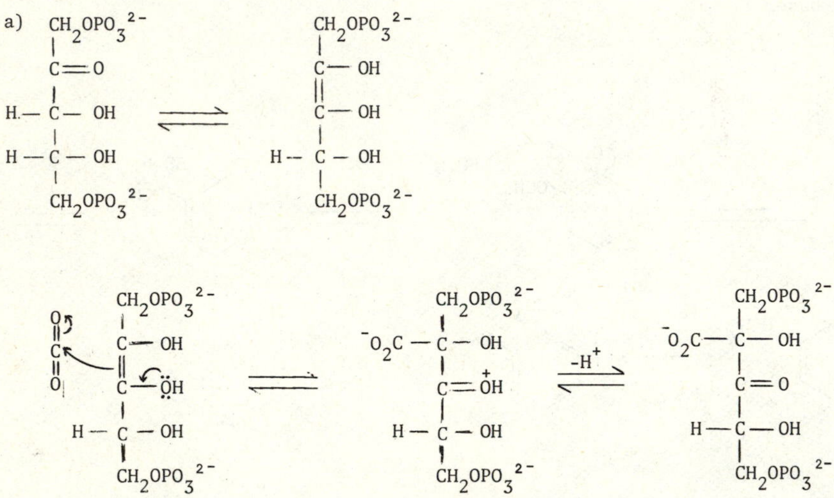

b)

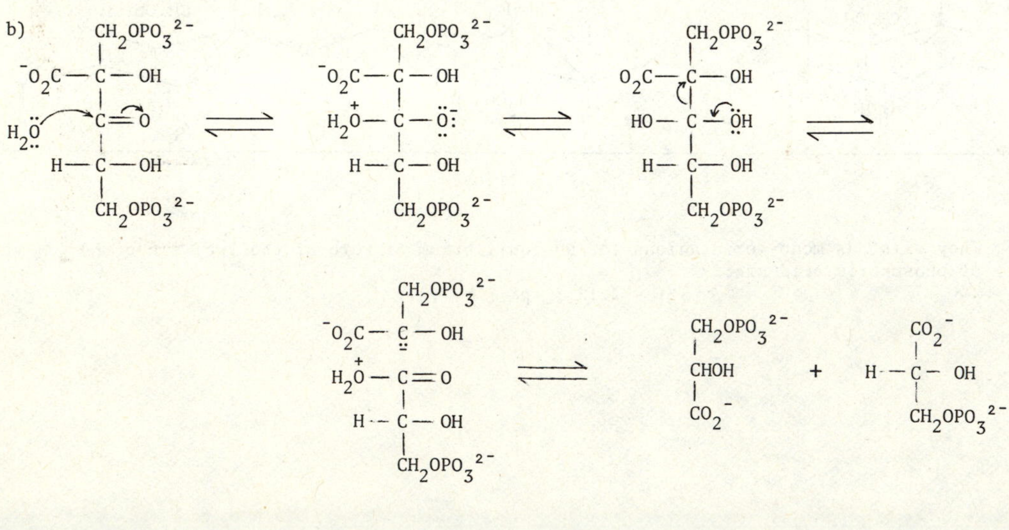

21-44 In one experiment a plant was fed $C^{18}O_2$ and in the other $H_2{}^{18}O$. In the first experiment no ^{18}O was found in the oxygen formed while the second experiment yielded labeled oxygen.

21-45 The 10 reaction of the Krebs cycle are summarized below:

1) The thioester acetyl-CoA enolizes and adds to the α-ketoacid in an aldol reaction. Subsequent hydrolysis of the thioester leads to a more stable carboxylate anion.

2) Elimination of water to give the conjugated double bond.

3) Readdition of water in a conjugate addition to produce a different isomer.

4) Dehydrogenation (oxidation) of the alcohol serves to reduce NAD^+ to NADH.

5) Decarboxylation of a β-ketoacid releases the first molecule of carbon dioxide.

6) An oxidative decarboxylation releases the second carbon dioxide molecule. This is a complex step in which NAD^+ is also reduced to NADH. A CoA thioester forms.

7) The hydrolysis of the thioester, like that of an acid chloride, is exothermic and here is coupled with phosphorylation of a nucleotide phosphate from the di- to the higher-energy triphosphate. The nucleotides here are guanosine phosphates (GDP ⟶ GTP) which serve the same function as the more common adenosine phosphates, ADP and ATP.

8) Dehydrogenation is accomplished by another reaction which reduces NAD^+ to NADH as in step (4).

9) Conjugate addition of water parallels that of reaction (3).

10) In the final oxidation, NAD^+ dehydrogenates the secondary alcohol to the ketone of oxaloacetate [like step (4)], preparing it for condensation with a new thiol-activated acetate molecule in reaction (1).

21-46 In order to follow the convention that the aldehyde group is drawn at the top of the structural formula, some of the original D-aldohexoses become L-aldohexoses in these proposed transformations.

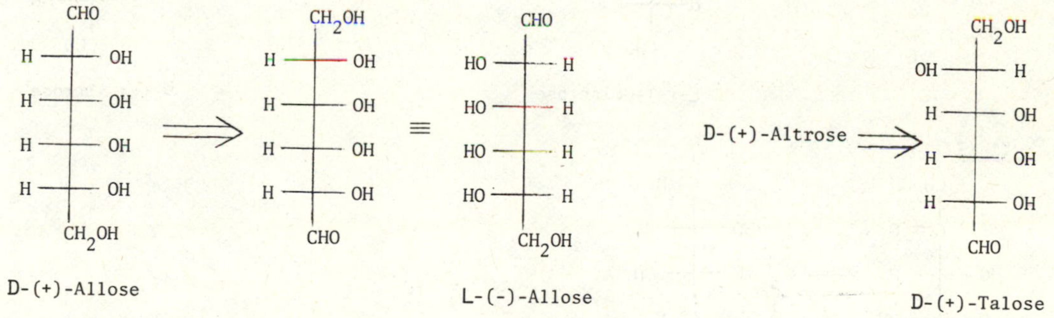

Contd...

21-46 contd...

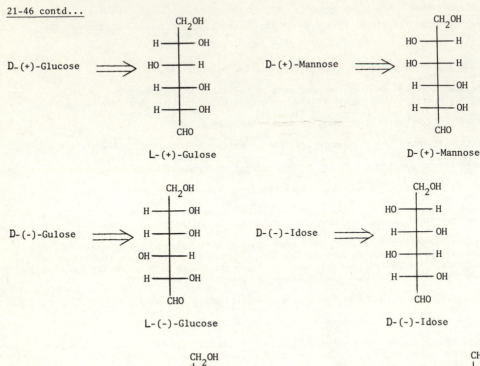

D-(+)-Glucose 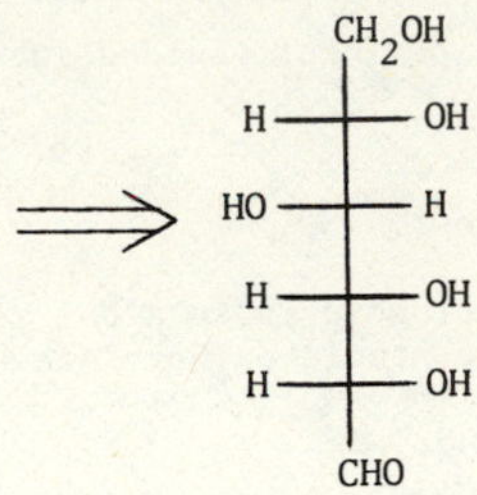

L-(+)-Gulose

D-(+)-Mannose 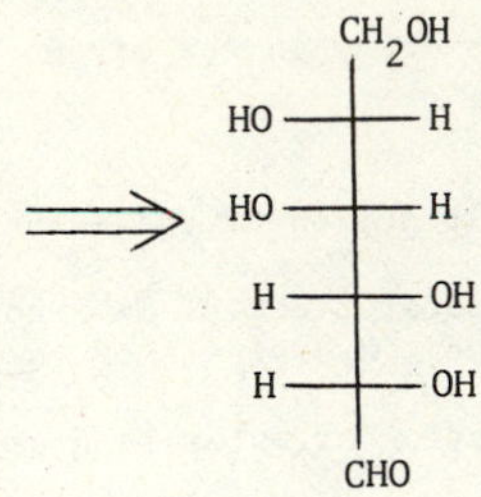

D-(+)-Mannose

D-(-)-Gulose 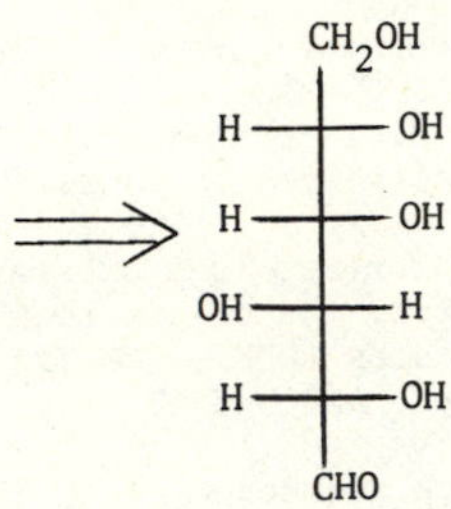

L-(-)-Glucose

D-(-)-Idose 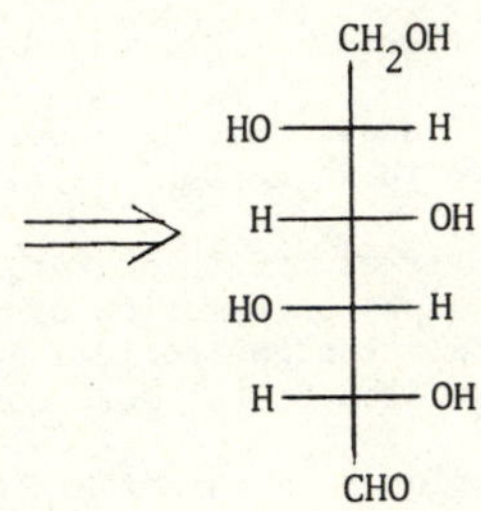

D-(-)-Idose

D-(+)-Galactose 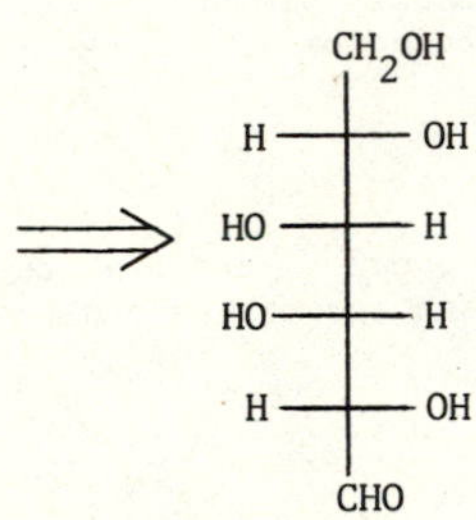

L-(-)-Galactose

D-(+)-Talose 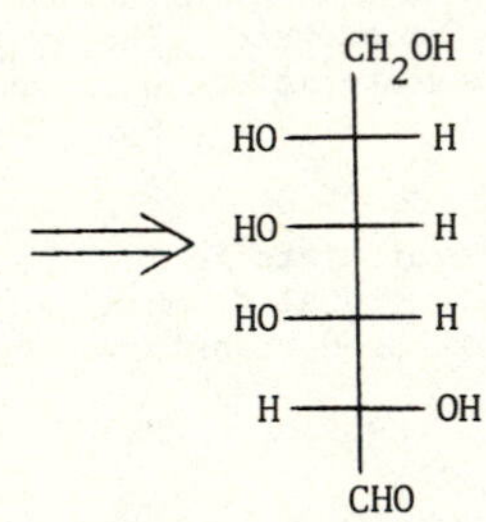

D-(+)-Altrose

21-47

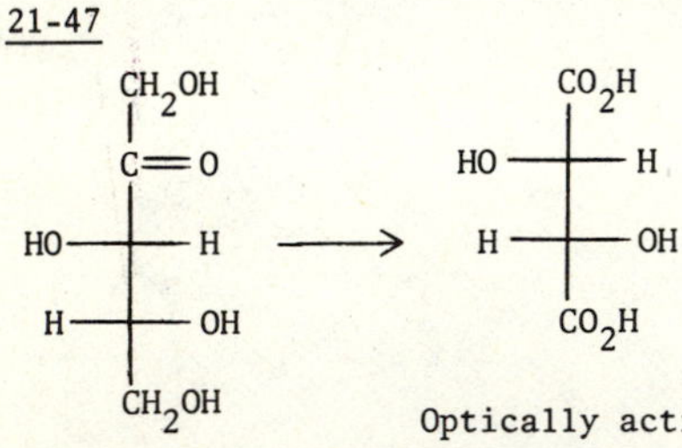

Optically active

D-Tartaric acid

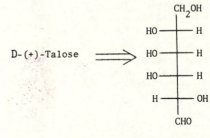

21-48

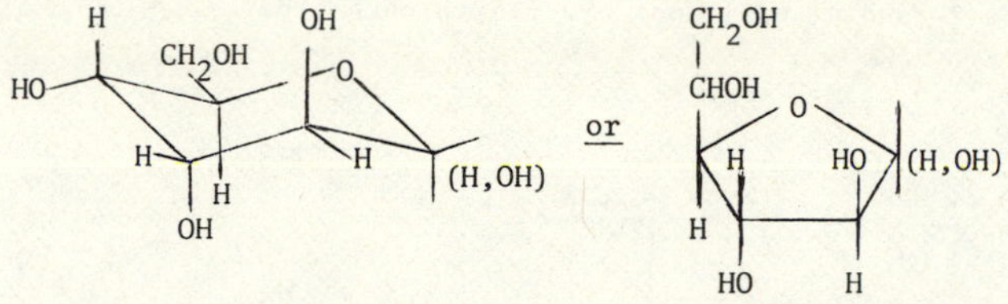

The furanose ring has the preferred all trans configuration. The pyranose ring has two unfavorable axial hydroxy groups.

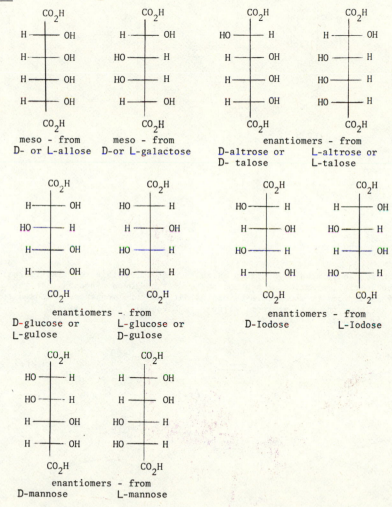

21-49

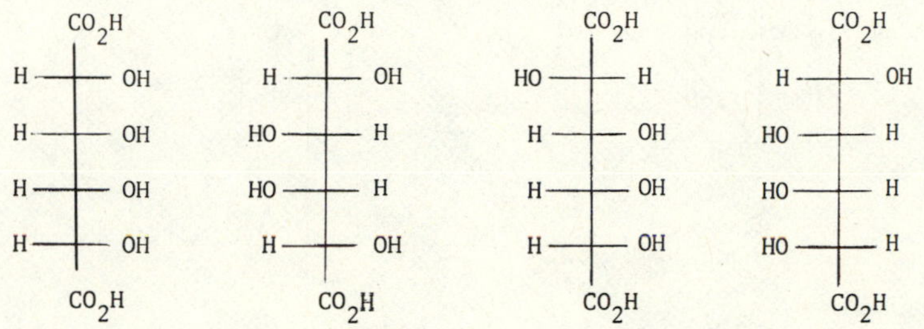

meso - from D- or L-allose; meso - from D-or L-galactose; enantiomers - from D-altrose or D- talose / L-altrose or L-talose

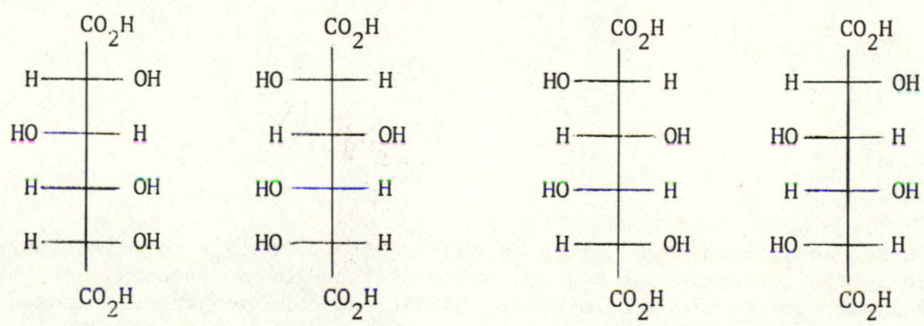

enantiomers - from D-glucose or L-gulose / L-glucose or D-gulose; enantiomers - from D-Iodose / L-Iodose

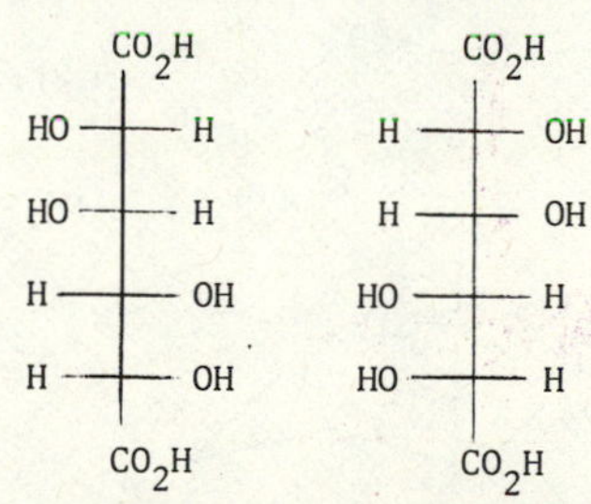

enantiomers - from D-mannose / L-mannose

21-50 Benzaldehyde normally forms 6-membered ring acetals with carbohydrates. In D-glucose only the hydroxy groups at the 2- and 4- positions are cis to each other.

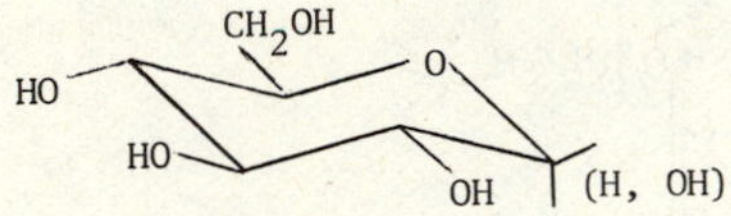

21-51

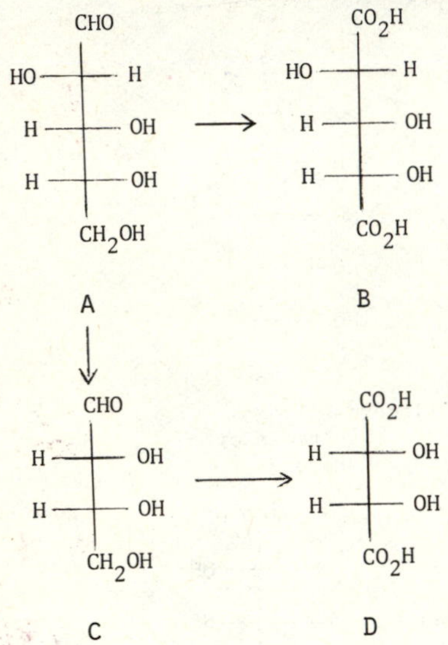

21-52 The α-anomer possesses an axial hydroxy group which is less accessible for reaction with bromine during the initial formation of a hypobromite. It has been proposed that the α-anomer actually isomerizes to the β form before oxidation. One mechanism proposed for the process also includes the anti stereoelectronic requirements favorable for elimination from the β-anomer intermediate.

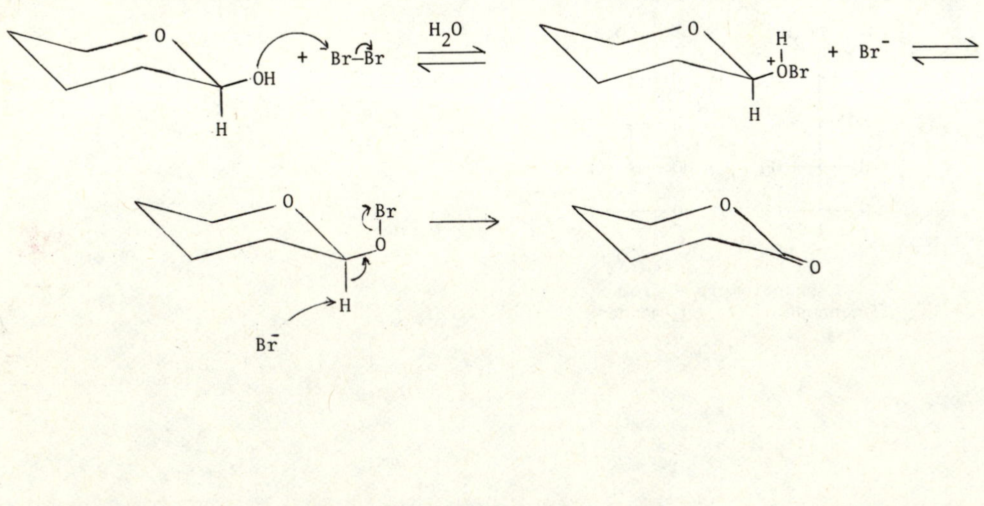

21-53 Formation of an acetone acetal (an isopropylidene derivative) requires that two hydroxy groups be relatively close together. Acetone typically gives a cyclic five-membered ring ketal when functioning as a protecting group.
The pyranose ring glucose can only form one 1,2-*o*-isopropylidene derivative. Normally the furanose is formed faster but rapidly interconverts to the more stable pyranose. As the very low concentration of the furanose is derivatized with two moles of acetone, equilibrium shifts to form more furanose, etc.

21-54

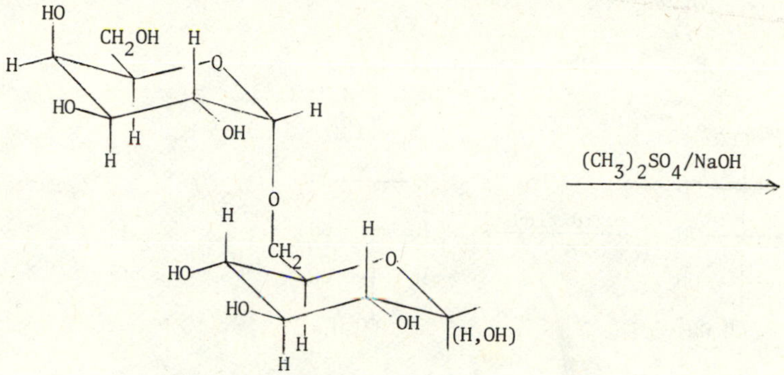

Melibiose (A reducing sugar must have at least one hemiacetal or hemiketal linkage.)

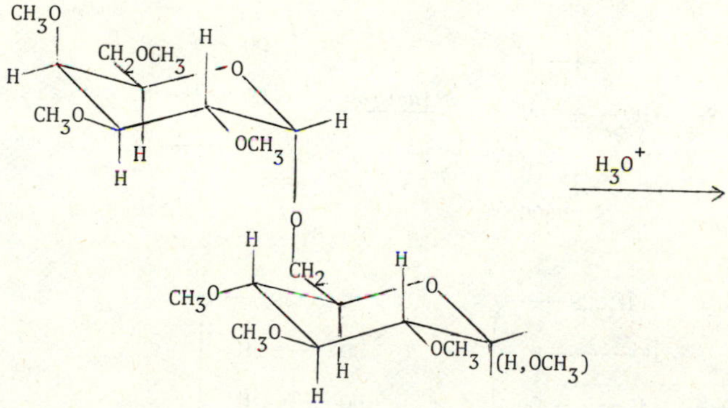

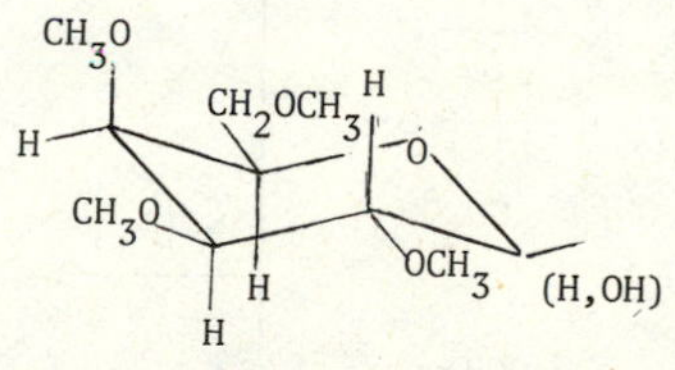

2,3,4,6-Tetramethyl-D-galactose

+

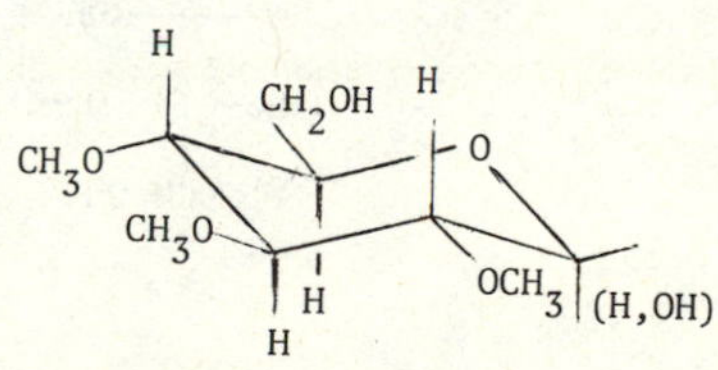

2,3,4-Trimethyl-D-glucose

Methylation labels the hydroxy groups of the disaccharide, then hydrolysis cleaves acetals to produce hydroxy groups at positions that were non-ether oxygen atom links. This defines ring size and the position of connection of the monosaccharides. One hydroxy must correspond to the hemiacetal that gives melibiose its reducing property. These data do not define configuration at the anomeric carbon atom or of the linkage between the monosaccharides.

21-55

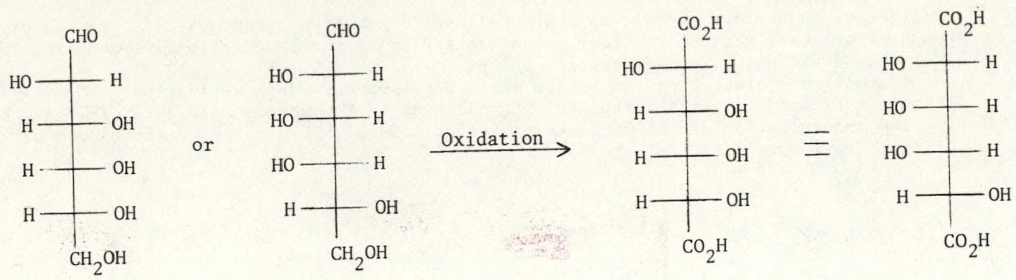

A B

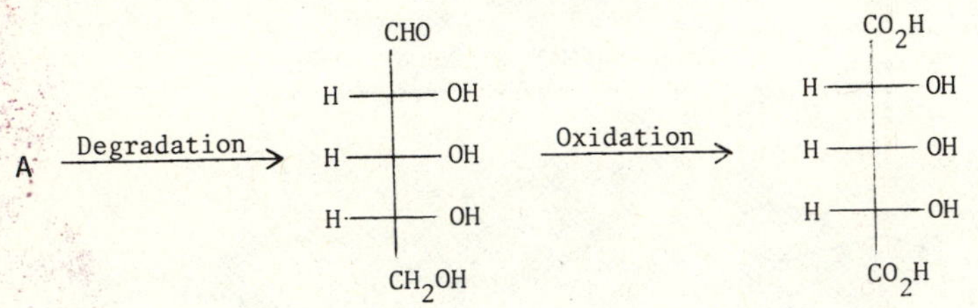

meso

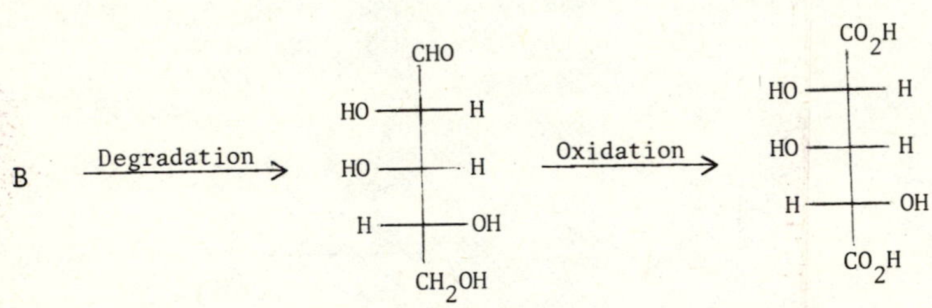

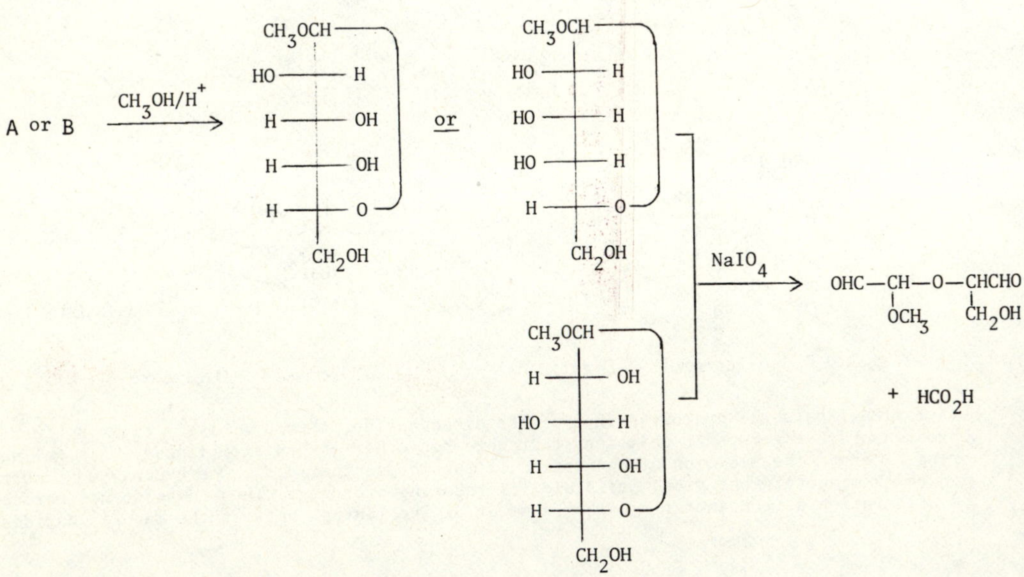

Methyl D-glucopyranoside

21-56

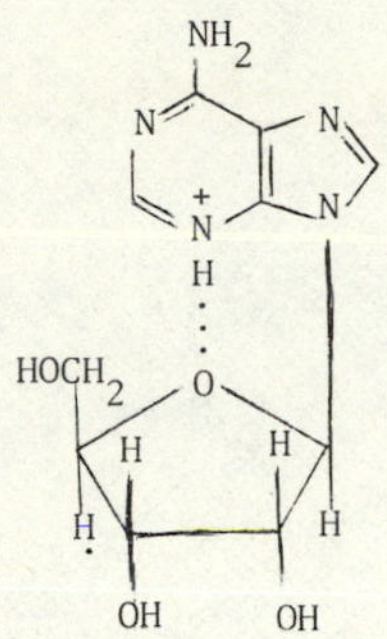

Protonation of the amino group is expected to be more rapid, but protonation of the heterocyclic nitrogen atom provides a more acidic intermediate which can transfer the proton in a more favorable conformation. No similar intramolecular interaction is possible with the pyrimidine base of cytidine.

21-57

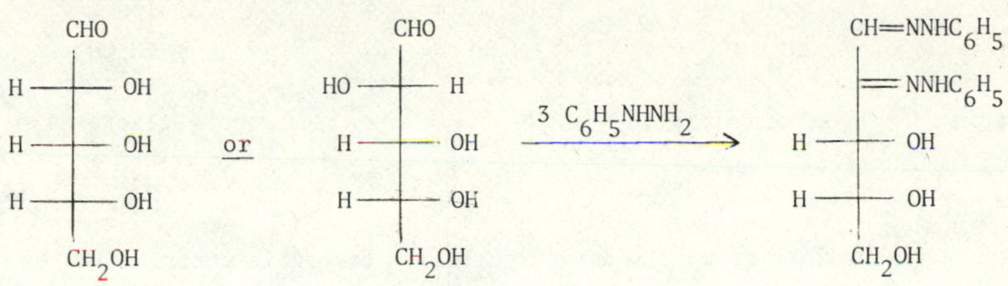

D-Ribose D-Arabinose

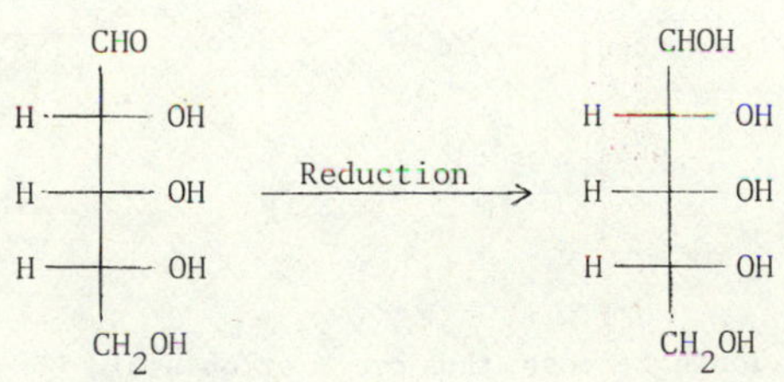

D-Ribose Ribitol (meso)

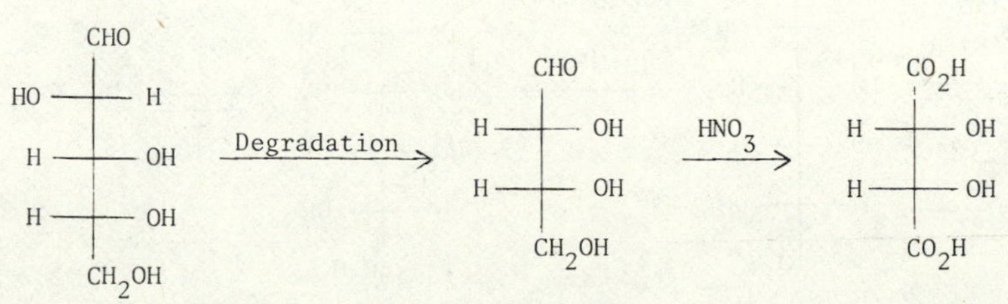

D-Arabinose D-Erythrose *meso*-Tartaric acid

21-58

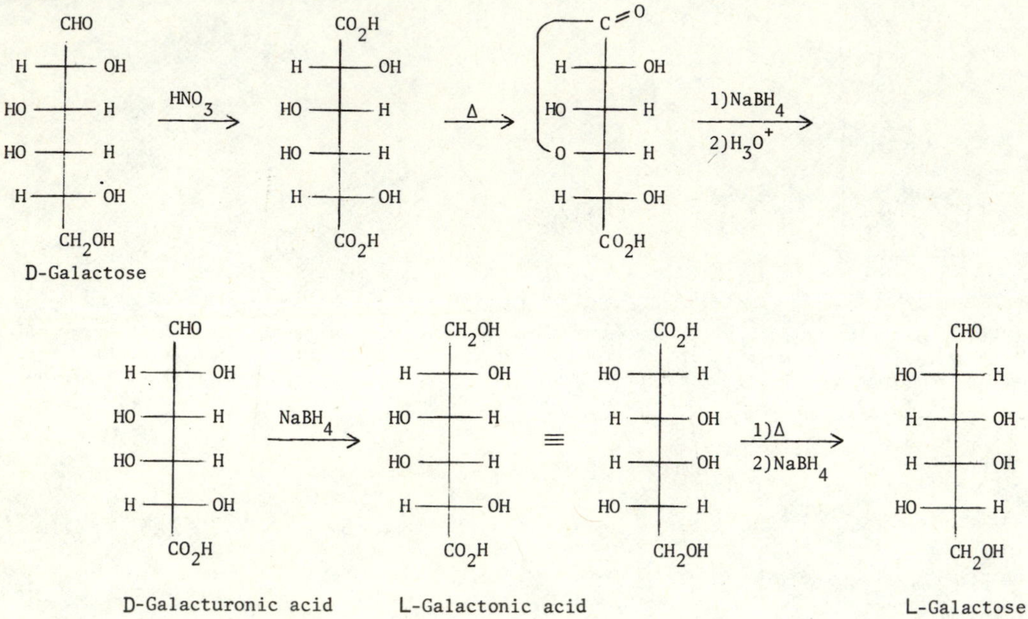

21-59 Hydrogen bonding between cis hydroxy groups is an energetically favorable contribution to the axial conformation.

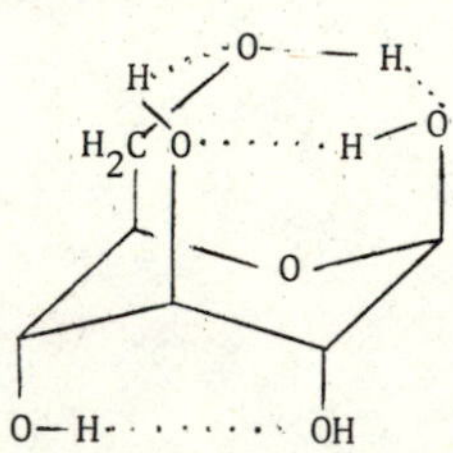

21-60 If the ring had opened the resulting aldaric acid would be meso, thus cyclization would produce the racemic lactone.

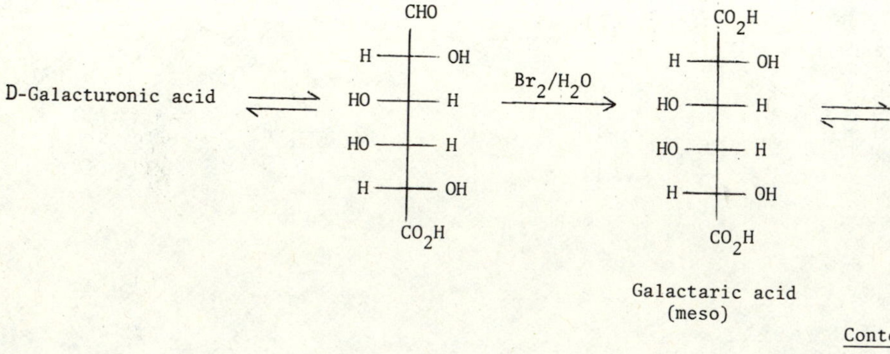

Contd...

21-60 contd...

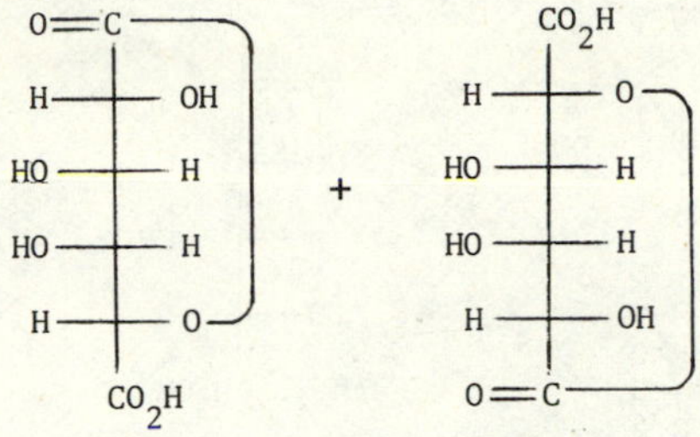

Galacturonic-1,5-lactone
(racemic)

21-61 The cyclodextrins are optically active chiral compounds since they are composed of D-glucose units. Inclusion of racemic sulfinate ester results in two diastereomers, one with (*R*) sulfinate and one with (*S*) sulfinate. Separation of the diastereomers and regeneration of the free sulfinate accomplishes resolution.

21-62 A positive benedicts test, mutarotation, and reaction with phenylhydrazine indicate that gentiobiose has a hemiacetal or hemiketal group. Cleavage by emulsion confirms a β-linkage between the two glucose units. (See also structure proof of melibiose; prob. 21-54.)

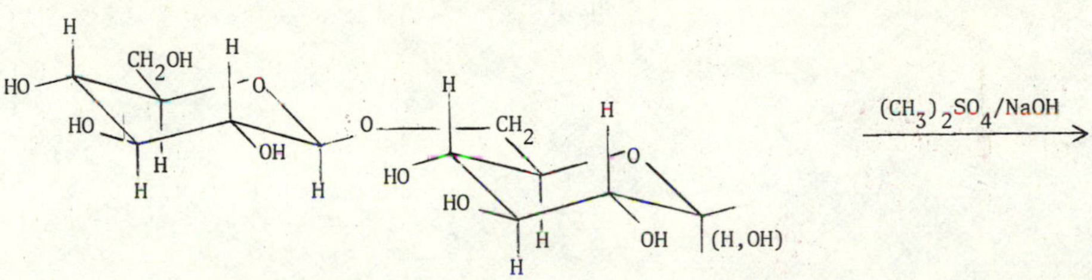

Gentiobiose

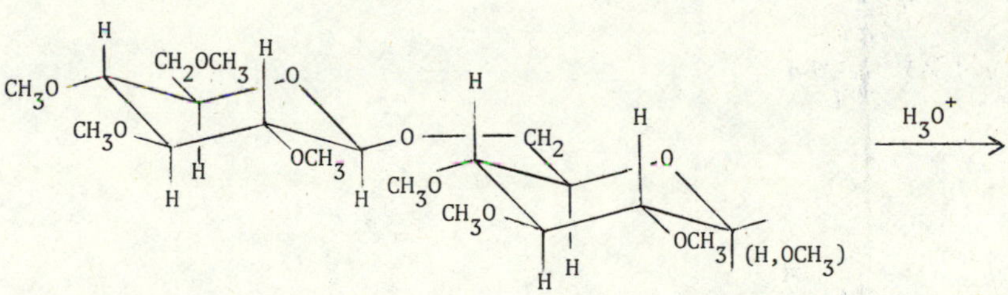

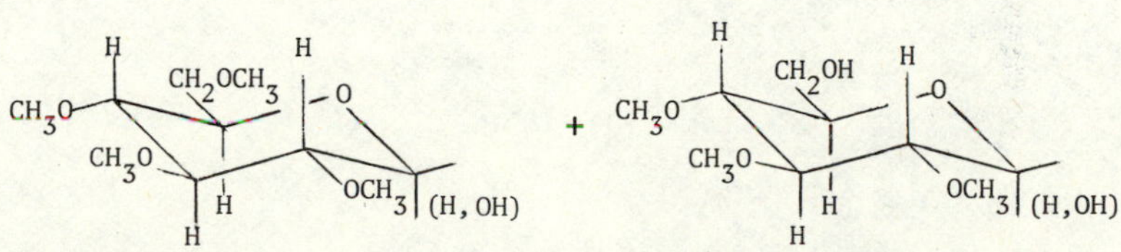

2,3,4,6-Tetramethyl-D-glucose + 2,3,4-Trimethyl-D-glucose

21-63

A ≡

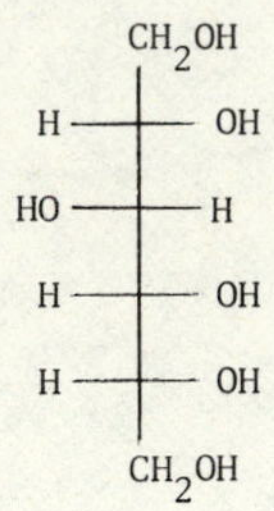
CH2OH
H
OH
HO
H
H
OH
H
OH
CH2OH

B ≡

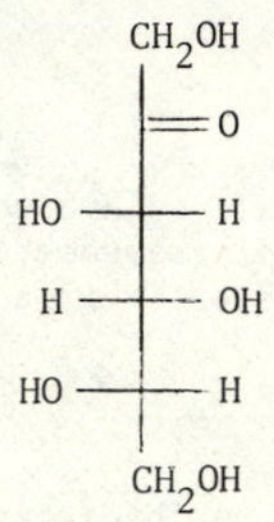
CH2OH
O
HO
H
H
OH
HO
H
CH2OH

C ≡

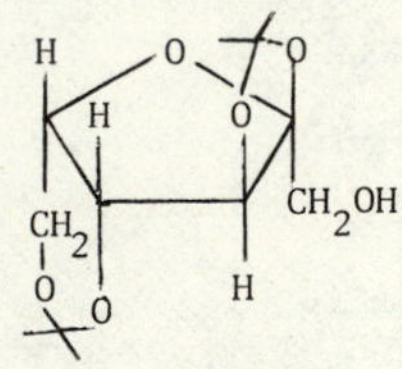
H
O
O
H
O
CH2
CH2OH
O
O
H

D ≡

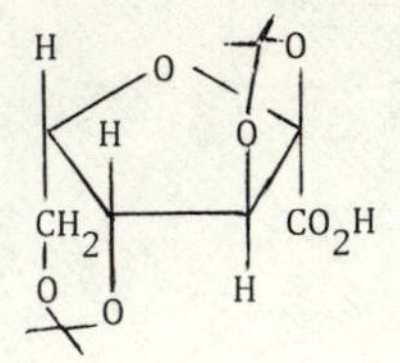
H
O
O
H
O
CH2
CO2H
O
O
H

E ≡

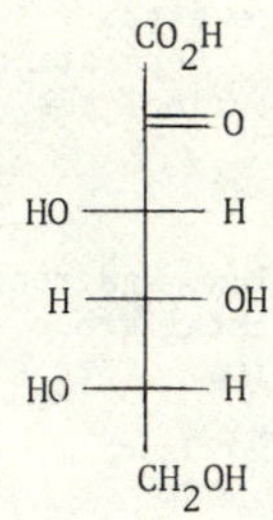
CO2H
O
HO
H
H
OH
HO
H
CH2OH

F ≡

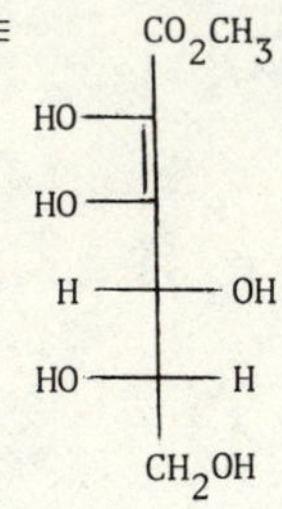
CO2CH3
HO
HO
H
OH
HO
H
CH2OH

21-64

A ≡ CH_3OH/H^+

B ≡ C_6H_5CHO/H^+

C ≡ 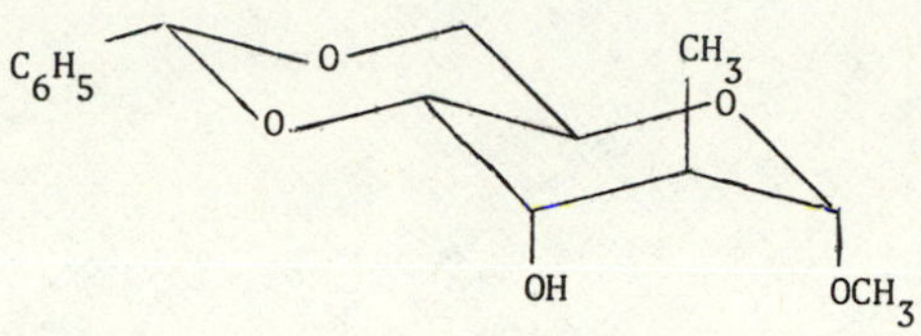

D ≡ 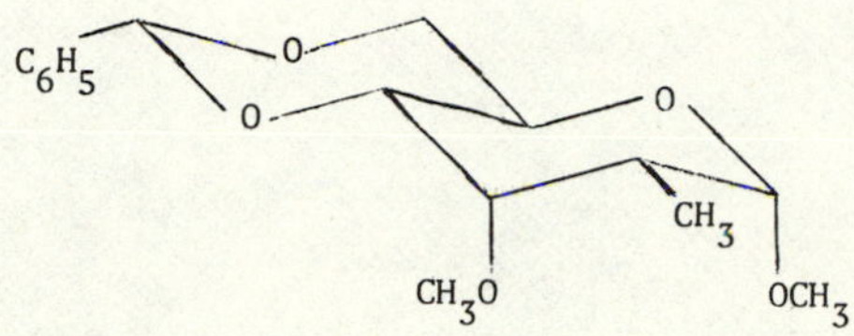

E ≡ CrO_3/pyridine

F ≡ 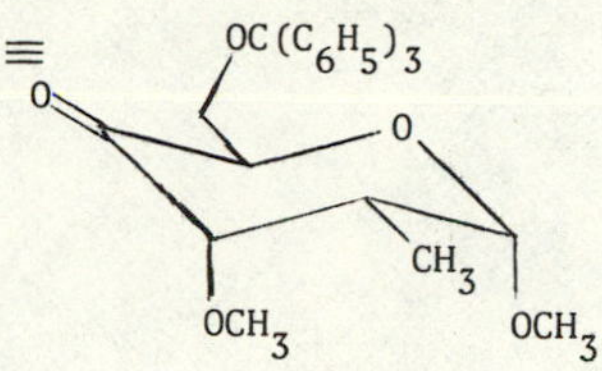

G ≡ 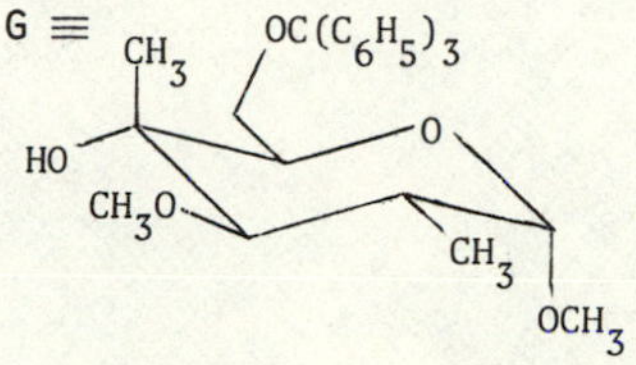

1) The methyl ketal gives principally the α-anomer (the anomeric effect - see problem 21-26).

2) Benzaldehyde acetal forms a six-membered ring in protecting hydroxy groups at the 4 and 6 positions.

3) The reagent adds a methyl group to the less-hindered side of the epoxide to give the trans-diaxial product.

4) Epimerization takes place alpha to the carbonyl group to give the more stable equatorial methyl.

5) Hydrogenolysis takes place at the benzylic protecting group.

6) This accomplishes protection of one of the hydroxy groups as triphenylmethyl (trityl) ether. The bulky reagent is selective for the primary alcohol.

7) The base promotes epimerization alpha to the carbonyl to give the equatorial methoxy group.

8) The addition actually leads to a mixture of axial and equatorial products, but the desired isomer crystallizes from the mixture.

22 Amino Acids and Proteins

22-1

a) 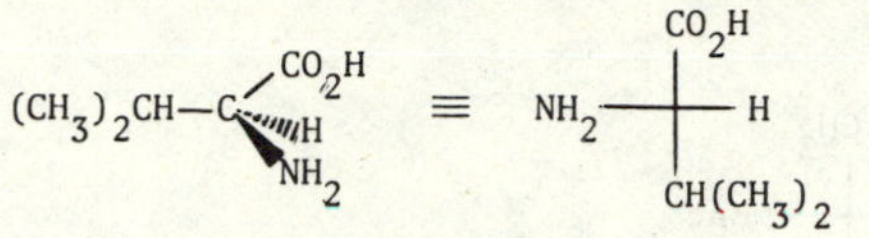

L-Valine

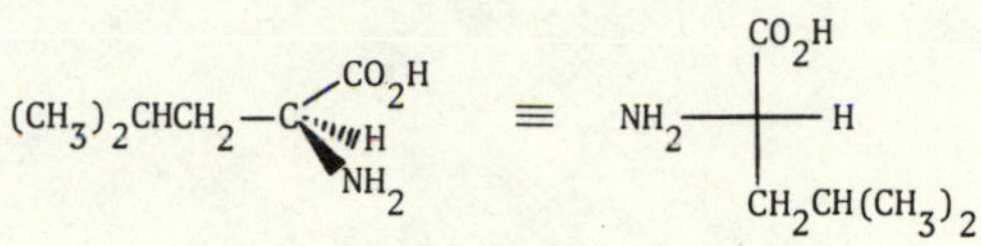

L-Leucine

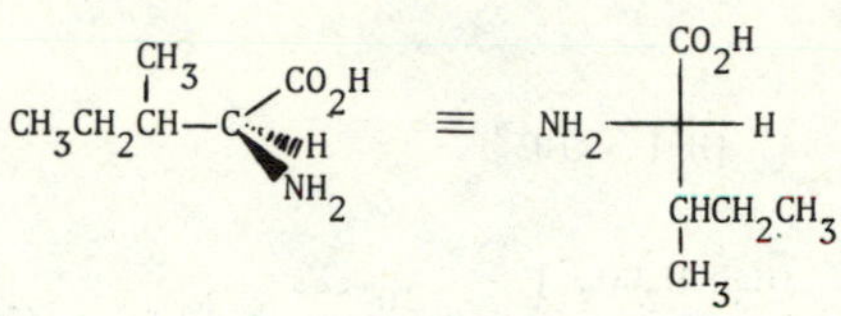

L-Isoleucine

b) Only one L-stereoisomer of valine and leucine are possible. Isoleucine can exist as two L-stereoisomers because the β-carbon is chiral also.

c) 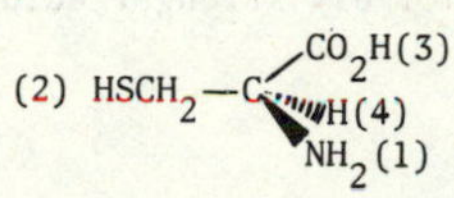

The thiomethylene group has priority over the carboxy group since the sulfur atom has a higher atomic number than the oxygen atom.

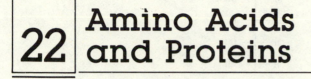

22-2

A ≡

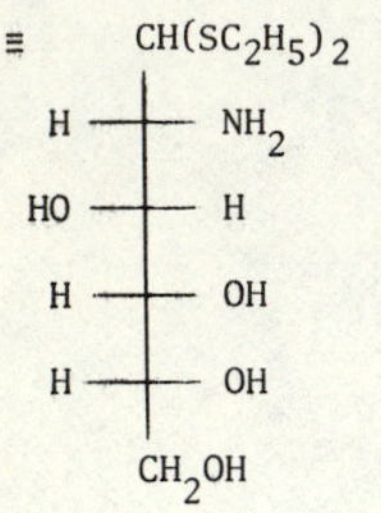

C ≡

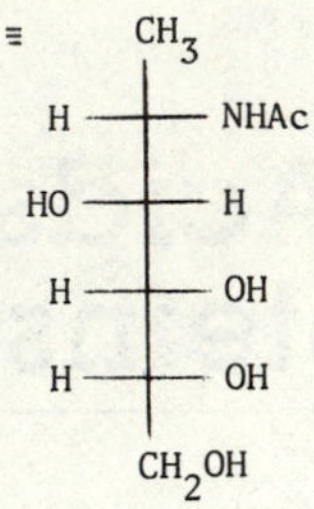

B ≡

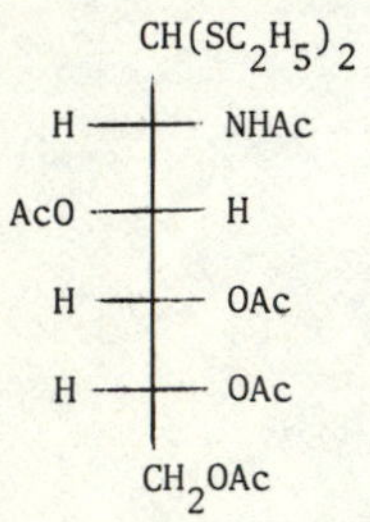

D ≡

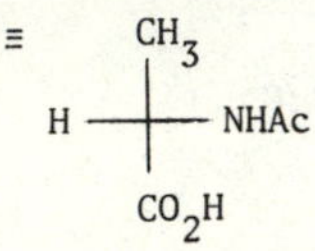

22-3

a) $$K_{a1} = \frac{[H^+][H_3\overset{+}{N}CH_2CO_2^-]}{[H_3\overset{+}{N}CH_2CO_2H]}$$

$$K_{a2} = \frac{[H^+][H_2NCH_2CO_2^-]}{[H_3\overset{+}{N}CH_2CO_2^-]}$$

b) At pH = 4 $[H^+] = 10^{-4}$

$$\frac{K_{a1}}{[H^+]} = \frac{[H_3\overset{+}{N}CH_2CO_2^-]}{[H_3\overset{+}{N}CH_2CO_2H]} = \frac{10^{-2.35}}{10^{-4}} = 10^{1.65} = 45$$

22-4 The observation indicates that the carboxylic acid group is a relatively stronger acid than the amino group is a base.

22-5 At pH values other than the isoelectric point most of the amino acid molecules in solution are of the same charge. They repel each other and are solvated by the water. At the electrically neutral p*I* value, charge repulsion is minimized. Molecules come together and precipitate (proteins coagulate). Solvation is less favorable because a solvent molecule must accommodate two different charges in the same amino acid molecule.

22-6

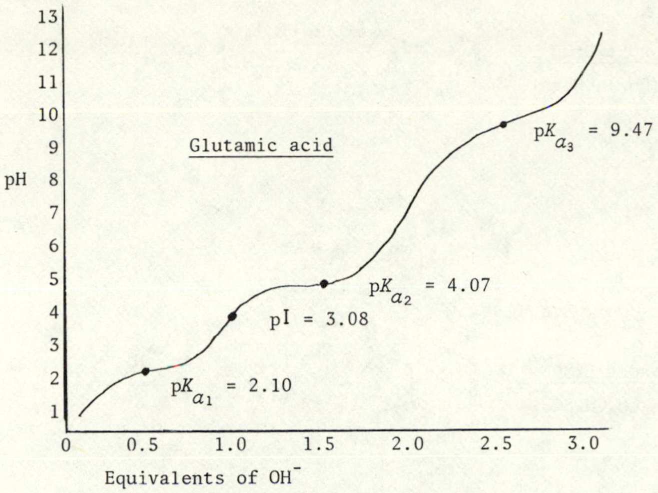

Species present at pH = 2.10:

$HO_2CCH_2CH_2CH(\overset{+}{N}H_3)CO_2H$ and

$HO_2CCH_2CH_2CH(\overset{+}{N}H_3)CO_2^-$;

at pH = 3.08:

$HO_2CCH_2CH_2CH(\overset{+}{N}H_3)CO_2^-$;

at pH = 4.07:

$HO_2CCH_2CH_2CH(\overset{+}{N}H_3)CO_2^-$ and

$^-O_2CCH_2CH_2CH(\overset{+}{N}H_3)CO_2^-$

at pH = 9.47:

$^-O_2CCH_2CH_2CH(\overset{+}{N}H_3)CO_2^-$ and

$^-O_2CCH_2CH_2CH(NH_2)CO_2^-$

22-7 The diethyl acetamidomalonate approach can also be used for each of the following.

$$CH_2(CO_2C_2H_5)_2 + Br_2 \xrightarrow{CCl_4} BrCH(CO_2C_2H_5)_2$$

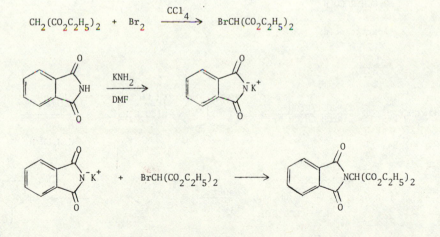

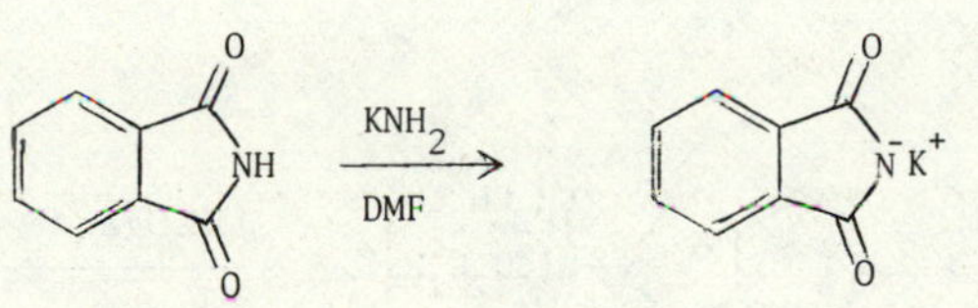

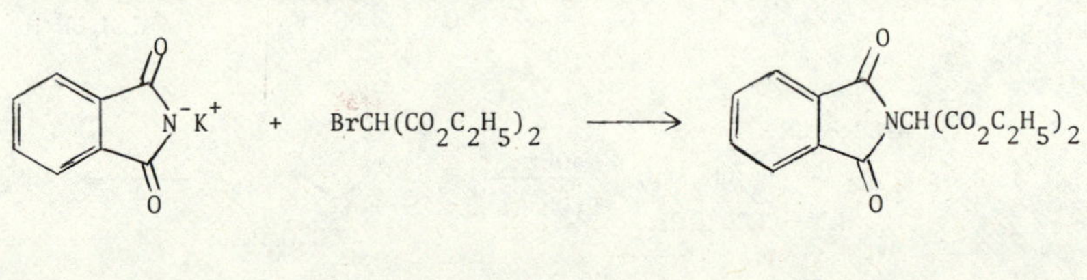

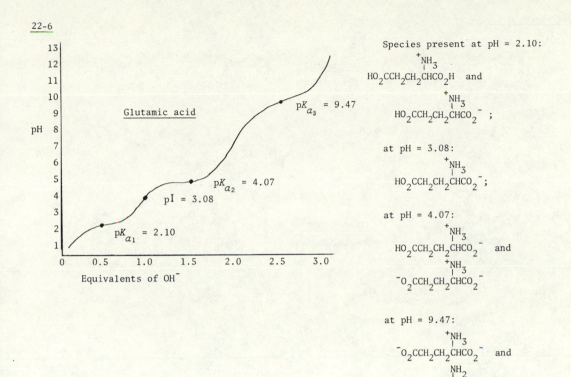

22-7 contd...

a)

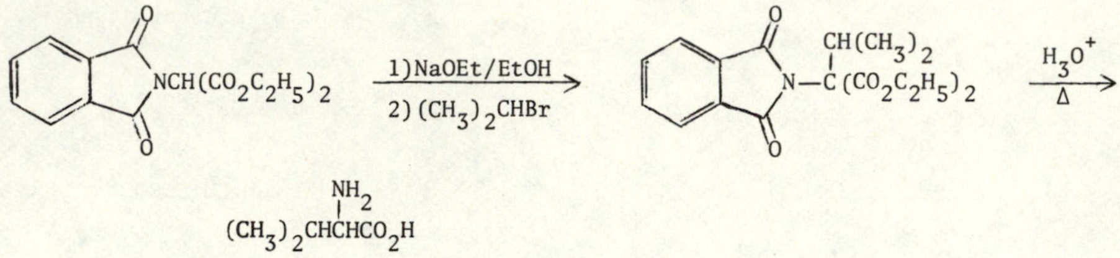

Valine

b)

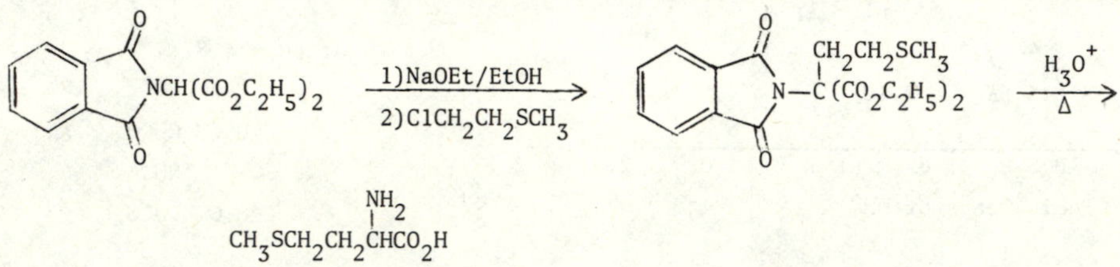

Methionine

c)

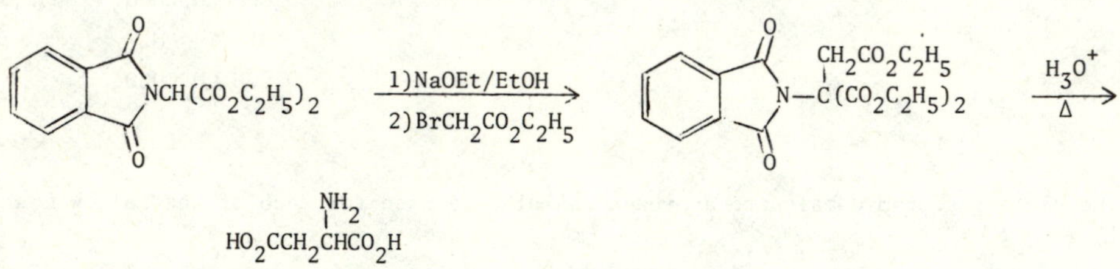

Aspartic acid

22-8

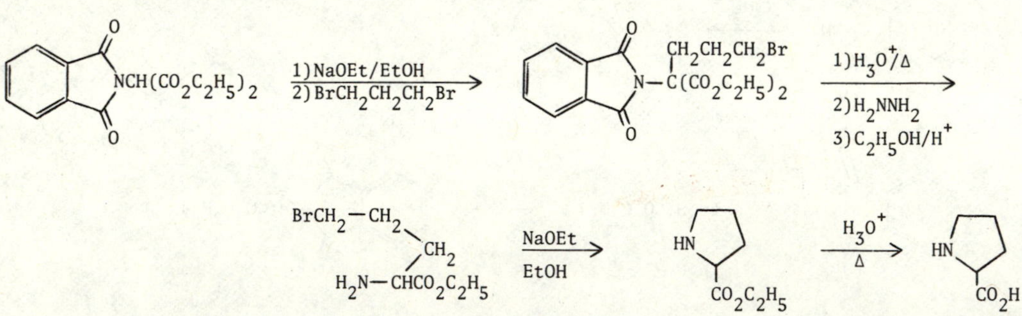

Proline

22-9

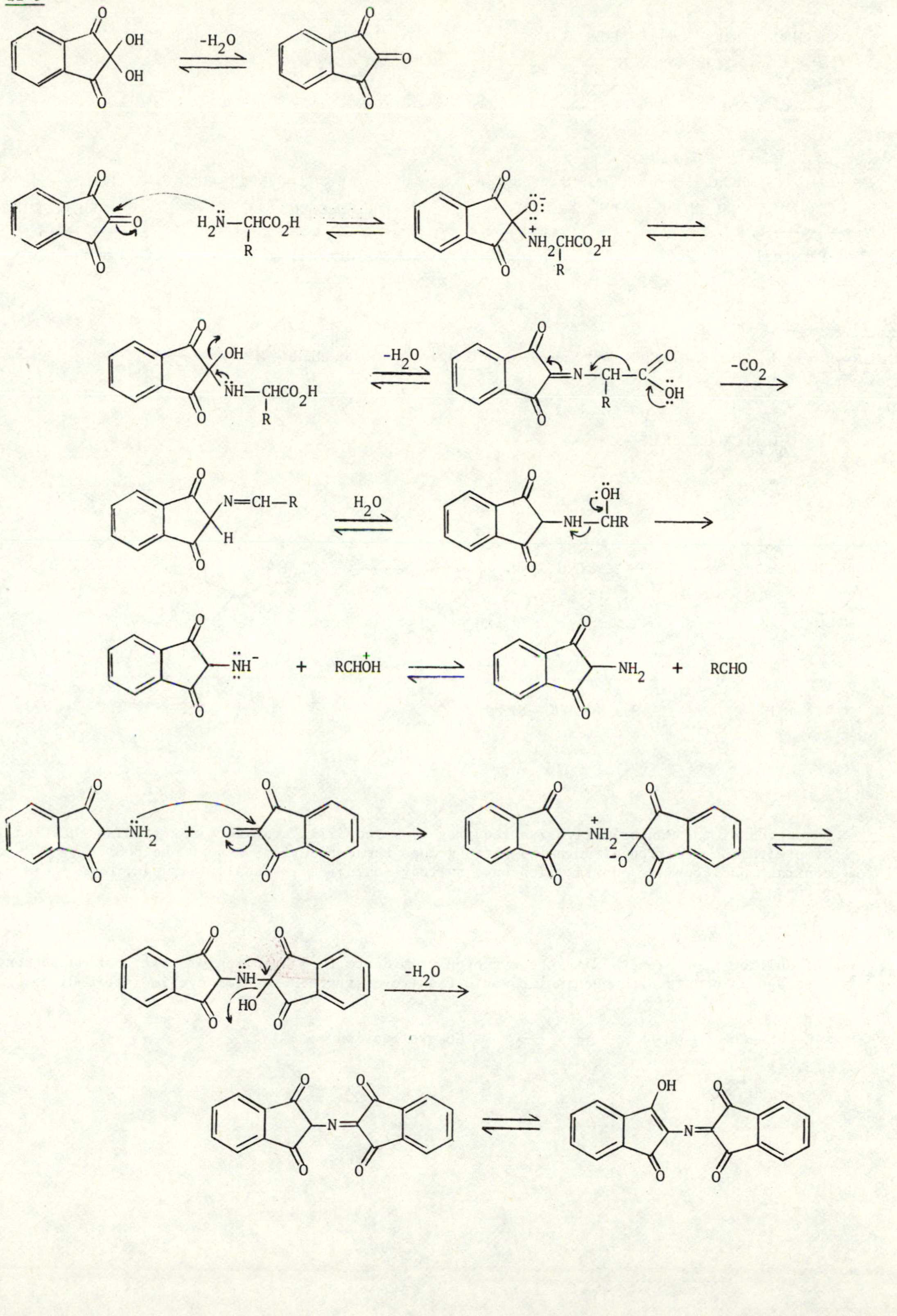

22-10

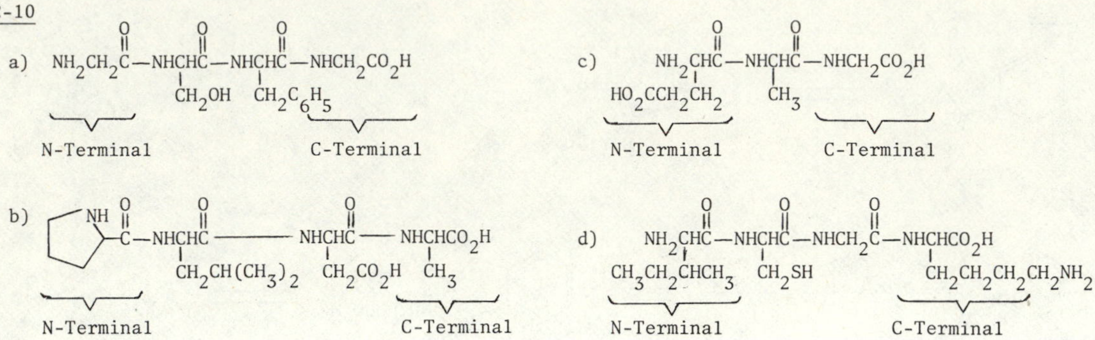

22-11

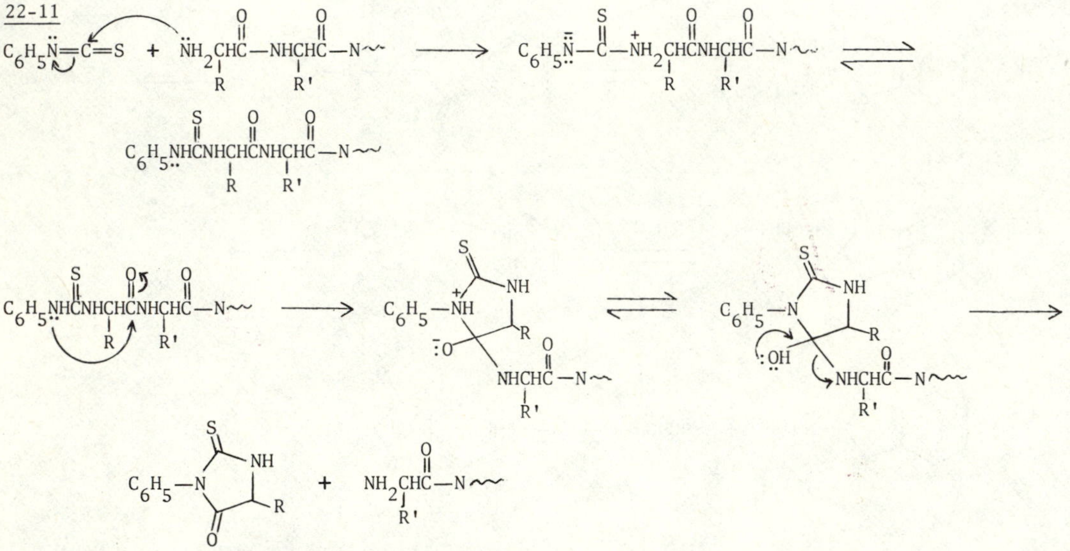

22-12 Formation of the Sanger derivative requires a nucleophilic aromatic substitution reaction. Even with the electron withdrawing nitro groups these reactions tend to be poor. By contrast nucleophilic substitution on a sulfonyl chloride is usually a rapid, high yield process.

22-13 The sulfide group reacts with the cyanogen bromide to form a sulfonium salt, but subsequent cleavage as occurs with methionine would require a strained, 4-atom cyclic intermediate.

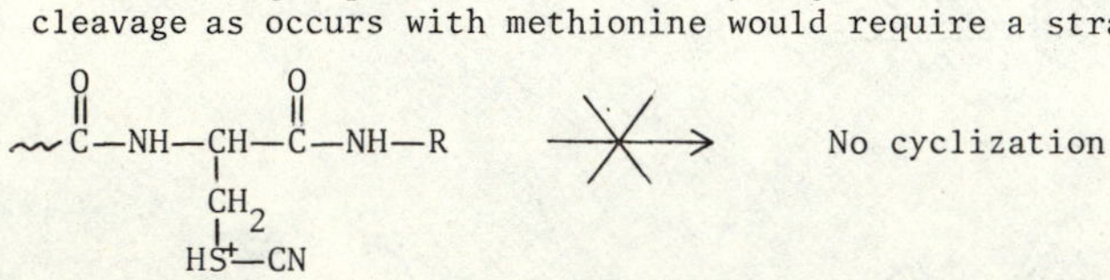

22-14 Since reaction with 2,4-dinitrofluorobenzene shows that one valine is the N-terminal amino acid, either fragment *3* or *5* must be at that end. The cyanogen bromide reaction indicates that methionine must be connected to glycine or valine by the methionine carboxylic acid group. Fragment *1* shows that it is the glycine that is connected to methionine and fragment *5* indicates that valine and glycine are connected. The composition of *1* also shows that phenylalanine must be connected to methionine. The valine of fragment *5* must be to the right (toward the C-terminal end) of the peptide.

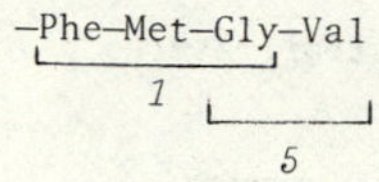

The other valine - the N-terminal amino acid - is that of fragment *3* and must be connected to leucine, while fragment *2* indicates that proline is attached to the leucine.

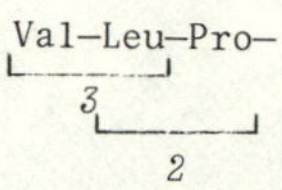

Fragment *4* confirms the relationship between our two partial structures. The heptapeptide structure can now be completed: Val—Leu—Pro—Phe—Met—Gly—Val.

22-15 The enzymes trypsin and chymotrypsin cleave the peptide at specific amino acids and at the carboxy group of those amino acids. Dansyl chloride identifies the N-terminal amino acid and carboxypeptidase the C-terminal amino acid. These data coupled with the results of hydrolysis lead to the amino acid sequence. In the following we show how each step in the analysis reveals more of the sequence. Each new datum must be consistent with previous data. The two pentapeptide fragments from chymotrypsin cleavage complete the analysis.

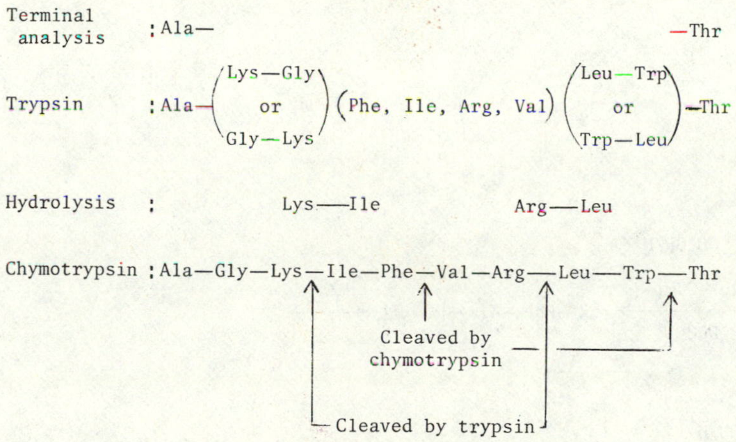

22-16

a) The terminal amino and carboxylic acid groups are so far from each other that they do not interact. Deviations in pK_a values of peptides from those of alkylamines and carboxylic are due principally to the influence of other adjacent groups in the peptide.

b) $pI = \dfrac{pK_{a_1} + pK_{a_2}}{2} = \dfrac{3.42 + 7.92}{2} = 5.68$

22-17 $(0.90)^{40} \times 100 = 1.5\%$

22-18

a) $(CH_3)_3COH + ClCCl \xrightarrow{Et_2O} (CH_3)_3COCCl$ (each C=O carbonyl drawn above the central C)

This reagent is unstable and is usually prepared and used at low temperatures as needed.

b) 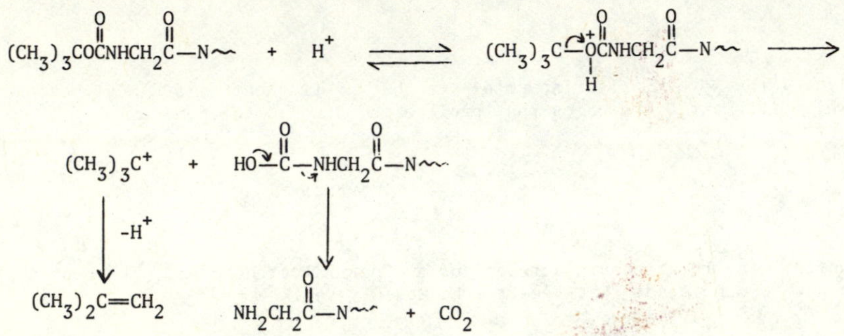

22-19 Four dipeptides would be expected to form: Ala—Ala; Gly—Gly; Ala—Gly; Gly—Ala.

22-20 Other protecting and activating groups can also be used in the following sequences. Abbreviations are:

Z ≡ Benzoxycarbonyl

Boc ≡ *t*-Butoxycarbonyl

NCA ≡ *N*-carboxyanhydride

a) 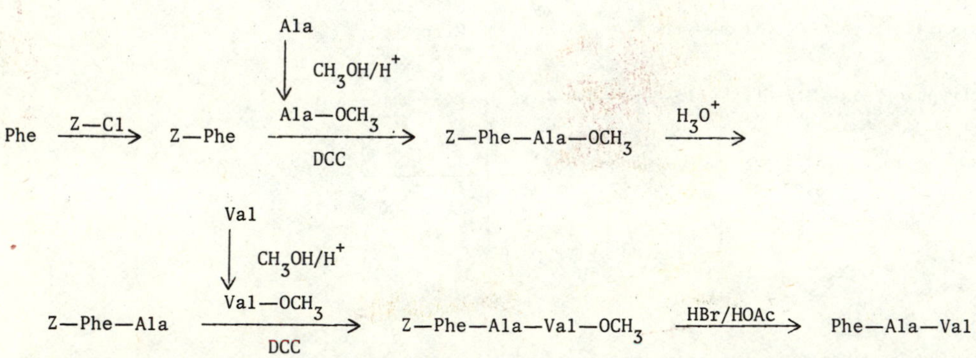

Contd...

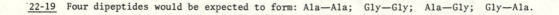

22-20 contd...

b) Ile $\xrightarrow{\text{Boc—Cl}}$ Boc—Ile $\xrightarrow{\text{ClCOCl}}$ Boc—Ile—NCA $\xrightarrow{\text{Met—OEt}}$ (Met $\xrightarrow{\text{EtOH/H}^+}$ Met—OEt)

Boc—Ile—Met—OEt $\xrightarrow{\text{H}_3\text{O}^+}$ Boc—Ile—Met $\xrightarrow{\text{Asp—(O-}t\text{-Bu)}_2}$ (Asp $\xrightarrow{t\text{-BuOH/H}^+}$ Asp—(O-t-Bu)$_2$)

Boc—Ile—Met—Asp—(O-t-Bu)$_2$ $\xrightarrow{\text{HBr/HOAc}}$ Ile—Met—Asp

c) Leu $\xrightarrow{\text{ClCOCl}}$ Leu—NCA $\xrightarrow{\text{Phe}}$ Leu—Phe $\xrightarrow{\text{MeOH/H}^+}$

Ala $\xrightarrow{\text{ClCOCl}}$ Ala—NCA

Leu—Phe—OMe $\xrightarrow{\text{Ala—NCA}}$ Ala—Leu—Phe—OMe $\xrightarrow[\text{DCC}]{\text{Boc}_2\text{Lys}}$

Boc$_2$Lys—Ala—Leu—Phe—OMe $\xrightarrow{\text{HBr/HOAc}}$ Lys—Ala—Leu—Phe

22-21 One step at a time: $(0.90)^5 \times 100 = 59\%$

Each tripeptide formed, then coupled: $(0.90)^2 \times 0.90 \times 100 = 73\%$

22-22

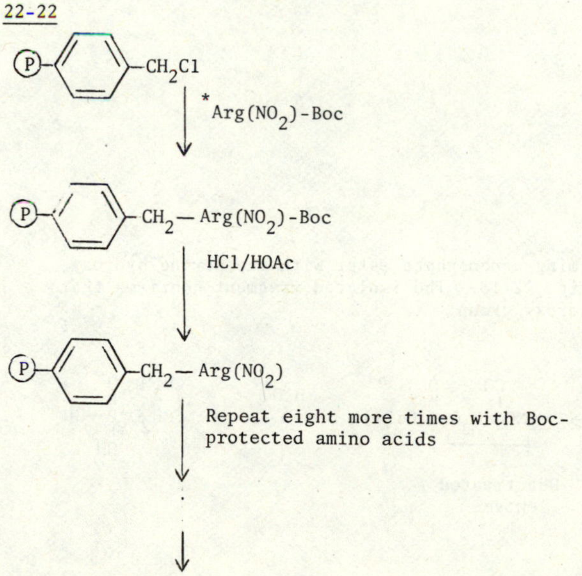

Contd...

22-22 contd...

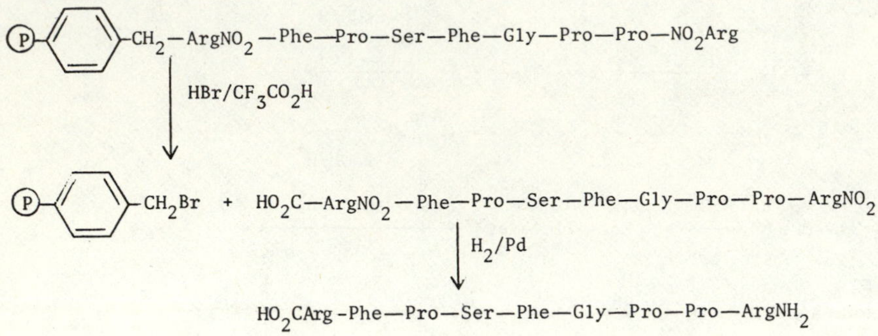

(The C-terminal amino acid is connected to the polymer so that the structure above is written opposite to the usual direction.)

*Arginine was used as a nitro derivative, then reduced to the free amine at the last step.

22-23 At pH 6-7 the side chain amino group of each lysine segment is protonated. Repulsions between the cations disrupt the α-helix conformation. At high pH each unit is electrically neutral so that the α-helix can form.

22-24 Two common mechanisms of enzyme inhibition are:
i) the inhibitor complexes with the active site and blocks the enzymes ability to complex with substrate;
ii) the inhibitor matches and complexes with the substrate and prevents approach of the enzyme.

22-25

a) 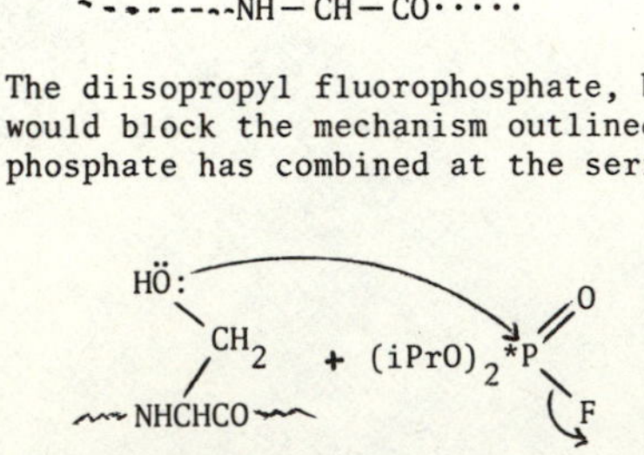

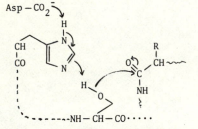

b) The diisopropyl fluorophosphate, by forming a phosphate ester with the serine hydroxy, would block the mechanism outlined in fig. 22-13. The isolated fragment confirms that phosphate has combined at the serine hydroxy group.

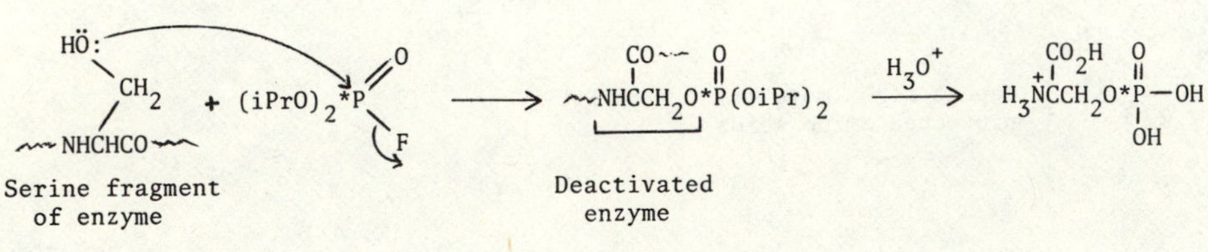

22-26

a) The anion of the conjugate base is stabilized inductively by the adjacent ammonium nitrogen atom and by d-orbital resonance with the adjacent sulfur atom.

b)

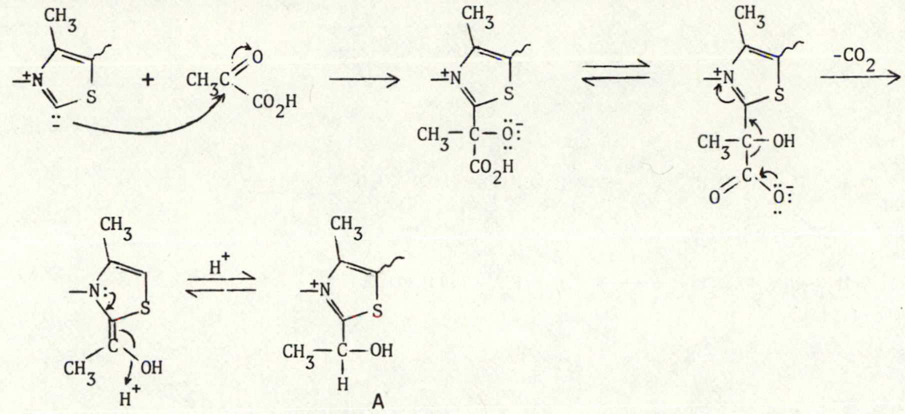

c) The lipoic acid is reduced as it participates in the formation of acetyl CoA.

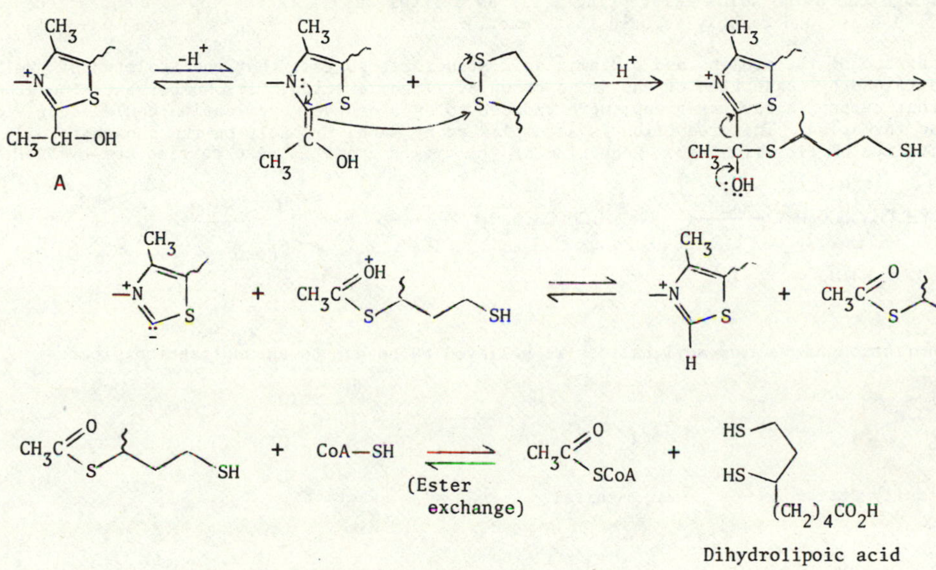

22-27

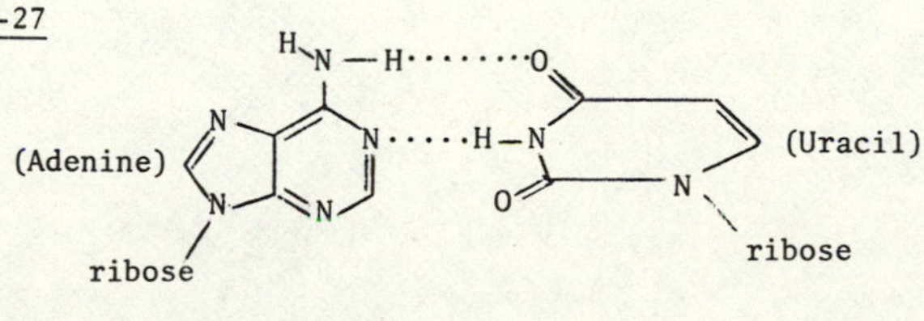

22-28

a)

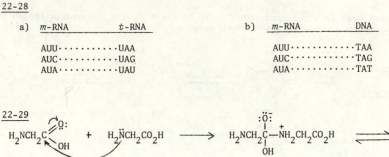

m-RNA	*t*-RNA
AUU	UAA
AUC	UAG
AUA	UAU

b)

m-RNA	DNA
AUU	TAA
AUC	TAG
AUA	TAT

22-29

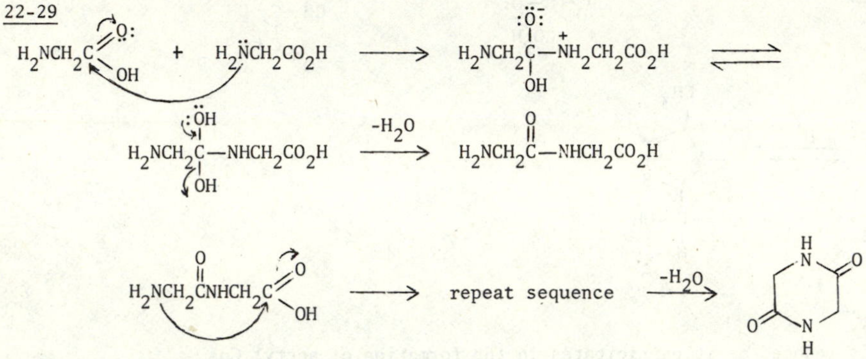

(Although the amino acids exist principally as zwitterions, it is the low concentrations of free amine and carboxylic acid that react.)

22-30 Comparison of the proline and glutamic acid structures suggest that the skeleton of proline be attained by connection of the amino group of glutamic acid to its number 5 carbon atom. If that carbon, a carboxy group, were reduced to an aldehyde, cyclization could occur via imine formation. That reduction is accomplished by NADH, probably on the phosphate derivative of glutamic acid. Reduction of the cyclic imine is also carried out by NADH.

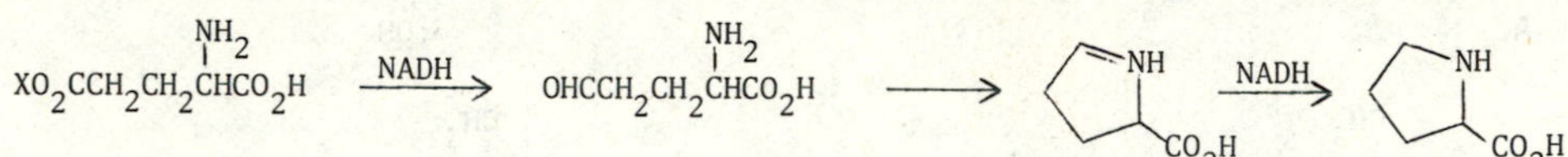

X = H or PO_3H_2

22-31 Racemization of α-amino acyl halides is believed to be due to an enolization process.

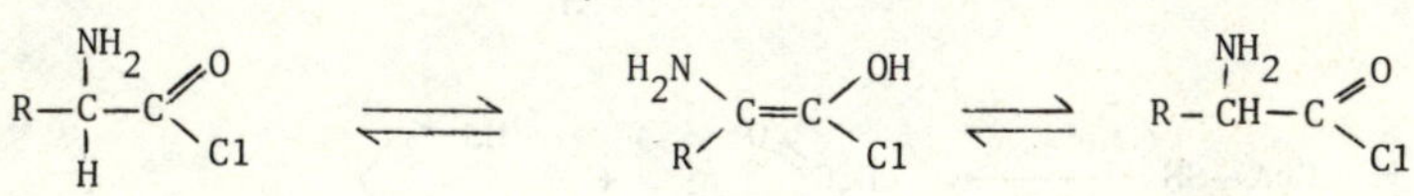

Optically active Achiral Racemic

22-32

a) Moles of glycine $= \frac{3.0g}{75g/mol} = 0.04$ mol

Moles of alanine $= \frac{0.9g}{89g/mol} = 0.01$ mol

Moles of valine $= \frac{3.7g}{117g/mol} = 0.03$ mol

Moles of proline $= \frac{6.9g}{115g/mol} = 0.06$ mol

Moles of serine $= \frac{7.3g}{105g/mol} = 0.07$ mol

Moles of arginine $= \frac{8.6g}{174g/mol} = 0.05$ mol

The ratio of amino acid units is:

Gly:Ala:Val:Pro:Ser:Arg = 4:1:3:6:7:5

b) The minimum molecular weight of the peptide is the sum of the molecular weights of all the amino acids less a mol of water for all the peptide bonds formed.

MW = (4 x 75) + (1 x 89) + (3 x 117) + (6 x 115) + (7 x 105) + (5 x 174) - (25 x 18) = 3,485

22-33

1) DNS—Gly shows that glycine is the N-terminal group.

2) DNS—Ser, Ser, Gly, Gly shows that two of the three glycine units are to the right (on the carboxyl side) of a serine. When combined with the information from (1), three sequences are possible that differ in the positions of the right three amino acids:
   a) Gly-Ser-Gly-Gly-Ser
   b) Gly—Ser—Gly—Ser—Gly
   c) Gly—Ser—Ser—Gly—Gly

3 & 4) Do not provide new useful structural information. However, the more careful analysis of 3 shows that the pentapeptide has structure a).

22-34

1) DNP—Val indicates that Val is the N-terminal amino acid and DNP—Val—Leu identifies the next unit.

2) Carboxypeptidase cleaves the C-terminal amino acid so that the result shows that Ala is that end unit and Glu is next to it.

3) The fragments recovered from partial hydrolysis identify Leu—Ile and Ala—Phe—Glu or Ale—Glu—Phe. The data from (2) support the Ala—Phe—Glu sequence. All of the above data lead to a structure for the heptapeptide of:

Val—Leu—Ile—Ala—Phe—Glu—Ala

22-35 The conversion of ornithine to arginine by reaction with H_2NCN is a key factor in establishing the ornithine structure. Those two amino acids differ by only that H_2NCN unit so that ornithine presumably has the structure:

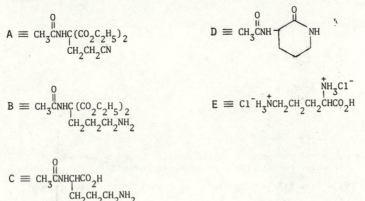

$$\underset{\text{Ornithine}}{H_2NCH_2CH_2CH_2\overset{NH_2}{\overset{|}{C}}HCO_2H} \xrightarrow[2)\,H_2NCN]{1)\,NaOH} \underset{\text{Arginine}}{H_2N\overset{NH}{\overset{\|}{C}}NHCH_2CH_2CH_2\overset{NH_2}{\overset{|}{C}}HCO_2H}$$

In the synthesis below note that the six-membered lactam readily forms on heating the δ-amino acid.

$$A \equiv CH_3\overset{O}{\overset{\|}{C}}NH\underset{CH_2CH_2CN}{\underset{|}{C}}(CO_2C_2H_5)_2$$

$$B \equiv CH_3\overset{O}{\overset{\|}{C}}NH\underset{CH_2CH_2CH_2NH_2}{\underset{|}{C}}(CO_2C_2H_5)_2$$

$$C \equiv CH_3\overset{O}{\overset{\|}{C}}NH\underset{CH_2CH_2CH_2NH_2}{\underset{|}{C}}HCO_2H$$

D ≡ 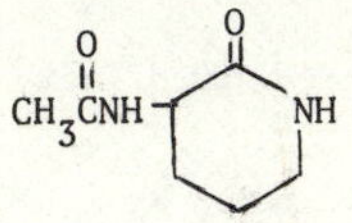

$$E \equiv Cl^-\,H_3\overset{+}{N}CH_2CH_2CH_2\overset{\overset{+}{N}H_3Cl^-}{\overset{|}{C}}HCO_2H$$

22-36 The fit into the enzyme is best for an L configuration of the benzyl, acetamido, and hydrogen groups. Methyl cannot replace benzyl or the hydrogen atom, nor can hydrogen replace the acetamido group. The cyclic substrate reacts rapidly in the configuration opposite to that of the natural L-phenylalanine. This suggests that tying back the benzyl and amido removes one restriction to a proper fit although the data do not indicate which group is more important. The position of the H-atom at the asymmetric carbon atom appears to be critical.

22-37 Fragments 4 and 5 indicate that the sequence in 5 must be Tyr—Ile—Glu. The data provided by fragment 2 extends the sequence to $CySO_3H$—Tyr—Ile—Glu (7). Similar reasoning using 3 and 6 leads to a tetrapeptide sequence of HO_3SCys—Pro—Leu—Gly (9). The cysteic acid unit in fragment 6 must be different from that in 2. The sequence of tetrapeptide 7 was established above and 8 can be deduced from 9 as being Asp—$CySO_3H$—Pro—Leu—Gly. Since the two cysteine units are connected together we have -

```
Glu—Ile—Tyr—Cys
              |
        Asp—Cys—Pro—Leu—Gly
```

Knowing that glycine is present as an amide and thus must be a C-terminal group, and that oxytocin is cyclic, provides the final structure.

```
Ile—Tyr—Cys
 |       |
Glu—Asp—Cys—Pro—Leu—Gly
```

22-38

a) $C_6H_6 + CH_3COCl \xrightarrow{AlCl_3} C_6H_5COCH_3 \xrightarrow[2)\ HCl/\Delta]{1)\ NaCN/NH_4Cl} C_6H_5C(NH_2)(CH_3)-CO_2H$

b) $CH_3(CH_2)_3CO_2H \xrightarrow{PBr_3} CH_3CH_2CH_2CH(Br)CO_2H \xrightarrow{NH_3} CH_3CH_2CH_2CH(NH_2)CO_2H$

c) $CH_3CH_2MgBr + {}^{14}CO_2 \xrightarrow[2)\ H_3O^+]{1)\ Et_2O} CH_3CH_2{}^{14}CO_2H \xrightarrow[2)\ LiAlH[OC(CH_3)_3]_3]{1)\ SOCl_2}$

$CH_3CH_2{}^{14}CHO \xrightarrow[2)\ HCl/\Delta]{1)\ NaCN/NH_4Cl} CH_3CH_2{}^{14}CH(NH_2)CO_2H$

d) 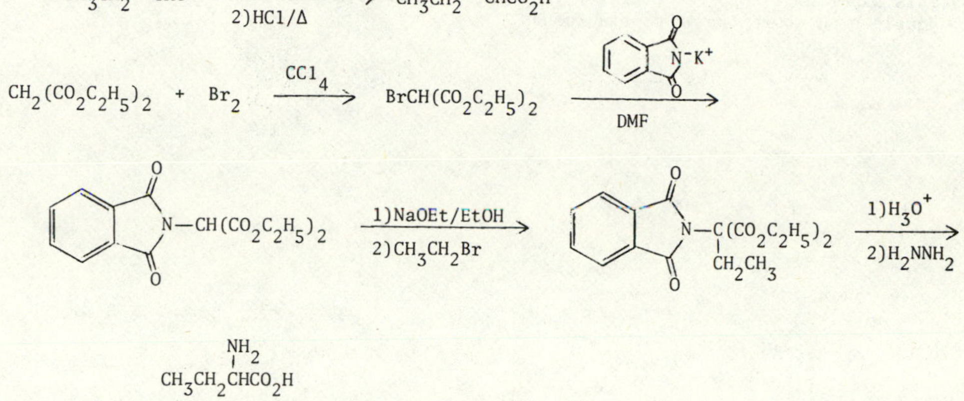

e) 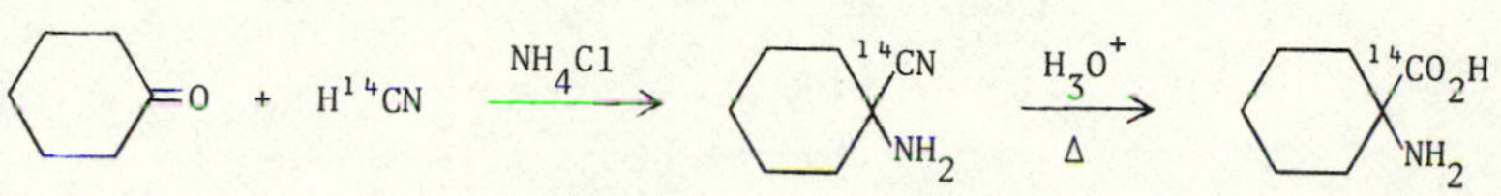

f)

Ala $\xrightarrow{\text{Boc—Cl}}$ Boc—Ala $\xrightarrow[\text{DCC}]{\text{Val—OEt}}$ Boc—Ala—Val—OEt $\xrightarrow{H_3O^+}$ (Val $\xrightarrow{EtOH/H^+}$ Val—OEt)

Boc—Ala—Val $\xrightarrow[\text{DCC}]{\text{Phe—OEt}}$ Boc—Ala—Val—Phe—OEt $\xrightarrow{H_3O^+}$ (Phe $\xrightarrow{EtOH/H^+}$ Phe—OEt)

Boc—Ala—Val—Phe $\xrightarrow[\text{DCC}]{\text{Met—OEt}}$ Boc—Ala—Val—Phe—Met—OEt $\xrightarrow{H_3O^+}$ (Met $\xrightarrow{EtOH/H^+}$ Met—OEt)

Boc—Ala—Val—Phe—Met $\xrightarrow[\text{DCC}]{\text{Ala—OEt}}$ Boc—Ala—Val—Phe—Met—Ala—OEt $\xrightarrow{H_3O^+}$ (Ala $\xrightarrow{EtOH/H^+}$ Ala—OEt)

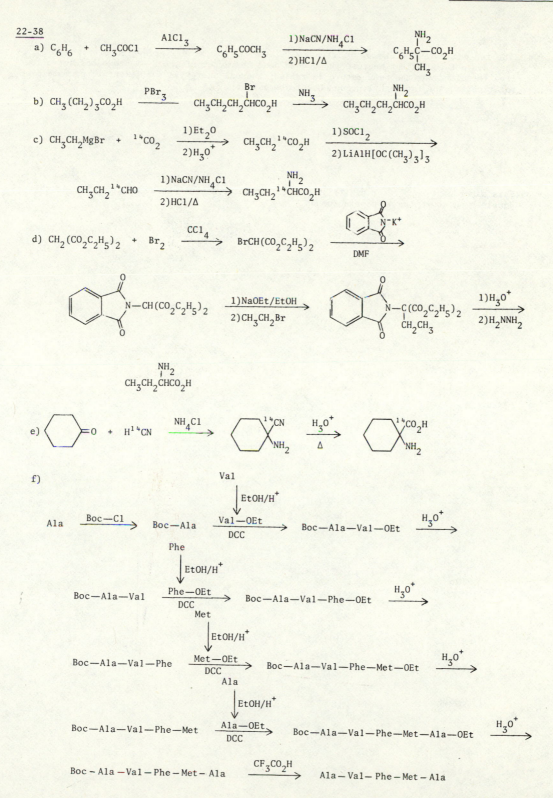

Boc–Ala–Val–Phe–Met–Ala $\xrightarrow{CF_3CO_2H}$ Ala–Val–Phe–Met–Ala

22-39

a) The two peaks in the ^{1}H-nmr spectrum at 25° are the two methyl groups on nitrogen. They are not equivalent because the double bond character of the C—N bond does not allow free rotation. As the temperature is raised, thermal energy is sufficient to promote free rotation and the two methyl groups become equivalent.

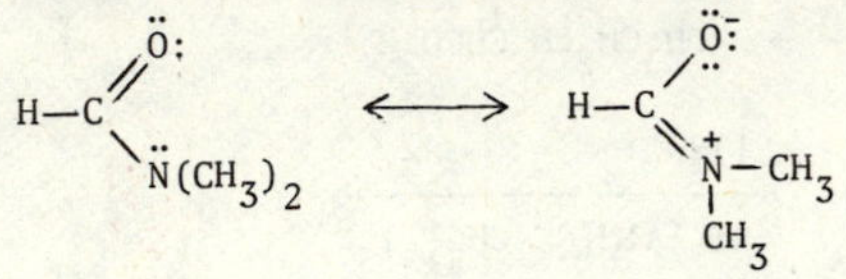

b) This experiment supports rigidity in peptide structures that has been attributed to the double bond character of peptide bonds.

23 Lipids

23-1

Lauric acid	-	Dodecanoic acid
Myristic acid	-	Tetradecanoic acid
Palmitic acid	-	Hexadecanoic acid
Stearic acid	-	Octadecanoic acid
Palmitoleic acid	-	*Z*-9-Hexadecenoic acid
Oleic acid	-	*Z*-9-Octadecenoic acid
Linoleic acid	-	*Z*,*Z*-9,12-Octadecadienoic acid
Linolenic acid	-	*Z*,*Z*,*Z*-9,12,15-Octadecatrienoic acid

23-2

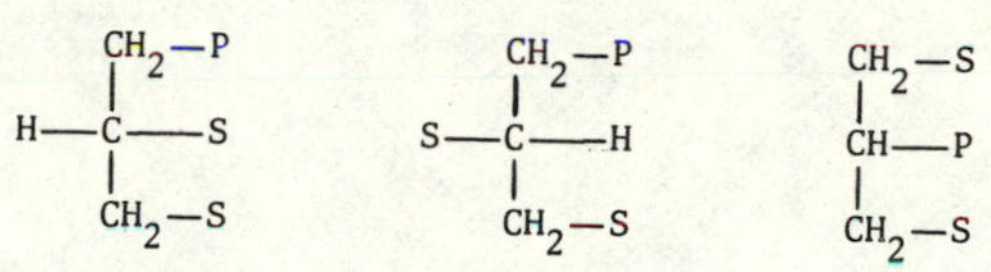

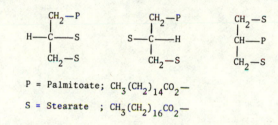

P = Palmitoate; $CH_3(CH_2)_{14}CO_2—$

S = Stearate ; $CH_3(CH_2)_{16}CO_2—$

23-3 Elaidic acid is the thermodynamically more stable *E*-configurational isomer.

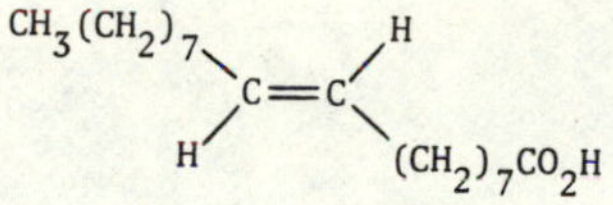

23-4 $CH_2O_2C(CH_2)_{12}CH_3$
|
$CHO_2C(CH_2)_{12}CH_3$
|
$CH_2O_2C(CH_2)_{12}CH_3$

23-5

A ≡ 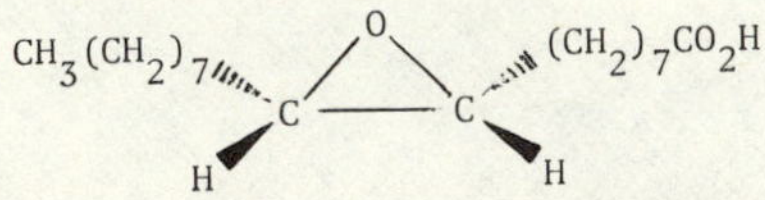

B ≡ 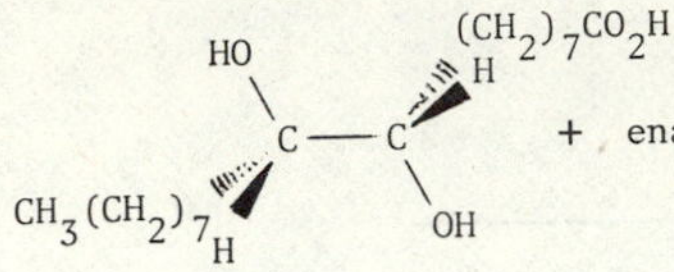 + enantiomer

threo

C ≡ 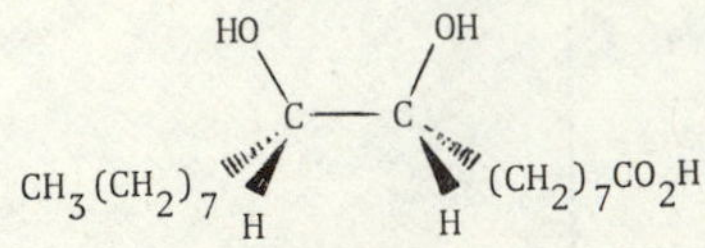

erythro

D ≡ $CH_3(CH_2)_7CHO$

E ≡ $OHC(CH_2)_7CO_2H$

23-6 Phloroacetophenone can be formed by an intramolecular "Claisen" reaction followed by aromatization.

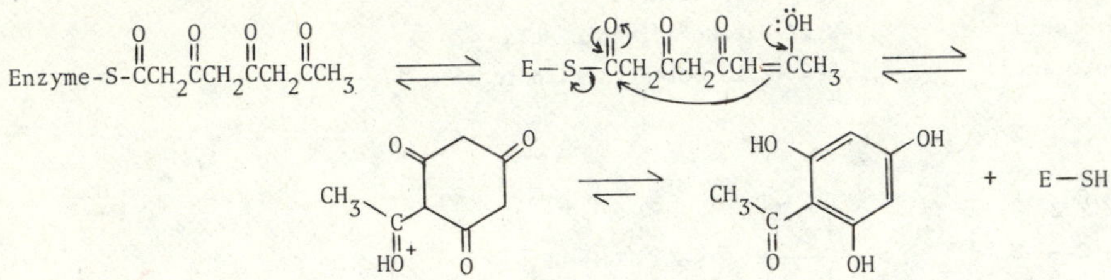

The formation of Orsellinic resembles an intramolecular "aldol" reaction.

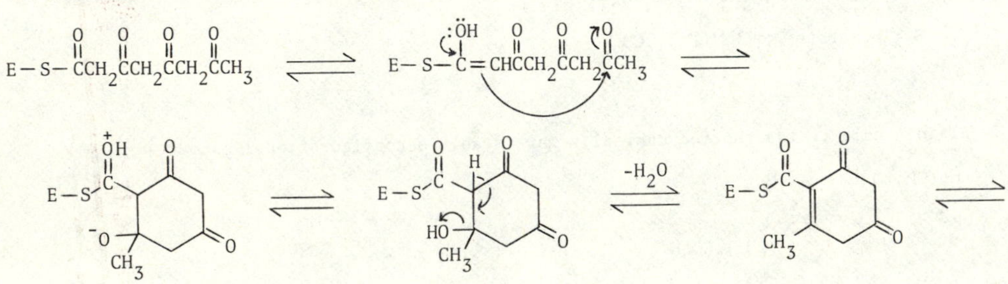

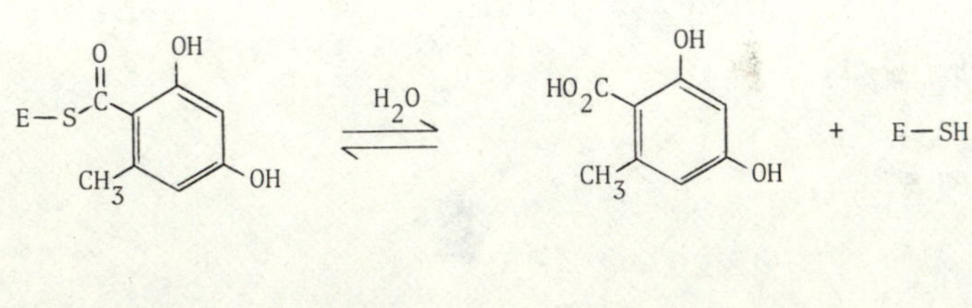

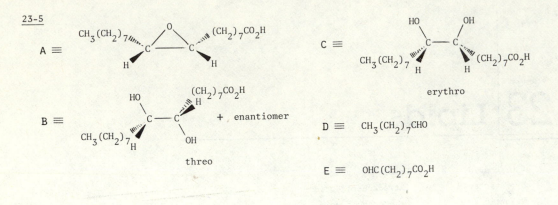

23-7

a) 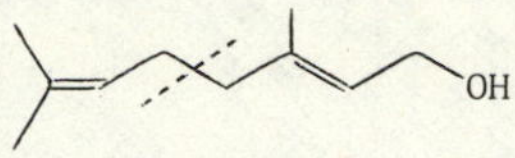

Geraniol

b) 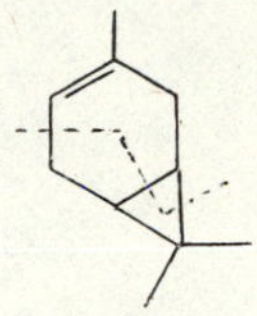

Carene

c) 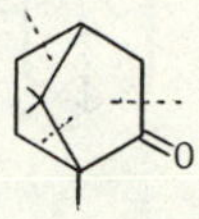

Camphor

d) 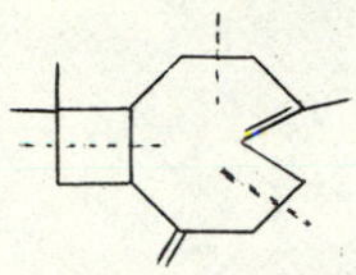

Caryophyllene

e) 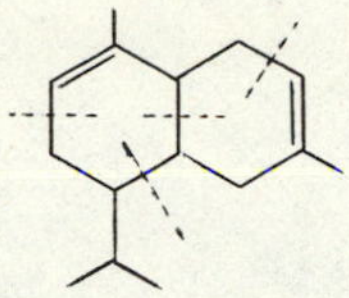

Cadinene

f) 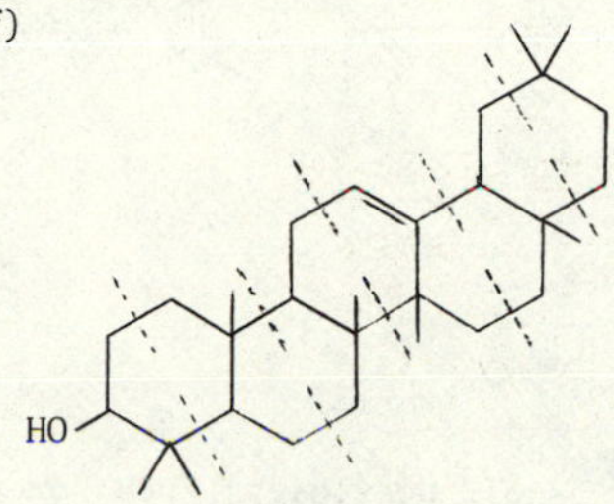

β-Amyrin

23-8

a) Phosphorylation converts the hydroxy to a good leaving group and is an energy transfer step (sec. 9-8B) in biological processes.

b) The good leaving group phosphate, promotes loss of CO_2 by accommodating the electron pair of the bond being cleaved.

c) 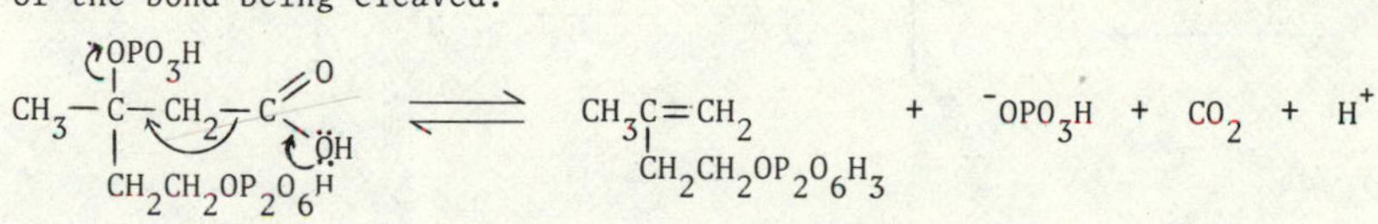

23-9 Acid-catalyzed cyclization of nerol forms an intermediate carbocation which is the precursor for the three terpenes.

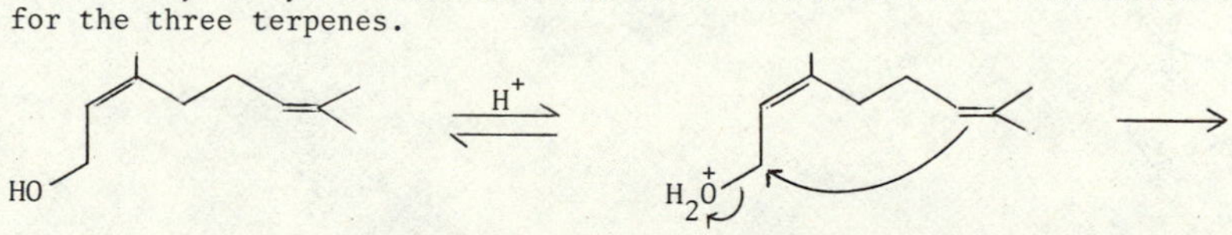

Contd...

23-9 contd...

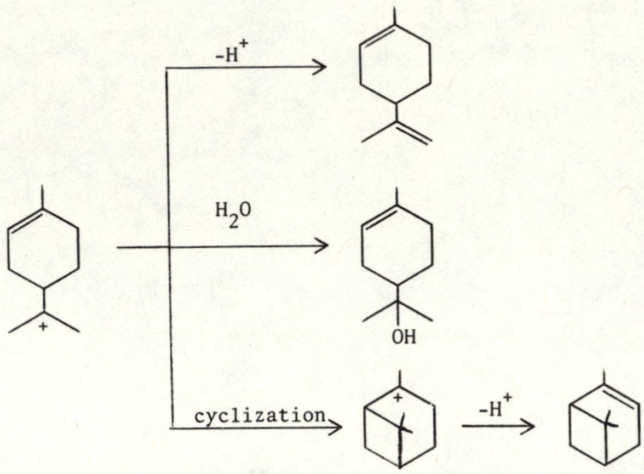

23-10 There are 2^9 = 512 possible stereoisomers of cholestanol (coprostanol).

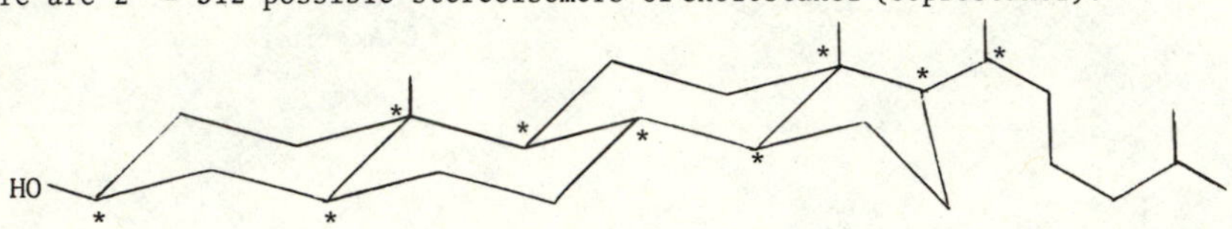

23-11 In each case the epimer with an equatorial hydroxy group is more favorable thermodynamically.

23-12

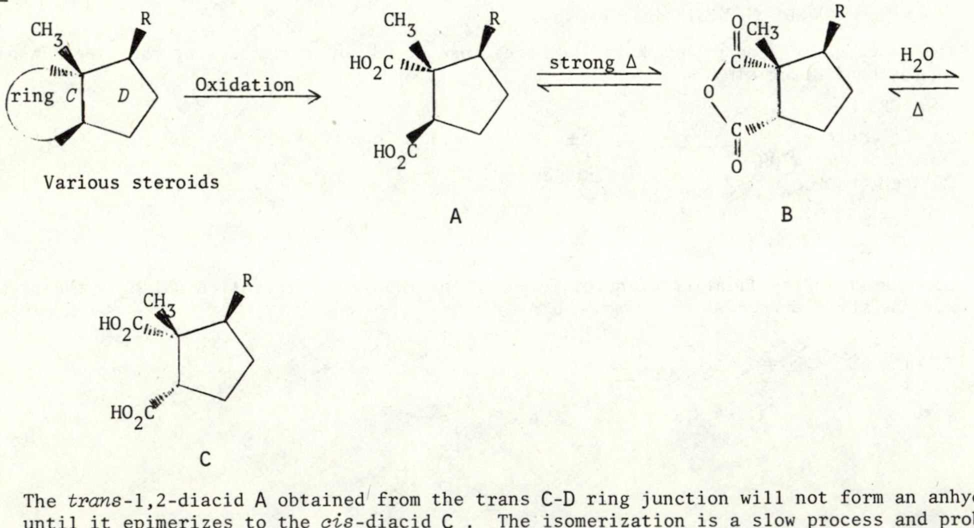

The *trans*-1,2-diacid A obtained from the trans C-D ring junction will not form an anhydride until it epimerizes to the *cis*-diacid C . The isomerization is a slow process and probably involves acid catalyzed enolization. The *cis*-diacid C readily interconverts with the anhydride B.

23-13 The sequence for the formation of labeled geranyl pyrophosphate from $^{14}CH_3CO_2H$ is depicted in fig. 23-4. A sequence to labeled farnesyl pyrophosphate and labeled squalene follows.

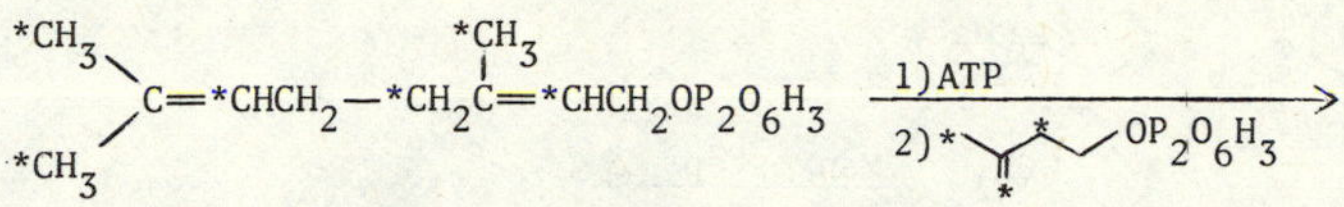

Geranyl pyrophosphate

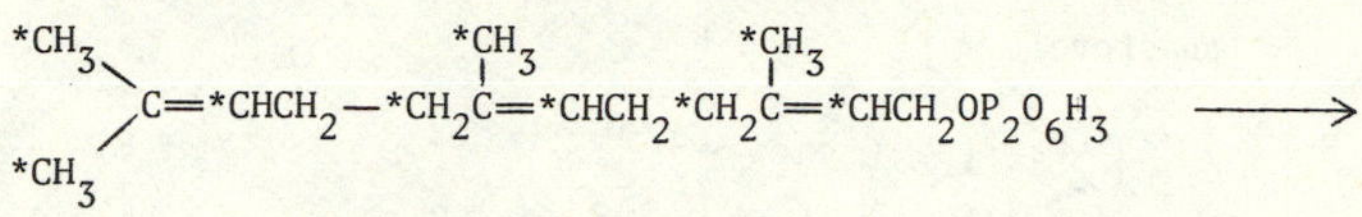

Farnesyl pyrophosphate
(Two molecules)

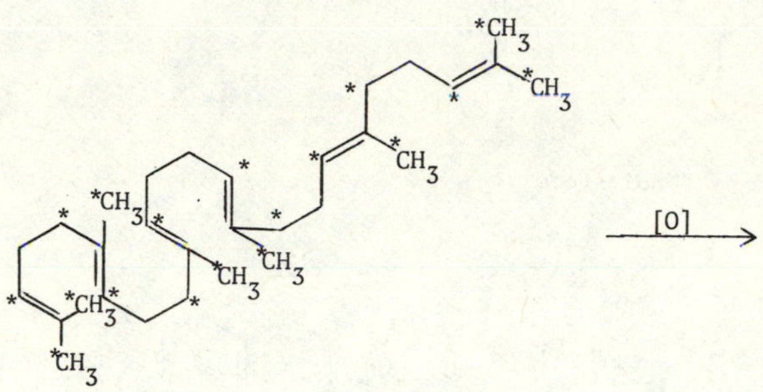

Squalene

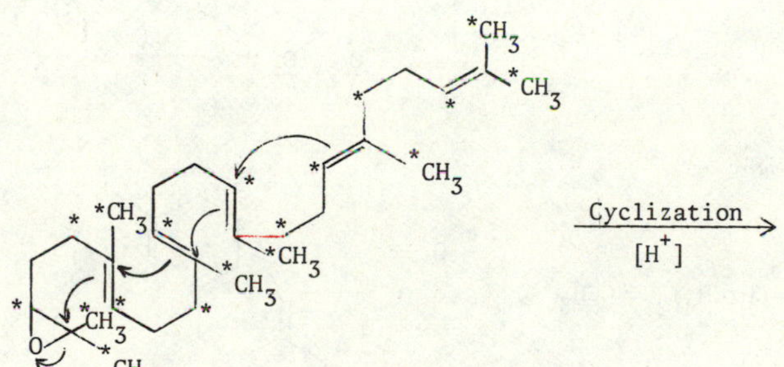

Squalene-2,3-epoxide

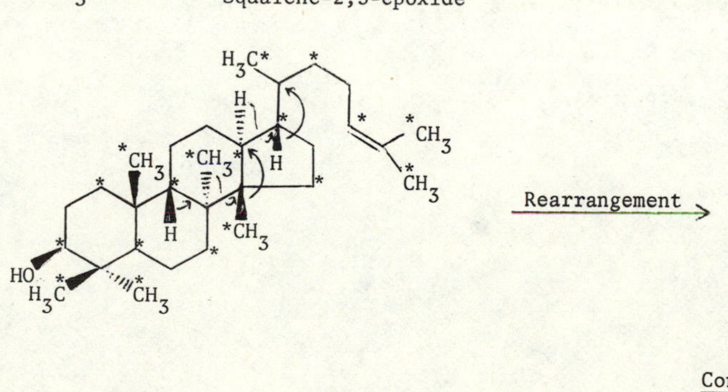

Contd...

23-13 contd...

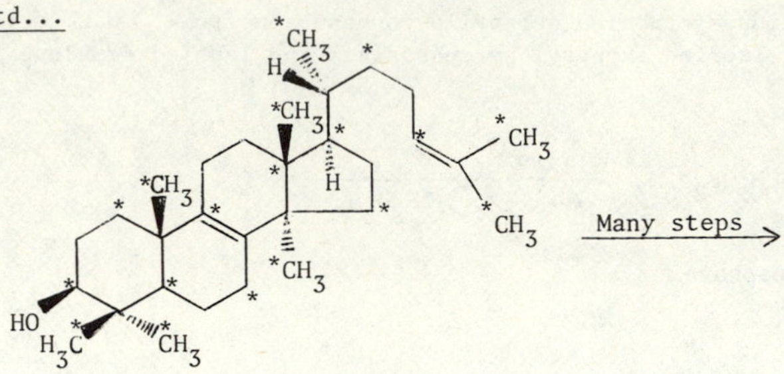

Lanosterol

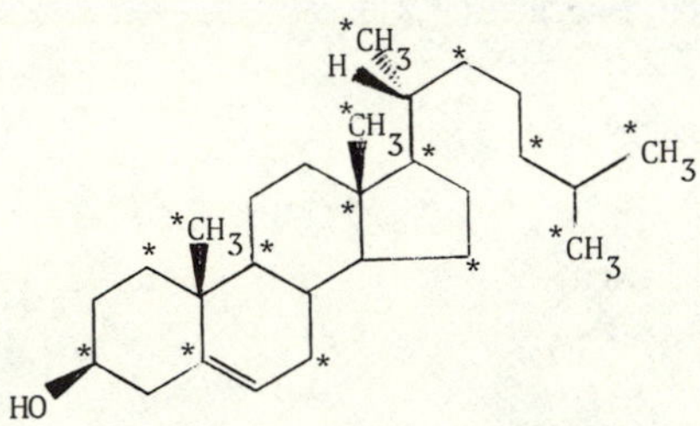

Cholesterol

23-14

A ≡ PBr_3

B ≡ $HOCH_2CH_2OH$/TsOH

C ≡ $(C_6H_5)_3P$

D ≡ $CH_3C{\equiv}C(CH_2)_2MgBr$

E ≡ $HOCH_2(CH_2)_2C(CH_3){=}CH(CH_2)_2C{\equiv}CCH_3$

F ≡ 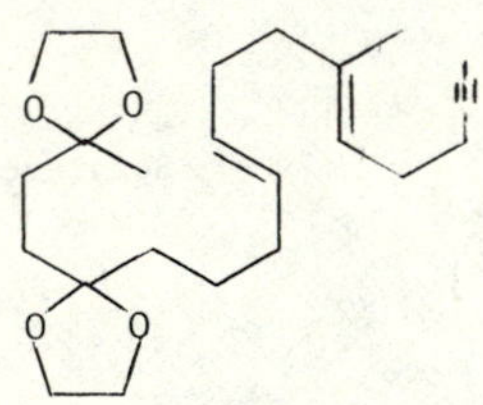

G ≡ 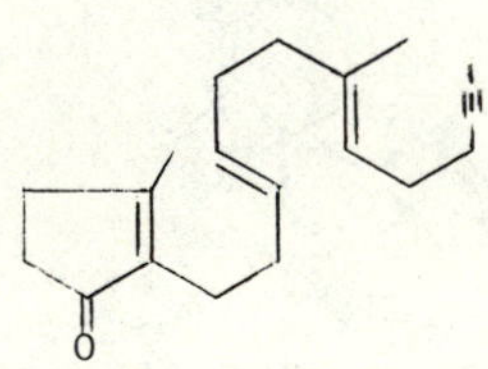

H ≡ O_3

I ≡ KOH/H_2O

23-15

a) Cholic acid has *cis*-fused A-B rings and the epicoprostanol configuration (prob. 23-11) in which the 3-hydroxy group is α and thus equatorial. The 7- and 12-hydroxy groups are also both α- but are more hindered by other parts of the steroid structure.

b) Yes, because the two functional groups are cis to each other.

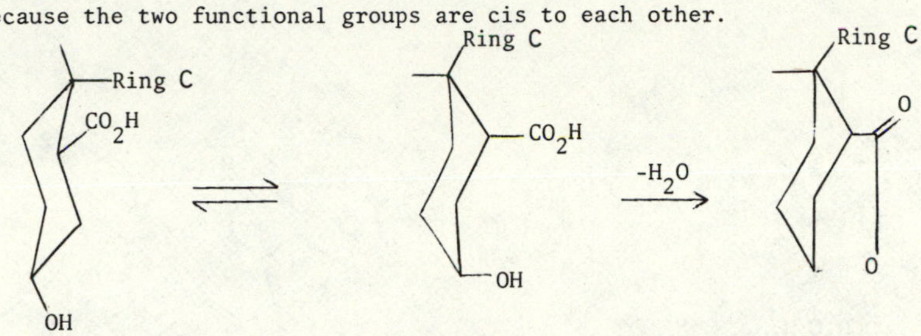

23-16

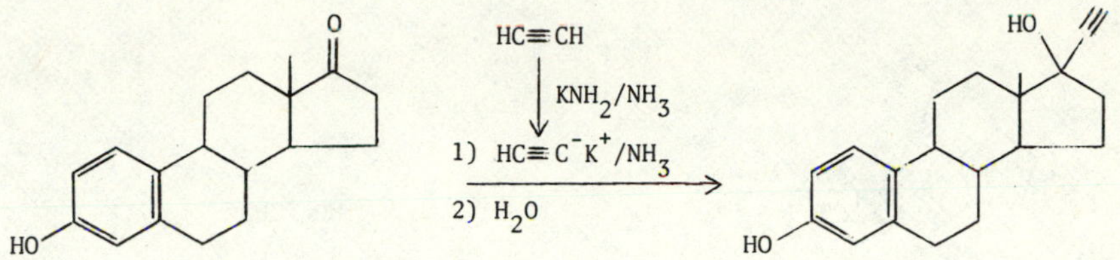

Estrone

23-17 $PGF_{1\beta}$ is the epimer of $PGF_{1\alpha}$ and differs in configuration at the position of the reduced carbonyl.

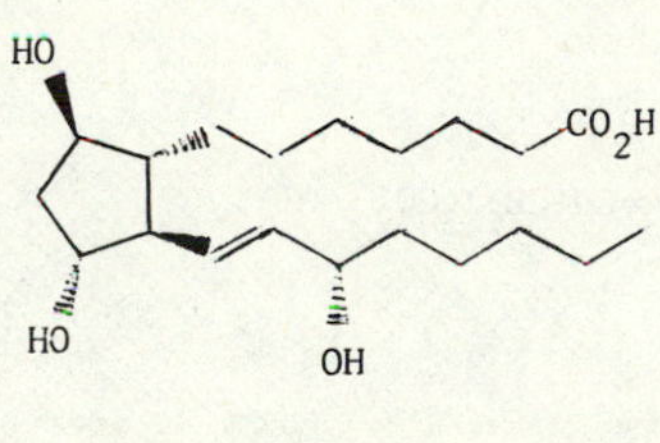

23-18

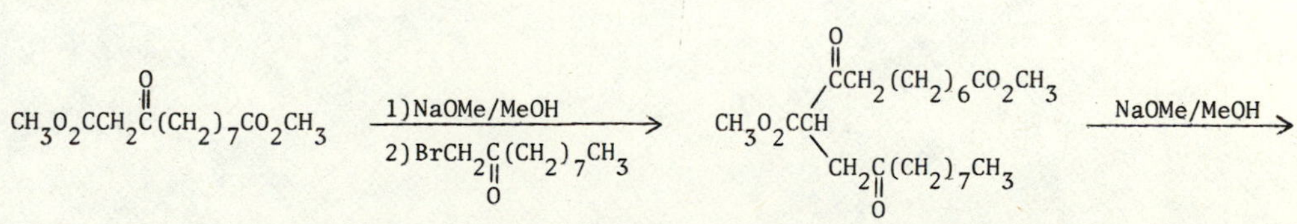

Contd...

23-18 contd...

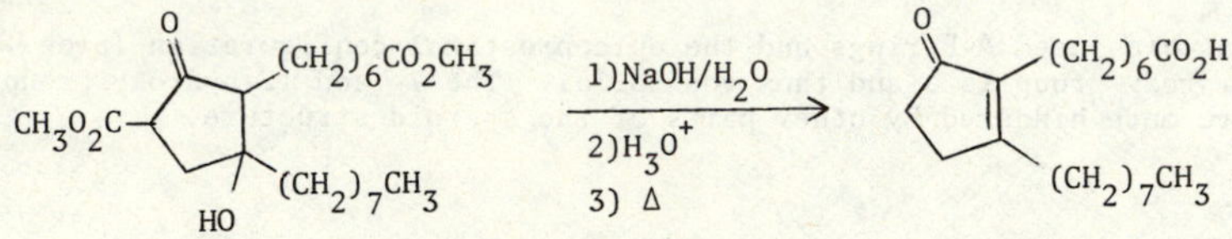

23-19

A ≡ $ClCH_2OCH_2C_6H_5$

B ≡

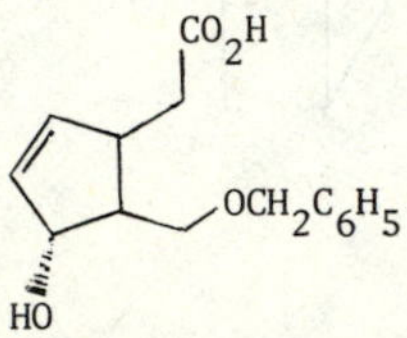

C ≡

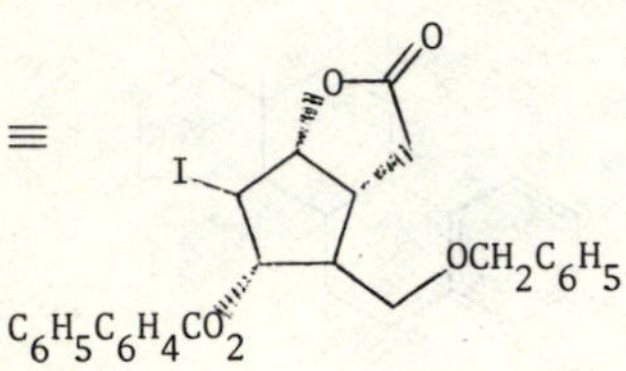

D ≡

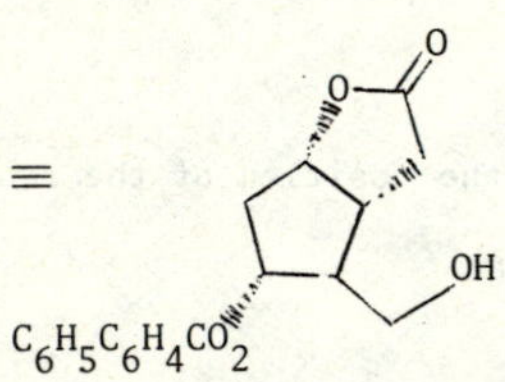

E ≡

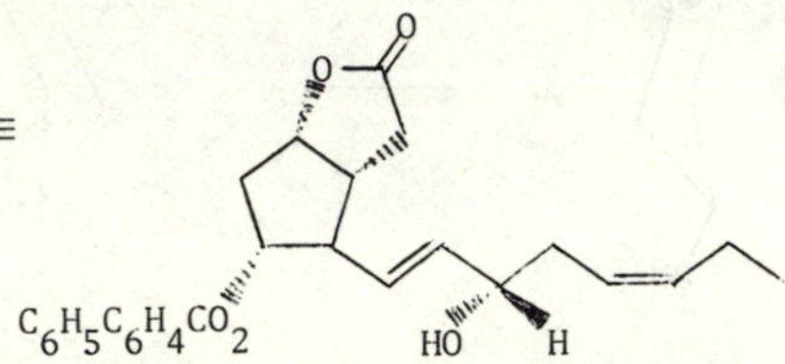

F ≡  , TsOH

G ≡

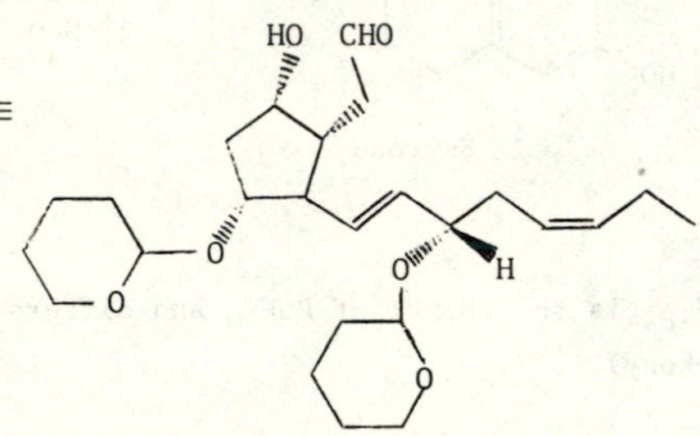

H ≡ $(C_6H_5)_3P{=}CH(CH_2)_3CO_2^-$/THF

23-20

A ≡ CH_3MgI/Et_2O

B ≡ 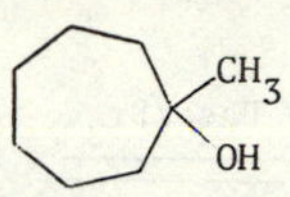

C ≡ $CH_3\overset{O}{\overset{\|}{C}}(CH_2)_5CHO$

D ≡ $OHC(CH_2)_3CHO$

E ≡ $HC{\equiv}CCH_2ZnBr/Et_2O$

F ≡ 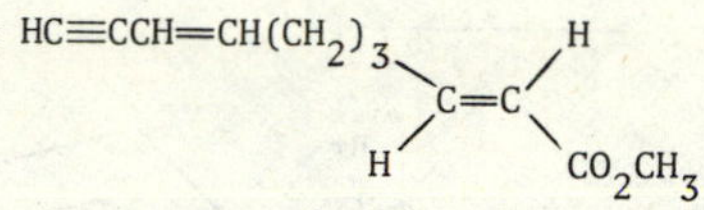

G ≡ $HgSO_4/H_3O^+$

H ≡ 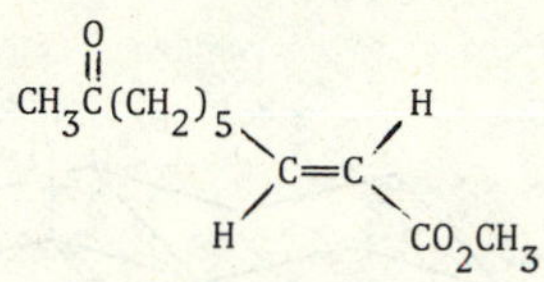

23-21

A ≡ 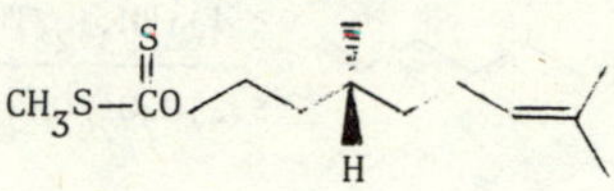

B ≡ H_3O^+/THF

C ≡ 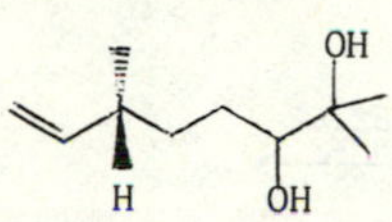

D ≡ $Pb(OAc)_4$

E ≡ 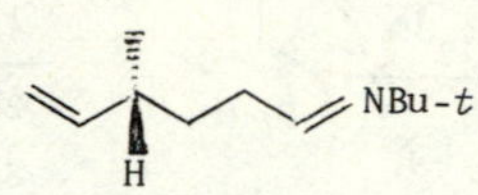

F ≡ $\xrightarrow[2)\,H_3O^+]{1)\,C_2H_5MgBr/Et_2O}$

G ≡ 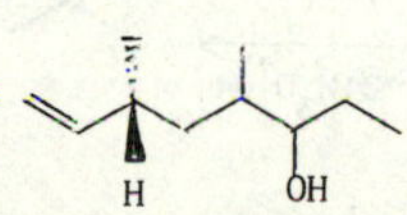

H ≡ m-$ClC_6H_4CO_3H$

23-22 Under the basic conditions of the Robinson annelation, both A and B undergo elimination to produce methyl vinyl ketone *in situ*.

23-23 The two 6-bromocholestanone isomers differ in configuration at the 6-position. The more stable equatorial (β) isomer is favored thermodynamically but addition of bromine to the enol intermediate forms the 6-α epimer more rapidly (kinetics).

Contd...

23-23 contd...

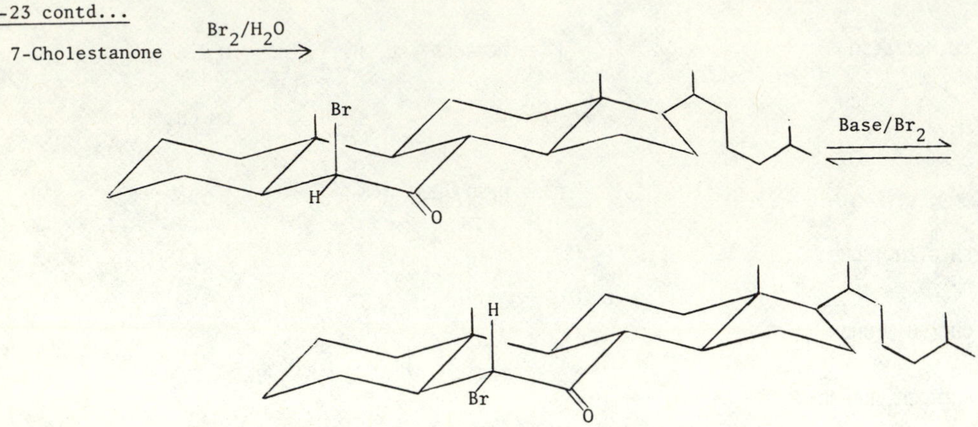

23-24

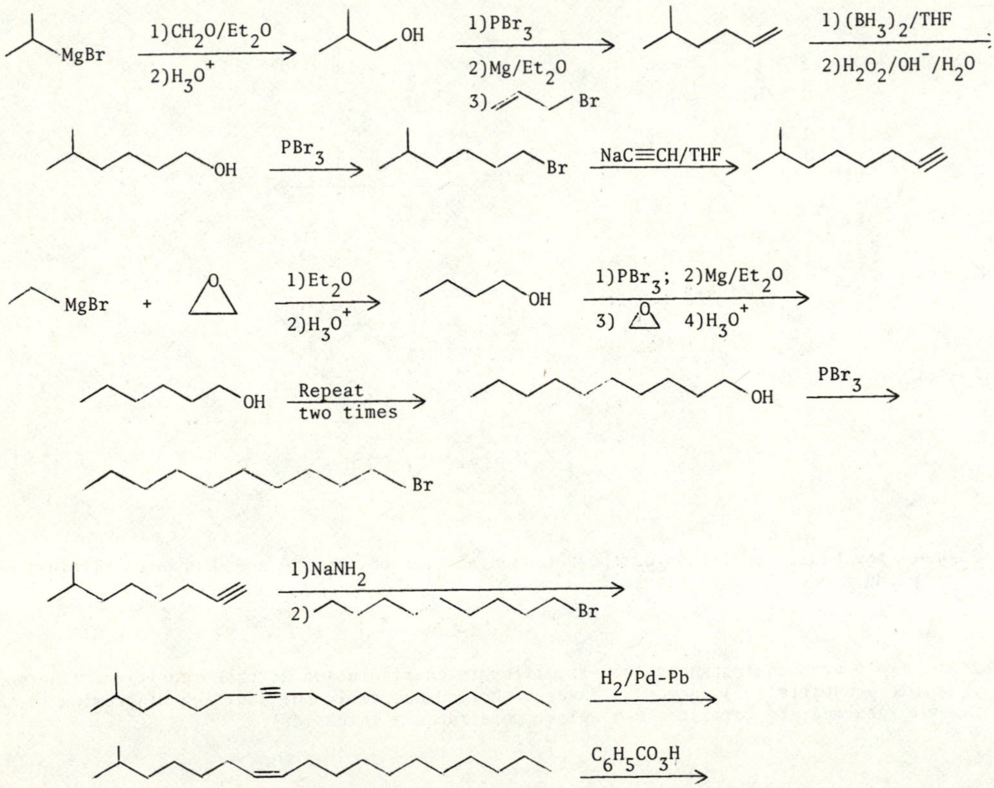

Contd...

23-24 contd...

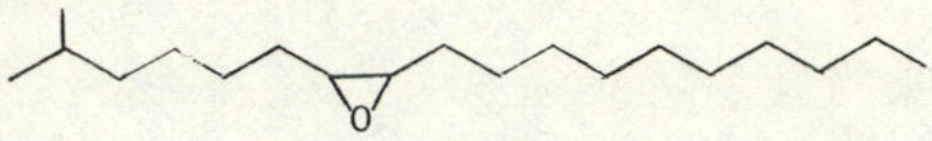

23-25

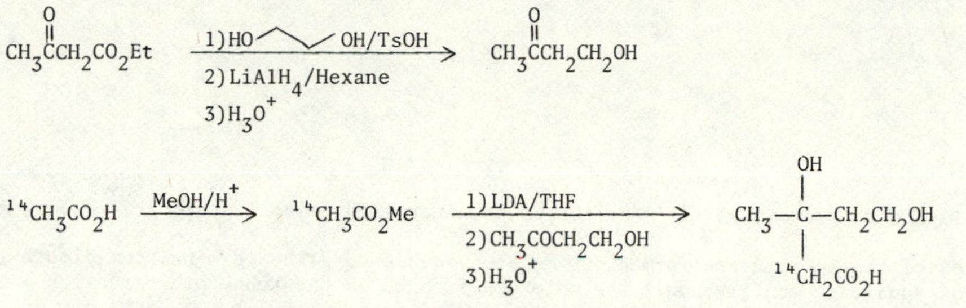

23-26

a + b) Epoxidation takes place from the side of the double bond further from the angular methyl group at C-10 (the back side in the drawing). Addition of acid opens the epoxide so as to give the more stable tertiary carbocation at C-9. The water adds to give the axial hydroxy at C-9 to maintain the chair conformation.

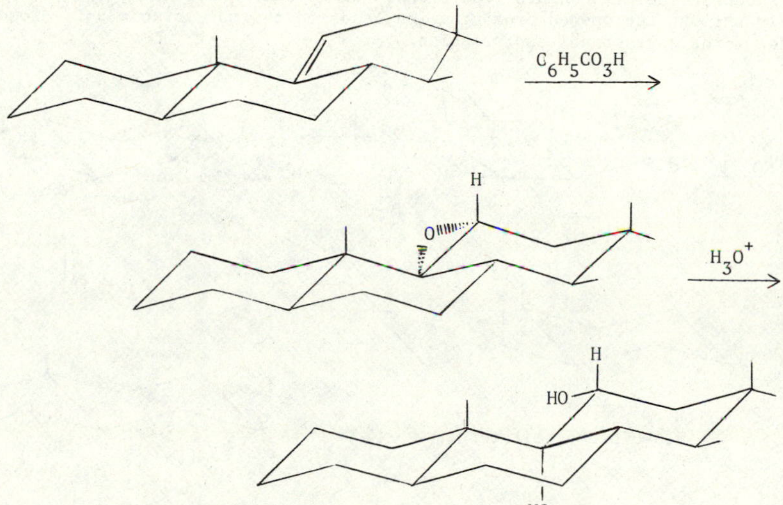

c) The *A*-ring has little effect on reaction of the 9,10-double bond. The bromonium ion intermediate follows a stereochemical course similar to that of the epoxide reaction in part (b).

Contd...

23-26 contd...

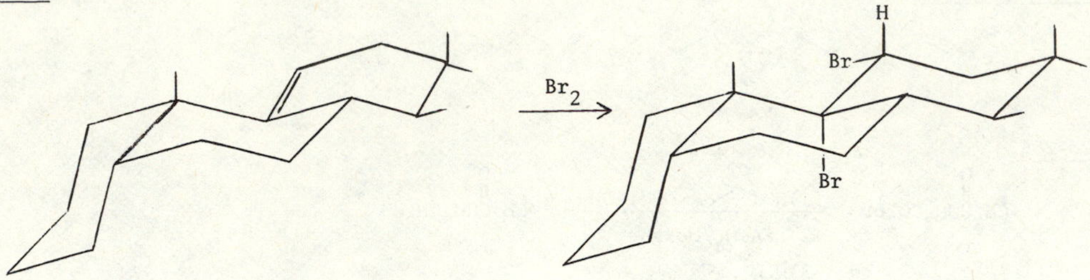

23-27 (See solution to prob. 23-13 for ^{14}C-labeled structural formulas.)

a) Six moles of acetic acid are formed per mole of squalene. (The terminal *gem*-dimethyl groups of squalene each give only one mole acetic acid in the Kuhn-Roth oxidation.) This accounts for 6/18 = 1/3 of the ^{14}C in labeled squalene.

b) Four moles of labeled acetic are recovered from cholesterol but one (that from the C-18 angular methyl group) will have both carbon atoms labeled. The proportion of ^{14}C recovered is thus 5/15 = 1/3.

23-28 Bromonium ion formation takes place from the bottom side, opposite to the angular methyl group, and leads to predominatnt anti addition. However, the axial bromine atoms encounter considerable nonbonded repulsions. Slow isomerization and epimerization, presumably through the opened bromonium ion leads to the more stable anti-diequatorial dibromide of the coprostanol conformation.

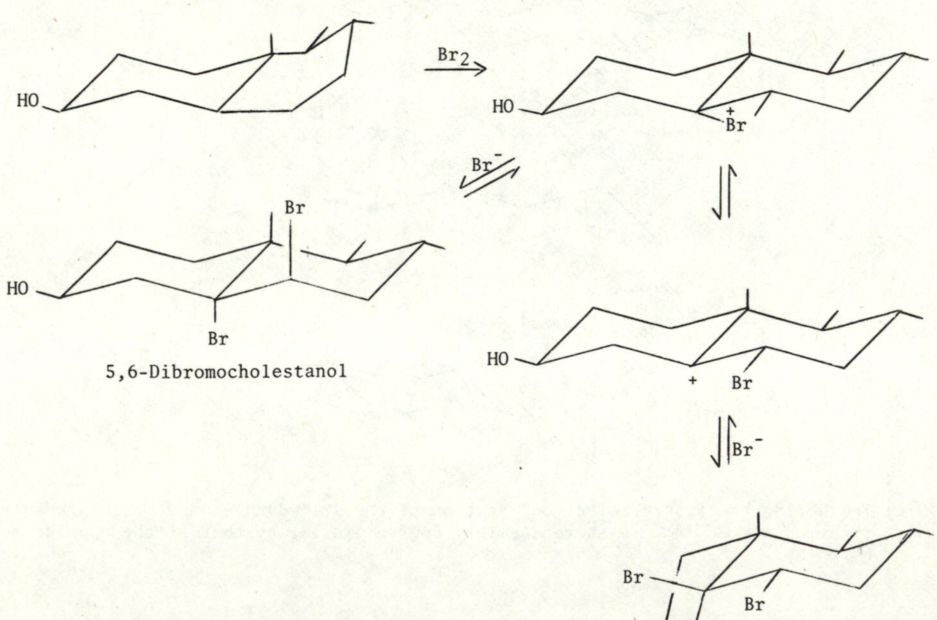

5,6-Dibromocholestanol

5,6-Dibromocoprostanol

23-29

A ≡ 2 C_6H_5MgBr/Et_2O

B ≡ H_3O^+/Δ

C ≡ 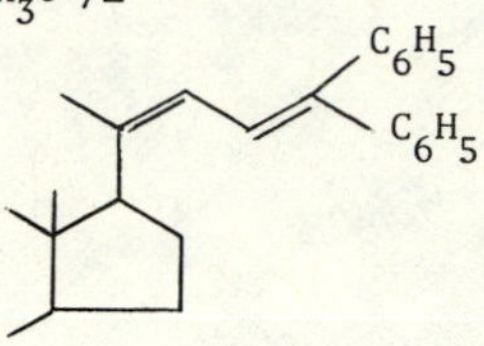

D ≡ 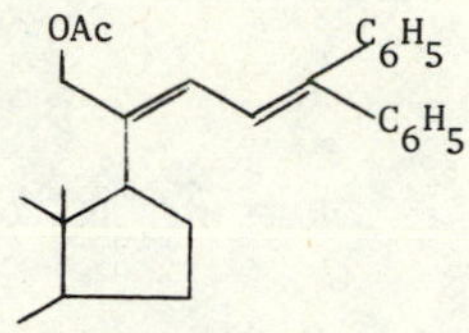

E ≡ O_3

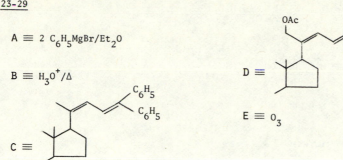

23-30 A double conjugate addition provides the product in one synthetic step.

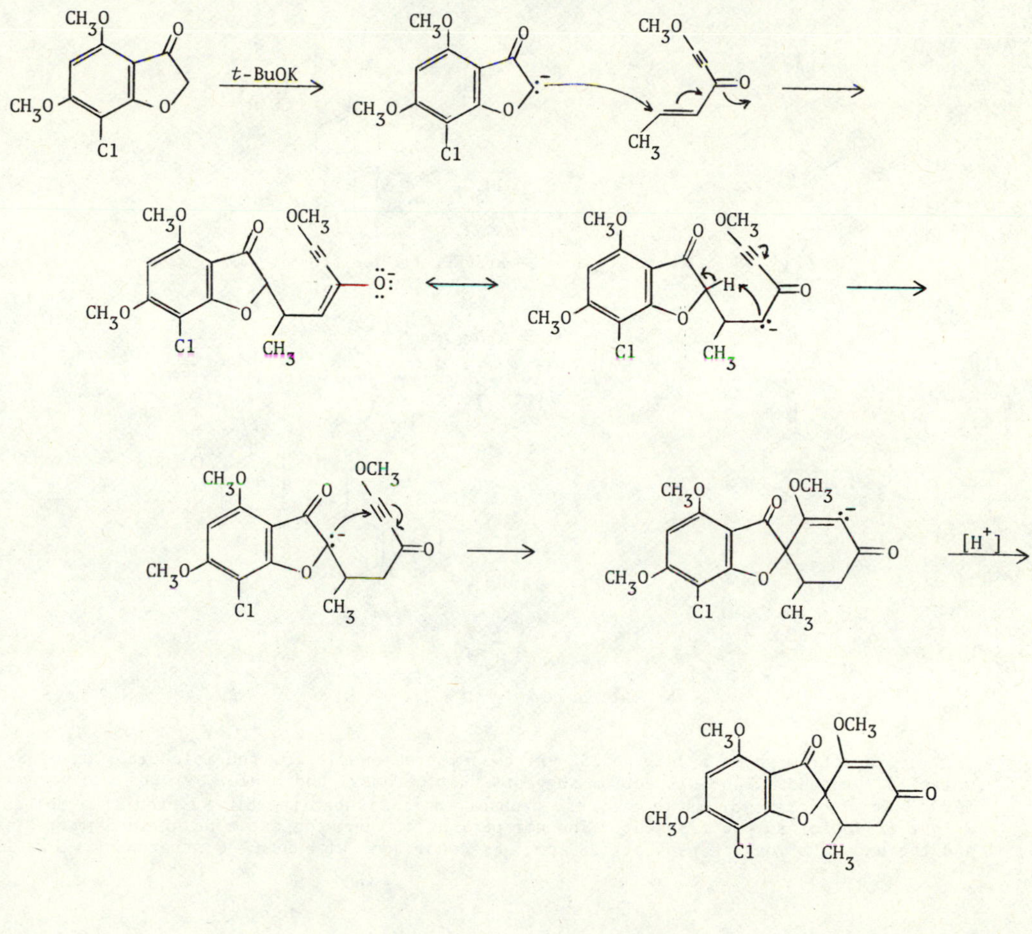

23-31

A ≡ 2 CH_3COCH_3/H^+

B ≡ 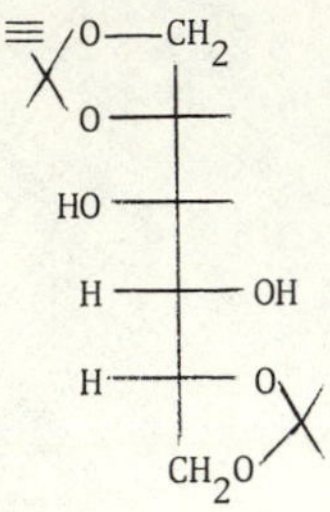

C ≡ $NaBH_4$

D ≡ *n*-$C_{15}H_{31}COCl$

E ≡ *n*-$C_{15}H_{29}COCl$

F ≡ H_2/Pd-C

23-32

A ≡ Ac_2O

B ≡ 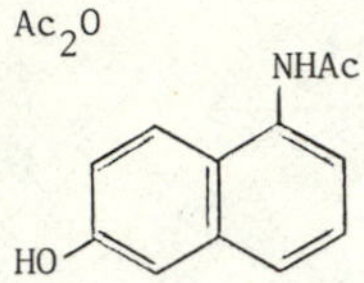

C ≡ $NaNO_2/H_2SO_4$

D ≡ KI

E ≡ 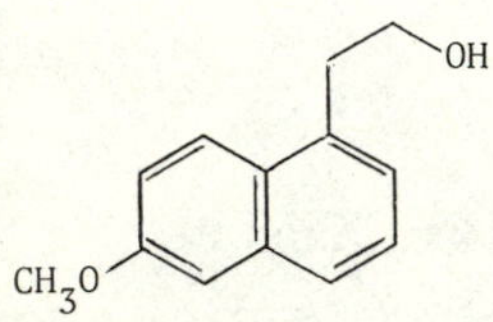

F ≡ 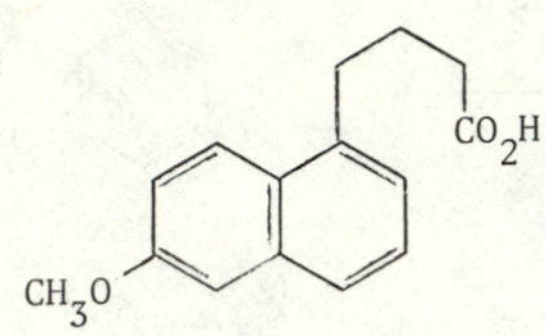

G ≡ $BrZnCH_2CO_2CH_3/Et_2O$

H ≡ NaOMe/MeOH

I ≡ 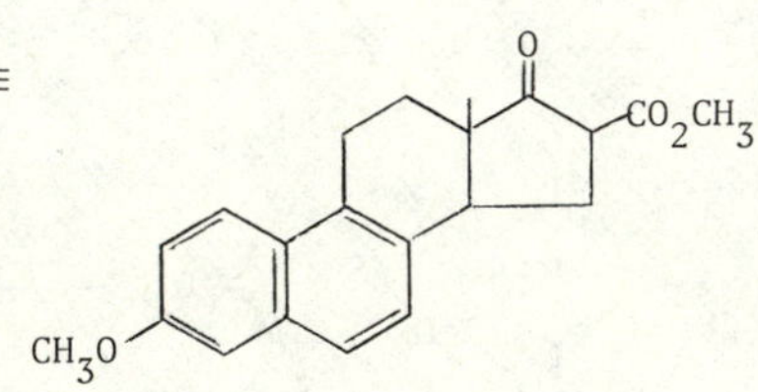

J ≡ H_3O^+/Δ

23-33 The mass spectrum shows a peak at 152 which seems reasonable for the molecular ion of a volatile compound. The IR spectrum suggests that carbonyl and hydroxy groups are present. The ^{1}H-nmr indicates aromatic protons, probably para disubstituted. We predict a molecular weight of 76 for such a fragment. The nmr peak at 4.0 ppm suggest a deshielded methyl group and the broad, strong IR peak at 1280 cm^{-1} is consistent with the C—O stretching vibration

Contd...

23-33 contd...

of an ester. A methyl ester is reasonable and the large peak in the mass spectrum at 121 is consistent with loss of 31 ($-OCH_3$) from the compound. A substituted methyl benzoate accounts for a molecular weight of 135 plus the substituent. The single proton deshielded to 10.5 ppm is consistent with a phenol so that the pheromone is identified as methyl *p*-hydroxybenzoate.

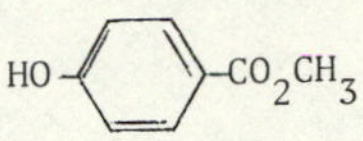

23-34 The IR shows —OH (3300 cm^{-1}) and C=C (1680 cm^{-1}). The peak at 1000 cm^{-1} is probably C—O.
The mass spectrum shows M^+ at m/z = 154.
The ^{13}C-nmr shows nine peaks, but since the compound is a terpene there are expected to be ten carbon atoms. A gem-dimethyl group could give a single peak for the two identical carbon atoms.
The ^{1}H-nmr shows 18 protons. This, plus data above, suggests a molecular formula of $C_{10}H_{18}O$ which is consistent with the M^+ = 154.
The IHD = 2.
The ^{1}H-nmr peaks at 1.6ppm suggest CH_3 and $(CH_3)_2$, both unsplit. The single broad peak at 3.2ppm is probably —OH and the two peaks at 5.0 - 5.5ppm are consistent with alkene protons. If these are indeed two double bonds, then two of those four carbon atoms have no H substituent. The peaks at 4.0 - 4.2 may be two singlets or a slightly distorted doublet. In any case the marked deshielding suggests CH—O
If we now make use of the isoprene rule to draw a basic terpene structure we can use the preceeding data to identify the compound as -

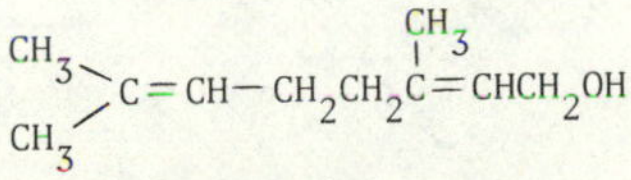

Geraniol

Note that the two CH_2 groups are close to being magnetically equivalent so a broad singlet is observed at 2.1 ppm.

24 Free Radicals

<u>24-1</u>

a) $$2\ (C_6H_5)_3C\cdot \longrightarrow (C_6H_5)_3C-C(C_6H_5)_3$$

b) Hexaphenylethane would be expected to show an approximate singlet in the aromatic region (six equivalent phenyl groups) and a UV spectrum characteristic of benzene (λ_{max} 254 nm).

The actual "Gomberg dimer" shows only protons for five aromatic rings in the nmr spectrum and a UV absorption typical of a conjugated alkene. The nmr spectrum also shows 4 alkene protons. The dimer is now known to be a *para*-disubstituted benzene with the structure indicated below. Steric hindrance apparently inhibits dimerization at two benzylic carbon atoms so that reaction takes place betweeen the tertiary carbon of one free radical and the *para* position of the second free radical.

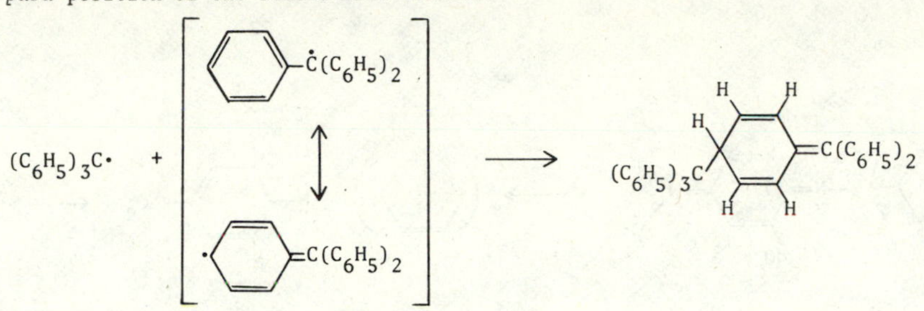

<u>24-2</u> Only a low concentration of free radicals is required to keep the chain reaction going. There is a higher probability that a free radical species will react with the high concentration of nonradical reactants to propagate the chain rather than with another radical in a termination step.

<u>24-3</u> The H—F bond energy is 136 kcal/mol (568 kJ/mol). The bond is stronger than those of HCl and HBr. A fluorine atom actually abstracts hydrogen atoms more rapidly than Cl· or Br·.

<u>24-4</u>

$$(CH_3)_2CHCH(CH_3)_2 + Cl_2 \xrightarrow[CCl_4]{h\nu} ClCH_2\overset{CH_3}{\overset{|}{C}}HCH(CH_3)_2 + (CH_3)_2\overset{Cl}{\overset{|}{C}}CH(CH_3)_2$$

There are 12 primary hydrogen atoms and 2 tertiary hydrogen atoms to account for the reactivity at each position:

$$\frac{0.60}{12} = 0.05 \qquad \frac{0.40}{2} = 0.20$$

$\therefore$ *prim* : *tert* reactivity = 0.05 : 0.20 = 1 : 4.

24-5

a) 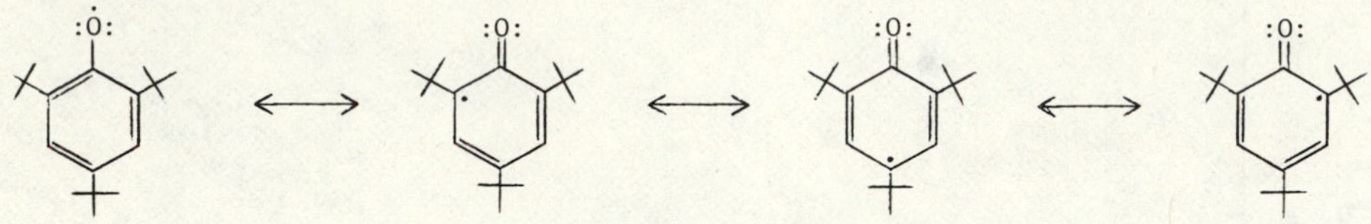

b) 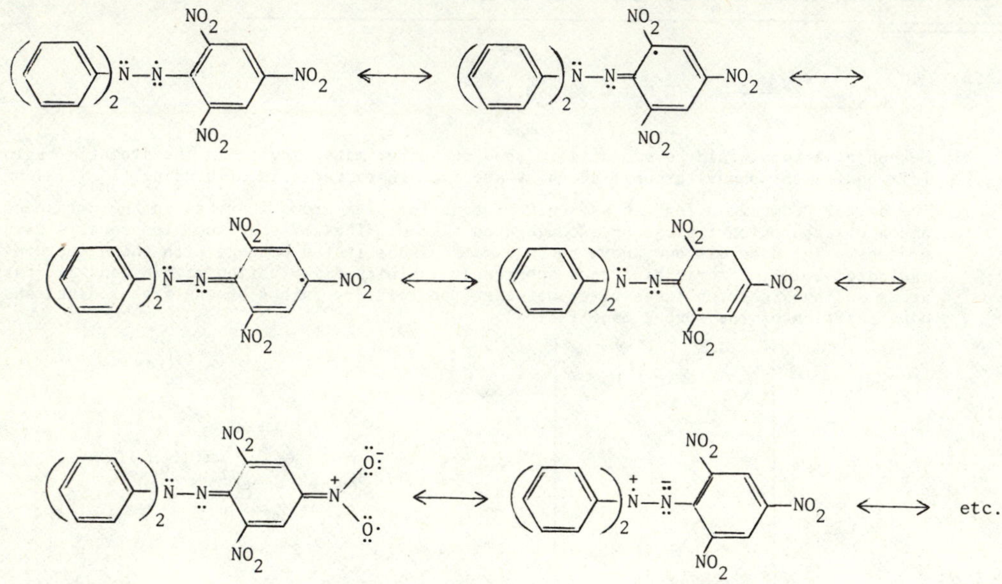

c) 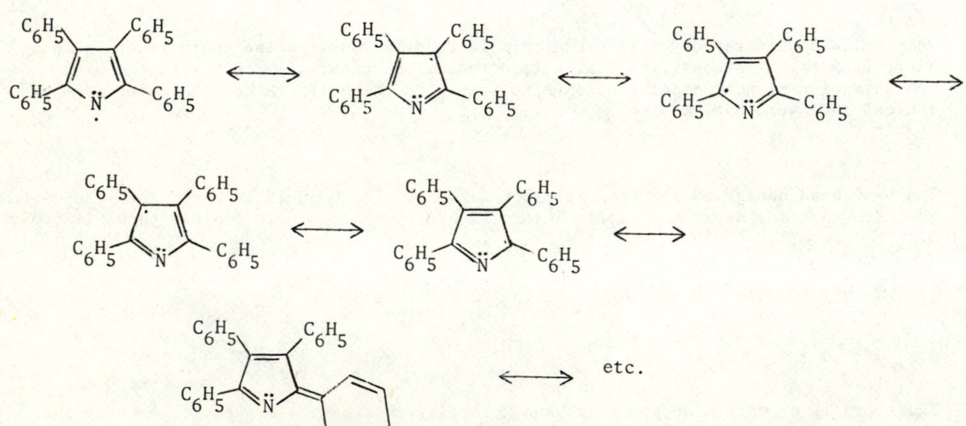

Contd...

24-5 contd...

d)

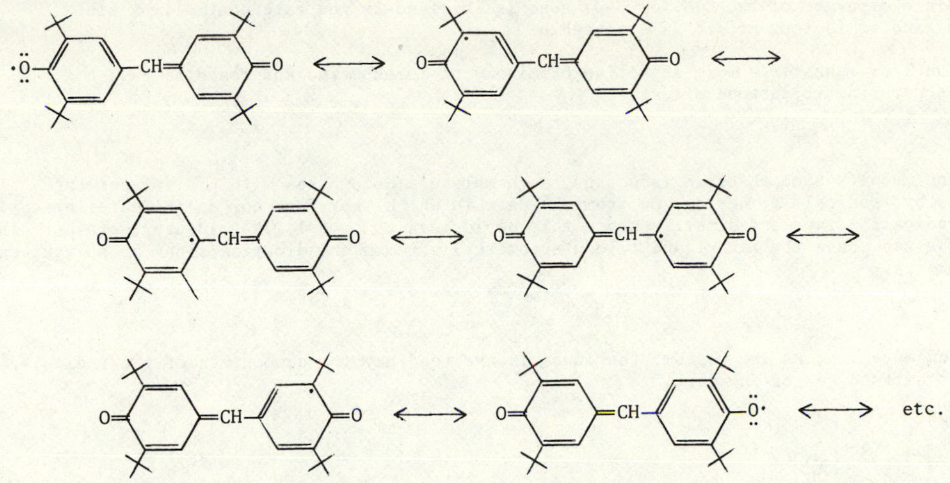

24-6

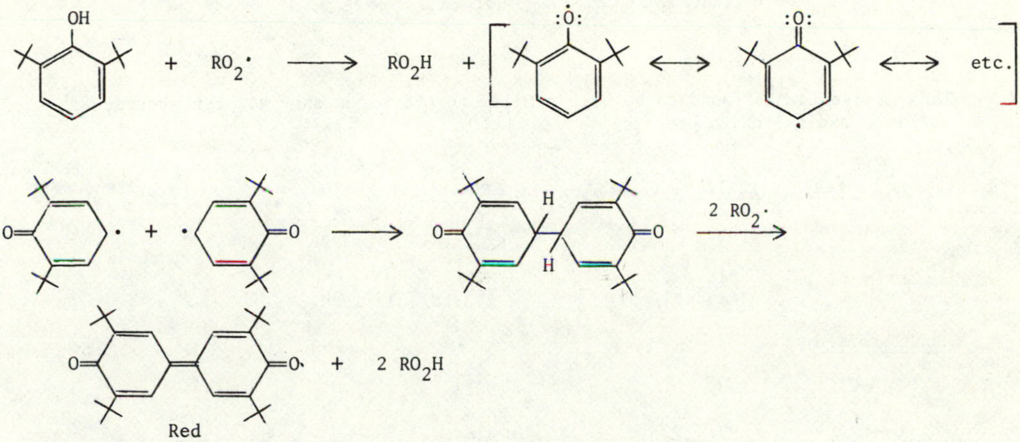

Dimerization takes place at the unhindered para position to give a highly conjugated product.

24-7

a) Since cleavage of the C—H (or C—D) bond is involved in the rate controlling step, a kinetic isotope effect is expected.

b) The less reactive - more selective bromine atom is expected and found to have a larger kinetic isotope effect.

24-8 The results show that, in isooctane, each molecule decomposes within a solvent cage. Methyl radicals do not escape from the cage in which they form nor do trideuteriomethyl radicals escape and dimerize since this would lead to some 1,1,1-trideuterioethane. In the gas phase the methyl and trideuteriomethyl scramble and dimerize since no solvent cage exists.

24-9 The more viscous the medium, the lower is the tendency for components of the radical pair to diffuse out of the cage.

24-10

a) $CH_3C(=O){-}O{-}O{-}C(=O)CH_3 \xrightarrow{\Delta} CH_3CH_3 + CH_3CO_2CH_3 + CO_2$

b) $C_6H_5C(=O){-}O{-}O{-}C(=O)CH_2CH_3 \xrightarrow{\Delta} C_6H_5CH_2CH_3 + C_6H_5CO_2CH_2CH_3 + CH_3CH_2CO_2C_6H_5 +$

$C_6H_5{-}C_6H_5 + CH_3CH_2CH_2CH_3 + CO_2$

24-11 These reagents are found to be very similar to NBS since they all can provide Br_2 within a radical chain process.

24-12

Bonds broken	kcal/mol	kJ/mol
Sn—H	80	334
C—Cl	81	339
	161 kcal/mol	673 kJ/mol

Bonds formed	kcal/mol	kJ/mol
Sn—Cl	120	502
C—H	99	414
	219 kcal/mol	916 kJ/mol

$\therefore \Delta H° = 161-219 = -58$ kcal/mol
(673-916 = -243 kJ/mol)

∴ Reaction is exothermic

24-13 Both reactions proceed with complete loss of optical activity as is typical of most radical reactions. The deuterated tin hydride produces deuterated product as is consistent with the chain mechanism.

24-14 The absence of a hydrogen-deuterium kinetic isotope effect indicates that hydrogen abstraction is not the rate controlling step. The addition step in the mechanism proposed is believed to be rate controlling.

24-15 Copper acts as the electron transfer agent. In the first step Cu(I) gives up an electron to the aromatic ring to form the phenyl radical and Cu(II). Abstraction of a bromine atom from $CuBr_2$ reduces the copper back to Cu(I) and the sequence repeats.

24-16 Addition of X• involves formation of a C—X bond and conversion of a C═C bond to a C—C• The energy for conversion of a double to a single bond is the same in all three cases.

>C═C< ⟶ >C—C< ΔH° = 146-83 = 63 kcal/mol
(610-347 = 263 kJ/mol)

∴ Cl• + >C═C< ⟶ Cl—C—C• ΔH° = 63-81 = -18 kcal/mol
(263-339 = -76 kJ/mol)

Br• + >C═C< ⟶ Br—C—C• ΔH° = 63-68 = -5 kcal/mol
(263-284 = -21 kJ/mol)

I• + >C═C< ⟶ I—C—C• ΔH° = 63-51 = 12 kcal/mol
(263-213 = 50 kJ/mol)

Abstraction of H• involves breaking an H—X bond and making a C—H bond. The energy of formation of the C—H bond is the same in all three cases.

C—C• + H• ⟶ C—C—H ΔH° =-99 kcal/mol
(-414 kJ/mol)

∴ Cl—C—C• + H—Cl ⟶ Cl—C—C—H + Cl• ΔH° = 103-99 = 4 kcal/mol
(431-414 = 17 kJ/mol)

Br—C—C• + H—Br ⟶ Br—C—C—H + Br• ΔH° = 87-99 = -12 kcal/mol
(365-414 = -49 kJ/mol)

I—C—C• + H—I ⟶ I—C—C—H + I• ΔH° = 71-99 = -28 kcal/mol
(299-414 = -115 kJ/mol)

Only the HBr addition has two exothermic steps which are presumed to be reflected in the activation energies.

24-17 At low HBr concentration, hydrogen atom abstraction is slow. The bridged bromine intermediate has a sufficient lifetime to equilibrate between stereoisomers.

24-18 The bromine atom of the reactant acts as a neighboring group to aid hydrogen abstraction and protect the back side of the chiral carbon atom at which reaction occurs. The new Br substitutes regiospecifically at the carbon atom expected to form the more stable radical.

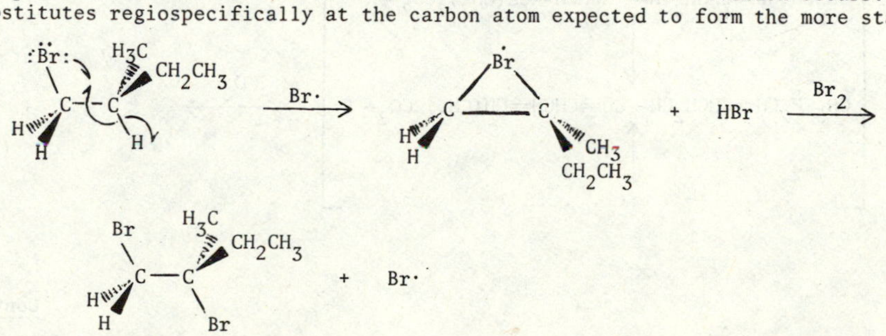

24-19 Addition proceeds to form the more stable benzylic free radical.

24-20

a) $CH_3(CH_2)_4COCH_3$

b) n-$C_6H_{13}CH(Cl)CH_2CCl_3$

c) $ICH_2CH_2CF_3$

d) n-$C_6H_{13}CH_2CH_2C(OH)(CH_3)_2$

e) $CH_3CH_2CH_2CH_2C(=O)C_4H_9$-$n$

24-21 An initial free radical and subsequent hydroperoxide can form at the number 1,2, or 3 carbon atom. Carbon-carbon bond cleavage then leads to formic, acetic, or propionic acids respectively. (Interestingly, butyric acid is not an important product from this process.)

24-22

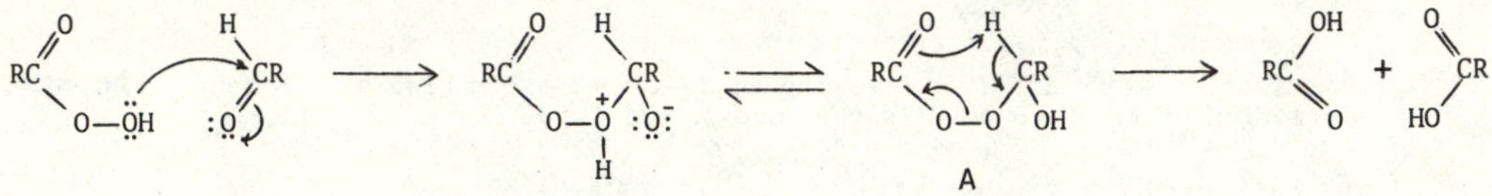

24-23 The three double bonds of the major fatty acid component of tung oil are conjugated, thus more readily lead to the stabilized radical intermediates which form dimers and polymers. The double bonds of the components of linseed oil are nonconjugated.

$CH_3(CH_2)_3CH{=}CHCH{=}CHCH{=}CH(CH_2)_7CO_2H$

9,11,13-Octadecatrienoic acid

24-24

$CH_3CH_2CH{=}CHCH_2CH{=}CHCH_2CH{=}CH(CH_2)_7CO_2{-}R \quad \xrightarrow{\text{Initiator}}$

Linolenic acid

$\left[\begin{array}{c} CH_3CH_2CH{=}CHCH_2CH{=}CH\overset{\bullet}{C}HCH{=}CH(CH_2)_7CO_2{-}R \\ \updownarrow \\ CH_3CH_2CH{=}CHCH_2\overset{\bullet}{C}H{-}CH{=}CHCH{=}CH(CH_2)_7CO_2{-}R \\ \updownarrow \\ \text{etc.} \end{array}\right] \quad \xrightarrow{O_2}$

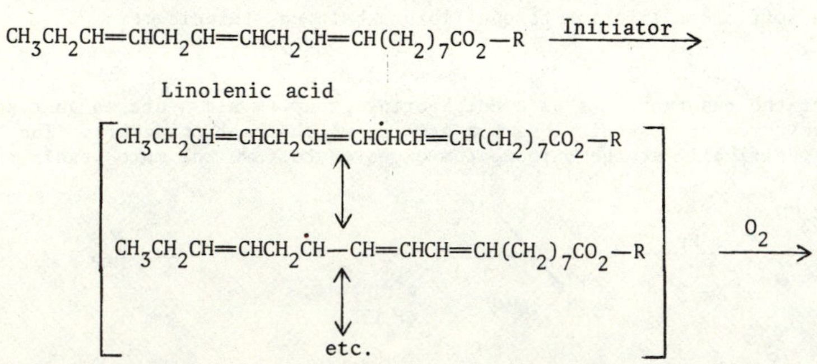

Contd...

24-24 contd...

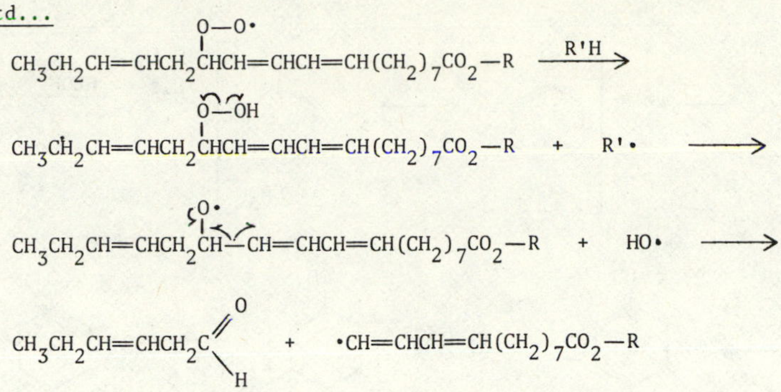

$$CH_3CH_2CH{=}CHCH_2C\underset{H}{\overset{O}{\diagup}} \quad \xrightarrow[\text{(Prob. 24-22)}]{\text{Further oxidation}} \quad CH_3CH_2CH{=}CHCH_2CO_2H$$

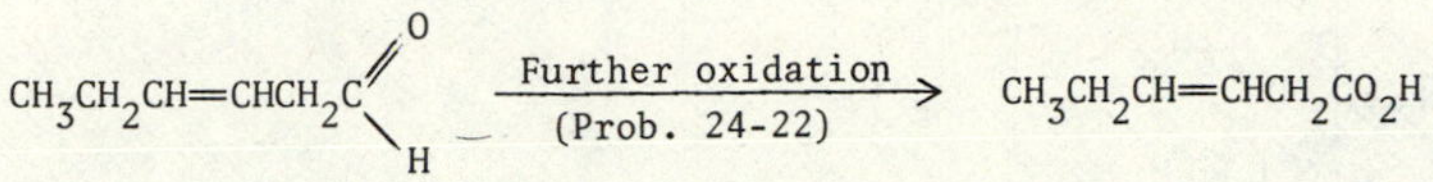

24-25 Abstraction of a hydrogen atom to form a free radical at a tertiary carbon atom of one isopropyl group is more favorable than the similar reaction at a secondary carbon atom of diethyl ether.

24-26

a) R $\equiv CH_3(CH_2)_7CH{=}CH(CH_2)_7$

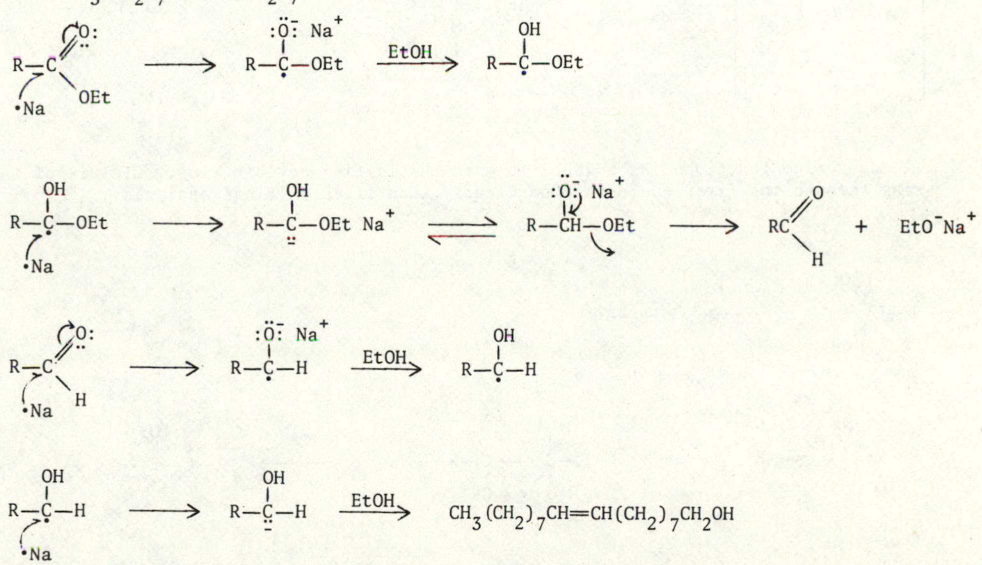

Contd...

24-26 contd...

b)

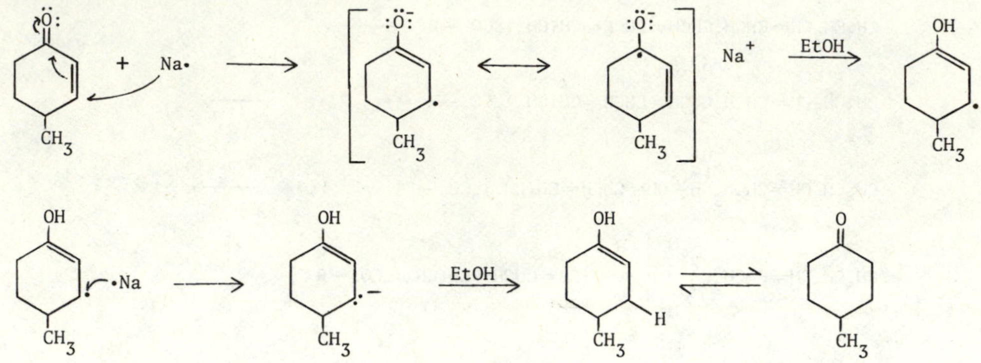

c)

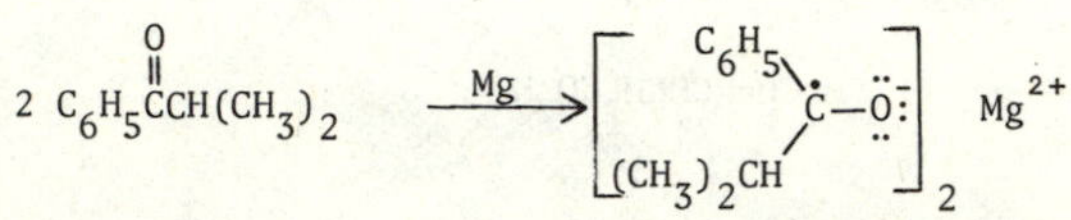

A magnesium ketyl

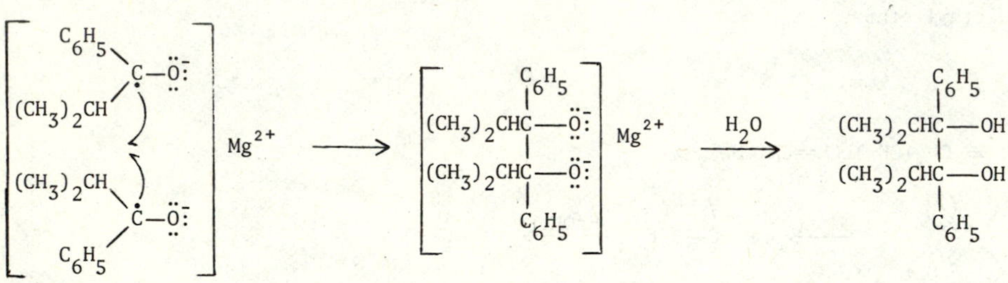

24-27 The deuterium functioned as a label to mark the initial acyloin ring. Closure of the second ring through the first could then be clearly identified on final analysis.

24-28

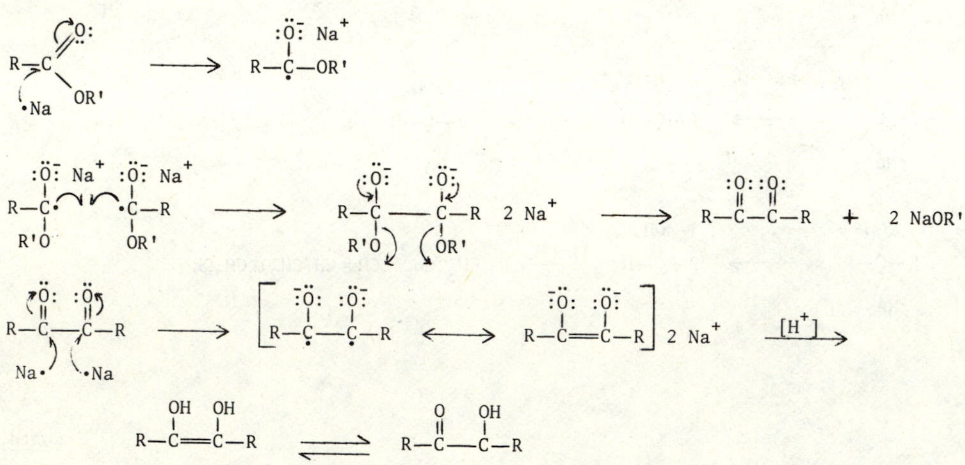

24-29

a) 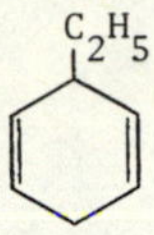

b) 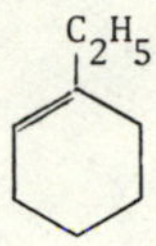

c) 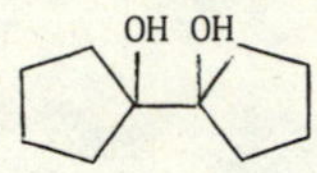

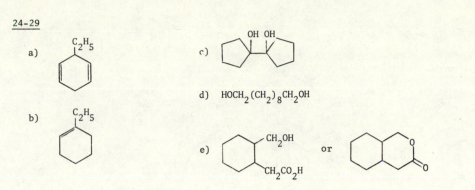

d) $HOCH_2(CH_2)_8CH_2OH$

e) 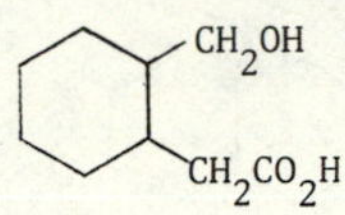 or 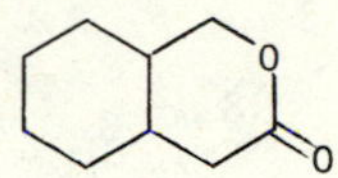

24-30

a) A doublet. The H nucleous splits the electron signal.

b) A singlet since no adjacent magnetic nuclei are encountered, even in the important resonance-delocalized structures.

c) Because the single electron is delocalized over six carbon atoms, only one-sixth of an electron (on the average) actually experiences splitting by a hydrogen atom.

d) The electron is actually on the atom which accounts for the splitting in the hydrogen atom. The magnitude of the coupling constant is therefore quite large.

24-31

a) $CH_3COCH_2CH_2CH_2O_2CCH_3$

b) 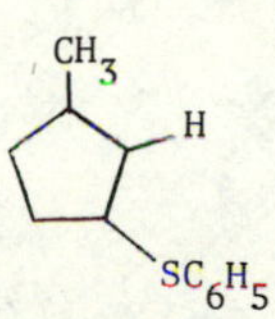

c) $C_6H_5\overset{Br}{C}HCH{=}CHCH_3$

d) 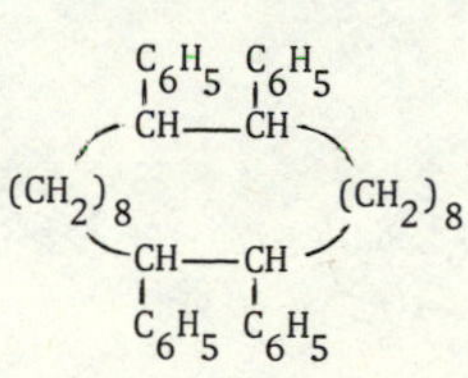

or 2 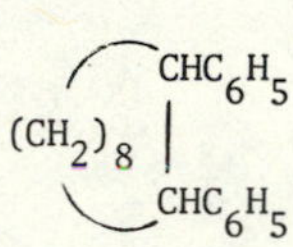

e) 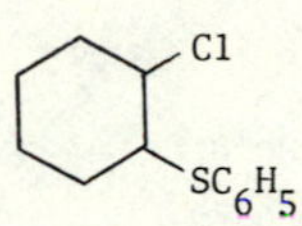

f) 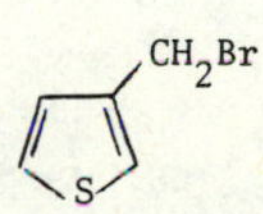

g) 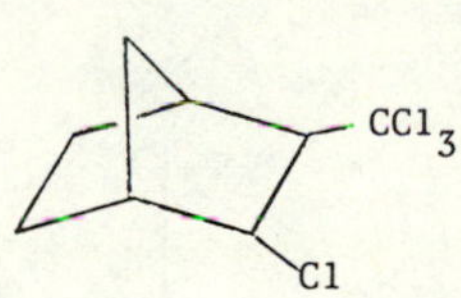

24-32

a)

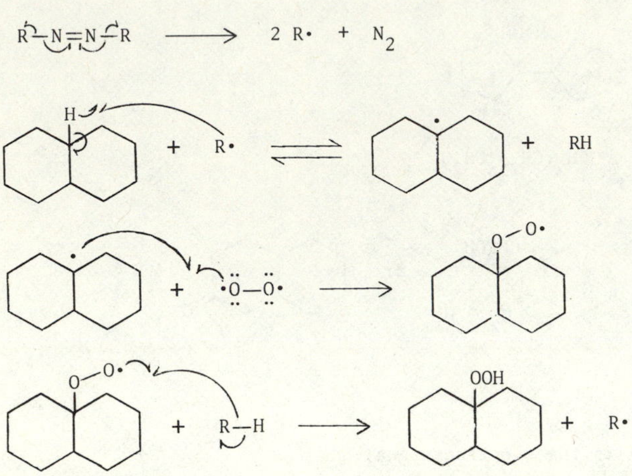

b)

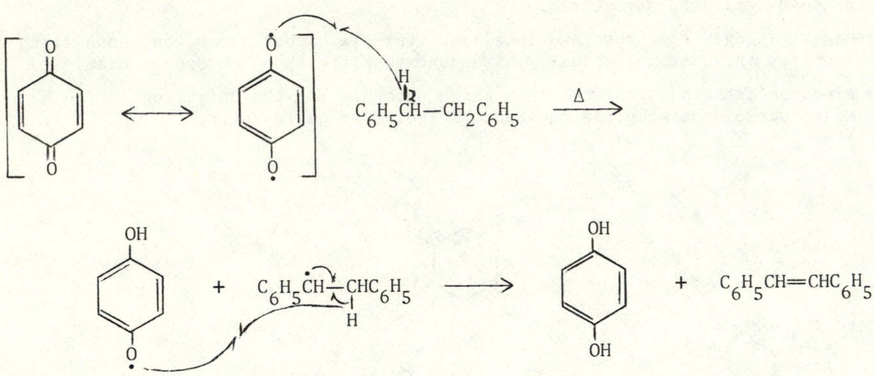

c)

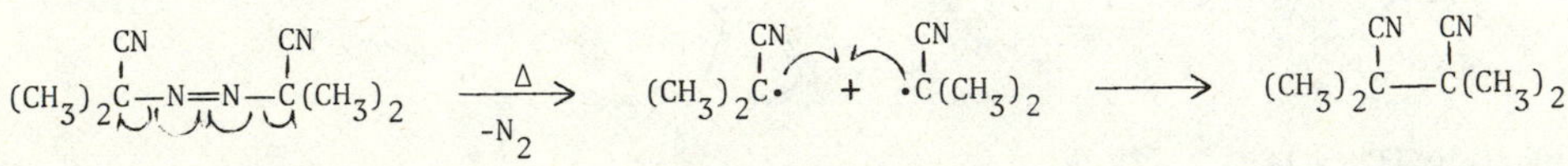

d)

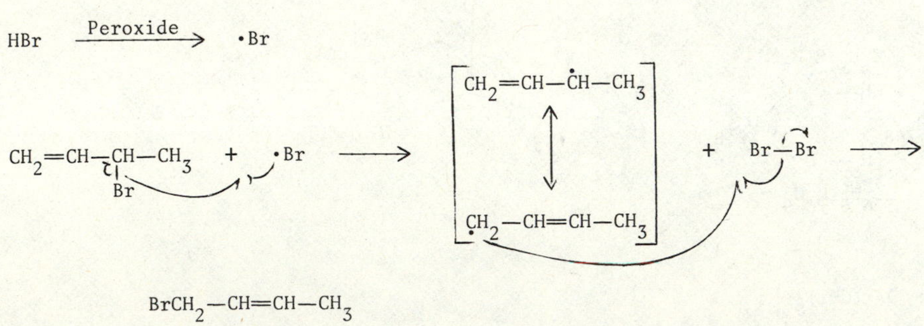

Contd...

24-32 contd...

e) (Only bridgehead position shown.)

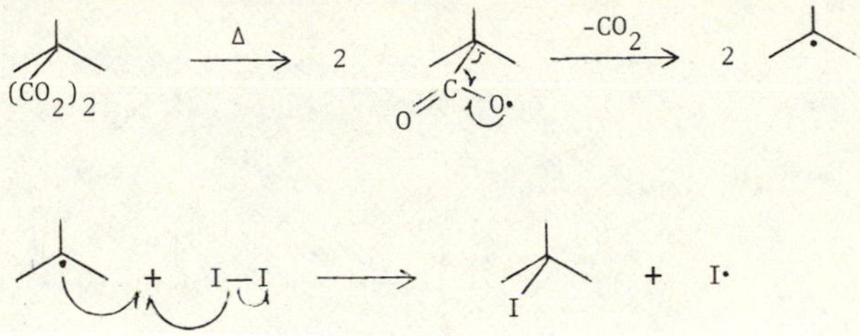

f)

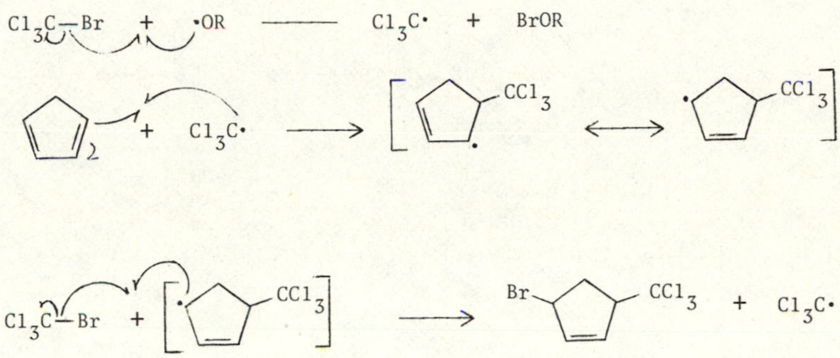

g)

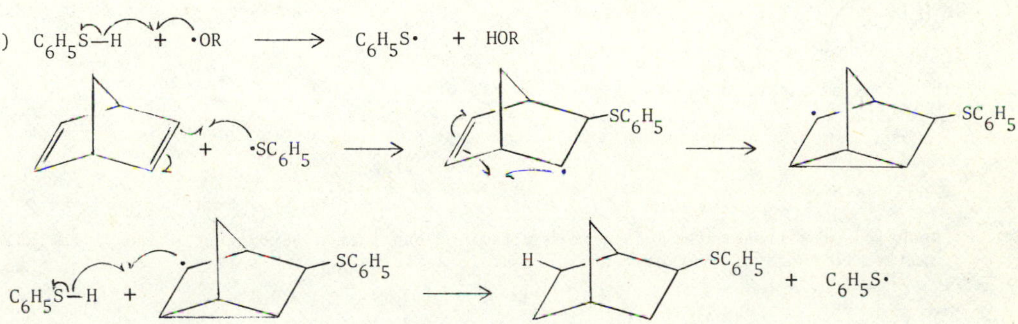

24-33 The order of reactivity reflects the relative stabilities of the intermediate free radicals formed on initial addition of a bromine atom. The tertiary free radical formed from 2-methylpropene is expected to be considerably more stable than the primary free radical formed from ethylene. The kinetic data support the suggestion that transition state stabilities are in the same order as stabilities of the intermediates.

24-34

a) The benzoyloxy radical pair could recombine without change (a), recombine at the opposite oxygen atoms from which cleavage occurrred (b), or rotate relative to each other, then combine (c).

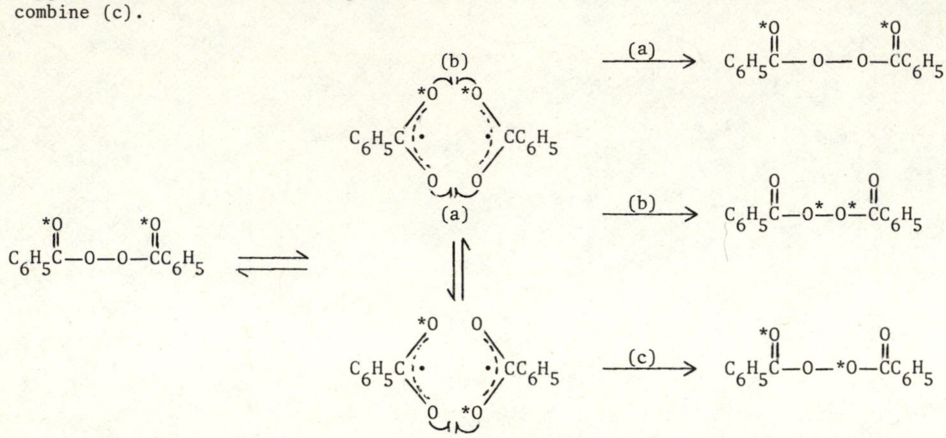

b)

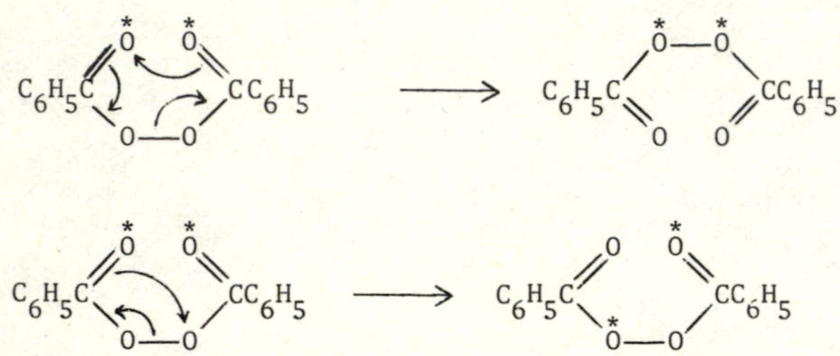

This four atom cyclic pathway is not very favorable

24-35 Reaction of a free radical (R•) with vitamin E can form a persistent phenoxyl radical similar to those of synthetic antioxidants. A radical chain is terminated.

a)

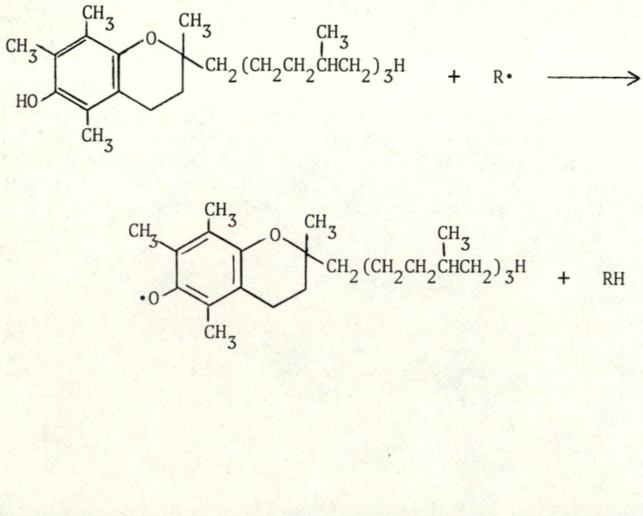

Contd...

24-35 contd...

b) Reduction of coenzyme Q, presumably by NADH, produces a diphenol which can function as an antioxidant as does vitamin E.

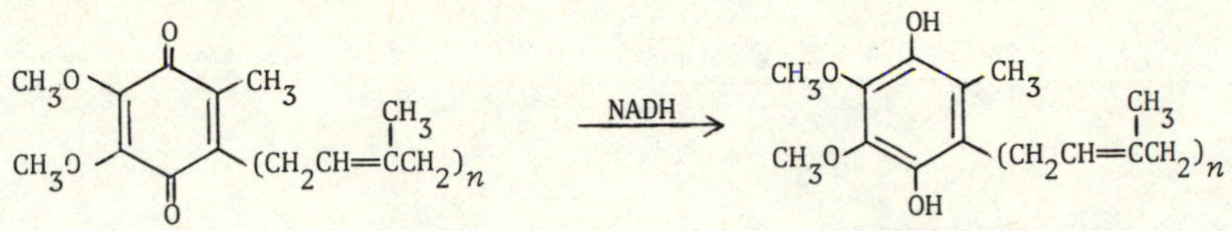

24-36

a)

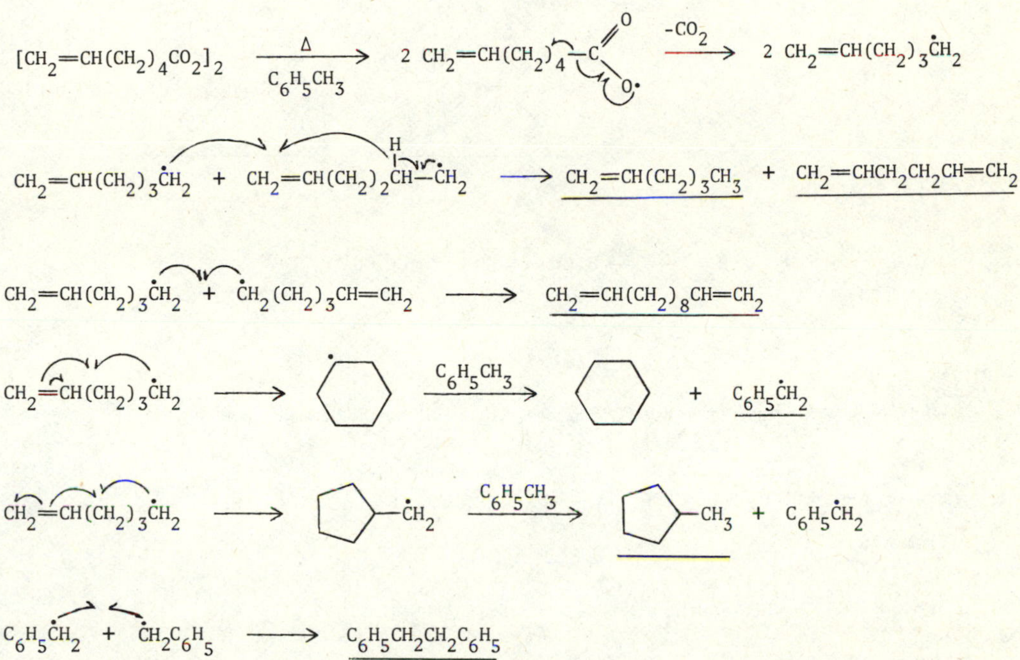

b)

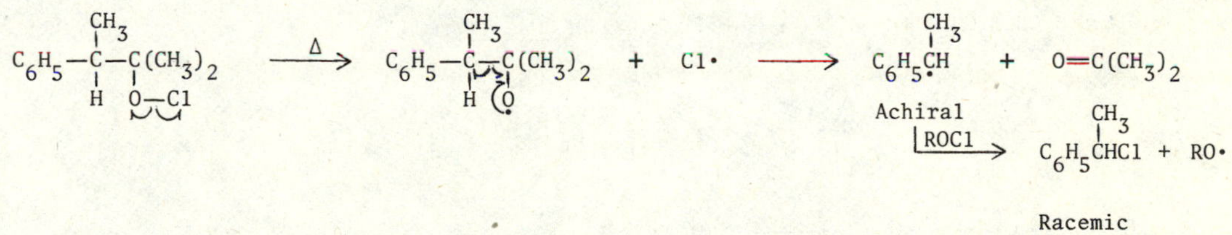

Contd...

24-36 contd...

c)

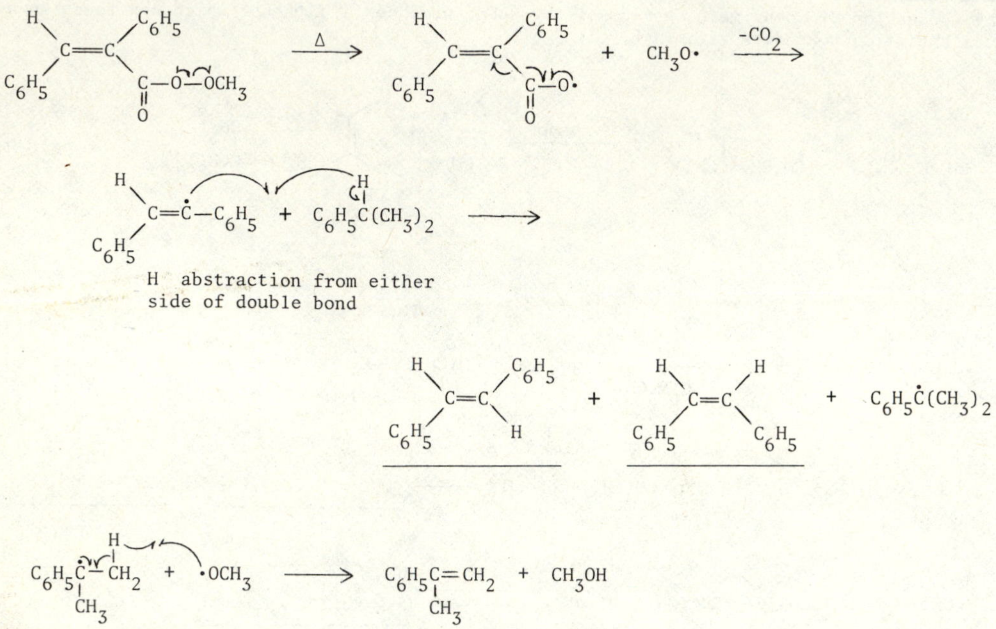

d)

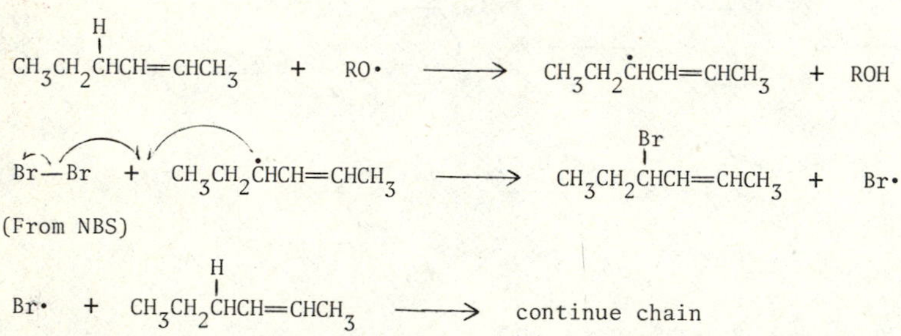

e)

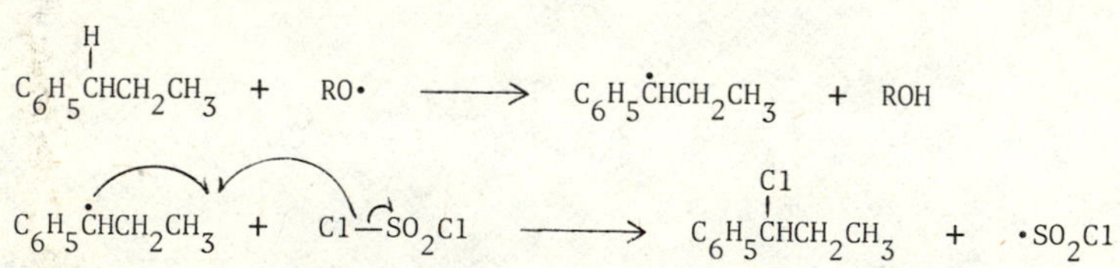

f)

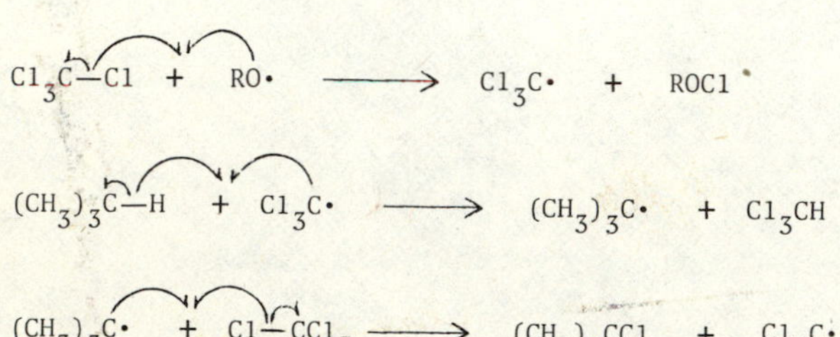

24-37 Thermal decomposition of A is expected to proceed by a radical pathway to produce N_2 and two cyano-stabilized cyclohexyl radicals. When radicals combine within or outside of the solvent cage, B is formed.
Bromine traps radicals that escape from the cage and produces C . Since the amount of cage escape is independent of added bromine, the 70% maximim yield of C indicates that 70% of the radicals escape from the cage, then react.

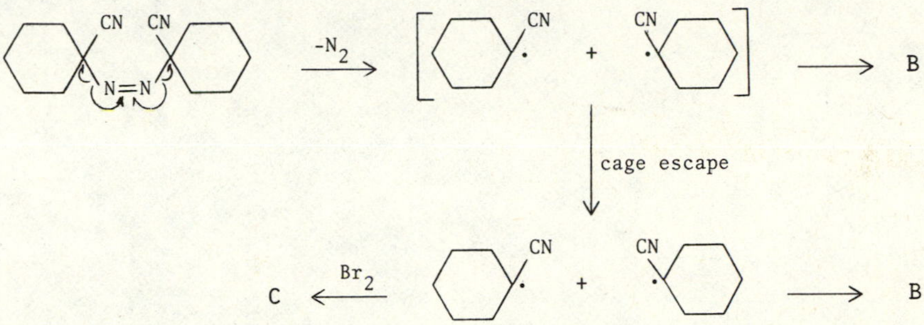

24-38 The doublet-triplet with an area ratio of 2:1 indicates two equivalent protons split by one proton and the one proton split by the other two. Since the doublet is deshielded the most, the protons it represents must be deshielded by a greater number of chlorine atoms than the proton which accounts for the triplet. The compound is

$Cl_2CHCHCHCl_2$ (with Cl on the central carbon)

24-39 Dimerization of the phenoxyl radical is sterically inhibited, but 1,4-addition to butadiene can take place through the electron-delocalized intermediate. If addition had occurred at either unsubstituted position of the aromatic rings more than two *tert*-butyl peaks would have shown in the nmr spectrum.

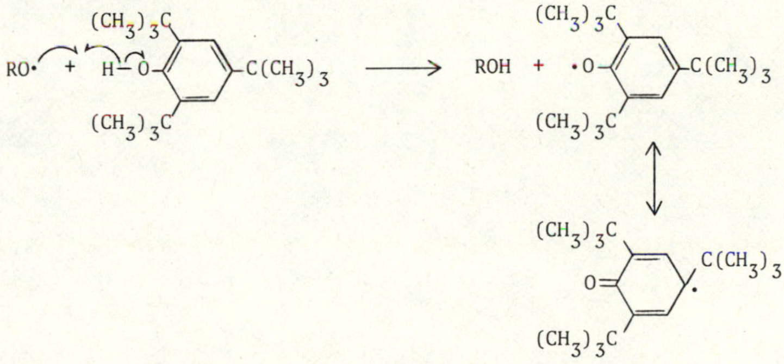

A

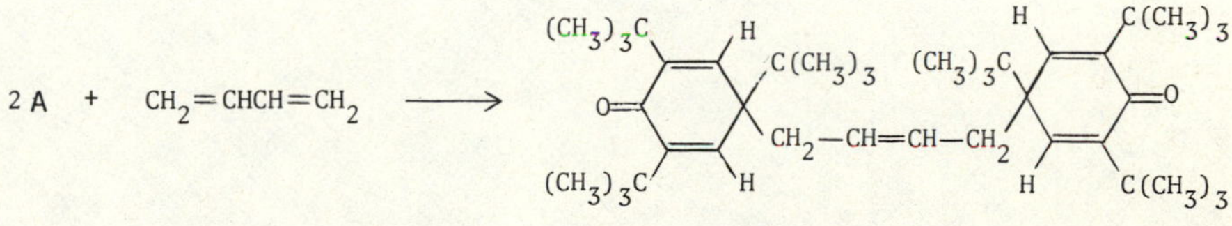

25 Natural and Synthetic Polymers

25-1

a) 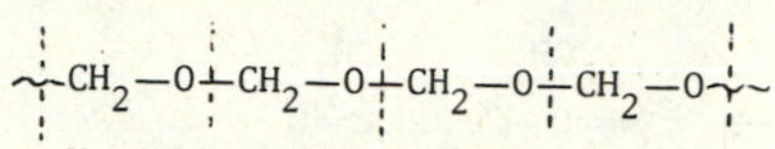

Homopolymer; no cross links.

b) 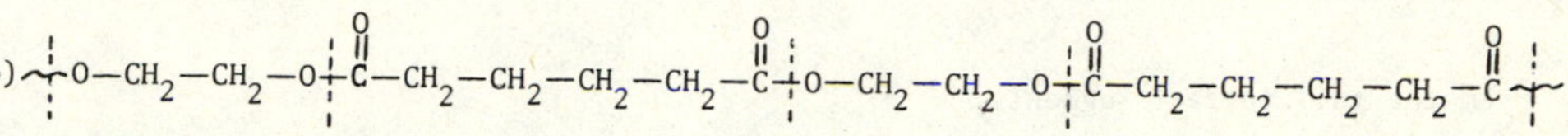

Copolymer; no cross links.

c) 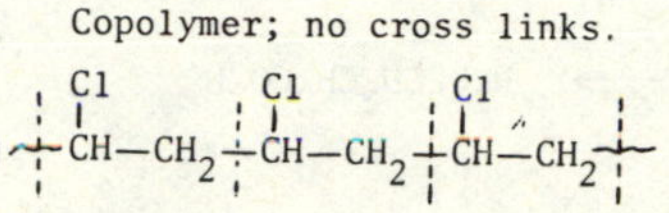

Homopolymer; no cross links

d) 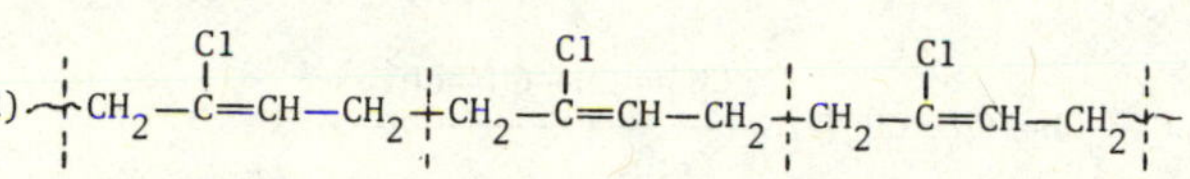

Homopolymer; no cross links.

e) 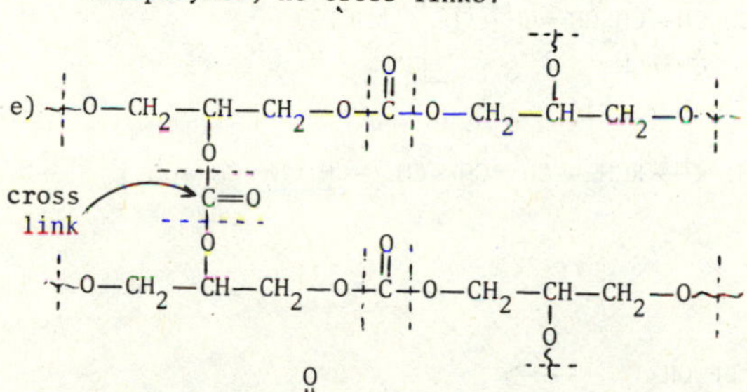

Copolymer ($Cl-C(=O)-Cl$ is one reactant)

f) 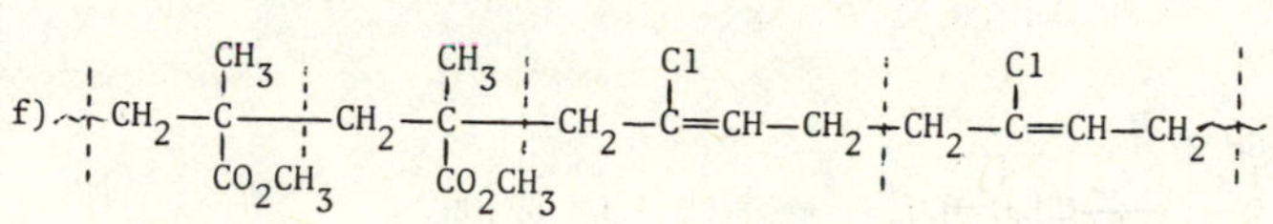

Copolymer; no cross links

25-2 Initiation probably involves abstraction of a hydrogen atom at the allylic carbon atoms of polyisoprene monomer units.

25-3

a) 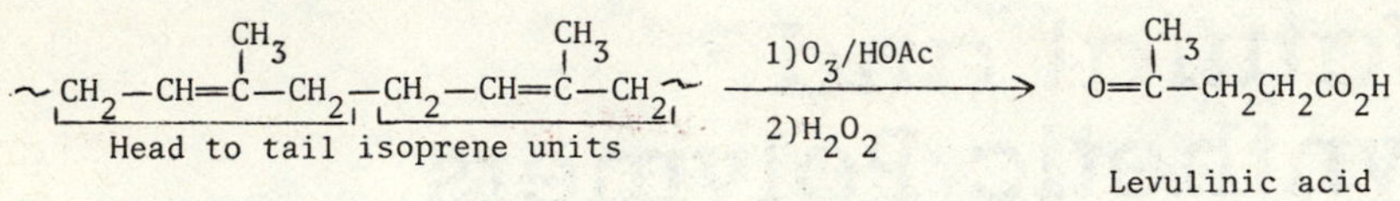

b) From the head to head segments:

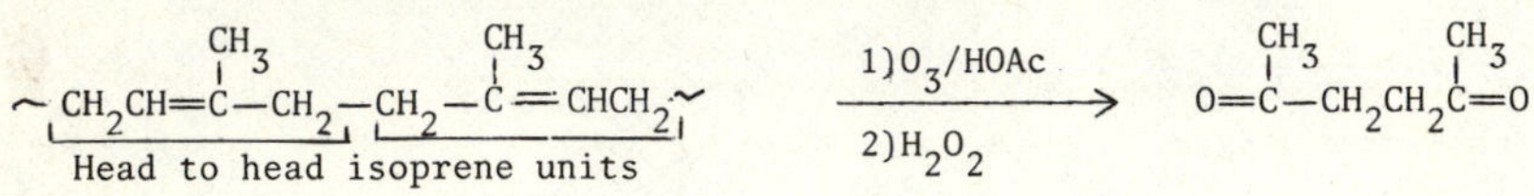

From the tail to tail segments:

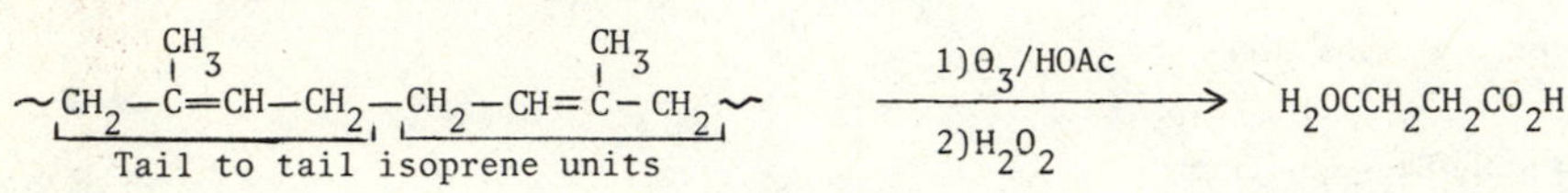

25-4

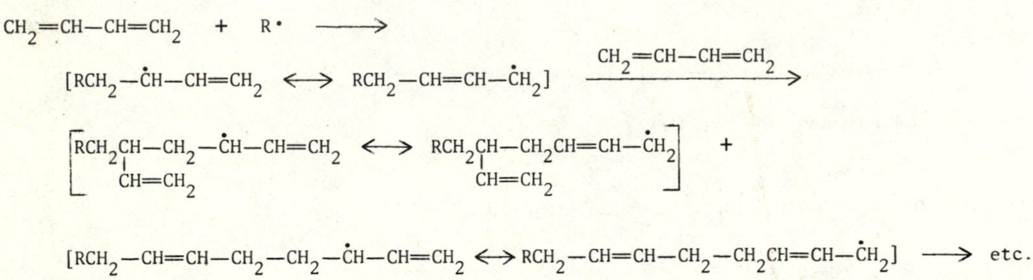

25-5

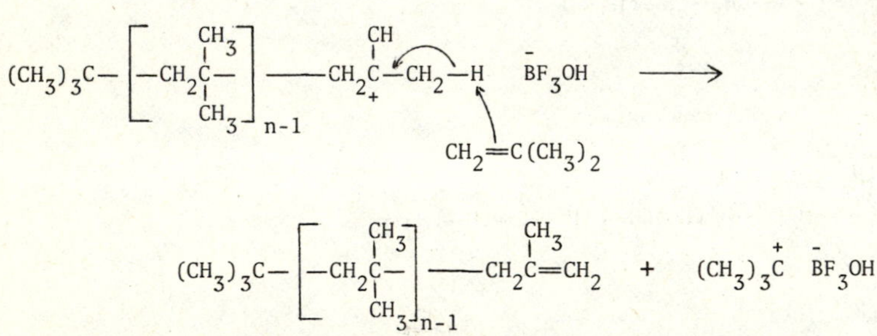

25-6

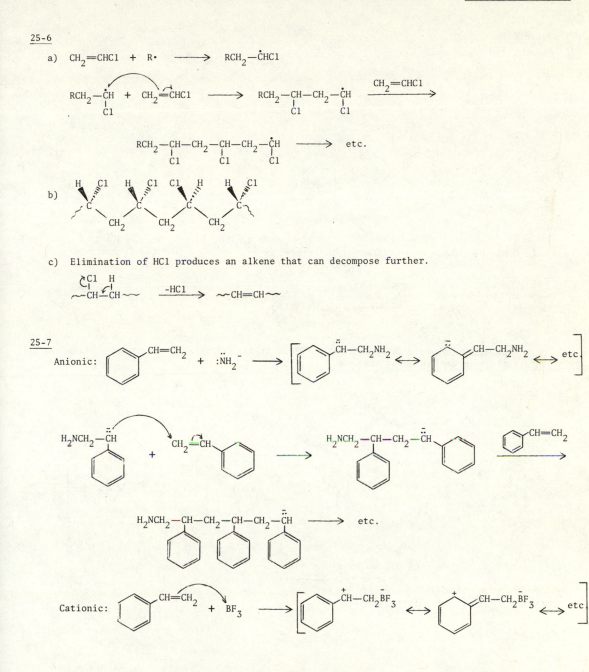

a) $CH_2{=}CHCl \; + \; R\cdot \;\longrightarrow\; RCH_2{-}\dot{C}HCl$

$$RCH_2{-}\underset{Cl}{\underset{|}{\dot{C}H}} \; + \; CH_2{=}CHCl \;\longrightarrow\; RCH_2{-}\underset{Cl}{\underset{|}{CH}}{-}CH_2{-}\underset{Cl}{\underset{|}{\dot{C}H}} \;\xrightarrow{CH_2=CHCl}$$

$$RCH_2{-}\underset{Cl}{\underset{|}{CH}}{-}CH_2{-}\underset{Cl}{\underset{|}{CH}}{-}CH_2{-}\underset{Cl}{\underset{|}{\dot{C}H}} \;\longrightarrow\; \text{etc.}$$

b)

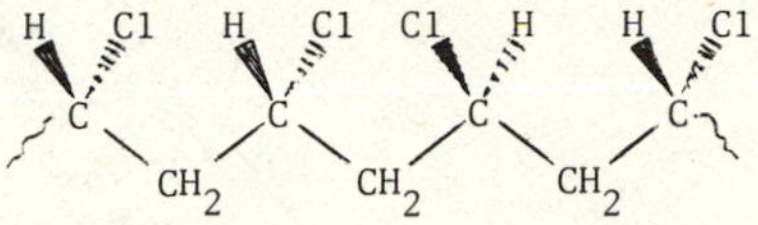

c) Elimination of HCl produces an alkene that can decompose further.

$$\sim\underset{}{\overset{Cl}{\overset{|}{CH}}}{-}\overset{H}{\overset{|}{CH}}\sim \;\xrightarrow{-HCl}\; \sim CH{=}CH\sim$$

25-7

Anionic:

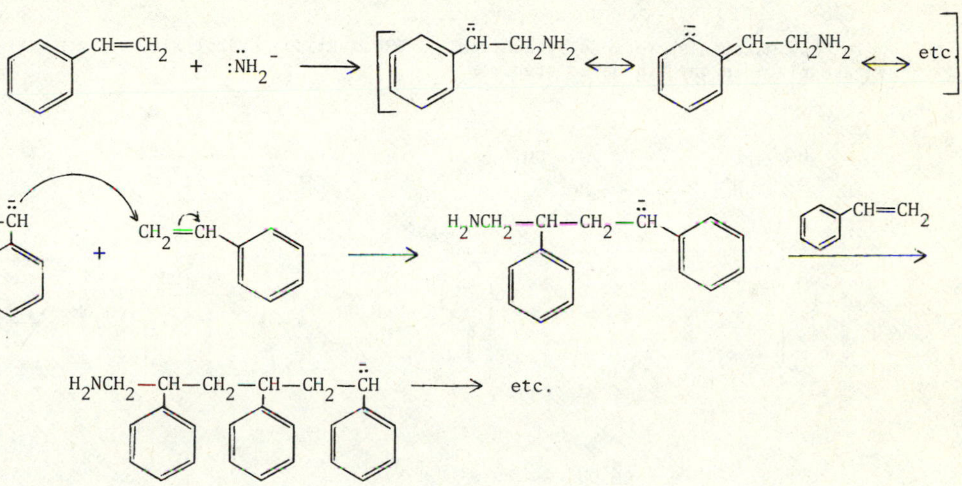

Cationic:

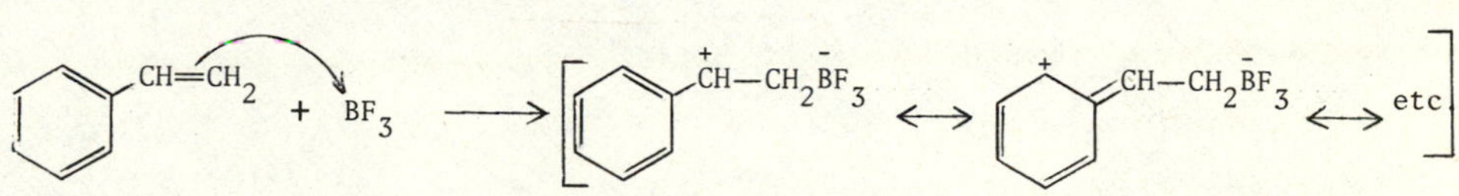

Contd...

25-7 contd...

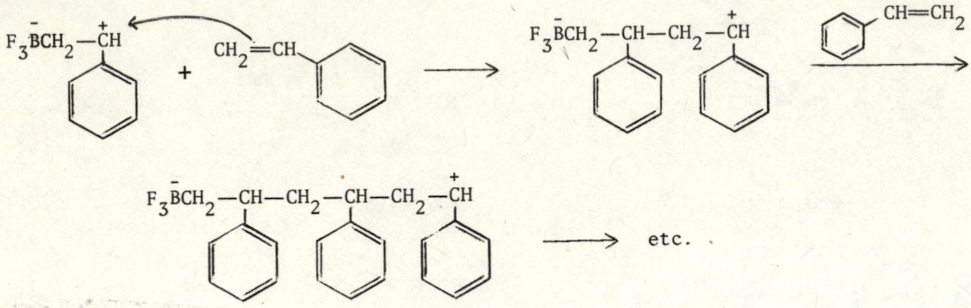

25-8

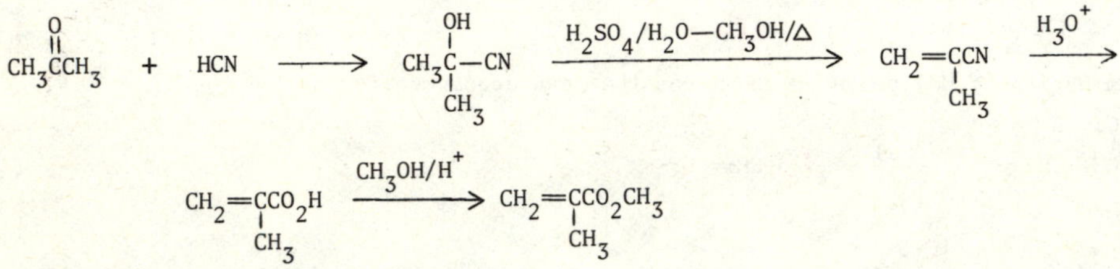

The raw materials are readily available. Elimination, hydrolysis, and esterification all take place in one synthetic step.

25-9

a)

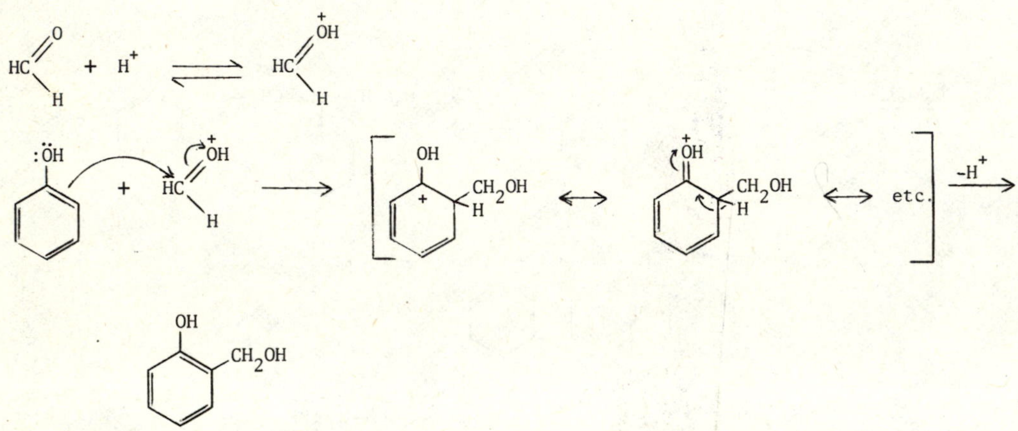

Contd...

25-9 contd...

b)

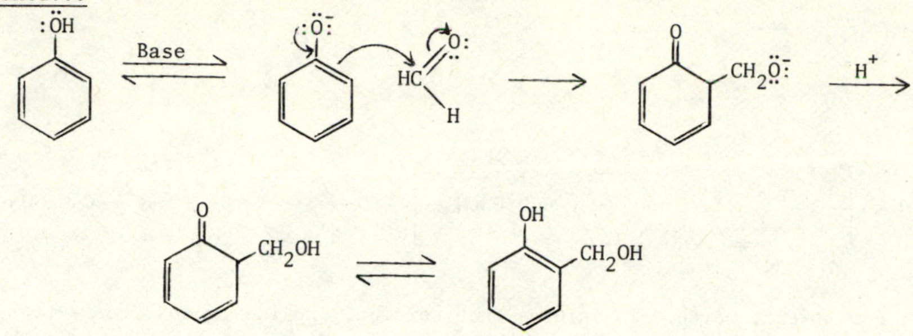

c)

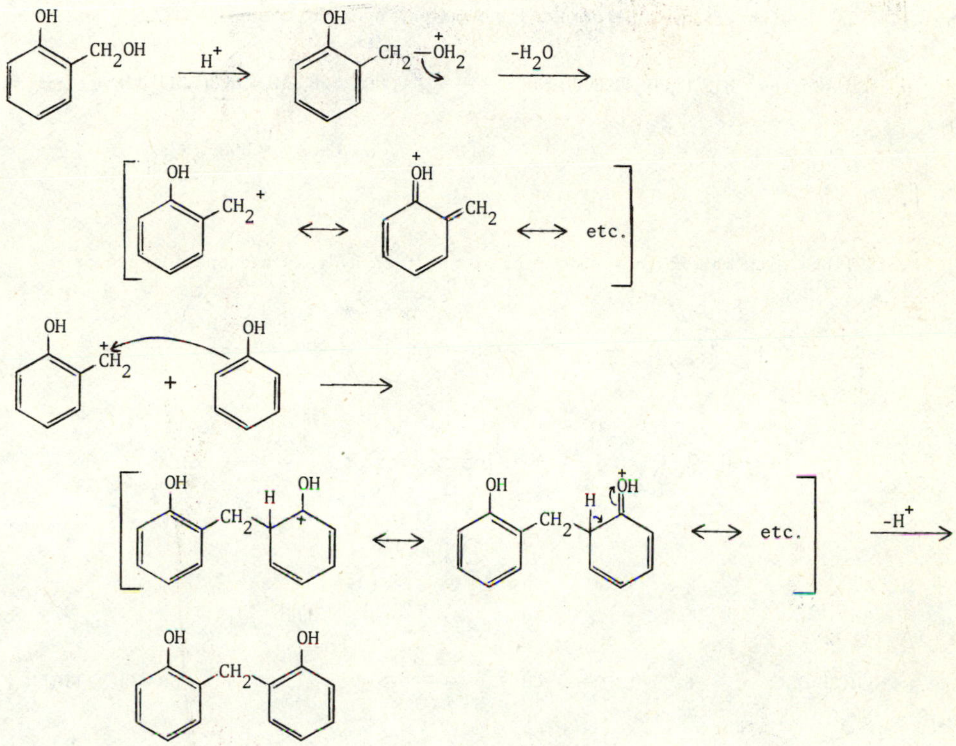

25-10

a) 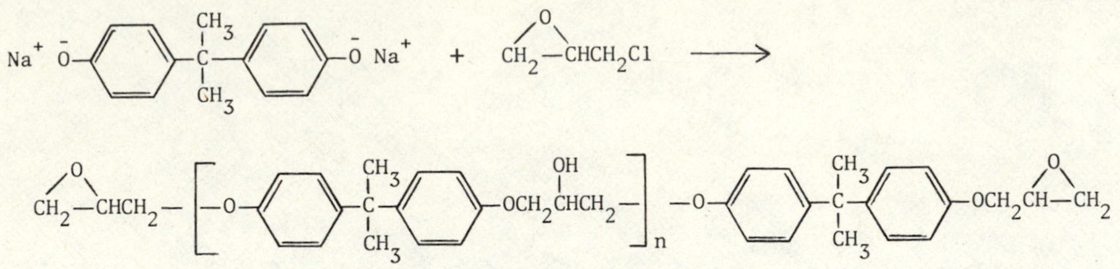

(The use of excess epichlorohydrin insures that terminal epoxide groups are present.)

b) Cross links with triamine B form at the terminal epoxide groups.

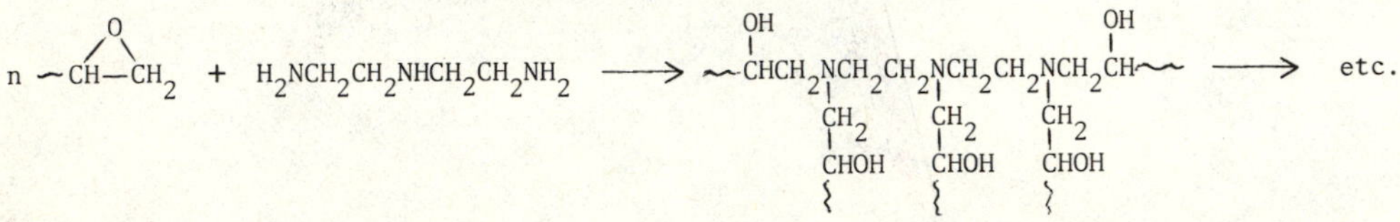

Cross links with maleic anhydride form at the free hydroxy and epoxide groups.

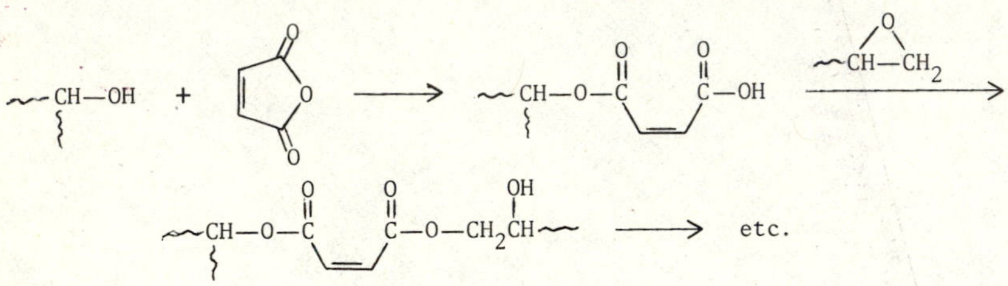

25-11

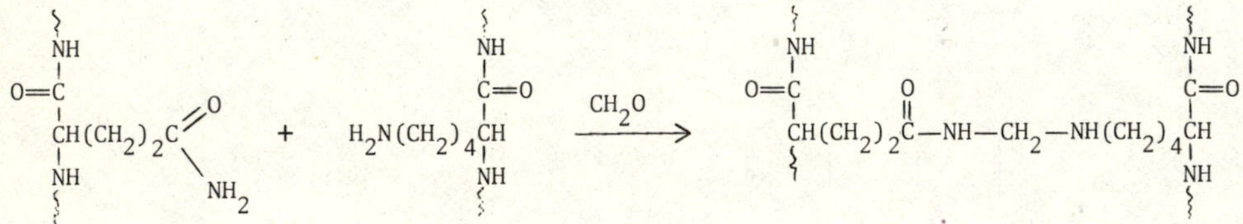

A glutamine unit A lysine unit

25-12

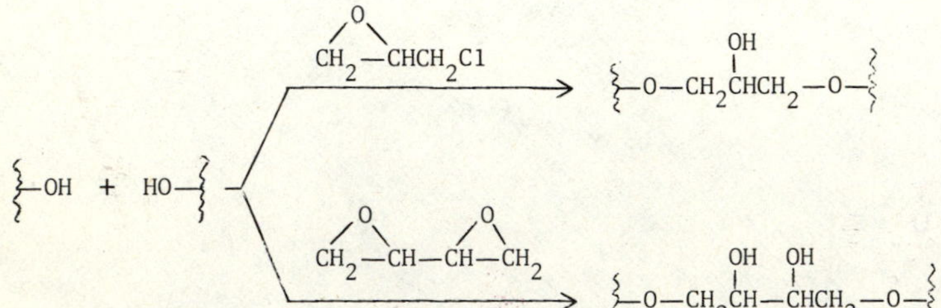

25-13

$$\underset{\text{"Cellulose"}}{ROH} + NaOH \longrightarrow RO^-Na^+ + H_2O$$

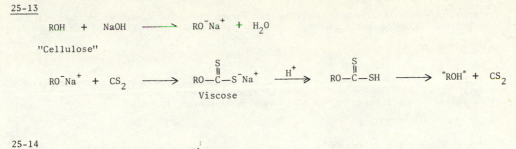

$$RO^-Na^+ + CS_2 \longrightarrow \underset{\text{Viscose}}{RO{-}\overset{S}{\overset{\|}{C}}{-}S^-Na^+} \xrightarrow{H^+} RO{-}\overset{S}{\overset{\|}{C}}{-}SH \longrightarrow \text{"ROH"} + CS_2$$

25-14

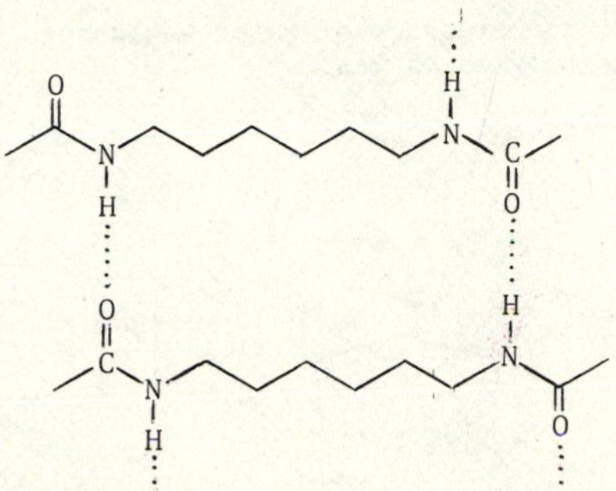

25-15

a) Fatty acids and their degradation products.

b) 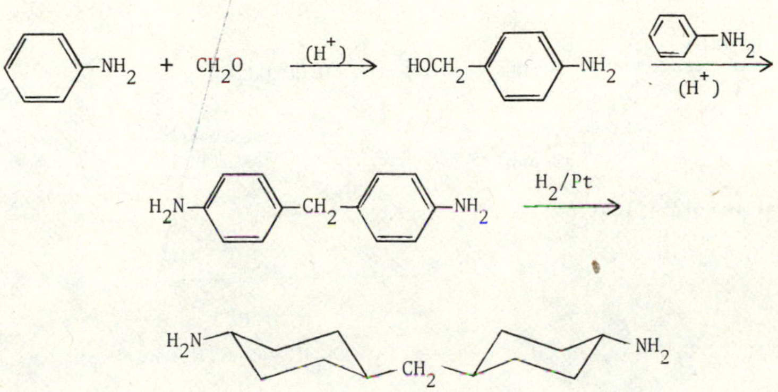

25-16

a) 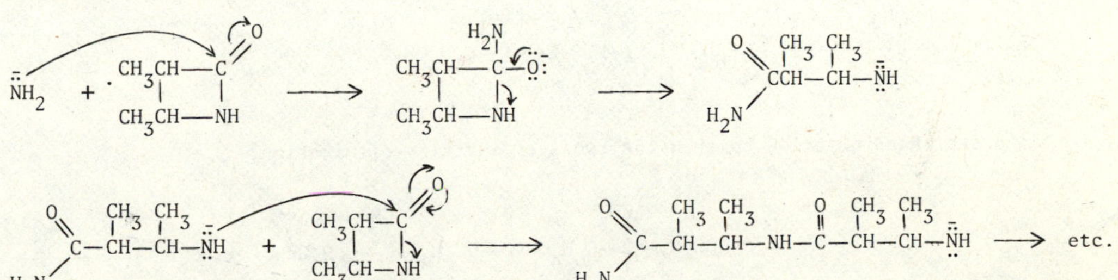

Contd...

25-16 contd...

b) The monomer used industrially has two asymmetric carbon atoms and exists as racemic cis and trans isomers. The subsequent polymer can have a combination of erythro and threo repeating units.

25-17 Polyesters are relatively nonpolar lipophilic (lipid-liking) materials because of their relatively high proportion of the nonpolar methylene and aryl groups. They tend to dissolve greases and resist penetration by the usual aqueous laundering agents.

25-18 At the high temperature of the melt-spin process, the presence of water would result in ester hydrolysis and consequent shortening of the polymer chains.

25-19

a)

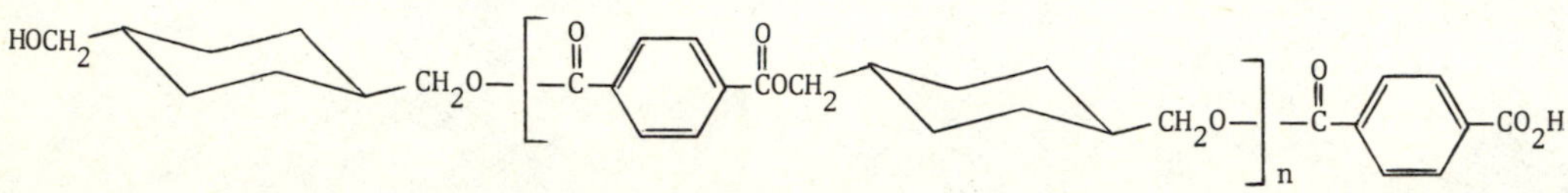

b) The large, nonplanar cyclohexane groups prohibit the polymer chains from packing as close together as is found in polyethylene terephthalate. Kodel fibers tend to be somewhat more labile thermally than the polyethylene terephthalates.

25-20

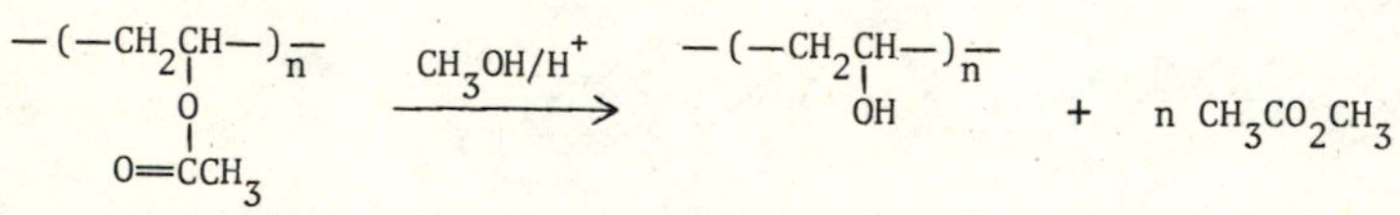

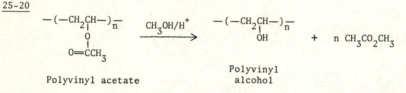

Polyvinyl acetate — Polyvinyl alcohol

The reaction is a transesterification.

25-21

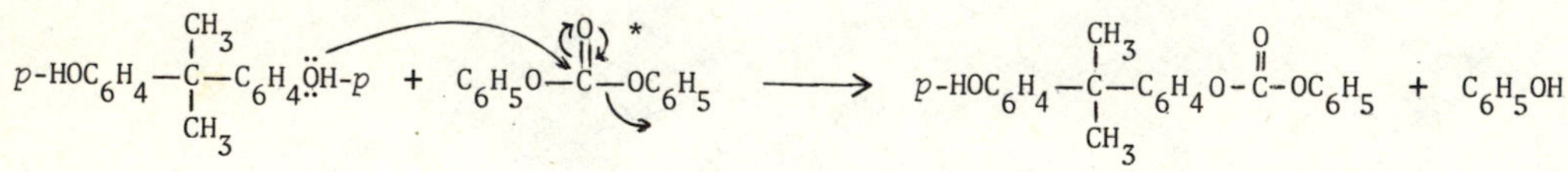

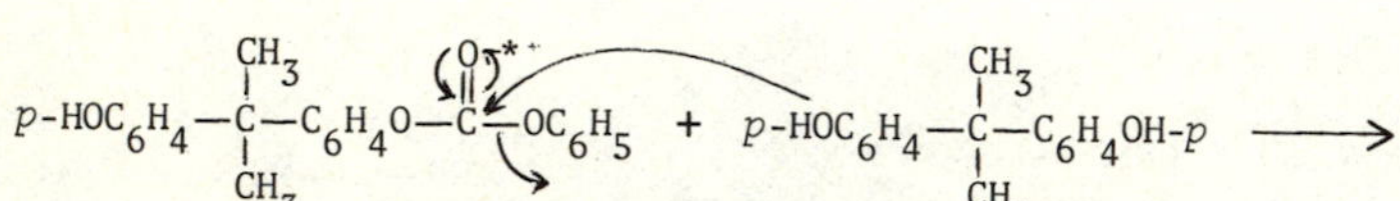

* A shorthand notation to show the two step addition-elimination.

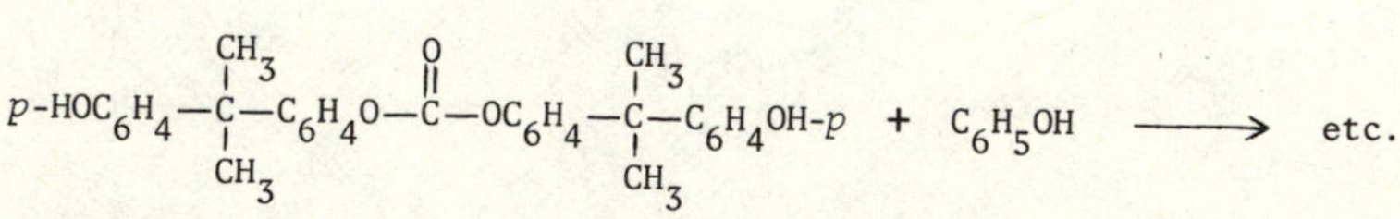

25-22 The low density (high pressure) polyethylene is formed by a relatively random free radical sequence. Branches can form from which new chains grow. The polymer is branched and randomly formed so that chains do not pack together well. The material is less dense and amorphorous.
The high density polyethylene is prepared by a stereoselective Zeigler-Natta catalyzed process. The polymer forms in long linear chains which fit together in a regular pattern and has considerable crystalline character.

25-23

a) Teflon is a linear perfluorinated polyethylene.

$-(-CF_2CF_2-)_n-$

b) Free radical chain polymerization catalyzed by peroxides or oxygen.

c) Thermal decomposition of the chlorodifluoromethane is believed to produce a carbene by α-elimination of HCl. Dimerization of the carbene gives the monomer.

$$HCClF_2 \xrightarrow{\Delta} :CF_2 + HCl$$

$$2\ :CF_2 \longrightarrow F_2C{=}CH_2$$

25-24 No, because no chiral centers are present in polyisobutylene.

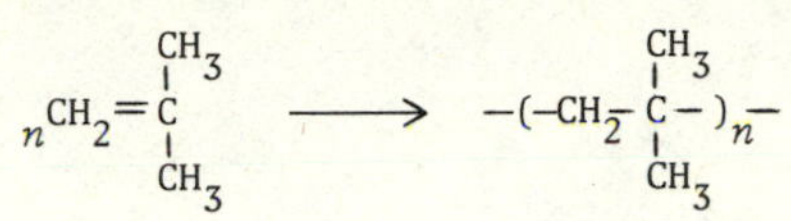

25-25

a)

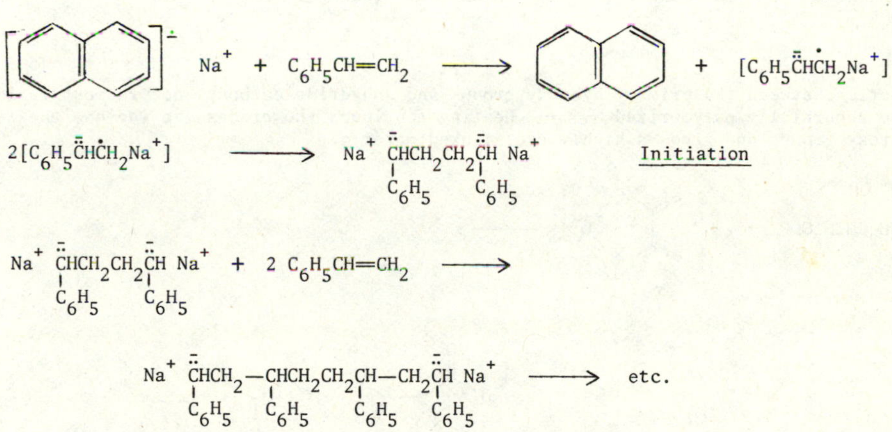

b) Anything that adds to an anion can be used to terminate the chain, i.e., H^+, $CO_2 + H^+$, etc.

25-26 The polymer end is a labile hemiacetal which readily begins the unzipping sequence. End-capping gives a more stable acetal or ester structure.

$$\sim CH_2OCH_2OCH_2{-}OH \xrightarrow[\text{or}\ R'CO_2X]{RX} \sim CH_2OCH_2OCH_2{-}OR \quad \text{or}\ {-}O_2CR'$$

The "unzipping" mechanism is: 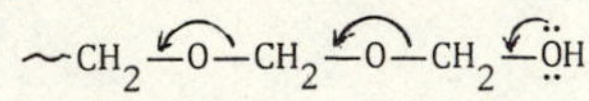

25-27 The polymer is a polyester formed in a chain process.

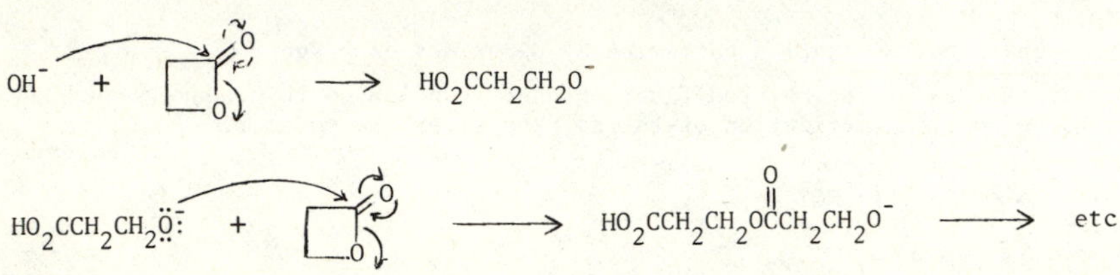

25-28 Polyethylene glycol is prepared commercially from ethylene oxide (formed from ethylene) by anionic chain polymerization. (Acid catalysis might also be used.)

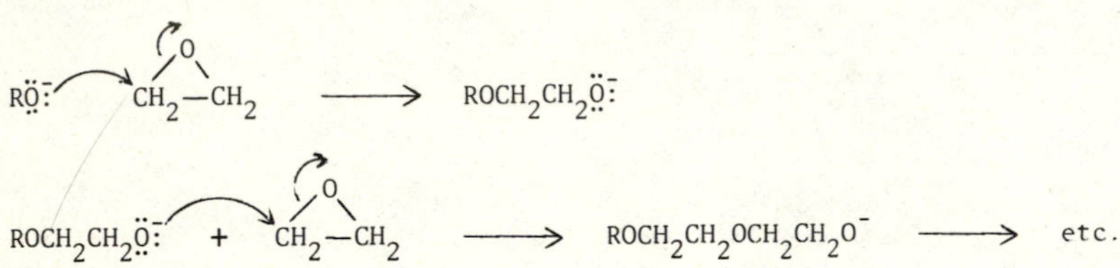

25-29 Reaction between the primary hydroxy groups and anhydride carbonyl occurs most rapidly to give a partially polymerized resin. Heating continues the process to include the secondary hydroxy groups and gives a highly crosslinked material.

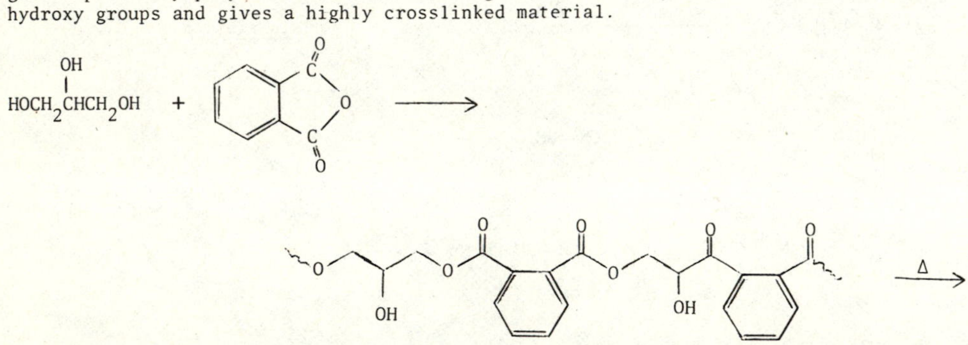

Contd...

25-29 contd...

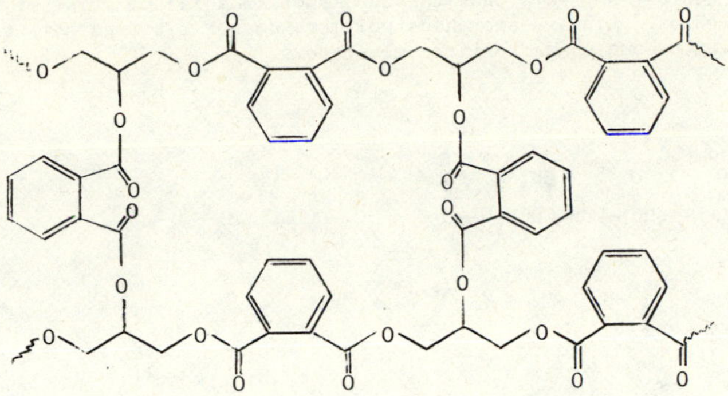

25-30

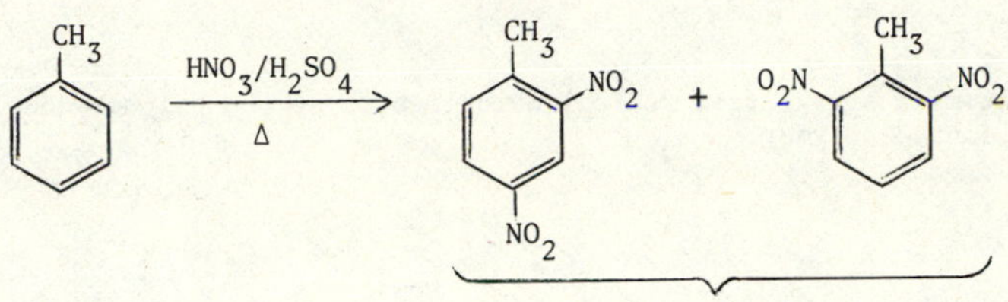

Separate isomers

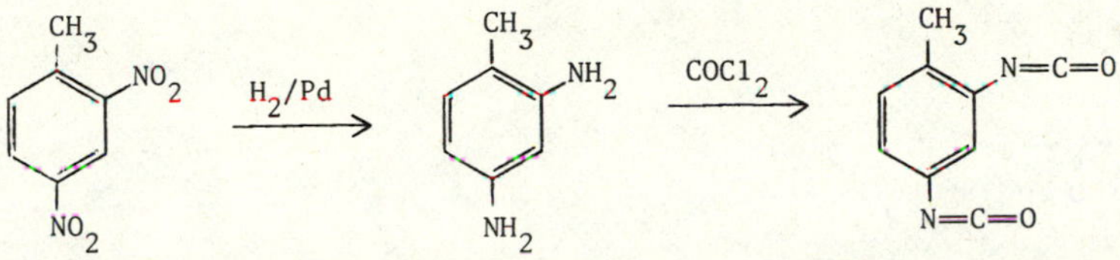

25-31

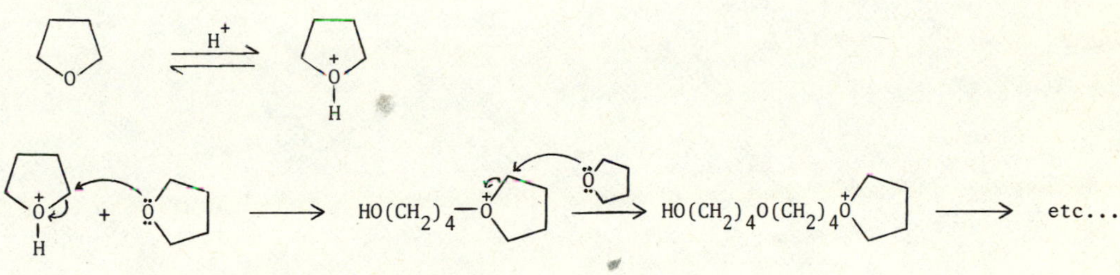

25-32 We expect this coordination vinyl type of polymerization to give a linear copolymer. The high ethylene content suggests that the copolymer is a random array of monomer units, thus is highly amorphous. Linear, amorphous polymers do not fit together well and on crosslinking can become characteristic elastomers.

25-33

a) 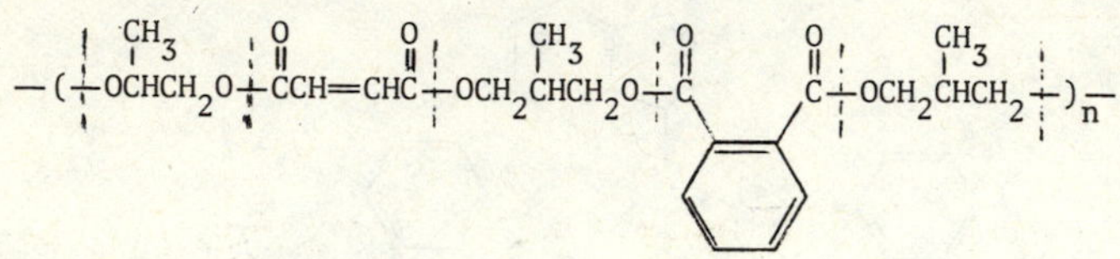

b) $HOCH(CH_3)CH_2OH$, 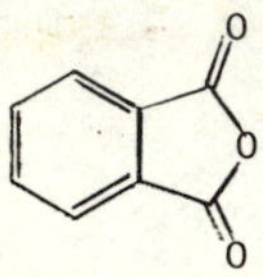 and 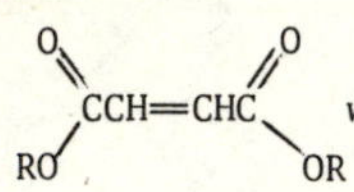 which actually comes from 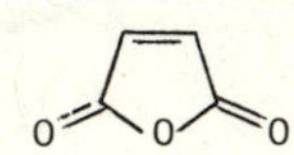

c) The radical initiator forms a styryl free radical which reacts at the polymer double bonds to form a highly cross linked polymer.

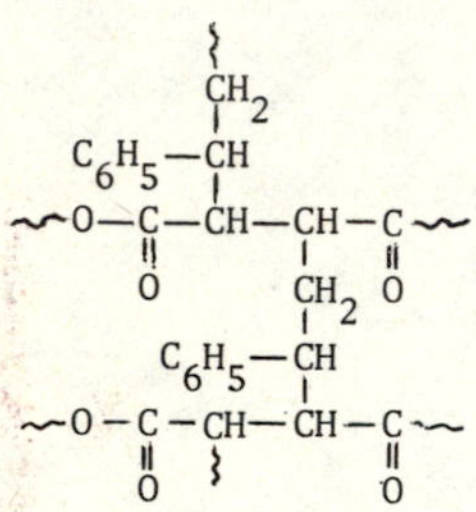

26 Molecular Rearrangements

26-1 For the reaction:

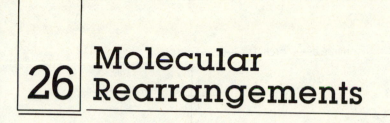

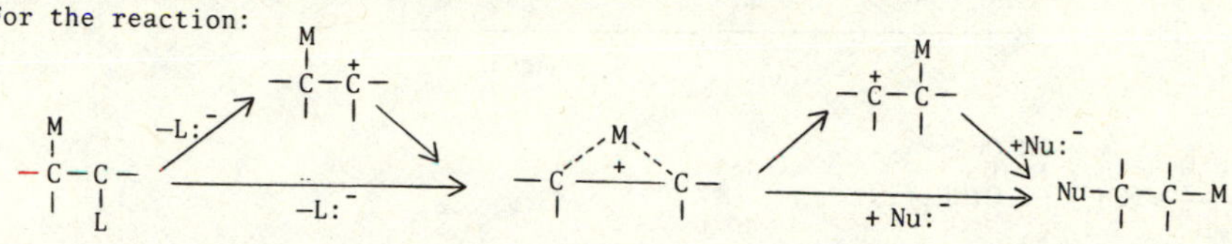

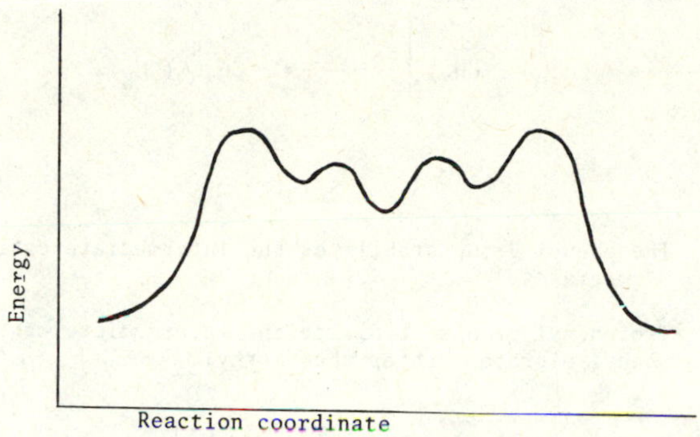

We assume that the bridged intermediate is more stable than the open carbocations. The relative heights of each transition state will depend on the specific structures involved.

26-2 In each case the same carbocation is formed which then reacts with water to form pinacol or rearranges to pinacolone.

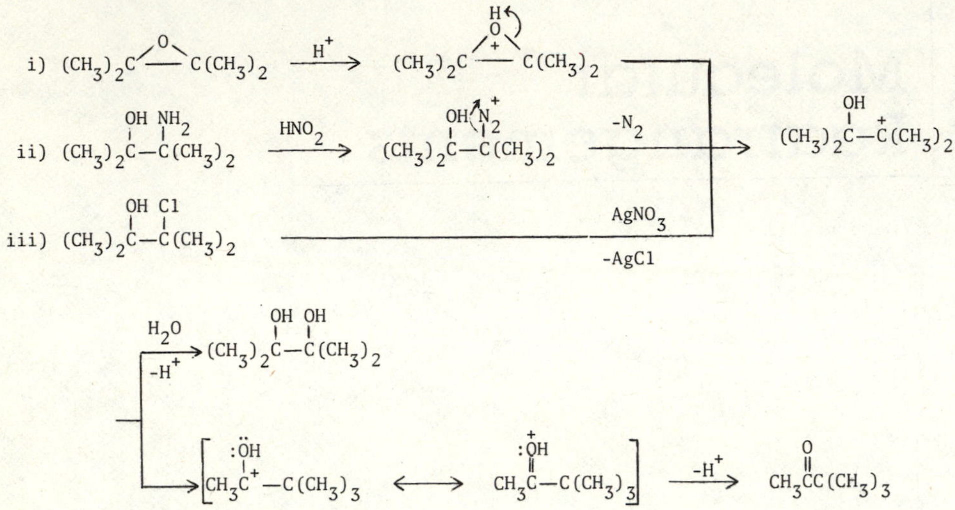

26-3

a) $C_6H_5CH_2CHO$ — The phenyl group stabilizes the intermediate cation and H^- migrates.

b) $(C_6H_5)_3C\overset{O}{\overset{\|}{C}}CH_3$ — Two phenyl groups stabilize the intermediate cation, then phenyl migrates better than methyl.

c) CH_3CH_2CHO — Methyl stabilizes the cation and H^- migrates.

d) $p\text{-}NO_2C_6H_4\overset{O}{\overset{\|}{C}}—C(C_6H_5)(C_6H_4OCH_3\text{-}p)_2$ — p-Anisyl stabilizes the cation; phenyl migrates better than $p\text{-}NO_2C_6H_4$-.

e) 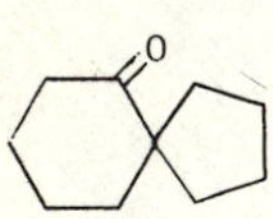 — Migration of a ring methylene (an alkyl group) leads to ring expansion.

26-4 The suggestion is based on the assumption that a concerted process would lead to 100% inversion as only the labeled phenyl group migrates. A front-sided concerted rearrangement is not normally considered to be a reasonable pathway for a nucleophilic displacement. Thus the results are consistent with formation of an intermediate cation.

26-5 In each case the most stable rotamers (the rotamers in which H is between the two aromatic groups) account for back sided migration. Migratory aptitudes have little effect on which group migrates in these deaminative-rearrangement sequences that do not involve a stabilized cation.

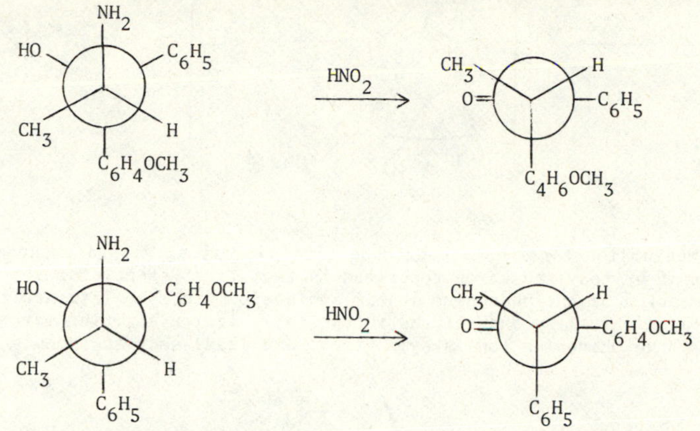

26-6 In this example deaminiation produces a phenyl stabilized carbocation which is sufficiently long-lived to partially rotate before migration takes place. The cationic intermediate leading to retention is less crowded than that leading to inverted product and therefore accounts for the predominant reaction pathway. The 2-amino-1,1-diphenyl-1-propanol forms a a non-stabilized carbocation which rapidly reacts by a predominant inversion pathway.

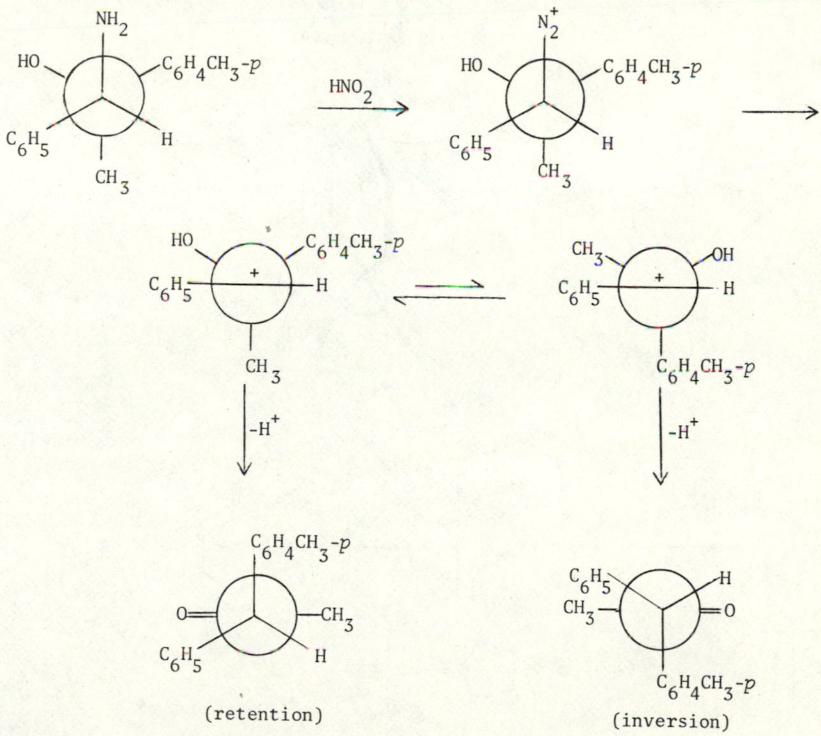

26-7 The electron donating *p*-methoxy group enhances formation of and stabilizes the intermediate phenonium ion.

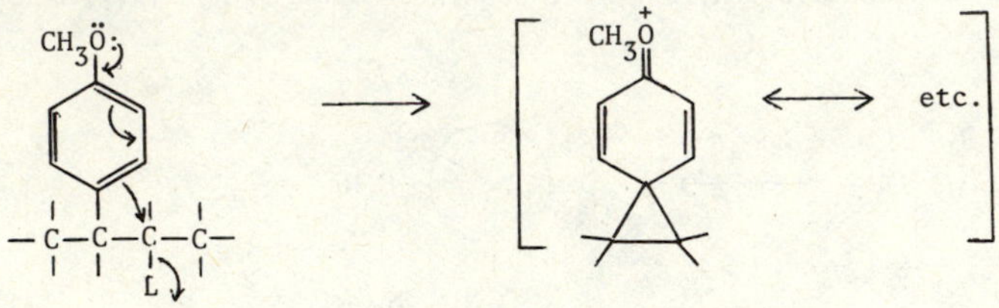

26-8 The intermediate phenonium ion forms by a stereospecific back sided displacement of tosylate. Back sided displacement by tosylate anion generated in that first step reforms starting material. Since formation of the phenonium ion is stereospecific, the intermediate from *erythro* reactant retains chirality and returns to optically active starting material. In the case of *threo*, the phenonium ion is symmetrical and leads back to racemic starting material.

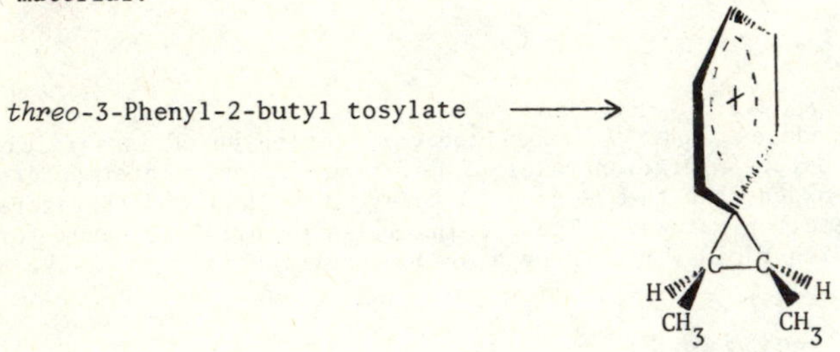

Achiral

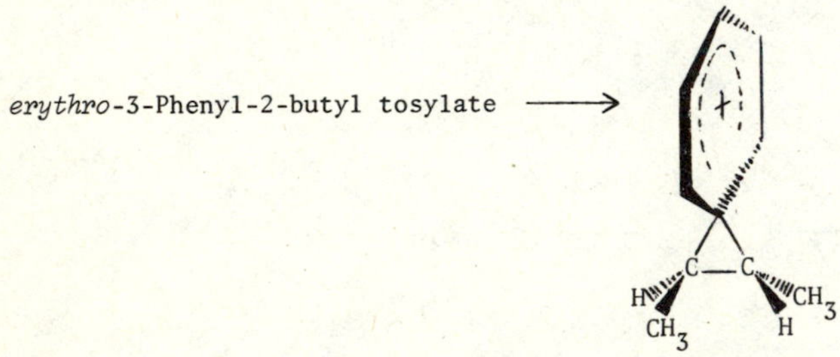

Chiral

26-9

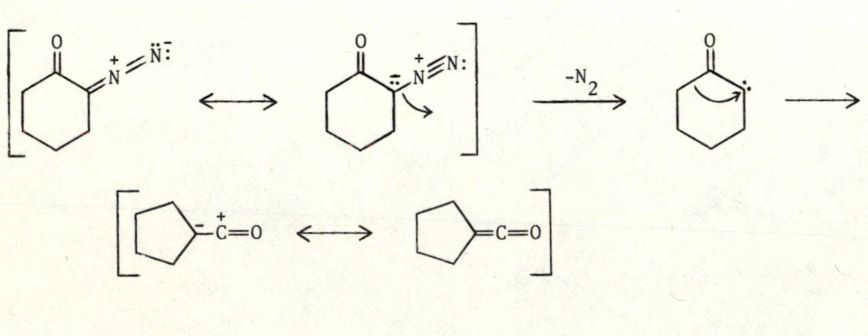

Contd...

26-9 contd...

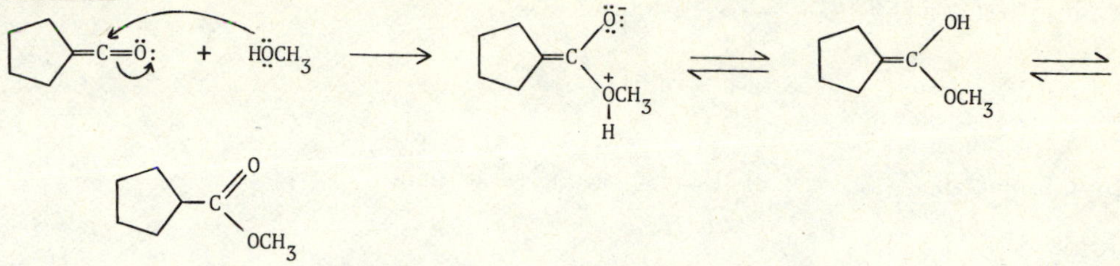

26-10

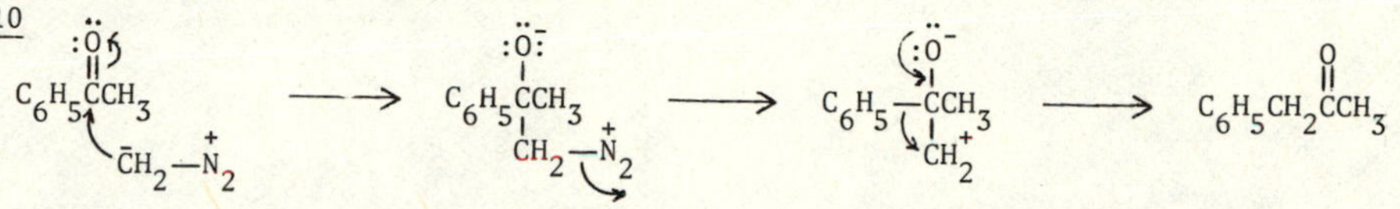

or

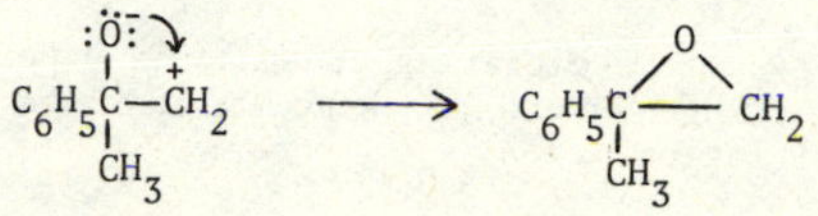

26-11

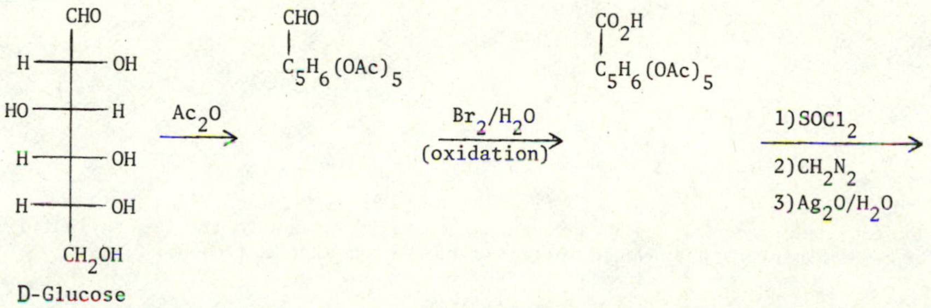

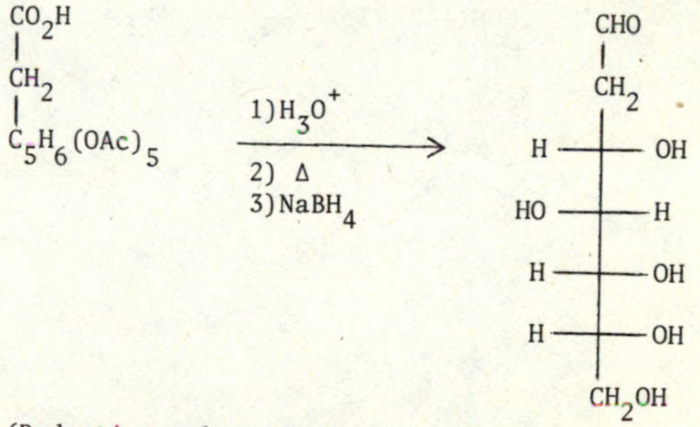

(Reduction takes place through the cyclic lactone sec. 21-2C).

26-12

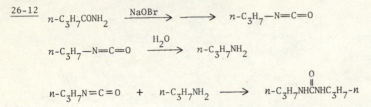

$$n\text{-}C_3H_7CONH_2 \xrightarrow{NaOBr} \longrightarrow n\text{-}C_3H_7\text{—}N\text{=}C\text{=}O$$

$$n\text{-}C_3H_7\text{—}N\text{=}C\text{=}O \xrightarrow{H_2O} n\text{-}C_3H_7NH_2$$

$$n\text{-}C_3H_7N\text{=}C\text{=}O + n\text{-}C_3H_7NH_2 \longrightarrow n\text{-}C_3H_7NH\overset{\overset{O}{\|}}{C}NHC_3H_7\text{-}n$$

26-13 The *N*-haloamide which is formed from an *N*-substituted amide has no hydrogen atom on nitrogen. The anion required for rearrangement cannot form.

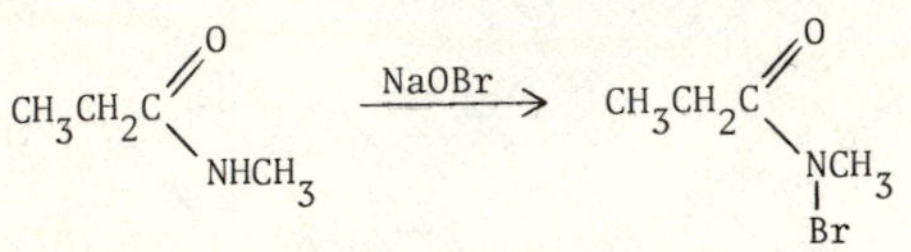

26-14 Migration is enhanced by increasing electron density at the migrating group as is expected if reaction is a nucleophilic substitution on the nitrogen atom. A phenonium ion-like intermediate has also been proposed for the aryl migration.

26-15 Neopentyl cation readily undergoes Wagner-Meerwein rearrangement to the 2-methylbutyl cation. The predominent pathway would therefore have been expected to be:

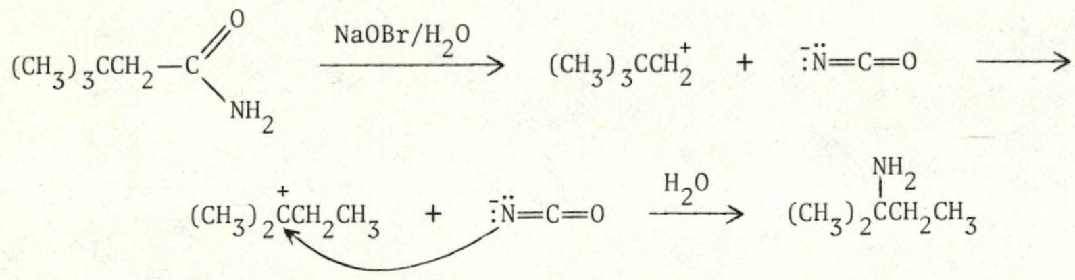

26-16

$(CH_3)_2CHCH_2-C\overset{O}{\underset{\overset{\cdot\cdot}{N}-N\equiv N:}{\diagdown}}$ \longrightarrow $(CH_3)_2CHCH_2-N=C=O$ + N_2

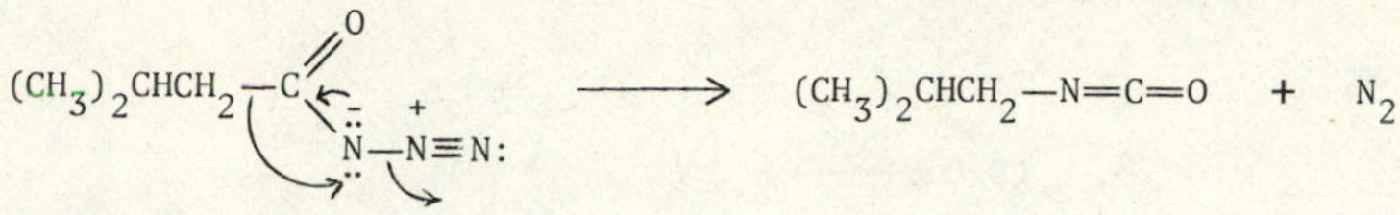

(The potential nitrene intermediate which would be formed if N_2 departed before rearrangement has never been detected.)

$(CH_2)_2CHCH_2-N=C=O$ + $H_2\ddot{O}$ \longrightarrow $(CH_3)_2CHCH_2-N=C\overset{\overset{\cdot\cdot}{\ddot{O}:}}{\underset{\overset{+}{O}H_2}{\diagdown}}$ \rightleftharpoons

$(CH_3)_2CHCH_2-N=C\overset{OH}{\underset{OH}{\diagdown}}$ \rightleftharpoons $(CH_3)_2CHCH_2NH-C\overset{O}{\underset{\ddot{O}H}{\diagdown}}$ \rightleftharpoons

$(CH_3)_2CHCH_2\overset{+}{N}H_2-C\overset{O}{\underset{\ddot{O}:}{\diagdown}}$ \longrightarrow $(CH_3)_2CHCH_2NH_2$ + CO_2

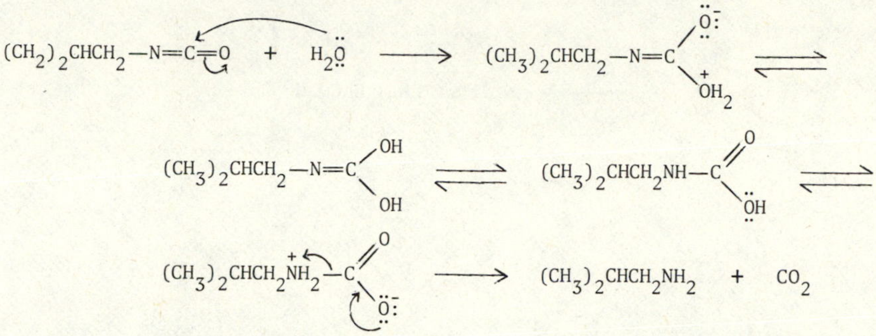

26-17

a) 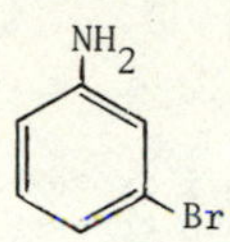

b) $CH_3(CH_2)_4NH_2$

c) 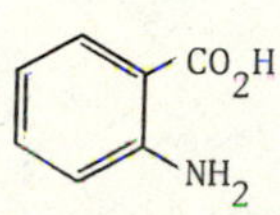

d) 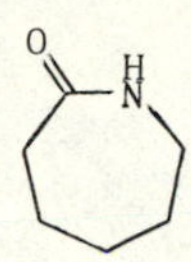

e) 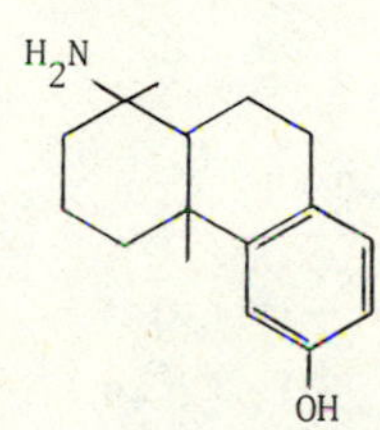

f) $H_2N(CH_2)_4NH_2$

g) $C_6H_5CH_2NHCO_2CH_3$

26-18

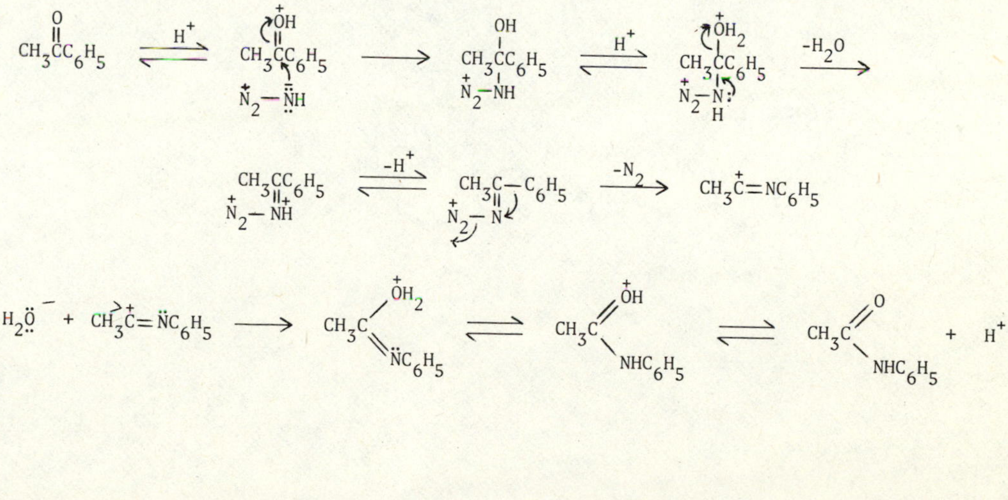

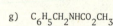

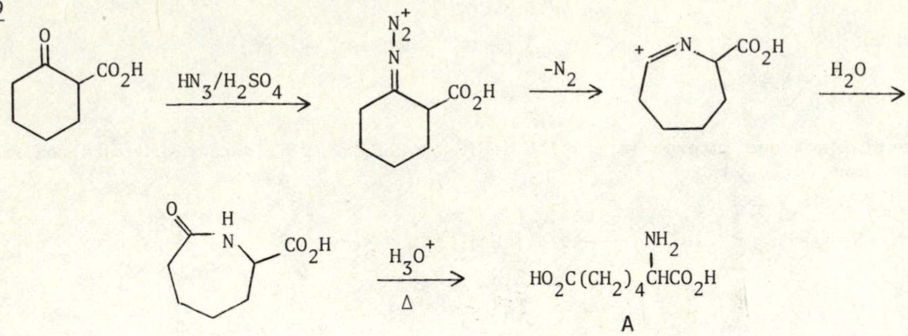

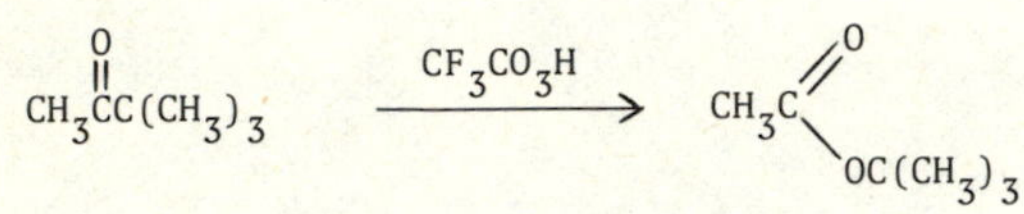

26-19

26-20

a) $CH_3\overset{O}{\overset{\|}{C}}C(CH_3)_3$ $\xrightarrow{CF_3CO_3H}$ $CH_3\overset{O}{\overset{\|}{C}}OC(CH_3)_3$ *t*-Butyl is the better migratory group because it is the more electron rich.

b) $p\text{-}CH_3OC_6H_4\overset{O}{\overset{\|}{C}}CH_2CH_3$ $\xrightarrow{CF_3CO_3H}$ $p\text{-}CH_3OC_6H_4O\overset{O}{\overset{\|}{C}}CH_2CH_3$ *p*-Methoxyphenyl is the better migrating group because it is the more electron rich.

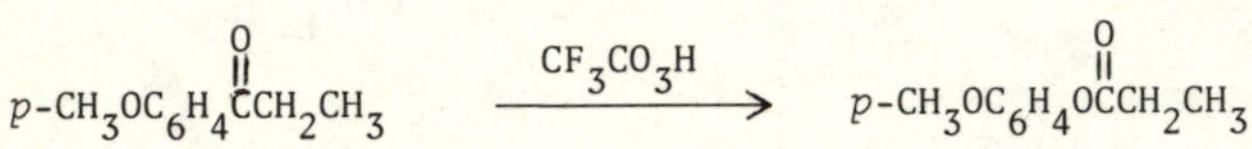

26-21

a) $HOO^-\,Na^+$ +

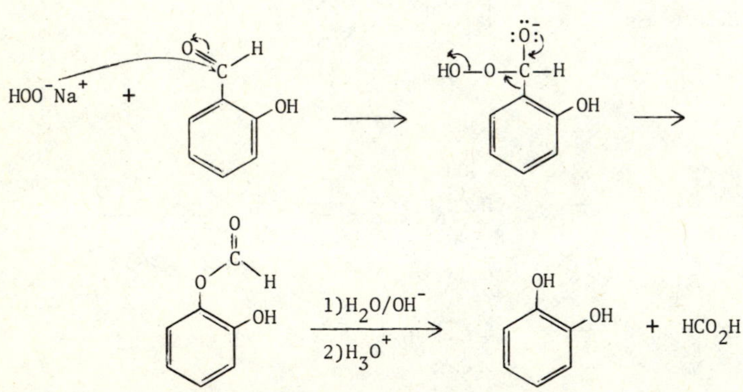

(This is a formate ester)

Contd...

26-21 contd...

b)

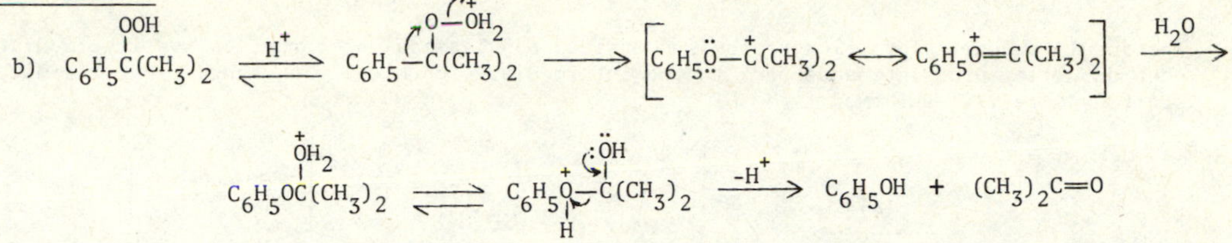

c)

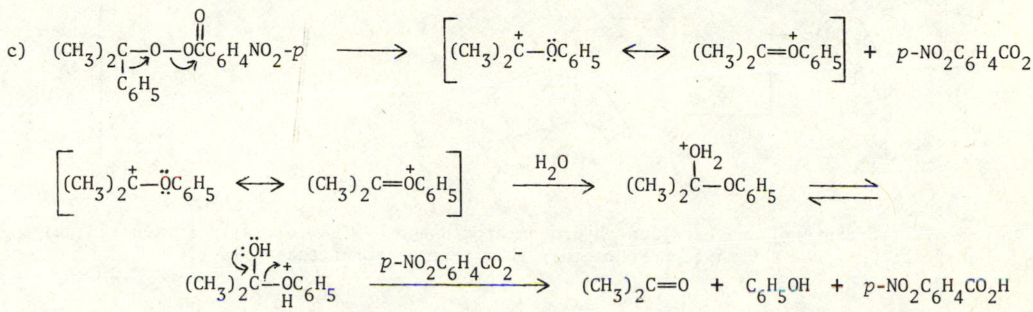

26-22 The observation that there is no scrambling of ethyl and butyl groups supports the proposed intramolecular nature of the rearrangement.

26-23

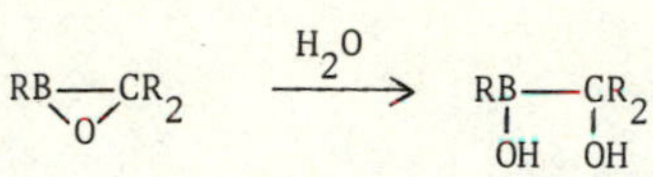

A 1,2-diol is presumed to be formed. That diol, by analogy with carbon analogs, would only be expected to rearrange under strongly acidic conditions similar to the pinacol rearrangement. By contrast epoxides readily open to provide a driving force for the third rearrangement that occurs under anhydrous conditions.

26-24

a) $3\ CH_3CH{=}CHCH_3 \xrightarrow[\substack{2)\,CO/H_2O\\3)\,NaOH/H_2O_2}]{1)\,(BH_3)_2/\text{Diglyme}} [CH_3CH_2CH(CH_3)]_2C{=}O$

b)

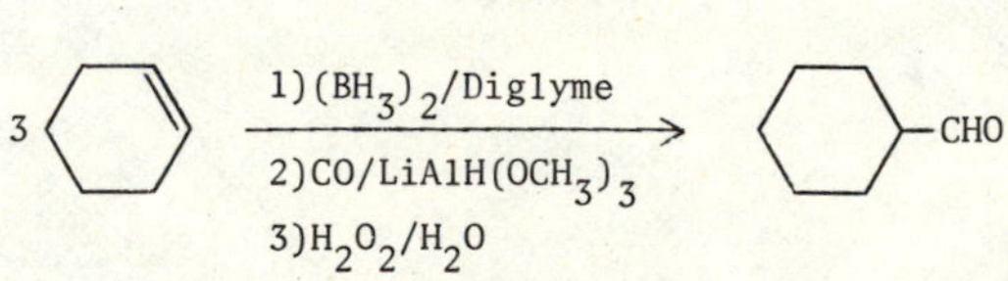

c)

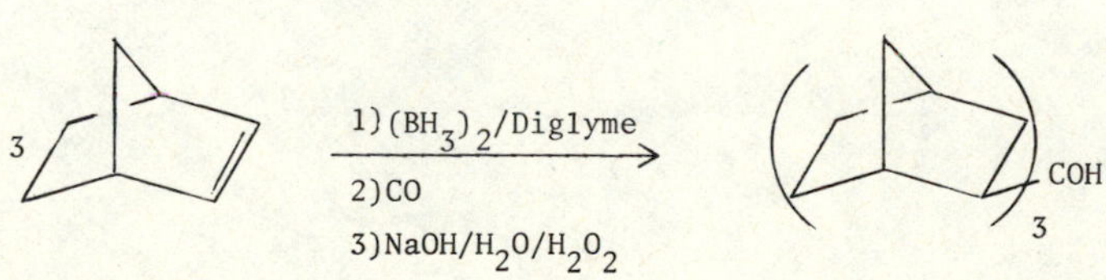

26-25

a) The result supports formation of the symmetrical nonclassical norbornyl cation intermediate. When departing brosylate adds back to the intermediate, chemically unchanged, but racemic, starting material is formed.

b)

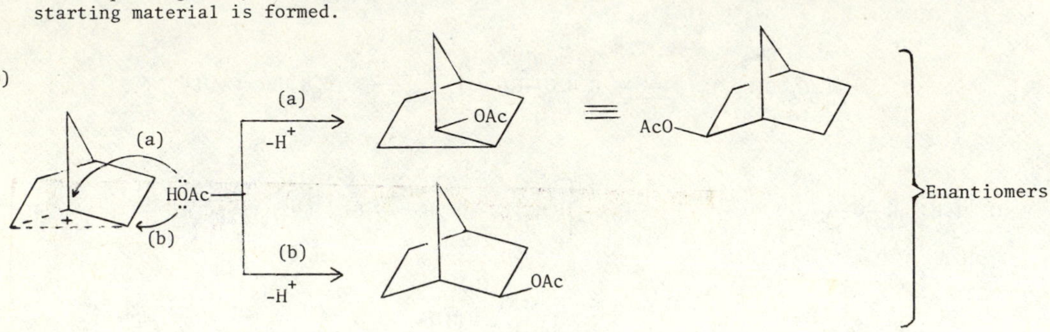

26-26 Thiols (mercaptans) are excellent hydrogen atom donors. The initially formed radical abstracts H• from the thiol more rapidly than rearrangement takes place.

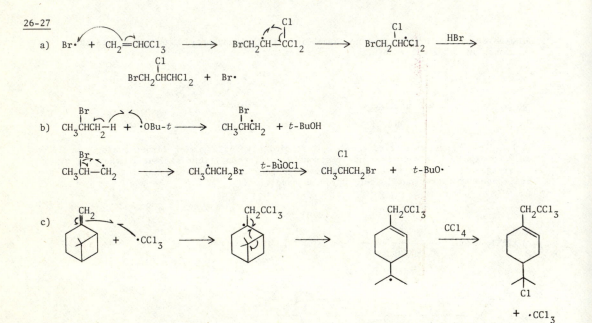

26-27

a) $Br\cdot + CH_2{=}CHCCl_3 \longrightarrow BrCH_2\dot{C}H{-}CCl_2$ (Cl on C) $\longrightarrow BrCH_2CH(Cl)\dot{C}Cl_2 \xrightarrow{HBr}$

$BrCH_2CH(Cl)CHCl_2 + Br\cdot$

b) $CH_3CH(Br)CH_2{-}H + \cdot OBu\text{-}t \longrightarrow CH_3CH(Br)\dot{C}H_2 + t\text{-}BuOH$

$CH_3CH(Br){-}\dot{C}H_2 \longrightarrow CH_3\dot{C}HCH_2Br \xrightarrow{t\text{-}BuOCl} CH_3CH(Cl)CH_2Br + t\text{-}BuO\cdot$

c)

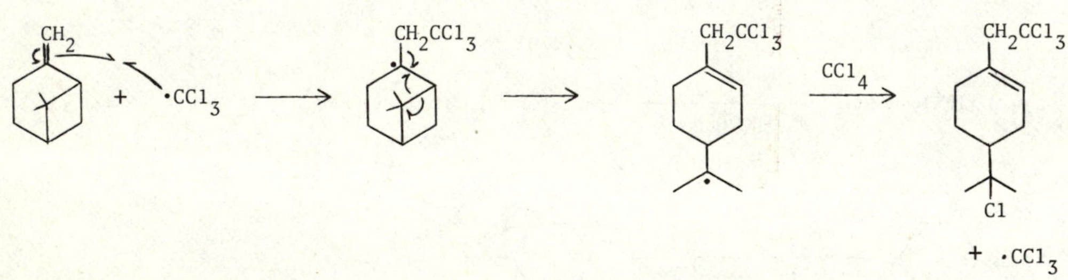

26-28 Both reactants form the same intermediate cyclopropanone.

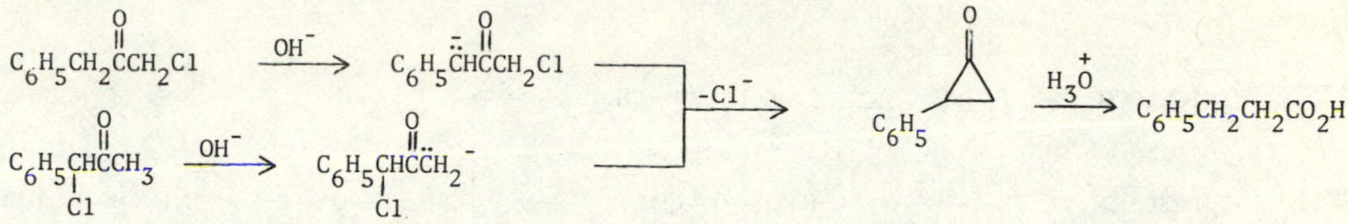

26-29 Back sided attack by the α-carbanion inverts configuration at the cyclohexane number one carbon atom as Cl^- is displaced.

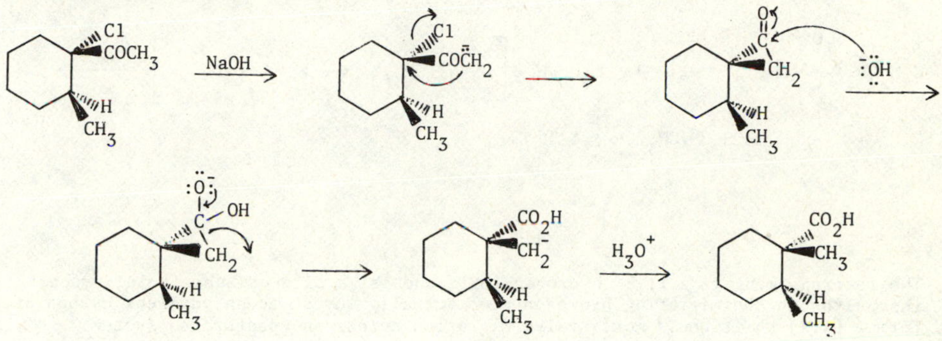

26-30

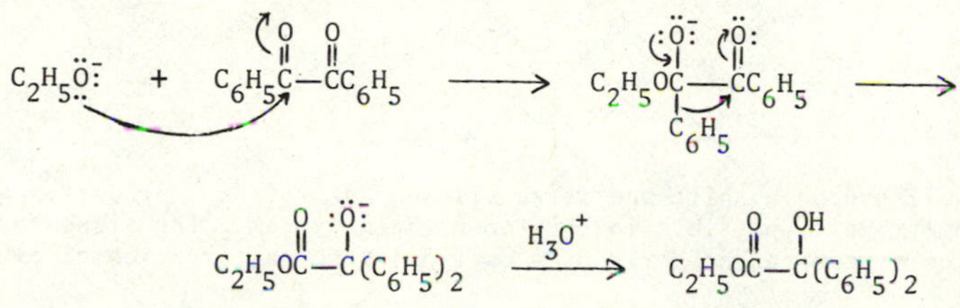

26-31

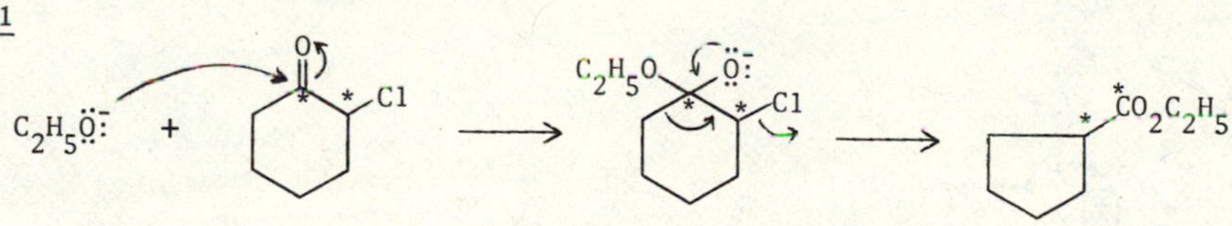

All labeled carbon would be at the ester carbonyl and the C—1 position of the cyclopentane ring if this mechanism were operating. We have already seen that the Favorskii rearrangement of this compound distributes the original C—2 label between the C—1 and C—2 carbons of the cyclopentane.

26-32 No atoms change positions; only an electron moves.

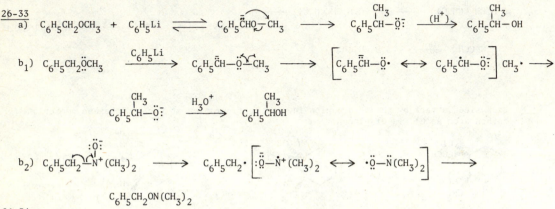

26-33

a) $C_6H_5CH_2OCH_3 + C_6H_5Li \rightleftharpoons C_6H_5\bar{C}HO-CH_3 \longrightarrow C_6H_5CH(CH_3)-\ddot{O}:^- \xrightarrow{(H^+)} C_6H_5CH(CH_3)-OH$

b$_1$) $C_6H_5CH_2\ddot{O}CH_3 \xrightarrow{C_6H_5Li} C_6H_5\bar{C}H-\ddot{O}-CH_3 \longrightarrow \left[C_6H_5\bar{C}H-\ddot{O}\cdot \longleftrightarrow C_6H_5\dot{C}H-\ddot{O}:^-\right] CH_3\cdot \longrightarrow$

$C_6H_5CH(CH_3)-\ddot{O}:^- \xrightarrow{H_3O^+} C_6H_5CH(CH_3)OH$

b$_2$) $C_6H_5CH_2-N^+(\ddot{O}:^-)(CH_3)_2 \longrightarrow C_6H_5CH_2\cdot \left[:\ddot{O}^--\dot{N}^+(CH_3)_2 \longleftrightarrow \cdot\ddot{O}-\ddot{N}(CH_3)_2\right] \longrightarrow$

$C_6H_5CH_2ON(CH_3)_2$

26-34

a) The rearrangement is a [1,5] hydrogen shift and is an allowed suprafacial process. Note that although the migrating hydrogen atom actually moves to an adjacent carbon atom, a formal [1,2] shift would not involve migration across an unsaturated system.

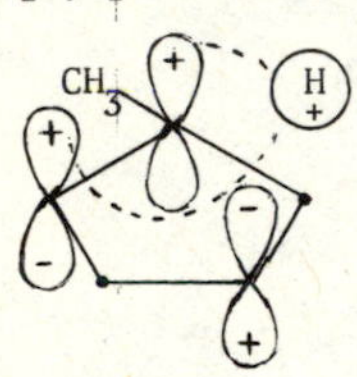

b) The rearrangement is a [1,7] hydrogen shift and is an allowed antarafacial process. A [1,7] antarafacial shift is geometrically possible in this open chain system. The signs for this ψ_4 orbital of a 7-atom system can be derived by continuing the alternation of orbital lobes observed in fig. 16-2.

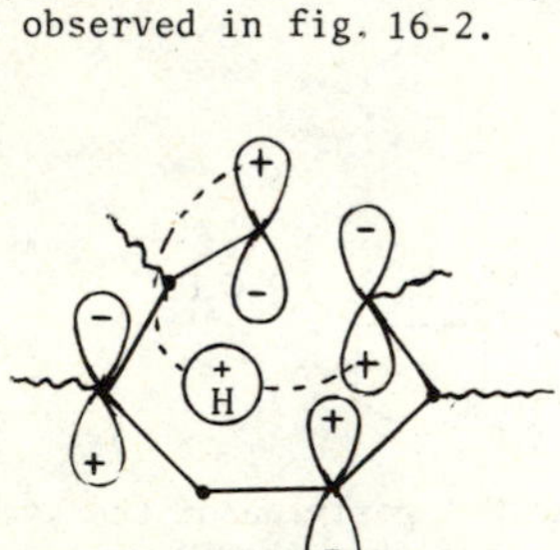

c) Migration of either deuterium atom involves a suprafacial [1,5] shift. Continual [1,5] shifts give only the two labeled materials. The orbital picture is the same pentadienyl ψ_3 used in part (a).

26-35 The reaction is a stereospecific [1,5] suprafacial shift of hydrogen. If we consider the *cis*-diene as rigid but the chiral atom freely rotating, then orbital symmetry analysis enables the hydrogen atom to move across the top or the bottom face of the pentadienyl system in an allowed suprafacial process. (Note that we can treat the hydrogen atom as if it were in a symmetrical 1*s* orbital of either + or - sign.)

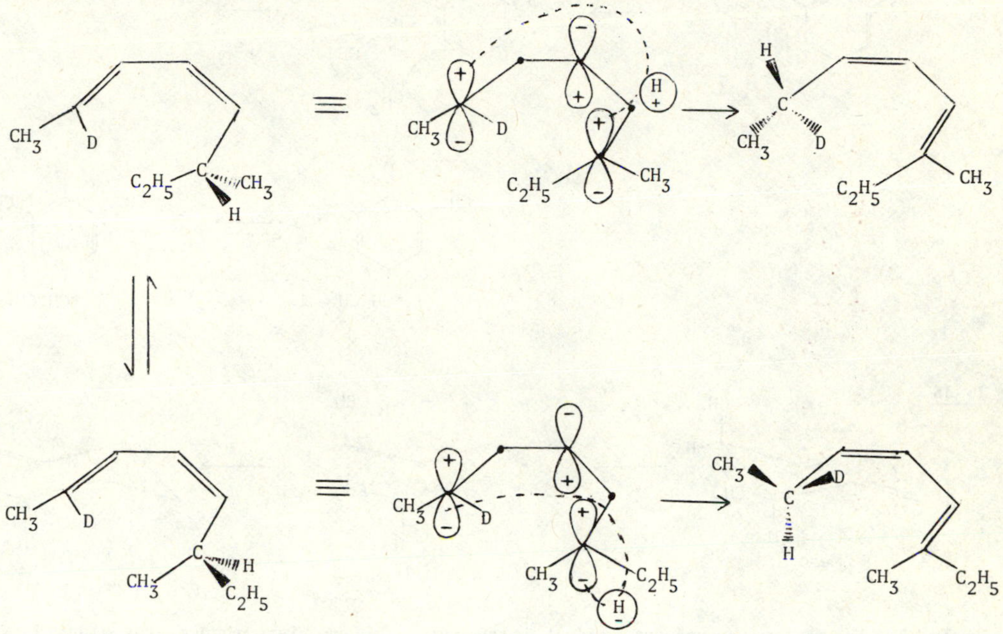

26-36 Each portion of the hexadiene system is considered as an allyl system with three electrons, two from the double bond and one from cleavage of the sigma bond. The HOMO's are ψ_2 of the allyl system.

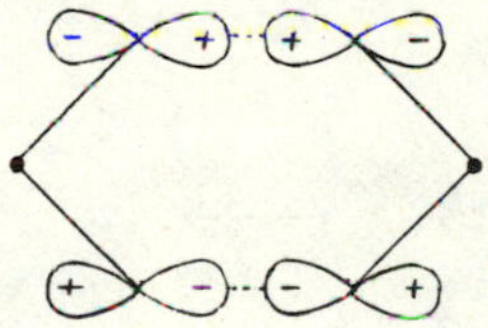

26-37 Two successive [3,3] shifts occur. The first is an *ortho*–Claison rearrangement and the second a Cope rearrangement.

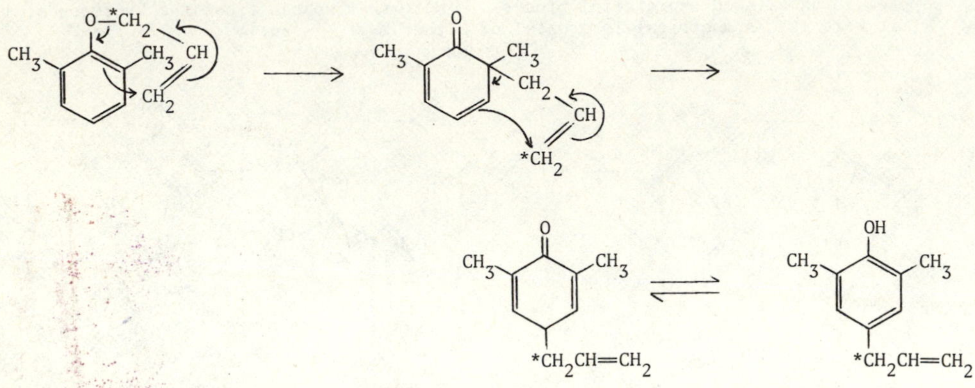

26-38

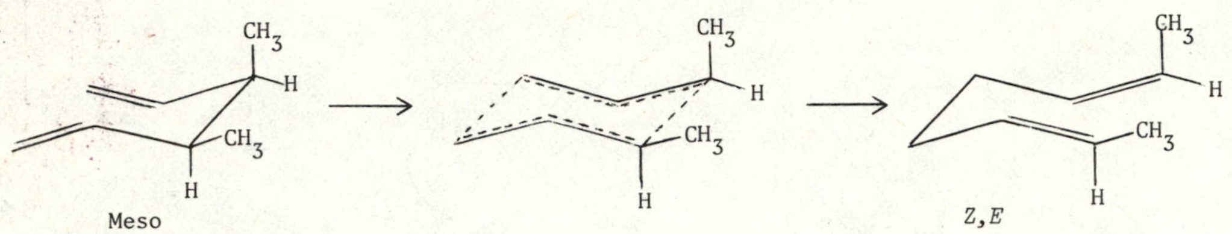

26-39 The initial rearrangement product is an enol. Tautomerism produces the aldehyde which cannot undergo a reverse Cope rearrangement.

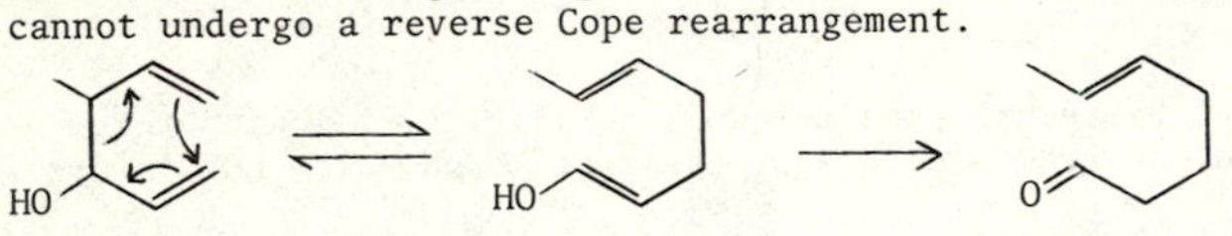

26-40

a)

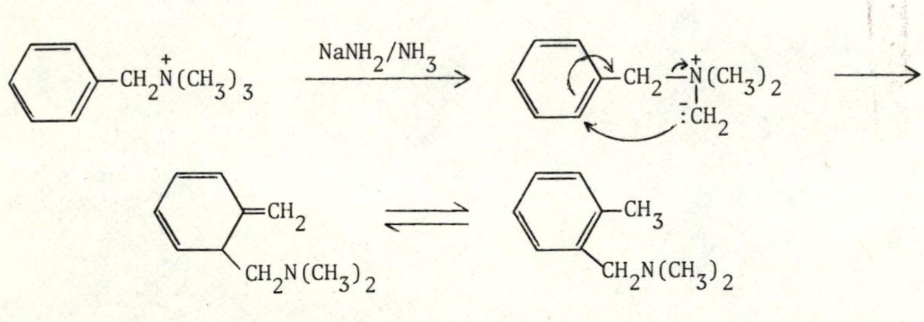

Contd...

26-40 contd...

b) 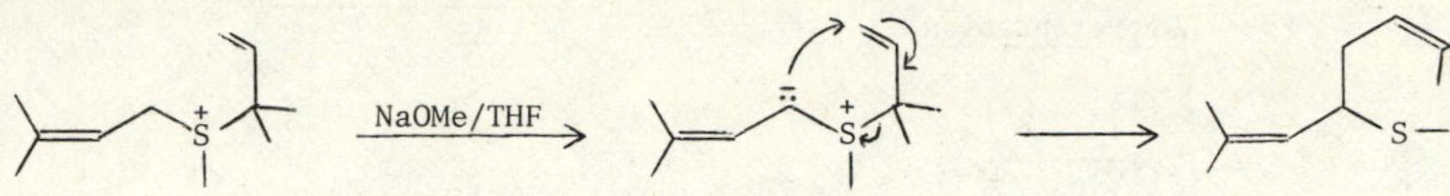

In each case formation of the ylid provides the precursor for an allowed, suprafacial [2,3] sigmatropic rearrangement. The HOMO of the C_3 fragment is ψ_2 and the LUMO for the C_2 fragment is ψ_2'.

(Use of the C_3 LUMO and C_2 HOMO gives identical results.)

∴ 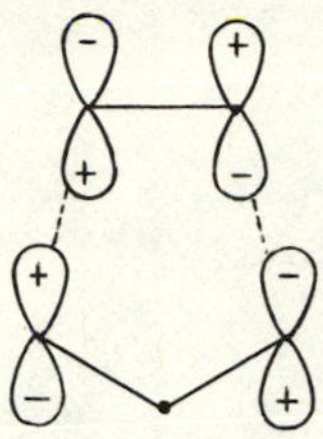

ψ_2' - LUMO

ψ_2 - HOMO

26-41 Symmetry properties of 1,3-cyclohexadiene relative to a plane and an axis of symmetry are:

Orbital	Representation	Symmetry element: Plane	Axis
σ^*	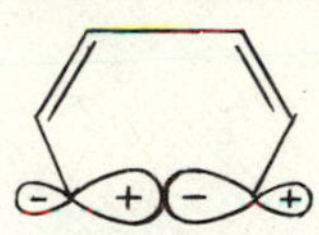	A	A
π_4^*	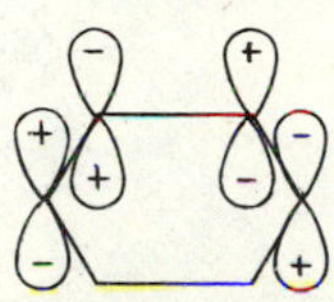	A	S
π_3^*	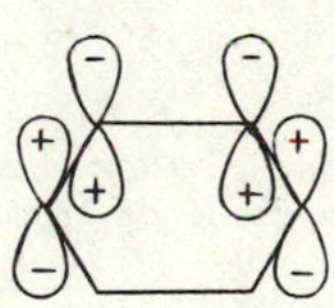	S	A

Contd...

26-41 contd...

Orbital	Representation	Symmetry element Plane	Symmetry element Axis
π_2	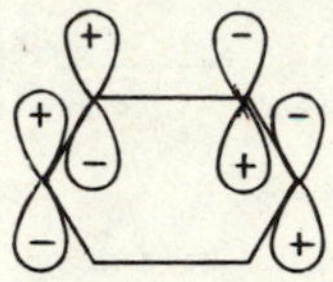	A	S
π_1	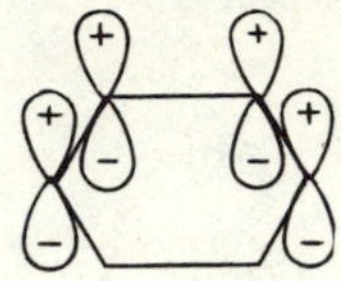	S	A
σ	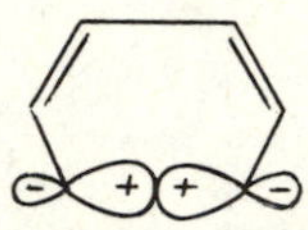	S	S

Symmetry properties for 1,3,5-hexatriene can be obtained from figure 16-2. The results are compiled below.

1,3,5-Hexadiene

Orbital	Symmetry element Plane	Symmetry element Axis
ψ_6	A	S
ψ_5	S	A
ψ_4	A	S
ψ_3	S	A
ψ_2	A	S
ψ_1	S	A

Correlation diagrams are constructed from these data. They show that correlation between ground state molecular orbitals is present when a plane of symmetry is maintained. A disrotatory reaction is therefore allowed.

a) Plane of symmetry (disrotatory)

(A)	ψ_6	—	—	σ^*	(A)
(S)	ψ_5	—	—	π_4^*	(A)
(A)	ψ_4	—	—	π_3^*	(S)
(S)	ψ_3	⇅	⇅	π_2	(A)
(A)	ψ_2	⇅	⇅	π_1	(S)
(S)	ψ_1	⇅	⇅	σ	(S)

b) Axis of symmetry (conrotatory)

(S)	ψ_6	—	—	σ^*	(A)
(A)	ψ_5	—	—	π_4^*	(S)
(S)	ψ_4	—	—	π_3^*	(A)
(A)	ψ_3	⇅	⇅	π_2	(S)
(S)	ψ_2	⇅	⇅	π_1	(A)
(A)	ψ_1	⇅	⇅	σ	(S)

26-42

a)

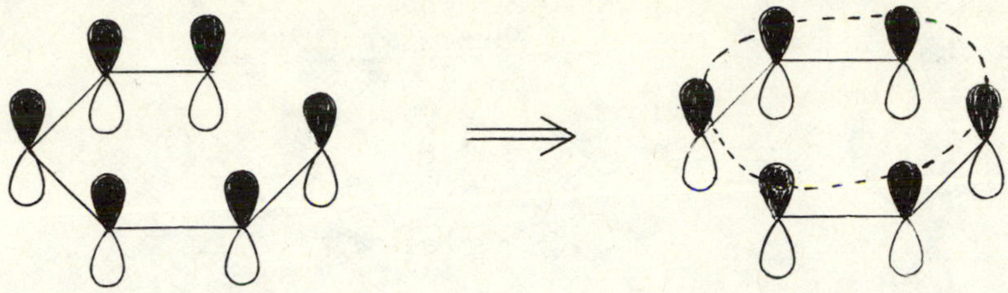

Zero phase changes; the pathway is Hückel. Since there are $4n + 2 = 6$ electrons, it is an allowed process.

b)

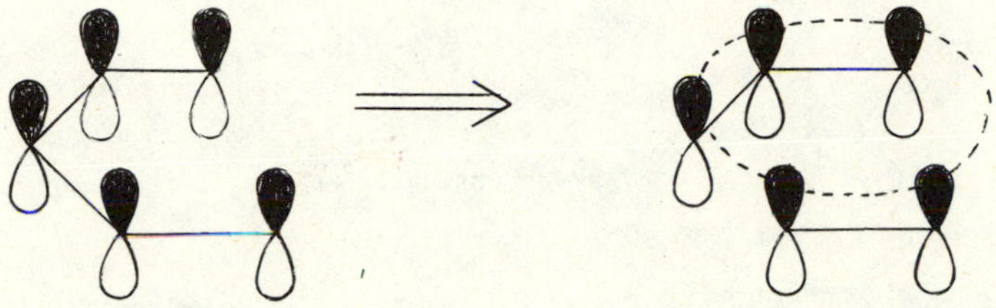

Zero phase changes; the pathway is Hückel. Since there are $4n + 2 = 6$ electrons (including the anion electrons) the reaction is allowed.

c)

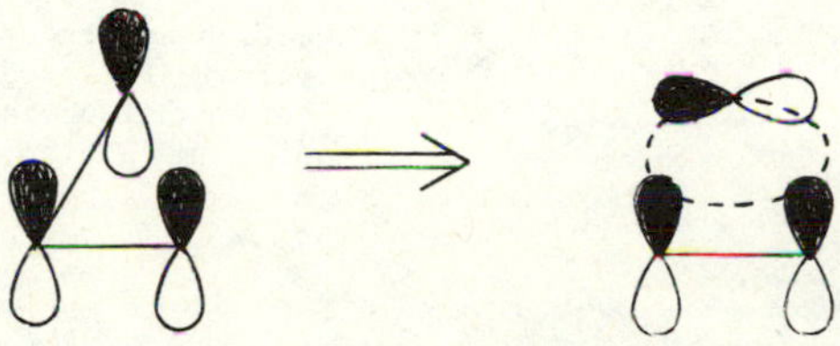

One phase change; the pathway is Möbius. Since there are $4n$ electrons (including the anion) the process is allowed.

d)

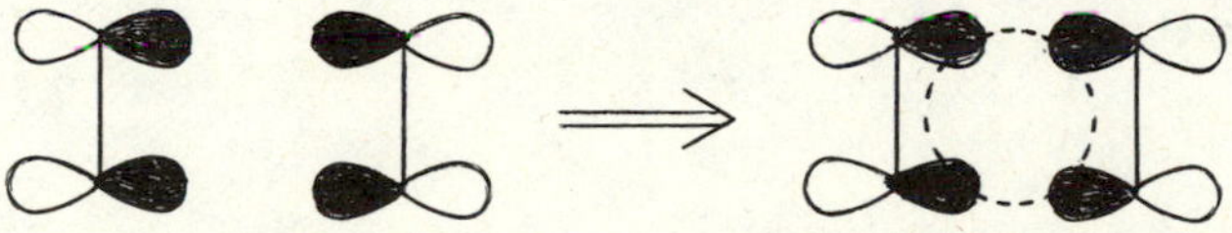

Zero phase change; the pathway is Hückel. Since there are $4n = 4$ electrons, the reaction is forbidden.

Contd...

26-42 contd...

e)

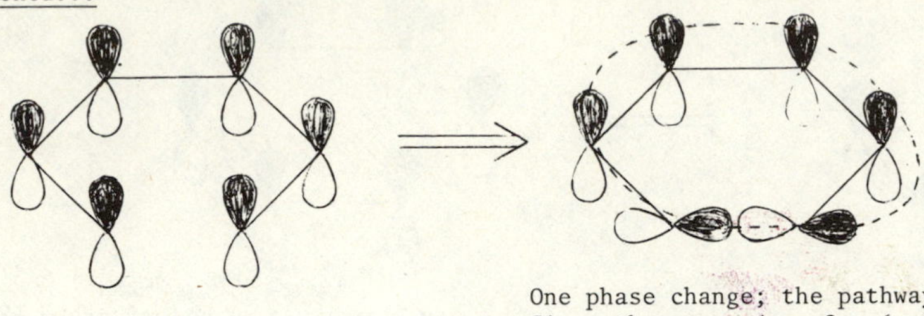

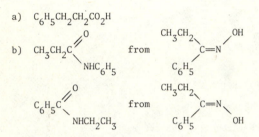

One phase change; the pathway is Möbius. Since there are $4n + 2 = 6$ electrons the process is forbidden.

26-43

a) $C_6H_5CH_2CH_2CO_2H$

b) 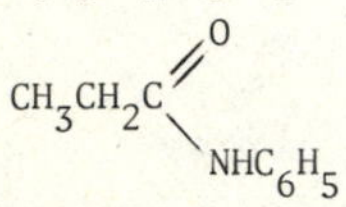 from 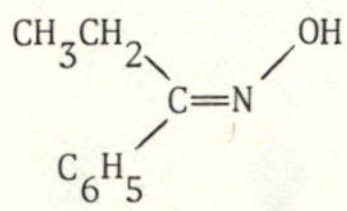

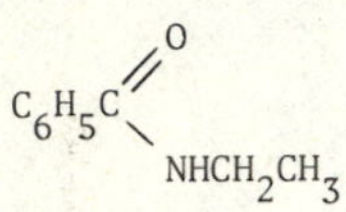 from 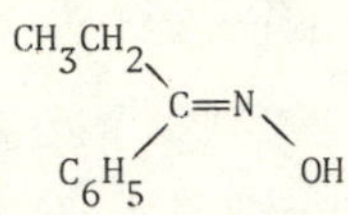

The first reaction which involves migration of phenyl is expected to be the faster of the two rearrangements.

c) 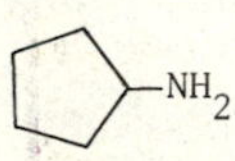

d) 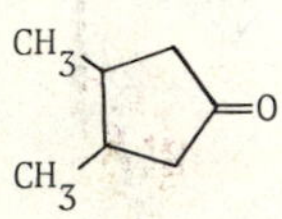

meso

e) 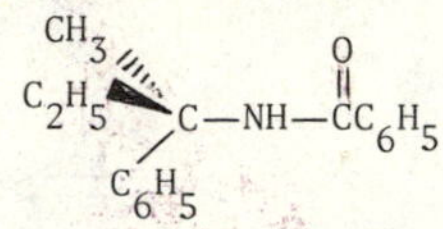

Optically active

f) 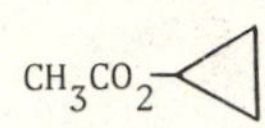

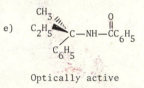

Cyclopropyl usually reacts as if it were more electron rich than an alkyl group, thus is favored in migration.

g) 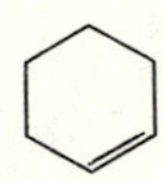

Reaction occurs by initial homolysis followed by ring expansion via a free radical mechanism.

26-44 Alkyl-oxygen cleavage would lead to the neopentyl carbocation. That would surely lead to alcohol with a rearranged skeleton.

$$(CH_3)_2C(OH)CH_2CH_3$$

26-45

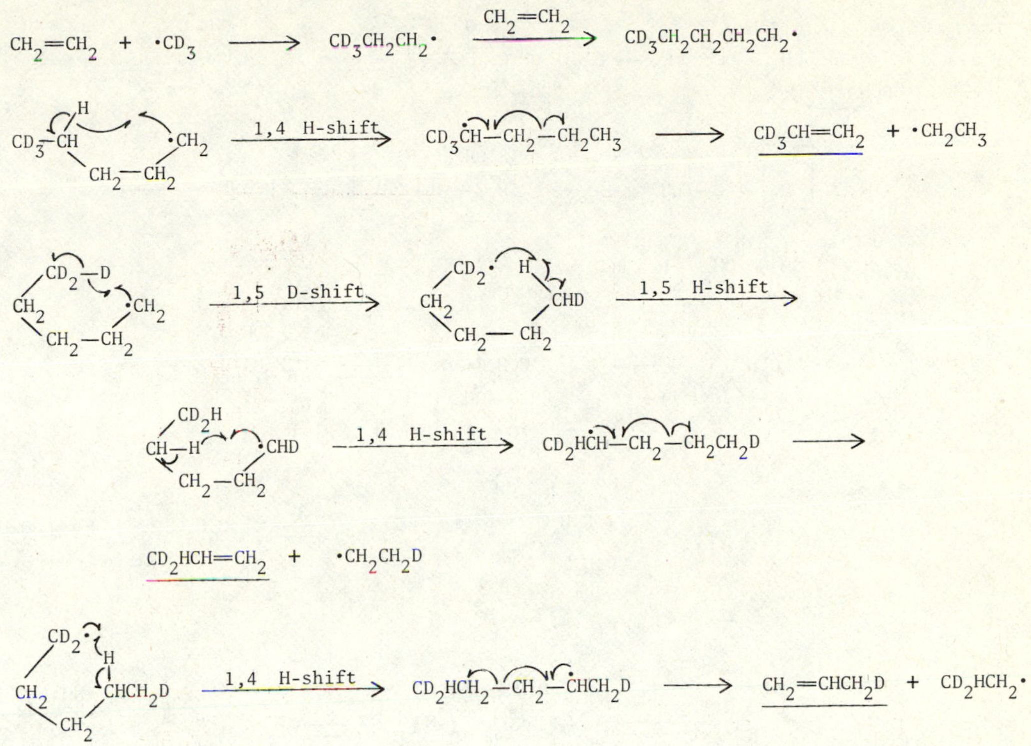

26-46 The cis isomer has methyl anti to the adjacent hydroxy group and leads to pinacol rearrangement with methyl migration.

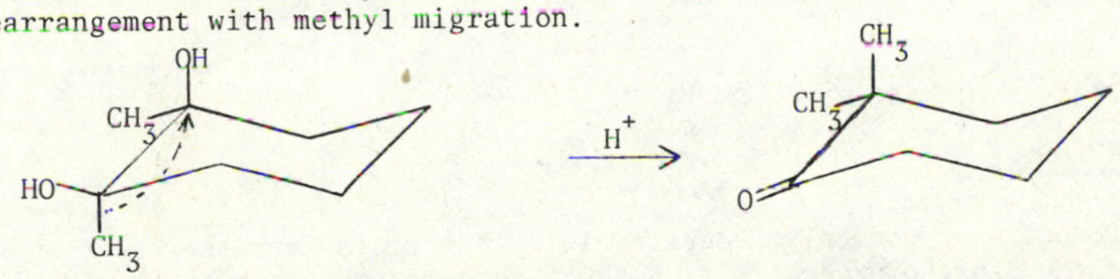

The more favorable conformation of the trans isomer has no easily accessible group for back-sided migration to a carbinol carbon atom. Ring flip provides a conformation in which a ring carbon atom can migrate.

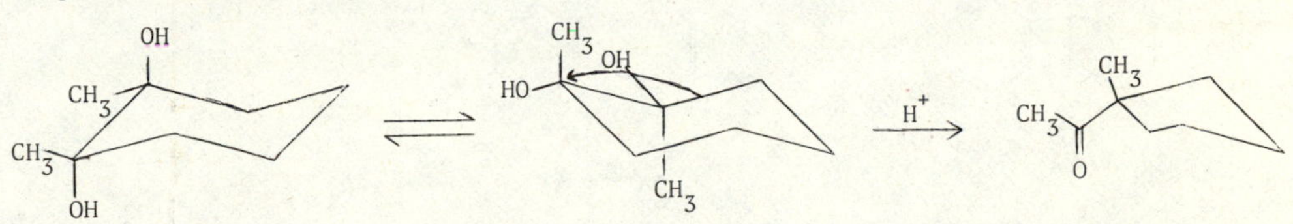

26-47

a) 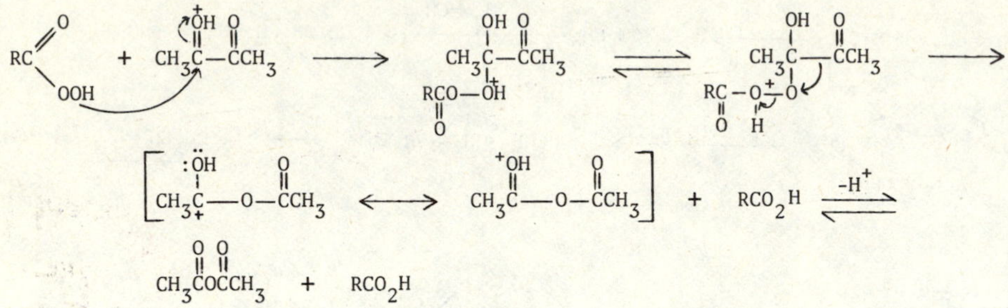

b) 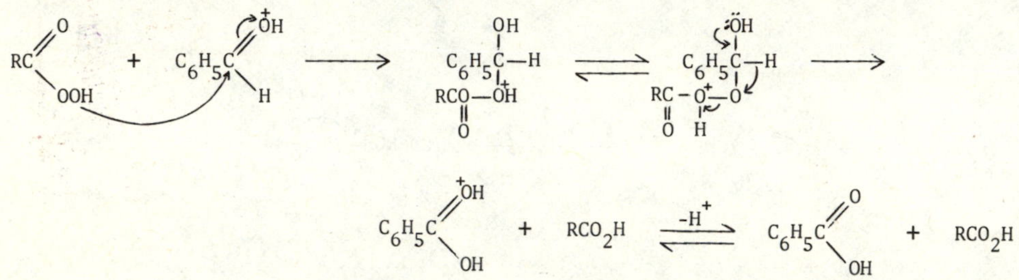

26-48

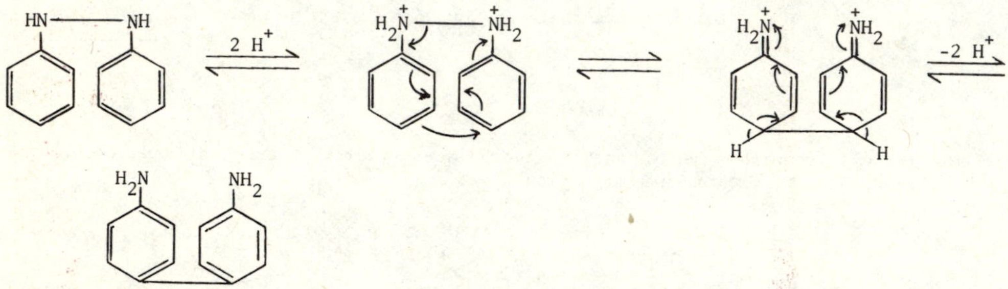

The ortho-para product A is commonly recovered in 20-30% yield from the reaction along with a small amount of the diortho product B . Both of these isomers can be accounted for by mechanisms similar to the above.

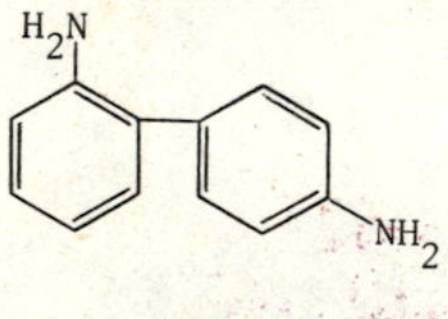

A

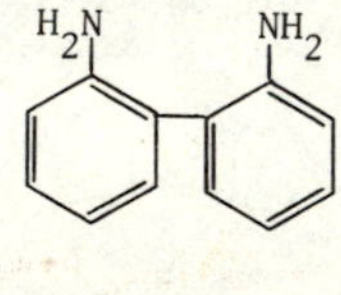

B

26-49 Most of the endo reactant looses brosylate and forms the nonclassical 2-norbornyl cation which leads only to racemic exo product. However, there is no backside participation in loss of the brosylate. A small amount of the acetate can approach the initial carbocation from the exo direction before formation of the nonclassical ion and thus form inverted product.

26-50

a)

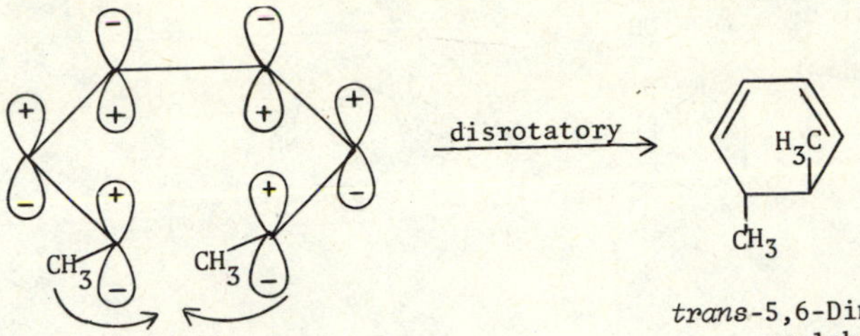

trans-5,6-Dimethyl-1,3-cyclohexadiene

b)

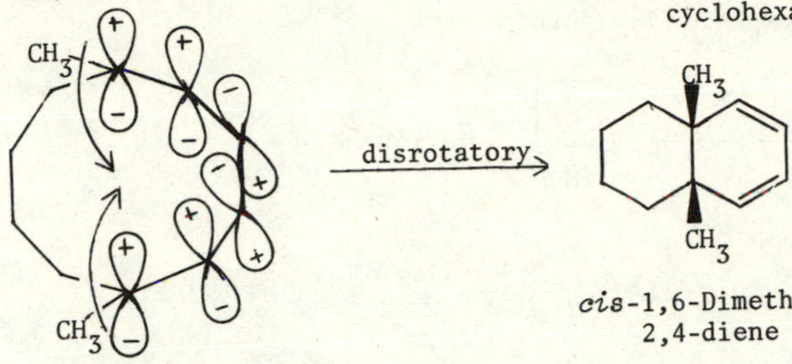

cis-1,6-Dimethylbicyclo[4.4.0]deca-2,4-diene

c)

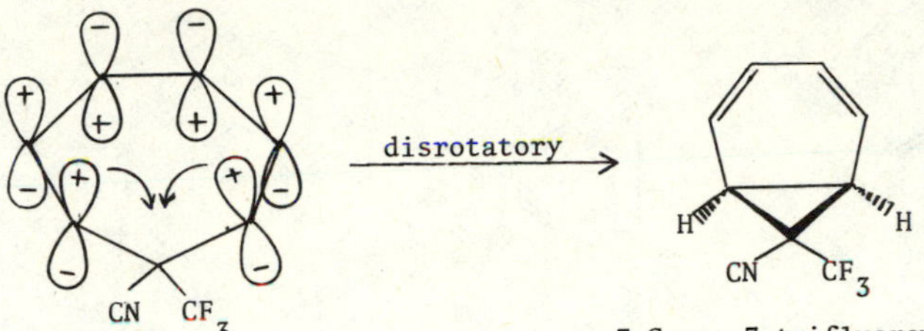

7-Cyano-7-trifluoromethylnorcaradiene

d)

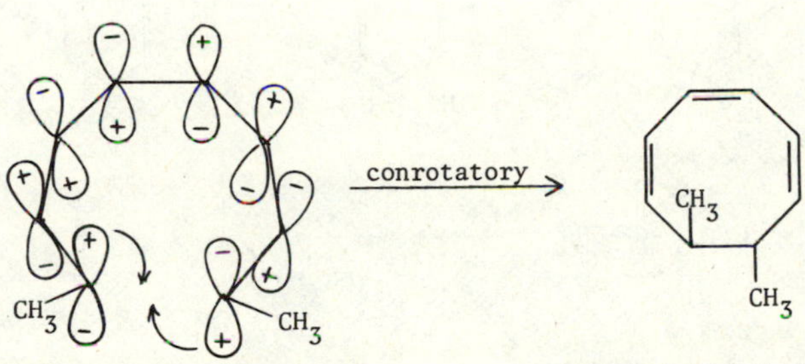

trans-7,8-Dimethyl-1,3,5-cyclooctatriene

26-51 A Beckmann rearrangement of cyclohexanone oxime provides a reasonable industrial route.

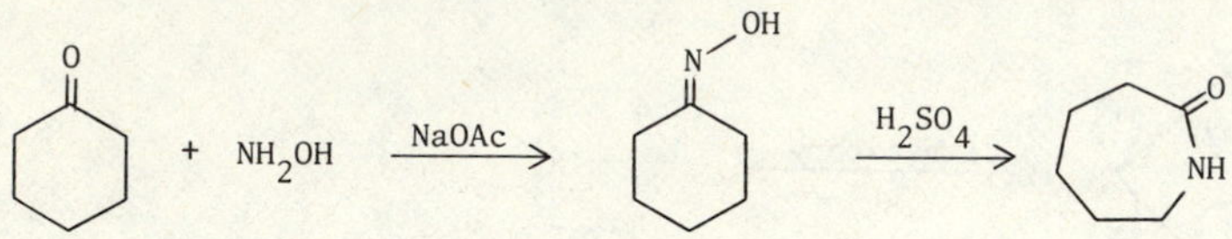

25-52

a) The dehalogenation is accompanied by rearrangement.

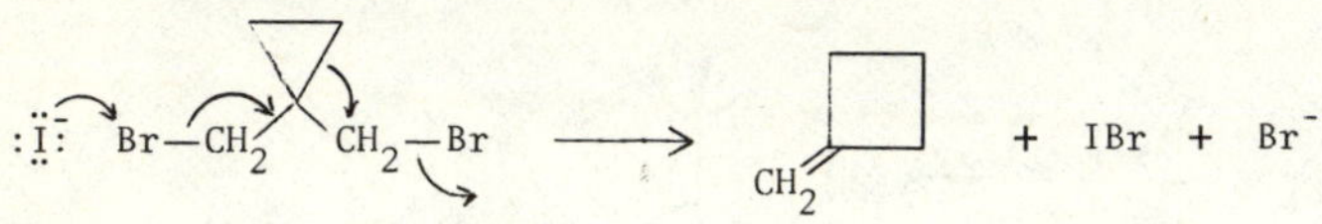

b)

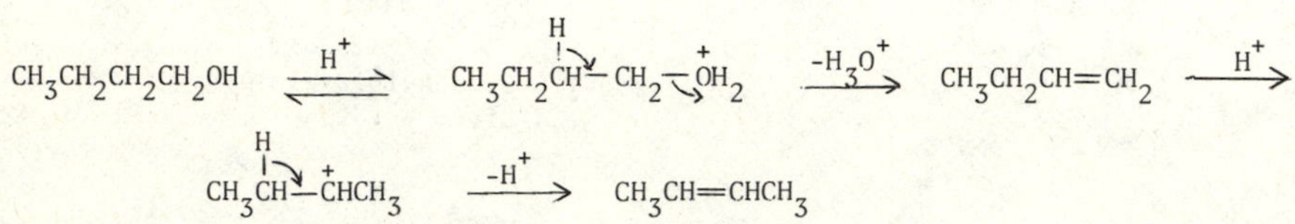

c)

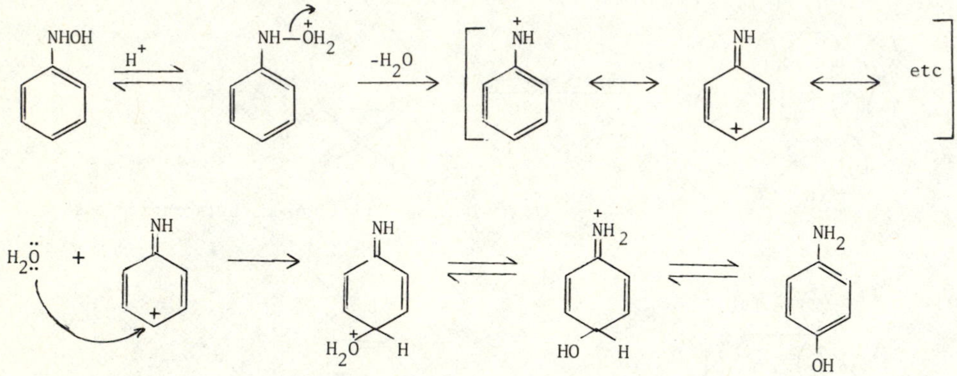

d)

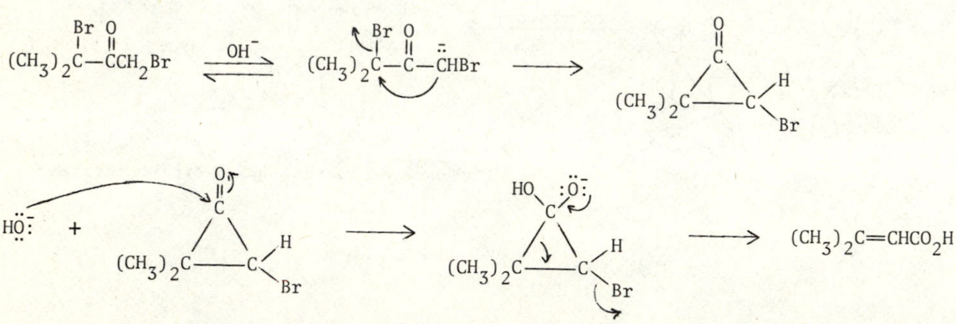

Contd...

26-52 contd...

e) Acyl-oxygen cleavage of the two esters gives acyl and phenacyl fragments and the corresponding phenoxy anions. Scrambling and recombination of all groups gives the product mixture.

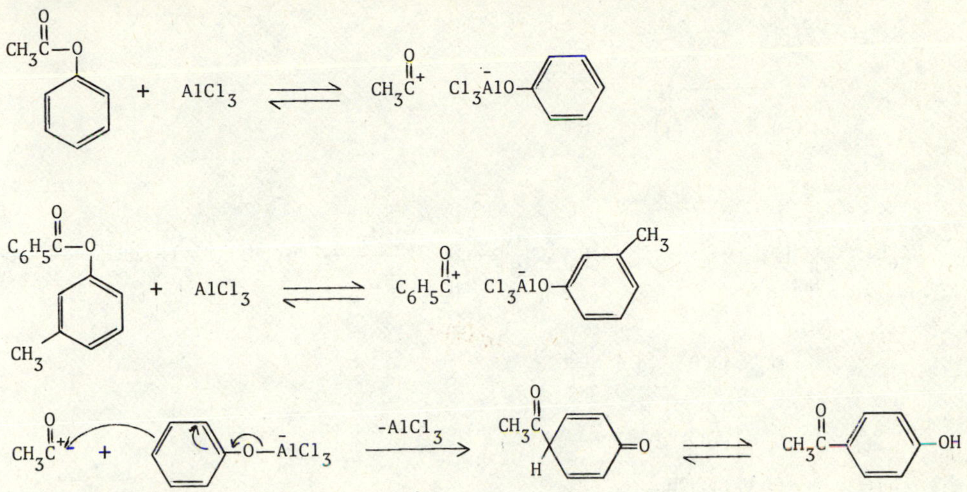

Same kind of sequences lead to the other products.

f) 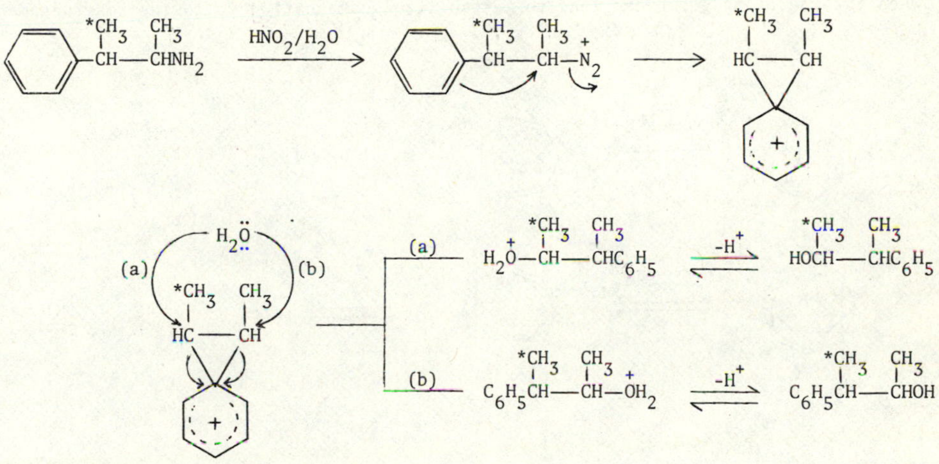

26-53 Orbital symmetry predicts inversion for such [1,2] sigmatropic processes which must, due to geometrical constraints proceed in a suprafacial manner. The orbital picture utilizes an ethylene orbital (the ylid bond) with three electrons thus the HOMO is ψ_2.

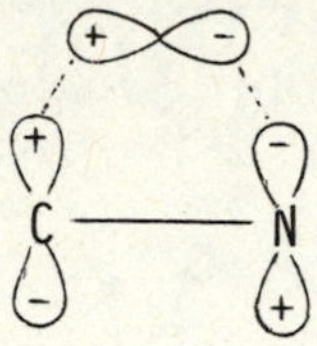

26-54

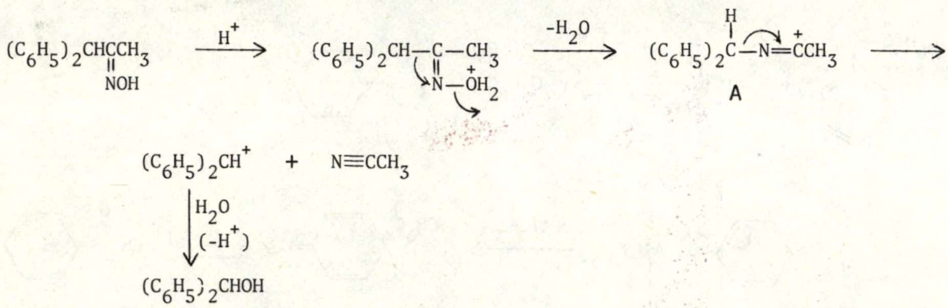

Both the Beckmann and abnormal Beckmann reactions are believed to proceed through the common intermediate A that results from migration of one group. Fragmentation is favored when the migrating group can lead to a stabilized carbocation following rearrangement.

26-55

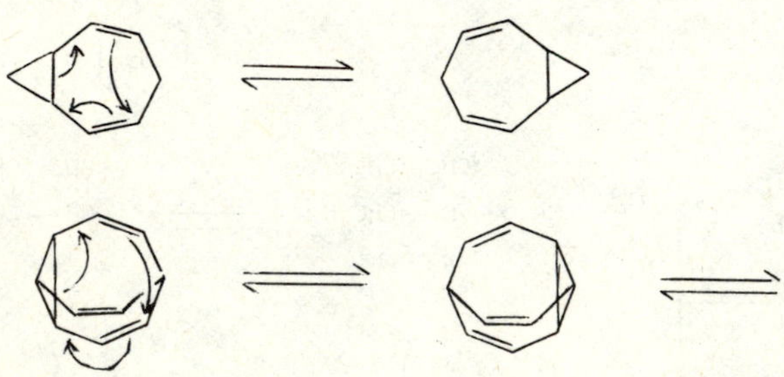

etc.

26-56

a)

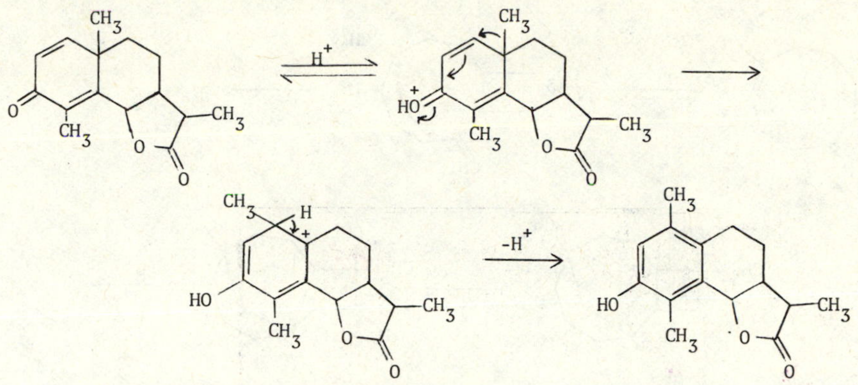

b)

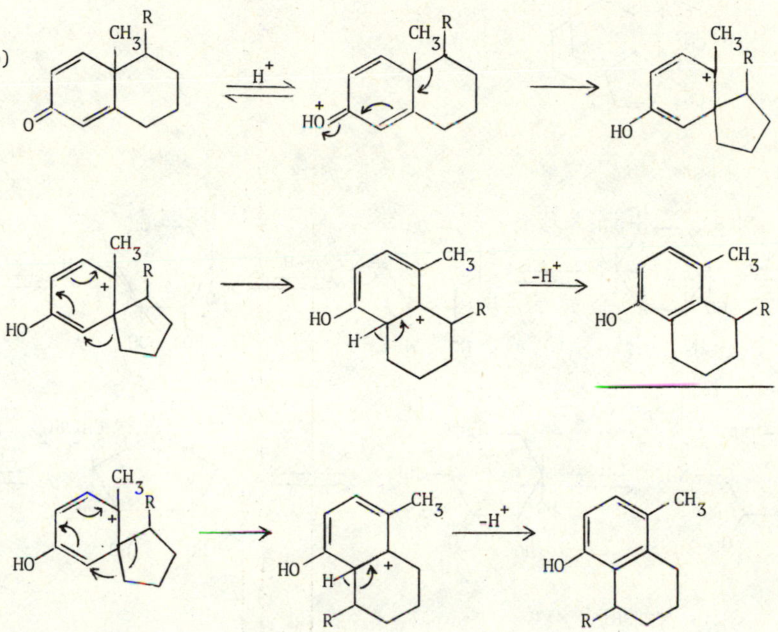

Contd...

26-56 contd...

c)

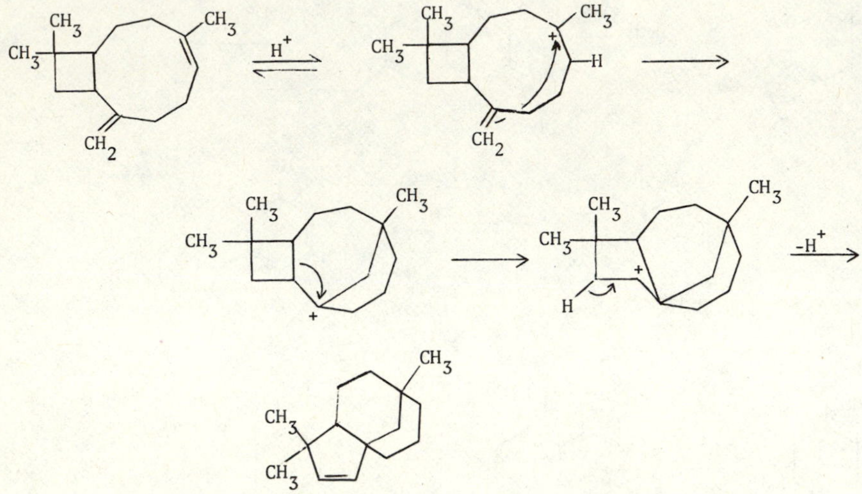

d)

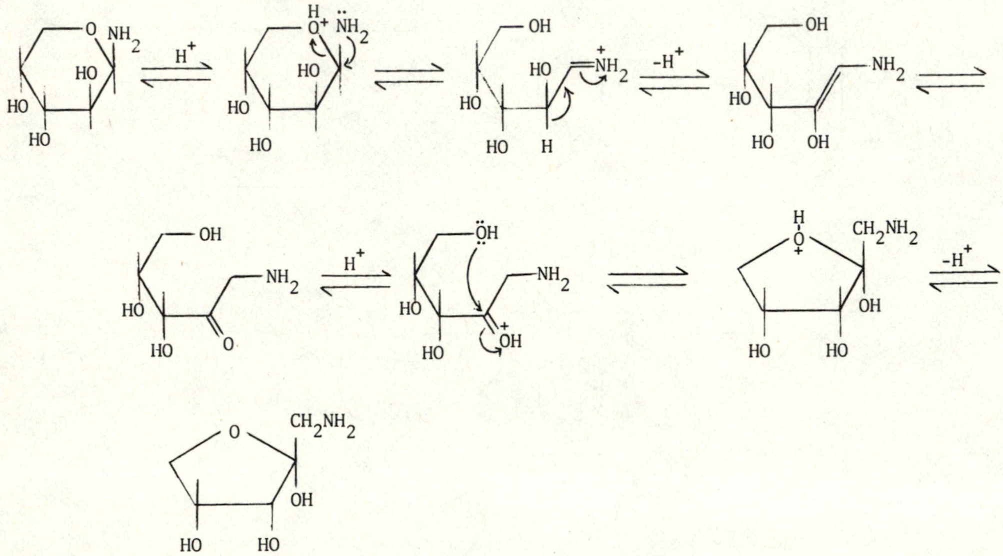

26-57

a) $(C_6H_5)_2CO \xrightarrow{NH_2OH} C_6H_5-\overset{\overset{NOH}{\|}}{C}-C_6H_5 \xrightarrow[H_2O]{H_2SO_4} C_6H_5CONHC_6H_5$

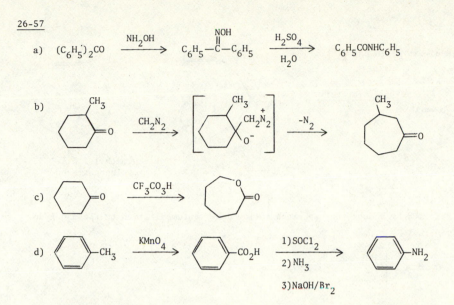

b)

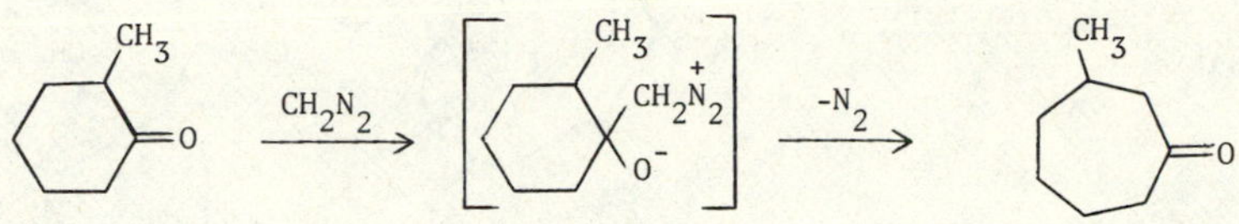

c)

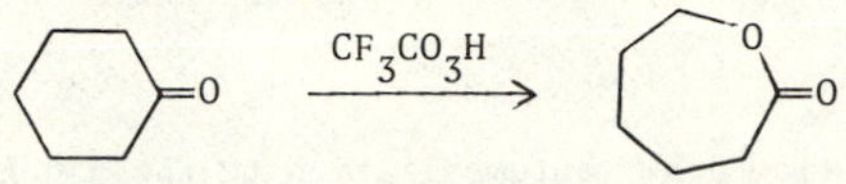

d)

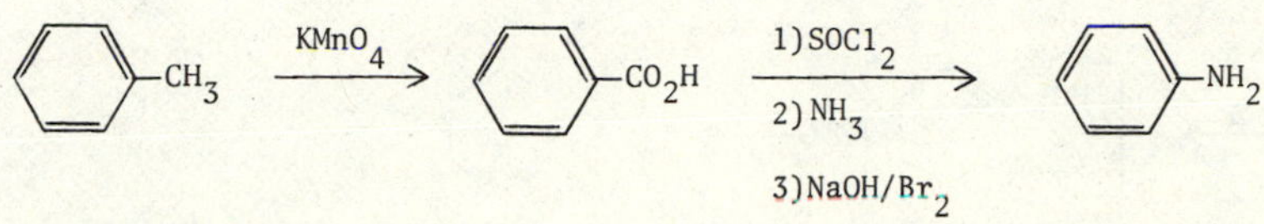

26-58 Refer to sec. 26-1F:
The first example confirms that the ketone carbonyl oxygen atom becomes the ester carbonyl oxygen atom.
The second example shows that the migrating group moves without loss of configuration; a process consistent with concerted migration by the bonding electrons.

26-59

a) The product butadiene HOMO(ψ_2) has identical end lobe signs on the opposite sides of the molecular plane. Conrotatory opening of the sigma bond is necessary to attain that configuration.

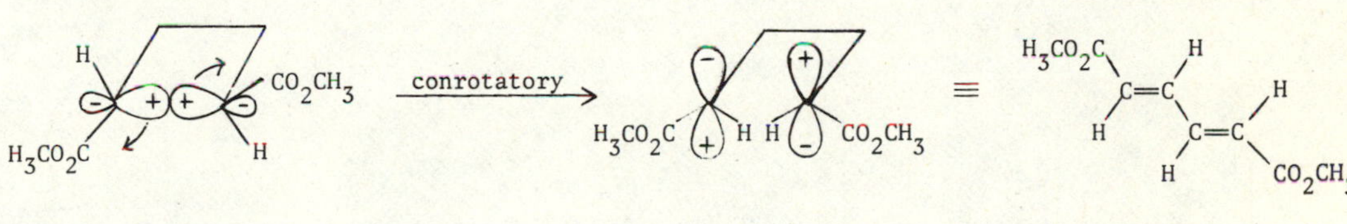

Dimethyl *E*,*E*-2,4-hexa-dienedioate

b) The hexatriene HOMO (ψ_3) has end orbital lobes of identical sign on the same side of the molecular plane; a disrotatory ring opening is required.

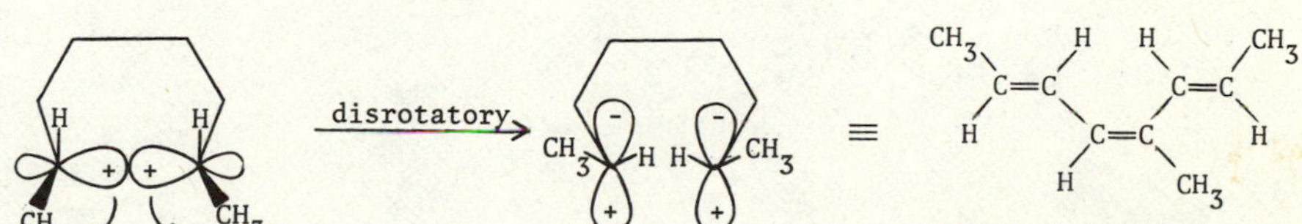

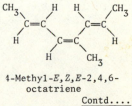

4-Methyl-*E*,*Z*,*E*-2,4,6-octatriene

Contd....

26-59 contd...

c) Same orbital analysis as in part (b).

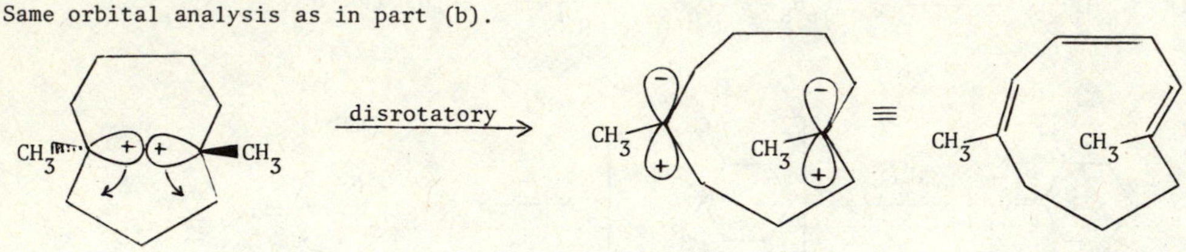

1,6-Dimethyl-*Z*,*Z*,*E*-1,3,5-cyclononatriene

26-60 The product is formed by a Cope rearrangement followed by tautomerization to the aldehyde.

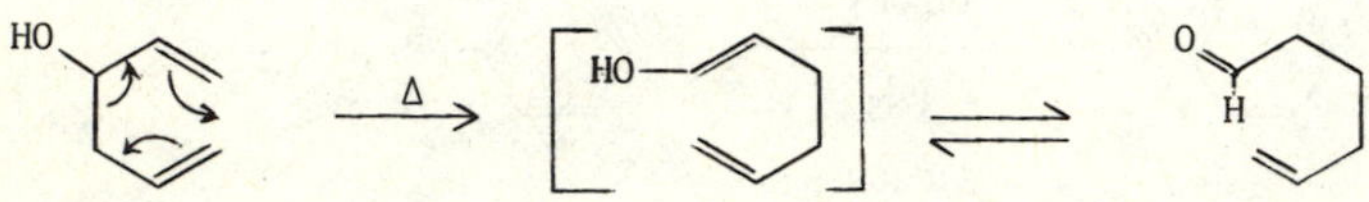

27 Photochemistry and Electrochemistry

27-1 The conjugated compound, 3-buten-2-one, has two ground state and two excited state π-molecular orbitals. Conjugation results in lowering the energies of the π orbitals relative to the nonconjugated π system of acetone. Little, if any change, takes place in the energy of the n-electrons. The $n \longrightarrow \pi^*$ transition therefore occurs at a lower energy (longer wavelength) for the conjugated molecule. The $\pi \longrightarrow \pi^*$ absorptions are at shorter wavelengths (higher energies) and 3-buten-2-one benefits again from conjugation.

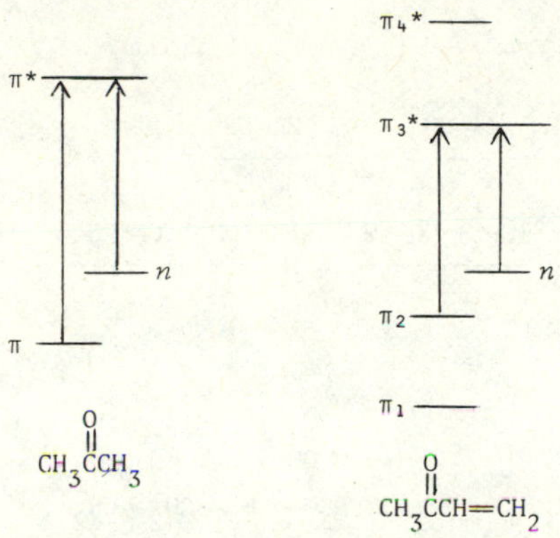

27-2 Some energy is lost from the excited state by vibrational relaxation. The emission transition is thus of lower energy (longer wavelength) than the original absorption.

27-3 An increase in temperature as energy is converted to molecular vibrations and rotations.

27-4 When the 2-hydroxy-2-propyl radical gives up a hydrogen atom to benzophenone, a stable molecule (acetone) is formed. Furthermore, the new radical (benzophenone ketyl) is more stabilized than is the 2-hydroxy-2-propyl radical. Similar favorable reaction energetics are not possible when the free radical is benzyl.

27-5

$$(C_6H_5)_2C{=}O \xrightarrow{h\nu} (C_6H_5)_2C{=}O^{*(3)}$$

$$(C_6H_5)_2C{=}O^{*(3)} + (C_6H_5)_2CHOH \longrightarrow 2\ (C_6H_5)_2\dot{C}{-}OH$$

$$2\ (C_6H_5)_2\dot{C}{-}OH \longrightarrow (C_6H_5)_2C(OH){-}C(OH)(C_6H_5)_2$$

The maximum quantum yield for the disappearance of benzophenone is expected to be 1 because only one required benzophenone ketyl is formed by absorption of one photon. Since formation of one excited benzophenone molecule produces one molecule of benzpinacol, the maximum quantum yield for formation of benzpinacol is 1.

27-6

$$(CH_3)_3CC(=O)C(CH_3)_3 \xrightarrow{h\nu} (CH_3)_3C\dot{C}{=}O + \cdot C(CH_3)_3 \quad (-CO)$$

$$\cdot C(CH_3)_3 + CCl_4 \longrightarrow \underline{ClC(CH_3)_3} + \cdot CCl_3$$

$$(CH_3)_2\dot{C}{-}CH_3 + \cdot CCl_3 \longrightarrow \underline{(CH_3)_2C{=}CH_2} + \underline{HCCl_3}$$

$$(CH_3)_3C\dot{C}{=}O + (CH_3)_2\dot{C}{-}CH_3 \longrightarrow \underline{(CH_3)_3CC(=O){-}H} + \underline{(CH_3)_2C{=}CH_2}$$

27-7

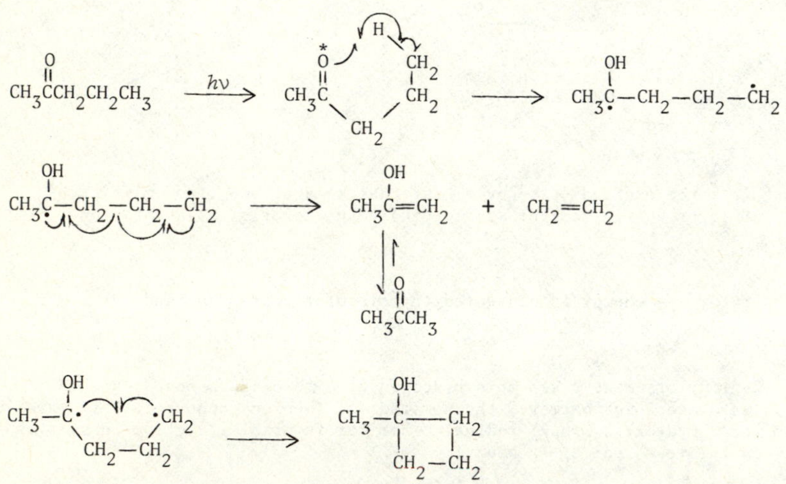

27-8

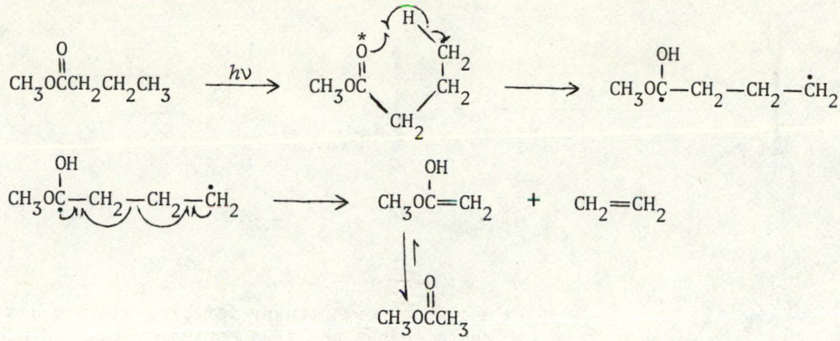

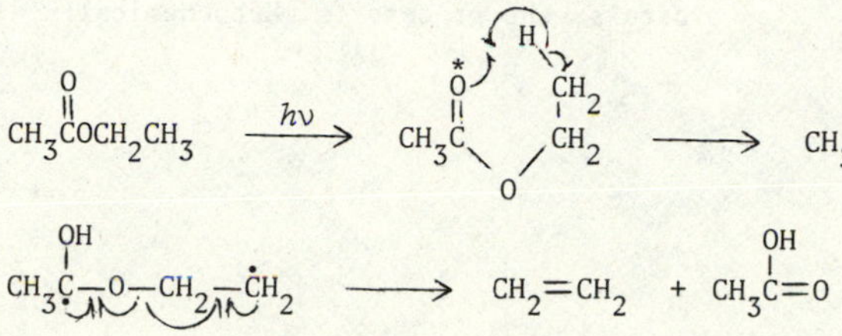

27-9

a) $(p\text{-}CH_3C_6H_4)_2C(OH)—C(OH)(C_6H_4CH_3\text{-}p)$ + $(CH_3)_2C{=}O$

b) $(CH_3)_3C—C(CH_3)_3$ + $(CH_3)_2C=CH_2$ + $(CH_3)_3CCHO$ + CO

c) 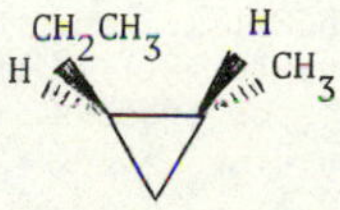

d) $C_6H_5C(=O)CH_3$ + $CH_2{=}CH_2$

e) 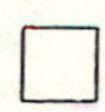 + CO

f) $C_6H_5CH_2CH_2C_6H_5$ + $C_6H_5CH_3$ + CO

27-10 The first synthesis of grandisol was carried out in the following way.

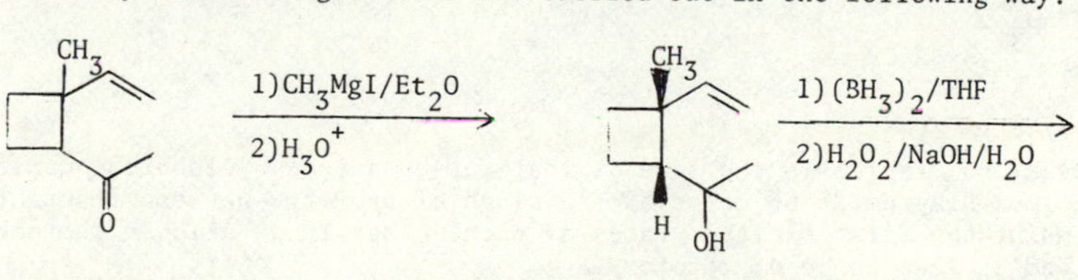

A mixture of stereoisomers

The isomeric mixture was separated by gas chromatography at this point.

Contd...

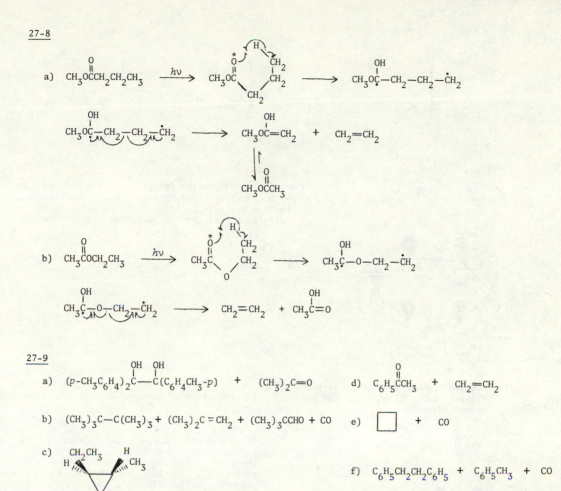

27-10 contd...

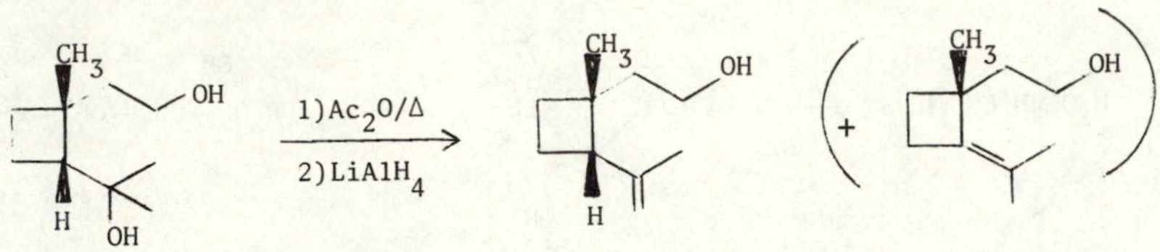

27-11

a) i)

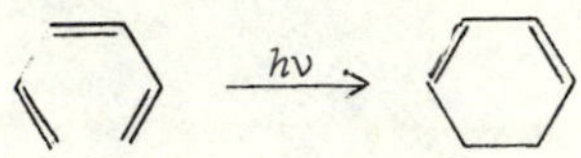

A conrotatory pathway for ring closure has one phase change and is a Möbius system. Since there are $4n + 2$ electrons ($n = 1$) in the interacting orbitals, the process is photochemically allowed

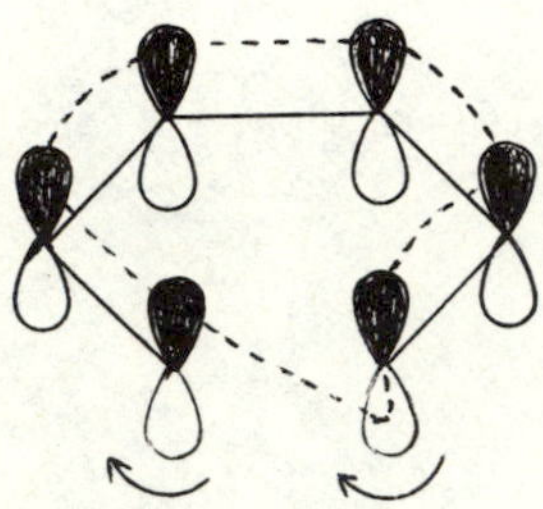

ii) The solution to problem 26-41 shows that an axis of symmetry correlates the first excited states of reactant and product. The photochemical process is thus predicted to be conrotatory.

b) i)

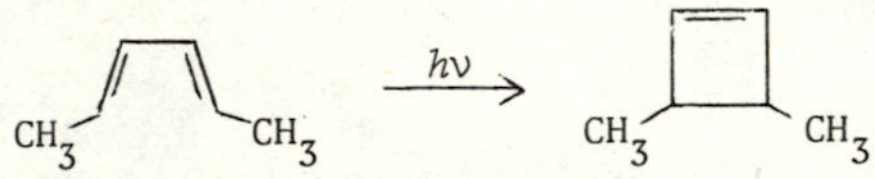

The disrotatory pathway has zero phase changes, thus is a Hückel system. Since there are $4n$ electrons ($n = 1$) in the interacting orbitals, the process is photochemically allowed.

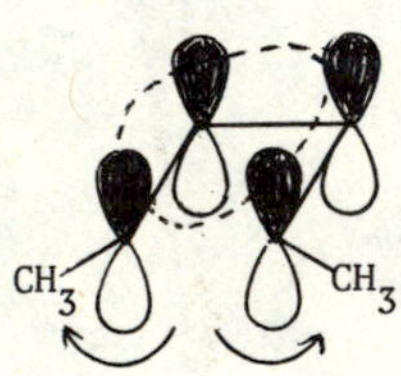

ii) The correlation diagram is the same as that for butadiene-cyclobutene depicted in figure 26-6. Diagram 26-6B correlates a plane of symmetry between reactant and product using the first excited states of each. That is an allowed photochemical process and is seen to be disrotatory.

27-12

a) 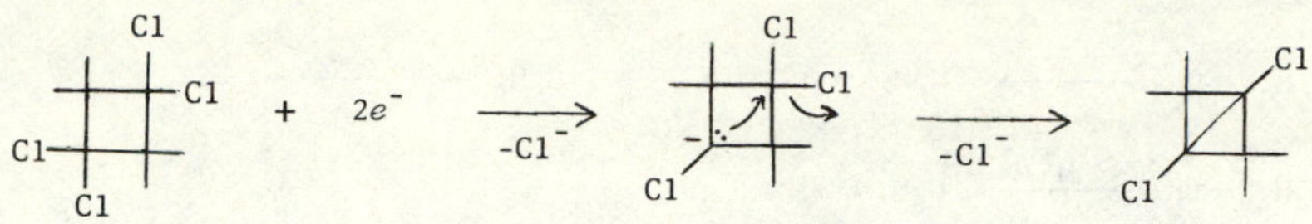

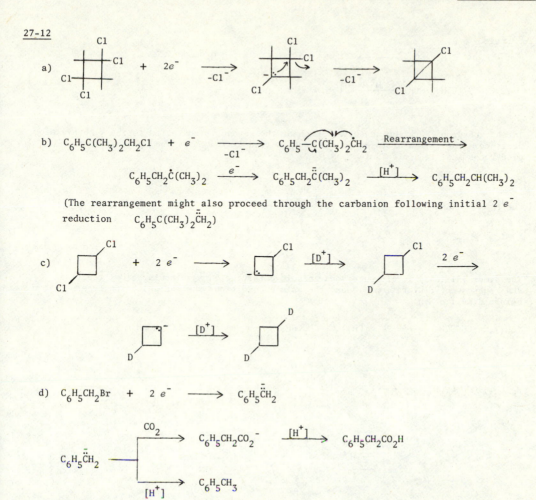

b) $C_6H_5C(CH_3)_2CH_2Cl + e^- \xrightarrow[-Cl^-]{} C_6H_5{-}C(CH_3)_2\dot{C}H_2 \xrightarrow{\text{Rearrangement}}$

$C_6H_5CH_2\dot{C}(CH_3)_2 \xrightarrow{e^-} C_6H_5CH_2\ddot{\bar{C}}(CH_3)_2 \xrightarrow{[H^+]} C_6H_5CH_2CH(CH_3)_2$

(The rearrangement might also proceed through the carbanion following initial 2 e^- reduction $C_6H_5C(CH_3)_2\ddot{\bar{C}}H_2$)

c) 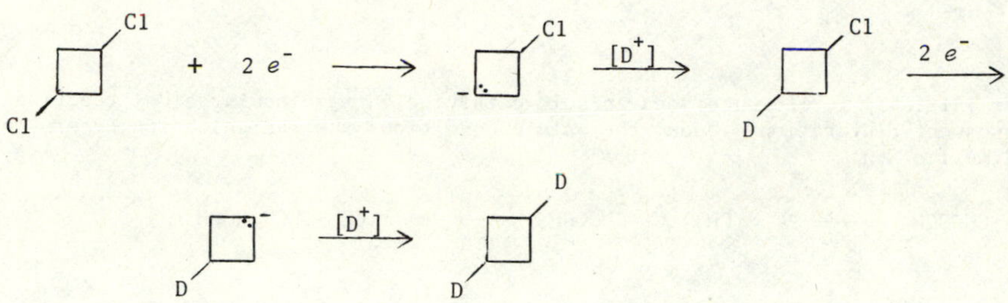

d) $C_6H_5CH_2Br + 2\,e^- \longrightarrow C_6H_5\ddot{\bar{C}}H_2$

$C_6H_5\ddot{C}H_2 \xrightarrow{CO_2} C_6H_5CH_2CO_2^- \xrightarrow{[H^+]} C_6H_5CH_2CO_2H$

$C_6H_5\ddot{C}H_2 \xrightarrow{[H^+]} C_6H_5CH_3$

27-13 The initial prediction about the structure of $\Delta^{1,4}$-bicyclo[2.2.0]hexane is that the molecule is held in a planar shape by the double bond and experiences considerable angle strain (possibly like cyclopropane). The ^1H nmr spectrum confirms this in exhibiting one sharp singlet peak (3.24 ppm) for the eight equivalent hydrogen atoms.

27-14 This sequence differs from that in aprotic media in that protonation can readily occur at each step.

$C_6H_5NO_2 + e^- \longrightarrow C_6H_5NO_2^{\bullet} \xrightarrow{H^+} C_6H_5\overset{+}{\dot{N}}(OH){-}\ddot{O}{:}^- \xrightarrow{e^-} C_6H_5\ddot{N}(OH){-}\ddot{O}{:}^- \xrightarrow{-OH^-}$

Contd...

27-14 contd...

$$C_6H_5\ddot{N}{=}\ddot{O} \xrightarrow{e^-} C_6H_5\dot{N}{-}\ddot{O}{:}^- \xrightarrow{H^+} C_6H_5\dot{N}{-}OH \xrightarrow{e^-}$$

$$C_6H_5\ddot{N}^{-}{-}OH \xrightarrow{H^+} C_6H_5NHOH$$

27-15 In the anhydrous methylamine solvent much of the ketone is converted to the *N*-methylimine derivate. The carbon-nitrogen double bond is readily reduced. The process is an electrochemical reductive amination.

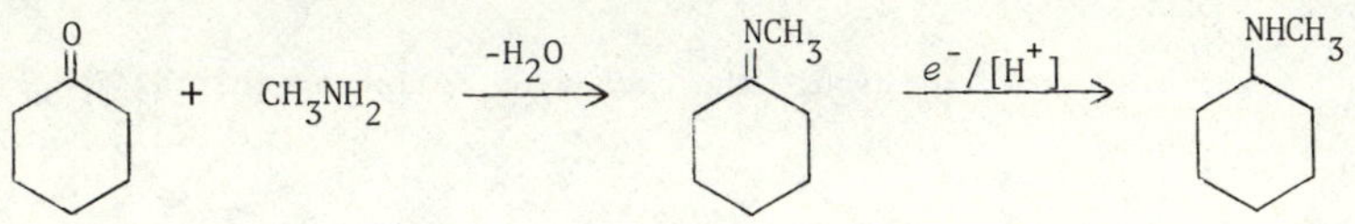

27-16 The dimer initially formed undergoes an acid catalyzed intramolecular aldol reaction and subsequent dehydration. Under the acidic conditions the radical anion is protonated before dimerization.

$$(CH_3)_2C{=}CHCOCH_3 \xrightarrow{e^-} (CH_3)_2\dot{C}{-}\ddot{C}HCOCH_3 \xrightarrow{H^+} (CH_3)_2\dot{C}CH_2COCH_3$$

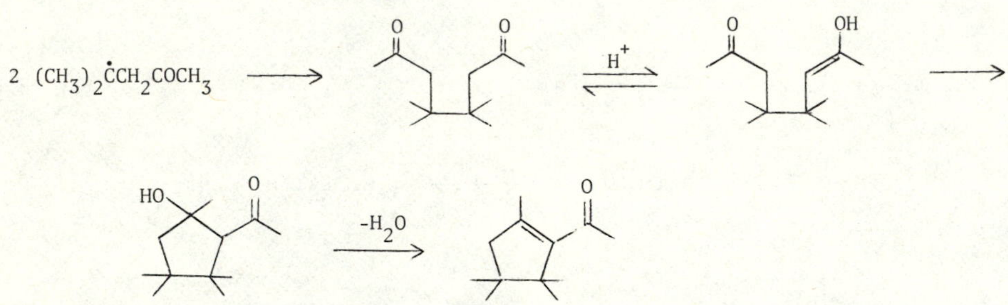

A B

27-17

a) 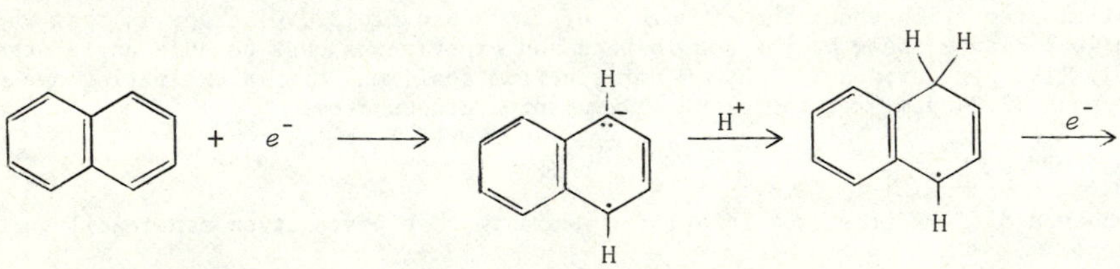

Contd...

27-17 contd...

a) Contd....

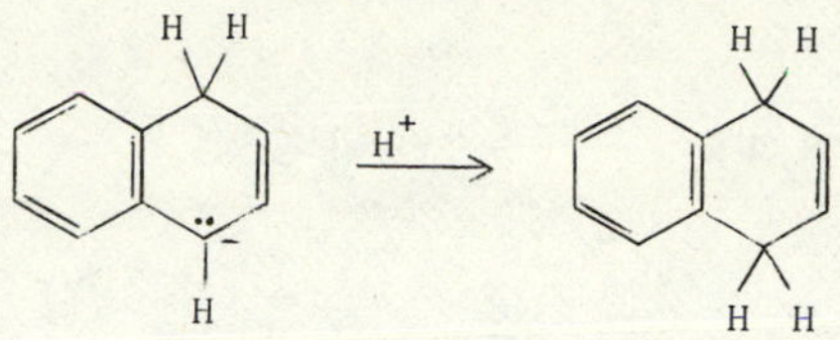

b) The reaction shown in part (a) has two distinct one-electron steps. In aprotic media the abstraction of H^+ from solvent is slow. The second wave in the voltammogram indicates that a second, more cathodic process takes place. This is probably formation of the naphthalene dianion.

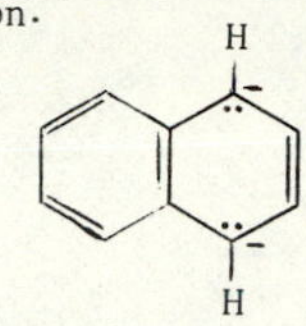

27-18

a) $EtO_2CCH_2CH(CH_3)\text{—}CH(CH_3)CH_2CO_2Et$

b)

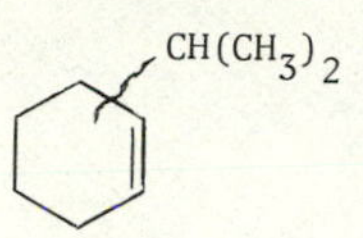

c)

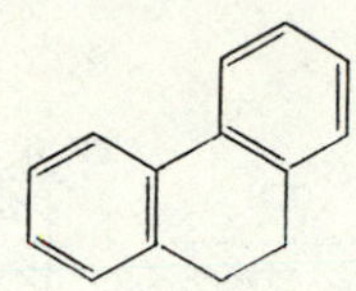

d) $EtO_2CCH_2CH(CO_2Et)CH(CO_2Et)CH_2CO_2Et$ + $EtO_2CCH_2CH(CO_2Et)CH_2CH_2CN$

(Only maleate forms a radical anion at -1.4V.)

27-19

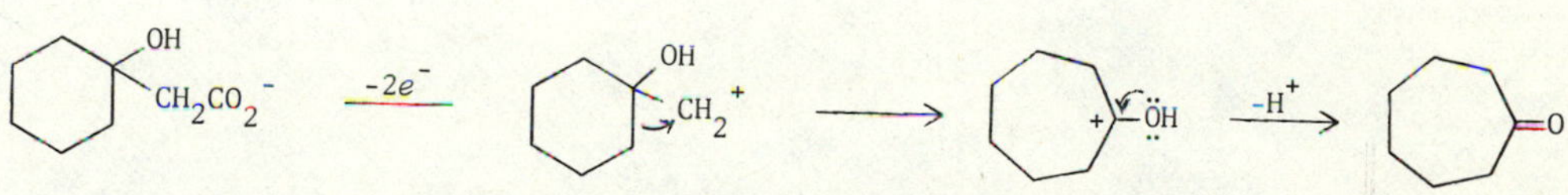

27-20

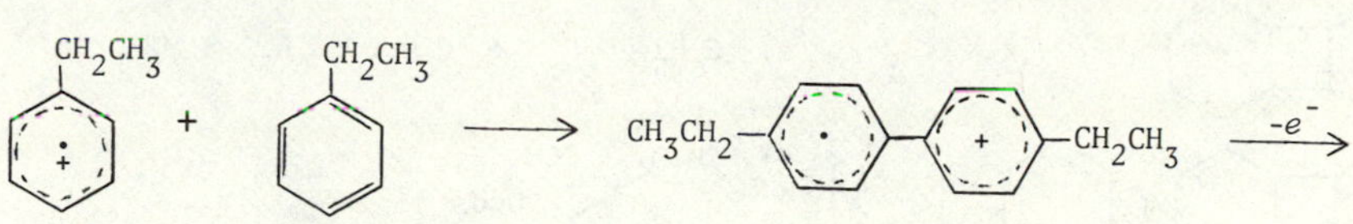

Contd...

27-20 contd...

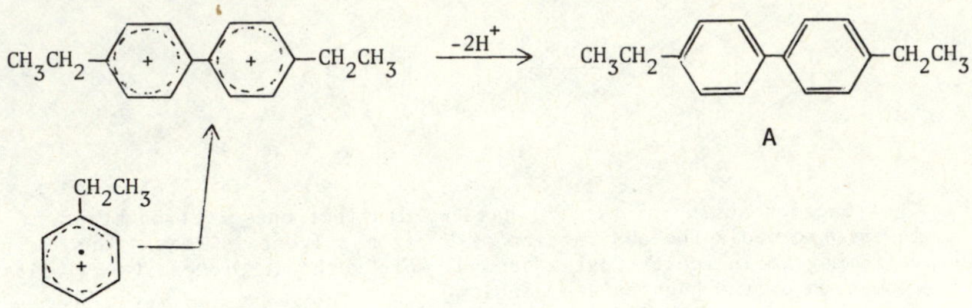

A

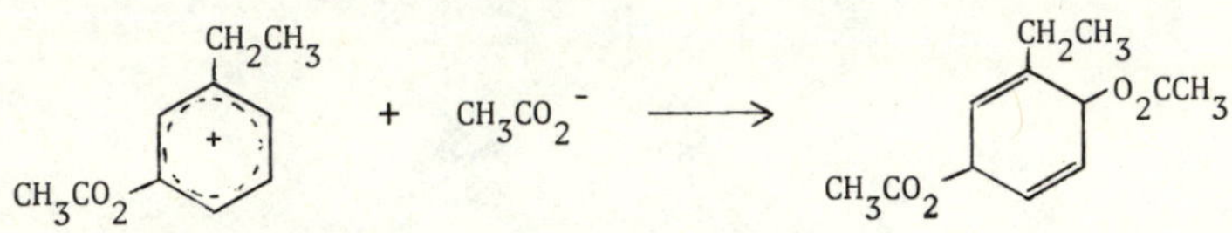

B

27-21

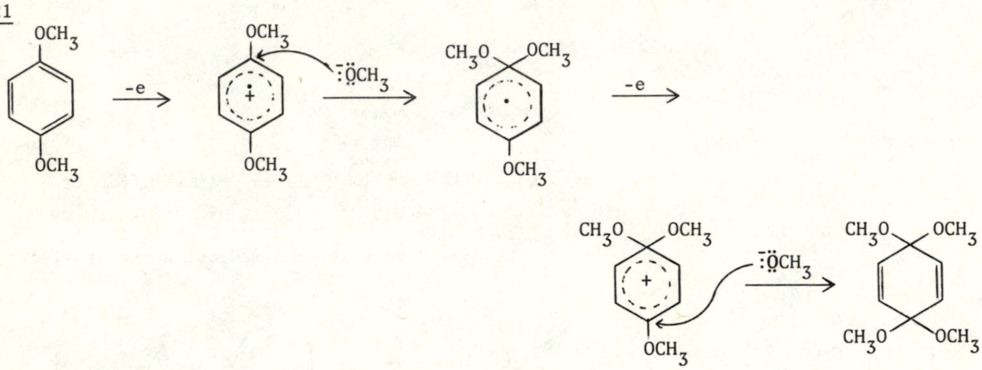

27-22

a) 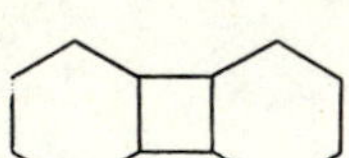

b) 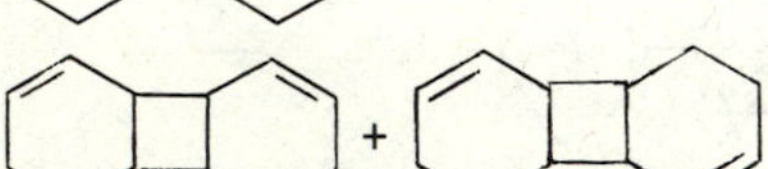

c) $EtO_2CCH_2CHCO_2Et$
$\quad\quad\quad\quad CH_2CH_2COCH_3$

d)  + CO

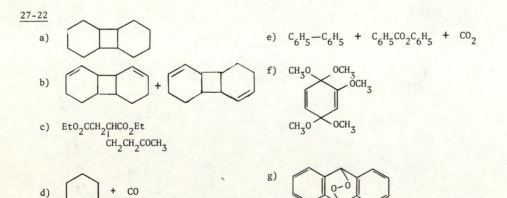

e) $C_6H_5—C_6H_5$ + $C_6H_5CO_2C_6H_5$ + CO_2

f) 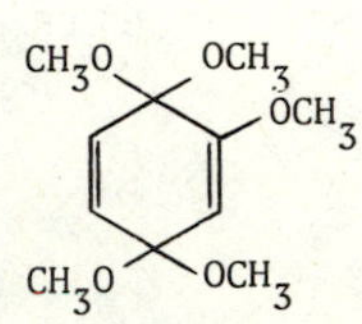

g) 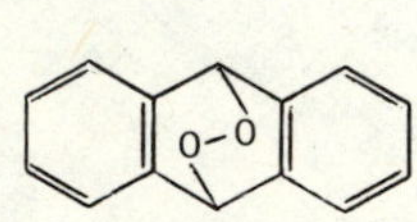

27-23 Only the carbonyl compounds absorb radiation above 200 nm. Excited acetophenone can transfer sufficient triplet energy to norbornene to promote cycloaddition. This is not possible with benzophenone so that excited benzophenone itself undergoes cycloaddition to norbornene.

27-24 Iodine is an excellent free radical trapping reagent.

$$CH_3COCH_3 \longrightarrow CH_3CO\cdot + CH_3\cdot \xrightarrow{I_2} CH_3COI + CH_3I$$

27-25

a) The major UV absorption of 1,3-butadiene comes at 217 nm so that visible light is not absorbed.

b) Direct absorption of UV light promotes butadiene to an excited singlet state from which a photochemically allowed disrotatory 4π-electron electrocyclic ring closure occurs.

c) Benzophenone absorbs the light due to its broad end absorption, then transfers triplet energy to butadiene. Triplet butadiene functions as a diradical (spins unpaired) and reacts stepwise with a molecule of ground state butadiene to give dimers which then cyclize.

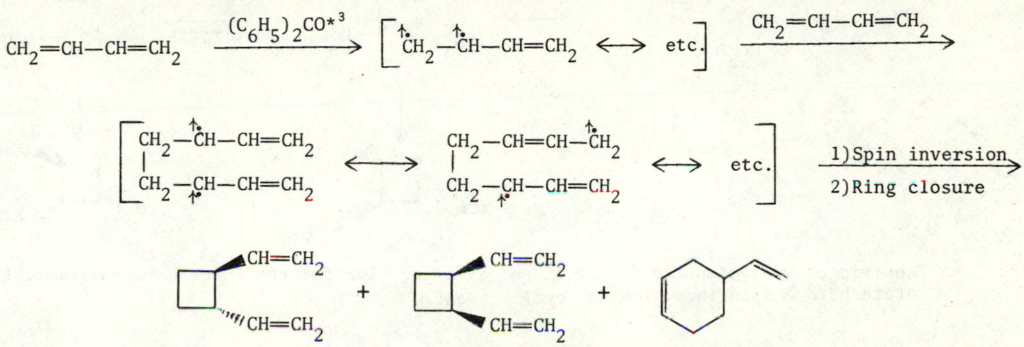

27-26

a) 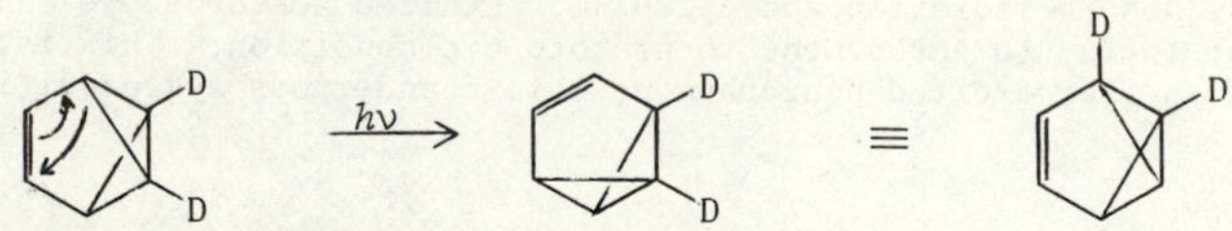

A photochemically allowed [1,3] sigmatropic rearrangement.

b) 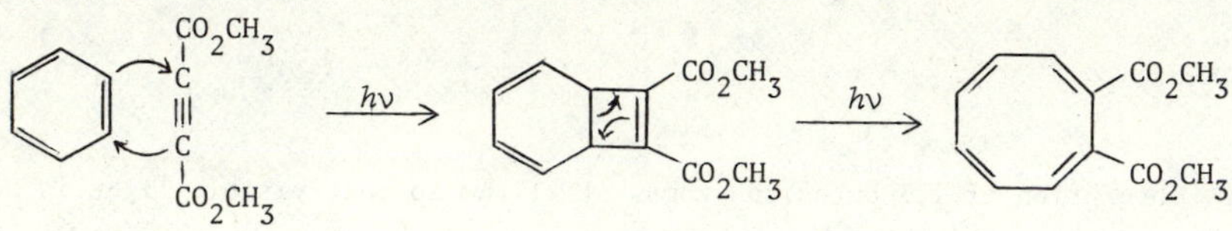

A photochemically allowed [2 + 2] cycloaddition followed by an allowed electrocyclic ring opening.

c) 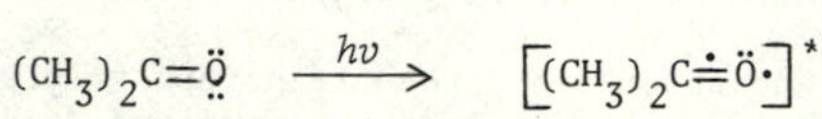 (This electron designation indicates that an *n* electron has been excited to a triplet π^* orbital.)

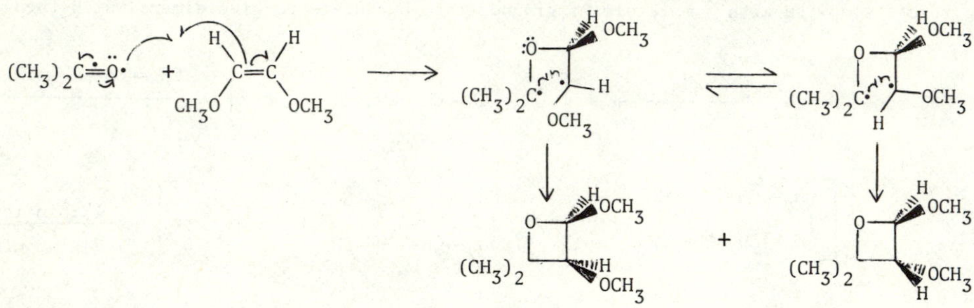

The triplet adds stepwise to the alkene allowing time for the intermediate diradical to rotate before spin inversion and cyclization.

d) 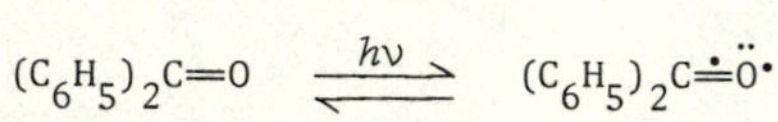

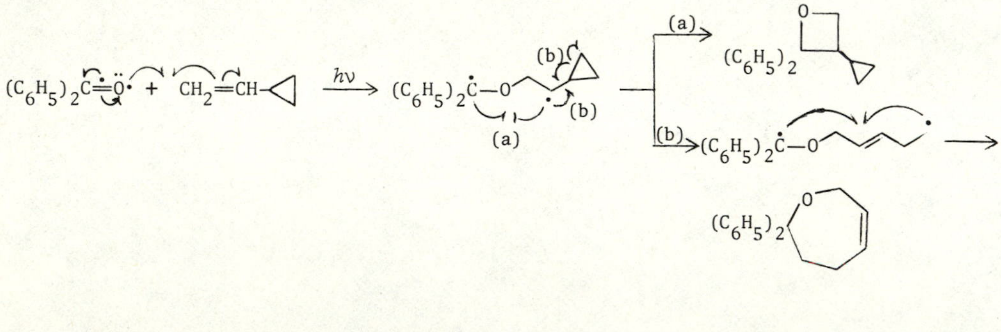

Contd...

27-26 contd...

e)

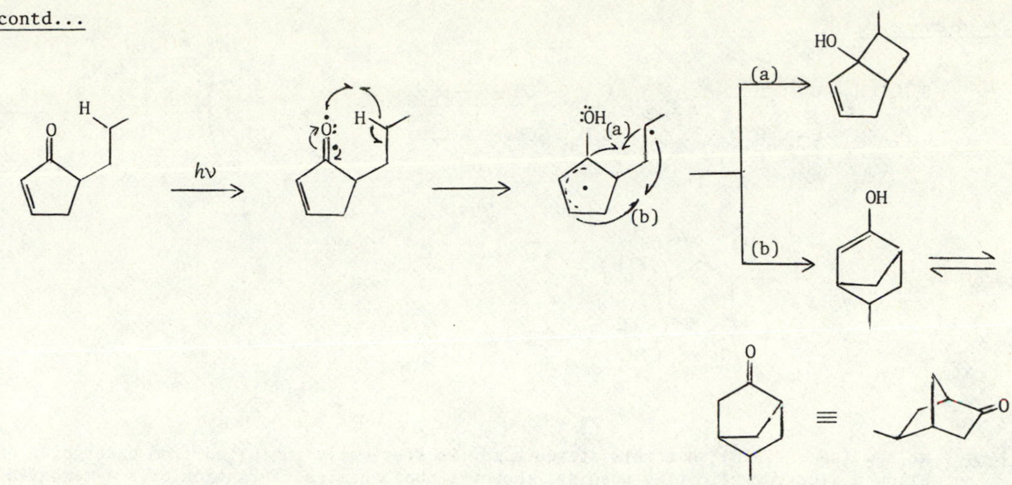

f) Both reactants produce common intermediates which then lead to common products.

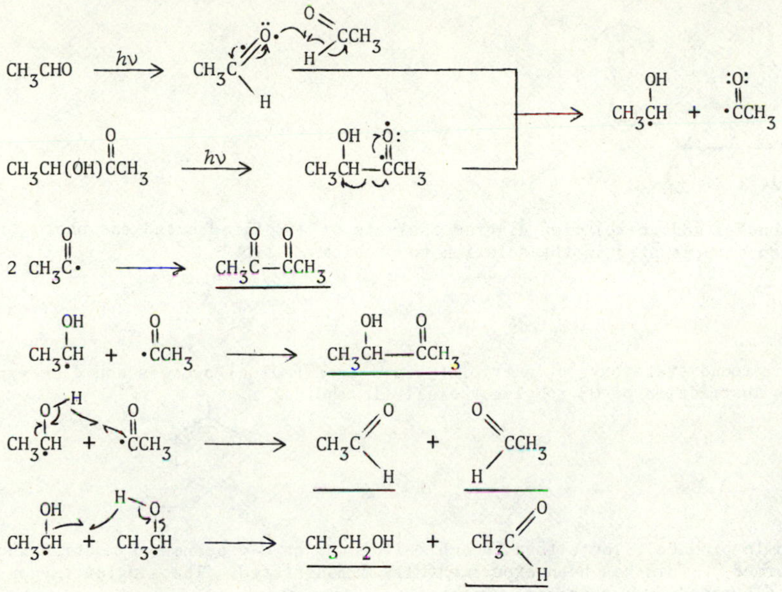

Contd...

27-26 contd...

g) 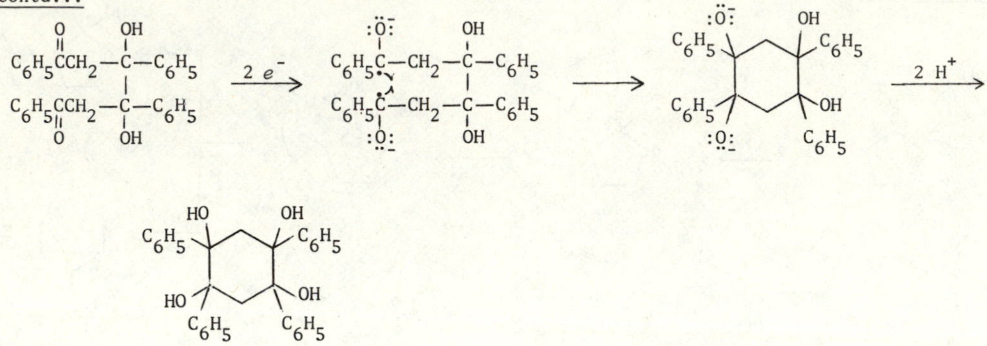

27-27 We saw (sec. 27-2D) that this stereoisomer is sterically inhibited from undergoing an allowed electrocyclic ring opening. However photochemical formation of a 4-membered ring can proceed by a symmetry and geometrically allowed cycloaddition. Orbital symmetry analysis by the HOMO method utilizes the ψ_3 M.O. of butadiene and predicts a disrotatory ring closure.

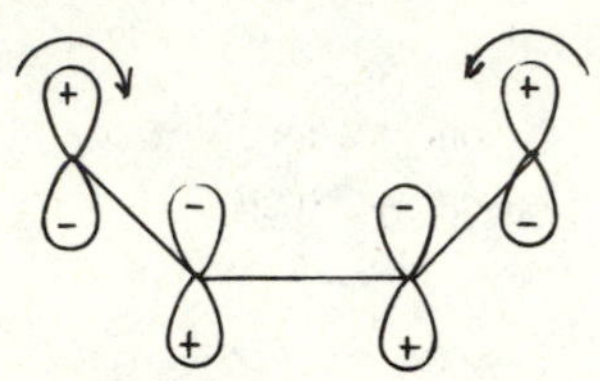

(The Mobius-Huckel and correlation diagram analysis of a related butadiene photo-induced cycloaddition is presented in the solution to problem 27-11b.)

27-28 Reaction with ground state oxygen (a triplet) in a free radical process would be expected to proceed by abstraction of H· to give the allylic radical

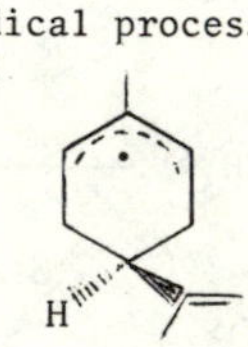

and thus racemic product. (Note that the chiral center at C-4 becomes symmetric when the radical is formed.) This has been experimentally demonstrated. The singlet oxygen reaction is believed to proceed by a concerted process.

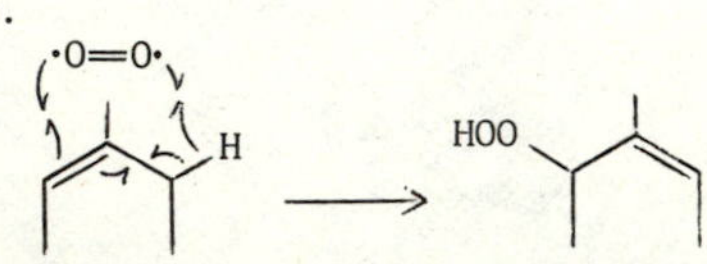

That pathway can be shown to be symmetry allowed by combining a C_2 M.O. having 2 electrons (the O_2) with a C_3 M.O. having three electrons (the allylic portion of limonene).

27-29

i) One photon is required for each benzyl free radical which is formed. Thus two photons are required for the two radicals which combine to give one molecule of bibenzyl. The maximum quantum yield is 0.5.

ii) In the formation of benzyldiphenyl carbinol only one photon is required for the generation of the the two precursor radicals. The maximum quantum yield is 1.

27-30 Reaction involves disproportionation of one-half of the ketyls to give acetone and dimerization of the other one-half to give pinacol. In the disproportionation step, H· is abstracted from a methyl group to give acetone enol which rapidly tautomerizes to acetone.

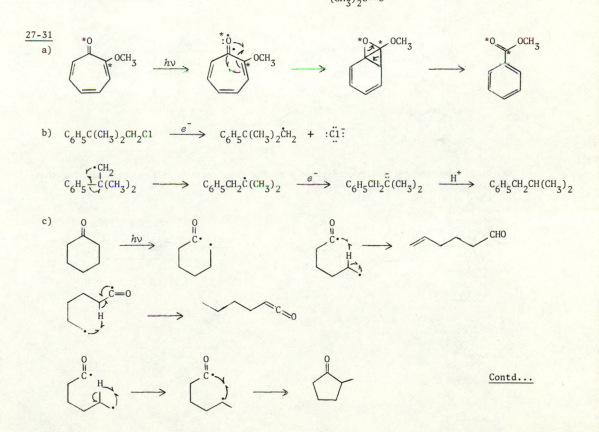

$$(CH_3)_2C{=}O \xrightarrow{h\nu} \longrightarrow (CH_3)_2C{=}O^{*3}$$

$$(CH_3)_2C{=}O^{*3} + (CH_3)_2CHOH \longrightarrow 2\ (CH_3)_2\dot{C}OH$$

$$2\ (CH_3)_2\dot{C}OH \xrightarrow{50\%} (CH_3)_2C(OH){-}C(OH)(CH_3)_2$$

$$2\ (CH_3)_2\dot{C}OH \xrightarrow{50\%} (CH_3)_2CHOH + \left[\cdot CH_2{-}\dot{C}(OH)CH_3 \equiv CH_2{=}C(OH)CH_3\right] \longrightarrow (CH_3)_2C{=}O$$

27-31

a)

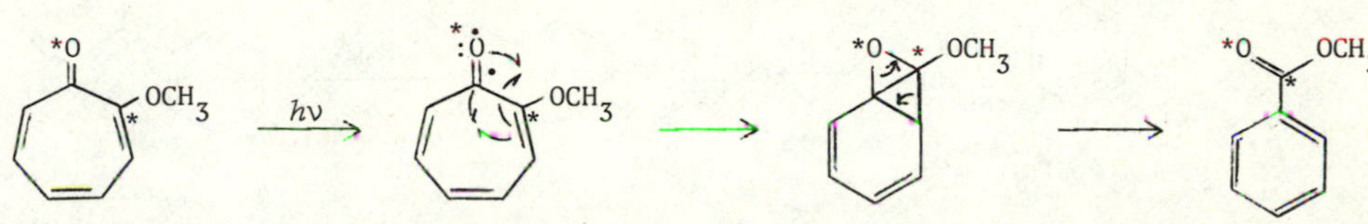

b) $$C_6H_5C(CH_3)_2CH_2Cl \xrightarrow{e^-} C_6H_5C(CH_3)_2\dot{C}H_2 + :\ddot{Cl}:^-$$

$$C_6H_5C(CH_3)_2\dot{C}H_2 \longrightarrow C_6H_5CH_2\dot{C}(CH_3)_2 \xrightarrow{e^-} C_6H_5CH_2\ddot{C}^-(CH_3)_2 \xrightarrow{H^+} C_6H_5CH_2CH(CH_3)_2$$

c)

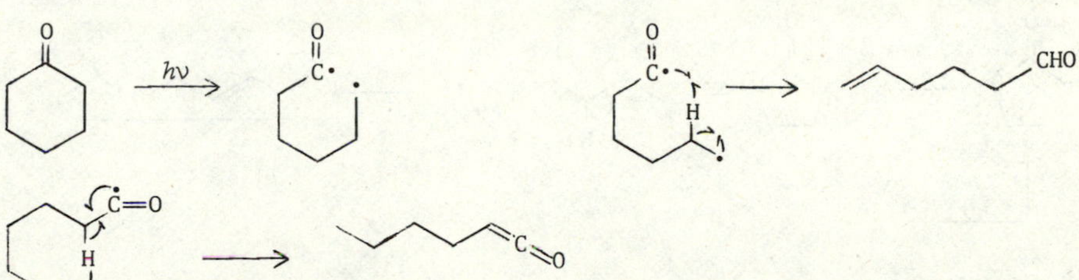

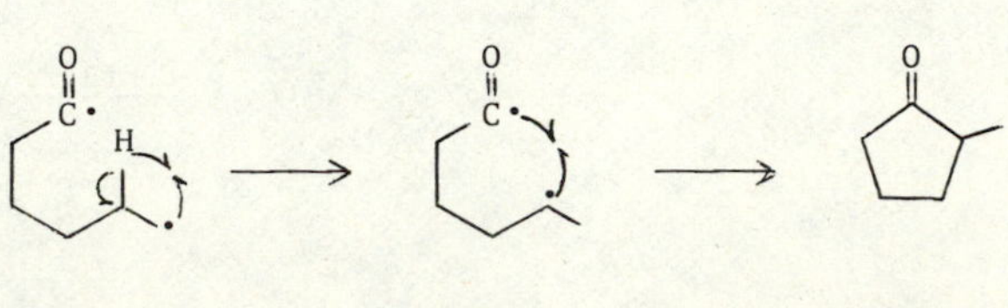

Contd...

27-31 contd...

d) 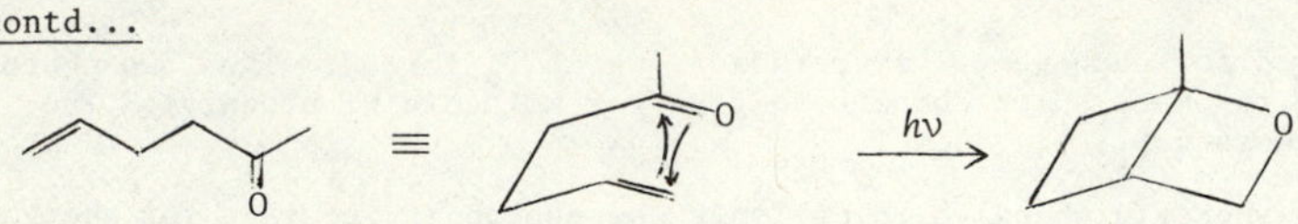

e) 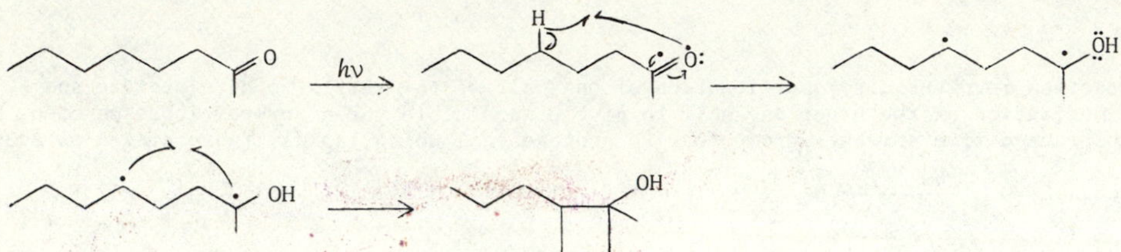

or

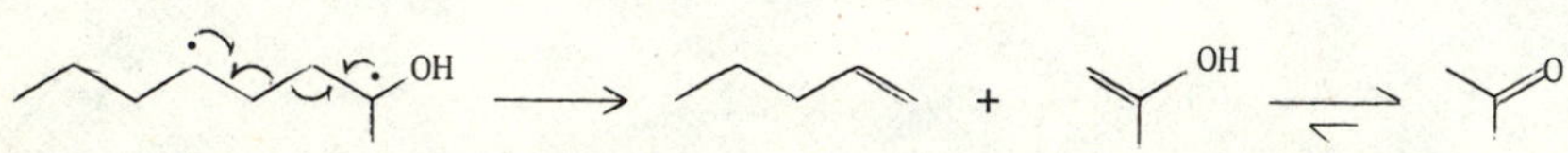

f) 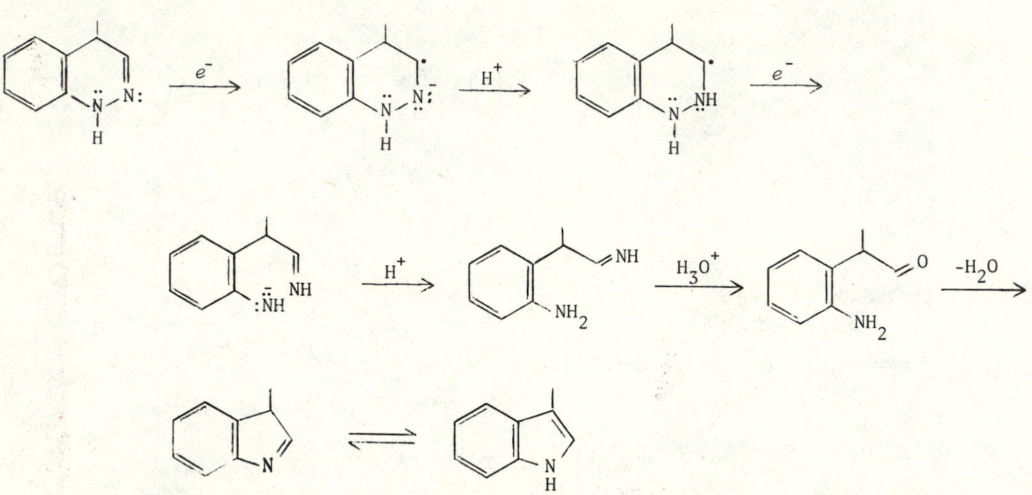

(The actual electrochemical process gives a single two-electron voltammogram showing that the second step is much faster than the first.)

g) 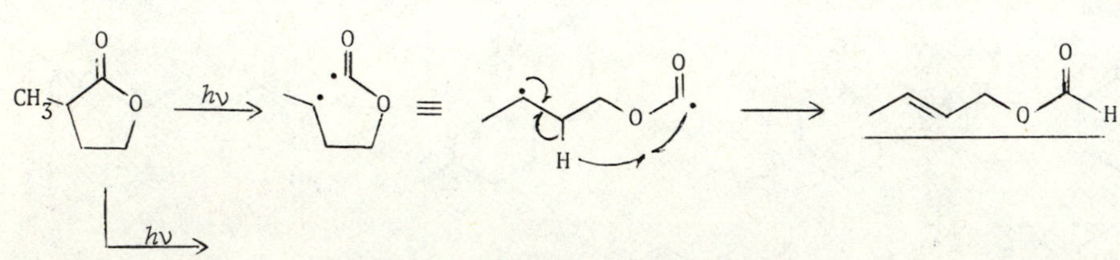

Contd...

27-31 contd...

g) contd...

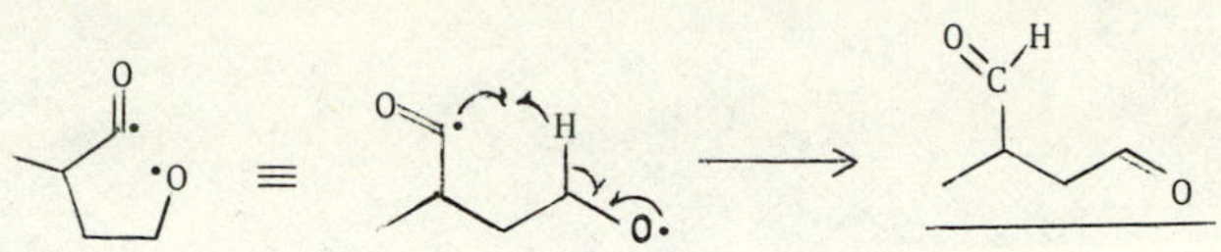

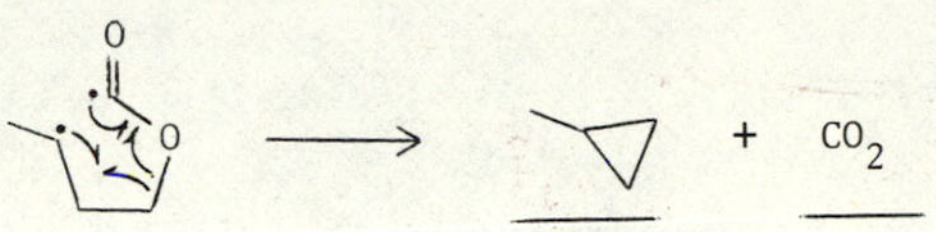

h)

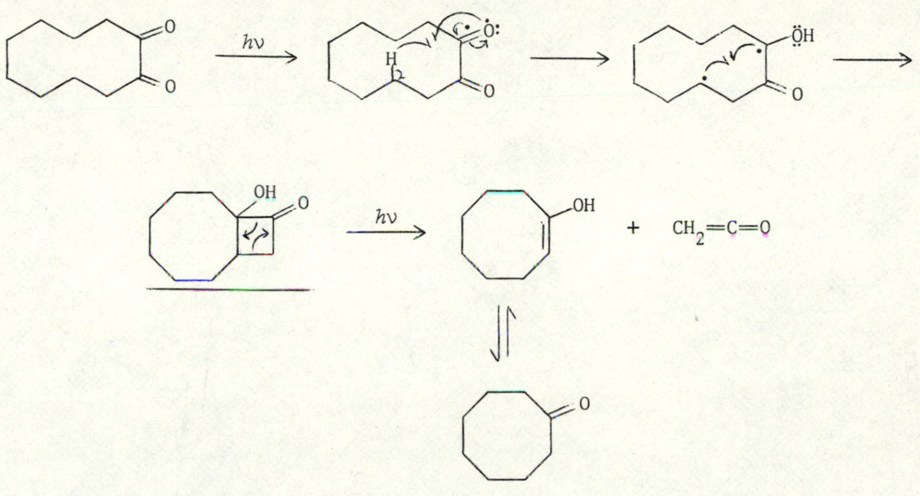

i)

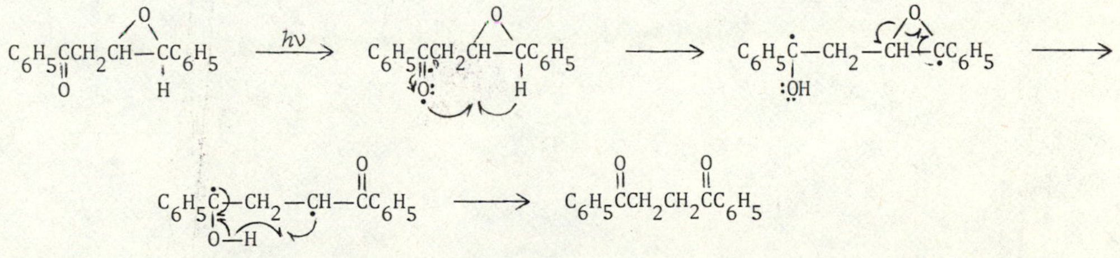

APPENDIX

A-1

i) $2.1 = \dfrac{\nu_A - \nu_{TMS}}{60 \times 10^6 \text{ Hz}} \times 10^6$

$\nu_A - \nu_{TMS} = 126$ Hz

ii) $2.1 = \dfrac{\nu_A - \nu_{TMS}}{100 \times 10^6 \text{ Hz}} \times 10^6$

$\nu_A - \nu_{TMS} = 210$ Hz

iii) $2.1 = \dfrac{\nu_A - \nu_{TMS}}{500 \times 10^6 \text{ Hz}} \times 10^6$

$\nu_A - \nu_{TMS} = 1{,}050$ Hz

A-2

a) δ:

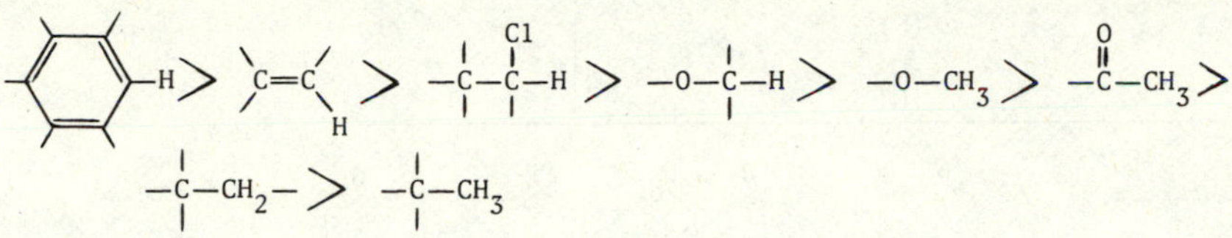

b)

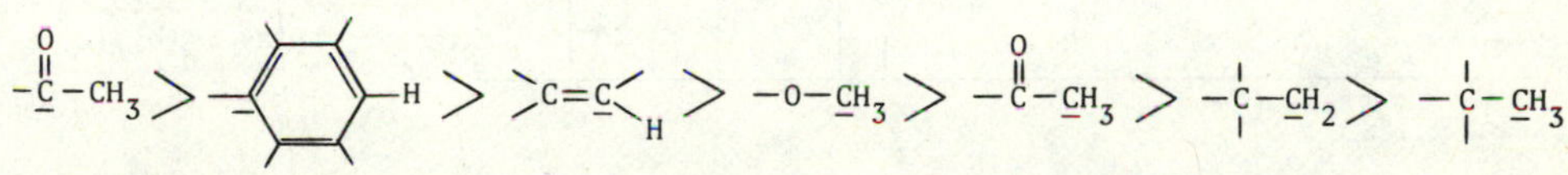

A-3 The chemical shift data listed below are experimental values.

		^{1}H-δ ppm	^{13}C-δ ppm
a)	CH_3OCH_3	3.2	60
b)	$ClCH_2C{\equiv}CH$ (a) CH_2, (b) CH; (x) CH_2, (y) C, (z) CH	a:4.1; b:2.4	x:44; y:83; z:68
c)	$(CH_3)_3N$	2.1	48
d)	$(CH_3)_4C$ (x) CH_3, (y) C	0.9	x:9; y:30

Contd...

A-3 contd...

		^1H-δ ppm	^{13}C-δ ppm
e)	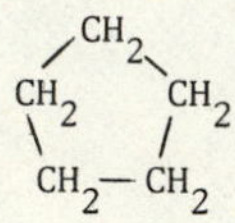	1.5	26
f)	(a) (b) $(CH_3)_3COH$ (x) (y)	a:1.2; b:OH variable	x:32; y:69
g)	(a) (b) $CH_3C(=O)OCH_3$ (x) (y) (z)	a:2.0; b:3.6	x:21; y:174; z:51
h)	$CH_3C{\equiv}CCH_3$ (x) (y)	1.8	x:3; y:79

A-4 The chemical shift data listed below are experimental values.

a) 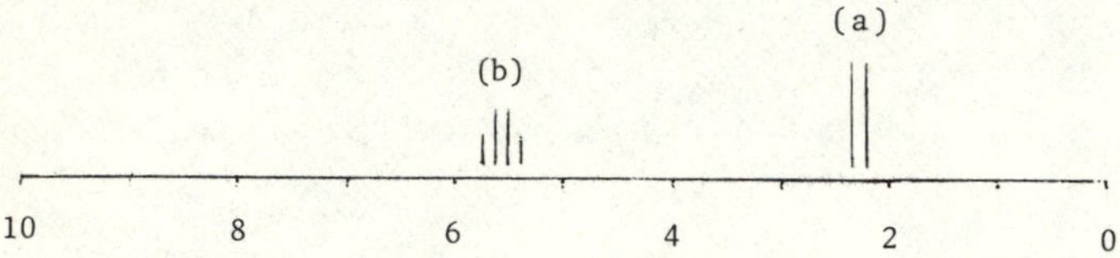

CH_3CHBr_2
(a) (b)

b)  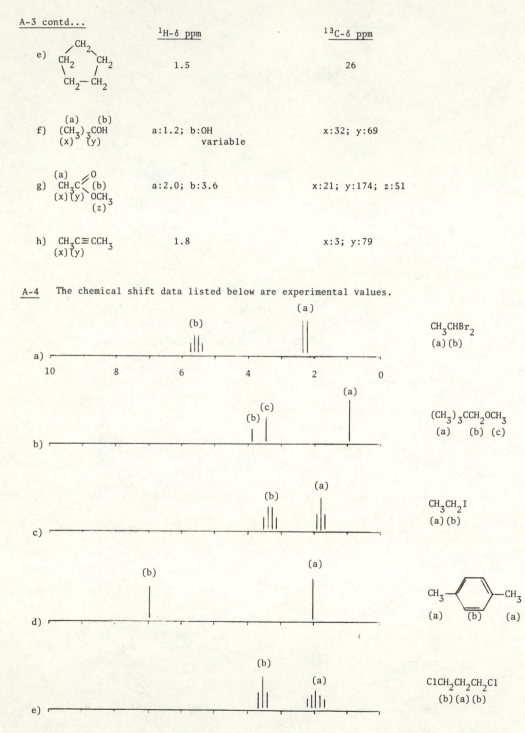

$(CH_3)_3CCH_2OCH_3$
(a) (b) (c)

c) 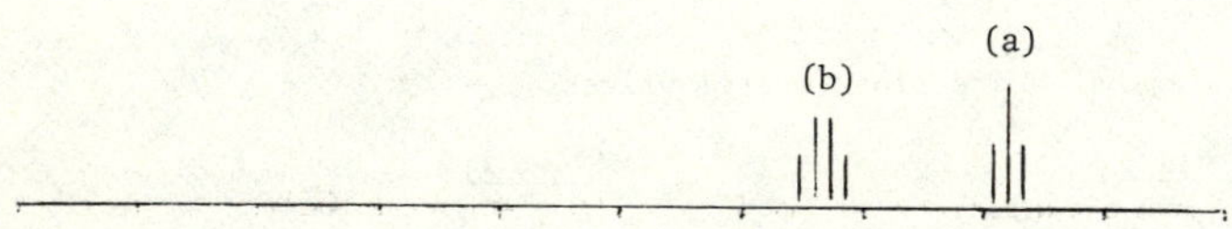

CH_3CH_2I
(a) (b)

d) 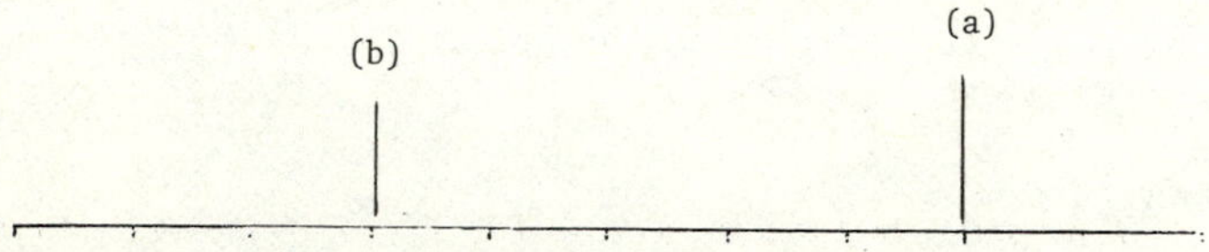

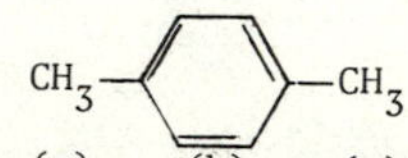

(a) (b) (a)

e) 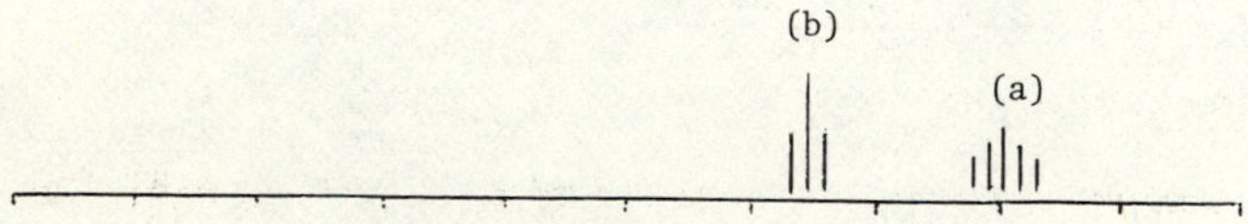

$ClCH_2CH_2CH_2Cl$
(b) (a) (b)

Contd...

A-4 contd...

f)

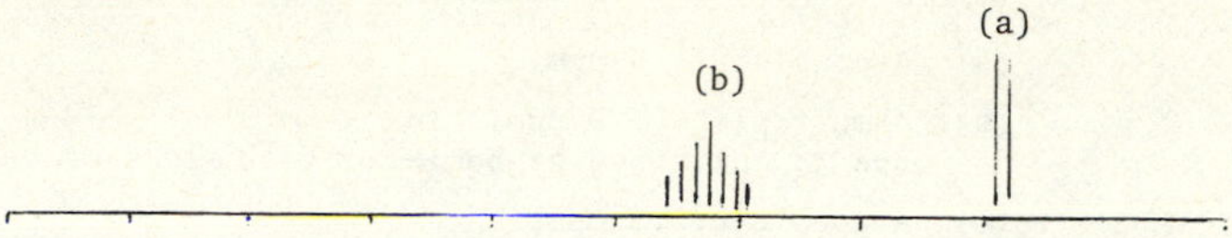

Br
CH_3CHCH_3
(a) (b) (a)

g)

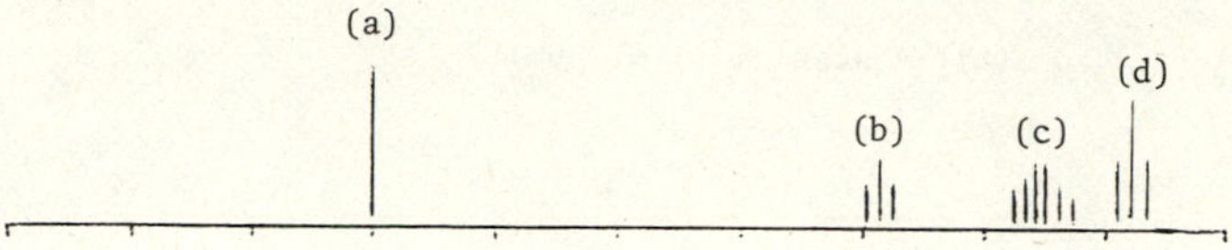

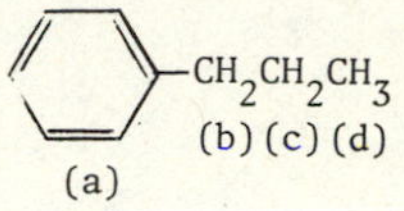

h)

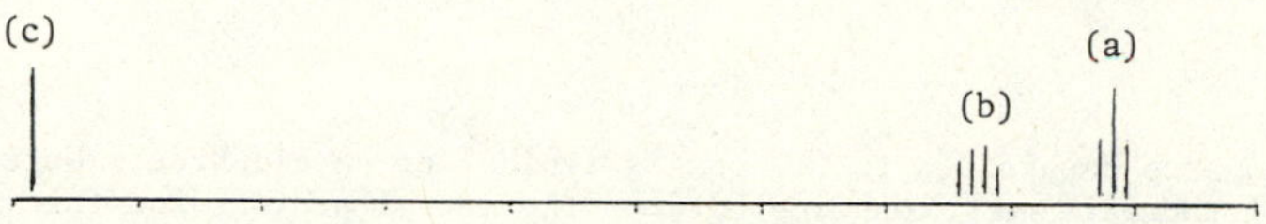

$CH_3CH_2C(=O)H$
(a) (b) (c)

(Coupling between (b) and (c) is very small in this example.)

A-5

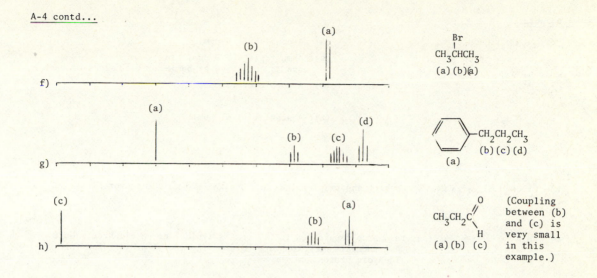

a) Under anhydrous conditions exchange of the hydroxy protons does not occur. They are not averaged so that splitting is observed.

i) The CH_3 resonance is a triplet due to splitting by the adjacent CH_2 group.

ii) The CH_2 resonance shows splitting into a quartet by the adjacent CH_3 group plus additional splitting of each part of the quartet into a doublet by the hydroxy H.

iii) The OH resonance is a triplet because of splitting by the adjacent CH_2 group. (Note that the —OH chemical shift is variable.)

b) The methyl group is a quartet (coupled to three equivalent hydrogen atoms) at 18 ppm and the methylene is a triplet (coupled to two equivalent hydrogen atoms) at 57 ppm.

A-6 The chemical shift data listed below are experimental values.

$CH_3CH_2CH_2CH_2Cl$
(a) (b) (c) (d)

There are four different kinds of protons. However the chemical shifts of (b) and (c) are similar and overlap to give a complex multiplet.

(a): A triplet (0.95 ppm)

(b) + (c): A multiplet (1.1-2 ppm) due to overlap of multiple splittings by nuclei on both sides of each carbon.

(d): A triplet (3.5 ppm)

Cl
$CH_3CH_2CHCH_3$
(a) (b) (c) (d)

(a): A triplet (1.0 ppm)

(b): A multiplet (1.7 ppm) due to splitting by nuclei on both sides of carbon

(c): A multiplet (3.9 ppm) due to splitting by nuclei on both sides of carbon

(d): A doublet (1.5 ppm)

Contd...

A-6 contd...

$(CH_3)_2CHCH_2Cl$
(a) (b)(c)

(a): A doublet (1.0 ppm)

(b): A multiplet (1.9 ppm)
due to splitting by nuclei on both sides of carbon.

(c): A doublet (3.3 ppm)

$(CH_3)_3CCl$
(a)

(a): A singlet (1.61 ppm)

A-7

a) Cl_2CHCH_2Cl - *IHD* = 0; doublet and triplet indicate one and two adjacent protons.

b) 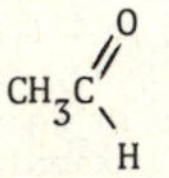 - *IHD* = 1; resonance deshielded to 9.8 ppm is typical of an aldehyde. Note that there is small coupling between the methyl protons and the aldehydic proton.

c) 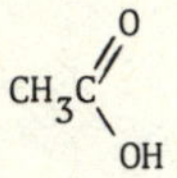 - *IHD* = 1; differs from (b) in having two oxygen atoms and no splitting since protons are not on adjacent atoms, and the hydroxy H exchanges.

d) 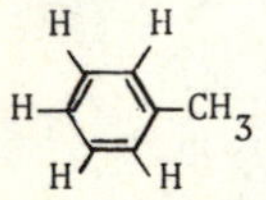 - *IHD* = 4; peak at 7.2 ppm is typical of aromatic compounds; the peak area ratio of 5:3 indicates monosubstitution; a singlet of area = 3 is typical of a methyl group. The methyl is deshielded by the aromatic.

A-8 The aromatic protons of benzene are chemically and magnetically equivalent, thus no spin-spin splitting occurs. The aromatic resonance of many aromatic compounds is a broadened singlet because of very small couplings between nonequivalent protons on the aromatic ring.

A-9

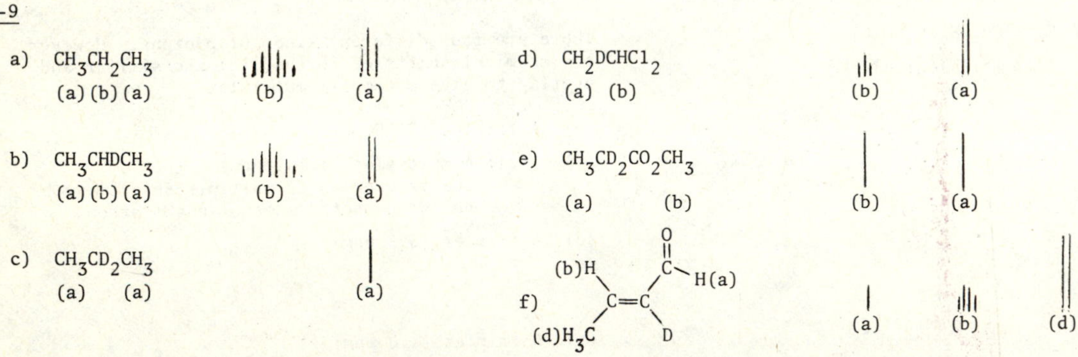

A-10 The probability of having two ^{13}C atoms next to each other is very low (.011x.011x100 = .012%)

A-11

a) $CH_3CO_2C_2H_5$ - *IHD* = 1. ^1H-NMR: A triplet-quartet pattern is typical of an ethyl group with the CH_2 deshielded by the ester oxygen atom.

Note the typical position of the resonance peak for an unsplit methyl group adjacent to C=O.

^{13}C-NMR: Shows four different carbons. Note the small strongly deshielded carbonyl carbon at 174 ppm.

b) 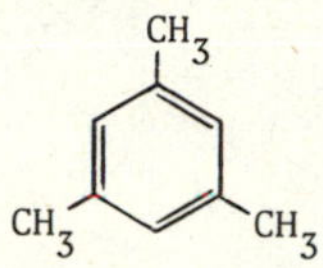

- *IHD* = 4. ^1H-NMR: A peak at 6.7 ppm suggests an aromatic group; the area ratio of 1:3 is actually 3:9 as per the molecular formula; the single aromatic peak is consistent with a symmetrical structure; the singlet of 9 protons is characteristic of three equivalent methyl groups.

^{13}C-NMR: Shows three different carbons. Note the two different kinds of aromatic carbons in the 120-140 ppm region.

c) $CH_3CH_2\overset{O}{\overset{\|}{C}}CH_3$ - *IHD* = 1. ^1H-NMR: The CH_2 group of ethyl adjacent to C=O is at 2.4 ppm in contrast to the CH_2 connected to an oxygen atom in (a).

^{13}C-NMR: Shows four different carbons. Note the strongly deshielded carbonyl carbon.

d) $CH_3C(=O)OCH_2C_6H_5$ - *IHD* = 5. ^1H-NMR: The chemical shifts and absence of splitting lead to the structure. Note that the CH_2 is deshielded to 5.1 ppm by a combination of the O atom and C_6H_5 group.

^{13}C-NMR: Here the *o,m,* and *p* aromatic carbons are almost equivalent. The aromatic carbon without hydrogens is very small and further deshielded. The carbonyl carbon at 170ppm is also very small.

A-12

a) Hydroxy; the spectrum is that of *m*-methylphenol. Also note the characteristic peaks of aromatic meta disubstitution in the 650-800 cm^{-1} region.

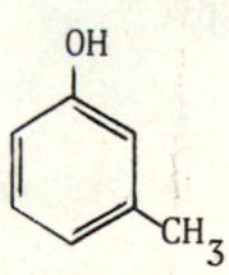

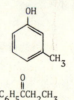

b) Carbonyl; the spectrum is that of 1-phenyl-1-propanone (Note that the conjugated ketone carbonyl is shifted to 1690 cm^{-1}.) $C_6H_5\overset{O}{\overset{\|}{C}}CH_2CH_3$

c) Hydroxy and carbon-carbon double bond; the spectrum is that of 2-propenol. $CH_2{=}CHCH_2OH$

d) Carbonyl; the spectrum is that of propyl propanoate. Also note the ester C—O stretching frequency near 1200 cm^{-1}. $CH_3CH_2CO_2CH_2CH_2CH_3$

Contd...

A-12 contd...

e) Primary amine; the spectrum is that of 1,2-diaminobenzene.

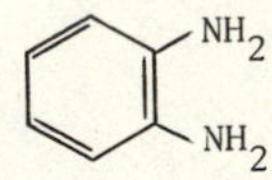

f) Carboxylic acid; the spectrum is that of butanoic acid. $CH_3CH_2CH_2CO_2H$

A-13

a) 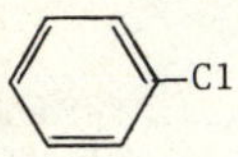

- *IHD* = 4 only fits an aromtic compound in this case; a weak C-H absorption is typical as are peaks at 1400-1600 cm^{-1}; the pattern at 600-800 cm^{-1} is typical of a monosubstituted aromatic with an additional peak at 680 cm^{-1} due to the C-Cl stretching absorption.

b) $CH_3\overset{O}{\overset{\parallel}{C}}CH_2CH_3$

- *IHD* = 1; the presence of O and a strong peak at 1710 cm^{-1} is typical of a ketone C=O; there is no indication of an aldehydic C-H stretching absorption; only one ketone has this molecular formula.

c) CH_3CO_2H

- *IHD* = 1; the broad peak at ≈ 3000 cm^{-1} with a strong absorption at 1710 cm^{-1} is typical of a carboxylic acid. Only one carboxylic acid has this molecular formula.

d) 1-Octene

- *IHD* = 1 for this hydrocarbon and a "relatively intense" peak at 1650 cm^{-1} is typical of a terminal alkene; the C–H stretching frequency of a terminal alkene is at 3100 cm^{-1}. Final analysis would depend on a match with the spectra of known compounds.

e) *Z*-2,5-Dimethyl-3-hexene

- *IHD* = 1 for this hydrocarbon fits an alkene (or carbocycle); the peak at 750 cm^{-1} is typical of a cis (*Z*) alkene and the very small peak at 1650 cm^{-1} suggests a relatively symmetrical alkene. Final analysis would depend on a match with the spectra of known compounds.

f) 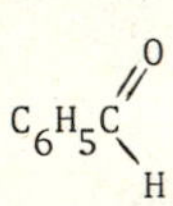

The strong absorption at 1710 cm^{-1} is typical of C=O while two peaks at 2700-2800 cm^{-1} suggest 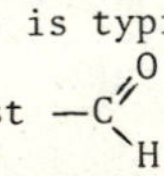 ; *IHD* = 5 (one for —C=O) and weak C-H stretching at 3050 cm^{-1} suggest an aromatic compound; a conjugated aldehyde accounts for the shift of the aldehyde carbonyl absorption.

A-14

a) Parent = 215 nm

α - C = 10 nm

β - C = 12 nm

1 exo = 5 nm

λ_{max} = 242 nm

b) Parent = 215 nm

3 C = 15 nm

1 exo = 5 nm

λ_{max} = 235 nm

c) Parent = 215 nm

β - C = 12 nm

β - Cl = 12 nm

λ_{max} = 239 nm

d) Parent = 250 nm

m - R = 3 nm

o - NR_2 = 20 nm

λ_{max} = 273 nm

e) Parent = 246 nm

o - OH = 7 nm

p - OH = 25 nm

2*m* - R = 6 nm

λ_{max} = 284 nm

A-15

a) The compound contains C, H, and O. If we assume a 100g sample, the g-atms of each atom can be calculated from the percentage composition. The percentage for oxygen is obtained by difference.

$$\text{g-atms C} = \frac{52.2}{12.011} = 4.35$$

$$\text{g-atms H} = \frac{4.38}{1.008} = 4.35$$

$$\text{g-atms O} = \frac{43.42}{16.00} = 2.71$$

The lowest common divisor for the g-atm values is 0.544 which leads to an empirical formula of $C_8H_8O_5$.

b) C_4H_5O

c) C_5H_9Br

d) C_4H_5NO

A-16

a) Empirical formula = molecular formula = $C_8H_8O_5$

b) Empirical formula ≠ molecular formula = $C_8H_{10}O_2$

c) Empirical formula = molecular formula = C_5H_9Br

d) Empirical formula ≠ molecular formula = $C_8H_{10}N_2O_2$

A-17

$CH_4 \xrightarrow{e^-} [CH_4]^{+\cdot}$ (m/z = 16)

The molecular ion is relatively stable and is the base peak. The peak at m/z = 17 is due to the presence of the ^{13}C isotope (1.1%).

$[CH_4]^{+\cdot} \longrightarrow [CH_3]^+ + H\cdot$ (m/z = 15)

Loss of a hydrogen atom is also relatively favorable.

The remaining processes do not lead to important fragments, thus have low relative intensities.

$[CH_3]^+ \longrightarrow [CH_2]^{+\cdot} + H\cdot$ (m/z = 14)

$[CH_4]^{+\cdot} \longrightarrow CH_2 + [H_2]^{+\cdot}$ (m/z = 2)

$[CH_2]^{+\cdot} \longrightarrow [CH]^+ + H\cdot$ (m/z = 13)

$[CH_4]^{+\cdot} \longrightarrow CH_3\cdot + [H]^+$ (m/z = 1)

$[CH]^+ \longrightarrow [C]^{+\cdot} + H\cdot$ (m/z = 12)

A-18

a) $CH_3CH_2CH_2CH_2CH_2CH_2CH_2CH_3$

Peaks come in fairly regular groupings, 14 units apart.

b) $CH_3C(CH_3)_2CH_2CH(CH_3)CH_3$

Fragmentation readily leads to the relatively stable tertiary carbocation.

$(CH_3)_3C^+$ (drawn as CH_3C^+ bearing two CH_3 groups) m/z = 57

c) $(CH_3)_2CHCH_2CH_2CH_2CH_2CH_3$

The most intense peak is fragmentation to give the 2-propyl cation.

$(CH_3)_2CH^+$ m/z = 43

A-19

a) i)

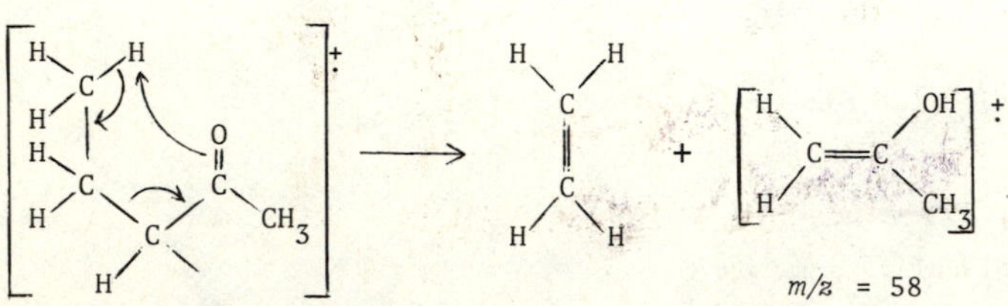

m/z = 58

Contd...

A-19 contd...

a) contd...

ii) 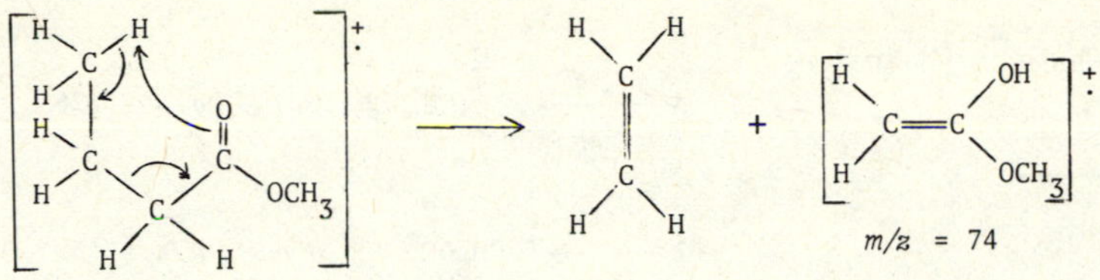

$m/z = 74$

iii) 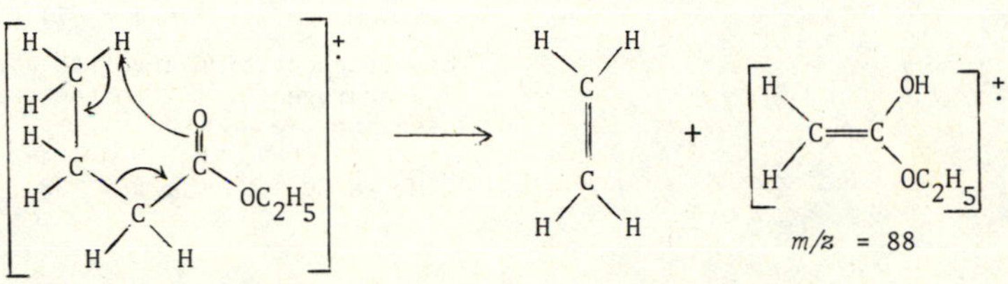

$m/z = 88$

b) The nitrogen atom is tricoordinate, thus amides lead to a fragment with an odd m/z value.

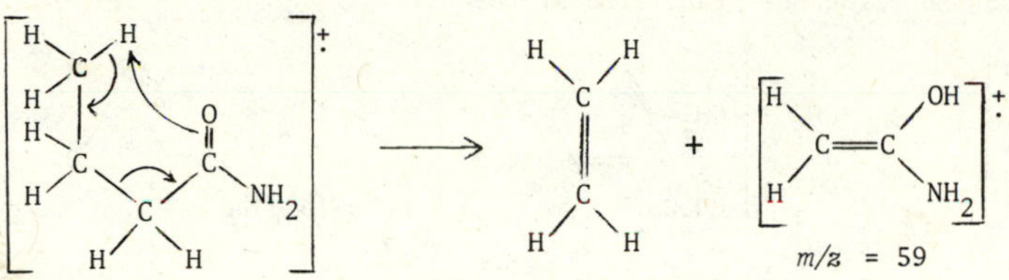

$m/z = 59$

A-20

a) 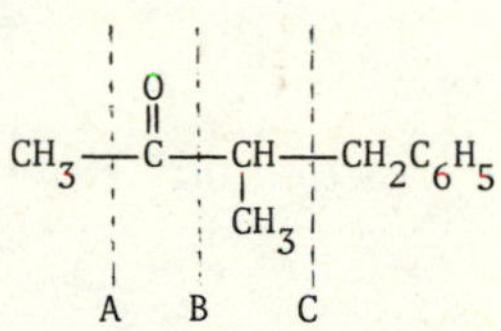

⟶ Molecular ion m/z = 162

A ⟶ $^{+}\overset{O}{\overset{\|}{C}}CH(CH_3)CH_2C_6H_5$ m/z = 147

B ⟶ $CH_3\overset{O}{\overset{\|}{C}}{}^{+}$ m/z = 43

C ⟶ $^{+}CH_2C_6H_5$ m/z = 91

Contd...

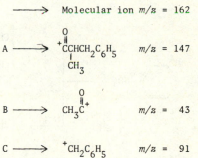

A-20 contd...

b)

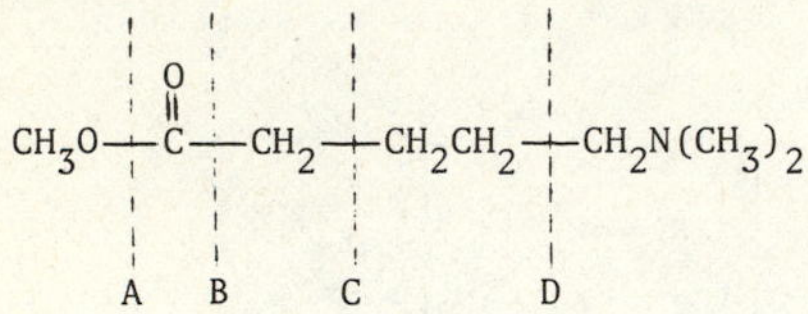

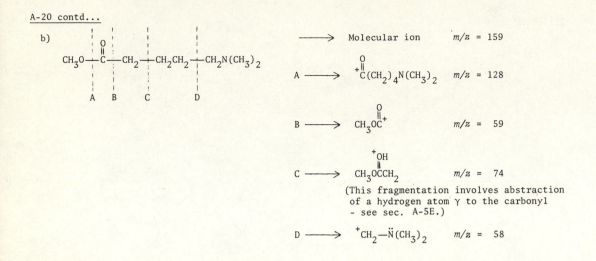

⟶ Molecular ion m/z = 159

A ⟶ $^{+}C(=O)(CH_2)_4N(CH_3)_2$ m/z = 128

B ⟶ $CH_3OC(=O)^{+}$ m/z = 59

C ⟶ $CH_3OC(=\overset{+}{O}H)CH_2$ m/z = 74

(This fragmentation involves abstraction of a hydrogen atom γ to the carbonyl - see sec. A-5E.)

D ⟶ $^{+}CH_2—\ddot{N}(CH_3)_2$ m/z = 58

A-21

a) Cleavage of a methyl free radical from the ethyl group at the four position produces a radical anion that can be resonance stabilized by the nitro group.

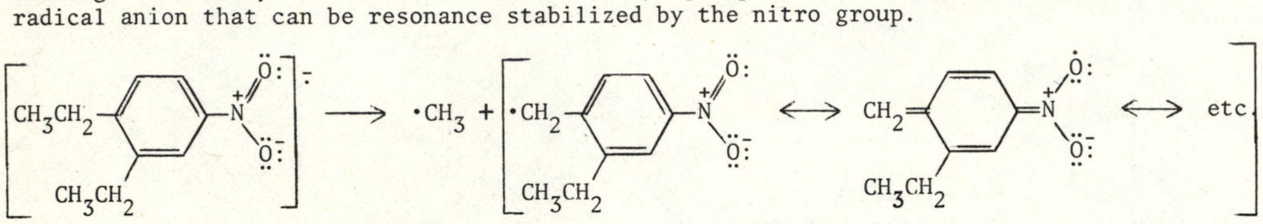

b) Deuterium could be used to label the methyl of the ethyl group at the four position as a CD_3. If this were the group which cleaved, the radical anion would differ in m/z value by three units relative to cleavage at the other position.

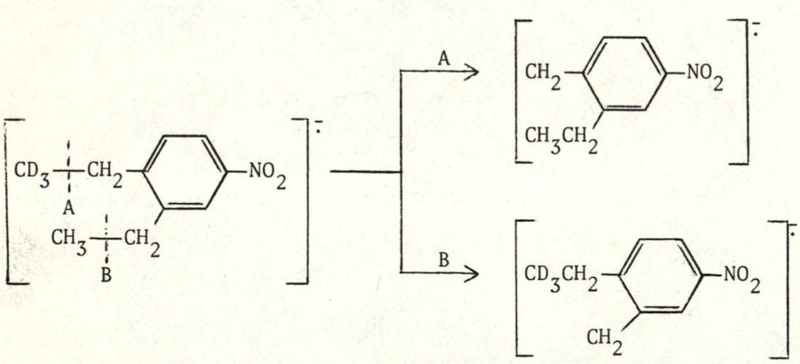

m/z = 164

m/z = 167

A-22

a) 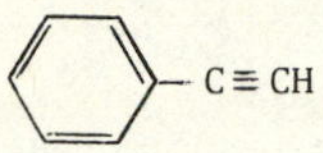

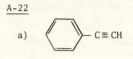

- The nmr spectrum suggest a monosubstituted aromatic (C_6H_5–) which accounts for a molecular weight contribution of 77. The presence of only one additional H in the spectrum suggests that C_2H remains to be accounted for to fit the molecular weight. The IR peak at 2210 cm^{-1} is due to a triple bond and the peak at 3310 cm^{-1} along with the position of the single H in the nmr spectrum suggest a terminal alkyne.

b) $CH_3CH(OH)CH_3$ (OH on the central carbon)

- The IR suggest a hydroxy group. Eight protons in the ^{1}H-nmr spectrum require at least three carbon atoms (C_nH_{2n+2}) while the low molecular weight and ^{1}H-nmr splitting patterns fit only one alcohol. The ^{13}C-nmr is consistent with two different kinds of carbon atoms.

c) CH_3CH_2I

- The typical nmr pattern for an ethyl group plus the low boiling point provide the structure.

d) $(CH_3)_3CCH{=}CH_2$

- The IR absorption at 1640 cm^{-1} suggests a nonsymmetrical alkene. The 12 hydrogen atoms in the nmr spectrum require at least six carbon atoms (C_nH_{2n}) and the low bp confirms that more are unlikely. A single nmr peak of nine H's at 1.0 ppm is typical of the *tert*-butyl group. The remaining nmr peaks fit a terminal alkene.

A-23 These two isomers can be differentiated by the different kinds of carbon atoms shown in the ^{13}C-nmr spectra. The 1,2-disubstituted benzene has three different kinds of carbon atoms while the 1,3-disubstituted benzene has four.

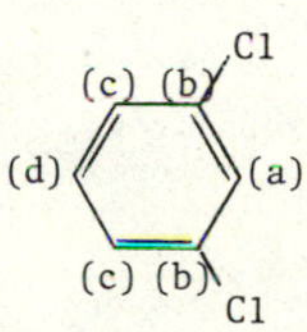

(a)

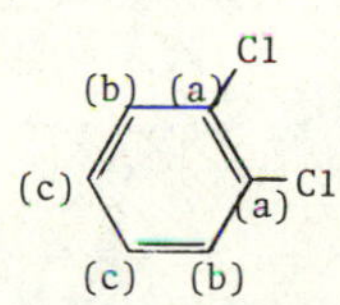

(b)

A-24

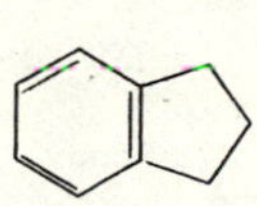

The *IHD* of 5 along with the nmr peak at 7.2 ppm indicate an aromatic ring (*IHD* = 4) plus another ring or multiple bond. No alkene H's are observed in the nmr spectrum, so that a second ring is suggested. The aromatic resonance peak of four protons and the triplet-pentet multiplets, when related to the molecular formula, provide the structure.

A-25 A - Fig. A-62d The IR shows an O—H stretching absorption but no peaks for the additional functional groups of the other alcohols, B and F.

B - Fig. A-62g A typical isopropyl doublet and heptet in the ^{1}H-nmr spectrum could fit compounds A and B. The rest of the nmr spectrum is not consistent with A. In particular note that the H atom on the carbon connected to OH (the carbinol carbon atom) is deshielded to 4.1 ppm and split by the adjacent H and long range by the terminal C—H. The terminal acetylenic H at 2.4 ppm also clearly shows this same long range splitting.

C - Fig. A-62e The ^{1}H-nmr spectrum shows only saturated hydrocarbon peaks, thus can fit none of the other compounds.

D - Fig. A-62c The only characteristic IR absorption is a weak peak at 2260 cm^{-1}. Though this could fit B or D, the O—H absorption characteristic of B is not present.

E - Fig. A-62f This is the only compound with an aromatic group (^{1}H-nmr multiplet at 7.1 ppm). The singlet at 2.3 ppm is characteristic of $CH_3-\overset{O}{\overset{\|}{C}}-$.

F - Fig. A-62h The ^{1}H-nmr spectrum clearly shows alkene protons between 5-6 ppm. The two broad peaks between 0.8-1.7 ppm are typical of straight chain alkanes, in this case $-CH_2CH_2CH_3$. The hydroxy H is an unsplit singlet (2.1 ppm) and carbinol H is at 4.0 ppm.

G - Fig. A-62a The odd molecular weight (73) is consistent with the N-containing compounds D and G, but only fits G. Fragmentation of the N—C bond accounts for m/z = 44.

H - Fig. A-62b Only A and H have four different kinds of carbon atoms. H has an ester carbonyl carbon which is strongly deshielded (175 ppm).

A-26 $ClCH_2CH_2CH_2Cl$ *IHD* = 0 indicates an acyclic saturated compound. The integral areas correspond to 2 (pentet) to 4 (triplet). The four hydrogens are deshielded and equivalent (to give the simple splitting pattern), thus must represent two $-CH_2-$ groups. The remaining $-CH_2-$ is slightly deshielded.

A-27 $N{\equiv}CCH_2CO_2CH_2CH_3$ *IHD* = 3 which can be accounted for by a triple bond and a carbonyl group as are indicated by the IR spectrum. The ^{1}H-nmr spectrum shows a 3:2 triplet-quartet characteristic of an ethyl group, with the CH_2 deshielded as is typical for an ethyl ester. Since there is no single H resonance peak in the nmr spectrum the triple bond cannot be due to a terminal acetylene, and there is no indication that the nitrogen atom is present as an amino group. The compound must be a nitrile. The singlet at 3.5ppm is consistent with deshielded $-CH_2-$.

A-28

$C_6H_5CH_2\overset{O}{\overset{\|}{C}}CH_3$

IHD = 5. The nmr shows a monosubstituted aromatic and the IR shows a carbonyl. These groups account for the *IHD* and C_7H_5O of the molecular formula. The nmr shows that the remaining C_2H_5 must be due to deshielded methyl and methylene groups that do not split each other.

A-29 The terminal methyl group is the least deshielded and thus is closest to TMS. The next two methylene groups are increasingly deshielded as they come closer to the nitrogen atom. Of the two methylene groups adjacent to nitrogen, the one closer to the oxygen is more deshielded. The methylene connected to oxygen is most deshielded because an oxygen atom is more electronegative than a nitrogen atom.

A-30

a) $(CH_3)_3CCH_2NH_2$ - A fishy smelling liquid containing nitrogen is typical of amines and the the IR double peak at 3300 cm^{-1} suggests a primary amine. From the elemental analysis we calculate:

$$\frac{68.93}{12.01} = 5.74; \quad \frac{15.04}{1.008} = 14.92; \quad \frac{16.08}{14.005} = 1.15$$

$\therefore$ the empirical formula is $C_5H_{13}N$. A *tert*-butyl and two unsplit groups of protons are evident from the nmr spectrum. (Remember that hydrogen atoms on N and O usually don't split adjacent protons.)

b) 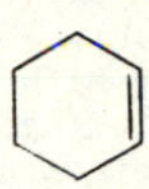

- *IHD* = 2. The molecular formula shows that the nmr peak areas correspond to 4:4:2 with the two H's at a chemical shift typical of alkenes. The four H's at 2.0 ppm are the allylic methylene groups. The ^{13}C-nmr shows three different kinds of carbon atoms with the alkene carbons markedly deshielded.

c) $C_6H_5CH_2\overset{O}{\overset{\|}{C}}CH_2\underset{CH_3}{CH}CO_2H$

or

$C_6H_5CH_2\overset{O}{\overset{\|}{C}}\underset{CH_3}{CH}CH_2CO_2H$

- A broad IR absorption at 2900-3300 cm^{-1} along with a C═O absorption and the nmr peak at 11.2 ppm suggest —CO_2H. The second carbonyl absorption cannot be an aldehyde because no aldehydic H is present in the nmr spectrum. It is presumed to be a ketone. The nmr also shows a monosubstituted benzene ring. These units (—CO_2H, C═O, and C_6H_5—) account for 150 MW units. The eight remaining H's require at least four carbon atoms and C_4H_8 fits the required MW. The singlet at 3.6 ppm is a deshielded and unsplit methylene. Chemical shifts and splitting patterns lead to two possible structures.

d) $N{\equiv}CCH_2CH_2CO_2CH_3$ - The IR peak at 2240 cm^{-1} in a nitrogen containing compound suggests a nitrile. The C═O peak at 1730 cm^{-1} is consistent with an ester and the nmr singlet at 3.8 ppm fits a methyl ester. The singlet at 2.7 ppm can only fit two magnetically equivalent CH_2 groups. In this compound, the nitrile and ester affect the CH_2's identically.

A-31 Integral areas would show zero, one and two, alkene protons respectively.

A-32

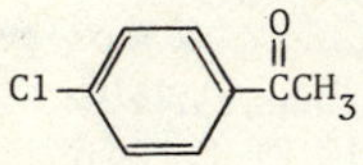

The dual molecular ion peaks are characteristic of a chlorine atom. The IR absorption and nmr singlet suggests a methyl ketone while the symmetrical aromatic multiplet with an area ratio of four is typical of para disubstitution.

A-33

A - $C_6H_5CH(OH)CH_3$

B - $C_6H_5CH_2CH_2OH$

An *IHD* = 4 is consistent with the presence of an aromatic ring as are the nmr spectra. The aromatic peak areas indicate monosubstitution. The IR spectra indicate a hydroxy group but both spectra are quite similar. These two groups account for C_6H_6O. The nmr splitting patterns and chemical shifts provide the information for the arrangements of the remaining C_2H_4 and for final structural assignments. The spectrum of A is fig. A-68 and that of B is fig. A-69.

A-34 $CH_3CH{=}CHCO_2C_2H_5$

The IR spectrum suggests a conjugated carbonyl group and possibly a nonsymmetrical alkene. Those two groups account for the *IHD* of 2. The second oxygen is consistent with an ester group and the 3:2 triplet-quartet with the deshielded quartet in the ^{1}H-nmr spectrum suggests an ethyl ester. The ^{1}H-nmr multiplets at 5.9 and 7.0 are indicative of conjugated alkene protons. The doublet at 1.9ppm is probably a methyl group adjacent to a C–H and further split by long range coupling with an alkene proton. The ^{13}C-nmr shows six different carbon atoms including a weakly deshielded peak at 166 ppm consistent with the ester carbonyl carbon. The two alkene carbons are at 124 ppm and 144 ppm. The $-CH_2-$ attached to the ester oxygen atom is deshielded to 60 ppm.

A-35 In each of the following ^{13}C-nmr spectra the chemical shift of each carbon atom is indicated. Resonance peaks move further downfield (to higher chemical shift values) when electronegative atoms or groups are near the atom.

A - Fig. A-71e - $\overset{(a)}{CH_3}\overset{(b)}{CN}$

(a) 1.3
(b) 117.7

B - Fig. A-71c -

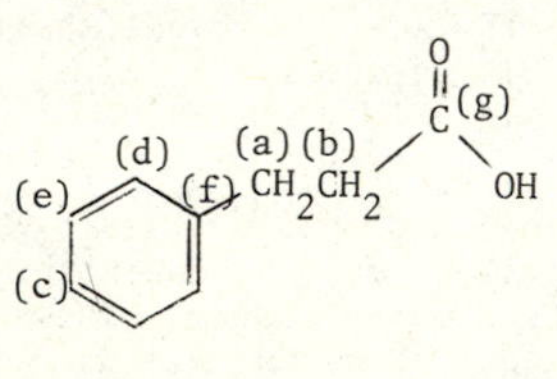

(a) 30.5
(b) 35.5
(c) 126.3
(d) 128.2
(e) 128.5
(f) 140.1
(g) 179.5

C - Fig. A-71b -

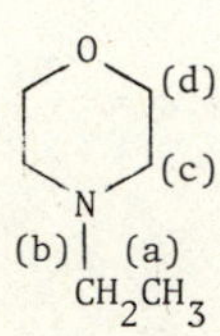

(a) 11.7
(b) 52.7
(c) 53.4
(d) 66.9

Contd...

A-35 contd...

D - Fig. A-71d - $\overset{(b)}{Cl_2CH}\overset{(a)}{CH_2Cl}$

(a) 50.1
(b) 70.4

E - Fig. A-71a - $\overset{(a)}{CH_3}\overset{(c)}{CH_2}\underset{(d)}{CH}(\overset{(b)}{CH_3})\overset{(c)}{CH_2}\overset{(a)}{CH_3}$

(a) 11.4
(b) 18.8
(c) 29.3
(d) 36.4

A-36

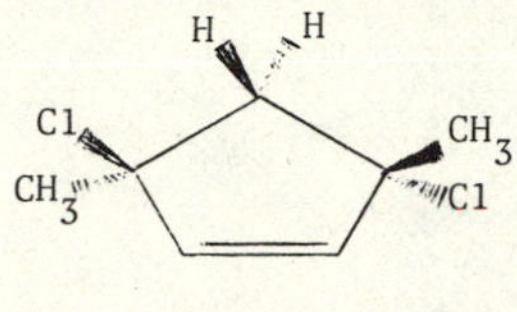

A

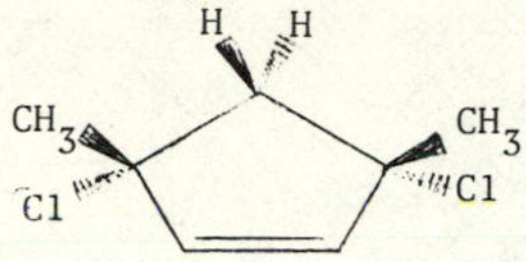

B

One diastereomer will have the methyl groups in a cis relation and the other will have trans methyl groups. The cis isomer has a plane of symmetry and one methylene hydrogen is cis to the methyl groups whereas the other is cis to the chlorine atoms. The two hydrogen atoms are magnetically nonequivalent and give an nmr multiplet. The cis isomer is compound B. The trans isomer is A and has a two fold axis of symmetry. The methylene hydrogen atoms are equivalent and give a singlet nmr peak.

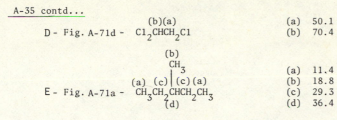